INTRODUCTION TO FERROIC MATERIALS

INTRODUCTION TO FERROIC MATERIALS

Vinod K. Wadhawan
Centre for Advanced Technology, Indore, India

CRC Press is an imprint of the
Taylor & Francis Group, an **informa** business

CRC Press
Taylor & Francis Group
6000 Broken Sound Parkway NW, Suite 300
Boca Raton, FL 33487-2742

First issued in paperback 2019

ISBN-13: 978-90-5699-286-6 (hbk)
ISBN-13: 978-0-367-39780-7 (pbk)

Visit the Taylor & Francis Web site at
http://www.taylorandfrancis.com

and the CRC Press Web site at
http://www.crcpress.com

Mere scholarship will not help you to attain the goal.
Meditate. Realise. Be free.

Swami Sivananda

DEDICATED TO MY PARENTS
Late SMT. KAMLA RANI WADHAWAN
AND
SH. ONKAR NATH WADHAWAN

Contents

FOREWORD

When Vinod Wadhawan mentioned to me a few years ago that he was considering writing a book on ferroic materials, I was immediately enthusiastic about the idea. Why? Well, this is a subject in which many people are working, often without realising it, and for which there is at the moment a distinct lack of material written at the level of a textbook. Most of this topic can be found in research publications as well as in bits in various textbooks, but hardly in a form that addresses the whole subject within one place. So the idea of having a book to describe this topic excited me from the start.

To understand why this should be, I have first to explain that the term 'ferroic' is all about the potential or actual ability of a material to switch some of its physical properties from one state to another. A moment's thought will demonstrate that this is a very wide remit indeed, and this is why many scientists engaged in materials research are often studying ferroic materials and ferroic properties without understanding that they actually form almost a complete scientific discipline. In the 19th century the ability of certain materials to switch the directions of their magnetisations under an applied field led to the classification known as ferromagnetism. The 'ferro' part simply referred to the fact that iron was the most common type of magnet showing such behaviour, but the term was adopted to cover all types of switchable magnetic materials. The switching ability was found to be described by a form of hysteresis curve between applied field and magnetisation. Then in the early part of the 20th century, it was discovered that some materials showed a similar sort of switchability, this time in their electrical polarisation when an electric field was applied. Such materials were termed ferroelectrics, simply by analogy with ferromagnets (nothing to do with iron !), and again these showed characteristic hysteresis curves between field and polarisation. This led to a new science and a new scientific community, that of ferroelectricity, which then developed largely independently from ferromagnetism.

More recently, it was realised that there was a third type of switchable behaviour to consider, this time in the elastic properties. Thus a plot of applied stress against strain in some crystalline materials again showed a hysteresis curve, and this then led to the term ferroelasticity being coined. It then became apparent that it would be possible to unify these three properties under the overall title of 'ferroics', and in so doing the classification was then easily extended to cover cross-terms in the above properties. Thus, for instance, the property of piezoelectricity in which an applied stress creates an electric charge, well known these days in the kitchen when one lights a modern gas cooker with a spark generated by a special battery-less lighter, is then thought of as indicative of ferroelastoelectricity (provided one can identify or imagine an acceptable 'prototype' symmetry for the material). You can see therefore that anyone working on just about any piezoelectric substance is in fact working on a ferroic material.

Another important aspect of ferroic materials is that they conform to definite symmetry principles. In general they have symmetries that are subgroups of others, which in crystal structural terms means that their structure consists of small deviations from a parent or prototypic structure - the concept of pseudosymmetry. This means that ferroic materials often undergo phase transitions as a function of temperature or other variables, and so the need for an understanding of ferroicity automatically leads one to the theory and experimental study of phase transitions.

It is important to understand that once one has a scheme to classify a large number of apparently disparate behaviours of a substance, one can then look for the relationships between the properties. The concept of ferroicity does this in a very successful way and this has enabled much to be learnt about how the different properties arise. Such knowledge is then invaluable in meeting the technological challenges of designing new materials. I believe therefore that the readership of this book will be those who are already informed about ferroics and those who know little or nothing about this concept. The first group will be pleased to have the essential ideas located within one text rather than having them scattered throughout the scientific literature; and the second are probably in for a surprise when they discover that all along they too have been working in this field, but were unaware of it.

Vinod Wadhawan is one of the world's leading authorities on ferroic materials, as well as being a personal friend for many years. I am very pleased that he has succeeded in writing such a book, and I think it will have a lasting influence on those who read it. This is a subject that spans all of materials science from theory to practice, and involves knowledge of many disciplines including physics, chemistry, crystallography and engineering. It is therefore a multidisciplinary subject, making it accessible to scientists from many backgrounds. Vinod has managed to put together a book that can be read at all levels, whether as a student or as an active research scientist. I am particularly delighted by the clarity and at the same time thoroughness with which the concepts are discussed here.

Clarendon Laboratory
Oxford

A. M. Glazer
President, B. C. A.
Professor, University of Oxford

PREFACE

Although the term 'ferroic materials' is not very widely familiar yet, ferroic materials themselves have been known for a long time, especially because of their device applications. Lithium niobate, the Ni-Ti alloy NITINOL, chromium dioxide, and quartz are examples of ferroic materials. Lithium niobate is a ferroelectric, finding a variety of applications in the manipulation and control of laser radiation. NITINOL is a ferroelastic, well known for its extensive shape-memory applications. Chromium dioxide is a ferromagnetic, used in the recording-tapes industry. Quartz is a ferrobielastic, having wide-ranging applications as a transducer material. The special properties of ferroic materials, making them important for device applications, arise from the presence of a *ferroic phase transition*. We call a 'nondisruptive' phase transition a ferroic phase transition if it entails a change of the point-group symmetry of the material with reference to a certain prototype symmetry. The loss of point-symmetry operators results in the occurrence of domain structure, and the domain structure of ferroic materials can be manipulated to advantage, at least in a certain temperature range. This is one of the salient features of ferroic materials.

Two other important features of ferroic materials arise in the vicinity of the ferroic-transition temperature. One is the enhancement of some macroscopic properties (e.g. certain generalized susceptibilities, and properties coupled to them). To mention a familiar example: in the vicinity of a 'proper' ferroelectric phase transition the polarization is a very sensitive and nonlinear function of electric field. This means that in the vicinity of the transition the material develops a large response function (susceptibility) for the property (polarization) corresponding to the primary instability driving the transition. And if there is a strong coupling of the spontaneous polarization with, for example, birefringence, then birefringence also will exhibit an enhanced sensitivity to electric fields, giving rise to large electro-optic effects.

The other important feature of ferroics in the vicinity of the ferroic phase transition is their amenability to certain field-induced phase transitions. A well-known manifestation of this is the occurrence of the shape-memory effect in many ferroelastics.

Several excellent texts are available on ferroelectric, ferroelastic, and ferromagnetic materials, as also on the industrial applications of transducer materials like quartz. But a single book putting the entire field of ferroics in an adequate, systematic, perspective, and written in a pedagogical manner, has not been available so far. The present book is intended to meet that requirement. It is the first of its kind, in that a substantial portion of it (in fact, much of Part A) deals with the *common* characteristics of this class of materials. And there is no better way of doing this than through symmetry considerations. A ferroic phase transition results from a spontaneous breaking of symmetry. Symmetry considerations, being largely system-independent, have a great unifying influence on the understanding of large classes of apparently unrelated phenomena. A wide range of properties of ferroic materials can be understood in terms of the symmetry-breaking ferroic transition. This book attempts to explain (at an elementary level) how this understanding is acquired, with emphasis on the utilitarian role of symmetry in materials science.

Over the last decade and a half, the field of *smart materials and structures* has been progressing at a good pace. A smart material is designed to alter its properties automatically to suit the changes in environmental conditions. In the so-called actively smart materials, the *in situ* fine-tuning of properties is effected through an external biasing mechanism. The highly nonlinear behaviour of ferroic materials in the neighbourhood of the ferroic transition makes them obvious candidates for the field-tuning of their relevant properties for smart-structure applications. NITINOL, lead lanthanum zirconate titanate (PLZT), and lead magnesium niobate (PMN) are some of the more popular ferroic materials being investigated and used for this purpose. TERFENOL – D, in many ways the magnetic analogue of NITINOL, is another promising material in this context. Deeper insights into why ferroic materials behave the way they do will surely lead to exciting new developments in smart-materials research.

The level of presentation of the book should make it useful to undergraduate and graduate students of materials science and physics. Basic physical principles are brought out. A reasonably self-contained and connected account of the subject is attempted. However, the level of presentation is not uniformly the same throughout. This is because I have tried to meet two somewhat disparate objectives. One is to expose the beginner to this field to practically the entire gamut of concepts, definitions, and jargon. The other is to let the experienced worker in one part of the field have a feel for what other experts have been doing. From the beginner's point of view the approach adopted is somewhat different from that of many of the existing texts on crystallography, crystal physics, and phase transitions, in that the Curie-Shubnikov principle of symmetry is used explicitly, time and again.

When I was more than half way through with the writing work, I came across an unusual new book: *Principles of Condensed Matter Physics* by Chaikin & Lubensky (1995). This is an excellent text at an advanced level. What is unusual about it is the subject matter covered under the title of condensed-matter physics. Practically the entire book is designed to deal with the consequences of spontaneous breaking of continuous and discrete symmetries in condensed-matter systems. The similarity of approach with the present book is obvious. Ferroic properties are a consequence of spontaneous breaking of a specific type of symmetry, namely crystallographic point-group symmetry. Although the present book has been written at a more elementary level, and covers very different ground from that of Chaikin & Lubensky, it reflects a shifting trend in the modern approach to condensed-matter physics, particularly after the formulation of K. G. Wilson's renormalization-group theory.

Acknowledgements

A large number of professional colleagues have influenced whatever little understanding I have acquired of the field of ferroic materials. I take this opportunity to put on record my sense of gratitude to them.

Dr. V. M. Padmanabhan, Dr. R. Chidambaram, and Dr. S. K. Sikka shaped my development as a crystallographer. Obeisance to the teacher.

I had the privilege of spending a year as a Nuffield Foundation Travelling Fellow at the Clarendon Laboratory, Oxford, and working with Mike Glazer. Although that was way back in 1979-80, Mike continues to be a great source of inspiration. His enthusiastic response to the news that I was planning to write a book on ferroic materials provided the initial impetus, and kept me going.

My one-year stay at the Materials Research Laboratory of the Penn State University was a great educational experience. I found the enthusiasm of Bob Newnham, Eric Cross, and Amar Bhalla for the application potential of ferroic materials very infectious indeed.

I got an opportunity to work for a month at the Institute of Physics, Prague. Vaclav Janovec, Jan Fousek, Vojtek Kopsky and Z. Zikmund were very generous in sharing with me their knowledge and understanding of the symmetry aspects of ferroic materials.

I am grateful to Keitsiro Aizu and Jinzo Kobayashi for the large number of reprints received from them over the years, and also for several personal communications. Jinzo explained to me the intricacies of the high-accuracy universal polarimeter (HAUP) developed in his laboratory. This technique is indeed a boon for studying ferrogyrotropic materials.

My tenure as a Visiting Professor at the University of Aix-Marseille in the laboratory of Claude Boulesteix exposed me to some of the fascinating complexities of nonstoichiometric and vacancy-ordered ferroics. Claude's contributions to this field are quite substantial, and discussions with him helped me in improving my understanding of this topic.

During the preparation of this book I had to send requests to several scientists all over the world for sending me reprints of papers published from their laboratories. The response was invariably generous, and I want to thank all of them, especially Vaclav Janovec, Bob Newnham, Eric Cross,

Ekhard Salje, Theo Hahn, Claude Boulesteix, Dorian Hatch, Dan Litvin, Mike Glazer, and Wilfried Schranz.

Mike Glazer, Satish Bhargava, Praveen Chaddah, Sindhunil Roy, Dorian Hatch, Dhananjai Pandey, Surya Gupta, Ekhard Salje, and Keshav Bhagwat went through portions of the book, and offered critical comments. I am thankful to them for sparing the necessary time for doing this. I am also thankful to my colleagues Vidya Sagar Tiwari, and Jagdish Sisodiya for their help.

A. K. Rajarajan was my guru for all questions related to the LaTeX coding of the manuscript. C. P. Verma, P. B. Kamble, G. R. Nair and R. S. Jain did the art work with great interest and skill. My sincere thanks to all of them.

I thank Dr. D. D. Bhawalkar, Director, Centre for Advanced Technology, Indore, for his support for this venture.

This book was written during evenings and weekends. Because of my official duties and other constraints, I had to spread out the writing work over a rather long period of five years. My wife Rajni, daughter Namrata, and son Anupam made their own contributions by not being too demanding on my time and attention. Thanks a lot, you silent sufferers !

Indore
November 1999

V. K. Wadhawan

Part A

GENERAL CONSIDERATIONS

Chapters 1 to 8 comprise Part A of the book. Chapters 1 to 5 deal with basic condensed-matter physics, covering topics such as crystallography, crystal physics, diffraction physics, and phase transitions. Chapter 6 describes a tensor classification of ferroic materials. It also deals with optical and acoustical ferrogyrotropy. These two topics are not considered in a separate chapter later because ferrogyrotropy is only an implicit form of ferroicity, which can occur in a material only as an adjunct to another, explicitly ferroic, property.

Chapters 7 and 8 deal with the general aspects of domain structure of ferroics.

Contents of Part A serve to highlight the *common* features of the wide variety of ferroic materials.

Chapter 1

INTRODUCTION

This chapter is in two parts. The first provides an overview of what this book is all about, and how the subject matter is divided among its chapters. In the second part we trace briefly the historical development of the subject of ferroic materials.

1.1 OVERVIEW

A large fraction of solids in the inanimate world have a crystalline atomic structure. This indicates that the most stable arrangement of atoms or molecules in a solid is frequently that which involves a periodic repetition of an atom or a group of atoms along three noncoplanar directions. This arrangement can be viewed as a repeated stacking of a certain building block (the *unit cell*) along the three dimensions in a space-filling manner. Because of their regular atomic structure, crystals usually display a rather high degree of symmetry, as also anisotropy, in their physical properties.

To understand and describe adequately the repetitive and symmetric internal structure of crystals, and also the properties resulting from such a structure, one must use an appropriate mathematical language. Group theory provides such a language (although not entirely adequately). We recapitulate in Appendix B the relevant basics of group theory, after listing some elementary terms and concepts of set theory in Appendix A.

The atomic structure of a crystal possesses symmetry. It always has at least the *translational symmetry*. What this means is that if someone were to translate all the atoms in the crystal by an integral number of repeat distances along one or more of the three directions of repetition, the atomic arrangement would look exactly as before. One expresses this fact by saying that the crystal structure, because of its translational symmetry, is invariant under translations corresponding to the repeat-distance vectors.

The connection between symmetry and invariance is of a general nature. It is known in physics as *Noether's theorem*, which can be stated in a highly simplified form as follows: Wherever there is a symmetry in Nature, there is a corresponding conservation law. Somewhat more rigorously: If there exists an infinitesimal transformation of the dynamical variables which leaves the Lagrangian invariant, then every parameter of such a transformation is associated with a conservation law (Noether 1918; also see Kimberling 1972).

In addition to translational symmetry, a crystal may possess *rotational symmetry*. That is, its structure and other properties may remain invariant under certain specific rotations (or rotations combined with inversion) performed about specific axes.

Another type of symmetry that can be possibly associated with a crystal is that involving *time inversion*. When a crystal possesses this symmetry, it is a nonmagnetic crystal, whereas its absence corresponds to the occurrence of magnetic properties. This is because we associate a magnetic moment with the flow of current in a loop, and time reversal and the concomitant reversal of the direction of flow of current corresponds to a flipping of the direction of the magnetic moment.

The various types of symmetry possessed by a crystal can occur together only in certain restricted combinations; the restrictions ensure self-consistency. The totality of symmetry operations applicable to a crystal defines its *crystallographic space group* (cf. Chapter 2).

Now the structure a crystal has at, say, room temperature may not be the most stable structure at other temperatures. The stable structure or phase a crystal can possess at a given temperature (and at given values for other control parameters like hydrostatic pressure, uniaxial stress, electric and magnetic fields etc.) can be understood in terms of its thermodynamic free energy. That phase is stable which has the lowest free energy. Fig. 1.1.1 shows schematically the temperature dependence of the free energies of two competing phases a crystal may exist in. For $T > T_c$ Phase I has a lower free energy than Phase II, and is therefore more stable. The two free-energy curves intersect at T_c, and for $T < T_c$ it is Phase II which has a lower value of the free energy. Consequently, as the crystal is cooled from a high temperature, it makes a (first order) transition from Phase I to Phase II at the temperature T_c.

When such a phase transition occurs, there is also a change of the space-group symmetry of the crystal (though not always a change of the space-group *type*; cf. §2.2.4 for a distinction between space groups and space-group types). Let this change of space group be from $S(I)$ to $S(II)$. What determines the symmetry group of the new phase after a phase transition ? This question was first tackled by Landau (1937a, b, c, d). In the formulation of his landmark theory of phase transitions, he implicitly used

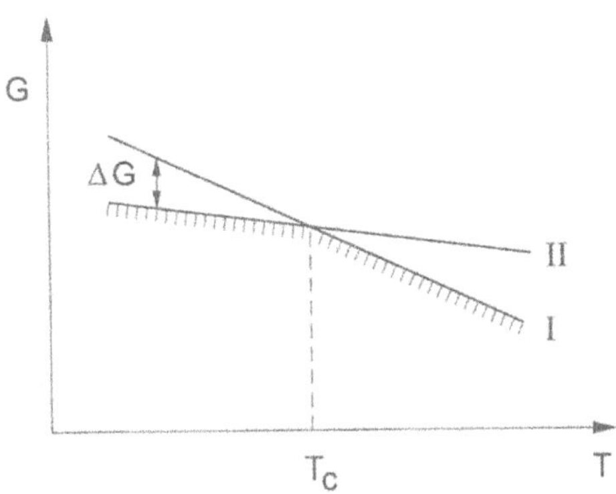

Figure 1.1.1: Temperature dependence of the Gibbs free energy G of a crystal for two different phases I and II. For $T > T_c$, $G(I) < G(II)$; and for $T < T_c$, $G(II) < G(I)$. For any T less than T_c, ΔG provides the driving force influencing the rate of transition from one phase to the other.

the *Curie principle* of superposition of symmetries. According to the Landau theory, a symmetry-lowering phase transition in a crystal is heralded by the emergence of a so-called *order parameter*. This order parameter has a certain symmetry of its own, and the symmetry of the resultant phase is the highest symmetry common to the parent phase and the order parameter (in accordance with the Curie principle).

The Curie principle was given its extended mathematical formulation by Shubnikov (see Shubnikov & Koptsik (1974)). We propose to call this generalized principle the *Curie-Shubnikov principle*. It is discussed in Appendix C. It is of fundamental importance, not only to crystal physics and phase transitions, but also to much else in physics. We shall have occasion to refer to it repeatedly in this book.

In Chapter 2 we recapitulate the basics of crystallography. How does a crystal form? What is its smallest size for which the symmetry is the same as that of the bulk crystal? Such questions are touched upon, and references are given for further reading.

The 'Suggested Reading' listed at the end of most of the sections is intended to serve two purposes. It leads the reader into areas which could not be covered in this book either for reasons of space or of scope. It also serves as an acknowledgement, in several cases, of the fact that the contents of the section have been drawn, in some way or the other, from the material originally published in these references.

Although in the ultimate analysis it is the atomic structure of a crystal which determines all its properties, including macroscopic physical properties, for purposes such as device applications it is usually advantageous to view the crystal as an *anisotropic continuum*. This is the underlying feature of what has come to be known as *crystal physics*, and Chapter 3 is devoted to some salient aspects of this subject. The importance of this branch of physics has grown enormously as crystals have come to be employed in ever-increasing and vitally important technology-related applications. Discussion of crystal physics requires the use of tensors, and we introduce them briefly in this chapter.

The transition from Chapter 3 to Chapter 4 can be viewed as that of going over from a macroscopic description of crystals to a microscopic or space-group level description. Basics of diffraction theory are described, leading to the introduction of the concept of the reciprocal lattice. Representations of translation groups and space groups are described briefly. This chapter provides the groundwork for Chapter 5, which deals with phase transitions.

It is instructive and useful to categorize phase transitions in crystals in terms of the changes of symmetry that accompany them. The primary subdivision can be on the basis of whether or not there is a change in the space-group type of the crystal at the phase transition. If the initial and the final phase have the same space-group type, the two phases are said to be isomorphous, and one speaks of an *isomorphous phase transition*. Those phase transitions for which there *is* a change of the space-group type are described as *nonisomorphous phase transitions*.

The symmetry operations comprising a crystallographic space group can involve lattice translations, fractional translations, rotational operations, time-reversal operations, and the allowed combinations of all these. For the theme of the present book, the most relevant question to ask here is: In a nonisomorphous phase transition, is there also a change in the point-group symmetry of the crystal? If the answer is yes, *and if this change of point symmetry occurs in a nondisruptive manner* (cf. §5.1 and 5.2), we are dealing with a ferroic phase transition.

> *A phase transition is called a ferroic phase transition if: (a) it can be viewed as a nondisruptive modification of a certain "prototypic phase", and (b) it involves a loss of one or more point-symmetry operators present in the prototype.*

By "nondisruptive modification" we mean that the new phase can be described by (i.e. its symmetry elements, Wyckoff positions, atomic parameters, etc. can be located) in the frame of reference of the other phase, after making (if necessary) continuous distortions (affine mappings) that do not themselves entail any additional change of symmetry.

> *A ferroic material is one which can, or can be conceived to, undergo one or more ferroic phase transitions.*

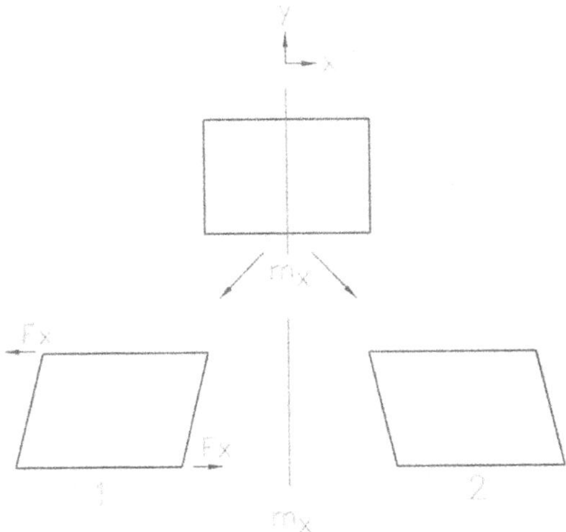

Figure 1.1.2: Illustration of how at least two distinct orientation states must arise in any ferroic phase transition.

(In certain situations a ferroic material may not actually be able to undergo a ferroic phase transition (for example because it decomposes before attaining the requisite temperature). Nevertheless, the postulation of a hypothetical ferroic phase transition may still be very useful for understanding several properties of the material (cf. §5.2).)

The motivation behind the above, rather abstract-looking, definition of ferroic phase transitions becomes clear when we look at the consequences of such a transition. As a result of the occurrence of a ferroic transition in a material, the material comes to possess four important features. The first is that in the phase in which a point-symmetry operator has been lost the crystal has at least two equivalent states which differ only in their orientation (either of structure, or of spontaneous magnetic moment, or both) and/or handedness. Aizu (1962, 1969) therefore spoke of these states as *orientation states*. Fig. 1.1.2 provides a simple illustration of why more than one orientation state must be possible in the ferroic phase. The prototype phase, comprised of schematic rectangle-shaped unit cells, undergoes a transition to a lower-symmetry phase represented by a unit cell that has the shape of a parallelogram. It is clear that, in this example, the ferroic phase can arise in two distinct orientations. The ferroic phase is said to possess *domain structure*: in the example depicted in Fig. 1.1.2, some regions of the crystal may exist in Orientation State 1 (or Domain 1), and some others in Orientation State 2 (or Domain 2), with *domain walls* separating the domains. Domain structure of ferroic crystals is considered in detail in Chapters 7 and 8.

The term "ferroic" was coined by Aizu (1969a, 1970a). He defined a crystal as being a ferroic "when it has two or more orientation states in the absence of magnetic field, electric field, and mechanical stress, and can shift from one to another of these states by means of a magnetic field, an electric field, a mechanical stress, or a combination of these". In the example of Fig. 1.1.2, Orientation State 1 can be changed to Orientation State 2 by a shear stress F_x applied as shown. Similarly, Orientation State 2 can be shifted to Orientation State 1 by an opposite stress. When orientation states are thus changed to one another under the action of external fields, the interfaces (domain walls) separating them move accordingly. Thus, under certain conditions, the domain structure of ferroics can be manipulated (to advantage) by applying suitable external fields. This is a central feature of ferroic materials.

The orientation states defined by Aizu (1970a) are identical or enantiomorphous in crystal structure. They can be mapped onto one another by appropriate transformation operations. In Fig. 1.1.2 the two orientation states can be mapped onto each other by a mirror operation (m_x) normal to the x-axis. As reflected above in our definition of a ferroic phase transition, the fact that a ferroic crystal can exist in two or more equivalent orientation states can be understood by postulating that the ferroic phase is the result of a transition from a real or hypothetical phase of the crystal in which all the orientation states of the ferroic structure lose their identity; that is, not only do the orientation states have the same crystal structure, they also have the same orientation in this reference phase. In other words, in the reference phase the whole specimen is just one single crystal, with no orientation states. Aizu (1970a) called such a reference structure the *prototype*. The concept of prototype symmetry plays a central role in the systematic description of ferroic materials, although Aizu's (1970a, 1975) definition of it was not rigorous enough. We introduce a new, and more rigorous, definition of prototype symmetry in §5.1.

Clearly, the prototype has a higher point-group symmetry than the ferroic phase under consideration. For example, in Fig. 1.1.2 the parent phase has an additional mirror plane of symmetry, m_x, which the ferroic phase does not have. We also note that the symmetry operation (m_x) lost on passing to the lower-symmetry phase is also the operation which can map Orientation State 1 to Orientation State 2. This is a general result, and such operations are called *F-operations* (F standing for "ferroic").

According to the extended Neumann theorem of crystal physics (cf. Chapter 3 and Appendix C), the space-time point-group symmetry possessed by any macroscopic physical property of a crystal cannot be lower than the point-group symmetry of the crystal. Because of this, some or all the components of one or more tensor properties can become forbidden for any particular point-group symmetry. A very simple example is that of spontaneous polarization, which cannot exist in any crystal that is centrosymmetric, because the symmetry elements possessed by spontaneous polarization do not include the inversion-symmetry operation present in a centrosymmetric crystal (inversion symmetry means that whatever occurs in a crystal at any point (x, y, z) also occurs at $(-x, -y, -z)$, with a suitable choice of the origin of the coordinate system). Now, if the crystal undergoes a (nondisruptive) phase transition which lowers its prototype point-group symmetry (and a ferroic transition is, by definition, such a transition), one or more of the forbidden components of the macroscopic tensor properties can acquire nonzero values if the point-symmetry operators forbidding their occurrence are lost at the ferroic transition. In our above-mentioned example, if inversion symmetry is lost at the phase transition, the lower-symmetry (or ferroic) phase can allow the occurrence of spontaneous polarization (if this phase belongs to a "polar" symmetry group), so that we have what is known as a *ferroelectric phase transition.* Ferroelectric phase transitions are a subset of ferroic phase transitions.

The spontaneous polarization, P_s, of a crystal which undergoes a ferroelectric transition at a certain *critical temperature*, T_c, is zero in the higher-symmetry (prototypic) phase, and steeply rises from zero as the temperature is lowered below T_c. Fig. 1.1.3(a) shows the temperature variation of P_s for the ferroelectric phase of $BaTiO_3$ having tetragonal symmetry; the crystal is centrosymmetric and cubic for a temperature range above $T_c = 393K$. We thus see here a second characteristic feature of certain ferroelectric and other ferroic transitions (the first being the occurrence of domain structure in the ferroic phase), namely a steep temperature dependence of a macroscopic property (spontaneous polarization in the present example) for temperatures just below T_c.

We should remember that not in all ferroic phase transitions a macroscopic tensor property is the order parameter. But since the transition is ferroic, there is at least one macroscopic tensor property which is either the order parameter itself, or couples with the order parameter. Only in the former case does a macroscopic property necessarily have a steep temperature dependence just below T_c. In the latter case the temperature dependence of the property coupled to the order parameter depends on the nature of the coupling.

The important point is that, in general, in the vicinity of a ferroic phase transition one or more macroscopic properties become large. For a "proper" ferroelectric, $\partial D/\partial T$ or pyroelectric response becomes large. So also does the dielectric susceptibility $\epsilon(= \partial D/\partial E)$. An example is shown in Fig. 1.1.3(b) for $BaTiO_3$.

It is in the nature of the ferroelectric transition in $BaTiO_3$ that the dielectric constant becomes large in the vicinity of T_c. In practical terms what this means is that, in the vicinity of T_c, application of even a small

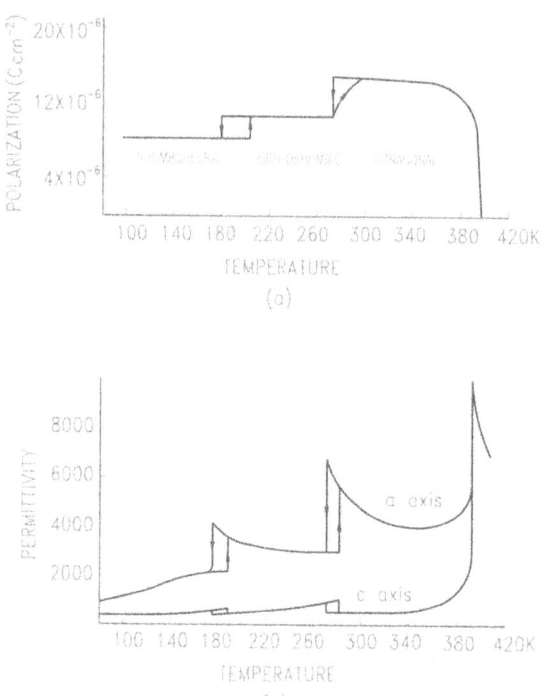

Figure 1.1.3: Temperature dependence of (a) spontaneous polarization, and (b) static permittivities ϵ_{11} and ϵ_{33} of $BaTiO_3$. [After Kay & Vousden (1949) and Merz (1949).]

electric field invokes a large response in the induced polarization. It is as if the material has become "electrically soft" in the vicinity of this ferroelectric transition. This is a common feature of certain ferroic phase transitions. However, we must emphasize that the response function corresponding to only the order parameter becomes large around T_c; other response functions do not necessarily blow up in the vicinity of T_c, although it is quite common that several of them do.

The various physical properties of a crystal are coupled to one another to a small or large extent. To consider the example of $BaTiO_3$ again, in the cubic, centrosymmetric, prototypic phase, both spontaneous polarization (P_s) and spontaneous birefringence (Δn) are zero, and they become nonzero in the tetragonal, noncentrosymmetric, ferroelectric phase. The temperature-variation or field-variation of P_s causes a corresponding variation of Δn through an internal electrooptic coupling. And since P_s has

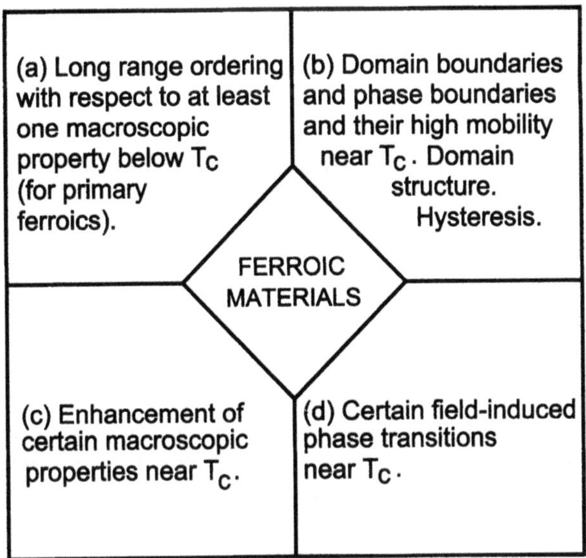

Figure 1.1.4: Salient features of ferroic materials. See text for details.

a strong temperature and electric-field dependence just below T_c, so also may Δn. This transmission of anomalous behaviour from the order parameter to other properties coupled to it is a fairly common feature of ferroic materials.

Another quite common feature of ferroics is their tendency to undergo field-induced phase transitions, particularly in the vicinity of T_c. In other words, application of certain fields may shift the effective T_c. This property finds applications, notably in smart structures (employing materials such as NITINOL and PMN).

Lastly, it is important not to forget that spontaneous breaking of directional symmetry at a ferroic transition amounts to long-range ordering with respect to at least one macroscopic tensor property. It is useful to count this as a distinct salient feature of ferroic materials.

The four characteristic features of ferroic materials are listed in Fig. 1.1.4, and can be summed up as follows:

(a) *Long-range ordering of at least one macroscopic tensor property* (for 'primary' ferroics; see below). If this property is, for example, spontaneous polarization, it may indicate larger than normal linear and nonlinear polarizabilities.

(b) *Occurrence of domain structure (domains and domain boundaries) in the ferroic phase, and the possibility of its modification by a field conjugate to the order parameter.* Domain-wall mobilities are high at temperatures just below T_c[1].

(c) *Large and nonlinearly varying response functions (or generalized susceptibilities) corresponding to certain properties, particularly in the vicinity of* T_c. This fact has several practical implications. For instance it makes it possible to fine-tune certain properties of ferroic materials by applying suitable biasing fields.

(d) *The possibility of field-induced phase transitions.*

Ferroic phase transitions involving the emergence of spontaneous polarization, or spontaneous strain, or spontaneous magnetization, result in phases called *primary ferroic phases*, with "primary" referring to the fact that the free-energy difference between at least one pair of orientation states of such a ferroic phase involves the first power of electric, magnetic, or mechanical field.

But ferroic phases are possible in which no two orientation states differ in either spontaneous polarization, or spontaneous strain, or spontaneous magnetization. The most notable example of these *higher-order ferroics* is quartz. At high temperatures this crystal has hexagonal symmetry, and is known as β-quartz. On cooling, it undergoes a ferroic transition at 846 K to α-quartz, which has trigonal symmetry. There is no emergence of 'relative' spontaneous strain at this transition because there is no shape-changing (in contrast to size- changing) distortion of the crystal lattice in going from the hexagonal crystal system to the trigonal system; so the α-phase of quartz is not a ferroelastic phase. Nor is it a ferroelectric or a ferromagnetic phase. α-quartz is thus not a primary ferroic. Yet its two orientation states differ in certain higher-rank tensor coefficients, notably some elastic stiffness and piezoelectric coefficients. Therefore, if the two orientation states are subjected to the same mechanical stress, their elastic and piezoelectric responses will be different. This is an example of *secondary ferroic* behaviour, which, in this particular case, is a manifestation of *ferrobielasticity* and *ferroelastoelectricity*. The term "secondary" refers to the fact that the driving fields involved are products (self products or cross products) of degree 2 of the primary fields. In the case of ferrobielasticity, for example, the driving field needed is the square of uniaxial mechanical stress. In Chapter 6 we discuss a thermodynamic classification of ferroic materials in terms of

[1] As we shall see in Chapter 5, for discontinuous phase transitions there is a *range* of temperatures in which the parent phase and the daughter phase are both stable; i.e. they coexist. In this temperature range, we not only have domain boundaries, but also *phase* boundaries.

macroscopic tensor properties of various ranks.

Also considered in Chapter 6 is the property of *ferrogyrotropy*. This refers to orientation states differing in optical activity (*optical ferrogyrotropy*) or acoustical activity (*acoustical ferrogyrotropy*).

In Chapters 7 and 8 we discuss the domain structure of ferroic materials, with special emphasis on the symmetry aspects of domains and domain walls. The common features of, as well as the dissimilarities between, domain structure and twinning are analyzed. Morphology of the ferroic phase, growing in the matrix of the surrounding parent phase, is described. The general atomic mechanism underlying the switching from one domain state to another is explained. Crystallographic shear planes, irrational shear planes, and chemical twin planes, all arising as compositional extended defects due to vacancy ordering, are also described briefly.

The next three chapters (9, 10 and 11) deal, respectively, with special features of the three types of primary ferroic materials, namely ferromagnetics, ferroelectrics, and ferroelastics. And Chapter 12 is about secondary and higher-order ferroics.

Chapter 13 deals with some materials-science aspects of ferroics, where we take cognisance of the fact that, although ferroic phase transitions are often discussed with reference to single crystals, actual ferroic materials may also be ceramics or composites. In single-crystal physics we usually assume that the specimen crystal is large enough to justify our neglect of surface effects. However, when crystal sizes are small, or when crystallites have dissimilar neighbours, dramatically new phenomena can emerge. Symmetry of composite ferroics is an important area of research, the fundamentals of which are also described in this chapter.

In Chapter 14 we discuss the applications of ferroic materials, highlighting the physical principles involved in these applications.

Over the last decade an increasing amount of research effort has been expended into the development of what are known as *smart materials and structures*. These are designed so as to change their properties in a preconceived "intelligent" manner to suit the changing environmental conditions in which they are employed. The need for a continuous fine-tuning of their properties by suitably configured external fields requires that the materials have large and nonlinear response functions at the temperature of application. Ferroic materials, by their intrinsic nature, are ideally suited for this purpose. In Chapter 14 we also describe the successes achieved, as also the promise of more to come, in this exciting field of applied research, where the device-application potential of ferroic materials may truly come of age.

SUGGESTED READING

K. Aizu (1970a). Possible species of ferromagnetic, ferroelectric, and ferroelastic crystals. *Phys. Rev. B*, **2**, 754.

V. Janovec, V. Dvorak & J. Petzelt (1975). Symmetry classification and properties of equitranslation structural phase transitions. *Czech. J. Phys.*, **B25**, 1362.

M. E. Lines & A. M. Glass (1977). *Principles and Applications of Ferroelectrics and Related Materials.* Clarendon Press, Oxford.

J. C. Toledano (1979). Symmetry-determined phenomena at crystalline phase transitions. *J. Solid State Chem.*, **27**, 41.

1.2 HISTORICAL

In this section we take a look at the major milestones in the history of ferroic materials. The account is bound to be somewhat sketchy because the range of materials and phenomena to be covered is very large and diverse, and only a limited amount of space can be given to this topic in this book. One way of limiting the requirement on space is by not covering history that is too recent. By and large we follow this precept here.

In 1970 Aizu introduced the term "ferroic", and presented a unified treatment of certain symmetry-dictated aspects of ferroelectric, ferroelastic, and ferromagnetic materials. Before the publication of this paper by Aizu (1970a), although the prefix "ferro" was borrowed from the field of ferromagnetism to take note of certain similarities of the properties of ferroelectrics and ferroelastics with those of ferromagnetics, notably hysteresis (Fig. 1.2.1) and the Curie-Weiss law, little common ground among them was emphasized, and these subjects grew more or less independently. We shall therefore divide our historical narrative into subsections dealing separately with each of the major categories of ferroic materials. Their common features are highlighted where appropriate.

1.2.1 Ferromagnetic Materials

Ferromagnetism is the oldest of the ferroic properties known to science. The element Fe is ferromagnetic at room temperature (as also are Co and Ni). And metallic meteorites consist mostly of pure iron. Thus, at least one ferroic material, iron, existed on earth millions of years before the emergence of *homo sapiens.*

Loadstone ($FeO.Fe_2O_3$), a ferrimagnetic material, is said to have been

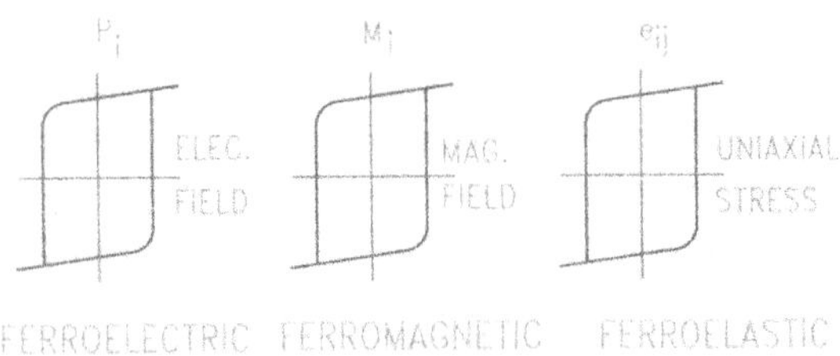

Figure 1.2.1: The hysteretic behaviour of ferroelectrics and ferroelastics is similar to that of ferromagnetics. M_i, P_i, and e_{ij} denote components of spontaneous magnetization, spontaneous polarization, and spontaneous strain. These, when plotted against the driving fields conjugate to them, are non-single-valued functions of the fields.

discovered by the Greeks in the island of Magnesia, over 3500 years ago. According to Chen (1986), the world's first compass, using a loadstone, was invented in China about 85 AD. The invention of the compass had far-reaching consequences for the history of mankind, as it made possible navigation on the high seas.

Apart from the use of the compass, there are at least three major areas of technology where ferromagnetic materials have played a vital role in human progress and welfare. These are: (i) generation and distribution of electric power; (ii) data storage and processing; and (iii) telecommunications. Most of these applications involve the use of materials (iron-silicon alloys, ferrites, garnets, compounds of rare-earths and transition metals, etc.) which are ferromagnetic (or ferrimagnetic) at room temperature, and have a substantial saturation magnetization.

In the early part of the 19th century Ampère put forward the hypothesis that the magnetic moment of a ferromagnetic material arises from internal electric currents in the molecules of the material. This idea was developed further by Weber, who could explain the occurrence of saturation magnetization as arising from a situation wherein the molecular magnets have been aligned almost along the same direction by the external field. A random distribution of the directions of the these magnets would then correspond to the unmagnetized state on a macroscopic scale.

Weiss (1907) invoked the presence of an internal *effective magnetic field* to explain the occurrence of spontaneous magnetization in ferromagnetic materials. This work of Weiss was preceded by that of Curie, and of Langevin, on the response of paramagnetic materials to magnetic fields.

Weiss used the effective-field approach to explain the ferromagnetic phase transitions in Fe, Co and Ni. He also introduced the concepts of magnetic domains and domain walls.

Iron-Silicon Alloys

The low cost of iron due to its abundance on earth, combined with its highly favourable magnetic properties, made possible the generation of electricity on a massive scale at low expense. The Westinghouse Electric Company set up the first AC generating station in 1886. This application of magnetic materials requires the largest possible saturation magnetization, and the least possible loss from causes such as eddy currents. The iron-silicon alloy has been meeting this need for a long time. Barrett et al. (1900) found that the presence of about 3% silicon in iron increased its resistivity and permeability, and reduced the coercive-field value, resulting in reduced hysteresis and eddy-current losses. The material was used in transformers in the form of hot-rolled polycrystalline sheets, initially with random orientation of grains (Gumlich & Goerens 1912).

The hysteresis loop of a single crystal of iron-silicon is nearly rectangular in shape, which means that application of a magnetic field only slightly higher than the coercive field is sufficient to drive the crystal to a state of saturated magnetization. This results in a high value for the maximum induction possible for the crystal. If high cost and other practical problems were not a consideration, one would like to choose for the core of a transformer single-crystal sheets oriented so as to achieve a closed rectangular flux path along the preferred directions, namely $< 100 >$. Goss (1935) developed a cold-rolling and annealing method for achieving a texture in the polycrystalline Fe-Si alloy such that the grains had their $\{110\}$ planes oriented preferentially along the plane of the sheet, with the $< 100 >$ directions of most of them nearly parallel to one another in this plane. Because of this grain orientation, the coercive field value dropped to about 0.1 Oe, the maximum permeability rose to 70,000, the core losses dropped to about 0.6 Watt/kg at 60 cps, and the magnetic induction rose to 10 kG ($1 \mathrm{Wb/m^2}$). Further advances continue to be made in this direction (see Enz (1982) for a review).

Ferrites

The relatively low resistivity of iron-silicon alloys results in large eddy-current losses, particularly in high-frequency applications, even when they are used in the form of sheets. Attempts to overcome this problem led to the development of magnetic oxides for high-frequency applications (Hilpert 1909).

Ferrites have emerged as the material of choice for applications in the frequency range 10^3 to 10^{11} Hz. They are now indispensable for several areas in telecommunication and electronics industries.

Ferrites have the general composition $MOFe_2O_3$, where M stands for divalent metal ions like Ni, Mn, Zn. The ore magnetite ($FeO.Fe_2O_3$) mentioned earlier is also a ferrite. The magnetic structure of ferrites was first explained by Néel (1948), who introduced the concept of partially compensated antiferromagnetic order, and called this property *ferrimagnetism.* Because of their ferrimagnetic nature, the saturation magnetization of ferrites is only a fraction of that of iron, but their most useful feature, namely high resistivity (10^2 to 10^{10} Ωcm), makes them the sole choice for a variety of technological applications.

The occurrence of nearly rectangular hysteresis loops in certain ferrites led to their extensive use as cores for computer memories. In fact, until 1970 practically all the main-frame computers had ferrite cores as memory elements.

Ferrites also played (and continue to play) a major role in the design considerations for several types of particle accelerators (Brockman et al. 1969). These machines use large transformers acting as resonance cavities, which accelerate the charged particles. The design is such that the circulating beam of charged particles acts as the secondary winding of the transformers. As the particles gain energy in successive cycles, the period per cycle decreases, which must therefore be continuously compensated for. This is done by controlling the self-inductance of the core of the transformer by using a bias field. Ferrites offer a low-eddy-current-loss solution to the problem.

Garnets

The ferrimagnetic properties of some rare-earth garnets were discovered by Bertaut & Forrat (1956), and Pauthenet (1956). The best known among these is yttrium iron garnet (YIG), $Y_3Fe_5O_{12}$. The most notable property of YIG is its extremely small ferromagnetic resonance linewidth. This was discovered by Spencer et al. (1956), and Dillon (1957). Soon, linewidths as small as 0.1 Oe (8 A/m) at 10 MHz were achieved by improving the quality of YIG crystals (LeCraw et al. 1958). This made YIG an excellent material for applications in microwave devices (see Wang (1973) for a review).

Following the work of Bobeck (1967), magnetic garnets were inducted into another major application, namely as the medium for magnetic-bubble devices for high-density storage of information. Magnetic bubbles are domains of cylindrical shape, occurring in magnetized thin films of the material. One associates a binary number 1 or 0 with the presence or absence of a bubble at a pre-defined location and time. Nielsen (1976), Malozemoff & Slonczewski (1979), and Leeuw et al. (1980) have reviewed the use of various garnets for this application.

Permanent Magnets

Permanent magnets find a large number of uses in modern science and technology. Carbon steel was widely used for making permanent magnets till the development of 'alnico' magnets, following the work of Mishima (1932) on the alloy AlNiFe. The coercive field for alnico is double that of the materials used earlier. It also has much better mechanical hardness. This was achieved by special thermal treatment, leading to the precipitation of a finely dispersed second phase.

Further research for the development of better permanent magnets, particularly at the Bell Laboratories (Nesbitt et al. 1966), led to the emergence of rare-earth transition-metal compounds (Strnat et al. 1966). The best known material in this category is $SmCo_5$.

Theory of Magnetic Symmetry

The notion of black-and-white symmetry was introduced by Speiser (1927) and Weber (1929). It was developed further by Heesch (1929, 1930), who derived the 122 point groups used at present for describing the magnetic symmetry of crystals (*Heesch groups*). The work of Heesch, however, was of a rather abstract nature, and it was left to Shubnikov (1951) to interpret these groups as involving the operations of *antiequality* or *antisymmetry*. In due course it was realized that the antiequality operation can also serve, for many purposes, as the time inversion operation, and the groups were rederived as magnetic point groups (Landau & Lifshitz 1958; Tavger & Zaitsev 1956).

The derivation (Zamorzaev 1953) of the 1651 space groups of magnetic symmetry (*Shubnikov groups*) owes much to the introduction of the time-inversion operation in the original edition of the book by Landau & Lifshitz (1958).

1.2.2 Critical-Point Phenomena

Processes occurring in the vicinity of the 'critical point' are called *critical-point phenomena*, or *critical phenomena*. The term *critical point* was introduced by Andrews (1869), who pioneered the study of critical phenomena by his extensive measurements on the carbon-dioxide system (see Stanley 1971). He established that there exists a critical point (T_c, P_c, ρ_c) in the phase diagram of this system at which the identities of the liquid phase and the gaseous phase merge into a single phase. The theoretical rationalization of the critical point was provided by van der Waals (1873) through his famous equation of state. The idea of *universal behaviour* also emerged from this work when it was found that a universal equation of state could be derived for practically all gases by working with reduced parameters

$(T/T_c, \rho/\rho_c)$.

In the vicinity of the critical point a fluid system undergoes large density fluctuations, giving rise to the phenomenon of *critical opalescence* (anomalous scattering of light). A quantitative explanation of this phenomenon was provided by Einstein (1910), who derived an expression for the mean-square density fluctuation in terms of the isothermal compressibility of the fluid.

The similarity of the critical phenomena in fluids and in ferromagnetic materials was recognized by Curie around 1895. If we take magnetic field as corresponding to specific volume V_c, then the paramagnetic and the ferromagnetic phases can be taken as corresponding to the gaseous and the liquid phases respectively. A common feature of critical phenomena is that a certain suitably defined quantity (now called the *order parameter*) is zero above the critical temperature, and nonzero below it. For a ferromagnetic system, the spontaneous magnetization is such a quantity normally, and for the liquid-gas system it is the density difference $\rho_L - \rho_G$.

The analogy between fluids and magnetic systems was carried further by Langevin, who derived a statistical-mechanical equation of state for magnetic systems which described the response of such systems to magnetic field.

As mentioned earlier, Weiss (1907) postulated the presence of an internal mean magnetic field (in analogy with the presence of an internal pressure in fluids). This field represents the long-range interaction among the magnetic spins. On substitution of this field into the magnetic equation of state derived by Langevin, the well-known Curie-Weiss law for magnetic susceptibility was obtained.

Apart from fluids and magnetic systems, other systems in which critical phenomena were investigated quite early were binary alloys like CuAu. In such systems there is a random mixing of the two metallic elements at temperatures above T_c, but a separation and ordering into two phases below T_c. Just like susceptibility for ferromagnets and isothermal compressibility for fluids, the specific heat of the alloy becomes anomalously large in the vicinity of T_c. Bragg & Williams (1934, 1935a,b) defined an order parameter as the difference of the concentrations of, say, Au in the two phases that exist below T_c. They formulated a mean-field theory similar to the Weiss model for ferromagnets, and showed that the order parameter should decrease continuously on raising the temperature towards T_c.

The concept of the order parameter was generalized by Landau (1937a, b, c, d) to cover all continuous phase transitions in solids. Landau wrote a Taylor expansion of the Gibbs free-energy density in powers of the order parameter. Such an expansion assumes analyticity of the free energy in the neighbourhood of T_c. However, it is exactly in this region that the critical fluctuations can be dominant. The Landau theory of continuous phase transitions is thus not always valid in a certain neighbourhood of the critical point, although outside this region, and especially for phase transitions mediated by long-ranged interactions like strain, the theory offers a good explanation of phase transitions.

The fact that interactions underlying critical phenomena are often not long-ranged was realized quite early. A simple statistical-mechanical model for ferromagnetic phase transitions involving only a nearest-neighbour classical interaction was proposed in 1925 by Ising. In this model one assumes a spin variable on each lattice site that can take only two values: +1 (up) or -1 (down). The spins were assumed to interact via an exchange interaction which gives a lower energy when the spins are parallel, than when they are antiparallel. A 1-dimensional solution of this model by Ising (1925) did not predict any phase transition. An exact solution of the Ising model in two dimensions was given by Onsager (1944). A model wherein spins on lattice sites were permitted freedom of 3-dimensional orientation was formulated by Heisenberg (1928).

In the vicinity of the critical point large and correlated fluctuations of the order parameter can occur. These are ignored by a mean-field theory like the Landau theory, which therefore predicts only a discontinuity in the specific heat at T_c for a system of any number of spatial dimensions. Experimentally, however, a divergence, or at least a logarithmic divergence, is observed, as predicted by the solution of the 2-dimensional Ising model by Onsager (1944). Experimental and theoretical evidence mounted in favour of the general conclusion that, because of the large correlation length of the order parameter in the vicinity of a continuous phase transition, the critical-point exponents for the various thermodynamic quantities are quite insensitive to the small-scale details of the system, but depend on features such as the effective spatial dimension and symmetry of the system, and on the dimensionality and symmetry of the order parameter. In particular, experimental data pointed to the existence of *scaling*, as was first conjectured by Widom (1965). It was observed that near the critical point the free-energy density and other thermodynamic functions like susceptibility are "homogeneous functions" of temperature and of the field conjugate to the order parameter. Consequently, not all the critical exponents can be independent. In particular, the correlation length provides a scale for distances. And its critical exponent is related to the critical exponents for other thermodynamic quantities, almost irrespective of the value of the critical temperature and the nature of the atomic interaction involved.

It was also observed, on the other hand, that for a given spatial dimension the critical exponents depend sensitively on the symmetry of the order parameter. And the critical exponents are practically the same (or universal) for a given symmetry of the order parameter. This is referred to as the phenomenon of *universality*.

A rationalization of scaling in critical phenomena was provided by Kadanoff (1966). Taking note of the fact that the correlation length becomes arbitrarily large at the critical point, and thus allows for averaging over smaller lengths, he introduced 'block spins' (for the case of the ferromagnetic transition) defined by the same Hamiltonian as the actual spins. This *coarse-graining* led to a scaling of the free energy. In the so-called *Kadanoff construction* the interaction of the actual and the block spins via the same Hamiltonian was assumed, but not proved. Moreover, Kadanoff did not describe the procedure for actually calculating the critical exponents.

It was against such a background that K. G. Wilson formulated, in the early 1970s, his celebrated renormalization-group (RG) theory for systems possessing a whole range of length scales or time scales, in particular for critical phenomena (see Wilson (1977) for a general description of this approach). He made use of Kadanoff's coarse-graining approach (but in momentum space), and demonstrated that descriptions differing in length scales possess a certain symmetry, namely the symmetry of the renormalization group.

The RG theory has been applied with great success, not only to critical phenomena, but also to several other fluctuation-dominated systems.

1.2.3 Ferroelectric Materials

The term "ferroelectric" was first used by Erwin Schrodinger in 1912, although the history of ferroelectricity (or what was earlier called *Seignette-electricity*) can be said to have started around 1665 when Elie Seignette of La Rochelle, France, created "sel polychreste", later known as Rochelle salt (see Busch 1991). However, it was not until 1920 that Joseph Valasek (1920, 1921) demonstrated that the direction of spontaneous polarization of Rochelle salt (sodium potassium tartrate tetrahydrate, $NaKC_4H_4O_6.4H_2O$) could be reversed by the application of an electric field. He also observed hysteretic behaviour between 255 K and 297 K, as well as very large dielectric and piezoelectric coefficients.

Pyroelectricity is a phenomenon quite closely connected to ferroelectricity; all ferroelectrics are pyroelectrics (although the converse is not necessarily true). In the 19th century several studies were carried out to understand pyroelectricity (e.g. Brewster (1824); Gaugain (1856a,b)). A historical review of this work has been given by Lange (1974). Such investigations led eventually to the discovery of piezoelectricity by Jacques and Pierre Curie (1880a,b).

For well over a decade, Rochelle salt remained the only known example of a ferroelectric. Ferroelectricity was therefore taken as a rare phenomenon, and (erroneously) attributed to dipolar interaction between water molecules present in Rochelle salt. However, this perception changed when, between 1935 and 1938, a whole series of new ferroelectrics, namely phosphates and arsenates of potassium, were produced (Busch & Scherrer 1935; Busch 1938). The best known member of this family of crystals is potassium dihydrogen phosphate, KH_2PO_4, better known as KDP. This family of ferroelectrics has a much simpler crystal structure than Rochelle salt, and does not have any water of crystallization. The crystal structure involves intermolecular O-H$\cdots$O hydrogen bonding, and the hopping or quantum tunneling of hydrogen between the donor and acceptor oxygens results in different orientations of the $(H_2PO_4)^-$ dipoles. This fact formed the basis of the first microscopic theory of ferroelectricity, given by Slater (1941).

Another decade was to pass before ferroelectricity was discovered in any other type of crystal structure, and this property therefore continued to be viewed as a rarity, perhaps requiring hydrogen bonding as an essential ingredient for its existence. This notion was discarded when $BaTiO_3$, which has a very simple structure (Megaw 1945) with no hydrogen bonding, was found to be a ferroelectric (Wainer & Solomon 1942; Wul 1945, 1946; Wul & Goldman 1945a,b, 1946). Several other members of this structural family were soon shown to be ferroelectric: $KNbO_3$ and $KTaO_3$ (Matthias 1949); $LiNbO_3$ and $LiTaO_3$ (Matthias & Remeika 1949); and $PbTiO_3$ (Shirane, Hoshino & Suzuki 1950).

On the theoretical front, four years before Slater (1941) gave his theory of ferroelectricity in KDP, Landau (1937a, b, c, d) had put forward his phenomenological theory of phase transitions. Landau's theory had a thermodynamic aspect and a symmetry aspect (see Landau & Lifshitz 1958). In fact, he was the first to apply group-theoretical ideas to thermodynamics. According to the Landau theory, a symmetry-lowering continuous phase transition is heralded by the emergence of an order parameter, such that the symmetry of the daughter phase is given by the intersection group formed from the symmetry group of the parent phase and the symmetry group of the order parameter (in accordance with the Curie principle). One can work out, among other things, the temperature dependence of several macroscopic properties like the susceptibility of the crystal undergoing the phase transition by expanding its free energy as a Taylor series in powers of the order parameter, and then minimizing the free energy with respect to the order parameter, assuming that the coefficient of only the leading term in the Taylor expansion has a significant temperature dependence. In the context of ferroelectric phase transitions, Hans Mueller (1935, 1940a, b, c, d), in his "interaction theory", applied a similar approach to Rochelle salt, including in the free energy expansion a term corresponding to the strain arising from electrostrictive coupling to the spontaneous polarization. This theory was developed further and applied to the phase transitions in $BaTiO_3$ by Ginzburg (1945, 1949) and Devonshire (1949, 1951,

1954); also see Megaw (1952, 1957). Ginzburg (1946) applied the Landau theory to KDP-type crystals as well. Kittel (1951) extended the ideas to antiferroelectrics.

A major breakthrough in the microscopic understanding of structural phase transitions came during 1958-1960 when Cochran (1959, 1960) and Anderson (1958, 1960) provided an interpretation for them in terms of lattice dynamics, putting forward the notion that the "softening" of a lattice mode of vibration was responsible for a structural phase transition. It is often quite straightforward to identify the order parameter of the Landau theory with an appropriate soft mode driving the phase transition. The soft-mode theory, when applied to the ferroelectric phase transitions in $BaTiO_3$, was a vast improvement over the "rattling titanium ion" model proposed earlier by Slater (1950).

Although the Landau order parameter of a phase transition is not always a lattice-dynamical soft mode, the overlapping physics of these two concepts is worth emphasizing. As we shall see in more detail in Chapter 5, the functional dependence of the free energy of a crystal on the order parameter develops a "flat bottom" at the transition temperature for a continuous phase transition (cf. Fig. 5.6.1). This implies that large excursions (critical fluctuations) of the order parameter around its mean value become possible because of the vanishingly small change of free energy entailed by them. Also implied in this is the fact that the restoring force for these fluctuations is very small in the vicinity of the phase-transition temperature. The order parameter corresponds to a particular "normal mode", and this mode may be either underdamped or overdamped. If it is underdamped, its frequency of vibration would be low near T_c. On the other hand, if it is overdamped, there would be a large increase in the relaxation time (also referred to as *critical slowing down*). The soft-mode concept has now been generalized to cover both these possibilities.

Raman & Nedungadi (1940) and Saksena (1940) were the first to report the occurrence, for the $\alpha \leftrightarrow \beta$ phase transition in quartz, of what is now called a *soft mode*. Mason (1947, 1949), who viewed a ferroelectric transition as an order-disorder transition, proposed a mean-field type of theory which, on hindsight, appears similar to the notion of critical slowing down put forward by Landau & Khalatnikov (1954). Mention must also be made here of the book by Frohlich (1949) (first edition), wherein the Lyddane-Sachs-Teller equation was invoked to argue that the divergence of the dielectric susceptibility can be linked to an approach to zero frequency of a transverse-optical mode of lattice vibration. However, although the academic air was surcharged with the soft-mode idea for quite some time, none of the workers before Anderson and Cochran emphasized adequately the importance of this concept.

A phase transition can be a ferroelectric phase transition even when spontaneous polarization is not the order parameter. This was first pointed out by Indenbom (1960a,b), and materials which undergo such transitions are now referred to as *improper ferroelectrics* or *faint ferroelectrics* (Levanyuk & Sannikov 1969, 1971, 1974, 1975; Pytte 1970; Dvorak, 1970, 1971, 1974; Aizu 1972b,c, 1973b; Kobayashi, Enomoto & Sato 1972). Gadolinium molybdate (GMO for short), $Gd_2(MoO_4)_3$, was probably the first improper ferroelectric to be investigated experimentally (Cross, Fouskova & Cummins 1968; Cummins 1970). Since polarization is not the order parameter for the ferroelectric transition in this crystal, its dielectric susceptibility does not obey the Curie-Weiss law in the paraelectric phase; in fact, it displays only a slight variation with temperature in the entire temperature range investigated. Several other materials are now known to be improper ferroelectrics. These include many rare-earth molybdates, boracites, dicadmium diammonium sulphate, and ammonium fluoroberylate (see Levanyuk & Sannikov (1974)).

1.2.4 Ferroelastic Materials

The word "ferroelasticity" has had a number of connotations, all similar, and yet not quite the same. It was used in physical metallurgy in the early 1950s by F. C. Frank for describing the rubber-like (martensitic) behaviour of $Au_{1.05}Cd_{0.95}$ and InTl alloys (see Lieberman, Schmerling & Karz (1975)).

In physics, in the pre-1969 period, ferroelastic strains were associated primarily with ferroelectricity, through electrostriction. (Magnetostriction also leads to strain in ferromagnetic materials, but the effect is usually much smaller than that of electrostriction.) Thus, before Aizu (1969a, 1970a) gave a formal definition of ferroelasticity as a property which can exist on its own, physicists' perception was that it occurs predominantly as an adjunct to ferroelectricity or ferromagnetism. In fact, even after the publication of the work of Aizu (1969a), reluctance to recognize ferroelasticity as an independent property persisted for some years among some physicists who were experts on ferroelectrics.

Mention must also be made here of the work of Indenbom (1960) and Boccara (1968) who, though not using the term "ferroelastic", investigated in substantial depth phase transitions having spontaneous strain as the order parameter.

Progress in the symmetry description of ferroelastic materials was preceded by that on ferroelectric materials. The domain structure of ferroelectrics had been understood in group-theoretical terms, making use of the Curie principle (Zheludev & Shuvalov 1956, 1957). Symmetry considerations are often system-independent, and were readily extended to ferroelastic materials (Aizu 1970a,b,c; Janovec 1972; Cracknell 1972, 1974).

Aizu (1969a) not only introduced a formal definition of ferroelasticity, but also presented a unified symmetry description of ferroelectrics, ferroelastics, ferromagnetics, and secondary and higher-order ferroics (Aizu 1970a, 1973a). He did this after introducing the all-important concept of *prototype symmetry* (Aizu 1970a, 1975, 1978) mentioned in §1.1. The earlier literature was rather imprecise on the meaning of the terms "initial phase" and "parent phase". Introduction of the formally defined notion of prototype symmetry should be regarded as one of the major contributions of Aizu to the field of ferroic materials. However, it is also necessary to mention here that Aizu's definition of prototype symmetry uses the phrase "slight distortion", which is not quantified adequately. For certain situations this leaves room for ambiguity and lack of rigour. A rigorous definition of the prototype in terms of the nondisruption condition is being introduced in this book in §5.1.

Aizu (1969b, 1970c) also introduced the concept of *antiferroelasticity*, by analogy with antiferroelectricity.

In the same year in which Aizu (1969a) published his first paper on ferroelasticity, Alefeld et al. (1969) published an analysis of the ferroelastic behaviour of hydrogen-containing metals. Alefeld et al. (1969, 1970, 1971) were investigating the Nb – H system, in which the hydrogen atoms occupy interstitial sites in the Nb lattice. The hydrogen atoms act like point defects, creating local strain fields ("elastic dipoles"), as well as a macroscopic strain field. These elastic dipoles can undergo discrete reorientations under the action of an external uniaxial stress. This process, known as *Snoek relaxation* (see, e.g., Nowick & Heller (1963)), results in stress-strain hysteresis that is characteristic of a ferroelastic material. Further, the interstitial hydrogens modify the elastic constants of the crystal, and a Curie-Weiss-type temperature dependence of the appropriate elastic-compliance coefficients ensues (Alefeld et al. 1969).

Obviously, if these point defects are not present, the related ferroelastic behaviour of the host crystal would disappear. Aizu's (1969a, 1970a) formalism of ferroelasticity is different from, and more general than, the treatment given by Alefeld, in that ferroelastic behaviour envisaged by Aizu exists even in the absence of point defects, as it is a consequence of the pseudosymmetry resulting from the phase transition from the prototype structure.

An interesting situation arises in certain crystals (e.g. $YBa_2Cu_3O_{7-x}$, popularly known as Y – Ba – Cu – O) in which Aizu's and Alefeld's formulations overlap. This happens when a part of the structure itself (e.g. the basal-plane oxygen atoms in Y-Ba-Cu-O) behaves like point defects (Wadhawan 1989; Wadhawan & Bhagwat 1989). Such crystals have been termed *nonstoichiometric ferroelastics* (Wadhawan 1991).

1.2.5 Secondary and Higher-Order Ferroics

During World War II, when mostly natural quartz had to be used for transducer applications, several attempts were made to obtain twinning-free crystals of quartz by detwinning its Dauphiné twins through the use of twisting forces and thermal gradients (Wooster & Wooster 1946; also see Klassen-Neklyudova (1964) for additional references). However, these attempts met with only partial success, the reason being that they did not have the most appropriate theoretical basis. The right theoretical basis was provided much later by Aizu (1970a, 1972a, 1973a), and that too in a perfectly general way, covering all ferroics, and not just the secondary ferroic quartz. He did so by introducing the concept of prototype symmetry. The ease and the systematic nature of the approach with which one can now calculate the most appropriate direction in which to apply a field for detwinning any ferroic crystal should be regarded as one of the triumphs of the modern theory of ferroic materials. The crux of the matter is the realization of the fact that (to consider the case of quartz) the Dauphiné twins arise from the breaking of a certain prototype symmetry. The knowledge of the prototype and the ferroic symmetries enables us to write an exact expression for the difference in the free energies of the two twin states in terms of the relevant compliance and piezoelectric coefficients, from which it is straightforward to work out the optimum configuration for applying electric and/or mechanical fields for effecting detwinning (Indenbom 1960a; Aizu 1972a, 1973a; Anderson, Newnham & Cross 1977; Bertagnolli, Kittinger & Tichy 1979; Laughner, Wadhawan & Newnham 1981; Wadhawan 1982).

1.2.6 Ferrogyrotropic Materials

Fig. 1.2.1 shows the typical hysteretic behaviour of, respectively, a first-rank polar, a first-rank axial, and a second-rank polar tensor property, corresponding to ferroelectricity, ferromagnetism, and ferroelasticity. Can there be a fourth primary ferroic property, for which a second-rank axial tensor property shows hysteresis? The answer involves both a 'yes' and a 'no'.

Let us begin by noting that the optical gyration tensor represents a candidate property for this possibility. Aizu (1970a) did not include it in his general formalism for ferroic materials. Gyrotropic phase transitions were analysed by Konak, Kopsky & Smutny (1978) and by Wadhawan (1979). Wadhawan's analysis was mainly from the point of view of ferroic switching between orientation states differing in their gyration tensors. In his manuscript, he proposed the term *ferroenantiomorphism* for the property whereby orientation states differing in their gyration tensors could be switched from one to another. However, he had to settle for the term *gy-*

rotropy, because the referee thought that the word 'ferroenantiomorphism' was too long for comfort ! Later (Wadhawan 1982) the term 'gyrotropy' was changed to *ferrogyrotropy* to emphasize the requirement of switchability between orientation states.

It was also pointed out by Wadhawan (1979, 1982) that optical ferrogyrotropy is only an *implicit* form of ferroicity. What this means is that although ferrogyrotropic state shifts are a reality (see Konak et al. (1978) for a review), there is no contribution from the gyrotropy term to the change of free energy accompanying such state shifts. Therefore, such state shifts can occur only as an adjunct to some other, *explicit*, type of ferroic state shifts.

Optical activity arises from certain peculiarities of the crystal structure (Devarajan & Glazer 1986) which result in a spatial dispersion of the dielectric tensor. Similarly, acoustical activity arises from the spatial dispersion of the elastic-stiffness tensor. The notion of optical ferrogyrotropy was therefore extended to that of *acoustical ferrogyrotropy* (Wadhawan 1982). Acoustical ferrogyrotropy, or *ferroacoustogyrotropy*, is described by a fifth-rank tensor. It was shown by Bhagwat et al. (Bhagwat, Subramanian & Wadhawan 1983; Bhagwat, Wadhawan & Subramanian 1986) that, because of its intrinsic symmetry, this tensor is completely equivalent to a fourth-rank tensor. It was also shown by these authors that an earlier description of a similar fourth-rank tensor was, in fact, in error.

In Chapter 6 we shall see that when it comes to the application of the Hermann theorem for transverse isotropy in crystals, it is important to work with the *lower*-rank tensor for drawing correct conclusions from this theorem about optical or acoustical activity.

SUGGESTED READING

V. A. Koptsik (1968). A general sketch of the development of the theory of symmetry and its applications in physical crystallography over the last 50 years. *Sov. Phys. - Cryst.*, **12**, 667.

H. E. Stanley (1971). *Introduction to Phase Transitions and Critical Phenomena.* Oxford University Press, Oxford.

M. E. Lines & A. M. Glass (1977). *Principles and Applications of Ferroelectrics and Related Materials.* Clarendon Press, Oxford.

U. Enz (1982). Magnetism and magnetic materials: Historical developments and present role in industry and technology. In E. P. Wohlfarth (Ed.), *Ferromagnetic Materials*, Vol. 3. North-Holland, Amsterdam.

S. Chikazumi (1991). Progress in the physics of magnetism in the past forty-five years. In Y. Ishikawa & N. Miura (Eds.), *Physics and Engineering Applications of Magnetism*. Springer-Verlag, Berlin.

Chapter 2

CRYSTALLOGRAPHY

> . . . *What immortal hand or eye,*
> *Could frame thy fearful symmetry ?*
>
> William Blake, *The Tiger*

Although a variety of noncrystalline solids exist (glasses, polymers), a basic knowledge of the crystalline state is necessary for a better understanding of the condensed state of matter, if only because the properties of crystals often serve as *benchmarks* for comprehending the properties of noncrystalline materials. In this chapter we review briefly the geometrical and symmetry aspects of the structure of perfect single crystals. It is also instructive to discuss the question of how a crystal is formed from its building blocks. Many of the concepts used in the theories of crystal growth are also relevant to the growth of a ferroic phase in the surrounding matrix of the prototypic phase or the parent phase. The switching of ferroic domains by driving fields is also akin to crystal growth in many respects. The chapter ends with a brief description of the structural features of incommensurately modulated crystals.

2.1 GROWTH OF A CRYSTAL

> *Here and elsewhere we shall not obtain the best insight into things until we actually see them growing from the beginning.*
>
> William Aristotle, *Politics*

In a crystal there is a regular and repetitive arrangement of atoms or molecules. One can identify a certain basic motif or building block, the *unit cell*, which occurs again and again in a space-filling fashion along three noncoplanar (and pairwise noncollinear) directions.

The nucleation and growth of a crystal is a process of phase transition. The transition can occur from a vapour phase to the solid phase, or from a liquid phase to the solid phase, or from one solid phase to another. The last-mentioned case will be taken up in detail in Chapter 5. We discuss here some basic features of the formation of crystals from the vapour state or the liquid state.

2.1.1 Nucleation

The creation of an ordered arrangement of atoms or molecules when crystallization occurs from a random or nearly random configuration of the liquid or vapour state entails a large change (lowering) of entropy. This is therefore necessarily a first-order phase transition.

Fig. 1.1.1 in Chapter 1 depicts a necessary condition for the occurrence of a phase transition, namely a lowering of the Gibbs free energy, G. In the absence of external electric, magnetic or uniaxial-stress fields, G is defined for a multicomponent system as

$$G = U - TS + pV + \mu_i n_i \qquad (2.1.1)$$

Here U denotes the internal energy, T the temperature, S the entropy, p the ambient hydrostatic pressure, V the volume, and n_i the number of moles of the ith component having the chemical potential μ_i (summation over the repeated index i is assumed). For highlighting the role of the entropy term in the crystallization process, we rewrite the above expression as

$$G = H - TS, \qquad (2.1.2)$$

with H denoting the enthalpy.

The coming together of atoms or molecules to form a crystal is a dynamic process. Let us call the crystalline phase as Phase 2, and the mother phase from which the crystal is to be grown as Phase 1. One can view the process as a two-way chemical reaction, with Fig. 2.1.1 depicting a typical free-energy vs. reaction-coordinate diagram. The reaction coordinate can be, for example, the radius (or some other typical size) of the growing crystal. The reaction $1 \to 2$ involves the surmounting of an activation-free-energy barrier ΔG^*_{12}, with ΔG^*_{21} playing a similar role for the process $2 \to 1$. Thermal fluctuations provide the necessary energy for overcoming these barriers.

The net rate J at which the atoms or molecules aggregate into the solid phase can be written as a Boltzmann relation:

$$J \sim N_1 \exp(-\Delta G^*_{12}/RT) - N_2 \exp(-\Delta G^*_{21}/RT) \qquad (2.1.3)$$

Here N_1 and N_2 are the numbers of molecules in Phase 1 and Phase 2 respectively.

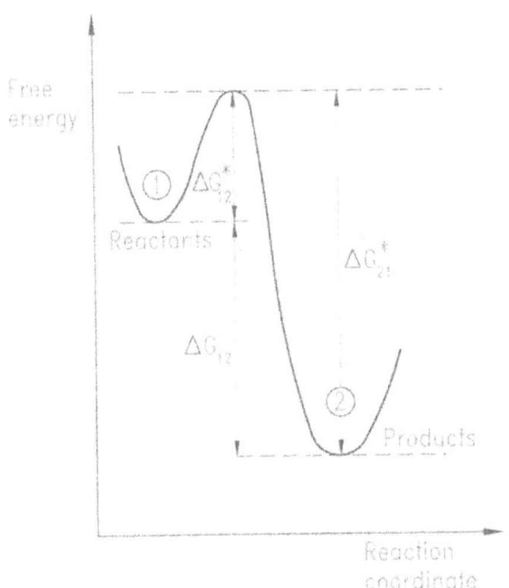

Figure 2.1.1: Illustration of the concept of activation free energy for a chemical reaction or phase transformation. [After Brophy, Rose & Wulff (1965).]

When thermodynamic equilibrium prevails, we have $J = 0$, giving

$$N_2/N_1 = \exp(\Delta G_{12}/RT), \tag{2.1.4}$$

where

$$\Delta G_{12} = \Delta G_{21}^* - \Delta G_{12}^* \tag{2.1.5}$$

An experiment designed for the crystallization of a material must create conditions (supersaturation, supercooling, etc.) so that the activation barrier ΔG_{12}^* can be overcome. Gibbs (1876-78) was the first to study such matters in detail, and his work led to the recognition of a very important aspect of first-order phase transitions, namely *nucleation.* We can understand the necessity for nucleation by rewriting Eq. 2.1.3 with $N_2 = 0$, and substituting for ΔG_{12}^* from Eq. 2.1.2:

$$J \sim N_1 \exp(\Delta S^*/R) \exp(-\Delta H^*/RT) \tag{2.1.6}$$

The activation entropy for the crystal-growth process, as also for any other first-order phase transition, is thus seen to be negative, and of extremely large magnitude. (This is in sharp contrast to the situation in a second-order phase transition, for which $\Delta S^* = 0$.)

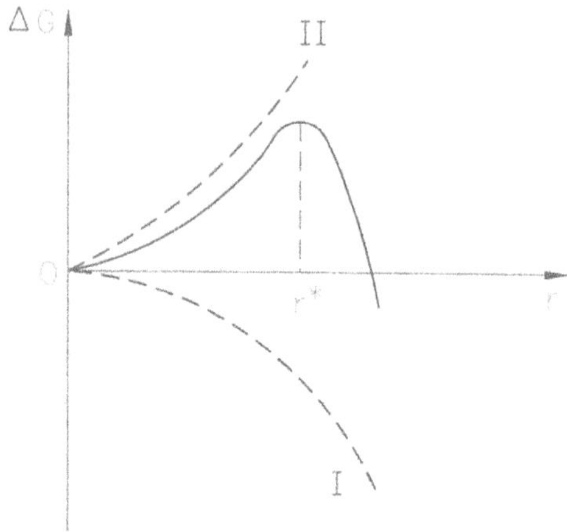

Figure 2.1.2: Dependence of the bulk free energy (Curve I) and the surface free energy (Curve II) on crystal radius r, and the resultant overall variation of ΔG described by Eq. 2.1.7.

Therefore the change of phase must begin on a very small scale, with the formation of small nuclei of the new phase, followed by the gradual growth of these nuclei (if the conditions are favourable).

Let us assume, for the sake of simplicity of further discussion, that the nucleus of the crystal to be grown is of spherical shape, with a radius r. The so-called *Gibbs work of nucleation* can then be written as

$$\Delta G = -4\pi r^3 \, \Delta\mu \,/\, (3v) \, + \, 4\alpha\pi r^2, \tag{2.1.7}$$

with

$$\Delta\mu = \mu_v - \mu_c \tag{2.1.8}$$

Here μ_v and μ_c are the chemical potentials for the vapour phase and the condensed phase, v is the molecular volume, and α the surface energy per unit area.

The two terms in Eq. 2.1.7 have opposite signs, and their variation with r goes as r^3 and r^2 respectively (Fig. 2.1.2). There is thus a critical value of r $(= r^*)$, below which $\partial(\Delta G)/\partial r$ is positive, implying that the system can lower its free energy by *redissolving* the newly formed nucleus. However, if somehow (for example through thermal fluctuations) the radius of the nucleus can be greater than r^*, it will grow into a larger and larger crystal because the $\Delta G(r)$ function decreases with increasing r in this regime.

As we shall see in more detail in Chapter 5, such considerations are applicable to practically all first-order phase transitions. Typical features of such transitions are:

- Coexistence of the two phases in a certain temperature region.
- The consequent occurrence of *interfaces* separating the two phases.
- The existence of a *nucleation barrier.*

It is rather easy to visualize these three features for a crystalline phase growing in a fluid matrix. But the three features may also be present in a first-order ferroic phase transition. The main difference is that in a ferroic transition the new phase is surrounded by a rigid and anisotropic crystalline matrix of the parent phase. This fact has several important consequences, including an alteration of the net symmetry of the new phase (cf. §8.1), as well as a volume-dependent restoring force tending to restrain the transition. This volume-dependent term is quite different from, and is in addition to, the surface-energy term included in Eq. 2.1.7.

2.1.2 The Cluster-to-Crystal Transition

At the beginning of the formation of the nucleus, i.e. when r in Eq. 2.1.7 is only a few atomic or molecular dimensions across, the surface-energy term is very dominant. In fact, for such small dimensions, $\Delta\mu$ and α in this equation cannot even be taken as well-defined constants (Mutaftschiev 1993). One also has to make a distinction between what may properly be called *large molecules* on the one hand, and what have come to be known as *clusters* on the other. The term clusters is used for aggregates of atoms that, unlike molecules, are not found in appreciable numbers in vapours in equilibrium (Martin 1988).

Small clusters undergo a process of *reconstruction*: Every time a unit (atom or molecule) of the crystallizing species attaches itself to the cluster, the units rearrange themselves completely. And the symmetries possessed by these changing clusters do not necessarily have any resemblance to the symmetry the bulk crystal would finally acquire. A stage comes in the gradually increasing size of the cluster when it no longer reconstructs drastically on the attachment of additional units. One then speaks of a *microcrystal*, or a *microcrystallite*, which now has the symmetry of the bulk crystal. A *cluster-to-crystal transition* can be said to have taken place at this stage (Sugano, Nishina & Ohnishi 1987; Jena, Rao & Khanna 1987; Benedeck, Martin & Pacchioni 1988; Multani & Wadhawan 1990; Haberland 1994). This is an example of a *size-induced phase transition.*

Clusters may consist of about 100-1000 growth units, or less. What is the structural symmetry possessed by them? A number of studies have been performed, mostly with spherical metallic and rare-gas atoms, to answer this question. One fact which emerges is that icosahedral symmetry is frequently favoured for stable configurations of small clusters (Hoare 1979). Gold clusters, for example, have icosahedral symmetry in the 4-15 nm size regime (Renou & Gillet 1981).

As we shall see later in this chapter, icosahedral symmetry is not compatible with the translational periodicity of a bulk single crystal. Therefore, at the microcrystallite stage a major transition must occur to bulk-crystalline symmetry, sometimes involving multiple twinning as an adjustment mechanism[1] (Mackay 1962; Senechal 1986; Ajayan & Marks 1990; Riley 1990). Gold clusters change from icosahedral symmetry to face-centered-cubic (fcc) symmetry on reaching a size of ~15 nm (Renou & Gillet 1981).

Certain *magic numbers* of atoms in some clusters make them exceptionally stable, and almost spherical in shape (Mackay 1962; Hoare & Pal 1972). The packing of atoms in many such cases has icosahedral symmetry, often changing to dodecahedral or cubic configurations on increase of the particle size (Kimoto & Nishida 1977; Echt, Sattler & Recknagel 1981; Mort la Brecque 1988).

We shall return to this topic when we discuss size effects in ferroic materials in Chapter 13.

2.1.3 Growth Mechanisms

How does the nucleus grow into a larger and larger crystal, once it has crossed the free-energy activation barrier corresponding to the critical size r^* (Fig. 2.1.2)? As mentioned in §2.1.1, the coexistence of the growing crystal with its mother phase involves the presence of an interface. This interfacial region is where the various crystal-growth processes occur. The growth units come and get attached on the interface, from where they may or may not get knocked off back into the mother phase, depending upon the sites where they got attached. The net rate at which the interface advances into the mother phase, and therefore the net growth rate of the crystal, depends on several parameters, including supersaturation or supercooling of the solution or melt. To explain the observed growth rates of crystals, the so-called *terrace-ledge-kink (TLK) model* was evolved by Stranski and Volmer during the 1920's (see, for example, Givargizov (1991) for a review). Fig. 2.1.3 shows the situation schematically, where the growth units are taken as cubes for simplicity of explanation. A cube attached to Site 1 (a *terrace* site) is bonded to the growing crystal across only one of its six faces, and is therefore quite likely to be blown off by the thermodynamic

[1] As we shall discuss in §11.6.5, in a realistic modeling of a ferroelastic phase transition, multiple twinning must be incorporated as an integral part of the phase-transition process itself.

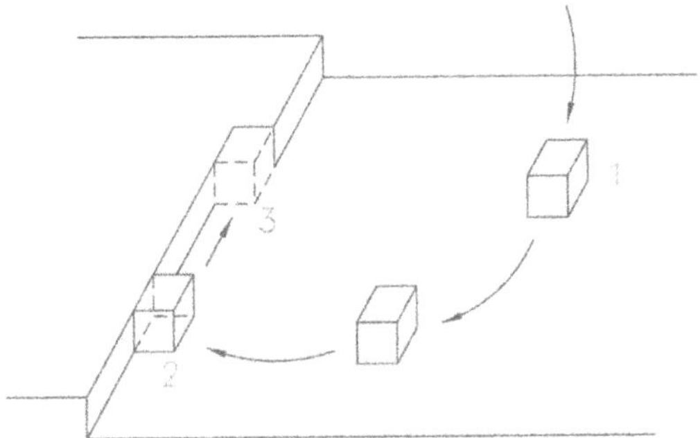

Figure 2.1.3: Illustration of the various types of attachment sites for the growth units (taken as cubelets). Site 1 is a terrace site, Site 2 a ledge site, and Site 3 a kink site. [After Sangwal & Rodriguez-Clemente (1991).]

fluctuations. But it may also wander on the terrace and succeed in reaching a *ledge* site (Site 2), where the bonding is stronger, being on two of the six faces. The strongest bonding, of course, will occur if it reaches a *kink* site (Site 3), where as many as three of the six faces are bonded to the growing crystal. An important feature of this model is that the kink sites are self-regenerative: after a unit attaches at Site 3, the resulting configuration is once again a kink site, ready to receive and bind strongly another growth unit.

What happens when the entire ledge has grown end to end, and so also the entire terrace by the successive addition of ledges (or *steps*) to it? The crystal will have to start and sustain another layer of growth by beginning with a terrace site. This would require a rather high degree of supersaturation or supercooling of the mother phase. The growth rates predicted by this model for given degrees of supersaturation were therefore found to be far too low compared to experimental observations.

This situation was remedied by the celebrated BCF theory (Burton, Cabrera & Frank 1951), which, among other things, explained the observed high rates of crystal growth by invoking the role of screw dislocations for providing a spiralling (and therefore never-ending) growth ledge. Crystal growth can advance indefinitely on the helicoidal surface of the screw dislocation by the addition of growth units at the never-ending source of kink sites. Thus the need for surface nucleation is eliminated or reduced. Because of the anchoring of the ledge at the emergence point of the screw dislocation, the ledge winds up into a growth spiral. Experimental confirmation of this model was provided by Verma (1951a, 1951b, 1953).

We shall see in §10.7.1 that the kinetics of domain switching in ferroelectrics has several features in common with crystal growth from a fluid phase.

The Roughening Transition

The phenomenon of the growth of a crystal involves a large ensemble of atoms or molecules. Extensive use of statistical mechanics for dealing with it has therefore been made from early times (see, for example, Chernov (1984) and van der Eerden (1993) for reviews). A variety of Ising models have been employed for describing the processes occurring in the interface between the mother phase and the growing crystal. The interface is partitioned into equal cells, each cell being either in the solid state or fluid state. One of the widely used models is the so-called *solid-on-solid (SOS) model* (Temkin 1964). In this model, the material below the interfacial region is assumed to be completely crystallized, and that above the interface is assumed to be completely fluid. Within the interface, solid cells can sit on top of solid cells, but overhangs (i.e. solid cells on top of fluid cells) and vacancies (i.e. occurrences of one state in the other) are disallowed. One can thus define a site variable, h_i, which can take all integral values between $-\infty$ and $+\infty$ with respect to a reference plane, and which is a measure of the heights of towers of solid cells.

At $T = 0$, $h_i = 0$ for all i, corresponding to a completely planar interface. As T increases, fluctuations of the site occupancy occur more and more, and the interface becomes increasingly rough. The general expression for the Hamiltonian has the form

$$\mathcal{H} = 2J \sum_{<i,j>} |h_i - h_j|, \tag{2.1.9}$$

where $< \cdots >$ denotes the thermodynamic expectation value, and the sum runs over all pairs of nearest-neighbour sites (in a two-dimensional lattice). For a two-level model ($h_i = \{0, 1\}$), the Hamiltonian becomes the same as that for a two-dimensional Ising model:

$$\mathcal{H} = -J \sum_{<i,j>} S_i S_j, \tag{2.1.10}$$

$$S_i = 2h_i - 1 \tag{2.1.11}$$

This model predicts a second-order phase transition at a temperature T_c, below which $< h_i >$ is less than $\frac{1}{2}$, and above which $< h_i >= \frac{1}{2}$. Above T_c, a central layer can be identified in the interface, which has 50% solid cells and 50% fluid cells, and the nucleation barrier for two-dimensional nucleation does not exist.

One speaks of a *roughening transition* occurring at a temperature T_R which is equal to, or slightly above the critical temperature T_c of the two-dimensional Ising model. Long-ranged fluctuations of the site variable h_i exist for temperatures above T_R, which decay very slowly, with a power law. This means that, above T_R, large numbers of kinks can exist on the crystal surface, even without screw dislocations, leading to a rapid and non-facetted growth of such surfaces.

The value of T_R is different for different habit planes of the crystal. If, for a given temperature T of crystal growth, $T < T_R$ for a particular habit face, such a face of the crystal will grow as a predominantly smooth face. If $T > T_R$ for a face, the face will be atomically rough, and will grow rapidly, with a rounded appearance.

The corresponding situation in a first-order crystal-to-crystal phase transition is more complex. We shall discuss some aspects of phase boundaries and polydomain phases in ferroelastics in §11.6.5. The equivalent of roughening transitions on different facets of a ferroic phase surrounded by the parent-phase matrix does not appear to have been thoroughly investigated. However, it is interesting to note that in several crystals a ferroic phase transition is intervened by the occurrence of an incommensurate phase in a narrow temperature range (§2.4, 4.1.5, 5.8). One of the best investigated systems of this type is quartz (see Saint-Gregoire (1995) and Dolino & Bastie (1995) for recent reviews). The ferrobielastic $\beta \rightarrow \alpha$ transition in quartz actually has the sequence $\beta \rightarrow$ *(incommensurate phase)* $\rightarrow \alpha$. A spectacular high-resolution electron microscopy picture, showing all three phases at slightly differing temperatures, was published by Amelinckx et al. (1989) (see Fig. 28 their paper). Although the occurrence of incommensurate phases is normally explained in terms of "competing interactions", one cannot help noticing a certain amount of similarity between a roughening transition and an incommensurate transition, particularly in the "discommensuration" regime (cf. §5.8).

Unlike a roughening transition at a solid-fluid interface, an incommensurate phase has a periodic crystal structure on both ends, a fact which forces it to adopt a somewhat regular-looking structure. The "triangular 3q structures" observed in quartz are an example of this (see Fig. 2 of Dolino & Bastie 1995).

2.1.4 Crystal Morphology

The morphology a crystal adopts is determined by the relative rates of growth of the various faces. Faces which grow too fast disappear eventually, and the final shape is determined by the slowest-growing faces.

A very fruitful concept for understanding the growth morphologies of crystals is that of *periodic bond chains* (PBCs) (see Hartman (1987) for a review of its applications). The growth of a crystal can be viewed as a process of formation of bonds between the growth units. Mainly strong bonds, usually those in the first coordination sphere, are relevant in this context. One can represent a bond between two growth units as a line segment joining them, with the length of the segment taken as proportional to the binding energy. Because of the periodic nature of the structure of the crystal, the line-segments of a specific length and orientation will lie on uninterrupted periodic chains, namely the PBCs.

In terms of PBCs, three distinct types of crystal faces can be identified: *F-faces* (or flat faces), with at least two nonparallel sets of PBCs running parallel to them; *S-faces* (or stepped faces), with one set of PBCs running parallel to them; and *K-faces* (or kinked faces), with no PBC parallel to them.

K-faces are atomically rough, with kink sites all over them. Therefore they have the largest growth rates, and are thus seldom important in determining the final morphology a crystal adopts. S-faces have, generally speaking, lower growth rates than K-faces, but, because of the presence of steps or ledges on them, they may still grow fast enough to eventually disappear. There are several important exceptions to this, however.

F-faces generally have the lowest growth rates, and therefore the highest morphological importance. An F-face growing below its roughening transition temperature, T_R, will be smooth, practically at the atomic level, except for the effects of defects like dislocations.

For computing the theoretical growth morphologies of crystals, several criteria are in use for determining the relative rates of growth of competing F-faces (cf. Bennema & van der Eerden 1987). According to one such criterion, the growth rate is directly proportional to the energy released when a slice of the crystal parallel to the face under consideration gets attached to the growing crystal. Another criterion takes the growth rate as inversely proportional to the value of T_R for that face.

The morphology of a crystal reflects its internal symmetry. In particular, it is in conformity with its point-group symmetry. Implicit in this statement is the assumption that the crystal grows in an isotropic environment, which is generally the case for crystals grown from a fluid phase. But when a crystalline phase is obtained as result of a solid-to-solid phase transition, the new phase has to grow in the anisotropic environment provided by the parent phase. Because of the strain fields, as well as other factors which vary with direction, the morphology of the new phase (e.g. a ferroic phase growing inside a prototypic phase) is modified. This modification can be determined by applying the Curie principle. We discuss this in §7.5 and 8.1.

SUGGESTED READING

P. Jena, S. N. Khanna & B. K. Rao (Eds.) (1992). *Physics and Chemistry of Finite Systems: From Clusters to Crystals*, Vol. 1. NATO ASI Series, Kluwer, Dordrecht.

J. P. van der Eerden (1993). Crystal growth mechanisms. In D. T. J. Hurle (Ed.), *Handbook of Crystal Growth.* North-Holland, Amsterdam. Vol. 1, Chap. 6.

P. Bennema (1993). Growth and morphology of crystals. Integration of theories of roughening and Hartman-Perdock theory. In D. T. J. Hurle (Ed.), *Handbook of Crystal Growth.* North-Holland, Amsterdam. Vol. 1, Chap. 7.

Y. Saito (1996). *Statistical Physics of Crystal Growth.* World Scientific, Singapore.

C. M. Pina, U. Becker, P. Risthaus, D. Bosbach & A. Putnis (1998). Molecular-scale mechanisms of crystal growth in barite. *Nature*, **395**, 483.

A. Pimpinelli & J. Villain (1998). *Physics of Crystal Growth.* Cambridge University Press, Cambridge.

A. A. Chernov (1998). Crystal growth and crystallography. *Acta Cryst.*, A**54**, 859.

2.2 SYMMETRY OF A CRYSTAL

2.2.1 The Symmetry Group of a Crystal

The symmetry of a crystal can be discussed in terms of a *density function* $\rho(x, y, z)$, which defines the electron density at any point (x, y, z) due to all the atoms in the unit cell. Certain coordinate transformations (translations, rotations, inversions, and their permitted combinations) leave the density function of a crystal invariant. The set of all such symmetry transformations for a crystal forms a group called the *symmetry group of the crystal* (cf. §B.1).

The symmetry of a crystal can be described at progressively increasing levels of detail. Crystals can be classified into 7 crystal systems, 14 Bravais lattices, 32 point groups, and 230 space groups. Magnetic crystals have additional features of symmetry, which we shall describe separately in §2.2.18.

2.2.2 Translational and Rotational Symmetry

The repetitive (periodic) arrangement of atoms in a crystal can be visualized in terms of a *lattice* of points, with an atom or a group of atoms (called the *basis*) placed at each lattice point in an identical manner. The crystal lattice is only a mathematical concept, with no physical identity. It is an *infinite* array of points in space with the property that all such points have identical surroundings. In three-dimensional space a lattice can be defined in terms of three *primitive translation vectors*, such that any lattice vector, $\mathbf{r}$, can be expressed as

$$\mathbf{r} = n_1\mathbf{a}_1 + n_2\mathbf{a}_2 + n_3\mathbf{a}_3, \qquad (2.2.1)$$

where n_1, n_2, n_3 can take all possible positive or negative integral values (including zero), and $\mathbf{a}_1$, $\mathbf{a}_2$, $\mathbf{a}_3$ are three non-coplanar, pairwise-noncollinear, basic vectors which are said to *span* the lattice.

Lattice translations defined by Eq. 2.2.1 generate *equivalent points*. The set of all such points defines a *lattice*. A lattice has translational symmetry: it is invariant under lattice translations defined by Eq. 2.2.1.

A given lattice may also have *rotational symmetry*; that is, it may be also invariant under specific rotations and reflections. The rotational symmetry of the lattice defines the *crystal system*.

Crystal systems possess specific *point-group symmetries*, that is; they are invariant under coordinate transformations related to specific symmetry axes and symmetry planes, applied so as to keep at least one point in space invariant. This invariant point is referred to as a *singular point*.

2.2.3 Crystal Structure

A crystal structure can be generated by choosing an appropriate set of atoms (called the basis), and placing it in a fixed orientation at the tip of each of the lattice vectors given by Eq. 2.2.1. This repetition of a pattern of atoms can be described mathematically as a *convolution* (cf. Appendix D) of a basis function $B(\mathbf{r})$ and a lattice function $L(\mathbf{r})$ (Lipson & Taylor 1958; Burns & Glazer 1990):

$$C(\mathbf{r}) = B(\mathbf{r}) * L(\mathbf{r}) = \int B(\mathbf{r} - \mathbf{r}')L(\mathbf{r}')\, d\mathbf{r}' \qquad (2.2.2)$$

The function $C(\mathbf{r})$ represents the infinite crystal structure.

2.2.4 Point Space

Mathematically, a crystal lattice can be visualized as a subset of an infinite *point space*. Those transformations or mappings in this point space which preserve distances between points are called *isometries*, or *isometric mappings*, or *rigid motions.*

A set of points is said to be *discrete* if the distance between any two of these points is always larger than some minimum distance r.

A set of points is *relatively dense* if any sphere (in three dimensions) or a circle (in two dimensions) that does not contain any point of the set (i.e. the largest "empty hole") has a radius (R) less than some fixed value.

An *(r,R) system of points* is that which is both discrete and relatively dense.

A pattern of points including translations as symmetry operations is said to be *periodic.*

When the points of an (r, R) system are such that their configuration looks the same from any point of the set, the system is said to be *regular.* A crystal lattice is an example of a regular (r, R) system of points. All points of a regular (r, R) system are *equivalent.* It can be proved that regularity implies periodicity.

2.2.5 Symmetry Elements in a Crystal

If an object possesses symmetry, it implies that it has 'parts' (or 'asymmetric units') which are equivalent. The equivalence of parts means that rigid motions exist which can map any part to the original position of any other part, with the appearance of the object remaining unchanged. Such motions are the *symmetry operations.*

Certain symmetry operations in a crystal leave one or more points fixed during their action. The set of all such points constitutes the *symmetry element* for that symmetry operation. In 3-dimensional space, the symmetry element of a rotational symmetry operation is a straight line (the *rotation axis*). The symmetry element of a reflection operation is a *mirror plane.* And the symmetry element of an inversion operation is just a point (rather than a line or a plane).

2.2.6. Orbits; Stabilizers

If we consider a point $\mathbf{r}$, and apply to it successively the various operations of the symmetry group G of the crystal, we generate a set of points equivalent to $\mathbf{r}$. This set of equivalent points defines the *orbit of* G with respect to the point $\mathbf{r}$. Naturally, the size and appearance of the orbit will depend on the location of $\mathbf{r}$.

Orbits of a finite group are finite sets of points. By contrast, a regular (r, R) system is an orbit of an infinite group.

A site in a crystal may be such that some operations of G leave it unmoved. The totality of all such symmetry operations constitutes a *stabilizer subgroup*, $S(\mathbf{r})$, of G, and is known variously as *isotropy group*, *little group*, or *site-symmetry group* (Evarestov & Smirnov 1993).

Suppose a symmetry operation g of the group G carries the point $\mathbf{r}$ to $\mathbf{r}'$: $\mathbf{r}' = g\mathbf{r}$. Then the stabilizer of $\mathbf{r}'$, namely $S(\mathbf{r}')$, is given by $gS(\mathbf{r})g^{-1}$; i.e. $S(\mathbf{r}')$ and $S(\mathbf{r})$ are equivalent or conjugate subgroups of G (cf. §B.1).

Thus if two or more points are members of the same orbit (as $\mathbf{r}$ and $\mathbf{r}'$ are), their stabilizer is the same, except for equivalence transformations. Therefore one can speak of *the* stabilizer of an orbit.

2.2.7 Attributes of Space

Four attributes of space are relevant for a discussion of the symmetry of crystals. The space may be: (i) homogeneous or inhomogeneous; (ii) infinite or finite; (iii) continuous or discrete; and (iv) isotropic or anisotropic.

The question of homogeneity of space is linked to the scale adopted for defining it. At the atomic-structure level, the space a crystal exists in is inhomogeneous, because not all its points are symmetrically equal. However, at a sufficiently crude or macroscopic level, the same crystal space may be regarded as homogeneous, though not necessarily isotropic.

The attribute of infiniteness is essential for the purpose of defining translational invariance or periodicity. Unless the space is infinite in a given direction, it is not possible to define translational invariance along that direction or dimension. For example, in the definition of translational periodicity of a crystal along a particular direction, it is implicitly assumed that the crystal is of infinite size along that direction; then only can all lattice translations qualify as symmetry operations. In real-life situations, of course, the crystal size is always finite, but one can often ensure that the size is still large enough to make the effect of finiteness negligible in a given context.

A 3-dimensional crystal also provides an example of discrete space. Not all its points are equivalent by symmetry. It has a discrete lattice; any lattice point is equivalent only to those that are related by the discrete lattice translations. By contrast, in a continuous, homogeneous, space any point is equivalent or identical to any other point.

The attributes of an m-dimensional space may not be the same along all its dimensions. Its different n-dimensional subspaces ($n < m$) may have a variety of combinations of attributes, each with its own group of symmetry. In the context of crystals and other periodic structures, the symbol G_n^m is frequently used for denoting these symmetry groups. The integer n indicates the number of dimensions in which the crystal is periodic and therefore infinite, as well as (discretely) homogeneous.

A space is said to be fully inhomogeneous if it has no homogeneous subspaces ($n = 0$).

2.2.8 Rational and Irrational Directions

In Eq. 2.2.1, n_1, n_2, n_3 are integers, so that $\mathbf{r}$ is a lattice vector. A vector $\mathbf{r}$ defines a rational direction in the crystal lattice when n_1, n_2, n_3, though not necessarily integers, are rational numbers. For example, if $n_1 = 1.1$, $n_2 = 2.3$, and $n_3 = 7.26$, we can write $100\mathbf{r} = 110\mathbf{a}_1 + 230\mathbf{a}_2 + 726\mathbf{a}_3$, so that although $\mathbf{r}$ is no longer a lattice translation, $100\mathbf{r}$ is still so.

If one or more of the coefficients n_1, n_2, or n_3 is an irrational number (e.g. $\sqrt{2}$), $\mathbf{r}$ will define an irrational direction. In this case no multiple or submultiple of $\mathbf{r}$ can be a lattice translation.

2.2.9 The Crystallographic Restriction on Axes of Symmetry

There are only seven distinct types of crystallographic unit cells that remain invariant under the rotational symmetry operations of the crystal lattice. These define the seven *crystal systems*: triclinic, monoclinic, orthorhombic, tetragonal, cubic, rhombohedral, and hexagonal. To see how these restrictions arise, consider a lattice vector $\mathbf{r}$, and a rotational symmetry operator $\mathbf{R}$. Under the action of $\mathbf{R}$, $\mathbf{r}$ changes to another lattice vector $\mathbf{r}'$:

$$\mathbf{r}' = \mathbf{R}\,\mathbf{r} \tag{2.2.3}$$

One can assume, without loss of generality, that the axis of rotation for the symmetry operator $\mathbf{R}$ is along one of the three basic vectors spanning the lattice, say along $\mathbf{a}_3$. If $\mathbf{R}$ denotes a proper rotation through an angle θ, it can be represented by the following matrix with respect to a cartesian frame of reference:

$$\mathbf{M} = \begin{bmatrix} \cos\theta & -\sin\theta & 0 \\ \sin\theta & \cos\theta & 0 \\ 0 & 0 & 1 \end{bmatrix}$$

We now carry out a coordinate transformation from the cartesian frame of reference (in which the above representation of the operator $\mathbf{R}$ is defined) to one based on basic vectors $\mathbf{a}_1$, $\mathbf{a}_2$, $\mathbf{a}_3$ spanning the lattice. The new coordinate axes need not, in general, be mutually orthogonal.

Since $\mathbf{r}'$and $\mathbf{r}$ in Eq. 2.2.3 are both lattice vectors, this equation can be rewritten as

$$n_1'\mathbf{a}_1 + n_2'\mathbf{a}_2 + n_3'\mathbf{a}_3 = \mathbf{R}(n_1\mathbf{a}_1 + n_2\mathbf{a}_2 + n_3\mathbf{a}_3) \tag{2.2.4}$$

Since $n_1', n_2', n_3', n_1, n_2, n_3$ are arbitrary integers, Eq. 2.2.4 can hold only if all the terms in the new matrix representation (say $\mathbf{M}'$) of $\mathbf{R}$ are integers. In particular, the sum of the diagonal elements, i.e. the trace, must be an integer. This trace is $(2\cos\theta + 1)$ in the cartesian frame of reference. However, since a distance-preserving transformation (i.e. a similarity transformation) does not change the trace of $\mathbf{R}$, the trace is $(2\cos\theta+1)$ even for the coordinate system defined by the basic vectors $\mathbf{a}_1$, $\mathbf{a}_2$, $\mathbf{a}_3$. Therefore, $(2\cos\theta + 1)$ must be an integer.

Since $|\cos\theta|$ cannot exceed unity, the following are the only possibilities: $2\cos\theta + 1 = 3, 2, 1, 0, -1$, corresponding to

$$\cos\theta = 1, \frac{1}{2}, 0, -\frac{1}{2}, -1 \tag{2.2.5}$$

This implies that the only allowed values of θ are 2π, $2\pi/6$, $2\pi/4$, $2\pi/3$, and $2\pi/2$. The corresponding rotation axes of symmetry are defined as 1-fold, 6-fold, 4-fold, 3-fold, and 2-fold respectively. Thus the only rotational axes of symmetry allowed in a three-dimensional crystal are those denoted by the symbols 1, 2, 3, 4, and 6. This is sometimes referred to as the crystallographic restriction.

Apart from these six *proper* rotation axes of symmetry possible in a crystal, there are the six *improper* symmetry axes. These are obtained by defining a composite symmetry operation consisting of a proper rotation followed (or preceded) by an inversion operation $\bar{1}$. [An inversion operation through the origin changes all the atomic coordinates (x, y, z) to $(-x, -y, -z)$.] The composite operation is denoted by putting a "bar" over the symbol of the proper rotation operator. One gets $\bar{1}$, $\bar{2}$, $\bar{3}$, $\bar{4}$, and $\bar{6}$. It is readily verified that a 2-fold rotation, followed by inversion through a point on the rotation axis, is equivalent to a reflection operation across a plane perpendicular to the 2-fold axis and passing through the inversion point. Therefore, $\bar{2}$ really represents the mirror-symmetry operation m.

Thus, a crystal in three dimensions can possess only the following ten rotational symmetries, or directional symmetries, and their self-consistent combinations (we use the term "rotational symmetry" to represent both proper and improper rotations):

$$1, 2, 3, 4, 6, \bar{1}, \bar{2}, \bar{3}, \bar{4}, \bar{6}.$$

2.2.10 Crystal Systems and Crystal Families

In the light of the above, the point-group symmetries possessed by unit cells of various shapes can be described as follows:

(i) *Arbitrary parallelepiped*, with edges parallel to the basic lattice vectors $\mathbf{a}_1$, $\mathbf{a}_2$, $\mathbf{a}_3$, such that $a_1 \neq a_2 \neq a_3$, and $\alpha \neq \beta \neq \gamma$. Here α is the angle between $\mathbf{a}_2$ and $\mathbf{a}_3$, β that between $\mathbf{a}_3$ and $\mathbf{a}_1$, and γ that between $\mathbf{a}_1$ and $\mathbf{a}_1$. Such a unit cell has two symmetry elements, namely the identity operation 1, and the inversion operation $\bar{1}$. Its point symmetry is thus described by a group of order 2.

Two main types of notation are in common use for describing symmetry in crystallography. One is the so-called *international* or *Hermann-Mauguin* notation, which for the point group of the present unit cell is $\bar{1}$.

The other notation is the *Schoenflies* notation; for the present case it is C_i, the two elements of the group being denoted by E and J. In the symbol C_i, i stands for the inversion operation, and C represents the fact that it is a cyclic group (cf. §B.1).

(ii) *Right parallelepiped*, with $a_1 \neq a_2 \neq a_3$, and $\alpha = \gamma = 90^o \neq \beta$. Such an object has a 2-fold axis of symmetry, C_2, parallel to the a_2-axis, as also a mirror plane of symmetry, σ_h, perpendicular to the a_2-axis. Its symmetry group has the elements E, C_2, J, and σ_h (or 1, 2, $\bar{1}$ and m_y), and is denoted by C_{2h} (or $2/m$).

(iii) *Rectangular parallelepiped*, with $a_1 \neq a_2 \neq a_3$, and $\alpha = \beta = \gamma = 90^o$. Its symmetry group is of order 8, with elements E, C_2, C_2', C_2'', J, σ_h, σ_v, σ_v' (or 1, 2_x, 2_y, 2_z, $\bar{1}$, m_x, m_y, m_z), and is denoted by D_{2h} (or mmm).

(iv) *Right square prism*, with $a_1 = a_2 \neq a_3$ and $\alpha = \beta = \gamma = 90^o$. Its symmetry group is denoted by D_{4h} or $4/mmm$, and it has 16 elements.

(v) *Cube*, with $a_1 = a_2 = a_3$ and $\alpha = \beta = \gamma = 90^o$. The symmetry group of a cube is of order 48, and is denoted by O_h or $m\bar{3}m$. The letter O denotes "octahedral"; a cube and an octahedron have the same symmetry elements. More details can be found in, for example, Burns & Glazer (1990) and Hahn (1992).

(vi) *Rhombohedron*, with $a_1 = a_2 = a_3$ and $\alpha = \beta = \gamma \neq 90^o$. Its symmetry group D_{3d} or $\bar{3}m$ has 12 elements.

(vii) *Regular hexagonal prism*, with $a_1 = a_2 \neq a_3$, $\alpha = \beta = 90^o$, and $\gamma = 120^o$. The point group for this case is denoted by D_{6h} or $6/mmm$, and has 24 elements.

Each of these seven types of unit cells, if stacked together and repeated in a space-filling manner along $\mathbf{a}_1$, $\mathbf{a}_2$ and $\mathbf{a}_3$, generates the entire crystal. Every vertex of every unit cell is identical in all respects: configuration of electrons, atoms, everything. The vertices therefore constitute a set of equivalent points, related to one another through lattice vectors. Such a set of points constitutes a lattice.

A crystal may have more equivalent points than those generated through lattice translations alone. For example, rotational symmetry operations can generate additional equivalent points.

> *A lattice is a set of equivalent points generated by lattice translations alone.*

The seven types of unit cells listed above generate seven of the crystal lattices, called *primitive lattices.* The symbol P is used for representing primitive crystal lattices, except that R is used for denoting the rhombohedral lattice.

These seven types of unit cells also define the seven *crystal systems*, each with a distinct *essential symmetry*, namely C_i (triclinic), C_{2h} (monoclinic), D_{2h} (orthorhombic), D_{3d} (rhombohedral), D_{4h} (tetragonal), D_{6h} (hexagonal), and O_h (cubic). The seven crystal systems represent the seven distinct point-group symmetries that crystal lattices can possess.

The concept of *crystal families* is particularly useful in the context of ferroelastic phase transitions. The 32 crystallographic point groups are divided into six crystal families. The crystallographic family of a point group is the same as the crystal system to which it belongs, except that point groups under the rhombohedral crystal system are taken as belonging to the hexagonal crystal system (Hahn 1992; Janovec, Richterova & Litvin 1993). Thus, whereas there are seven crystal systems, there are only six crystal families.

2.2.11 Primitive and Nonprimitive Bravais Lattices

Of all the lattices which can be generated in 3-dimensional space by three basic lattice vectors, only 14 are unique or distinct. These are referred to as the 14 Bravais lattices.

In a *primitive Bravais lattice* only one lattice point is associated per crystallographic unit cell. (A *crystallographic unit cell* is that which possesses the full point-group symmetry of the crystal.) Thus there are seven primitime Bravais lattices, one for each crystal system.

There also exist seven *nonprimitive Bravais lattices*, in which the number of lattice points associated per crystallographic unit cell is more than one. (Such a unit cell is referred to as a *nonprimitive unit cell.*)

An example of a nonprimitive Bravais lattice is the body-centered cubic (bcc) lattice, in which one can define a unit cell that is a cube, for which the body centre, i.e. the point $(\frac{1}{2}, \frac{1}{2}, \frac{1}{2})$ in units of the basic vectors, is also a lattice point completely equivalent to the lattice points at the corners of the cube. The symbol I is used for denoting such a Bravais lattice.

Similarly, face-centered cubic (fcc) is another nonprimitive Bravais lattice (denoted by the symbol F), with lattice points at the centres of the six faces of the cubic unit cell, as well as at the corners of the cube.

One can always choose a *primitive* unit cell (with only one lattice point associated with every such cell) even for a nonprimitive Bravais lattice.

But, unlike the nonprimitive or crystallographic unit cell, such a cell does not possess the full point-group symmetry of the lattice.

It is readily verified that the set of symmetry operations defining any Bravais lattice constitutes a group, *the Bravais group.* The basic vectors $\mathbf{a}_1$, $\mathbf{a}_2$, $\mathbf{a}_3$ are the generators of this group.

2.2.12 Screw Axes and Glide Planes

We conclude our brief introduction to crystal lattices by mentioning the possibility of occurrence in them of symmetry operations that are combinations of rotations or reflections with *fractional translations.* There are two types of such symmetry elements: screw axes and glide planes.

A crystal lattice is said to have a screw axis of order n if it remains unchanged on rotation through an angle $2\pi/n$ followed by a translation through a vector $p\mathbf{a}/n$ $(p = 1, 2, \ldots n - 1)$, where a is the smallest period of the lattice along the direction of the axis. The symbol used for a screw axis is n_p. For example, a screw axis of order 3 can be of two kinds, involving either a fractional translation of $\mathbf{a}/3$, or of $2\mathbf{a}/3$. n successive applications of a screw operation of order n simply move the lattice by a nonfractional distance a.

A glide-plane of symmetry is said to exist in a crystal lattice if the lattice is invariant to reflection through such a plane combined with a fractional translation $\mathbf{A}/2$ along a specific lattice direction in this plane, A being the smallest period of the lattice in the direction of the fractional translation. Two successive operations of this type simply translate the lattice by a lattice vector $\mathbf{A}$. The repeat vector $\mathbf{A}$ may be either a basis vector $\mathbf{a}_1$, $\mathbf{a}_2$, or $\mathbf{a}_3$, or, for certain lattices, a vector like $\mathbf{a}_1 + \mathbf{a}_2$, or, for cubic lattices, even $\mathbf{a}_1 + \mathbf{a}_2 + \mathbf{a}_3$.

We emphasize that the additional equivalent points of a crystal lattice arising from the presence of any screw axes or glide planes, being not the points generated by full lattice translations, do not enter the definition of the corresponding Bravais lattice.

2.2.13 Wigner-Seitz Cell

This unit cell is constructed as follows: One chooses any of the lattice points as the origin O, and draws all the planes that are perpendicular bisectors of the lines joining O with the nearest (and sometimes also the next nearest) neighbours. Enough number of such planes are drawn to enclose a polyhedron, which is called the *Wigner-Seitz cell.* Such a unit cell contains only one lattice point, and has the same volume as the primitive unit cell.

With the crystallographic unit cell it shares the property that both display the full point symmetry of the lattice.

2.2.14 The Various Types of Unit Cells

We have described three types of unit cells in this section.

The *crystallographic unit cell* (also called the Bravais unit cell, or the *conventional unit cell*) possesses the full point-group symmetry of the crystal, but it may contain more than one lattice point in it.

The *primitive* cell may or may not display the full point symmetry of the crystal, but it has only one lattice point associated with it.

The *Wigner-Seitz* cell displays the full point symmetry, and there is also only one lattice point per cell of this type.

2.2.15 Crystallographic Point Groups

For macroscopic properties like mechanical deformation, thermal expansion and optical birefringence, crystals, in spite of their discrete atomic structure, can be treated as homogeneous continuous media. Such properties, however, need not be the same along all directions in the crystal. On the other hand, any rotational symmetry possessed by the atomic structure of a crystal results in a corresponding directional symmetry of its macroscopic physical properties. Thus, so far as the macroscopic properties are concerned, the crystal behaves as an anisotropic continuum, its directional symmetry determining the symmetry of these properties. The symmetry of directions in crystals can be described by one or the other of 32 crystallographic point groups. These are symmetry groups, the elements of which are derived from the 10 proper or improper rotational symmetries possible in crystals, including their allowed, self-consistent, combinations.

In contrast to rotational symmetry, the translational symmetry of the crystal does not lead to any symmetry of directions. However, it puts severe restrictions on the symmetry of directions a crystal can possess. Also, the fractional translations involved in the operations of screw axes and glide planes can be ignored when specifying the symmetry of directions of a crystal. The directional symmetry is thus determined only by the 10 proper and improper rotational symmetries, with screw axes and glide planes (if any) replaced by the corresponding simple rotation axes and mirror planes. The 32 groups of symmetry elements so obtained correspond to the 32 *crystal classes.*

Derivation of Point Groups

We begin by noting that the symmetry of a crystal class cannot be higher than that of the crystal system it belongs to; it can at the most be equal to it. This is because, when we generate a crystal structure by associating a basis (an atom, or a bunch of atoms) with each lattice point, the symmetry cannot possibly be enhanced by such a process. This suggests a way of deriving the various crystallographic point groups: We take each of the seven point groups corresponding to the seven crystal systems, and derive all their trivial and nontrivial subgroups.

On the application of such a procedure, it is bound to turn out that some of the subgroups so obtained occur in more than one crystal system. For example, since all the seven primitive Bravais lattices are centrosymmetric, the group C_i will occur as a subgroup in all the seven crystal systems. Now, it is physically highly unlikely that a crystal with a given point-group symmetry should belong to a crystal system of symmetry higher than the minimum necessary. For instance, it is very improbable that a crystal with point-group symmetry C_2 should belong to the cubic system (symmetry O_h), rather than belonging to the monoclinic system (symmetry C_{2h}). One can therefore impose the following physically reasonable condition for deriving the various crystallographic point groups from the symmetry groups of the seven crystal systems:

> *Each crystal class shall be assigned to the lowest-symmetry crystal system compatible with it.*

We start with the triclinic system, and determine the subgroups of C_i. There are two of them: C_1 and C_i.

The next higher system, monoclinic, contains three crystal classes: C_2, C_s, C_{2h}, not counting the classes C_1 and C_i already encountered in the triclinic system. And so on.

We list below the 32 crystallographic point groups derived in this way. The international notation is given in brackets.

Triclinic : C_1 (1), C_i ($\bar{1}$)

Monoclinic : C_s (m), C_2 (2), C_{2h} ($2/m$)

Orthorhombic : C_{2v} ($2mm$), D_2(222), D_{2h} (mmm)

Tetragonal : C_4 (4), S_4 ($\bar{4}$), C_{4h} ($4/m$), C_{4v} ($4mm$), D_{2d} ($\bar{4}2m$), D_4 (422), D_{4h} ($4/mmm$)

Rhombohedral : C_3 (3), S_6 ($\bar{3}$), C_{3v} ($3m$), D_3 (32), D_{3d} ($\bar{3}m$)

Hexagonal : C_{3h} ($\bar{6}$), D_{3h} ($\bar{6}m2$), C_6 (6), C_{6h} ($6/m$), C_{6v} ($6mm$), D_6 (622), D_{6h} ($6/mmm$)

Cubic : T (23), T_h ($m3$), T_d($\bar{4}3m$), O (432), O_h($m\bar{3}m$)

A word about the symbols used here. As mentioned before, C stands for *cyclic*; for example, C_3 is a cyclic group of order 3.

D denotes *dihedral*. A dihedral group D_n can be split into a direct product as follows:

$$D_n = C_n \otimes \{E,\ C_2[100]\} \tag{2.2.6}$$

Here the n-fold axis is taken along [001], and the 2-fold axis C_2 is along [100]. The orders of the groups C_n and D_n are n and $2n$ respectively.

T stands for "tetrahedron"; this group comprises the rotational (only the proper rotational) symmetry operations of a tetrahedron. The group T_d describes the full symmetry of a tetrahedron.

O denotes the proper-rotational part of the symmetry of an octahedron or a cube, and O_h their full symmetry.

There are other ways of deriving the 32 crystallographic point groups. We mention one more approach here, which helps bring out some geometrical aspects.

We have already come across the result that only the following 10 rotation and roto-inversion axes are possible in a crystal: $1, 2, 3, 4, 6, \bar{1}, \bar{2}, \bar{3}, \bar{4}, \bar{6}$. These account for 10 of the 32 crystallographic point groups; i.e. a crystal may have any of these as the only element of directional symmetry. Such point groups are called *monoaxial point groups*. In fact, three additional distinct monoaxial point groups are also possible, namely $2\bar{2}$ (or $2/m$), $4\bar{4}$ (or $4/m$), and $6\bar{6}$ (or $6/m$). The corresponding Schoenflies symbols are C_{2h}, C_{4h}, and C_{6h}. It is readily verified that $3\bar{3}$ is not distinct from $\bar{3}$ itself. The notation $n\bar{n}$ means that $\bar{n}$ is parallel to the n-fold axis, and n/m means that m is perpendicular to n.

Having listed the 13 monoaxial point groups for crystals, we consider next the *polyaxial point groups*. Suppose we have two nonparallel but intersecting symmetry axes n_1 and n_2. The net effect of rotation by an angle $2\pi/n_1$ about the axis n_1, followed by a rotation of $2\pi/n_2$ about n_2, is a rotation of $2\pi/n_3$ about a third axis n_3. There are three restrictions on n_3.

Restriction 1. n_3 can be 1, 2, 3, 4, 6, $\bar{1}$, $\bar{2}$, $\bar{3}$, $\bar{4}$, or $\bar{6}$ only.

Restriction 2. n_3 must be a proper rotation axis if the product $n_1 n_2$ is so; improper if the product is so. This means that only the following three types of combinations are possible:

$$n_1 n_2 n_3; \quad n_1\bar{n}_1\ n_2\bar{n}_2\ n_3\bar{n}_3; \quad n_1\bar{n}_2\bar{n}_3\ \bar{n}_1 n_2\bar{n}_3\ \bar{n}_1\bar{n}_2 n_3$$

Restriction 3. The interaxial angles among the symmetry axes n_1, n_2, n_3 must obey the Euler formula. For example, the angle $\phi_{\alpha\beta}$ between the axes n_1 and n_2 is restricted to the value given by

$$\cos\phi_{\alpha\beta} = \frac{\cos\gamma/2\ +\ \cos\alpha/2\ \cos\beta/2}{\sin\alpha/2\ \sin\beta/2} \tag{2.2.7}$$

Here $\alpha = 2\pi/n_1$, $\beta = 2\pi/n_2$ and $\gamma = 2\pi/n_3$.

There are six polyaxial point groups of the type $n_1n_2n_3$. These are: 222 (D_2), 32 (D_3), 422 (D_4), 622 (D_6), 23 (T), and 432 (O).

Six additional polyaxial point groups are of the type $n_1\bar{n}_1\ n_2\bar{n}_2\ n_3\bar{n}_3$. These are: $mmm(D_{2h})$, $\bar{3}m(D_{3d})$, $4/mmm(D_{4h})$, $6/mmm(D_{6h})$, $m3(T_h)$ and $m\bar{3}m(O_h)$.

The symbol $\bar{3}m$ is a short form for the full symbol $3\bar{3}\ 2\bar{2}\ 2\bar{2}$. This shortening is possible because $3\bar{3} = \bar{3}$, and $2\bar{2} = 2/m$. The 3-fold axis is taken as 'vertical', and therefore the 2-fold axes are horizontal. The m in $2/m$, being perpendicular to the horizontal 2-fold axis, becomes vertical and hence parallel to the $\bar{3}$-axis. Moreover, the presence of one such vertical mirror plane leads to the presence of two more such planes, obtained by the operation of the $\bar{3}$-axis; it is therefore enough to indicate the presence of one such mirror plane by writing the short symbol $\bar{3}m$. In the Schoenflies symbol D_{3d} for this point group, d stands for "diagonal"; the vertical m-planes bisect the angles between the 2-fold axes or *diads*.

The following relationships involving direct products of groups are instructive:

$$D_{nh} = D_n \otimes \{E, \sigma_h\} = D_n \otimes \{E, J\} \tag{2.2.8}$$

$$T_h = T \otimes \{E, \sigma_h\} = T \otimes \{E, J\} \tag{2.2.9}$$

$$O_h = O \otimes \{E, \sigma_h\} = O \otimes \{E, J\} \tag{2.2.10}$$

The remaining seven polyaxial crystallographic point groups are of the type $n_1\bar{n}_2\bar{n}_3$, $\bar{n}_1n_2\bar{n}_3$, $\bar{n}_1\bar{n}_2n_3$. These are: $2mm(C_{2v})$, $3m(C_{3v})$, $4mm(C_{4v})$, $\bar{4}2m(D_{2d})$, $6mm(C_{6v})$, $\bar{6}m2(D_{3h})$, and $\bar{4}3m(T_d)$.

Laue Classes

Centrosymmetric crystal classes (i.e. those possessing the inversion operation as a symmetry element) are called Laue classes. There are 11 Laue classes, and they include the seven classes corresponding to the highest symmetry of the seven crystal systems:

$$\bar{1}(C_i),\ 2/m(C_{2h}),\ mmm(D_{2h}),\ 4/m(C_{4h}),\ 4/mmm(D_{4h}),$$

$$\bar{3}(C_{3i}\ \mathit{or}\ S_6),\ \bar{3}m(D_{3d}),\ 6/m(C_{6h}),\ 6/mmm(D_{6h}),\ m3(T_h),\ m\bar{3}m(O_h)$$

Polar Groups

A direction in a crystal the two ends of which are not related by any symmetry operation of the point group of the crystal is called a *polar direction.*

Obviously, such a direction cannot exist for any of the 11 Laue groups. Out of the 21 noncentrosymmetric crystal classes, only the following 10 can possess such a direction:

$$1(C_1),\ 2(C_2),\ 3(C_3),\ 4(C_4),\ 6(C_6)$$

$$m(C_s),\ mm2(C_{2v}),\ 3m(C_{3v}),\ 4mm(C_{4v}),\ 6mm(C_{6v})$$

These are the 10 crystallographic polar groups, and crystals having any of these symmetries are said to belong to a *polar class.* These groups are, naturally, subgroups of the limit group ∞m (cf. §2.2.19 below).

2.2.16 Simple Forms

Any plane or line in a crystal is repeated by the operations of its point group into equivalent planes or lines. For example, if we consider a plane (100) in a crystal having point symmetry $m\bar{3}m$, identical planes $(\bar{1}00)$, (010), $(0\bar{1}0)$, (001), $(00\bar{1})$, will also occur because of the directional symmetry defined by $m\bar{3}m$. The entire set of these six planes is denoted by {100}, and together they enclose a polyhedron or form (a cube in this case), called a simple form.

A simple form is a polyhedron the faces of which are related by the symmetry operations of the crystal.

In our example of the simple form (cube) generated by {100}, the initial face (100) is in a special orientation, namely parallel to (in fact coinciding with) the mirror plane m_x of the group $m\bar{3}m$. If we choose an initial face in the most general orientation, so that it does not coincide with any symmetry element of the crystal, and is also inclined to the three coordinate axes at different angles, we obtain a *general simple form.* It is a hexoctahedron (48 faces) for the case of the crystal class $m\bar{3}m$ (Sirotin & Shaskolskaya 1982). The general simple form has the largest number of faces (or *hedra*) for a given crystal class.

Special simple forms, because of the special choice of the initial face, have less faces than the general simple form. For the example of $m\bar{3}m$ symmetry considered here, the number of faces of simple forms is 48 $\{hkl\}$, 24 $\{hhl\}$, 12 {110}, 8 {111}, and 6 {100}.

Simple forms may be either space-enclosing (e.g. a cube or an octahedron), or open (e.g. prisms, pyramids, or pinacoids). Crystals have forms which are usually combinations of several simple forms. Open forms can occur in combinations only.

Holohedry and Merohedry

The number of faces in a general simple form is equal to the order of the point group of the crystal. Special simple forms have symmetries of subgroups of the full point-group symmetry.

The 32 crystallographic classes are divided among the 7 crystal systems. The point-group symmetries of the Bravais lattices underlying the 7 crystal systems are the highest a crystal belonging to a particular crystal system can have. These are called *holohedral symmetries* or *holohedries.* The 7 holohedral symmetry classes are:

$$\bar{1},\ 2/m,\ mmm,\ 4/mmm,\ \bar{3}m,\ 6/mmm,\ m\bar{3}m$$

The other 25 crystal classes have point-group symmetries that are subgroups of the corresponding holohedral symmetry, and are called *merohedral classes*, or *merohedries.*

The possible merohedral forms are: *hemihedral* (subgroup(s) of index 2), *tetartohedral* (subgroup(s) of index 4), and *ogdohedral* (subgroup(s) of index 8). Hemihedral simple forms can have half the number of faces possible for the full holohedral symmetry. Tetartohedry amounts to cutting the possible number of faces by 4, and ogdohedry by 8.

A discussion of simple forms in terms of point-group symmetry is adequate for most purposes. If needed, the full space-group symmetry can be invoked for dealing with them. A total of 1403 types of simple forms are possible at the crystallographic space-group level (Shafranovsky 1968).

2.2.17 Crystallographic Space Groups

The full, microscopic, symmetry of a crystal is represented by its space group. A crystallographic space group is a group, the elements of which are all the symmetry operations (lattice translations, rotations, reflections, screw axes, glide planes) that map an infinite crystal onto itself.

The only translational symmetry a crystal can have is that described by one of the 14 Bravais groups. Therefore, to specify the space-group symmetry of a crystal, we have to identify its Bravais lattice, as well as the symmetry operations involving rotations and reflections (including screw axes and glide planes, if any). It is also necessary to identify the relative positions of the symmetry elements.

Seitz Operator

Space-group operations can be defined in terms of the widely used Seitz operator $\{\mathbf{R}|\mathbf{t}\}$ (Seitz 1936), where $\mathbf{R}$ denotes a point-group operation and $\mathbf{t}$ a translation. The Seitz operator, acting on a position vector $\mathbf{r}$, carries out the following transformation:

$$\{\mathbf{R}|\mathbf{t}\}\,\mathbf{r} = \mathbf{R}\mathbf{r} + \mathbf{t} \tag{2.2.11}$$

Successive application of two such operators, $\{\mathbf{Q}|\mathbf{u}\}$ and $\{\mathbf{R}|\mathbf{t}\}$, has the following effect:

$$\{\mathbf{R}|\mathbf{t}\}\,\{\mathbf{Q}|\mathbf{u}\}\mathbf{r} = \{\mathbf{R}|\mathbf{t}\}\,(\mathbf{QR} + \mathbf{u}) = \mathbf{RQr} + \mathbf{Ru} + \mathbf{t} \tag{2.2.12}$$

$$= \{\mathbf{RQ}|\mathbf{Ru} + \mathbf{t}\}\,\mathbf{r} \tag{2.2.13}$$

The inverse of an operator $\{\mathbf{R}|\mathbf{t}\}$ is given by

$$\{\mathbf{R}|\mathbf{t}\}^{-1} = \{\mathbf{R}^{-1}\,|\,-\mathbf{R}^{-1}\mathbf{t}\} \tag{2.2.14}$$

This is easily verified from Eq. 2.2.12 by identifying $\{\mathbf{Q}|\mathbf{u}\}$ with the inverse operator and demanding that the result of the product be $\{\mathbf{1}|0\}$.

The lattice of a crystal is characterized by the Seitz operator $\{\mathbf{1}|\mathbf{t}_n\}$ (or $\{\mathbf{E}|\mathbf{t}_n\}$), with $\mathbf{t}_n$ given by

$$\mathbf{t}_n = n_1\mathbf{a}_1 + n_2\mathbf{a}_2 + n_3\mathbf{a}_3 \tag{2.2.15}$$

However, the space-group operations of a crystal can also involve *essential translations* $\boldsymbol{\tau}$ which are *fractions* of the lattice translations. This comes from the possible presence of screw axes and/or glide planes of symmetry. Thus, in general, in Eq. 2.2.11

$$\mathbf{t} = \mathbf{t}_n + \boldsymbol{\tau} \tag{2.2.16}$$

The set of all lattice translations forms a group (the Bravais group or the translation group), T. It is a subgroup of the full space group, S. Each element of T forms a class by itself:

$$\{\mathbf{1}|\mathbf{t}_m\}^{-1}\,\{\mathbf{1}|\mathbf{t}_n\}\,\{\mathbf{1}|\mathbf{t}_m\} = \{\mathbf{1}|\mathbf{t}_{-m}\}\,\{\mathbf{1}|\mathbf{t}_{n+m}\} = \{\mathbf{1}|\mathbf{t}_n\} \tag{2.2.17}$$

Thus:

> *The Bravais group or the translation group of a crystal consists of complete classes, and is therefore a normal subgroup of the space group of the crystal.*

Space Groups and Space-Group Types

Consider the symmetry of Si and Ge crystals. They both have the 'diamond structure', described by space-group symmetry $Fd\bar{3}m$. It is often stated that they both belong to the same space group. Strictly speaking this is a somewhat loose, though widely prevalent, statement. They have different lattice parameters, and therefore their Bravais groups are not the same. However, their Bravais groups (as also their space groups) are *isomorphic.*

Two groups are said to be isomorphic if their elements display the same relationships. This property of groups allows us to classify them into *isomorphism classes*, or *abstract groups*, or *group types.*

In our above example, Si and Ge do not belong to the same space group, but to the same space-group *type*: There is a one-to-one correspondence between the symmetry elements of the two crystal structures.

Although this distinction between space groups and space-group types has already appeared in the new *International Tables for Crystallography* (see page 718 of Hahn (1992)), some experts strongly argue in favour of not making this distinction in most practical situations, pointing out that it would be desirable to state that the number of crystallographic space groups is just 230, rather than infinite (Glazer; personal communication).

Symmorphic and Nonsymmorphic Space Groups

In a three-dimensional lattice, apart from the lattice translations, there are, say, g point-group operations that can transform the contents of the primitive unit cell onto themselves. The g space-group operations obtained by combining these with the identity element $\{\mathbf{1}|\mathbf{0}\}$ of the translation group are called *essential space-group operations.*

There are two types of space groups: symmorphic and nonsymmorphic. Symmorphic space groups can be entirely specified in terms of rotational symmetry operators, all acting around a *common point.* Also, they do not have to involve any essential fractional translation $\boldsymbol{\tau}$. In other words, for them an appropriate choice of origin can make all the translations primitive lattice translations. Their Seitz operators can therefore all be written in the form $\{\mathbf{R}|\mathbf{t}_n\}$, with $\{\mathbf{1}|0\}$ necessarily a member of this set (with a suitable choice of origin). There are 73 symmorphic space groups in all.

For nonsymmorphic space groups, on the other hand, no matter what point is chosen as the origin, it is necessary to specify at least one operation involving a fractional translation $\boldsymbol{\tau}$, so that a general Seitz operator has the form $\{\boldsymbol{R}\,|\,\boldsymbol{t_n} + \boldsymbol{\tau}\}$. There are 157 nonsymmorphic crystallographic space groups.

Because of the essential presence of fractional translations, the primitive unit cell of a crystal having a nonsymmorphic space group must have at least two identical sites for every atom in it.

We consider two space groups, $P2$ and $P2_1$, to illustrate the difference between symmorphic and nonsymmorphic space groups. $P2$ is symmorphic because its two essential space-group operations are $\{\mathbf{1}|0\}$ and $\{\mathbf{2}|0\}$, whereas $P2_1$ is nonsymmorphic because its essential space-group operations are $\{\mathbf{1}|0\}$ and $\{\mathbf{2}\,|\,0\,\frac{1}{2}\,0\}$.

Point Group of a Space Group

If we replace all the symmetry operations $\{\boldsymbol{R_i}\,|\,\boldsymbol{\tau_i}+\boldsymbol{t_n}\}$ of a space group S by $\{\mathbf{R}_i|0\}$, that is if we set all the fractional as well as lattice translations to zero, the resulting set constitutes a group, G_p, called the point group of the space group. For example, the point group of the space group *Pbca* is *mmm*.

If S is a symmorphic space group, the g operations $\{\mathbf{R_i}|0\}$ of G_p are always symmetry operations of S also. The term 'symmorphic' is used in the sense that the space group has a structure similar to that of its point group G_p.

If S is nonsymmorphic, at least one of the operators $\{\mathbf{R_i}|0\}$ of G_p is not a symmetry operator of S. For example, suppose 4_1 is a symmetry operation of S. Its Seitz operator is $\{\mathbf{4}|\boldsymbol{\tau}(0,0,\frac{1}{4})\}$. The corresponding operator in G_p is $\{\mathbf{4}|0\}$, and this is *not* a symmetry operator of S.

Derivation of Space Groups

Crystallographic space groups can be derived by forming semi-direct products of the Bravais groups T and appropriate point groups P, and considering the trivial and nontrivial subgroups of these product groups. All the trivial subgroups correspond to symmorphic space groups, and certain nontrivial subgroups correspond to nonsymmorphic space groups.

Consider the following semi-direct product:

$$S_s = T \circledS P \tag{2.2.18}$$

Geometrically, this amounts to placing the symmetry elements of the point group P at all the sites of the Bravais lattice described by T; the group P may either have the full rotational symmetry of T, or a lower symmetry. T is a normal subgroup of S_s, and P is not. What this means, in effect, is that whereas T can be identified everywhere, P can be recovered in full only at those points where all its symmetry elements intersect.

If P has only proper rotations as its symmetry elements, so also will S_s. Proper rotations are sometimes referred to as *transformations of the first kind*, and improper rotations as those of the *second kind*; superscripts I and II are used for denoting the respective groups involving them. Thus, if $P = P^I$, then $S_s = S_s^I$. Similarly, $S_s = S_s^{II}$ if $P = P^{II}$. All the 73 symmorphic space groups can be obtained by this procedure.

To understand how nonsymmorphic space groups arise, we consider a simple one-dimensional example in three-dimensional space. Let us assume that T is the translation group $\{0, x, 2x, 3x, \dots\}$, and $P = \{e, 2\}$. The

2-fold axis in P is taken as parallel to the x-direction. We construct the semi-direct product S_s:

$$\begin{aligned} S_s &= T\,P = \{0, x, 2x, \ldots\}\ \{e, 2\} \\ &= \{0.e, x.e, 2x.e, \ldots, 0.2, x.2, 2x.2, 3x.2, \ldots\} \end{aligned} \qquad (2.2.19)$$

This product group contains a new symmetry element, $x.2$, which corresponds to a screw operation: We can isolate from the symmorphic group S_s a nonsymmorphic subgroup S_n:

$$S_n = \{0.e, (2x).e, \ldots, 0.2_1, (2x).2_1, \ldots\}, \qquad (2.2.20)$$

where the translation group now has the basis $\{0, 2x\}$ (instead of $\{0, x\}$ for the original translation group T), and $2_1 \equiv x.2 = 2.x$, is a screw axis.

Thus nonsymmorphic space groups can be derived as proper subgroups, S_n, of larger symmorphic groups, S_s, obtained by semi-direct product of a Bravais group and a point group ($S_n \subset S_s = T ⓢ P$). The groups S_n have primitive-unit-cell volumes which are some integral multiple of those for the corresponding supergroups S_s. Both S_n and S_s are homomorphic to the same point group P, which can now be identified as the point group G_p underlying the two space groups.

Nonsymmorphic space groups can be of two types: *hemisymmorphic*, and *asymmorphic*. A hemisymmorphic space group, S_h^{II}, arises if, during the process of isolating it as a subgroup of the larger group S_s^{II}, we reject all those rotational symmetry elements of the second kind which intersect at the same point where their axes intersect. For such space groups the highest site symmetry is that of a point group P^I of index 2 of the parent point group P^{II}.

On the other hand, if, during the process of selecting a nonsymmorphic subgroup of S_s, we select only those symmetry operations which have no common point of intersection of the axes along different directions, the result is an asymmorphic space group S_a.

Out of a total of 157 nonsymmorphic space groups, 54 are hemisymmorphic (S_h^{II}), and 103 are asymmorphic (S_a). There are 41 asymmorphic space groups of the first kind (S_a^I), and 62 of the second kind (S_a^{II}).

Wyckoff Positions

Consider a point $\mathbf{r}$ in a crystal. The set of all symmetry operations of the space group S of the crystal which leave the point $\mathbf{r}$ invariant forms a group called the *site-symmetry group* $S(\mathbf{r})$ of $\mathbf{r}$ with respect to S.

A Wyckoff position is the set of all points $\mathbf{r}$ for which the site-symmetry groups $S(\mathbf{r})$ are conjugate subgroups (§B.1) of the space group S.

The crystallographic community has found it very convenient to label each Wyckoff position by a specific letter (Hahn 1992), called the *Wyckoff letter*, or the *Wyckoff notation*.

A point **r** in a crystal is said to be a *Wyckoff point of general position* if no symmetry operation of S (other than the identity operation) leaves it fixed. If there is at least one such operation, the point is called a *Wyckoff point of special position.*

Two distinct types of Wyckoff positions are those *without* a variable parameter (e.g. $0, \frac{1}{2}, 0$; $\frac{1}{2}, 0, 0$), and those *with* a variable parameter (e.g. $0, \frac{1}{2}, z$; $\frac{1}{2}, 0, z; \ldots$).

Crystallographic Orbit, or Lattice Complex

For a crystallographic space group S, the set of all points generated by the application of its symmetry operations on a point **r** is called the crystallographic orbit or the lattice complex of **r** with respect to S.

The site-symmetry groups of the various points of a crystallographic orbit are conjugate subgroups of S. Therefore the crystallographic orbit consists of either points of general position (*general crystallographic orbits*), or points of special position (*special crystallographic orbits*). All points of a crystallographic orbit belong to the same Wyckoff position of S.

2.2.18 Magnetic Symmetry of Crystals

To provide an adequate description of the symmetry of magnetic crystals it is necessary to introduce an additional concept, that of *antiequality.* This concept was introduced by Shubnikov (1951). Any two figures or entities which have the same dimensions and other characteristics, but are mutually opposite in one property, can be considered as antiequal. This property can be colour (e.g. black and white), sign of electric charge, direction of magnetic moment, or the direction of flow of time. Time reversal is particularly relevant for discussing the symmetry of magnetic structures, because with time reversal the directions of all currents are reversed, and consequently the signs of magnetic moments, including spins parallel or antiparallel to a given direction, are reversed.

In this section we make no distinction between the various types of *antioperations* (colour reversal, time reversal, etc.), and use the same symbol, $1'$, for all of them. The antisymmetry operator, being a totally different type of operator, commutes with all orthogonal transformations p:

$$p.1' = 1'.p = p' \tag{2.2.21}$$

Two successive applications of the antioperator result in an identity operation:

$$1'.1' = 1 \tag{2.2.22}$$

Because of Eqs. 2.2.21 and 2.2.22, if $p_i p_j = p_k$ and $p_j p_i = p_l$, then:

$$p_i p'_j = p'_i p_j = p'_k \tag{2.2.23}$$

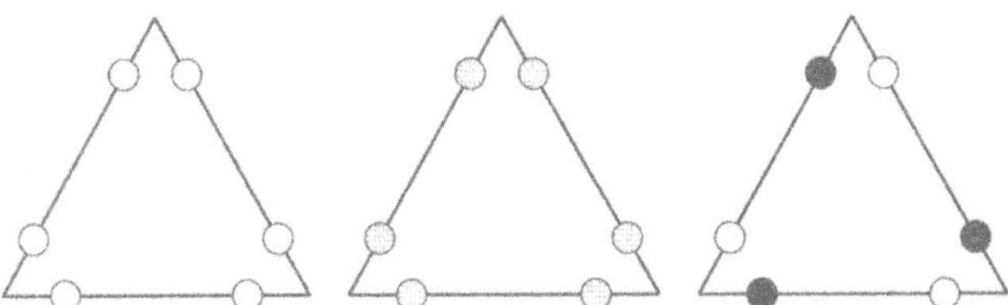

Figure 2.2.1: Magnetic point groups of the I, II, and III kind, exemplified for the case of the crystallographic point group C_{3v} (cf. Fig. B.1 of Appendix B). In (a) (C_{3v}^{I}), all the entities have the same colour (say white). In (b) (C_{3v}^{II}), they are grey (due to superposition of white and black colours occurring at the same locations). And in (c) (C_{3v}^{III}), half the entities are white, and half are black. [After Ludwig & Falter 1988.]

$$p_i' p_j' = p_k \tag{2.2.24}$$

$$p_j' p_i = p_j p_i' = p_l' \tag{2.2.25}$$

$$p_j' p_i' = p_l \tag{2.2.26}$$

Magnetic Point Groups

The 32 ordinary or *chemical* point groups can be derived from the following 10 generating elements:

$$1,\ 2,\ 3,\ 4,\ 6,\ \bar{1},\ \bar{2},\ \bar{3},\ \bar{4},\ \bar{6}$$

We can refer to these as the 10 'rotations' (both proper and improper). Inclusion of another symmetry operation, namely antiequality, gives rise to the following 10 additional generating elements called *antirotations*:

$$1',\ 2',\ 3',\ 4',\ 6',\ \bar{1}',\ \bar{2}',\ \bar{3}',\ \bar{4}',\ \bar{6}'$$

The 20 generating elements listed here give rise to a total of 122 distinct magnetic point groups. These can be split into three categories, say I (with 32 point groups), II (also with 32 point groups), and III (with 58 point groups). Fig. 2.2.1 explains the distinction between the three kinds with the help of an example.

I. *Magnetic Point Groups of the 1st Kind.* For groups in this category, also known as *white*, or *polar*, or *trivial magnetic*, groups (Kopsky 1976), all the antirotations (including 1′) are absent:

$$M_I = P; \quad 1' \notin M_I \tag{2.2.27}$$

II. *Magnetic Point Groups of the 2nd Kind.* These are also known as *neutral*, or *grey*, or *paramagnetic*, groups, as they possess $1'$ as an element of symmetry. This implies that if an element p ('white') is present, so is p' (black) at the same location. Hence the name 'grey groups' (combination of white and black). They can always be decomposed into cosets as follows:

$$M_{II} = P + P'; \quad 1' \in M_{II} \tag{2.2.28}$$

The group P is a normal subgroup of M_{II} of index 2, and M_{II} can therefore be written as the following direct product:

$$M_{II} = P \otimes \{e, 1'\} \tag{2.2.29}$$

Fig. 2.2.1(b) shows an example of a group of this kind:

$$C_{3v}^{II} = C_{3v} + r\, C_{3v}, \tag{2.2.30}$$

with C_{3v} having the symmetry elements

$$C_{3v} = (e, C_3, C_3^2, \sigma_v, \sigma_v', \sigma_v'') \tag{2.2.31}$$

Here r stands for the time-inversion operator $1'$, and the superscripts $'$ and $''$ on σ_v are part of the standard Schoenflies notation for denoting different vertical mirror operations (not to be confused with the time-inversion operation !).

III. *Magnetic Point Groups of the 3rd Kind.* These are also known as *black-white*, or *mixed polarity*, or *nontrivial-magnetic*, groups. Here the symmetry operation $1'$ is not present by itself, but at least one other antirotation is present. There are two types of point operations in this kind of magnetic point groups:

$$M_{III} = N + 1'\,(P - N), \tag{2.2.32}$$

$$N = h_1,\, h_2,\, \ldots h_k, \tag{2.2.33}$$

$$P - N \;=\; p_1', p_2', \ldots p_k' \tag{2.2.34}$$

If $(p_r)^n = E$ for odd n, then p_r' is not present. This is because

$$(p_r')^n = (1')^n\, (p_r)^n = 1'E = 1', \tag{2.2.35}$$

and $1'$ is not present in this category of groups. It follows that the symmetry operation $3'$ (or C_3' in Schoenflies notation) is excluded.

Eq. 2.2.32 can be rewritten as follows:

$$M_{III} = N + 1'p'N; \quad p' \in (P - N) \tag{2.2.36}$$

N is a normal subgroup of M_{III}. This fact presents a method for constructing the magnetic point groups of this kind: One takes a chemical point group, identifies its normal subgroups of index 2, and replaces the rest of the symmetry operations by their antioperations. For example, consider the point group $(2/m)$:

$$(2/m) = (1, 2, \bar{1}, m) \;=\; (1, 2) + (\bar{1}, m) \tag{2.2.37}$$

The corresponding black-white point group is $(2/m')$:

$$(2/m') = (1, 2, \bar{1}', m') \tag{2.2.38}$$

If a point group has more than one normal subgroups of index 2, we obtain as many black-white point groups.

Fig. 2.2.1(c) is an illustration of another example of this kind:

$$C_{3v}^{III} = (e, C_3, C_3^2, r\sigma_v, r\sigma_v', r\sigma_v'') \tag{2.2.39}$$

Shubnikov Groups

Space groups for the conventional magnetic or black-white symmetry of crystals are known as Shubnikov groups. Just as each of the 230 ordinary crystallographic space groups (*Fedorov groups F*) can be associated with an underlying point group (from among the 32 ordinary (or chemical) point groups, P), each of the 1651 Shubnikov groups, Ш, can be associated with one of the 122 crystallographic magnetic point groups, M. This association also provides a scheme for classifying the Shubnikov groups into three categories:

I. *White or polar Shubnikov groups.* These correspond to the 230 ordinary space groups, or Fedorov groups. They do not contain any antioperations:

$$Ш_I = F; \quad 1' \notin Ш_I \tag{2.2.40}$$

The underlying point groups for any of these are from among the 32 white or polar magnetic groups, M_I.

II. *Neutral or grey Shubnikov groups.* In this category of 230 Shubnikov groups, the presence of any symmetry operation is invariably accompanied by the presence of the corresponding antioperation at the same location:

$$Ш_{II} = F + 1'F; \quad 1' \in Ш_{II} \tag{2.2.41}$$

The underlying point group is one of the neutral or grey groups, M_{II}.

III. *Black-white or mixed-polarity Shubnikov groups.* In this category, time inversion, $1'$, is not a symmetry operation by itself, but some other antioperations are always present. There are 1191 such groups, in which no rotation or translation is accompanied by the corresponding antirotation or antitranslation. If all the antioperations of such a group are replaced by the corresponding ordinary operations, one obtains a Fedorov group, F. With respect to this group, there exists a Fedorov group F_N which is a normal subgroup of F of index 2. [The subscript N stands for 'normal', not for 'nonsymmorphic'.] This fact provides a method for constructing the Shubnikov groups of this category. One scans all the subgroups of index 2 of all the Fedorov groups F. The operations corresponding to F_N are left unchanged, and the rest are replaced by the corresponding antioperations; the result is a Shubnikov group of the third kind.

Suppose P is the point group underlying the Fedorov group F. The point group underlying F_N can be either H, a subgroup of P of index 2 (case a), or it can be the group P itself (case b). In case b, both F_N and F have the same rotational symmetry operations, but the translational symmetry of F_N is half that of F; i.e. the volume of the primitive unit cell of F_N is twice that of F. In case a, F_N and F have the same translational symmetry, but F_N has only half the number of rotational symmetry operations.

For case a, the black-white Shubnikov group III_{IIIa} has a black-white underlying point group M_{III}. For case b, this point group is a grey group M_{II}. That is,

$$\text{III}_{IIIa} = F_N + 1' \, (F - F_N); \quad 1' \notin M_{IIIa} \tag{2.2.42}$$

Here the set $(F - F_N)$ does not contain pure lattice translations; all colour reversal or time inversion operations are effected only through rotational operations.

For the category IIIb, the antiequality or time inversion operations are all associated with lattice translations only; i.e. we have a black-white lattice in this case. Such a lattice can be defined in terms of the ordinary Bravais lattice T by augmenting it with a set of antitranslations:

$$T_{IIIb} = T + 1' \, \{\mathbf{E}|\mathbf{t}\}, \tag{2.2.43}$$

where $\mathbf{t}$ is a Bravais lattice translation. Accordingly,

$$\text{III}_{IIIb} = F_N + 1' \, \{\mathbf{E}|\mathbf{t}\} \, F_N \tag{2.2.44}$$

Here F_N does not contain the element $\{\mathbf{E}|\mathbf{t}\}$. It is a normal subgroup of F of index 2 with respect to translations:

$$F = F_N + \{\mathbf{E}|\mathbf{t}\} \, F_N \tag{2.2.45}$$

2.2.19 Limit Groups

Nonmagnetic Curie Groups

Point groups involving at least one axis of ∞-fold symmetry are called limit groups, or Curie groups.

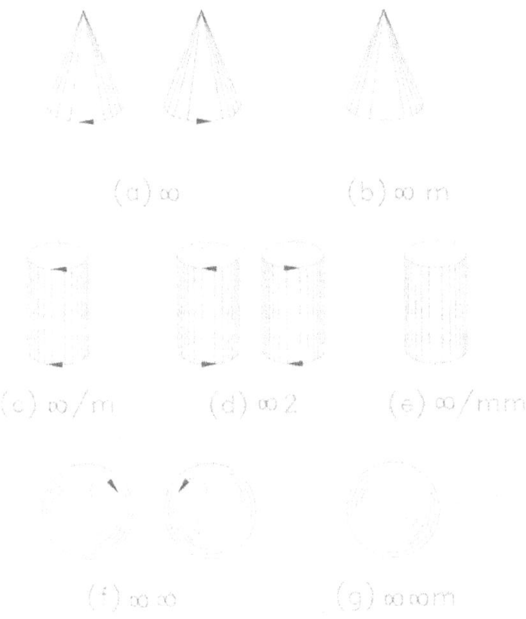

Figure 2.2.2: Geometrical objects having the point-group symmetries of the seven nonmagnetic Curie groups. The arrows show the directions of rotation or twisting of the objects.

There are seven nonmagnetic Curie groups in all: ∞, ∞m, ∞/m, $\infty 2$, ∞/mm, $\infty\infty$, and $\infty\infty m$.

It is convenient and instructive to visualize the symmetry represented by these groups in terms of geometrical objects. Fig. 2.2.2 shows the objects possessing the point-symmetries corresponding to these groups. Each of the 32 crystallographic chemical point groups is a proper subgroup of at least one of the seven nonmagnetic Curie groups.

A cone rotating about its central axis (Fig. 2.2.2(a)) has the symmetry group ∞. The ∞-fold axis coincides with the central axis, and there is no other symmetry element present.

A non-rotating cone (Fig. 2.2.2(b)) has a higher symmetry, ∞m. This group comprises an ∞-fold axis, and an infinite number of planes of mirror symmetry parallel to this axis. A constant, uniform, electric field has the symmetry of this limit group. By a process called poling, a ferroelectric ceramic can be made to acquire a preferred orientation or *texture* by cooling it from the paraelectric phase to the ferroelectric phase while under the action of a sufficiently strong electric field. Such a poled ceramic acquires, on an appropriately macroscopic scale, the same symmetry as the electric field, namely ∞m.

A uniform magnetic field has the *spatial* symmetry ∞/m (in fact, it has an additional symmetry involving time inversion, which we shall describe presently). Here the mirror plane is not parallel, but perpendicular, to the axis of ∞-fold symmetry. This symmetry can be visualized as that of a cylinder rotating about its central axis (Fig. 2.2.2(c)).

The limit group $\infty 2$ corresponds to the symmetry of a cylinder twisted as shown in Fig. 2.2.2(d). There are an infinite number of axes of 2-fold symmetry *transverse* to the ∞-fold axis, the latter coinciding with the central axis of the cylinder. [The symbol for this group should really be $\infty/2$, but $\infty 2$ is used conventionally].

If a cylinder at rest has no twisting or other forces acting on it (except hydrostatic pressure), it has the symmetry ∞/mm (Fig. 2.2.2(e)). There is one mirror plane of symmetry perpendicular to the ∞-fold axis and passing through the centre of gravity, and an infinite number of mirror planes coinciding with the ∞-fold axis. A force field generated by a uniform tensile or compressive mechanical stress has this symmetry.

The limit group $\infty\infty$ (Fig. 2.2.2(f)) can be visualized as that of a sphere with all its radii rotating, so that there is an infinite number of ∞-fold axes, but no mirror planes of symmetry. This is called the *group of rotations*. An alternative symbol for this group is SO(3) (§B.4).

The *orthogonal group*, $\infty\infty m$, describes the symmetry of an ordinary, nonrotating, sphere, with an infinite number of ∞-fold axes, all intersecting at a point (the centre of the sphere), and an infinite number of mirror planes passing through the centre (Fig. 2.2.2(g)). An alternative symbol for this group is O(3) (§B.4).

Limit groups ∞, $\infty/2$, and $\infty\infty$, which lack a mirror plane of symmetry, are *enantiomorphous groups*. This means that objects having these symmetries can be right-handed or left-handed. Those crystallographic point groups which are subgroups of these groups are therefore also enantiomorphous groups.

All the limit groups described above are subgroups of $\infty\infty m$

Magnetic Curie Groups

So far we have assumed implicitly the existence of time-inversion symmetry. In other words, we have assumed that for every spatial symmetry element present in the group, there is another one at the same location for which time t is replaced by $-t$. This means that the seven limit groups described above, and depicted geometrically in Fig. 2.2.2, are really *grey* groups, or magnetic point groups of the second kind (cf. §2.2.18). To take cognisance of this, we must attach the symbol $1'$ to the symbols of the seven nonmagnetic Curie groups considered above. For example, we should really be writing $\infty\infty m1'$, rather than $\infty\infty m$, for the group which represents the symmetry of an ordinary sphere. Similarly for the other six grey groups.

The magnetic Curie groups are, naturally, proper subgroups of these seven grey groups. We list them below. The subgroups given in any particular line are only those which have not appeared in the line(s) above them.

$$\infty\infty m1' \supset \infty\infty m', \infty\infty m \tag{2.2.46}$$

$$\infty/mm1' \supset \infty/m'm, \infty/mm', \infty/m'm', \infty/mm \tag{2.2.47}$$

$$\infty m1' \supset \infty m', \infty', \infty m \tag{2.2.48}$$

$$\infty 21' \supset \infty 2', \infty 2 \tag{2.2.49}$$

$$\infty\infty 1' \supset \infty\infty \tag{2.2.50}$$

$$\infty/m1' \supset \infty/m \tag{2.2.51}$$

$$\infty 1' \supset \infty \tag{2.2.52}$$

There are thus 14 magnetic Curie groups in all. Since the 7 grey or non-magnetic Curie groups are all subgroups of $\infty\infty m1'$, the 14 magnetic Curie groups are also subgroups of $\infty\infty m1'$.

Coming back to the question of the full symmetry of a uniform magnetic field, this symmetry is ∞/mm', and not ∞/m.

2.2.20 Layer Groups and Rod Groups

Domain structure is a characteristic feature of ferroic materials. Any two contiguous domains are separated by a usually planar interface called the domain wall. Each of the domains separated by the wall has the same triperiodic crystal structure defined by the space group of the ferroic phase. Since two such periodic structures meet at the domain wall, the wall itself will have a periodic structure. But the wall can have periodicity in only two directions or dimensions. What kind of groups are appropriate for describing this periodicity? We deal with this question in this section.

Crystallographic groups possible in three dimensions are: G_3^3 (space groups), G_2^3 (layer groups), G_1^3 (rod groups), and G_0^3 (point groups) (cf. §2.2.7 for notation).

In the symmetry operations of layer groups, at least one plane remains invariant. This plane is therefore unique, or *singular*; there is no other plane obtainable from it through any of the symmetry operations of the group G_2^3.

Similarly, the rod groups G_1^3 involve a *singular line*, and the point groups G_0^3 a *singular point.*

A singular plane or a singular line may be either *polar* or *nonpolar*, depending on whether or not its two sides are different or identical.

Types of groups possible in 2-dimensional space are G_2^2, G_1^2, and G_0^2.

The G_2^2 are *plane groups*, and are to be distinguished from the layer groups G_2^3. The latter describe 2-dimensionally periodic (or "diperiodic") objects existing in 3-dimensional space (e.g. domain walls in ferroics). By contrast, plane groups are defined in a 2-dimensional space only. The total number of groups of type G_2^2 is 17, whereas there are 80 G_2^3 groups. The 17 2-dimensional space groups (or plane groups) are included in the 80 layer groups, and are derived by confining our attention to only those symmetry operations which involve motions in strictly two dimensions.

In 1-dimensional space, the possible group types are G_1^1 and G_0^1.

Layer groups are of interest, not only for describing the symmetry of domain walls in ferroic materials, but also for dealing with certain phase transitions (Hatch & Stokes 1986; Litvin & Wike 1991).

Although diperiodic systems have periodicity in only two dimensions, additional symmetry may still be present in the third dimension. For example, the singular plane may be a plane of mirror symmetry, making it a nonpolar plane.

The 80 layer groups (or diperiodic space groups) have been derived and tabulated by Wood (1964), and Hatch & Stokes (1986). One can associate a 3-dimensional space group with each layer group. The latter can be obtained from the former by removing the z-component from all translations.

2.2.21 Colour Symmetry

The antiequality operator introduced in §2.2.18 allows only two states, say up and down, or black and white. Some crystal configurations (e.g. helical magnetic structures, or certain incommensurate phases) require the use of a more general description of the symmetry involved. This is done by introducing the concept of colour symmetry (Naish 1963; Koptsik 1975; Opechowsky 1977).

In addition to the operators which act on atomic coordinates, one introduces a set of rotations ϕ_i of the atomic spins. The 1651 Shubnikov groups then become special cases of this dispensation, with $\phi = 0$ and $\phi = 180^o$.

Three types of colour groups have been introduced.

In the so-called *Q-groups*, a symmetry element can be written in the form

$$g_Q = g(\mathbf{r}, \mathbf{S})\ \phi(\mathbf{S}) \tag{2.2.53}$$

The first factor on the right acts on the spatial coordinates of atoms and inverts their spins. The second factor further reorients the spins.

For the so-called *P-groups* or *permutational colour groups* the symmetry elements can be expressed as

$$g_P = g(\mathbf{r})\ \phi(\mathbf{S}) \tag{2.2.54}$$

The first factor, called a *base element*, acts only on the spatial coordinates. The second factor is the *colour load* (a permutation). Permutational colour groups are the simplest crystallographic colour groups. Several examples of them have been discussed by Litvin, Kotzev & Birman (1982).

The third type of colour groups are called *W-groups.* In the Q-groups and the P-groups all the spins are rotated by the same angle ϕ. In W-groups individual angles of rotation have to be assigned to each spin in the unit cell (see Izyumov & Syromyatnikov (1990) for references to original work.

Recently, Lifshitz (1997) has formulated a new, and comprehensive, theory of colour symmetry, which encompasses quasicrystals also.

SUGGESTED READING

A. V. Shubnikov, N. V. Belov, and others (1964). *Colored Symmetry.* Pergamon Press, Oxford.

E. M. Lifshitz & L. P. Pitaevskii (1980). *Statistical Physics*, third edition. Pergamon Press, Oxford. Chap. XIII.

B. K. Vainshtein (1981). *Modern Crystallography I. Symmetry of Crystals: Methods of Structural Crystallography.* Springer-Verlag, Heidelberg.

Yu. I. Sirotin & M. P. Shaskolskaya (1982). *Fundamentals of Crystal Physics.* Mir Publishers, Moscow.

G. Venkataraman, D. Sahoo & V. Balakrishnan (1989). *Beyond the Crystalline State: An Emerging Perspective.* Springer-Verlag, Berlin.

G. Burns & A. M. Glazer (1990). *Space Groups for Solid State Scientists*, second edition. Academic Press, Boston.

M. Senechal (1990). *Crystalline Symmetries: An Informal Mathematical*

Introduction. Adam Hilger, Bristol. A compact, highly readable book. Strongly recommended.

T. Hahn & H. Wondratschek (1994). *Symmetry of Crystals: Introduction to International Tables for Crystallography Volume A*. Heron Press, Sofia.

R. Lifshitz (1997). Theory of colour symmetry for periodic and quasiperiodic crystals. *Rev. Mod. Phys.*, **69**, 1181.

R. Mirman (1999). *Point Groups, Space Groups, Crystals, Molecules*. World Scientific, Singapore.

2.3 CRYSTAL SYMMETRY AND THE CURIE SHUBNIKOV PRINCIPLE

All actions take place in time by the interweaving of the forces of nature.

Bhagavad Gita

... nature does not simply find symmetry a convenient feature in building physical structures, nature absolutely demands it.

Kaku & Thompson (1997)

Why do crystals possess the symmetries they do? This question does not yet have an entirely satisfactory answer, but certain apparently reasonable conclusions have emerged from the work of some investigators (Sheftal 1966a, b, 1976; Vainshtein 1981, 1988).

A crystal is said to have symmetry because, when certain transformations are applied to it, it transforms back into itself. For the sake of concreteness, we shall restrict ourselves to coordinate transformations only.

The effect of any nontrivial coordinate transformation is to carry, or map, one part of the crystal to another. The invariance of the crystal under a symmetry transformation implies that the crystal consists of *equal parts*, which are mapped onto one another under the action of the symmetry transformations.

However, it is not sufficient that a symmetric object be composed of equal and identical parts. There must also be an *equal (or identical) placement* of these equal parts. Any arrangement other than this, for example a random aggregate of equal parts, cannot allow symmetry transformations to be applied; such an aggregate can possibly have symmetry only in a statistical sense. Thus:

The symmetry of a crystal can be regarded as synonymous with an identical or equal placement of equal parts.

The 230 crystallographic space groups define exhaustively all possible site symmetries in any crystal. The symmetry of any site in a crystal has to be from among the 32 crystallographic point-group symmetries. Any "particle" (atom or molecule) occupying a given site in the crystal can only have a point symmetry that is compatible with the site symmetry. If it had more symmetry than that of the site it occupies, the environment around that site would tend to distort it. As a result, the symmetry of the particle would decrease to achieve compatibility with the site symmetry. If it has less symmetry than the site symmetry, one of two things can happen. Either the lower symmetry of the particle would distort the crystalline environment, so that the overall space-group symmetry is reduced; or, else, the particle would be forced to move to a neighbouring, less symmetric, site.

It follows that high-symmetry particles have only a few allowed modes of packing in a crystal. In contrast to this, low-symmetry particles can crystallize in a richer variety of packing arrangements. High-symmetry particles can, of course, form low-symmetry *combinations* for qualifying for occupying low-symmetry sites. The most familiar example of this are molecules, which have lower symmetries than the atoms from which they are formed, the latter having a very high (spherical) symmetry in the isolated state. This is the reason crystals can even have symmetries as low as triclinic, even though they are composed of atoms which had spherical symmetry to start with.

2.3.1 The Asymmetric Unit

We can identify the equal parts that a crystal is composed of as those (smallest) regions which do not get transformed into themselves under the operations of the space group of the crystal. Crystallographers refer to such a region of the unit cell as the asymmetric unit.

It is worthwhile mentioning here that the asymmetric unit, though devoid of any symmetry described by the space group of the crystal, may sometimes have (exact or approximate) *noncrystallographic* symmetry (Fichtner 1986). This usually happens when certain high-symmetry subunits of the asymmetric unit come together to form an appropriately asymmetric combination. The meaning of "appropriate" will become clear presently.

2.3.2 Interplay between Dissymmetrization and Symmetrization

The occurrence of symmetry in crystals can be rationalized in terms of the laws of thermodynamics. The most important postulate in this regard appears to be the following:

> *A crystal acquires the symmetric configuration it does because that amounts to a state of least energy for the system.*

If an asymmetric unit finds for itself a certain least-energy configuration, other asymmetric units will also usually occur in the same configuration, with the same energy. Furthermore, normally the coming together (bonding) of all the asymmetric units resulting in the growth of the complete crystal must also occur in an identical or quasi-identical manner, everywhere, because only this can ensure that the crystal as a whole also has the least energy. If it occurred in nonidentical ways in different parts, then either the subunits (asymmetric units) are not identical (which is not possible), or there is inequality of interaction among identical subsystems (which is absurd). In this sense, formation of a crystal is a process of *symmetrization* (mentioned in the statement of the Curie-Shubnikov principle), i.e. development of higher symmetry by the equal placement of lower-symmetry subsystems, namely the asymmetric units (cf. Appendix C).

The opposite process of *dissymmetrization*, i.e. lowering of symmetry (cf. Appendix C), also occurs in the total process of crystal formation, if we trace the chain of events from an earlier stage, namely the formation of molecules or other growth units from the spherically symmetric atoms. Atoms undergo chemical reactions to form molecules, reducing the total energy of the system by an amount equal to the binding energy. The molecules have a lower symmetry compared to the spherically symmetric atoms, so this is dissymmetrization. A large number of identical or equal molecules (or asymmetric units) assemble into a crystal in such a way that they all have identical surroundings. This happens because the overall energy is minimized by this process. But this also happens to be a process of symmetry enhancement (symmetrization), because at least one new type of symmetry arises, namely translational symmetry, which was not present to start with. New rotational symmetries also arise often.

Apart from the equality of subsystems, another factor which contributes to the occurrence of symmetry in crystals is the fact that there is only a *finite* number of *types* of equal subsystems. There is a finite number of types of 'elementary' particles[2], atoms, and molecules. When they assemble under the restriction of "equal placement of equal parts" (to ensure minimization of energy), the possible configurations are not only finite in number, they also develop additional symmetry. Had the number of available options been unlimited, crystallographic symmetry need not arise.

[2]The superstring theory notwithstanding !

SUGGESTED READING

Sheftal, N. N. (1976). A crystal as a medium that orders phenomena. In N. N. Sheftal (Ed.), *Growth of Crystals*, Vol. 10. Consultants Bureau, New York.

B. K. Vainshtein (1981). *Modern Crystallography. I. Symmetry of Crystals: Methods of Structural Crystallography.* Springer-Verlag, Berlin.

2.4 INCOMMENSURATELY MODULATED CRYSTALS

One can introduce a regular modulation in a periodic signal by superimposing on it another periodic signal with a different period. Something similar happens in certain phases of crystals. The structure $C(\mathbf{r})$ of a normal crystal is a convolution of the lattice function $L(\mathbf{r})$ with the basis function $B(\mathbf{r})$ (Eq. 2.2.2). The presence of an additional modulation, say $M(\mathbf{r})$, over the lattice function $L(\mathbf{r})$ leads to the following mathematical description of the modulated crystal structure:

$$C_{mod}(\mathbf{r}) = B(\mathbf{r}) * [L(\mathbf{r}) . M(\mathbf{r})] \tag{2.4.1}$$

It is a convolution of $B(\mathbf{r})$ with a product of two functions in real (or direct) space.

If it is a so-called *deformation-type* or *displacement-type* modulation (Amelinckx et al. 1989), the average positions of atoms of the basic structure described by Eq. 2.2.2 are not affected. In other words, the vector sum of the displacements of all the atoms equivalent under the space-group operations of the basic structure is zero. Therefore, the gross macroscopic (or point-group) symmetry of the crystal is not changed by such a modulation.

Such modulations can arise when there are two *competing interactions*, each tending to lower the overall free energy, but each leading to a different periodicity. The crystal structure settles for a compromise between the two, with one dominant interaction defining the basic lattice function $L(\mathbf{r})$, and the other causing the modulation $M(\mathbf{r})$ of the basic structure. An example is provided by the Peierls distortion (Peierls 1955), wherein the total electronic energy can be decreased by the opening up of energy gaps along certain reciprocal-lattice planes parallel to flat parts of the Fermi surface, and the resulting mechanical deformation which tends to oppose this change of structure.

Charge-density waves can also lead to a similar situation sometimes (Overhauser 1978).

Typically, such a modulation occurs along some specific crystallographic direction, say along the lattice vector **a**. If the periodicity (say d) of the modulation function $M(\mathbf{r})$ is such that d/a is not a ratio of two integers (i.e. if it is an *irrational* number), the crystal structure is said to be incommensurately modulated.

As is the case when two periodic signals are superimposed, the overall system is very sensitive to any changes in their periods (like in Moiré fringes). In the case of a modulated crystal structure, even small changes of, for example, temperature can cause large changes in the relationship of the modulation to the basic structure. In particular, outside a certain temperature interval, the ratio d/a may become a *rational* number, leading to a structure which is *commensurate* to the basic structure. Such a structure has a larger period of repetition, and is called a *superstructure.*

The symmetry of incommensurate phases of crystals can be described in terms of "superspace groups" (Janner & Janssen 1979; de Wolff, Janssen & Janner 1981).

SUGGESTED READING

G. Burns & A. M. Glazer (1990). *Space Groups for Solid State Scientists*, second edition. Academic Press, Boston.

Chapter 3

CRYSTAL PHYSICS

The practical importance of ferroic crystals stems from the occurrence of ferroic phase transitions in them. We have defined a ferroic phase transition as one which entails a nondisruptive change of the point-group symmetry of the prototype of the crystal. A reduction of the point-group symmetry is necessarily accompanied by the emergence of at least one macroscopic property coefficient, which was forbidden by the point-group symmetry of the crystal from being nonzero in the parent phase. The symmetry restrictions on physical properties of crystals come from the *Neumann theorem* (cf. §C.1 of Appendix C).

Since macroscopic physical properties of a crystal are translation invariant, it is sufficient and appropriate to consider only its point-group symmetry for determining the restrictions on these properties, the relevant point group being the *point group underlying the space group* of the crystal (§2.2.17).

Thus the atomic structure of a crystal is taken as having only an indirect bearing on the symmetry of its macroscopic properties. For example, the translational periodicity of the structure requires that the directional symmetry of the macroscopic properties be from among the 32 crystallographic point-group symmetries only (§2.2.15). For the discussion in this chapter, we therefore take the crystal to be a *homogeneous continuum*, which may, however, be *anisotropic*, in general.

By replacing the actual structure of a crystal by an anisotropic continuum, we are able to determine the effect of external fields on macroscopic properties in a form that can be related directly to experimental results. This is an important feature of crystal physics.

3.1 TENSOR PROPERTIES

Macroscopic physical properties are specified in terms of relations between measurable quantities. For example, the density ρ of a material is defined in terms of mass and volume:

$$m = \rho V \tag{3.1.1}$$

Since both V and m can be defined without reference to any direction in space, ρ also does not depend on direction, and can be specified by a single number. This is an example of a *scalar* macroscopic property.

Unlike m and V, there are physical quantities, like the electric field, for which one must specify not only the magnitude but also the direction. These are called *vector* quantities. One way of specifying them is by choosing three mutually perpendicular coordinate axes (the Cartesian axes) Ox_1, Ox_2, Ox_3, and stating the components along these axes:

$$\mathbf{E} = (E_1, E_2, E_3) = (E_i) \tag{3.1.2}$$

Thus three numbers are required to specify a vector quantity like $\mathbf{E}$ in 3-dimensional space.

Electric field is an example of a vector quantity, but it is not a *vector property* of a crystal. An example of the latter is provided by the *pyroelectric effect* exhibited by crystals belonging to any of the 10 polar classes (§2.2.15). In such crystals there can exist spontaneously an electric polarization $\mathbf{P}$ (dipole moment per unit volume) which changes with temperature:

$$P_i = p_i \, \Delta T, \quad i = 1,2,3 \tag{3.1.3}$$

How do scalar and vector properties behave under transformations of coordinate axes? A transformation from one Cartesian set of axes to another can be defined as follows:

$$x'_i = a_{ij} \, x_j \tag{3.1.4}$$

Here we have adopted the convention that when a subscript is repeated (e.g. j in the above equation), a sum over it is implied ($j = 1,2,3$). There are three such equations, one for each i.

Under the above transformation, a scalar quantity ϕ undergoes no change at all:

$$\phi' = \phi \tag{3.1.5}$$

A vector property, like the pyroelectric vector, transforms as follows:

$$p'_i = a_{ij} \, p_j \tag{3.1.6}$$

After considering a property (density) which relates two scalars, and a property (pyroelectric vector) which relates a scalar (ΔT) to a vector (polarization), we consider a property which relates two vectors. An example of this is provided by the dielectric response of a crystal. When an electric field (E_j) is applied to a crystal along a general direction, an electric displacement (D_i) is induced:

$$D_i = \epsilon_{ij}\, E_j, \quad i = 1, 2, 3 \tag{3.1.7}$$

Each of these three equations has three terms on the right-hand side, one each for $j = 1, 2, 3$, so that there are, in all, nine coefficients ϵ_{ij} involved in the relationship between the vectors $\mathbf{D}$ and $\mathbf{E}$.

Eqs. 3.1.6 can be solved for $\mathbf{p}$ in terms of $\mathbf{p}'$. A similar equation for electric field can be solved for $\mathbf{E}$ in terms of $\mathbf{E}'$, to yield

$$E_i = a_{ji}\, E'_j \tag{3.1.8}$$

How does the set of quantities (ϵ_{ij}) transform under a transformation of coordinate axes specified by the matrix (a_{ij})?

Let us suppose that, under this transformation, (D_i) changes to (D'_i), and (E_i) to (E'_i). Then

$$D'_i \;=\; a_{ij}D_j \;=\; a_{ij}\epsilon_{jk}E_k \;=\; a_{ij}\epsilon_{jk}a_{lk}E'_l \tag{3.1.9}$$

Here we have made use of Eqs. 3.1.7 and 3.1.8.

We rewrite Eq. 3.1.9 as follows:

$$D'_i = \epsilon'_{il}E'_l, \tag{3.1.10}$$

where

$$\epsilon'_{il} = a_{ij}a_{lk}\epsilon_{jk} \tag{3.1.11}$$

A physical quantity which transforms according to Eq. 3.1.11, i.e. as a product of two vectors (or two coordinates), is called a *tensor of rank 2*, or a *second-rank tensor*. The dielectric behaviour of a crystal is thus described by a second-rank tensor. Its components are represented by two indices.

In the same spirit, a vector quantity is a tensor of rank 1, and a scalar quantity is a tensor of rank zero. We can generalize these considerations to define tensors of various ranks:

$$\begin{aligned}
\text{Zero}-\text{rank tensor} &: \quad \phi' = \phi && (3.1.12)\\
\text{First}-\text{rank tensor} &: \quad p'_i = a_{ij}p_j && (3.1.13)\\
\text{Second}-\text{rank tensor} &: \quad \epsilon'_{ij} = a_{ik}a_{jl}\epsilon_{kl} && (3.1.14)\\
\text{Third}-\text{rank tensor} &: \quad d'_{ijk} = a_{il}a_{jm}a_{kn}\, d_{lmn} && (3.1.15)\\
\text{Fourth}-\text{rank tensor} &: \quad M'_{ijkl} = a_{im}a_{jn}a_{ko}a_{lp}\, M_{mnop} && (3.1.16)
\end{aligned}$$

3.1.1 Symmetrized and Alternated Tensors

A tensor **B** is said to be an *isomer* of tensor **A** (and vice versa) if **B** has been obtained from **A** through any permutation(s) of its indices.

Consider a tensor **A**, and form all its isomers. A tensor **B** obtained as the arithmetic mean of all the isomers of **A** is called a symmetrized form of tensor **A**.

If in the above arithmetic mean, the isomers obtained from **A** by an even number of permutations are taken with a plus sign, and those by an odd number of permutations with a minus sign, we obtain an alternated form of tensor **A**.

Symmetrization and alternation of tensors can be defined, not only for the entire set of indices, but also for any subset of these indices.

3.1.2 Polar Tensors and Axial Tensors

Let $\det \mathbf{A}$ denote the determinant of the transformation matrix in Eq. 3.1.4. For orthogonal transformations not involving a reflection or inversion operation, $\det \mathbf{A} = +1$, and for those involving such an operation, $\det \mathbf{A} = -1$.

Tensors described by Eqs. 3.1.12 to 3.1.16 are examples of polar tensors: Their transformation properties do not depend on whether $\det \mathbf{A} = +1$ or $\det \mathbf{A} = -1$. By contrast, the transformation properties of an axial tensor (or a pseudotensor) **A** are defined by the following equation:

$$A_{ijkl\ldots} = (\det \mathbf{A})\, a_{im}\, a_{jn}\, a_{ko}\, a_{lp}\, \ldots\, A_{mnop\ldots} \tag{3.1.17}$$

3.1.3 Matter Tensors and Field tensors

Tensors which describe properties of crystals and other materials are called matter tensors. Such tensors must conform to the symmetry of the material.

By contrast, a tensor describing a field applied to a material (electric field, stress field, etc.) does not represent a property of the material. It can have any orientation inside the material, and its symmetry is quite independent of the symmetry of the system on which it is applied. Such a tensor is called a field tensor.

3.1.4 Intrinsic Symmetry of Tensors; the Jahn Symbol

A tensor is said to be symmetric with respect to two or more of its indices if all its isomers differing only by permutations of these indices are equal among themselves. For example, if for a third-rank tensor (d_{ijk}) we have

$$d_{ijk} = d_{ikj} \tag{3.1.18}$$

for all values of indices j and k, it is said to be *symmetric* with respect to its last two indices.

If, on the other hand,

$$d_{ijk} = -d_{ikj}, \qquad (3.1.19)$$

the tensor is said to be *antisymmetric* (or *skew-symmetric*) with respect to its last two indices.

The property of a tensor to be symmetric or antisymmetric with respect to some or all indices, or groups of indices, can be shown to be independent of the coordinate system used for defining its components. It is therefore described as its *intrinsic symmetry*, or *internal symmetry*. This symmetry usually has its origins in the symmetry of the space the tensor is defined in; it may also result from thermodynamic considerations.

Jahn (1949) introduced an elegant and compact scheme of notations for representing the intrinsic symmetry of tensors. In this scheme, the symmetry of a polar vector (first-rank polar tensor) is denoted by the letter V. Since a polar tensor of rank r transforms as the product of the components of r polar vectors, its intrinsic symmetry is denoted by V^r.

A tensor of rank r symmetric with respect to all its indices is denoted by $[V^r]$. If it is symmetric with respect to only q of its indices, the symbol used is $[V^q]V^{r-q}$. For example, the intrinsic symmetry of a tensor satisfying Eq. 3.1.18 is denoted by $V[V^2]$.

Let us consider a fourth-rank tensor $\mathbf{T}$, such that

$$T_{ijkl} = T_{klij} \neq T_{jikl} \neq T_{ijlk} \qquad (3.1.20)$$

The intrinsic symmetry of such a tensor is denoted by $[(V^2)^2]$, since it is symmetric with respect to permutation of only certain pairs of indices, and not for permutations within a pair.

By contrast, a tensor with the Jahn symbol $[[V^2]^2]$ is a fourth-rank tensor symmetric with respect to the permutation of the first and the second pair of indices, and also with respect to permutation within a pair.

An extra symbol ϵ is used for representing the intrinsic axial or "pseudo" nature of a tensor. Thus, the intrinsic symmetry of a pseudoscalar is denoted by ϵ, that of a pseudovector by ϵV, that of a symmetric pseudotensor of rank 2 by $\epsilon[V^2]$, and so on.

The intrinsic symmetry of an ordinary scalar (like density) is denoted by the symbol 1.

For representing the intrinsic symmetry of antisymmetric tensors, square brackets are replaced by braces. For example, the Jahn symbol for a tensor $\mathbf{T}$, such that $T_{ij} = -T_{ji}$ is $\{V^2\}$.

3.1.5 Extrinsic Symmetry of Tensors

The intrinsic symmetry of tensors is the same in all coordinate systems of reference. In addition, properties of a tensor may also exhibit invariance of all its components under certain specific coordinate transformations. This is referred to as the extrinsic symmetry of tensors.

In this section we focus attention on those point groups G_0^3 of orthogonal transformations under which a tensor exhibits invariance of all its components. Any such transformation, (a_{ij}), is a symmetry transformation for a polar tensor **T** if, for all its components $T_{ijk...}$,

$$a_{mi}\, a_{nj}\, a_{ok} \cdots T_{ijk...} = T'_{mno...} = T_{ijk...} \tag{3.1.21}$$

We consider here the example of inversion symmetry, which is particularly easy to visualize. For an inversion operation, $a_{11} = a_{22} = a_{33} = -1$, and all other elements of the transformation matrix are zero. Therefore, if **T** is a polar tensor of rank r, Eq. 3.1.21 reduces to

$$T'_{mno...} = (-1)^r\, T_{ijk...} \tag{3.1.22}$$

It follows that if r is even, **T** possesses inversion symmetry.

Similarly, if **P** is an axial tensor of rank r, then, under an inversion operation,

$$P'_{mno...} = (-1)^{r+1}\, P_{ijk...} \tag{3.1.23}$$

This implies that all odd-rank axial tensors possess inversion symmetry.

Thus, spatial inversion symmetry is possessed by all scalars, axial vectors, second-rank polar tensors, third-rank axial tensors, fourth-rank polar tensors, and so on.

We now consider some general features of the extrinsic symmetry of tensors of various ranks.

A zeroth-rank polar tensor, i.e. a scalar, has the symmetry of the full orthogonal group $\infty\infty m$, or $O(3)$ (§2.2.19). And a pseudoscalar has the extrinsic symmetry $\infty\infty$, or $SO(3)$.

A first-rank polar tensor, i.e. a polar vector, possesses the symmetry of a non-rotating cone, namely ∞m.

A first-rank axial tensor, i.e. an axial vector, is centrosymmetric. It has the symmetry of a cylinder rotating about its central axis, the corresponding point-group being ∞/m.

A second-rank tensor may be polar or axial, and for each of these cases one has to consider the possibility of the tensor being symmetric, antisymmetric, or nonsymmetric. There are thus six different possibilities to consider.

The case of the second-rank polar tensor is particularly important. It can be given a geometrical interpretation in terms of a second-degree surface (or a *quadric*). Such a surface can be defined by the equation

$$S_{ij}\, x_i\, x_j = 1, \tag{3.1.24}$$

with $S_{ij} = S_{ji}$. If we carry out a coordinate transformation defined by

$$x_i = a_{ki}\, x'_k \quad x_j = a_{lj}\, x'_l, \tag{3.1.25}$$

Eq. 3.1.24 changes to

$$S_{ij}\, a_{ki}\, a_{lj}\, x'_k\, x'_l = 1 \tag{3.1.26}$$

We can rewrite this as

$$S'_{kl}\, x'_k\, x'_l = 1, \tag{3.1.27}$$

with

$$S'_{kl} = a_{ki}\, a_{lj}\, S_{ij} \tag{3.1.28}$$

This transformation law for the coefficients S_{ij} of the quadric is identical to that of a polar second-rank tensor, *provided that the tensor is a symmetric tensor.* Eq. 3.1.24 therefore defines a representation surface (the *representation quadric*) for such a tensor.

A quadric has the important property that it possesses *principal axes.* These are three mutually perpendicular axes, such that when Eq. 3.1.24 is referred to them, it assumes the form

$$S_1\, x_1^2 + S_2\, x_2^2 + S_3\, x_3^2 = 1 \tag{3.1.29}$$

Accordingly, S_1, S_2, S_3 correspond to the *principal components* of the tensor (S_{ij}).

If S_1, S_2, S_3 are all positive, Eq. 3.1.29 defines an ellipsoid. If two of them are positive and one negative, the quadric is a hyperboloid of one sheet, and if one is positive and two negative, it is a hyperboloid of two sheets. If all three are negative, the surface is an imaginary ellipsoid. In all these cases, so long as the three principal components have different magnitudes the symmetry of the quadric is mmm (D_{2h}). This is therefore also the extrinsic symmetry of the symmetric second-rank polar tensor represented by the quadric.

If the magnitudes of any two of the principal components are equal, the representation quadric acquires a circular cross-section perpendicular to the third axis, and therefore its symmetry is enhanced to that of a spheroid of revolution, namely ∞/mm.

When all three components are equal in magnitude, the symmetry of the tensor becomes that of a sphere, namely $\infty\infty m$, and the tensor reduces to a scalar.

We consider next the extrinsic symmetry of an antisymmetric second-rank polar tensor. The most familiar example of this is the tensor defined by the vector product of two polar vectors. Let the two polar vectors be **p** and **q**. Their vector product is a second-rank tensor **V** defined by

$$V_{ij} = p_i\, q_j - q_i\, p_j \tag{3.1.30}$$

Therefore, $V_{ii} = 0$, and $V_{ji} = -V_{ij}$. Consequently, this second-rank tensor has only three nonzero, independent components: V_{12}, V_{23}, and V_{31}. As is well known (see, for example, Nye 1957), the result of the vector product is an axial vector. A polar second-rank antisymmetric tensor is thus *dual to* an axial vector. We have seen above that the extrinsic symmetry of the latter is ∞/m. The same is therefore the symmetry of the former also.

It can be shown likewise that an antisymmetric second-rank axial tensor is dual to a polar vector, and therefore has the symmetry ∞m.

What is the extrinsic symmetry of a symmetric second-rank axial tensor **P** ? Such a tensor can be reduced to the diagonal form, and can be represented by a second-degree surface. However, unlike a symmetric second-rank polar tensor, this tensor is not centrosymmetric. Consequently, its representation quadric lacks a centre of symmetry. We can deduce its symmetry properties by removing the inversion operation as a generator from the symmetry group of the symmetric second-rank polar tensor.

The case $P_1 \neq P_2 \neq P_3$ therefore corresponds to extrinsic symmetry 222 (obtained from the group mmm by removing the symmetry operations generated by the inversion operation).

The case $P_1 = P_2 = P_3$ corresponds to the symmetry of a pseudoscalar, namely $\infty\infty$ (instead of $\infty\infty m$ for a true scalar).

Lastly, the case $|P_1| = |P_2| \neq P_3$ corresponds to the symmetry group $\infty 2$.

We now consider *nonsymmetric* second-rank tensors. To determine their extrinsic symmetries we invoke the following general result: Any second-rank tensor can be written as the sum of a symmetric part and an antisymmetric part:

$$T_{ij} = S_{ij} + A_{ij}, \tag{3.1.31}$$

where

$$S_{ij} = \frac{1}{2}(T_{ij} + T_{ji}), \tag{3.1.32}$$

$$A_{ij} = \frac{1}{2}(T_{ij} - T_{ji}) \tag{3.1.33}$$

If **T** is a polar tensor, so also are **S** and **A**.

The extrinsic symmetry of **T** can thus be determined from that of **S** and **A** by applying the Curie principle of superposition of dissymmetries (§C.1). A number of possibilities can arise depending on the mutual orientations of the principal axes of **S** and **A**.

If each of them possesses an ∞-fold axis, and these are parallel, the intersection group determined by the Curie principle is ∞/m, which is therefore the extrinsic symmetry of the nonsymmetric second-rank polar tensor.

If the ∞-fold axes are mutually perpendicular, the intersection group obtained is $2/m$. The group $2/m$ is also obtained if the planes and axes of symmetry of the two tensors coincide.

Lastly, if there is no coincidence of axes or planes of symmetry, the net symmetry is just $\bar{1}$.

Similar considerations lead to the result that the extrinsic symmetry of a nonsymmetric second-rank *axial* tensor can be ∞, 2, 1, $mm2$, or m (see Shuvalov 1988).

The extrinsic symmetry of tensors becomes increasingly difficult to determine as the rank increases beyond 2. One can often determine a coordinate system in which the matrix of the tensor components assumes the simplest form. And the symmetry elements of the tensor can be usually associated with the principal axes.

3.1.6 Tensor Invariants

A change of coordinate system can, in general, result in a change of the values of the components of a tensor. Tensor invariants are those linear combinations of the components of a tensor that remain invariant under all coordinate transformations.

A tensor component transforms as a product of coordinates, or, what is the same thing, as a product of vector components. We assign to each index of the tensor a different vector $\mathbf{V}$, and, for determining the transformation properties, use the component $(\mathbf{V})_1$ or $(\mathbf{V})_2$ or $(\mathbf{V})_3$ of the vector depending on whether the corresponding index labelling the tensor is 1 or 2 or 3.

For example, to a second-rank tensor (T_{ij}) we assign two vectors $\mathbf{V}_1$ and $\mathbf{V}_2$. Then, for example, T_{13} will transform as $(\mathbf{V}_1)_1(\mathbf{V}_2)_3$, and T_{21} as $(\mathbf{V}_1)_2(\mathbf{V}_2)_1$, etc.

This rule, which is a part of the definition of any tensor, can be used for constructing the invariants of the tensor. We simply have to construct those products of the vectors assigned to the tensor that remain constant under all coordinate transformations. In other words, we look for all possible scalars that can be constructed from $\mathbf{V}_1$, $\mathbf{V}_2$, $\mathbf{V}_3$, etc. We consider some simple examples.

Only one scalar can be constructed from two vectors $\mathbf{V}_1$ and $\mathbf{V}_2$ representing the two indices of a second-rank tensor (T_{ij}). This is the scalar product $\mathbf{V}_1.\mathbf{V}_2$. Since

$$\mathbf{V}_1.\mathbf{V}_2 = (\mathbf{V}_1)_x(\mathbf{V}_2)_x + (\mathbf{V}_1)_y(\mathbf{V}_2)_y + (\mathbf{V}_1)_z(\mathbf{V}_2)z, \tag{3.1.34}$$

the invariant of the tensor is $T_{11} + T_{22} + T_{33}$.

Similarly, a third-rank tensor (T_{ijk}) has only one invariant under a pure rotation of the coordinate system, corresponding to the scalar triple product $\mathbf{V}_1 \times \mathbf{V}_2 \cdot \mathbf{V}_3$. This product is:

$$\begin{array}{lclcl}(\mathbf{V}_1)_y(\mathbf{V}_2)_z(\mathbf{V}_3)_x & - & (\mathbf{V}_1)_z(\mathbf{V}_2)_y(\mathbf{V}_3)_x & + & (\mathbf{V}_1)_z(\mathbf{V}_2)_x(\mathbf{V}_3)_y \ - \\ (\mathbf{V}_1)_x(\mathbf{V}_2)_z(\mathbf{V}_3)_y & + & (\mathbf{V}_1)_x(\mathbf{V}_2)_y(\mathbf{V}_3)_z & - & (\mathbf{V}_1)_y(\mathbf{V}_2)_x(\mathbf{V}_3)_z\end{array}$$

corresponding to the tensor invariant

$$T_{231} - T_{321} + T_{312} - T_{132} + T_{123} - T_{213}$$

For higher-rank tensors, usually more than one scalars can be constructed from their assigned vectors. A set of *linearly independent invariants* has then to be selected (arbitrarily) from among the available invariants.

3.1.7 Equilibrium Properties and Transport Properties

The tensor properties discussed by us so far in this chapter are *equilibrium properties.* A tensor property $\mathbf{T}$ was defined above as relating a force $\mathbf{X}$ to a response $\mathbf{Y}$ through a linear constitutive relation:

$$\mathbf{Y} = \mathbf{T}\mathbf{X} \qquad (3.1.35)$$

There is no restriction here on how slowly $\mathbf{X}$ must be applied on the crystal. This ensures that the system can move from one equilibrium state to the next, till the final equilibrium state is reached. It also means that the change of state is of a thermodynamically reversible nature.

Transport properties of crystals (like thermal conductivity, electrical conductivity, thermoelectric power, diffusivity), though tensor properties, are of a fundamentally different nature in that they involve thermodynamically irreversible processes (entailing an overall increase of entropy). This is so even when steady-state conditions prevail (see Nye (1957)). By and large, we shall not be dealing with such properties in this book. However, the concept of time-reversal symmetry (or the lack of this symmetry) is relevant for a discussion of magnetic properties. The irreversible nature of transport properties implies that they cannot possess time-reversal symmetry.

3.1.8 i-Tensors and c-Tensors

The tensors in Eq. 3.1.35 may or may not possess time-reversal symmetry. Magnetic field $\mathbf{H}$, viewed as arising from electric currents, changes its sign on time reversal:

$$\mathbf{H}(-t) = -\mathbf{H}(t) \qquad (3.1.36)$$

Electric field E, on the other hand, is time-symmetric:

$$\mathbf{E}(-t) = \mathbf{E}(t) \qquad (3.1.37)$$

The same is true about dielectric permittivity:

$$\epsilon(-t) = \epsilon(t) \qquad (3.1.38)$$

Tensors invariant under time reversal are called i-tensors, and those which change their sign under time reversal are called c-tensors (Birss 1964). The magnetic-field tensor is a c-tensor, and the electric field an i-tensor.

3.1.9 Special Magnetic Properties

Consider the basic equation, namely Eq. 3.1.35, defining a matter tensor **T** in terms of a force tensor **X** and a response tensor **Y**.

When both **X** and **Y** are i-tensors, so is **T**.

When both **X** and **Y** are c-tensors, again **T** is an i-tensor.

When either **X**, or **Y**, but not both are c-tensors, **T** is a c-tensor. In this case **T** is a *special magnetic property* (Nowick 1995), provided it is an equilibrium or static property, and not a transport property.

For transport properties, because of the irreversible thermodynamics involved, time-symmetry is not even conceivable, and it is meaningless to think of a situation where **X** is an i-tensor and **Y** a c-tensor, or vice versa.

The magnetic permeability tensor μ provides a good example of what is *not* a special magnetic property:

$$B_i = \mu_{ij} H_j \tag{3.1.39}$$

Since both **H** and **B** are c-tensors, μ is not a special magnetic property. In fact it is an i-tensor.

One can generalize and say that in a constitutive relation like Eq. 3.1.39 involving c-tensor components, if these components occur an even number of times, then the matter tensor involved is not a special magnetic property. If they occur an odd number of times, the matter tensor represents a special magnetic property.

An example of the latter type is the magnetoelectric tensor (α_{ij}):

$$B_i = \alpha_{ij} E_j \tag{3.1.40}$$

One has to use one of the 90 magnetic point groups when dealing with the symmetry properties of (α_{ij}) (§2.2.18). By contrast, the 32 chemical or ordinary point groups are appropriate for dealing with (μ_{ij}), in spite of the fact that the latter connects two magnetic quantities (**H** and **B**).

SUGGESTED READING

J. F. Nye (1957). *Physical Properties of Crystals: Their Representation by Tensors and Matrices.* The Clarendon Press, Oxford.

S. Bhagavantam (1966). *Crystal Symmetry and Physical Properties.* Academic Press, London.

H. J. Juretschke (1974). *Crystal Physics: Macroscopic Physics of Anisotropic Solids.* Benjamin, London.

Yu. I. Sirotin & M. P. Shaskolskaya (1982). *Fundamentals of Crystal Physics.* Mir Publishers, Moscow.

Shuvalov, L. A. (Ed.) (1988). *Modern Crystallography IV. Physical Properties of Crystals.* Springer-Verlag, Berlin.

3.2 RESTRICTIONS IMPOSED BY CRYSTAL SYMMETRY ON TENSOR PROPERTIES

3.2.1 Neumann Theorem

All properties of a crystal result from its atomic structure. In accordance with the Curie principle (Appendix C), the elements of symmetry in the *cause* (namely the atomic structure) must be present in the *effects* (the properties, including tensor properties).

Since the repeat distances characteristic of translational symmetry of crystals are in the subnanometer range, they are not considered important in determining the point-group symmetry of macroscopic tensor properties of crystals. Therefore the restrictions imposed by crystal symmetry on the tensor properties can be adequately determined by considering merely the point-group symmetry of the crystal.

Thus, all the symmetry elements of the point group of the crystal must be present in the totality of (extrinsic) symmetry elements possessed by any macroscopic tensor property exhibited by the crystal. There are no restrictions on a tensor property displaying *higher* symmetry than that embodied in the point group of the crystal. The Neumann theorem of crystal physics expresses this as follows (Nye 1957):

> *The symmetry elements of any physical property of a crystal must include the symmetry elements of the point group of the crystal.*

If $G_{crystal}$ is the point-symmetry group of the crystal, and $G_{property}$ that of the property under consideration, then the former either coincides with the latter or is its subgroup:

$$G_{property} \supseteq G_{crystal} \tag{3.2.1}$$

In other words, a tensor property can exist in a crystal only if the point group of the crystal is a subgroup of the symmetry group of the property.

The Neumann theorem must hold even when c-tensors are involved. The symmetry elements of any physical property of a crystal must include *all* the symmetry elements of the point group of the crystal, even those in space-time, rather than those in space alone.

This statement, however, is valid only for static or equilibrium tensor properties. It does not hold for transport properties. For such properties time-symmetry is not an allowed symmetry operation (because of the preference for that direction of time in which overall entropy increases).

An obvious consequence of the validity of the Neumann theorem in space-time is that if the point-group symmetry of the crystal includes time reversal, i.e. if it is a nonmagnetic crystal, then all the matter tensors that are c-tensors must be identically zero for such a crystal. For example, all components of the magnetoelectric and piezomagnetic tensor must be identically zero for such crystals. But the magnetic permeability tensor, being an i-tensor, is not zero.

3.2.2 Crystallographic System of Coordinates

The symmetry elements of the extrinsic symmetry of a tensor property must coincide with the respective symmetry elements of the point group of the crystal. However, the point symmetry of the crystal is often lower than that of the tensor property under consideration. The crystal classes 1 and $\bar{1}$ do not provide any internal or natural choice of coordinate axes for relating to the symmetry elements of their tensor properties. The classes 2, 3, 4, 6, m, $\bar{3}$, $\bar{4}$, $\bar{6}$, $2/m$, $4/m$ and $6/m$ provide only one axis of symmetry which may serve as a principal axis for tensor properties. The principal axes become available more easily as we consider higher point-group symmetries.

When the point-group symmetry of a crystal does not define uniquely all the three principal axes for its tensor properties and other anisotropic properties, it becomes necessary to follow some universally adopted convention and define the so-called *crystallographic system of coordinates* (*Standards on Piezoelectric Crystals* 1949). This system of coordinates not only has an arbitrary orientation of axes for crystals of low symmetry, it is also orthogonal (whereas the angles between basis vectors in some low-symmetry crystals are not).

We mention here only one example of how such a system $Ox_1x_2x_3$ is defined with respect to the crystal axes $Oxyz$: For crystals belonging to the monoclinic system (point groups 2, m, $2/m$), Ox_2 is taken either along the 2-fold axis, or perpendicular to the mirror plane m. And Ox_1 and Ox_3 are defined in the plane perpendicular to the 2-fold axis, or in the mirror plane m (if present), such that Ox_1, Ox_2, Ox_3 are mutually orthogonal;

further, Ox_1 coincides with O_x.

3.2.3 Some Consequences of the Neumann Theorem

The Neumann theorem (Eq. 3.2.1) puts severe and definite restrictions on the macroscopic tensor properties a crystal can possess. As a consequence of this theorem, one can make categorical statements about the tensor coefficients which must be zero in a crystal of a given point-group symmetry (magnetic or nonmagnetic).

Any polar vector property (e.g. the pyroelectric tensor) possesses the extrinsic symmetry $\infty\infty m$. It follows from Eq. 3.2.1 that only those crystallographic point groups are compatible with the occurrence of a first-rank polar tensor property which are subgroups of the group $\infty\infty m$. There are 10 such groups, and they are called polar groups (cf. §2.2.15):

$$1, 2, 3, 4, 6, m, mm2, 3m, 4mm, 6mm \tag{3.2.2}$$

Since a ferroelectric phase of a crystal is characterized by the occurrence of a spontaneous polarization vector, it follows that only these 10 crystal classes can allow ferroelectric behaviour.

Similarly, first-rank axial tensor properties can occur in only those crystal classes the point-symmetry of which is a subgroup of the limit group ∞/m. There are 13 such classes:

$$1, 2, 3, 4, 6, \bar{1}, m, \bar{3}, \bar{4}, \bar{6}, 2/m, 4/m, 6/m \tag{3.2.3}$$

For dealing with magnetic macroscopic symmetry, we must use magnetic point groups. Ferromagnetic crystals are characterized by the occurrence of spontaneous magnetization, the extrinsic symmetry of which is described by the limit group ∞/mm'. Ferromagnetism can therefore occur in only those crystal classes which are subgroups of this limit group describing the symmetry of an axial magnetic vector. There are 31 such classes (Tavger 1958):

$$\begin{gathered} 1, \bar{1}, m', 2', 2'/m', m, mm2', 2, m'm'2, 22'2', 2/m, m'm'm, \\ \bar{4}, \bar{4}2'm', 4, 4m'm', 42'2', 4/m, 4/mm'm', 3, 3m', 32', \bar{3}, \bar{3}m', \\ 6, 62'2', 6m'm', 6/m, 6/mm'm', \bar{6}, \bar{6}m'2' \end{gathered} \tag{3.2.4}$$

Fumi's Method of Direct Inspection

Continuing our illustration of the applications of the Neumann theorem, we now consider a more analytical and exhaustive approach to the question of symmetry of tensor properties. The method involves a transformation of the axes of reference of the matrix representation of the tensor by one of the symmetry operations possessed by the crystal, and to enforce the condition that the tensor coefficients must be the same before and after the coordinate transformation. The algebra involved can be reduced very substantially by applying Fumi's method of direct inspection (Fumi 1952a, b, c; Fieschi & Fumi 1953). Fumi's method is applicable most conveniently for symmetry elements 2, m, and 4, and for the 3-fold axes occurring in the cubic system. It can be applied to practically all the crystallographic point groups, with the exception of point group 3.

Fumi's method is based on the fact that tensor components transform as products of components of vectors. It also makes use of the fact that components of matter tensors are defined with reference to the *crystallophysical system of coordinates.* This system of coordinates is based on the symmetry elements of the crystal: the coordinate axes are taken to coincide with symmetry axes, mirror-plane normals, or with directions bisecting angles between these axes or plane-normals.

We consider the operation of a 4-fold symmetry axis on a second-rank polar tensor as an example.

Let us consider a position vector $\mathbf{r}(x_1, x_2, x_3)$ referred to the mutually orthogonal basis vectors $\mathbf{e}_1$, $\mathbf{e}_2$, $\mathbf{e}_3$. Under a symmetry operation of the 4-fold axis the basis vectors change to $\mathbf{e}_1'$, $\mathbf{e}_2'$, $\mathbf{e}_3'$, and the vector can be represented as $\mathbf{r}\,(x_1', x_2', x_3')$:

$$x_1\,\mathbf{e}_1 + x_2\,\mathbf{e}_2 + x_3\,\mathbf{e}_3 = x_1'\,\mathbf{e}_1' + x_2'\,\mathbf{e}_2' + x_3'\,\mathbf{e}_3' \qquad (3.2.5)$$

It is conventional to take the 4-fold axis along $\mathbf{e}_3$. After performing one symmetry operation with this axis, i.e. after a rotation of 90° about $\mathbf{e}_3$, the new basis vectors have the following relationship with the old basis vectors: $\mathbf{e}_3' = \mathbf{e}_3$; $\mathbf{e}_1' = \mathbf{e}_2$; and $\mathbf{e}_2' = -\mathbf{e}_1$. Therefore, Eq. 3.2.5 can be rewritten as

$$x_1\,\mathbf{e}_1 + x_2\,\mathbf{e}_2 + x_3\,\mathbf{e}_3 = x_1'\,\mathbf{e}_2 - x_2'\,\mathbf{e}_1 + x_3'\,\mathbf{e}_3 \qquad (3.2.6)$$

Since the unit basis vectors are linearly independent, Eq. 3.2.6 can be satisfied only if

$$x_1' = x_2; \;\; x_2' = -x_1 \;\; x_3' = x_3 \qquad (3.2.7)$$

We notice that the components of $\mathbf{r}$ either remain unchanged ($x_3' = x_3$), or change into one of the other components, with ($x_2' = -x_1$) or without ($x_1' = x_2$) a change of sign. This is because all the direction cosines of the symmetry transformation under consideration are either 1, or -1, or 0. Fumi's method of direct inspection applies only to symmetry operations defined by integral direction cosines. This is the case for symmetry elements $\bar{1}$, 2_x, 2_y, 2_z, 2_{xy}, m_x, m_y, m_z, m_{xy}, 4_z, $\bar{4}_z$, and 3_{xyz}.

To consider how the components of a second-rank polar tensor $\mathbf{T}$ transform under the operations of a 4-fold symmetry axis 4_z, we make use of the fact that any component of this tensor will transform as the product of two vector components. For example, the tensor component with indices 11 will transform as the product x_1x_1 or x_1^2, and the component 12 as the product x_1x_2. It is convenient to denote the tensor component ij by x_ix_j. Then, for the 90^0 rotation operation, $x_1' = x_2$, $x_2' = -x_1$, and $x_3' = x_3$. It follows that $[x_1x_1]' = [x_2x_2]$, implying that $T_{11}' = T_{22}$. But since the considered coordinate transformation is a *symmetry operation*, we must also have $T_{11}' = T_{11}$. It follows that $T_{11} = T_{22}$.

One can show similarly that $T_{22}' = T_{11}$, and $T_{33}' = T_{33}$.

Considering next the nondiagonal coefficient T_{12}, we find that $[x_1x_2]' = -[x_2x_1]$, implying that $T_{12}' = T_{12} = -T_{21}$. Similarly $T_{23} = -T_{23}$, (implying that $T_{23} = 0$), $T_{32} = -T_{32}$ (or $T_{32} = 0$), $T_{31} = T_{32}$ $(= 0)$, and $T_{13} = T_{23}$ $(= 0)$.

A second-rank polar tensor for a crystal having point-group symmetry 4 thus has the form

$$(T_{ij}) = \begin{bmatrix} T_{11} & T_{12} & 0 \\ -T_{12} & T_{11} & 0 \\ 0 & 0 & T_{33} \end{bmatrix} \tag{3.2.8}$$

If this tensor is symmetric (i.e. $T_{ij} = T_{ji}$), then its only nonzero independent components are T_{11} and T_{33}.

If a crystallographic point group has more than one generators, we work with them successively, applying its restrictions on the tensor components obtained by the application of the previous generator(s).

SUGGESTED READING

J. F. Nye (1957). *Physical Properties of Crystals: Their Representation by Tensors and Matrices.* The Clarendon Press, Oxford.

H. J. Juretschke (1974). *Crystal Physics: Macroscopic Physics of Anisotropic Solids.* Benjamin, London.

Yu. I. Sirotin & M. P. Shaskolskaya (1982). *Fundamentals of Crystal Physics.* Mir Publishers, Moscow.

L. A. Shuvalov (Ed.) (1988). *Modern Crystallography IV. Physical Properties of Crystals.* Springer-Verlag, Berlin.

A. S. Nowick (1995). *Crystal Properties via Group Theory.* Cambridge University Press, Cambridge.

3.3 THE HERMANN THEOREM OF CRYSTAL PHYSICS

Apart from the Neumann theorem, another theorem of basic importance is the Hermann theorem of crystal physics (Hermann 1934). According to it:

If we consider an r-rank tensor with reference to a material having an N-fold axis of symmetry, and $r < N$, then this tensor property effectively conforms to an ∞-fold symmetry axis parallel to the N-fold axis.

We present here an abridged proof of the theorem. The textbook by Sirotin & Shaskolskaya (1982) should be consulted for more details.

We begin by introducing the notion of cyclic coordinates.

3.3.1 Cyclic Coordinates

One can define cyclic basis vectors, $\mathbf{j}, \bar{\mathbf{j}}, \mathbf{e}$, in terms of cartesian basis vectors, $\mathbf{e}_1, \mathbf{e}_2, \mathbf{e}_3$, as follows:

$$\mathbf{j} = \frac{1}{2}(\mathbf{e}_1 + i\mathbf{e}_2) \tag{3.3.1}$$

$$\bar{\mathbf{j}} = \frac{1}{2}(\mathbf{e}_1 - i\mathbf{e}_2) \tag{3.3.2}$$

$$\mathbf{e} = \mathbf{e}_3 \tag{3.3.3}$$

The converse equations are:

$$\mathbf{e}_1 = \mathbf{j} + \bar{\mathbf{j}} \tag{3.3.4}$$

$$\mathbf{e}_2 = -i(\mathbf{j} - \bar{\mathbf{j}}) \tag{3.3.5}$$

$$\mathbf{e}_3 = \mathbf{e} \tag{3.3.6}$$

For a real position vector $\mathbf{r}(x, y, z)$ the following equality holds in terms of the two sets of basis vectors:

$$x\mathbf{e}_1 + y\mathbf{e}_2 + z\mathbf{e}_3 \;=\; \xi\mathbf{j} + \bar{\xi}\bar{\mathbf{j}} + z\mathbf{e} \tag{3.3.7}$$

It follows from Eqs. 3.3.4-7 that

$$\xi = x - iy \tag{3.3.8}$$

$$\bar{\xi} = x + iy \tag{3.3.9}$$

$$z = z \tag{3.3.10}$$

Let us perform a rotation ϕ of the coordinate system about the principal axis X_3. The new basis vectors are given by:

$$\mathbf{e}_1' = \mathbf{e}_1 \cos\phi + \mathbf{e}_2 \sin\phi \tag{3.3.11}$$

$$\mathbf{e}_2' = -\mathbf{e}_1 \sin\phi + \mathbf{e}_2 \cos\phi \tag{3.3.12}$$

$$\mathbf{e}_3' = \mathbf{e}_3 \tag{3.3.13}$$

$$\mathbf{j}' = \frac{1}{2}(\mathbf{e}_1' + i\mathbf{e}_2') \tag{3.3.14}$$

$$\bar{\mathbf{j}}' = \frac{1}{2}(\mathbf{e}_1' - i\mathbf{e}_2') \tag{3.3.15}$$

$$\mathbf{e}' = \mathbf{e}_3' \tag{3.3.16}$$

These equations can be solved to yield the following results (with the help of Eqs. 3.3.4-6):

$$\mathbf{j}' = e^{-i\phi}\mathbf{j}, \quad \bar{\mathbf{j}}' = e^{i\phi}\bar{\mathbf{j}}, \quad \mathbf{e}' = \mathbf{e} \tag{3.3.17}$$

Since the magnitude and direction of the position vector $\mathbf{r}$ do not change as a result of the coordinate transformation, we can write

$$\xi\mathbf{j} + \bar{\xi}\bar{\mathbf{j}} + z\mathbf{e} \;=\; \xi'\mathbf{j}' + \bar{\xi}'\bar{\mathbf{j}}' + z'\mathbf{e}' \tag{3.3.18}$$

Substitution from Eqs. 3.3.17 yields

$$\xi' = e^{i\phi}\,\xi \;\; \bar{\xi}' = e^{-i\phi}\,\bar{\xi} \;\; z' = z \tag{3.3.19}$$

We are now ready to derive the law of transformation of the cyclic components $\mathbf{A}_{l_1 l_2 \ldots l_r}$ of a tensor $\mathbf{A}$ of rank r under a rotation ϕ about the axis X_3. For this it is convenient to give the position vector component ξ an index 1, $\bar{\xi}$ an index -1, and z an index 0. Then the direction cosines of the transformation can be written as

$$c_{k'l} \;=\; e^{ik'\phi}\,\delta_{k'l} \;=\; e^{il\phi}\,\delta_{k'l} \tag{3.3.20}$$

The following result is then evident:

$$A_{k_1' \ldots k_r'} \;=\; \exp[i(l_1 + \cdots + l_r)\phi]\,\delta_{k_1' l_1} \cdots \delta_{k_r' l_r}\,A_{l_1 \ldots l_r} \tag{3.3.21}$$

If the rotation ϕ is a symmetry operation about an N-fold axis, then $\phi = 2\pi/N$, and Eq. 3.3.20 becomes

$$c_{k'l} \;=\; \exp(\frac{2\pi i}{N}\,l)\,\delta_{k'l} \tag{3.3.22}$$

The tensor **A**, specified by its components $A_{l_1 l_2 \cdots l_r}$, must be invariant under the symmetry operation defined by $\phi = 2\pi/N$. Eq. 3.3.21 yields the following condition for this:

$$[\exp\{\frac{2\pi i}{N}(l_1 + \cdots l_r)\} - 1]\, A_{l_1 \cdots l_r} = 0 \qquad (3.3.23)$$

Only those cyclic tensor components will be nonzero for which the expression in the parenthesis is zero. This means that, if a tensor component $A_{l_1 l_2 \cdots l_r}$ is to be nonzero for a crystal having a point-group symmetry with an N-fold axis as a generator, the following condition must be satisfied:

$$l_1 + l_2 + \cdots + l_r = 0 \pmod N \qquad (3.3.24)$$

3.3.2 Proof of the Hermann Theorem

Since the cyclic-coordinate indices only take values 1, 0, or -1, their absolute sum for any tensor component cannot exceed the rank r of the tensor. If $r < N$, the only situation in which Eq. 3.3.24 can be satisfied is that when this sum is zero (only zero, and not zero modulo N). But when this is the case, Eq. 3.3.23 will be satisfied (for nonzero $A_{l_1 l_2 \cdots l_r}$) for all values of N ($N > r$), including $N = \infty$. Q.E.D.

3.3.3 Importance of the Hermann Theorem

The Hermann theorem defines a sufficient condition for *transverse isotropy*, i.e. isotropic behaviour of a tensor property of rank r in a plane perpendicular to an N-fold symmetry axis of the crystal. We shall come across applications of this theorem in later chapters. By way of illustration, we apply it here to the birefringence behaviour of crystals of various point-group symmetries.

Birefringence of a crystal is determined by its dielectric permittivity tensor (ϵ_{ij}), a second-rank polar tensor, i.e. $r = 2$. Crystals belonging to triclinic, monoclinic, and orthorhombic systems do not possess any symmetry axis of order greater than 2, i.e. for them $N \leq 2$. Thus there is no symmetry axis for them for which $N > r$, and therefore the provision $N = \infty$ does not exist for them, so far as the dielectric permittivity tensor is concerned. Such crystals are *optically biaxial*: One can find two directions in them (namely the two *optic axes*) for which there is transverse isotropy. That is, the crystal is optically isotropic (or the birefringence is zero) in planes transverse to the two optic axes. But these optic axes do not coincide with any symmetry element of the crystal, and their directions change with temperature, wavelength etc.

Crystals belonging to trigonal, tetragonal, and hexagonal systems do possess one (and only one) symmetry axis of order 3 or more, so that, for $r = 2$, we have $N > r$ for them. The Hermann theorem predicts transverse isotropy for them with respect to dielectric permittivity, and thence birefringence, i.e., effectively $N = \infty$ so far as the birefringence behaviour of such crystals is concerned. As is well known, the $N = \infty$ axis is nothing but the optic axis for these crystals. There is only one such axis, and the crystals are described as *optically uniaxial.*

Lastly, we consider the cubic system. In this case there are four 3-fold axes of symmetry, directed along the body-diagonals of the cubic unit cell. The Hermann theorem demands isotropy of dielectric permittivity (i.e. zero birefringence) in planes normal to each of these four 3-fold axes. The four conditions can be satisfied simultaneously only if the crystal is optically isotropic in planes normal to all directions. Cubic crystals are therefore *optically anaxial.*

To discuss another important feature of the Hermann theorem, we begin by reminding ourselves that if the application of the Neumann theorem tells us that a physical phenomenon (represented by an appropriate tensor component) is not forbidden from occurring, there is still no guarantee that the phenomenon would indeed occur; it still may not occur, say, for thermodynamic reasons. However, the converse situation, whereby a phenomenon or property is forbidden by symmetry, corresponds to absolutely certain knowledge (regarding the nonoccurrence of the phenomenon).

We notice next the inverse relationship between the order of the point group of a crystal and the number of independent components of a given tensor property that the crystal can exhibit. On one extreme we have the situation wherein the crystallographic point-group symmetry is just 1 (i.e. no rotational symmetry at all). In this case all the tensor components, which are nonzero in spite of the intrinsic and extrinsic symmetries of the tensor, are allowed by the crystal symmetry to be nonzero. The number of these nonzero independent components decreases as we consider point groups of higher and higher order. What amounts to the same thing, the number of zero (or forbidden) tensor components increases as we go up the symmetry scale.

Now, all the 32 crystallographic point groups are proper subgroups of one or more of the 7 limit groups. Accordingly, for a particular physical property, the number of tensor components that is zero by symmetry is smaller for any crystallographic point group than it is for any of its limiting supergroups. The Hermann theorem, by stating that if $N > r$, then $N = \infty$, increases the domain of certain knowledge (by giving a larger number of zero components, i.e. certain knowledge), and a correspondingly smaller number of *possibly* nonzero components (uncertain knowledge). This is shown schematically in Fig. 3.3.1. In part (a) of Fig. 3.3.1, the domain of knowledge about a tensor property of rank r is divided into certain

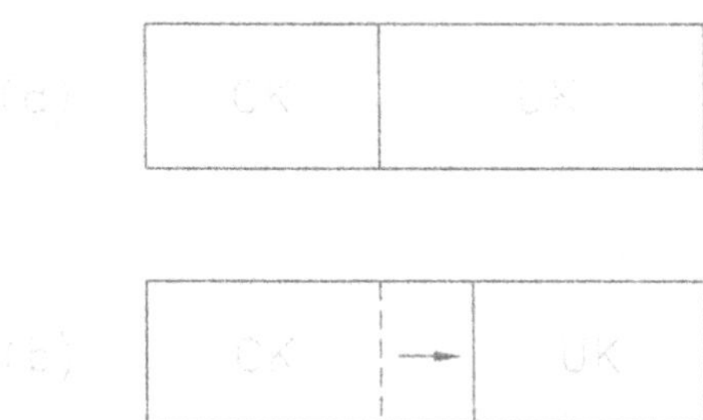

Figure 3.3.1: Depiction of how applicability of the Hermann theorem to a given situation amounts to an effective increase of the domain of *certain knowledge (ck)*, and a corresponding decrease of the domain of *uncertain knowledge (uk)*.

knowledge (ck), and uncertain knowledge (uk). The size of the domain ck is specified by the fraction of tensor components that are zero by crystal symmetry. If $r < N$, so that the Hermann theorem applies, then the fraction of tensor components that is zero becomes larger, and the domain ck becomes larger (part (b) of Fig. 3.3.1).

The properties of a crystal (the *effects*) arise from its structure and symmetry (the *causes*). The "minimalistic use" of the symmetry principle (§C.1) provides a lower bound on the symmetry of an effect. The Hermann theorem, by requiring that $N = \infty$ if $N > r$, provides an effective raising of the lower bound on the symmetry of the effect, namely the tensor property of rank r.

We note in passing that the Hermann theorem applies only for $r < N$. It says nothing about the case when $r = N$. If it can be shown, by going back to the arguments leading to the proof of the theorem, that, for some situations $N = \infty$ even for $r = N$, then once again ck, the domain of certain knowledge, would expand.

SUGGESTED READING

Yu. I. Sirotin & M. P. Shaskolskaya (1982). *Fundamentals of Crystal Physics*. Mir Publishers, Moscow.

3.4 REPRESENTATIONS OF CRYSTALLOGRAPHIC POINT GROUPS

Representations of crystallographic point groups can be obtained either as subduced representations (§B.3) of continuous rotation groups (see, for example, Ludwig & Falter 1988), or as induced representations (§B.3) of low-order crystallographic point groups. We outline the latter approach here, the basis of which is the fact that if a group can be expressed as the direct product of two groups, its character table can be obtained from those of the latter groups (§B.3). We shall discuss only characters of representations, and not actual representations.

The class structure of a group determines its irreducible representations (IRs). We summarize here the main results (described in more detail in §B.3 for some cases) which are needed for determining the characters of the IRs of a group:

(A). *The number (n_r) of the inequivalent IRs (IIRs) a group can have is equal to the number (n_c) of classes in it* (Eq. B.3.15):

$$n_r = n_c \tag{3.4.1}$$

(B). *The sum of the squares of the dimensions of all the IIRs of a group is equal to the order (g) of the group* (Eq. B.3.18):

$$\sum_{\mu=1}^{n_c} n_\mu^2 = g \tag{3.4.2}$$

(C). *The first orthogonality theorem for characters* (Eq. B.3.11):

$$\sum_{C_s=1}^{n_c} g_s \, \chi_\nu^*(C_s) \, \chi_\mu(C_s) = g \, \delta_{\nu\mu} \tag{3.4.3}$$

It is useful to write down two special cases of Eq. 3.4.3. When $\mu = \nu$ this equation becomes

$$\sum_s [\chi_\mu(s)]^2 = g \tag{3.4.4}$$

In Eq. 3.4.3 g_s is the number of elements in the class C_s, and the summation is over all the classes. In Eq. 3.4.4 we carry out the summation over all the elements of the group, which is equally valid. Eq. 3.4.4 states that *the sum of the squares of the characters of any IR of a group is equal to the order of the group.*

The other special case of Eq. 3.4.3 is when $i \neq j$:

$$\sum_s \chi_\mu(s) \, \chi_\nu(s) = 0 \tag{3.4.5}$$

In words: *Vectors (in the g-dimensional space defined in §B.3) the components of which are characters of two different IRs are orthogonal to one another.*

(D). *The second orthogonality theorem for characters* (Eq. B.3.12):

$$\sum_{\mu=1}^{n_r} \chi_\mu^*(C_a)\,\chi_\mu(C_b) \;=\; \frac{g}{g_a}\,\delta_{ab} \tag{3.4.6}$$

(E). *If H is a normal subgroup of G, and thence G/H the factor group of G with respect to H (§B.1), the representations of G/H are also (unfaithful) representations of G.*

(F). *The direct product of representations of two commuting groups is a representation of the direct-product group.*

(G). *Direct products of the IRs of two commuting groups exhaust all the IRs of the direct-product group.* In other words, there is no IR of the product group which cannot be expressed as a direct product of an IR of the first constituent group and an IR of the second constituent group.

(H). Sometimes it becomes necessary to use the following result, which can be derived from class-multiplication relationships:

$$g_i\, g_j\, \chi_\mu(C_i)\,\chi_\mu(C_j) \;=\; n_\mu \sum_k c_{ij}^k\, g_k\, \chi_\mu(C_k) \tag{3.4.7}$$

Here c_{ij}^k are nonnegative integers called *class constants.* They are defined by the equation

$$C_i\, C_j \;=\; \sum_k c_{ij}^k\, C_k \tag{3.4.8}$$

which describes the fact that the class C_k appears c_{ij}^k times in the product of the classes C_i and C_j. (The product of two classes of a group always consists of whole classes.) The class constants are symmetric in the indices i and j.

As an illustration of the use of some of these results, we consider the point group C_{3v}. For this group $g = 6$ and $n_c = 3$. It follows that there are three IIRs (Eq. 3.4.1). Therefore Eq. 3.4.2 has a unique solution: $1^2 + 1^2 + 2^2 = 6$. This means that the group must have two 1-dimensional IRs and one 2-dimensional IR. The character table for this group is shown in Table 3.4.1.

Table 3.4.1: Character table for the point group C_{3v}.

C_{3v}		E	$2C_3 = \{C_3, C_3^2\}$	$3\sigma_v = \{\sigma_v, \sigma_v', \sigma_v''\}$
Γ_1	A_1	1	1	1
Γ_2	A_2	1	1	-1
Γ_3	E	2	-1	0

We first describe the notation used in Table 3.4.1 for labelling the IRs.

Several systems of notation are in use in the literature. Two of them, namely the *Mulliken notation* and the *Bethe notation*, are the most common.

In the Mulliken notation, the symbols used for the IRs are A, B, E, T and F, alongwith subscripts and superscripts where necessary. 1-dimensional IRs are denoted by A or B, 2-dimensional by E, and 3-dimensional by T (or sometimes by F). In the Bethe notation the labels used are simply $\Gamma_1, \Gamma_2, \cdots$ etc.

1-dimensional real IRs which are symmetric with respect to rotation by $2\pi/n$ about the principal axis C_n are labelled as A, and those antisymmetric with respect to this rotation are labelled B. 'Symmetric' here means $\chi(C_n) = 1$, and 'antisymmetric' means $\chi(C_n) = -1$.

If there is a 2-fold symmetry axis perpendicular to C_n, or if there is a 'vertical' mirror plane σ_v containing the axis C_n, subscripts 1 or 2 are attached to A and B to represent symmetry and antisymmetry with respect to these additional symmetry elements.

If a 'horizontal' mirror-plane of symmetry, σ_h, is present, primes and double primes are attached to A, B, E, T etc. to indicate symmetry and antisymmetry with respect to operations of σ_h.

For groups that include inversion symmetry, the subscript g (from the German *gerade*, meaning even) is attached for denoting symmetry under inversion. The subscript u (for *ungerade*, meaning uneven or odd) is used for denoting antisymmetry with respect to the inversion operation.

For 1-dimensional complex IRs the Mulliken and Bethe approaches are different. On account of time-reversal symmetry a 1-dimensional representation having a complex character is degenerate with its complex-conjugate representation. In the Mulliken system it is viewed as a 2-dimensional representation in the sense that it is physically irreducible. In the Bethe notation, on the other hand, the pair is regarded as two independent 1-dimensional representations.

Several variations of the Mulliken and Bethe notations are also used for dealing with, for example, IRs of space groups. [see Ludwig & Falter (1988), page 406, for a comparative description.]

Continuing with our description of Table 3.4.1, we note that the IR in the first line, namely Γ_1 or A_1, is a property of every group, since the simplest representation of any group is that homomorphic mapping in which all the elements of the group are mapped to a unit matrix of order 1 (§B.3). This is the so-called *identity representation*, with $\chi_1(s) = 1$ for all s.

For the group C_{3v} there must be two 1-dimensional representations and one 2-dimensional representations in all. We first look for the second 1-dimensional IR.

The IR Γ_1 corresponds to a vector χ_1 in 6-dimensional space (§B.3) with all its components as +1. A second vector in this space, corresponding to a 1-dimensional representation and orthogonal to the vector for Γ_1 (Eq. 3.4.5) must have its components as either +1 or -1, because the sum of the squares of these components must equal 6 (Eq. 3.4.4). For this vector $\chi_2(E) = +1$. For satisfying Eq. 3.4.5 three components must be +1 and the other three -1. Since we already have $\chi_2(E) = +1$, this is possible only if the characters χ_2 for the two elements in the class (C_3, C_3^2) are +1, and those in the class $(\sigma_v, \sigma_v', \sigma_v'')$ are -1. The second 1-dimensional IR (labelled Γ_2 or A_2) therefore has the characters shown in the second line of Table 3.4.1.

The remaining IR is 2-dimensional. For it the label used in Table 3.4.1 is Γ_3 or E, and $\chi_3(E) = 2$. To find $\chi_3(C_3)$ and $\chi_3(\sigma_v)$ we invoke the orthogonality relationships expressed by Eq. 3.4.5:

$$\sum_s \chi_1(s)\,\chi_3(s) \;=\; 0 \;=\; 1\times 2 + 2[1\times\chi_3(C_3)] + 3[1\times\chi_3(\sigma_v)] \qquad (3.4.9)$$

$$\sum_s \chi_2(s)\,\chi_3(s) \;=\; 0 \;=\; 1\times 2 + 2[1\times\chi_3(C_3)] + 3[(-1)\times\chi_3(\sigma_v)] \qquad (3.4.10)$$

These two equations, when solved, yield $\chi_3(\sigma_v) = 0$ and $\chi_3(C_3) = -1$. This completes the explanation of the last line of entries in Table 3.4.1. Eq. 3.4.4 provides a further check on the correctness of the characters determined for the IR E $[2^2 + 2(-1)^2 + 3(0)^2 = 6]$.

We have so far considered only *irreducible* representations in this section. When dealing with a reducible representation (with characters, say, $\chi(s)$) it is possible to determine readily the number (q_μ) of times the various IIRs with characters $\chi_\mu(s)$ are contained in it by using the following very important result, which we state without proof:

$$q_\mu \;=\; \frac{1}{g}\sum_s \chi_\mu(s)^*\,\chi(s) \qquad (3.4.11)$$

Thus, knowing only the characters of each IR (without having to know the actual representations) it is possible to calculate the number of times each IR is contained in a given reducible IR.

Table 3.4.2: Character table for the point group C_i.

C_i	E	J
A_g	1	1
A_u	1	-1

Table 3.4.3: Character table for the point group D_{3d}.

D_{3d}	E	$2C_3$	$3C_2$	J	$2S_6$	$3\sigma_d$
A_{1g}	1	1	1	1	1	1
A_{1u}	1	1	1	-1	-1	-1
A_{2g}	1	1	-1	1	1	-1
A_{2u}	1	1	-1	-1	-1	1
E_g	2	-1	0	2	-1	0
E_u	2	-1	0	-2	1	0

Any reducible representation can then be written as a direct sum of the IRs D_i:

$$D(s) = q_1 D_1(s) \oplus q_2 D_2(s) \oplus \cdots \tag{3.4.12}$$

Having constructed all the IRs for the group C_{3v} (or $3m$), we now illustrate the construction of the IRs of a group which can be obtained as a direct product of C_{3v} and another group commuting with this group.

C_i (or $\bar{1}$) is a group of order 2 (having elements E, J) which commutes with C_{3v}, as well as with other point groups. Table 3.4.2 is its character table, which does not require any additional explanation. The point group D_{3d} (or $\bar{3}m$) can be obtained as the following direct product:

$$D_{3d} = C_{3v} \otimes C_i \tag{3.4.13}$$

Since C_{3v} has three IIRs (Table 3.4.1) and C_i has two IIRs (Table 3.4.2), D_{3d} must have six, and only six, IIRs. Table 3.4.3 lists these. Half of them are 'g' IIRs, and the other half are 'u' IIRs.

SUGGESTED READING

F. A. Cotton (1971). *Chemical Applications of Group Theory*, second edition. Wiley, New York.

W. A. Wooster (1973). *Tensors and Group Theory for the Physical Properties of Crystals.* Clarendon Press, Oxford.

A. W. Joshi (1982). *Elements of Group Theory for Physicists*, third edition. Wiley Eastern, New Delhi.

A. S. Nowick (1995). *Crystal Properties via Group Theory.* Cambridge University Press, Cambridge.

3.5 EFFECT OF FIELDS ON TENSOR PROPERTIES

When we speak of the point-group symmetry of a crystal, we implicitly presume that the crystal exists in an isotropic environment. If the same crystal is acted upon by a field that is not isotropic, the net symmetry of the composite system (crystal plus field) is that given by the Curie-Shubnikov principle of superposition of symmetries (§C.2).

A related question is that of the mutual interaction of the tensor properties of the crystal. Application of an external force field $\mathbf{X}$ leads to a *primary response* $\mathbf{Y}_1$:

$$\mathbf{Y}_1 = \mathbf{A}\mathbf{X} \tag{3.5.1}$$

The same external field can produce another, different, response via a matter tensor other than $\mathbf{A}$:

$$\mathbf{Y}' = \mathbf{A}'\mathbf{X} \tag{3.5.2}$$

And the response $\mathbf{Y}'$ can, in turn, act as an internal field, producing an additional (secondary) response via another tensor property $\mathbf{B}$:

$$\mathbf{Y}_2 = \mathbf{B}\mathbf{Y}' = \mathbf{B}\mathbf{A}'\mathbf{X} \tag{3.5.3}$$

Combining Eqs. 3.5.1 and 3.5.3 we get

$$\mathbf{Y} = \mathbf{Y}_1 + \mathbf{Y}_2 = (\mathbf{A} + \mathbf{B}\mathbf{A}')\mathbf{X} \tag{3.5.4}$$

Experimental conditions determine the presence or absence of the second term on the right-hand side of this equation.

We illustrate such a possibility with an example from nonlinear optics.

Example of Pockels Effect

Application of an electric field to a crystal changes its charge distribution, with a concomitant change of the refractive indices. If the crystal is noncentrosymmetric, there can always be a change of refractive index proportional to the first power of the electric field. This linear electrooptical effect is called the Pockels effect. In centrosymmetric crystals the Pockels effect is absent, but a change of refractive index proportional to the square of the electric field (the Kerr effect) can occur in all crystals.

Electrooptic effects of various orders are described by the following equation:

$$D_i = \epsilon_{ij}E_j + r_{ijk}E_jE_k + R_{ijkl}E_jE_kE_l + \cdots \tag{3.5.5}$$

The first term on the right-hand side describes the linear part of the dielectric response. The rest of the terms describe nonlinear effects of various orders. (r_{ijk}) is a third-rank, polar, i-tensor governing the Pockels effect. The Kerr effect is determined by the fourth-rank tensor (R_{ijkl}).

For relating the tensor (r_{ijk}) to the optical indicatrix, we bring in the *relative dielectric impermeability tensor* (B_{ij}), which is defined by (Nye 1957):

$$B_{ij} = \partial E_i/\partial D_j \tag{3.5.6}$$

In terms of this tensor the equation of the optical indicatrix (in the absence of external electric and mechanical fields) takes the form

$$B_{ij}x_ix_j = 1 \quad (E = 0;\ \sigma = 0) \tag{3.5.7}$$

Application of an external electric field **E** leads to a change (δB_{ij}) in the tensor (B_{ij}):

$$\delta B_{ij} = r_{ijk}E_k \quad (\sigma = 0) \tag{3.5.8}$$

This equation is valid for a crystal that is free to expand or contract $(\sigma = 0)$. However, such an unclamped crystal will also exhibit a piezoelectric strain when the electric field is applied:

$$e_{ij} = d_{kij}\,E_k \tag{3.5.9}$$

This strain causes a further change in (B_{ij}) for the unclamped crystal via the elastooptical effect governed by the fourth-rank tensor (P_{ijlm}), so that the total effect is

$$\delta B_{ij} = r^*_{ijk}\,E_k + P_{ijlm}\,e_{lm}, \tag{3.5.10}$$

which on substitution from Eq. 3.5.8 becomes

$$\delta B_{ij} = (r^*_{ijk} + P_{ijlm}\,d_{klm})\,E_k \tag{3.5.11}$$

The first term on the right-hand side describes the primary Pockels effect, observable if the crystal is prevented from expanding or contracting when the electric field is applied. The second term defines the *secondary* Pockels effect, arising because the strain produced by the electric field via the (converse) piezoelectric effect causes a further change in (B_{ij}) via the elasto-optic effect.

Thus, if only the primary Pockels effect is to be observed, conditions amounting to effective clamping must be created. In practice, this can be realized when the electric field applied is an ac field of a frequency much higher than the natural frequency of vibrations of the crystal.

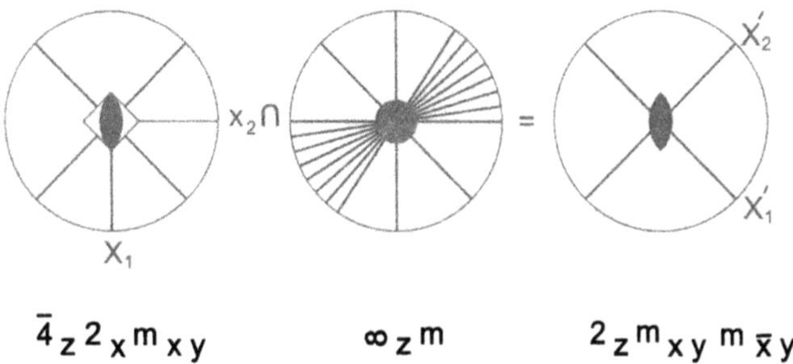

Figure 3.5.1: Direct application of the Curie principle for determining the change in the symmetry of the optical indicatrix of a KDP crystal when an electric field is applied along its optic axis.

Direct Application of the Curie Principle

KDP (KH_2PO_4), or its deuterated analogue, is a crystal commonly used for making electro-optic modulators, exploiting the Pockels effect mentioned above. The point-group symmetry of this crystal at room temperature is $\bar{4}2m$. In the so-called longitudinal-geometry mode of application, the electric field is applied along the same direction as that of the propagating laser beam. This crystal is optically uniaxial, and the laser beam is usually made to propagate along the optic axis. It can be shown by detailed calculation that when the electric field is switched on, the crystal becomes optically biaxial, so that a progressively increasing phase difference gets introduced between the ordinary wave and the extraordinary wave as the light propagates through the crystal (see, e.g., Yariv 1985). Variations in the applied electric field thus get translated into corresponding variations in the phase of the light beam emerging from the crystal.

We show here how a direct application of the Curie principle of superposition of symmetries (§C.1) leads qualitatively (and quickly) to the same result.

In terms of the standard coordinate axes shown in Fig. 3.5.1, the point-group symmetry of the KDP crystal can be expressed as $\bar{4}_z 2_x m_{xy}$. The electric field, which has the symmetry of a cone, is applied along the z-axis, so that its point-group symmetry can be written as $\infty_z m$. According to the Curie principle, the net symmetry of the composite system (crystal plus electric field) consists of symmetry elements common to the crystal and the electric field.

Since the symmetry axis $\bar{4}_z$ includes the symmetry axis 2_z, the intersection symmetry along the z-axis is that of the diad 2_z.

In addition, there are an infinite number of mirror planes of symmetry (containing the z-axis) in the symmetry group of the electric field. But there are only two such planes in $\bar{4}_z 2_x m_{xy}$ (see Fig. 3.5.1). What survives in the intersection group from this set are thus these two planes only.

The complete intersection group is thus $2_z m_{xy} m_{\bar{x}y}$, or $2mm$. This point group belongs to the orthorhombic system, which is optically biaxial.

We also note from Fig. 3.5.1 that the mirror planes of $2mm$ are inclined at $\pm 45^o$ to the x-axis of the unbiased KDP crystal. This again is in conformity with experiment, as well as with results of detailed calculation.

SUGGESTED READING

A. Yariv & P. Yeh (1984). *Optical Waves in Crystals.* Wiley, New York.

A. Yariv (1985). *Optical Electronics.* Holt, Rinehart & Winston, Holt-Suanders Japan.

Chapter 4

CRYSTALS AND THE WAVEVECTOR SPACE

To "see" an object we shine radiation on it. The object scatters this radiation, some of which is received and processed by the eye and the brain, or by some other recording and interpreting device. Determining the internal structure of a crystal amounts to "seeing" it on an atomic scale. Since the sizes of atoms, as well as their typical separations, are in the Ångstrom range, the radiation employed should also have a wavelength in this range. X-rays, thermal neutrons, and electron beams are commonly used for this purpose.

The radiation waves scattered by an object interfere among themselves to produce the diffraction pattern of the object. The description of the diffraction pattern involves introduction of the concept of *reciprocal space* or *wavevector space.* The wavevector $\mathbf{k}$ also serves as a natural 'label' for the irreducible representations of space groups of crystals.

4.1 DIFFRACTION BY A CRYSTAL. THE RECIPROCAL LATTICE

4.1.1 Diffraction by a General Distribution of Scatterers

Let $\rho(\mathbf{r})$ denote the density function for a distribution of scatterers. In a typical X-ray diffraction experiment, one shines a collimated, monochromatic beam of wavelength λ on the specimen. Let $\mathbf{k}_0$ be a vector of magnitude $2\pi/\lambda$ along the direction of the incident beam, and $\mathbf{k}_1$ a similar vector along any direction of interest. Let us consider scattering from a point $P(\mathbf{r})$ (Fig. 4.1.1). The path difference between the wave scattered from P,

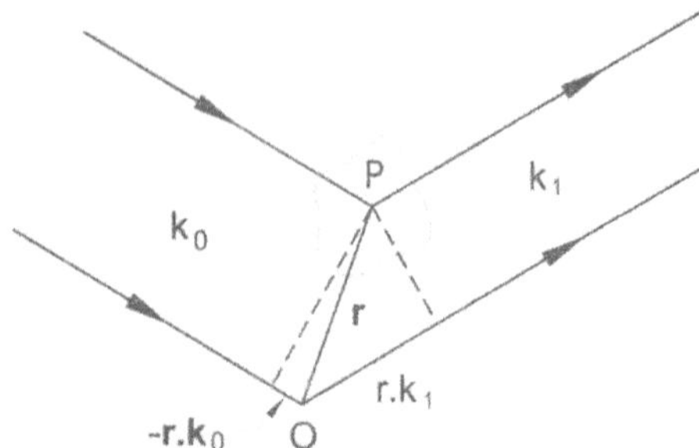

Figure 4.1.1: Scattering of radiation from a point P, with reference to that from the origin O.

relative to that scattered from the origin O, is $(\mathbf{r}\cdot\mathbf{k}_1 - \mathbf{r}\cdot\mathbf{k}_0)\lambda/(2\pi)$ (Fig. 4.1.1). And therefore we have the phase difference, δ, given by

$$\delta = \mathbf{r}\cdot(\mathbf{k}_1 - \mathbf{k}_0) = \mathbf{r}\cdot\mathbf{k}, \tag{4.1.1}$$

where $\mathbf{k}$ $(= \mathbf{k}_1 - \mathbf{k}_0)$ is called the *scattering vector*, or the *wavevector*. It is readily verified that

$$k = |\mathbf{k}| = 4\pi\sin\theta/\lambda, \tag{4.1.2}$$

where θ is half the angle between $\mathbf{k}_1$ and $\mathbf{k}_0$.

We now consider an element of volume $d\mathbf{r}$ at the point $P(\mathbf{r})$. The wave scattered from this volume of scatterers is given by $[\rho(\mathbf{r})\,d(\mathbf{r})]\ \exp(i\delta)$, with δ given by Eq. 4.1.1. Integrating over all space, we get the total wave scattered along $\mathbf{k}_1$:

$$f(\mathbf{k}) = \int \rho(\mathbf{r})\ \exp(i\mathbf{r}\cdot\mathbf{k})\,d\mathbf{r} \tag{4.1.3}$$

This is just the Fourier transform of $\rho(\mathbf{r})$ (cf. Appendix D).

As a special case, if we identify the above general distribution of scatterers with that in a single atom, then $f(\mathbf{k})$ given by Eq. 4.1.3 defines the scattering factor for the atom.

Unit-Cell Transform

Let us consider a set of N atoms, which may either constitute a molecule in the usual sense, or, else, may comprise the contents of the unit cell of a crystal. The N atoms are characterized by position vectors $\mathbf{r}_n$, and scattering factors $f_n(\mathbf{k})$ $(n = 1, 2, \cdots N)$. The position vectors are, naturally, referred to a *common* origin. The position of any general point, P, in, say, the nth atom can be expressed as

$$\mathbf{r}' = \mathbf{r}_n + \mathbf{r}, \tag{4.1.4}$$

where $\mathbf{r}$ is the position vector of the same point, referred to the local origin for that atom, and $\mathbf{r}_n$ is the position vector of this local origin, referred to the common origin.

The amplitude of waves scattered from the nth atom is given by

$$A_n = \int [\rho(\mathbf{r})d\mathbf{r}] \exp(i\mathbf{r}' \cdot \mathbf{k}) \tag{4.1.5}$$

On substituting from Eq. 4.1.4, and using Eq. 4.1.3, we get

$$A_n = f_n(\mathbf{k}) \exp(i\mathbf{r}_n \cdot \mathbf{k}) \tag{4.1.6}$$

The total wave scattered by all the N atoms is

$$D_0(\mathbf{k}) = \sum_{n=1}^{N} f_n(\mathbf{k}) \exp(i\mathbf{r}_n \cdot \mathbf{k}) \tag{4.1.7}$$

$D_0(\mathbf{k})$ is referred to as the unit-cell transform.

4.1.2 Diffraction by a Crystal

We first consider scattering of radiation from two unit cells of the crystal stacked along the direction of the basic translation vector $\mathbf{a}_1$. For the first unit cell, the cell transform is $D_0(\mathbf{k})$, given by Eq. 4.1.7. The second cell is displaced from the first cell by the vector $\mathbf{a}_1$, so its cell transform is

$$D_1(\mathbf{k}) = \sum_{n=1}^{N} f_n(\mathbf{k}) \exp(i(\mathbf{r}_n + \mathbf{a}_1) \cdot \mathbf{k}) = D_0(\mathbf{k})\,[1 + \exp(i\mathbf{a}_1 \cdot \mathbf{k})] \tag{4.1.8}$$

The total wave scattered from these two cells is the sum of $D_0(\mathbf{k})$ and $D_1(\mathbf{k})$. The total wave scattered by M_1 such cells in a row is

$$\begin{aligned} D_{x_1} &= \sum_{m=0}^{M_1-1} D_m(\mathbf{k}) \\ &= D_0(\mathbf{k}) \sum [1 + \exp(i\mathbf{a}_1 \cdot \mathbf{k}) + \exp(2i\mathbf{a}_1 \cdot \mathbf{k}) + \\ &\quad \cdots + \exp((M_1 - 1)i\mathbf{a}_1 \cdot \mathbf{k})] \end{aligned} \tag{4.1.9}$$

This equation can be simplified to yield

$$D_{x_1}(\mathbf{k}) = D_0(\mathbf{k}) \frac{\sin(M_1\,\mathbf{a}_1 \cdot \mathbf{k}/2)}{\sin(\mathbf{a}_1 \cdot \mathbf{k}/2)} \frac{\exp(i(M_1\,\mathbf{a}_1 \cdot \mathbf{k}/2))}{\exp(i\mathbf{a}_1 \cdot \mathbf{k}/2)} \tag{4.1.10}$$

The last factor in Eq. 4.1.10 is a phase factor, which drops out when we compute the intensity of the diffracted beam, which is proportional to $|D(\mathbf{k})^2|$:

$$|D_{x_1}(\mathbf{k})|^2 = |D_0(\mathbf{k})|^2 \sin^2(M_1\,\mathbf{a}_1 \cdot \mathbf{k}/2) \,/\, \sin^2(\mathbf{a}_1 \cdot \mathbf{k}/2) \tag{4.1.11}$$

For bulk single crystals, M_1 is usually a very large number, say of the order of 10^6. Eq. 4.1.11 therefore represents a very rapidly oscillating function. Its numerator is zero whenever $M_1\mathbf{a}_1 \cdot \mathbf{k}/2$ is an integer. The intensity of the diffracted beam is then zero, unless $\mathbf{a}_1 \cdot \mathbf{k}/2\pi$ is also an integer for that value of $\mathbf{k}$. Thus, nonzero diffracted intensity is observed only along those directions for which $\mathbf{a}_1 \cdot \mathbf{k}/2\pi$ is equal to an integer, say h:

$$\mathbf{a}_1 \cdot \mathbf{k}/2\pi \;=\; h \tag{4.1.12}$$

The whole row of M_1 cells can be regarded as a single unit, repeated, say, M_2, times along the lattice vector $\mathbf{a}_2$. The full 3-dimensional crystal can be constructed by repeating the entire net of M_1M_2 cells M_3 times along $\mathbf{a}_3$. The amplitude of the radiation diffracted from such a crystal can be written by inspection from Eq. 4.1.10:

$$D_x(\mathbf{k}) \;=\; D_0(\mathbf{k})\,\frac{\sin(M_1\,\mathbf{a}_1\cdot\mathbf{k}/2)}{\sin(\mathbf{a}_1\cdot\mathbf{k}/2)}\,\frac{\sin(M_2\,\mathbf{a}_2\cdot\mathbf{k}/2)}{\sin(\mathbf{a}_2\cdot\mathbf{k}/2)}\,\frac{\sin(M_3\,\mathbf{a}_1\cdot\mathbf{k}/2)}{\sin(\mathbf{a}_3\cdot\mathbf{k}/2)} \tag{4.1.13}$$

We have dropped the phase factor from this equation, as it does not enter the expression for the intensity of the diffracted beam.

Laue Equations

By an extension of the reasoning by which Eq. 4.1.12 was arrived at, we write the following three equations, called the Laue equations, which must be satisfied simultaneously if nonzero diffracted intensities are possibly to be observed along directions defined by the scattering vector $\mathbf{k}$:

$$\mathbf{a}_1 \cdot \mathbf{k}/2\pi = h, \tag{4.1.14}$$

$$\mathbf{a}_2 \cdot \mathbf{k}/2\pi = k, \tag{4.1.15}$$

$$\mathbf{a}_3 \cdot \mathbf{k}/2\pi = l \tag{4.1.16}$$

Here h, k, l can take all possible positive and negative integral values, including zero. Even when the three Laue equations are satisfied, the diffracted intensity would be zero if $D_0(\mathbf{k})$ in Eq. 4.1.13 is zero.

Bragg Law

Often it is more convenient to combine the three Laue equations into a single diffraction condition, the Bragg law. Subtracting Eq. 4.1.15 from 4.1.14 we get

$$(\mathbf{a}_1/h - \mathbf{a}_2/k)\cdot\mathbf{k} \;=\; 0 \tag{4.1.17}$$

This equation implies that the vector $(\mathbf{a}_1/h - \mathbf{a}_2/k)$ is perpendicular to $\mathbf{k}$. This vector lies in a plane which is a member of a family of parallel equidistant planes dividing the cell length a_1 into h equal parts and the cell length a_2 into k equal parts. Similarly, $\mathbf{k}$ is perpendicular to the vector $(\mathbf{a}_2/k - \mathbf{a}_3/l)$ which lies in a plane belonging to a set dividing the basic repeat distance a_2 into k equal parts, and a_3 into l equal parts.

A family of crystallographic planes which divide a_1 into h, a_2 into k, and a_3 into l equal parts is said to have *Miller indices* h, k, l, and such planes are denoted by (hkl). [The Miller index k is not to be confused with the magnitude k of the scattering vector $\mathbf{k}$.]

Thus, for a 3-dimensional crystal, when the three diffraction conditions are satisfied, $\mathbf{k}$ is perpendicular to the planes (hkl). But, since $\mathbf{k} = \mathbf{k}_1 - \mathbf{k}_0$ and $|\mathbf{k}_1| = |\mathbf{k}_0|$ ("elastic scattering"), $\mathbf{k}$ has the direction of the bisector of the angle between k_1 and k_0. Thus the direction of this bisector of the incident and diffracted-beam directions can be identified with the normal to the planes (hkl).

Let d_{hkl} denote the spacing between the planes (hkl). It is equal to the projection of $\mathbf{a}_1/h$, or $\mathbf{a}_2/k$, or $\mathbf{a}_3/l$, along $\mathbf{k}$:

$$d_{hkl} = (\mathbf{a}_1/h) \cdot \mathbf{k}/k = 1/k = \lambda/(2 \sin \theta_{hkl}), \tag{4.1.18}$$

or,

$$2 d_{hkl} \sin \theta_{hkl} = \lambda \tag{4.1.19}$$

We have made use of Eq. 4.1.14 in the derivation of Eq. 4.1.18. Eq. 4.1.19 is known as the Bragg law, or the Bragg condition for constructive interference. When this diffraction condition is satisfied, one also speaks of Bragg *reflection* because the incident and the diffracted beams make the same angle, θ_{hkl}, with the planes (hkl).

4.1.3 The Reciprocal Lattice

Each of the Laue equations (Eqs. 4.1.14-16) defines a family of parallel equidistant planes in the space of $\mathbf{k}$. [This space is called the *wavevector space*, or *Fourier space*, or *reciprocal space* because k $(= 4\pi \sin\theta/\lambda)$ has the dimensions of L^{-1}.]

For example, the first Laue equation (4.1.14) defines, for $h = 0, 1, 2, \cdots$, a family of planes normal to $\mathbf{a}_1$ (Fig. 4.1.2).

Suppose the spacing between the consecutive planes is d^*. Then, as is clear from Fig. 4.1.2,

$$d^* = \mathbf{k} \cdot (\mathbf{a}_1/a_1) = 2\pi/a_1 \tag{4.1.20}$$

(Eq. 4.1.14 has been used here.)

Thus, if we have a system of scatterers which is periodic only in one direction (that of a_1), the diffraction condition for it will be satisfied for all scattering vectors, $\mathbf{k}$, which terminate on any of the parallel and equidistant planes having a spacing $2\pi/a_1$ (Fig. 4.1.2).

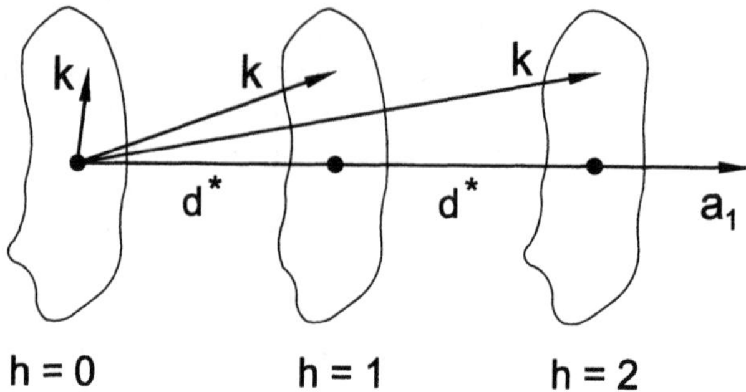

Figure 4.1.2: A family of parallel equidistant planes in reciprocal space, normal to the a_1 axis.

Let us now consider a system for which the second Laue condition is also satisfied. That is, the system has lattice periodicity in two dimensions. There will now be a second set of parallel equidistant planes in reciprocal space, with a spacing $2\pi/a_2$, in addition to the first set shown in Fig. 4.1.2. The two Laue equations will be simultaneously satisfied only for $\mathbf{k}$ terminating on the lines of intersection of the two families of planes in reciprocal space. These lines are sometimes referred to as *rods*: the diffraction pattern of a planar net of scatterers is a set of rods in reciprocal space.

Finally, all the three Laue equations must be satisfied for a 3-dimensional crystal lattice, and this will happen when $\mathbf{k}$ terminates on points (in fact, a lattice of points) where all the three families of parallel, equidistant, planes in reciprocal space, perpendicular respectively to $\mathbf{a}_1$, $\mathbf{a}_2$, and $\mathbf{a}_3$, intersect. This lattice of points in reciprocal space is called *the reciprocal lattice.*

Thus appreciable diffracted radiation can possibly be observed in only those directions, and for only those orientations of the crystal with respect to the incident collimated monochromatic radiation, which make the scattering vector $\mathbf{k}$ $(= \mathbf{k}_1 - \mathbf{k}_0)$ equal to a reciprocal-lattice vector. Let us denote by $\mathbf{a}_1^*, \mathbf{a}_2^*, \mathbf{a}_3^*$ the fundamental translation vectors which span the reciprocal lattice. Any vector $\mathbf{K}$ in this lattice can then be expressed as

$$\mathbf{K} = h\mathbf{a}_1^* + k\mathbf{a}_2^* + l\mathbf{a}_3^* \tag{4.1.21}$$

The diffraction condition contained in the Laue equations implies that $\mathbf{k}$ must be equal to one of the $\mathbf{K}$s:

$$\mathbf{k} = \mathbf{K} \tag{4.1.22}$$

$$= h\mathbf{a}_1^* + k\mathbf{a}_2^* + l\mathbf{a}_3^*, \tag{4.1.23}$$

h, k, l being arbitrary integers.

It is straightforward to derive $\mathbf{a}_1^*, \mathbf{a}_2^*, \mathbf{a}_3^*$ in terms of $\mathbf{a}_1, \mathbf{a}_2, \mathbf{a}_3$. From Eqs. 4.1.23 and 4.1.14,

$$\mathbf{a}_1 \cdot \mathbf{k} \;=\; 2\pi h \;=\; h\mathbf{a}_1 \cdot \mathbf{a}_1^* \,+\, k\mathbf{a}_1 \cdot \mathbf{a}_2^* \,+\, l\mathbf{a}_1 \cdot \mathbf{a}_3^* \tag{4.1.24}$$

Since h, k, l can take arbitrary (integral) values, this equality can hold only if

$$\mathbf{a}_1 \cdot \mathbf{a}_1^* \;=\; 2\pi, \;\; \mathbf{a}_1 \cdot \mathbf{a}_2^* \;=\; 0, \;\; \mathbf{a}_1 \cdot \mathbf{a}_3^* \;=\; 0 \tag{4.1.25}$$

Similarly,

$$\mathbf{a}_2 \cdot \mathbf{a}_1^* \;=\; 0, \;\; \mathbf{a}_2 \cdot \mathbf{a}_2^* \;=\; 2\pi, \;\; \mathbf{a}_2 \cdot \mathbf{a}_3^* \;=\; 0 \tag{4.1.26}$$

$$\mathbf{a}_3 \cdot \mathbf{a}_1^* \;=\; 0, \;\; \mathbf{a}_3 \cdot \mathbf{a}_2^* \;=\; 0, \;\; \mathbf{a}_3 \cdot \mathbf{a}_3^* \;=\; 2\pi \tag{4.1.27}$$

These equation imply that $\mathbf{a}_1^*$ is perpendicular to both $\mathbf{a}_2$ and $\mathbf{a}_3$. It can therefore be expressed as

$$\mathbf{a}_1^* \;=\; C\,\mathbf{a}_2 \times \mathbf{a}_3 \tag{4.1.28}$$

Since $\mathbf{a}_1 \cdot \mathbf{a}_1^* \;=\; 2\pi$, the proportionality constant C must be

$$C \;=\; 2\pi/(\mathbf{a}_1 \cdot \mathbf{a}_2 \times \mathbf{a}_3) \;=\; 2\pi/V, \tag{4.1.29}$$

V being the volume of the unit cell. The same proportionality constant applies for $\mathbf{a}_2^*$ and $\mathbf{a}_3^*$ also, and we finally arrive at the following equations:

$$\mathbf{a}_1^* \;=\; 2\pi\mathbf{a}_2 \times \mathbf{a}_3/(\mathbf{a}_1 \cdot \mathbf{a}_2 \times \mathbf{a}_3), \tag{4.1.30}$$

$$\mathbf{a}_2^* \;=\; 2\pi\mathbf{a}_3 \times \mathbf{a}_1/(\mathbf{a}_1 \cdot \mathbf{a}_2 \times \mathbf{a}_3), \tag{4.1.31}$$

$$\mathbf{a}_3^* \;=\; 2\pi\mathbf{a}_1 \times \mathbf{a}_2/(\mathbf{a}_1 \cdot \mathbf{a}_2 \times \mathbf{a}_3) \tag{4.1.32}$$

For any translation $\mathbf{t}_n$ ($= n_1\mathbf{a}_1 + n_2\mathbf{a}_2 + n_3\mathbf{a}_3$) of the direct lattice, and any reciprocal-lattice vector $\mathbf{K}$ (defined by Eq. 4.1.23), the following relation holds because of Eqs. 4.1.25-27:

$$\mathbf{K} \cdot \mathbf{t}_n \;=\; 2\pi\,\mathcal{M}, \tag{4.1.33}$$

where $\mathcal{M}$ is an integer.

Diffraction Pattern as a Sample of the Unit-Cell Transform

Eq. 4.1.13 for the amplitude of the wave diffracted by a crystal has factors of the form $\sin Mx/\sin x$. Now

$$\frac{\sin Mx}{\sin x} \;=\; \frac{\sin Mx}{Mx} \cdot M \cdot \frac{x}{\sin x} \;\to\; M \tag{4.1.34}$$

for $x = n\pi$ and large M.

Thus for those values of **k** for which the Laue conditions are satisfied, i.e. at the reciprocal-lattice points, Eq. 4.1.13 reduces to

$$D_x(\mathbf{k}) = M_1 M_2 M_3 D_0(\mathbf{k}) \tag{4.1.35}$$

This means that, at the reciprocal-lattice points, the crystal transform is simply $M_1 M_2 M_3$ times the unit-cell transform. When molecules or unit cells come together to form a large crystal, the cell transform gets wiped out at all points in reciprocal space except near the reciprocal-lattice points, and at these points it is simply scaled up by a (usually large) factor $M_1 M_2 M_3$. It follows that the recording of the diffraction pattern of a crystal may be thought of as amounting to *sampling* the unit-cell transform at the reciprocal-lattice points. The crystal not only "anchors" the unit-cell contents in position and orientation, it also provides a large amplification of the diffracted intensities.

The peak of a Bragg reflection varies as M^2, and the peak width varies as M^{-1}, so that the integrated intensity varies as M^2/M, or M.

Symmetry of the Reciprocal Lattice

Consider an element g of the point group of the real-space lattice or the direct lattice. For any lattice translation $\mathbf{t}_n$ of this lattice, $g\mathbf{t}_n$ is also a lattice translation and therefore satisfies Eq. 4.1.33:

$$\mathbf{K} \cdot (g\mathbf{t}_n) = 2\pi\mathcal{M} \tag{4.1.36}$$

Operations of the point group of the direct lattice are orthogonal transformations. Therefore the scalar product of two vectors, such as that in Eq. 4.1.36, must remain invariant under them. It follows that, if we apply the orthogonal transformation g^{-1} to both $\mathbf{K}$ and $g\mathbf{t}_n$ in Eq. 4.1.36, we must get

$$g^{-1}\mathbf{K} \cdot g^{-1}(g\mathbf{t}_n) = \mathbf{K} \cdot (g\mathbf{t}_n), \tag{4.1.37}$$

which, on using Eq. 4.1.36 again, yields

$$g^{-1}\mathbf{K} \cdot \mathbf{t}_n = 2\pi\mathcal{M} \tag{4.1.38}$$

Since this is true for any $\mathbf{t}_n$, $g^{-1}\mathbf{K}$ must be a reciprocal-lattice vector. And since this is true for any g, any symmetry operation of the direct lattice must also be a symmetry operation of the reciprocal lattice.

The converse is also true because the direct lattice is the reciprocal of the reciprocal lattice.

We are therefore led to the following result:

The reciprocal lattice has the same point-group symmetry as the direct lattice.

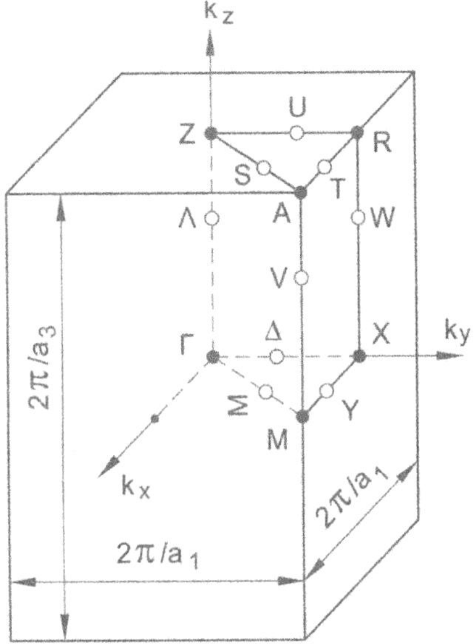

Figure 4.1.3: The Brillouin zone of a simple tetragonal lattice.

4.1.4 The Brillouin Zone

The Brillouin zone is the Wigner-Seitz cell (§2.2.13) of reciprocal space or **k**-space. It is defined by all those points of reciprocal space that are closer to the point $k = 0$ than to any other reciprocal-lattice point. Like the Wigner-Seitz cell it is a closed polygon bounded by planes that are perpendicular bisectors of lines joining the point $k = 0$ to the nearest and sometimes to the next nearest reciprocal-lattice points. It exhibits the point symmetry, G_p, of the reciprocal lattice.

Two vectors $\mathbf{k}_1$ and $\mathbf{k}_2$ in the reciprocal-lattice space are said to be *equivalent* if

$$\mathbf{k}_2 = \mathbf{k}_1 + \mathbf{K}, \tag{4.1.39}$$

where $\mathbf{K}$ is any reciprocal-lattice vector.

No two points in the interior of the Brillouin zone can be equivalent. However, for every point on the *surface* of the Brillouin zone there is at least one point equivalent to it, also on the surface.

Points, Lines, and Planes of Symmetry of the Brillouin Zone. Lifshitz Points

Fig. 4.1.3 shows the Brillouin zone for a crystal with a simple-tetragonal lattice.

Different points in the Brillouin zone have different directional symmetry $P_{\mathbf{k}}$. The point group $P_{\mathbf{k}}$ is defined as comprising all those transformations which either map $\mathbf{k}$ back into itself, or change it into an equivalent vector in reciprocal space. Obviously, $P_{\mathbf{k}} \subseteq G_p$.

A point $\mathbf{k}$ in reciprocal space is said to be a *symmetry point* (or a *point of symmetry*) if there does not exist in the neighbourhood of $\mathbf{k}$ any point $\mathbf{k}'$ which has a point-group symmetry $P_{\mathbf{k}'}$ equal to or higher than $P_{\mathbf{k}}$.

Symmetry points of the Brillouin zone are also called *Lifshitz points.*

The wavevector $\mathbf{k}$ in the Brillouin zone is said to lie on a *line (plane) of symmetry* if in the neighbourhood of $\mathbf{k}$ there is a line (plane) of points that passes through $\mathbf{k}$ and has the same point-group symmetry $P_{\mathbf{k}}$ as the point $\mathbf{k}$.

A characteristic feature of Lifshitz points is that they have numerically fixed coordinates [e.g. $(0, \frac{1}{2}, 0)$]. By contrast, for non-Lifshitz points some or all coordinates are running variables.

4.1.5 Diffraction by an Incommensurately Modulated Crystal

The structure of an incommensurately modulated crystal can be described by the following function (cf. Eq. 2.4.1):

$$C_{inc}(\mathbf{r}) = B(\mathbf{r}) * [L(\mathbf{r}) \cdot M(\mathbf{r})], \tag{4.1.40}$$

where the ratio of the periods of $L(\mathbf{r})$ and $M(\mathbf{r})$ is an irrational fraction. The diffraction pattern of the crystal is determined by the Fourier transform of $C_{inc}(\mathbf{r})$.

Let $C_{inc}(\mathbf{k})$, $B(\mathbf{k})$, $L(\mathbf{k})$ and $M(\mathbf{k})$ denote, respectively, the Fourier transforms of $C_{inc}(\mathbf{r})$, $B(\mathbf{r})$, $L(\mathbf{r})$ and $M(\mathbf{r})$. According to the convolution theorem (Appendix D), $C_{inc}(\mathbf{k})$ is equal to the product of $B(\mathbf{k})$ with the convolution of $L(\mathbf{k})$ and $M(\mathbf{k})$ (Burns & Glazer 1990):

$$C_{inc}(\mathbf{k}) = B(\mathbf{k}) \cdot [L(\mathbf{k}) * M(\mathbf{k})] \tag{4.1.41}$$

$L(\mathbf{k})$ is a set of regular reciprocal-lattice points. Let us assume for simplicity that $M(\mathbf{r})$ is a sinusoidal modulation along the lattice vector $\mathbf{a}_3$. $M(\mathbf{k})$ will then be two delta functions (Appendix D), symmetrically displaced from the neighbouring basic reciprocal-lattice peak. The convolution of $M(\mathbf{k})$ with $L(\mathbf{k})$ (Eq. 4.1.41) means that this pair of *satellite diffraction peaks* repeats around each reciprocal-lattice point (provided, of course, that $B(\mathbf{k})$ is nonzero for that point).

The modulation function is seldom a pure sine wave. As a result, one obtains a series of satellite peaks, instead of just one pair, around the main diffraction peaks.

SUGGESTED READING

H. Lipson & I. Taylor (1958). *Fourier Transforms and X-Ray Diffraction.* G. Bell, London.

G. Burns & A. M. Glazer (1990). *Space Groups for Solid State Scientists,* second edition. Academic Press, London.

4.2 REPRESENTATIONS OF CRYSTALLOGRAPHIC TRANSLATION GROUPS

A crystallographic translation group T is a subgroup of the general translation group $T(3)$ (§B.4), and is characterized by the Seitz operator $\{\mathbf{E}|\mathbf{t}_n\}$, with $\mathbf{t}_n$ given by (Eq. 2.2.15):

$$\mathbf{t}_n = n_i\,\mathbf{a}_i \qquad (4.2.1)$$

Since $T(3)$, and therefore T, is a commutative group (§B.4), one can write the following direct product:

$$T = T_1 \otimes T_2 \otimes T_3, \qquad (4.2.2)$$

with $\{\mathbf{E}\,|\,n_i\,\mathbf{a}_i\}$ as elements of the group T_i.

The definition of the translational symmetry of a crystal requires that the space occupied by the crystal be infinite in all the dimensions; then only can all lattice translations qualify as symmetry operations. Real crystals, however, are finite objects, bounded by surfaces. Mathematically, the problem of the presence of surfaces can be recognized (and overcome) by introducing the so-called *cyclic boundary conditions*:

$$\{\mathbf{E}\,|\,M_1\,\mathbf{a}_1\} = \{\mathbf{E}\,|\,M_2\,\mathbf{a}_2\} = \{\mathbf{E}\,|\,M_3\,\mathbf{a}_3\} = \{\mathbf{E}\,|\,0\} \qquad (4.2.3)$$

Here M_1, M_2, M_3 are sufficiently large integers (cf. Eq. 4.1.13). The crystal is thus taken as composed of $M_1M_2M_3$ unit cells. This finite, though large, number is also the order of the translation group T.

The cyclic boundary conditions also imply that T_1, T_2, T_3 are cyclic groups (§B.1).

A cyclic group has the property that it can be generated by a single generating element (§B.1). In particular, T_1 is a cyclic group of order M_1, with lattice translation $\mathbf{a}_1$ as the generating element ($\mathbf{a}_1^{M_1+n} = \mathbf{a}_1^n$). All cyclic groups are Abelian or commutative.

Because of the commutative nature of the group T_1, it follows from Eq. B.1.3 that each of the elements of this group constitutes a class by itself (cf. Eq. 2.2.17). Therefore the number of classes in the group T_1 is equal to M_1 (cf. Eq. 3.4.1).

It follows then from Eq. 3.4.2 that cyclic groups can have only 1-dimensional irreducible representations (IRs). The characters of the group T_1 must therefore satisfy the equation

$$[\chi(\{\mathbf{E}|\mathbf{a}_1\})]^{M_1} = \chi(\{\mathbf{E}|M_1\mathbf{a}_1\}) = 1, \tag{4.2.4}$$

and can thus be written as

$$\chi_{m_1}(\{\mathbf{E}|\mathbf{a}_1\}) = \exp(2\pi i m_1/M_1) \tag{4.2.5}$$

The integer m_1 characterizes the various IRs of T_1.

Similarly, the IRs of T_2 and T_3 can be labelled in terms of integers m_2 and m_3, and the set of integers (m_1, m_2, m_3) labels the IRs of the translation group T.

We can define a vector $\mathbf{k}$ in the reciprocal space as follows:

$$\mathbf{k} = m_1\mathbf{a}_1^*/M_1 + m_2\mathbf{a}_2^*/M_2 + m_3\mathbf{a}_3^*/M_3 \tag{4.2.6}$$

This vector serves as a label for the IRs of the translation group T, with characters defined by

$$\chi_{\mathbf{k}}(\{\mathbf{E}|\mathbf{a}_n\}) = \exp(i\mathbf{k}\cdot\mathbf{a}_n) \tag{4.2.7}$$

Because of Eqs. 4.1.25-27 $\chi_{\mathbf{k}}$ has the following property:

$$\chi_{\mathbf{k}+\mathbf{K}} = \chi_{\mathbf{k}}, \tag{4.2.8}$$

where $\mathbf{K}$ is any reciprocal-lattice vector (defined by Eq. 4.1.21). Because of this property it is sufficient to consider $\mathbf{k}$ confined to the Brillouin zone.

Bloch Functions

The energies and wavefunctions of an electron in a crystal are determined by the Schrodinger equation, $\mathcal{H}\psi(\mathbf{r}) = \mathcal{E}\psi(\mathbf{r})$. Since the Hamiltonian $\mathcal{H}$ is invariant under the operations of the crystallographic translation group, the eigenfunctions $\psi(\mathbf{r})$ can be used for constructing an IR of the translation group. The action of any element of the translation group on $\psi(\mathbf{r})$ is simply to multiply it by a scalar; this scalar is then the representation of the group element, with $\psi(\mathbf{r})$ as the basis.

We write Eq. B.2.28 of Appendix B for the case when the Seitz operator is for just a lattice translation, $\{\mathbf{E}|\mathbf{t}_n\}$, and the function $f(\mathbf{r})$ is the wave function $\psi(\mathbf{r})$:

$$\{\mathbf{E}|\mathbf{t}_n\}\psi(\mathbf{r}) = \psi(\mathbf{r}-\mathbf{t}_n) \tag{4.2.9}$$

The IRs of the translation group are specified by the wavevector $\mathbf{k}$. Let $\psi_{\mathbf{k}}(\mathbf{r})$ be the basis functions for the IR labelled by $\mathbf{k}$. Since all the IRs here are 1-dimensional, the matrices representing the symmetry operators are of order 1, and are identical to the characters $\chi_{\mathbf{k}}$. The Seitz operator for a lattice translation therefore has the following effect on $\psi_{\mathbf{k}}$ (cf. Eq. 4.2.7):

$$\{\mathbf{E}|\mathbf{t}_n\}\,\psi_{\mathbf{k}} = \chi_{\mathbf{k}}(\{\mathbf{E}|\mathbf{t}_n\})\psi_{\mathbf{k}} = \exp(i\mathbf{k}\cdot\mathbf{t}_n)\psi_{\mathbf{k}} \tag{4.2.10}$$

Identifying $\psi(\mathbf{r})$ in Eq. 4.2.9 with $\psi_{\mathbf{k}}(\mathbf{r})$ of Eq. 4.2.10 we arrive at the so-called *Bloch theorem:*

$$\psi_{\mathbf{k}}(\mathbf{r}-\mathbf{t}_n) = \exp(i\mathbf{k}\cdot\mathbf{t}_n)\,\psi_{\mathbf{k}}(\mathbf{r}) \tag{4.2.11}$$

The functions $\psi_{\mathbf{k}}(\mathbf{r})$ satisfying Eq. 4.2.11 are called *Bloch functions.*

Bloch functions are the basis functions of the IRs of the translation group T.

We can introduce functions $u_{\mathbf{k}}(\mathbf{r})$ through the relationship

$$\psi_{\mathbf{k}}(\mathbf{r}) = \exp(-i\mathbf{k}\cdot\mathbf{r})\,u_{\mathbf{k}}(\mathbf{r}) \tag{4.2.12}$$

Eq. 4.2.11 then yields

$$u_{\mathbf{k}}(\mathbf{r}-\mathbf{t}_n) = u_{\mathbf{k}}(\mathbf{r}) \tag{4.2.13}$$

In other words, $u_{\mathbf{k}}(\mathbf{r})$ is a periodic function.

An alternative expression for Bloch functions is

$$\psi_{\mathbf{k}} = \sum_n \exp(-i\mathbf{k}\cdot\mathbf{t}_n)\,a(\mathbf{r}-\mathbf{t}_n), \tag{4.2.14}$$

implying that $\mathbf{k}$ represents the wavevector of Bloch waves propagating in the periodic potential of the crystal.

4.3 THE GROUP OF THE WAVEVECTOR, AND ITS REPRESENTATIONS

Group of the Wavevector, and Star of the Wavevector

Consider a wavevector $\mathbf{k}$, and apply on it all the operations s $(=\{\mathbf{R}|\mathbf{t}\})$ of the space group S of the crystal. The resultant set of wavevectors can be divided into two categories: those equivalent to $\mathbf{k}$, and those not equivalent to $\mathbf{k}$.

The symmetry operators that generate the set of wavevectors equivalent to $\mathbf{k}$ constitute a group (a space group) ($S_{\mathbf{k}}$) ($\subseteq S$) called the *group of the wavevector* $\mathbf{k}$, or the *little group.*

$S_{\mathbf{k}}$ is a subgroup of the full space group S, and consists of elements $\{A|\mathbf{t}\}$ taken from the group of elements $\{\mathbf{R}|\mathbf{t}\}$ such that

$$\{\mathbf{A}|\mathbf{t}\}\,\mathbf{k} = \mathbf{k} + \mathbf{K}, \tag{4.3.1}$$

where $\mathbf{K}$ is any vector of the reciprocal lattice of the crystal. $S_\mathbf{k}$ describes the inherent symmetry of the wavevector $\mathbf{k}$.

Tables for $S_\mathbf{k}$ for all distinct (inequivalent) $\mathbf{k}$-vectors can be found in Kovalev (1993) and Miller & Love (1967).

The set of nonequivalent wavevectors $\mathbf{k}_L$ obtained by the action of all the operators of the space group of the crystal constitutes what is called the *star of the wavevector* $\mathbf{k}$, and is denoted by $\{\mathbf{k}\}$ or $^*\mathbf{k}$. Members (or *arms*, or *prongs*) of the star of $\mathbf{k}$ are wavevectors given by the relation

$$\mathbf{k}_L = s_L \, \mathbf{k}, \tag{4.3.2}$$

where the operators s_L are representative elements of the coset decomposition

$$S = \sum_{L=1}^{l_\mathbf{k}} s_L \, S_\mathbf{k} \tag{4.3.3}$$

The s_Ls are symmetry operators of the type $\{\mathbf{R}|0\}$, and are nothing but some or all the elements of the point group G_p underlying the space group of the crystal (cf. §2.2.17).

If all the $\mathbf{k}_L$s are inequivalent, $\mathbf{k}$ must be a general point (§4.1.4) of the Brillouin zone. In such a case the number of arms or prongs of $\{\mathbf{k}\}$ is equal to the order of the point group G_p. This number can be only 48 at the most.

If $\mathbf{k}$ is invariant under n of the n_p elements of G_p, i.e. if $\mathbf{k}$ corresponds to a point, line, or plane of symmetry in the Brillouin zone (§4.1.4), its star has n_p/n arms ($l_\mathbf{k} = n_p/n$ in Eq. 4.3.3). Fig. 4.3.1 shows a couple of examples of the star of a wavevector. The usage of the term 'star' becomes obvious from these examples.

If $\mathbf{k}$ is a point of symmetry, i.e. if $\mathbf{k}$ is a Lifshitz point (§4.1.4), its star is conveniently referred to as the *Lifshitz star* (Lifshitz 1941).

To understand better the nature of the little group $S_\mathbf{k}$, we examine its relationship with the translation group T. T being a subgroup of the full space group S, we can write the following coset decomposition:

$$S = \sum_{i=1}^{n_p} s_i \, T, \quad s_1 = \{\mathbf{E}|0\} \tag{4.3.4}$$

The coset representatives s_i can involve, in general, both a rotation and a fractional translation:

$$s_i = \{\mathbf{R}_i|\mathbf{T}_i\} \tag{4.3.5}$$

We have seen in §4.2 that the IRs of the group T have Bloch functions $\psi_\mathbf{k}$ as the basis. Let us consider a function obtained by the operation of

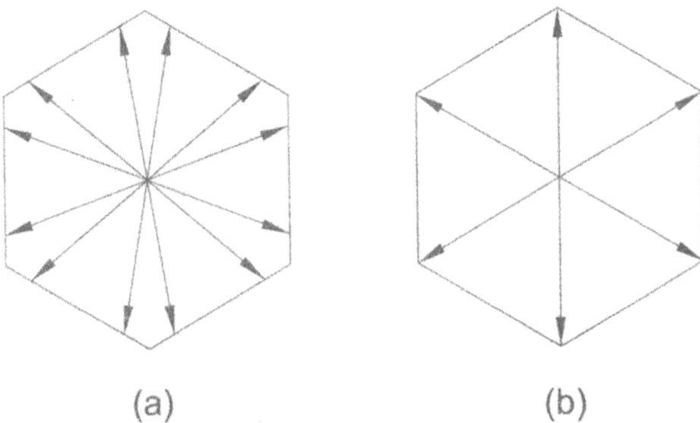

Figure 4.3.1: Stars of wavevectors for a 2-dimensional hexagonal lattice. (a) Star of a general point in the Brillouin zone. (b) Star of a symmetry point.

any of the s_is on a Bloch function:

$$\phi_{\mathbf{k}} = s_i \psi_{\mathbf{k}} = \{\mathbf{R}_i|\mathbf{T}_i\}\,\psi_{\mathbf{k}} \qquad (4.3.6)$$

Any lattice translation operator $\{\mathbf{E}|\mathbf{t}_n\}$ has the following effect on this function:

$$\begin{aligned} \{\mathbf{E}|\mathbf{t}_n\}\,\phi_{\mathbf{k}} &= \{\mathbf{E}|\mathbf{t}_n\}\,\{\mathbf{R}_i|\mathbf{T}_i\}\,\psi_{\mathbf{k}} && (4.3.7)\\ &= \{\mathbf{R}_i|\mathbf{T}_i+\mathbf{t}_n\}\,\psi_{\mathbf{k}} \text{ (cf. Eq. 2.2.13)} && (4.3.8)\\ &= \{\mathbf{R}_i|\mathbf{T}_i\}\,[\{\mathbf{E}|\mathbf{R}_i^{-1}t_n\}\psi_{\mathbf{k}}] \text{ (cf. Eq. 2.2.13)} && (4.3.9) \end{aligned}$$

Using Eq. 4.2.10 for the expression in square brackets we get

$$\{\mathbf{E}|\mathbf{t}_n\}\,\phi_{\mathbf{k}} = \{\mathbf{R}_i|\mathbf{T}_i\}\,\exp(i\mathbf{k}\cdot(\mathbf{R}_i^{-1}))\psi_{\mathbf{k}} = \exp(i\mathbf{k}\cdot(\mathbf{R}_i^{-1}\mathbf{t}_n))\phi_{\mathbf{k}} \quad (4.3.10)$$

Since a scalar product of vectors remains invariant under the orthogonal transformation defined by $\mathbf{R}_i^{-1}$, we must have

$$\boldsymbol{k}\cdot\boldsymbol{R}_i^{-1}\boldsymbol{t}_n = \boldsymbol{R}_i\boldsymbol{k}\cdot\boldsymbol{R}_i\boldsymbol{R}_i^{-1}\boldsymbol{t}_n = \boldsymbol{R}_i\boldsymbol{k}\cdot\boldsymbol{t}_n \qquad (4.3.11)$$

We thus arrive at the following result:

$$\{\mathbf{E}|\mathbf{t}_n\}\,\phi_{\mathbf{k}} = \exp(i\mathbf{R}_i\mathbf{k}\cdot\mathbf{t}_n)\,\phi_{\mathbf{k}}; \qquad (4.3.12)$$

that is, *the function $\phi_{\mathbf{k}}$ defined by Eq. 4.3.6 is the basis function for the IR* $\mathbf{R}_i\mathbf{k}$ *of the translation group.*

The inequivalent wavevectors in the set $\mathbf{R}_i\mathbf{k}$ $(i = 1, 2, ..n_p)$ constitute the star of $\mathbf{k}$, and the number of arms of the star is called the *order* of the star.

If $\mathbf{k}$ corresponds to a symmetry point of the Brillouin zone, some or all the operators s_i either leave it unchanged, or change it by the addition of a reciprocal-lattice vector $\mathbf{K}$. Let $\{\mathbf{A}_i|\mathbf{T}_i\}$ be one such operation. For the rotations $\mathbf{A}_i$ we have

$$\exp(i\mathbf{A}_i\mathbf{k}\cdot\mathbf{t}_n) = \exp(i\mathbf{k}\cdot\mathbf{t}_n) \tag{4.3.13}$$

Therefore, from Eq. 4.3.12

$$\{\mathbf{E}|\mathbf{t}_n\}\{\mathbf{A}_i|\mathbf{T}_i\}\psi_{\mathbf{k}} = \exp(i\mathbf{k}\cdot\mathbf{t}_n)\{\mathbf{A}_i|\mathbf{T}_i\}\psi_{\mathbf{k}} \tag{4.3.14}$$

This implies that $\{\mathbf{S}_i|\mathbf{T}_i\}\psi_{\mathbf{k}}$ belongs to the same IR $\mathbf{k}$ of T as $\psi_{\mathbf{k}}$ does. As stated above, the set of all such operations $\{\mathbf{S}_i|\mathbf{T}_i\}$ constitutes the group $S_{\mathbf{k}}$ of the wavevector $\mathbf{k}$. When $\mathbf{k}$ is a general point of the Brillouin zone, no point-group operations will leave it unchanged, and $S_{\mathbf{k}}$ will be nothing but the translation group T.

In general, $S_{\mathbf{k}}$ can consist of several of the cosets in Eq. 4.3.4, and T is a normal or invariant subgroup of $S_{\mathbf{k}}$.

$S_{\mathbf{k}}$ is a factor group of S with respect to T, with T playing the role of the identity element (cf. Eqs. B.1.11-13).

The following result is readily verifiable:

> *The order of the point-group underlying $S_{\mathbf{k}}$, multiplied by the number of arms in the star of $\mathbf{k}$, is equal to the order of the point-group underlying the space group of the crystal.*

Small Representations

IRs of the group of the wavevector are called small representations. Small representations can be built up from IRs of point groups (§3.4) and of translation groups.

Since the basis functions of the IRs of translation groups are Bloch functions satisfying Eq. 4.2.10, we can write down the representation matrix for a lattice translation t_n as

$$d_{\mathbf{k}}(\{\mathbf{E}|\mathbf{t}_n\}) = \exp(i\mathbf{k}\cdot\mathbf{t}_n)\,1_d, \tag{4.3.15}$$

where 1_d denotes a unit matrix having an order equal to the dimension d of the representation.

Let Γ be an IR of the point group $P_{\mathbf{k}}$ of the wavevector. It is pertinent to see whether the matrices $d_{\mathbf{k}}(\{\mathbf{A}|\mathbf{t}\})$, defined by

$$d_{\mathbf{k}}(\{\mathbf{A}|\mathbf{t}\}) = \exp(i\mathbf{k}\cdot\mathbf{t})\,\Gamma(\mathbf{A}) \tag{4.3.16}$$

constitute an IR of $S_{\mathbf{k}}$.

Since $\Gamma(\mathbf{A})$ is taken as irreducible, the irreducibility of $d_{\mathbf{k}}$ can be taken as granted.

We next compute the matrix products. The product of two such matrices is

$$d_{\mathbf{k}}(\{\mathbf{A}_2|\mathbf{t}_2\})\, d_{\mathbf{k}}(\{\mathbf{A}_1|\mathbf{t}_1\}) = \exp[i\mathbf{k}\cdot(\mathbf{t}_1+\mathbf{t}_2)]\,\Gamma(\mathbf{A}_2\mathbf{A}_1) \tag{4.3.17}$$

And the matrix for the product operation $\{\mathbf{A}_2|\mathbf{t}_2\}\,\{\mathbf{A}_1|\mathbf{t}_1\}$ is

$$d_{\mathbf{k}}(\{\mathbf{A}_2\mathbf{A}_1|\mathbf{A}_2\mathbf{t}_1+\mathbf{t}_2\}) = \exp[i\mathbf{k}\cdot(\mathbf{A}_2\mathbf{t}_1+\mathbf{t}_2)]\,\Gamma(\mathbf{A}_2\mathbf{A}_1) \tag{4.3.18}$$

The matrices occurring in Eqs. 4.3.17 and 4.3.18 can be identical only if

$$\exp[i(\mathbf{A}_2^{-1}\mathbf{k}-\mathbf{k})\cdot\mathbf{t}_1] = 1 \tag{4.3.19}$$

If $\mathbf{k}$ is inside the Brillouin zone, $\mathbf{A}_2^{-1}\mathbf{k}-\mathbf{k} = 0$ always, and Eq. 4.3.19 is satisfied, implying that the matrices defined by Eq. 4.3.16 constitute a valid small representation.

If $\mathbf{k}$ is a point on the zone boundary,

$$\mathbf{A}_2^{-1}\mathbf{k}-\mathbf{k} = \mathbf{K} \tag{4.3.20}$$

where $\mathbf{K}$ is some reciprocal-lattice vector. In this case Eq. 4.3.19 will be satisfied if $\mathbf{t}_1$ is a primitive lattice translation, i.e. if there is no fractional lattice translation involved ($\mathbf{T}_{n1}=0$ in $\mathbf{t}_1=\mathbf{t}_{n1}+\mathbf{T}_{n1}$). Such is the case when the little group $S_{\mathbf{k}}$ is a symmorphic space group.

The case when $S_{\mathbf{k}}$ is a nonsymmorphic space group presents special problems, and calls for special solutions. One solution is the use of 'ray representations'. Another is the application of Herring's method (Herring 1942).

SUGGESTED READING

W. A. Wooster (1973). *Tensors and Group Theory for the Physical Properties of Crystals.* Clarendon Press, Oxford.

T. Inui, Y. Tanabe & Y. Onodera (1990). *Group Theory and Its Applications in Physics.* Springer-Verlag, Berlin.

4.4 REPRESENTATIONS OF SPACE GROUPS

A detailed discussion of the theory of representations of space groups falls outside the scope of this book. The reader is advised to consult, for example, Koster (1957), Heine (1960), Ludwig & Falter (1988), and Kovalev

(1993) for a formal treatment of the subject. We list here some important information, mainly relating to the Landau theory of phase transitions (Chapter 5).

A basic result of the theory of irreducible representations of space groups is the following:

> *An IR of a space group can be obtained from that of the group of the wavevector. By letting the wavevector run throughout the inside and the surface of the Brillouin zone one can obtain all the IRs of the space group.*

The translation group T is an invariant subgroup of the space group S. Therefore an IR of S can be characterized by the group $S_{\mathbf{k}}$ of $\mathbf{k}$ and the star of $\mathbf{k}$.

We recall that $T \subseteq S_{\mathbf{k}} \subseteq S$. Similarly, in terms of point groups, $1 \subseteq P_{\mathbf{k}} \subseteq G_p$. And a typical element of $S_{\mathbf{k}}$ has the form $\{\mathbf{A}|\mathbf{t}\}$, where $\mathbf{t} = \mathbf{t}_n + \mathbf{T}$.

We can rewrite Eq. 4.3.16 as follows:

$$d_{\mathbf{k},\nu}(\{\mathbf{A}\,|\,\mathbf{t}_n + \mathbf{T}\}) = \exp(i\mathbf{k}\cdot\mathbf{t})\, d_{\mathbf{k},\nu}(\{\mathbf{A}\,|\,\mathbf{T}\}) \tag{4.4.1}$$

Here $d_{\mathbf{k},\nu}(\{\mathbf{A}\,|\,\mathbf{T}\})$ is a representation matrix for representation number ν of the point group $P_{\mathbf{k}}$. These matrices have been derived and tabulated by Kovalev (1993) for all the inequivalent values of $\mathbf{k}$ for all space groups.

Eq. 4.3.3 enables us to derive matrices $D_{\{\mathbf{k}\},\nu}(s)$ of the IRs of S from the representation matrices of the group $S_{\mathbf{k}}$. The following relation can be shown to hold between the two:

$$D^{L\lambda,M\mu}_{\{\mathbf{k}\},\nu}(s) = d^{\lambda,\nu}_{\mathbf{k},\nu}(s_L^{-1} s s_M) \quad \text{for } s_L^{-1} s s_M \in S_{\mathbf{k}} \tag{4.4.2}$$

$$D^{L\lambda,M\mu}_{\{\mathbf{k}\},\nu}(s) = 0 \quad \text{for } s_L^{-1} s s_M \notin S_{\mathbf{k}} \tag{4.4.3}$$

Here L and M are labels of the star arms involved, and $\lambda, \mu = 1, 2, \ldots l_\nu$ are the indices of the matrices of the representation $d_{\mathbf{k},\nu}$.

The choice of the basis functions and the representation determines the form taken by the various invariants, as also the detailed expression for physical quantities like strain. Stokes & Hatch (1988), in Table 7 of their book, list the correspondence between five major space-group representation tables.

SUGGESTED READING

W. Ludwig & C. Falter (1988). *Symmetries in Physics.* Springer-Verlag, Berlin. An excellent book at an advanced level.

Yu. A. Izyumov & V. N. Syromyatnikov (1990). *Phase Transitions and Crystal Symmetry.* Kluwer, Dordrecht. For the mathematically-minded reader this could be the book of choice for a large number of topics which are only touched upon in the present book.

Chapter 5

PHASE TRANSITIONS IN CRYSTALS

Broken symmetry is actually the basic underlying concept of solid state physics.

P. W. Anderson

Consider a crystal under the action of external influences (temperature, pressure, directional force fields). When any of these parameters is varied, the thermodynamic state of the crystal changes accordingly, tending to keep the free energy at the minimum value under all conditions. At certain specific values of the external parameters (or even internal parameters like composition), the crystal makes a transition to another phase with a lower free energy than the existing phase (cf. Fig. 1.1.1). We discuss phase transitions in this chapter.

As described in §1.1, only those phase transitions which are nondisruptive and which involve a change of the point-group symmetry of the crystal are relevant for discussing ferroic materials. However, for putting ferroic phase transitions in a proper perspective, we adopt here a somewhat more general approach to the description of phase transitions in crystals.

5.1 PROTOTYPE SYMMETRY

5.1.1 Guymont's Nondisruption Condition

A phase transition involves a "new" phase and an "old" phase, and there is, in general, a loss or a gain of symmetry operators in going from one phase to the other. As rightly emphasized by Guymont (1981), it is only when the following nondisruption condition is satisfied that we can speak of lost or gained symmetry operators at phase transitions.

> *A phase transition is said to satisfy the nondisruption condition if the new structure arising from this transition can be described (i.e. its symmetry elements, Wyckoff positions, etc. can be located) in the frame of reference of the old structure, after making the necessary continuous distortions applied under the proviso that they themselves do not entail any additional change of symmetry.*

The "continuous distortions" mentioned above are affine mappings (cf. §B.2), required for annulling the effects of factors such as thermal expansion or contraction, and possible volume changes at first-order phase transitions, etc.

We use the term *nondisruptive phase transitions* (NDPTs) for transitions which obey the nondisruption condition.

Martensitic transitions are examples of situations where the nondisruption condition is usually *violated.*

5.1.2 Parent-Clamping Approximation

For NDPTs the emergence of the new phase from the old can be viewed as a two-step process.

In Step 1 the symmetry group of the new phase is taken as a (hypothetical) strictly isometric subgroup of the old symmetry group, with lattice parameters (lengths) either exactly the same as, or multiples of, those of the old phase.

In Step 2 the hypothetical new structure is allowed to undergo a continuous affine distortion so as to become identical to the new structure. No additional changes of symmetry accompany Step 2.

If we stop at Step 1, and assume that the resultant structure is the same as the actual structure obtained at the completion of Step 2, it amounts to making the so-called parent-clamping approximation (PCA) (Zikmund 1984; Janovec et al. 1989).

The PCA amounts to neglecting distortions that either accompany the NDPT (e.g. a change of volume at a first-order transition), or have features that depend on ambient temperature or pressure (e.g. isotropic and anisotropic thermal expansion or contraction).

5.1.3 Definition of Prototype Symmetry

The concept of prototype symmetry (Aizu 1970a, 1978; Levanyuk & Sannikov 1971) is of central importance for a systematic symmetry-based description of ferroic phase transitions and ferroic materials.

A symmetry analysis of a symmetry-lowering phase transition, typically in the spirit of the Landau theory (§5.3), requires that the symmetry group of the phase after the transition (usually the lower-temperature phase) be a subgroup of the symmetry group of the phase before the transition. However, it happens often that this condition is not satisfied. $BaTiO_3$, for example, has the tetragonal symmetry $P4mm$ at room temperature, and, on cooling, it makes a transition to an orthorhombic phase of symmetry $Amm2$:

$$BaTiO_3 : \; P4mm \stackrel{273K}{\longrightarrow} Amm2 \tag{5.1.1}$$

The space group $Amm2$ is not a subgroup of $P4mm$.

It so happens that $BaTiO_3$ undergoes several NDPTs as a function of temperature:

$$BaTiO_3 : \; Pm\bar{3}m \stackrel{393K}{\longrightarrow} P4mm \stackrel{273K}{\longrightarrow} Amm2 \stackrel{183K}{\longrightarrow} R3c \tag{5.1.2}$$

One can give a satisfactory symmetry-based explanation of this entire sequence of phase transitions by taking the cubic phase of symmetry $Pm\bar{3}m$ as the *prototypic phase*, and assuming that each of the other phases shown in Eq. 5.1.2 is derived from this phase, rather than from the next neighbouring phase in the sequence shown. The space-group symmetry of each of these phases is indeed a subgroup of $Pm\bar{3}m$.

Aizu (1970a, 1975, 1977, 1978, 1979) defined prototype symmetry as follows: "In general, a ferroic crystal may be regarded as a slight distortion (lowering of symmetry) of a certain nonferroic ideal crystal, which is referred to as the prototype of that ferroic crystal."

If a crystal has a phase which, in the absence of external directional influences, has the same symmetry as the prototype, such a phase is called a prototypic phase. The prototypic phase is not always the next neighbouring higher-symmetry phase in a sequence of phase transitions.

In Eq. 5.1.2 the space group $Pm\bar{3}m$ serves as the prototype symmetry for all the other space groups listed.

Aizu (1978) made the following stipulations for the definition of prototypic and ferroic phases:

- Suppose a crystal has three phases designated I, II and III. If phase I is a slight distortion of phase II, phase I is said to be ferroic. If II cannot be regarded as a slight distortion of any other phase (including phase III), then phase II is the prototype for phase I.

- If, on the other hand, both II and I are slight distortions of phase III, and III is not a slight distortion of any higher-symmetry phase, then III, *and not phase II*, is prototypic to phase I. Phase III is prototypic to phase II as well, and both I and II are ferroic phases.

- No phase can be both ferroic for one phase, and prototypic for another.

Some crystals [e.g. $BaCl_2.2H_2O$ (Wadhawan 1978a, 1982)] melt or decompose before reaching a temperature at which the prototypic phase may exist. In such a case we say that the prototypic phase does not exist. However, a prototype symmetry can still be defined.

According to Aizu (1978), the prototypic phase, being an actual physical entity, varies with temperature and pressure (as also composition). The prototype symmetry, on the other hand, is a concept and an idealization, independent of temperature and pressure. All components of spontaneous magnetization, spontaneous polarization[1], and spontaneous strain are zero for the prototype at all temperatures.

The prototype is an idealized *reference phase.*

We return to the example of $BaTiO_3$ to illustrate the point that a crystal can have more than one prototype symmetries. Apart from the perovskite structure which its cubic prototypic phase has, this crystal exists in another polymorphic form, of symmetry $P6_3/mmc$ (Sawaguchi, Akishige & Kobayashi 1985). This phase is also prototypic, though not to the tetragonal, orthorhombic and rhombohedral phases listed in Eq. 5.1.2. It appears that, on cooling, this hexagonal form of $BaTiO_3$ undergoes a phase transition at 222 K, and also perhaps at 60 K (Sawaguchi et al. 1985). The new phase(s), when identified unambiguously, will perhaps be said to have a hexagonal prototype of space-group symmetry $P6_3/mmc$, quite independent of the cubic prototype $Pm\bar{3}m$.

The concept of orientation states and F-operations was mentioned briefly in §1.1, and will be discussed in greater detail in Chapter 6. According to Aizu (1970a) an F-operation from an orientation state S_1 to another orientation state S_2 of a ferroic phase is any point-group operation of the crystal which can map S_1 to S_2. Aizu (1970a) imposed the following two restrictions for a correct identification of prototype symmetry:

(a) The point-group symmetry of the prototype should contain all the F-operations from all to all the orientation states of the ferroic.

(b) If any symmetry operation of the prototype point group is performed on any orientation state, the result must be one of the possible orientation states only, <u>and none other</u>.

Condition (b) serves to define a *complete set* of orientation states, and at the same time ensures that the chosen prototype does not have superfluous

[1]As we shall see in §10.1.7, what Aizu meant by spontaneous polarization is what we define in this book as *relative* spontaneous polarization. Similarly for spontaneous strain.

or higher point-group symmetry than what is sufficient for an adequate explanation of all the information about the observed orientation-state structure of the ferroic material.

The "slight distortion" mentioned in Aizu's definition of prototype symmetry has to be usually such that there is no rupture and reconstructive rearrangement of the chemical bonds, or a drastic change of the coordination numbers of the various atoms of the crystal.

Aizu's formal definition of prototype symmetry, though a substantial improvement over the somewhat unclearly defined terms such as "initial phase", "paraphase", " parent phase" used prior to his work, still has the limitation of not specifying the term "slight distortion" with mathematical rigour.

We introduce here a *rigorous* definition of prototype symmetry (Wadhawan 1998):

> *The prototype symmetry for a phase transition, or for a sequence of phase transitions, in a crystal is the highest space-group (Fedorov, Shubnikov, or colour) symmetry attainable by, or conceivable for, that crystal by an affine mapping of the structure that does not violate the nondisruption condition.*

Affine mappings are coordinate transformations which do not necessarily preserve distances, but which preserve parallelism of straight lines (Hahn & Wondratschek 1994). Deformation of a cube to a square prism or a rhombohedron (of same or different volume) is an example of an affine transformation.

Invocation of the nondisruption condition in our definition of prototype symmetry ensures that statements about lost or gained symmetry operators in going from a ferroic phase to the prototypic phase, or vice versa, have meaning. This condition also puts a natural upper limit on the "slight distortions" mentioned in Aizu's definition of prototype symmetry; beyond this limit a correspondence between the Wyckoff positions of the prototypic phase and a ferroic phase does not exist.

SUGGESTED READING

K. Aizu (1978). The concepts "prototype" and "prototypic phase" - their differences and others. *J. Phys. Soc. Japan*, **44**, 683.

V. Dvorak (1978). On structural phase transitions from a hypothetical phase. Czech. *J. Phys. B*, **28**, 989.

M. Guymont (1981). Symmetry analysis of structural phase transitions between phases not necessarily group-subgroup related. Domain structures. *Phys. Rev. B*, **24**, 2647.

Yu. A. Izyumov & V. N. Syromyatnikov (1990). *Phase Transitions and Crystal Symmetry.* Kluwer, Dordrecht.

T. Hahn & H. Wondratschek (1994). *Symmetry of Crystals: Introduction to International Tables for Crystallography Vol. A.* Heron Press, Sofia.

V. K. Wadhawan (1998). Towards a rigorous definition of ferroic phase transitions. *Phase Transitions*, **64**, 165.

5.2 A CRYSTALLOGRAPHIC CLASSIFICATION OF PHASE TRANSITIONS

We introduce here a comprehensive classification of phase transitions (Wadhawan 1997) that takes due note not only of the metrical relationship between the two phases, but also of the presence or absence of a structural correspondence in terms of crystallographic symmetric elements.

The primary subdivision of phase transitions in this scheme is into the *disruptive* and *nondisruptive* categories (Fig. 5.2.1) A disruptive phase transition violates the nondisruption condition, and a nondisruptive phase transition (NDPT) does not.

NDPTs can be either *ferroic*, or *nonferroic-nondisruptive.* Ferroic transitions are NDPTs entailing a change of point symmetry. And nonferroic NDPTs are NDPTs in which there is no change of point symmetry.

Ferroic phase transitions can be either *ferroelastic*, or *nonferroelastic-ferroic.* Ferroelastic transitions are ferroic transitions involving a spontaneous distortion of the crystal lattice that entails a change of the crystal family. By 'distortion' of the lattice we mean a *change of shape* of the crystallographic or conventional unit cell (e.g. a cube changing into a rhombus, but not a cube changing into a larger or smaller cube).

Nonferroelastic-ferroic transitions are NDPTs with a change of point symmetry, but no distortion of the crystal lattice.

Nonferroic NDPTs can be either *translational NDPTs*, or *isostructural* phase transitions.

5.2.1 Disruptive Phase Transitions

The conventional definition of crystallographic space groups is in terms of a set of symmetry operators which carry out a permutation of equivalent points in a self-consistent manner. A phase transition in a crystal results in a change of not only the space group, but also usually of the space-group *type* (§2.2.17) to which the crystal belongs. In disruptive transitions the

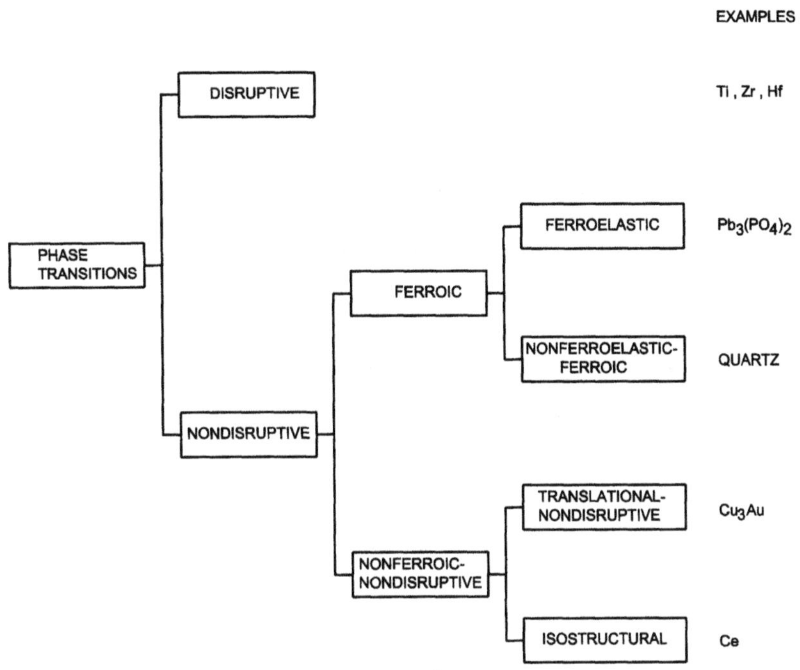

Figure 5.2.1: A crystallographic classification of phase transitions.

change of space-group type is so drastic that the nondisruption condition is violated. Both the phases across the transition have at least one symmetry element not present in the other phase. Classical Landau theory is not applicable to such transitions. Although there is a loss or gain of point-symmetry operators, such transitions are not ferroic transitions because a supergroup prototype symmetry cannot be defined for them.

The transition $\beta \leftrightarrow \omega$ in Ti, Zr and Hf is an example of a disruptive phase transition (Dmitriev & Toledano 1994). All *hcp* $\leftrightarrow$ *bcc* "martensitic" transitions are disruptive transitions.

What makes a phase transition disruptive ? In other words, what makes a phase transition violate the nondisruption condition ? To answer this question, we recall from §2.2.5 that a crystal can have certain special points, lines, or planes (called the symmetry elements of the crystal) which remain fixed under the operations of inversion symmetry, rotational symmetry, and reflection symmetry respectively. Wyckoff positions (which feature so importantly in the statement of the nondisruption condition) are determined directly by the locations of these symmetry elements in the crystal. If the displacements of all the atoms due to a phase transition are so small that all the crystallographic symmetry elements maintain their identities, the nondisruption condition is not violated, and a group-subgroup relationship between the symmetry groups of the two phases can exist. On the other hand, if the displacements of some or all atoms are comparable to the distances between the crystallographic symmetry elements, a number of possibilities may arise. For example, some or all atoms may sit on symmetry elements, giving rise to additional symmetry. Alternatively, old symmetry elements may disappear and new ones may appear. Any of these possibilities is enough to violate the nondisruption condition, making the phase transition disruptive.

5.2.2 Nondisruptive Phase Transitions

Nondisruptive phase transitions (NDPTs) may occur with or without a change of point symmetry. In the former case we call them ferroic phase transitions.

Ferroic Phase Transitions

Ferroic phase transitions can be further classified as *ferroelastic* and *nonferroelastic ferroic.* This division is based on whether or not a spontaneous distortion of the crystal lattice, entailing a change of crystal family, accompanies the transition. Ferroelastic transitions involve such a distortion of the crystal lattice, and nonferroelastic transitions do not.

The cubic-to tetragonal-transition in $BaTiO_3$ is an example of a ferroelastic transition. It is an NDPT, and it entails a distortion of the shape of the unit cell from a cube to a square prism.

Nonferroic Nondisruptive Phase Transitions

Nonferroic NDPTs do not involve a change of point symmetry. There are two possibilities. There can be a change of translational symmetry (*translational NDPTs*), as in the alloy Cu_3Au ($Fm\bar{3}m \leftrightarrow Pm\bar{3}m$). Alternatively, there may be a change of space group *without a change of space-group type.* Such transitions are called *isostructural transitions.* The phase transition that occurs in Ce under high pressure is an example of this type. Under the influence of high hydrostatic pressure the atoms of Ce simply collapse to a state of lower volume, with no change in the symmetry elements of the crystal structure. There is a change of lattice parameters, but no change of the space-group type to which Ce belongs.

Fig. 5.2.2 shows the set-theoretic relationships among the various types of phase transitions. Nondisruptive and disruptive transitions constitute disjoint sets.

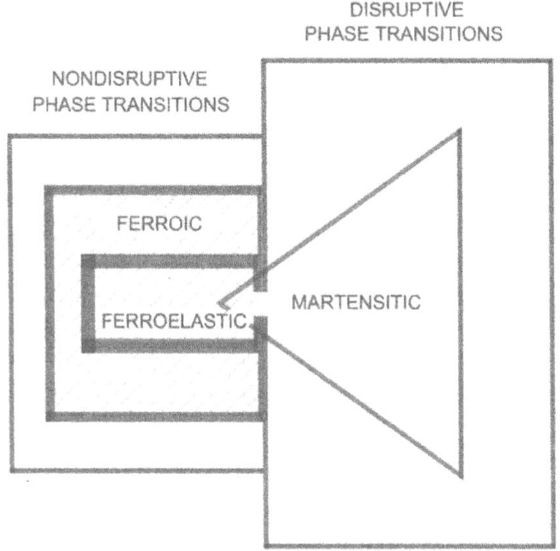

Figure 5.2.2: A Venn-Euler diagram for phase transitions.

Martensitic transitions (indicated by the triangular region in this figure) are generally of the disruptive type, but some are mild enough to qualify as NDPTs. The latter are the so-called Type M_1 transitions (see Izyumov, Laptev & Syromyatnikov 1994). We shall consider martensitic transitions in Chapter 11.

5.3 EXTENDED LANDAU THEORY OF CONTINUOUS PHASE TRANSITIONS

The Landau theory of continuous phase transitions (Landau 1937a, b, c, d) embodies the first application of group-theoretical ideas to thermodynamics. Landau (1937a) emphasized at the outset that transitions between phases of different symmetry of a crystal are fundamentally different from those between liquids and gases in that, in the case of crystals, there is necessarily a disappearance or appearance of symmetry elements, and that symmetry elements are either present or absent; no intermediate situations are possible.

Several developments of the Landau theory have taken place after its original formulation. We shall touch on some of these at appropriate places.

The book by Izyumov & Syromyatnikov (1990) provides a comprehensive account of these advances (also see Stokes & Hatch (1984)).

In effect, what we discuss in this section is an *extended* Landau theory. And we restrict ourselves to *continuous* phase transitions here. Discontinuous transitions are taken up separately in §5.7.

The Landau theory introduces the concept of the order parameter, a thermodynamic quantity the emergence of which at the phase transition results in a lowering of the symmetry of the parent phase. The Gibbs free energy of the crystal is expressed as a Taylor series in powers of the order parameter, and minimized with respect to it. The occurrence of the phase transition corresponds to the taking of a nonzero average value by the order parameter at a minimum of the free energy.

The setting up of the "Landau expansion" of the free energy in powers of the order parameter, and its subsequent minimization (§5.3.12) can be a tedious exercise, especially when one is looking for all possible phase transitions which can ensue from a given initial symmetry. A number of 'direct' group-theoretical conditions have therefore been formulated by several workers, which can shortlist the possible number of Landau expansions to be considered (Birman 1966; Goldrich & Birman 1968; Jaric & Birman 1977; Birman 1978; Jaric & Birman 1981, 1982a, b). We describe these conditions, criteria and conjectures here, and consider the thermodynamic aspects in §5.3.12.

5.3.1 Subgroup Criterion

Let the space group symmetries of the two phases across a phase transition in a crystal be S_0 and S. The Landau theory assumes that S is a proper subgroup of S_0:

$$S \subset S_0 \tag{5.3.1}$$

If this is not so for a given situation, it is necessary to choose a suitable prototype symmetry group S_0 so that Eq. 5.3.1 is satisfied.

5.3.2 Order Parameter

In the Landau theory the reduction of the symmetry of the crystal from S_0 to S is taken as resulting from the emergence of an order parameter, $\boldsymbol{\eta}$, with components $(\eta_1, \eta_2, \cdots \eta_m)$.

The emergence of the order parameter is an example of spontaneous breaking of symmetry, leading to a phase transition (§5.6).

It is postulated in the theory that space-group symmetry S results as an intersection group between S_0 and the symmetry group of $\boldsymbol{\eta}$, in accordance with the Curie principle (Eq. C.1.3).

Suppose $\rho_0(\mathbf{r})$ is the density function of the phase of the crystal in the phase of symmetry S_0. Naturally, $\rho_0(\mathbf{r})$ is invariant under the symmetry operations of S_0. The emergence of the order parameter at the phase transition temperature (the *Curie point*, T_c) lowers the space-group symmetry to S, with $\rho(\mathbf{r})$ as the new density function. One can write, for a temperature T near T_c,

$$\rho(\mathbf{r}) = \rho_0(\mathbf{r}) + \Delta\rho(\mathbf{r}), \tag{5.3.2}$$

where $\Delta\rho(\mathbf{r}) \to 0$ as $T \to T_c$.

A *continuous phase transition* is defined as one for which $\Delta\rho \to 0$ *continuously* as the transition point is approached.

The subgroup criterion puts the restriction that S cannot include any symmetry operators not present in S_0.

According to the group-theoretical completeness theorem (Wigner 1959; Lomont 1964), the physical distortion function, $\Delta\rho(\mathbf{r})$, can always be expressed in terms of a complete set consisting of basis functions from all IRs of S_0:

$$\Delta\rho(\mathbf{r}) = \sum_n \sum_{i=1}^{m} \eta_{ni}\phi_{ni}(\mathbf{r}) \tag{5.3.3}$$

Here the index n runs over the various IRs of S_0, and i runs over the basis functions of the nth IR. The summation over n does not include the identity representation. This is because a function transforming as the identity representation cannot lead to any symmetry change, whereas $\Delta\rho$ is a physical distortion responsible for the change of symmetry from S_0 to S.

The coefficients η_i $(i = 1, 2, \ldots m)$ are the components of an m-dimensional vector, namely the order parameter, in the space spanned by the basis vectors of the relevant IR. For continuous phase transitions (i.e. for phase transitions in which $\boldsymbol{\eta}$ acquires a nonzero value continuously on changing the temperature across T_c), only certain specific directions of $\boldsymbol{\eta}$ are possible.

5.3.3 Isotropy Subgroups

The space group S which results from the application of the distortion $\Delta\rho(\mathbf{r})$ on $\rho_0(\mathbf{r})$ comprises all elements $s \in S_0$ which leave $\Delta\rho(\mathbf{r})$ invariant. Such a group is called an isotropy subgroup of S_0.

The choice of basis functions ϕ_{ni} in Eq. 5.3.3 is not unique. And even for a given choice of basis functions, the components η_{ni} may have different sets of values, corresponding to different directions of $\boldsymbol{\eta}$. However, a particular direction of $\boldsymbol{\eta}$ in the configuration space always generates the same isotropy subgroup, irrespective of the choice of basis functions.

The number of isotropy subgroups corresponding to all possible directions of $\boldsymbol{\eta}$ is finite for finite representations of S_0.

One can list all possible symmetry groups S which can arise from S_0 by considering all possible IRs of S_0 and all possible directions of $\boldsymbol{\eta}$. However, the number of IRs of S_0 is infinite, and for purposes of tabulation some restrictive choice has to be made. One is frequently interested only in *physically irreducible* representations of S_0 (see below), and only in Lifshitz points (§4.1.4) (Bradley & Cracknell 1972; Stokes & Hatch 1988). Application of these two restrictions limits the possible number of IRs of the 230 crystallographic space groups to 4777, and the total number of nonequivalent isotropy groups associated with these IRs to 15,239. These have been derived and tabulated by Stokes & Hatch (1988).

An extension of the isotropy subgroups to IRs wherein these two restrictions are not imposed has been made in the computer code ISOTROPY developed by Stokes & Hatch (1988). This code was made available on the web in 1998, and has been frequently expanded and updated since then. It can be accessed at
http://www.physics.byu.edu/˜ stokesh/isotropy.html

5.3.4 Physically Irreducible Representations

The physical distortion function $\Delta\rho(\mathbf{r})$ in Eq. 5.3.3 is a real function. Therefore, so also are the order-parameter components η_{ni}. This can be possible only if we choose the basis functions $\phi_{ni}(\mathbf{r})$ as real. This requires that we deal only with real representations of S_0. When a particular IR is complex we can work with real combinations of basis functions, viz. $\phi'_{ni} = \phi_{ni} + \phi^*_{ni}$ and $i\phi''_{ni} = \phi_{ni} - \phi^*_{ni}$ in place of ϕ_{ni} and ϕ^*_{ni}. This is possible because the basis functions ϕ^*_{ni} are linearly independent of the basis functions ϕ_{ni}. A real representation $\bar{D}_n = D_n + D^*_n$, with $2m$ real basis functions $\phi'_{ni}(i = 1, 2, \ldots m)$ and $\phi''_{ni}(i = 1, 2, \ldots m)$ is called a physically irreducible representation. Since such representations can always be defined, we shall assume henceforward that this has been done before writing Eq. 5.3.3. In other words, we assume that the coefficients ϕ_{ni} are real.

5.3.5 Single-IR Criterion; Active IR

Usually the transformation properties of the order parameter correspond to a particular, nonidentity, physically irreducible, IR of S_0. Such an IR is called the *active IR* (Lyubarskii 1960). Only rarely can it happen that more than one order parameters (governed by different IRs of S_0) set in at exactly the same temperature (however, see Janovec (1975)). One can therefore normally assume that the so-called single IR criterion is obeyed, and in Eq. 5.3.3 the summation over n can be dropped:

$$\Delta\rho(\mathbf{r}) = \sum_{i=1}^{m} \eta_i \phi_i(\mathbf{r}) \tag{5.3.4}$$

Thus we implicitly assume in the Landau theory that all the components of the order parameter belong to the same IR of S_0.

5.3.6 Subduction Criterion; Subduction Frequency

The density function $\rho(\mathbf{r})$, as well as the distortion $\Delta\rho(\mathbf{r})$, are invariant under operations of their own symmetry group S. This fact was exploited by Birman (1966, 1978) to device the so-called subduction criterion.

We have $s\rho(\mathbf{r}) = \rho(\mathbf{r})$, and $S \subset S_0$. Therefore, the operations $s \in S$ also leave $\rho_0(\mathbf{r})$ invariant: $s\rho_0(\mathbf{r}) = \rho_0(\mathbf{r})$. According to the subduction criterion, the IR of S_0 which drives the phase transition, i.e. the active IR $D(^*\mathbf{k}, \Gamma)$ (§4.4), must subduce (§B.3) the unit representation, or the identity representation, $(0, 1)$, of the group S.

In other words, the linear combination $\sum_{i=1}^{m} \eta_i \phi_i(\mathbf{r})$ of the basis functions of the active IR should be a basis function of the unit IR of S, implying that the (reducible) representation subduced from S_0 to S must contain the unit representation of S.

The notion of subduction frequency, $i(S)$, is relevant in this context. It is the number of times the unit IR is subduced in S by the active IR of S_0:

$$i(S) = \frac{1}{|S|} \sum_{s \in S} \chi^{(^*\mathbf{k},\Gamma)}(s) \tag{5.3.5}$$

Here the summation is taken over all elements of S.

The subduction frequency can be identified with the number of independent order-parameter vectors $\boldsymbol{\eta}$ which are invariant under the operations of S.

The *subduction criterion* simply requires that the subduction frequency be nonzero:

$$i(S) \neq 0 \tag{5.3.6}$$

5.3.7 Chain Subduction Criterion

The space group S is an isotropy subgroup of S_0 ($S \subset S_0$). All subgroups of S will also be isotropy subgroups of S_0. The chain-subduction criterion (Goldrich & Birman 1968; Cracknell 1974; Hosoya 1977; Deonarine & Birman 1983) provides a reasonable-looking recipe for deciding whether or not a continuous phase transition can occur to any such subgroup of S.

Suppose a space group S' exists such that $S' \subset S \subset S_0$. We compute the subduction frequencies $i(S)$ and $i(S')$. The chain subduction criterion states that if $i(S) = i(S') = 1$, then whereas a continuous phase transition $S_0 \to S$ may occur (if other conditions are satisfied), the transition $S_0 \to S'$ cannot occur.

A similar criterion was formulated by Jaric & Birman (1977) for the multiplicity situation, i.e. when $i(S) = i(S') > 1$. In this case again, the transition $S_0 \to S$ may occur, but $S_0 \to S'$ cannot occur as a continuous transition.

The case $i(S') > i(S)$ has also been analysed (Birman 1982), and the following chain subduction criterion with increasing multiplicity has been proposed:

If $i(S') > i(S)$, then $S_0 \to S$ may occur as a simple second-order transition, and $S_0 \to S'$ may occur as a "higher-order critical transition".

The last case needs further investigation and confirmation.

If $i(S') < i(S)$, then only S must figure in the list of relevant subgroups of S_0 for further shortlisting.

5.3.8 Landau Stability Condition

The subduction criterion and the chain subduction criterion are not only necessary but also sufficient for specifying the isotropy subgroups of S_0 for any of its IRs. The Landau condition (see Landau & Lifshitz 1980), for a phase transition to be continuous, narrows down the choice still further, and thus helps further shortlist the specification of the possible active IRs of S_0. It takes cognisance of the stability of the parent phase of symmetry S_0. It states that the decomposition of the symmetrized triple Kronecker product, $[D]^3$, into IRs must not contain the identity representation of S_0, i.e. we must have

$$\frac{1}{|S_0|} \sum_{s_0 \in S_0} \{\frac{1}{3}\chi(s_0^3) + \frac{1}{2}\chi(s_0)\,\chi(s_0^2) + \frac{1}{6}[\chi(s_0)]^3\} = 0, \qquad (5.3.7)$$

where the summation is taken over all elements of S_0. When Eq. 5.3.7 is satisfied, third-order invariants do not occur in the Landau expansion for the free energy (§5.3.12).

5.3.9 Lifshitz Homogeneity Condition

Like the Landau condition, the Lifshitz condition (cf. Landau & Lifshitz 1980) also helps narrow down the possible choice of (physically irreducible) active IRs. It stipulates that the lower-symmetry phase be *commensurate* with the higher-symmetry phase. It states that the antisymmetric part $\{D\}^2$ of the product representation D^2 must not contain the IR $\chi^\nu(s_0)$ by which the components of a vector transform. That is, we must have

$$\frac{1}{|S_0|} \sum_{s_0 \in S_0} \frac{1}{2}\{\chi^2(s_0) - \chi(s_0^2)\}\,\chi^\nu(s_0) = 0 \qquad (5.3.8)$$

If this condition is violated, we may get a phase *incommensurate* with the parent phase (Dzyaloshinskii 1964; Hass 1965; Janovec et al. 1975; Levanyuk & Sannikov 1976; Kopsky & Sannikov 1977). Alternatively, the phase transition may become discontinuous.

General points in the Brillouin zone, as well as lines and planes of symmetry in it, fail the Lifshitz condition. Thus only the *points* of special symmetry (cf. §4.1.4) are relevant for discussing the Lifshitz condition.

However, not all points of symmetry of the Brillouin zone satisfy the Lifshitz or Landau conditions. Stokes & Hatch (1988) considered all such points for all the space groups. In Table 1 of their book, the indices resulting from Eqs. 5.3.7 and 5.3.8 are listed for all subgroups which satisfy the subduction criterion.

5.3.10 Maximality Conjecture

The subduction criterion and the chain subduction criterion are relevant even for discontinuous transitions, provided such transitions obey the subgroup criterion. These criteria select the isotropy subgroups of S_0 corresponding to specific orientations of the order-parameter vector in the representation space. The Landau condition (for 3-dimensional systems) and the Lifshitz condition provide additional necessary restrictions for a phase transition to be continuous and commensurate. A further selection is made finally by looking for the *absolute* minima of the free-energy function. Ascher (1966a, b, c, 1977) conjectured that the isotropy subgroups of S_0 corresponding to absolute minima of the free energy are *maximal* subgroups of S_0, irrespective of the subduction frequency (also see Ascher & Kobayashi 1977). This is commonly referred to as the maximality conjecture.

The maximality conjecture appears reasonable from energy considerations. It implies that the number of domain walls (cf. Chapter 7) in the lower-symmetry phase is the minimum necessary. Detailed arguments in favour of this postulate have been given by Sutton & Armstrong (1982).

An example violating this conjecture has been described by Mukamel & Jaric (1984). However, the conjecture is expected to be valid in most situations.

5.3.11 Tensor Field Criterion

So far no information about the crystal structure, or about the physical nature of the order parameter, has entered our description of the Landau theory. The tensor field criterion makes use of this information. According to it: The active IR of S_0 must be contained in a "tensor field representation" of S_0 (Birman 1966; Litvin 1982; Litvin, Kotzev & Birman 1982). The tensor field representation, by definition, is a direct product of a tensor representation of S_0 and a "permutation representation" of the atoms of the crystal.

In a series of papers, Hatch and coworkers have shown how the crystal structure, and hence the tensor-field criterion, can be taken into account systematically (Hatch, Stokes & Putnam 1987; Hatch, Stokes, Aleksandrov & Misyul 1989; Stokes, Hatch & Wells 1991). A variety of local (micro) site distortions, for example atomic displacements, rigid-unit tilting modes, and atomic ordering, can be used for inducing the global (field) distortions. This information is now systematically contained in the computer code ISOTROPY (Stokes & Hatch 1998).

5.3.12 The Landau Expansion

In the Landau theory the thermodynamic-potential density g (Gibbs free energy per unit volume of the crystal) is assumed to depend not only on temperature (and pressure and composition), but also on the order parameter η. For a continuous phase transition the order parameter is zero above T_c, and rises continuously from zero as the temperature is decreased below T_c. Therefore, temperature ranges exist on both of T_c in which the order parameter can be taken to be sufficiently small to ensure the validity of the following *Landau expansion:*

$$g = g_0 + \alpha\eta + \frac{a}{2}\eta^2 + \beta\eta^3 + \frac{b}{4}\eta^4 + \cdots \tag{5.3.9}$$

Here g_0 is the value of the thermodynamic potential in the parent phase or the *disordered* phase, and $\alpha, a, \beta, b, \ldots$ are functions of temperature, as well as of functionals of various orders constructed from the basis functions ϕ_i of the active IR $D(^*\mathbf{k}, \Gamma)$ of the space group S_0 of the disordered phase (cf. Eq. 5.3.4). This active IR is instrumental in inducing the phase change from symmetry S_0 to symmetry S. Since g is invariant under the operations of S_0, the only allowed forms of the functionals are those that are invariant under the operations of S_0.

The equilibrium value of the order parameter is determined by minimizing the free energy with respect to it. That is, the first derivative of g with respect to it must be zero, and the second derivative positive:

$$\frac{\partial g}{\partial \eta} = \alpha + a\eta + 3\beta\eta^2 + b\eta^3 + \cdots = 0, \tag{5.3.10}$$

$$\frac{\partial^2 g}{\partial \eta^2} = a + 6\beta\eta + 3b\eta^2 + \cdots > 0 \tag{5.3.11}$$

If Eq. 5.3.11 is not satisfied, the system is unstable. It can therefore be regarded as a *stability condition.*

Since Eq. 5.3.10 must hold even when η is zero, we must have

$$\alpha = 0 \tag{5.3.12}$$

Eq. 5.3.10 can therefore be rewritten as

$$\eta\,(a + 3\beta\eta + b\eta^2 + \cdots) = 0 \tag{5.3.13}$$

This equation has two solutions. One of them, namely $\eta = 0$, corresponds to the disordered phase, and the other to the ordered phase.

Suppose X is the field conjugate to the order parameter. It is defined by

$$\frac{\partial g}{\partial \eta} = X \tag{5.3.14}$$

The stability condition (Eq. 5.3.11) can therefore be reexpressed as follows:

$$\frac{\partial^2 g}{\partial \eta^2} = \frac{\partial X}{\partial \eta} \equiv \chi^{-1} > 0, \tag{5.3.15}$$

where χ is a *generalized susceptibility*. Thus, for a phase to be stable, its inverse generalized susceptibility must be positive.

It follows from Eqs. 5.3.11 and 5.3.15 that for the disordered phase ($\eta = 0$) to be stable, we must have

$$\chi^{-1} = a > 0 \tag{5.3.16}$$

5.3.13 Stability Limit of a Phase

We can assume here, without loss of generality, that the transition to the lower-symmetry or ordered phase occurs on cooling the crystal, rather than on heating.

The inverse susceptibility is a function of temperature, and it must be positive for a phase to be stable. For the disordered phase, the *stability limit* is the temperature (T_0) below which its inverse susceptibility with respect to the order parameter is no longer positive:

$$\chi_T^{-1}(T = T_0) = 0 \tag{5.3.17}$$

In the vicinity of T_0, Eq. 5.3.16 for the disordered phase can be written as

$$\chi_T^{-1} = a = \frac{\partial a}{\partial T}(T - T_0) = a'(T - T_0), \tag{5.3.18}$$

or

$$\chi_T = \frac{C}{T - T_0}, \quad T > T_0, \tag{5.3.19}$$

where

$$C = \frac{1}{a'} \tag{5.3.20}$$

Eq. 5.3.19 has the form of the well-known Curie-Weiss law for the generalized susceptibility of the disordered phase.

We consider the lower-symmetry or ordered phase next, for which $\eta \neq 0$.

The Landau condition for a phase transition to be continuous (§5.3.8) requires that $\beta = 0$ in Eq. 5.3.9, and therefore in Eq. 5.3.9. Further, in Eq. 5.3.11, the coefficient a is negative for $T < T_0$ (cf. Eq. 5.3.18). It follows that the crystal cannot be stable below T_0 if the coefficient b is negative; therefore

$$b > 0 \quad (T < T_0) \tag{5.3.21}$$

With these stipulations the second solution of Eq. 5.3.13 is the root of the equation

$$b\eta^2 + a = 0 \quad (T < T_0), \tag{5.3.22}$$

where we have ignored the higher-order terms in the Landau expansion.

Eq. 5.3.22 has the solution (on using Eq. 5.3.21)

$$\eta = (-a/b)^{1/2} = ((T_0 - T)a'/b)^{1/2} \tag{5.3.23}$$

To determine the stability limit of the ordered phase, we substitute the equilibrium value of the order parameter given by Eq. 5.3.23 into Eq. 5.3.11 to get

$$\chi_T^{-1} = a - 3a = -2a = 2a'(T_0 - T) \tag{5.3.24}$$

The stability limit for the ordered phase is defined by $\chi_T^{-1} = 0$, and is therefore given by $T = T_0$. T_0 is also the stability limit of the disordered phase. Thus:

For a continuous phase transition, the stability limits of the disordered phase and the ordered phase coincide.

There is thus a temperature T_c, the temperature of the continuous phase transition, below which the disordered phase is not stable, and above which the ordered phase is not stable:

$$T_c = T_0 \tag{5.3.25}$$

We also note from Eq. 5.3.23 that the order parameter, which is zero for $T > T_c$, increases continuously as the temperature is decreased below the transition temperature. Hence the name "continuous phase transition".

In view of the above results we can rewrite the Landau expansion (Eq. 5.3.9) for a continuous phase transition as

$$g = g_0 + \frac{a'}{2}(T - T_c)\eta^2 + \frac{b}{4}\eta^4 \qquad (5.3.26)$$

And the order parameter is predicted by the Landau theory to have the following temperature dependence (Eq. 5.3.23):

$$\eta = (a'/b)^{1/2}(T_c - T)^{\beta}, \quad \beta = 1/2 \qquad (5.3.27)$$

We can also rewrite Eq. 5.3.24 as follows:

$$\chi_T(T < T_c) = \frac{C'}{T_c - T}, \qquad (5.3.28)$$

$$C' = \frac{1}{2a'} \qquad (5.3.29)$$

Comparison of Eqs. 5.3.20 and 5.3.29 shows that the Curie-Weiss constants above and below T_c are related by

$$\frac{C}{C'} = 2 \qquad (5.3.30)$$

Lastly we write Eq. 5.3.28 as follows:

$$\chi_T = C'(T_c - T)^{\gamma}, \quad \gamma = -1 \qquad (5.3.31)$$

The exponents γ in Eq. 5.3.31 and β in Eq. 5.3.27 are examples of *critical exponents*. According to the Landau theory, $\beta = \frac{1}{2}$ and $\gamma = -1$ for all continuous phase transitions.

Another important critical exponent is δ, which determines the field dependence of the order parameter at the critical point. To derive its Landau-theory value, we include a field term $-X\eta$ in the Landau expansion (Eq. 5.3.26). Then the minimization condition for the free energy reads

$$\eta[a'(T - T_c) + b\eta^2] = X \qquad (5.3.32)$$

Therefore, at $T = T_c$,

$$\eta = (X/b)^{1/3} \equiv (X/b)^{1/\delta} \qquad (5.3.33)$$

Thus, according to the Landau theory, $\delta = 3$ for all continuous phase transitions.

Critical-point exponents (or critical exponents) define the behaviour of thermodynamic variables in the vicinity of the critical point. Although their importance was highlighted by the Landau theory, critical fluctuations were neglected by it. We shall return to this topic in §5.5.4.

5.3.14 Tricritical Points

Eq. 5.3.26 defines the Landau potential for a continuous phase transition. [It is convenient to refer to it as the *2-4 potential* because it includes contributions from the 2nd and the 4th powers of the order parameter.] The stability condition for this situation is determined by Eq. 5.3.11 (with $\beta = 0$ because of the Landau condition (§5.3.8)). It follows from this equation that the 2-4 potential is adequate provided $b > 0$. We now discuss the situation when $b = 0$ (the case when $b < 0$ will be taken up in §5.7.1).

When $b = 0$ we must include the next (i.e. 6th power) term in the Landau expansion. In other words we must now work with the 2-6 potential:

$$g = g_0 + \frac{a'}{2}(T - T_c)\eta^2 + 0 + \frac{c}{6}\eta^6 \qquad (5.3.34)$$

The minimization condition for this potential with respect to the order parameter is

$$\eta\,[a'(T - T_c) + c\eta^4] = 0 \qquad (5.3.35)$$

Therefore, assuming that $c > 0$, the order parameter is predicted to have the following temperature dependence:

$$\eta = \frac{a'}{c}(T_c - T)^{\beta}, \quad T < T_c, \qquad (5.3.36)$$

where

$$\beta = \frac{1}{4} \qquad (5.3.37)$$

Thus this critical exponent is expected to have the value $\frac{1}{4}$, rather than $\frac{1}{2}$ (cf. Eq. 5.3.27).

Similarly, for $b = 0$ Eq. 5.3.30 is replaced by

$$\frac{C'}{C} = 4 \qquad (5.3.38)$$

As we shall see in §5.7, $b < 0$ corresponds to *discontinuous* or first-order phase transitions. The point $b = 0$ thus represents a *crossover* from continuous to discontinuous phase-transition behaviour.

How is this crossover effected experimentally ? There are several examples. The uniaxial antiferromagnet dysprosium aluminum garnet (DAG) undergoes a temperature-induced continuous phase transition in the absence of an external magnetic field, which changes to a discontinuous transition when a sufficiently high magnetic field is applied along its easy axis (Landau, Keen, Schneider & Wolf 1971; Giordano & Wolf 1977). Similarly, application of a hydrostatic pressure of about 2 kbar to KDP changes its discontinuous phase transition to a continuous one (Schmidt, Western & Baker 1976). PZT ($PbZr_xTi_{1-x}O_3$) is a crystal in which the order of the ferroelectric phase transition changes from first to second at a specific value of the composition x (Clarke & Glazer 1974, 1976; Whatmore, Clarke & Glazer 1978). Additional examples have been discussed by Dattagupta (1981).

Two kinds of external fields must be distinguished when discussing situations in the vicinity of $b = 0$. One is the field conjugate to the order parameter (e.g. electric field in the case of the ferroelectric transition in KDP). This is referred to as the *ordering field.* The other field (e.g. pressure in the KDP example), called the *nonordering field,* plays, among other things, the role of driving the coefficient b from a positive to a negative value. Griffiths (1970) argued that in the 3-dimensional phase diagram in which the nonordering field constitutes one of the axes, the point $b = 0$ corresponds to the intersection of two lines of continuous or second-order phase transitions and one line of first-order transitions. Therefore the name tricritical point was suggested for such a point.

SUGGESTED READING

Landau, L. D. & E. M. Lifshitz (1958). *Statistical Physics.* Pergamon Press, Oxford. Third edition published as Lifshitz & Pitaevsky (1980), Part 1.

R. Bausch (1972). Ginzburg criterion for tricritical points. *Z. Physik*, **254**, 81.

S. Dattagupta (1981). The tricritical point - a qualitative review. *Bull. Mater. Sci.*, **3**, 133.

D. B. Litvin, J. N. Kotzev & J. L. Birman (1982). Physical applications of colour groups: Landau theory of phase transitions. *Phys. Rev. B*, **26**, 6947.

Yu. A. Izyumov & V. N. Syromyatnikov (1990). *Phase Transitions and Crystal Symmetry.* Kluwer, Dordrecht.

P. Toledano (1992). Phenomenological approach to structural phase transitions. *Key Engg. Materials*, **68**, 1. This issue is also available in book form as C. Boulesteix (Ed.) (1992), *Diffusionless Phase Transitions and Related Structures in Oxides.* Trans Tech Publications, Switzerland.

E. K. H. Salje (1992). Application of Landau theory for the analysis of phase transitions in minerals. *Phys. Reports*, **215**, 49.

H. T. Stokes, D. M. Hatch & H. M. Nelson (1993). Landau, Lifshitz, and weak Lifshitz conditions in the Landau theory of phase transitions in solids. *Phys. Rev. B*, **47**, 9080.

A. Saxena, G. R. Barsch & D. M. Hatch (1994). Lattice dynamics representation theory versus isotropy subgroup method, with application to $M_{\bar{5}}$ mode instability in CsCl structure. *Phase Transitions*, **46**, 89.

5.4 LATTICE DYNAMICS, SOFT MODES

The order parameter in the Landau theory of phase transitions is a measure of the extent to which the atomic configuration of the lower-symmetry phase (or the ordered phase) has departed from the configuration of the parent phase. Atoms in a crystal execute vibrations, the eigenfrequencies and eigenvectors of which are determined by interatomic potentials. The displacement $\mathbf{u}_{ls}$ of an atom of mass m_s in the lth unit cell of a crystal consisting of N unit cells can be expressed as a Fourier series, the coefficients of which are the *normal coordinates* $Q_j(\mathbf{k})$ (Cochran 1973):

$$\mathbf{u}_{ls} = (Nm_s)^{-1/2} \sum_{\mathbf{k},j} Q_j(\mathbf{k}) \mathbf{e}_{sj} \cos[\mathbf{k} \cdot \mathbf{r}_{ls} - \omega_j(\mathbf{k})t + \alpha_j(\mathbf{k})] \qquad (5.4.1)$$

Here $\mathbf{e}_{sj}$ is the polarization eigenvector of a lattice-vibrational mode of wavevector $\mathbf{k}$ and eigenfrequency $\omega_j(\mathbf{k})$.

Since N is usually a very large number, the displacement $\mathbf{u}_{ls}$ of an atom due to any particular normal mode is quite small in general. However, this is no longer the case if one of the eigenfrequencies tends to zero. To see how this happens, we quote from lattice dynamics the standard result that the mean-squared amplitude $Q_j(\mathbf{k})$ is determined by the equation

$$\frac{1}{2}\omega_j^2(\mathbf{k})\, |Q_j(\mathbf{k})|^2 = (\bar{n}_j(\mathbf{k}) + \frac{1}{2})\, \hbar\omega(\mathbf{k}), \qquad (5.4.2)$$

where

$$\bar{n}_j(\mathbf{k}) = \frac{1}{\exp(\hbar\omega_j(\mathbf{k})/k_B T) - 1} \qquad (5.4.3)$$

Now suppose one of the eigenfrequencies for a particular wavevector $\mathbf{k}$ tends to zero as the critical temperature is approached: $\omega_j(\mathbf{k}) \to 0$ as $T \to T_c$.

To see the effect of this, we first substitute Eq. 5.4.3 into 5.4.2:

$$\frac{1}{2}\omega_j^2(\mathbf{k})\, |Q_j(\mathbf{k})|^2 = \frac{\hbar\omega_j(\mathbf{k})}{e^{\hbar\omega_j(\mathbf{k})/k_B T} - 1} + \frac{1}{2}\hbar\omega_j(\mathbf{k}) \qquad (5.4.4)$$

We use the result that $e^x \simeq 1 + x$ for small x. Therefore, for small $\omega_j(\mathbf{k})$, Eq. 5.4.4 becomes

$$\frac{1}{2}\omega_j^2(\mathbf{k})|Q_j(\mathbf{k})|^2 = k_B T + \frac{1}{2}\hbar\omega_j(\mathbf{k}) \qquad (5.4.5)$$

As $\omega_j(\mathbf{k}) \to 0$, the right-hand side of this equation approaches the constant value $k_B T$. The amplitude $Q_j(\mathbf{k})$ of vibration will therefore tend to become arbitrarily large, till it is restrained by the anharmonic character of the interatomic potential.

The reduction of the frequency of a vibration mode implies a reduction of the effective force constant controlling that mode. Such a mode is referred to as a *soft mode.*

We also note from Eq. 5.4.1 that when $\omega_j(\mathbf{k}) = 0$, the atomic displacements caused by this (soft) mode are static, and not vibratory any more. This freezing of displacements amounts to a phase transition, as it leads to a reduction of the crystal symmetry. The static displacements are identified with the order parameter in a structural phase transition. (A structural phase transition can be defined as one entailing a change of crystal structure, as opposed to, say, change of magnetic structure.)

5.4.1 Ferrodistortive Transitions

According to Eq. 5.4.1, if the structural distortion introduced by the softening of a vibrational mode is to be the same in every unit cell of the parent phase (i.e. if it is to be independent of unit-cell label l), we must have $\mathbf{k} = 0$ for this mode. Phase transitions occurring as a result of the softening of such zero-wavevector (or Brillouin-zone-centre) modes are called ferrodistortive phase transitions (Indenbom 1960; Granicher & Muller 1971; Aubry & Pick 1971). For such transitions there is no change in the number of formula units in the primitive unit cell.

The notion of (optical) soft modes was actually introduced in the context of such transitions (Anderson 1958; Cochran 1959). It was stipulated that the LST relation (Lydanne, Sachs & Teller 1941) provided the connection between the approach to zero of the transverse optic phonon frequency, ω_T, and the anomalous increase near T_c of the low-frequency dielectric constant $\epsilon(0)$:

$$\frac{\epsilon(0)}{\epsilon(\infty)} = \frac{\omega_L^2}{\omega_T^2} \tag{5.4.6}$$

Here $\epsilon(\infty)$ is the high-frequency dielectric constant, and ω_L is the long-wavelength longitudinal optical phonon frequency. As $\epsilon(0) \to \infty$ in a second-order phase transition, we get $\omega_T^2 \to 0$.

5.4.2 Antiferrodistortive Transitions

Phase transitions accompanied by a change in the number of formula units in the primitive unit cell of the crystal, or phase transitions driven by soft modes with $|\mathbf{k}| \neq 0$, are called antiferrodistortive phase transitions (Granicher & Muller 1971).

Typically, a Brillouin zone boundary mode goes soft, entailing a doubling of the smallest lattice translation in the direction of **k**. Consequently, the zone boundary jumps to a position midway to its position for the phase before the transition. The result is a "folding in" of the mode which softened, and a soft mode at the *zone centre* of the new phase with the lower translational symmetry.

In §5.3.6 we stated Birman's (1966) subduction criterion, according to which the IR of the parent symmetry group S_0 which drives a continuous transition, i.e. the active IR, is such that the reducible representation subduced from S_0 to the daughter symmetry group S always contains the unit representation of S. Since the order-parameter concept and the soft-mode concept cover overlapping grounds, the subduction criterion was restated by Worlock (1971) as follows:

> *The soft mode in the daughter phase is a totally symmetric mode, and is therefore Raman active.*

This Birman-Worlock statement was extended to the case of antiferrodistortive transitions by Lavrencic & Shigenari (1973). The symmetry of the soft mode driving a continuous phase transitions must be such that Eq. 5.3.5 is satisfied (irrespective of whether $|\mathbf{k}|$ is zero or not). The procedure described by Lavrencic & Shigenari (1973) makes it possible not only to identify the wavevector of the soft mode, but also to determine the characters that enter Eq. 5.3.5. Thus the candidate active IRs are identified, which can then be tested against the Landau condition and other necessary conditions described in §5.3.

If the wavevector of the soft mode is nonzero, the ordered phase has lower translational symmetry than the disordered phase. In other words, some of the lattice translations present in the disordered or parent phase are not allowed in the ordered or daughter phase.

Let T_0 and T denote the translation groups underlying the space groups S_0 and S of the parent phase and the daughter phase respectively. And let $\mathbf{t}_0$ denote the primitive lattice translations of T_0, and $\mathbf{t}$ those of T. The allowed $\mathbf{k}$ points of the Brillouin zone of the parent phase are those for which

$$e^{-i\mathbf{k}\cdot\mathbf{t}} = 1, \tag{5.4.7}$$

and

$$e^{-i\mathbf{k}\cdot\mathbf{t}_0} \neq 1, \tag{5.4.8}$$

for those $\mathbf{t}_0$ which are not present in T.

The case of the phase transition in GMO provides a simple illustration of this. Here $S_0 = D_{2d}^3\,(P\bar{4}2_1m)$, and $S = C_{2v}^8\,(Pba2)$. If we denote the primitive translation vectors of T_0 by $\mathbf{a}_1^0, \mathbf{a}_2^0, \mathbf{a}_3^0$, and those of T by $\mathbf{a}_1, \mathbf{a}_2, \mathbf{a}_3$, then

$$\mathbf{a}_1 = \mathbf{a}_1^0 - \mathbf{a}_2^0, \;\; \mathbf{a}_2 = \mathbf{a}_1^0 + \mathbf{a}_2^0, \;\; \mathbf{a}_3 = \mathbf{a}_3^0 \tag{5.4.9}$$

This lattice correspondence can be expressed in terms of reciprocal-lattice vectors as follows:

$$\mathbf{b}_1 = \frac{1}{2}(\mathbf{b}_1^0 - \mathbf{b}_2^0), \quad \mathbf{b}_2 = \frac{1}{2}(\mathbf{b}_1^0 + \mathbf{b}_2^0), \quad \mathbf{b}_3 = \mathbf{b}_3^0 \tag{5.4.10}$$

Use of Eqs. 5.4.7 and 5.4.8 then gives the result that the only allowed $\mathbf{k}$ for the soft mode is that corresponding to the $M = (\frac{1}{2}\,\frac{1}{2}\,0)$ point of the Brillouin zone of the parent phase:

$$\mathbf{k} = \frac{1}{2}\mathbf{b}_1^0 + \frac{1}{2}\mathbf{b}_2^0 \tag{5.4.11}$$

Using this information one can construct, from the character table for the small representation of $D_{2d}^3(M)$, one 2-dimensional real representation and two 2-dimensional physically irreducible representations (Lavrencic & Shigenari 1973).

5.4.3 Displacive vs. Order-Disorder Type Phase Transitions

So far in this section we have implicitly assumed that the dielectric function and the frequencies of vibration are real, rather than complex, quantities. For explaining certain properties of ferroelectrics and other materials it is found expedient to treat them as complex quantities:

$$\epsilon(\omega) = \epsilon'(\omega) + i\,\epsilon''(\omega), \tag{5.4.12}$$

$$\omega = \omega' + i\,\omega'' \tag{5.4.13}$$

The imaginary parts of the dielectric function and the frequency provide a measure of the damping and other loss mechanisms.

In particular, for a ferroelectric with no damping ($\omega'' = 0$), i.e. for the 'pure resonance' case, the LST relation (Eq. 5.4.6) can be recast as follows:

$$\frac{\epsilon(0)}{\epsilon(\infty)} = \frac{\omega_L'^2}{\omega_T'^2} \tag{5.4.14}$$

The longitudinal optical phonon frequency ω_L' is normally found to be quite independent of temperature, so that Eq. 5.4.14 gives

$$\epsilon(0) \sim \frac{A}{\omega_T'^2}, \tag{5.4.15}$$

where the constant A is fairly independent of temperature.

The Curie-Weiss law for the temperature dependence of $\epsilon(0)$ is

$$\epsilon(0) = \frac{C}{T - T_0}, \tag{5.4.16}$$

where T_0 is the stability limit of the prototypic phase. Comparing Eqs. 5.4.15 and 5.4.16 we get the following temperature dependence for the soft-mode frequency:

$$\omega_T'^2 = B(T - T_0) \tag{5.4.17}$$

This is the situation for an extreme case in which there is no damping of the soft mode. The other extreme is that of a *purely relaxational* ferroelectric, characterized by $\omega' = 0$, $\omega'' \neq 0$. In the simple *Debye model* (which we shall describe in Chapter 10), the following relationship is assumed:

$$\epsilon(\omega) = \epsilon(\infty) + \frac{\epsilon(0) - \epsilon(\infty)}{1 + i\omega\tau}, \tag{5.4.18}$$

with $\tau \equiv 1/\omega_p''$ for the purely relaxational situation.

The LST relation for such a system is linear (rather than quadratic) in the frequencies:

$$\frac{\epsilon(0)}{\epsilon(\infty)} = \frac{\omega_p''}{\omega_L''} \tag{5.4.19}$$

Detailed considerations, taking note of the temperature dependence of ω_L'', lead to the following equation (which is the counterpart of Eq. 5.4.17):

$$\tau = \frac{1}{\omega_p''} \sim \frac{1}{T - T_0} \tag{5.4.20}$$

As $T \to T_0$, $\tau \to \infty$. This is referred to as the *critical slowing down* of the Debye relaxation time τ (see Blinc & Zeks 1974).

In the pure resonance case, on the other hand, the critical slowing down of the fluctuations of the order parameter on approaching the phase-transition temperature T_0 appears directly through the *frequency* of the soft mode, rather than through a divergence of the Debye relaxation time.

This distinction between resonance behaviour and relaxation behaviour has its manifestation in the distinction between what are called *displacive* phase transitions and *order-disorder* phase transitions. Real systems usually are mixtures of the two, but it is instructive to consider the extreme model systems. The $BaTiO_3$ crystal is closer to the former model, and KDP is closer to the latter.

A theoretical exposition of this many-body problem is a highly complicated task. Some of the simplifications usually introduced are as follows (see Thomas 1971; Blinc & Zeks 1974; Lines & Glass 1977).

One takes note of the fact that structural phase transitions in a large number of crystals involve a restructuring of only a small fraction of the total structure, with the overall edifice remaining intact. To make the theory tractable, one therefore deals with the movements of only this fraction of atoms in the unit cell, treating the rest of the edifice as providing a 'heat bath'. The Hamiltonian for such a model system can be written as a sum of two parts:

$$\mathcal{H} = \sum_l \mathcal{H}_l^{(S)} + \mathcal{H}^{int} \tag{5.4.21}$$

Here the summation index l runs over all the N unit cells of the crystal. $\mathcal{H}_l^{(S)}$ is the 'single-particle' part of the Hamiltonian, describing the contribution from the 'local' normal coordinate Q_l and the corresponding momentum P_l. $\mathcal{H}^{int}$ denotes the 'interaction' part of the total Hamiltonian.

A simple form assumed for the single-particle Hamiltonian is

$$\mathcal{H}_l^{(S)} = \frac{1}{2M} P_l^2 + V(Q_l) \tag{5.4.22}$$

Here M is an effective mass, and $V(Q_l)$ the single-particle potential: it is the total potential energy corresponding to the normal coordinate Q_l, with all other normal coordinates taken as zero.

The distinction between displacive and order-disorder models of structural phase transitions can be made in terms of the nature of the potential $V(Q_l)$.

For the displacive case $V(Q_l)$ is only slightly anharmonic, and has a single minimum (at $Q = 0$, cf. Fig. 5.4.1(a)). Mode softening and condensation in such a system amounts to a shifting of the minimum of the potential from $Q = 0$ to a neighbouring nonzero value of Q. By contrast, the anharmonicity in an order-disorder system is very large, and the atomic configuration corresponding to $Q = 0$ is unstable. In the simple example depicted in Fig. 5.4.1(b), $V(Q)$ has two minima, at $Q = \pm Q_0$. Typically this corresponds to a situation wherein the order parameter, e.g. the spontaneous polarization in the case of a proper ferroelectric phase transition, becomes zero in the prototypic phase only in a statistically averaging sense, but in reality, at a microscopic level, there exist *permanent* dipoles. These dipoles get spontaneously aligned along a specific direction in the ferroelectric phase, giving rise to a macroscopic polarization in a given domain. In the paraelectric or prototypic phase, their orientation becomes random, averaging to a value zero. An example is that of ordering of hydrogen atoms in KDP crystals into one of the wells of the double-well potential shown in Fig. 5.4.1(b) in the ferroelectric phase. In the paraelectric phase the two well sites are occupied randomly.

Blinc & Zeks (1974) have pointed out another difference between these two models of phase transitions. The transition entropy is small ($<< k_B \ln 2$) for predominantly displacive phase transitions, and large ($\sim k_B \ln 2$) for predominantly order-disorder transitions.

Real systems are normally a combination of the two extreme models. The single-particle potential can then be expressed as

$$V(Q) = aQ^2 + bQ^4, \tag{5.4.23}$$

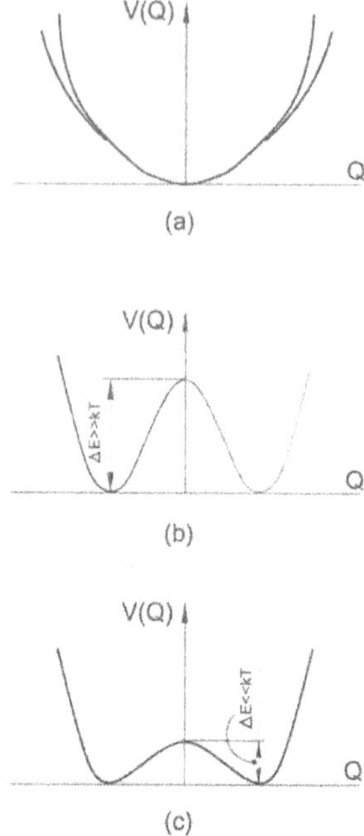

Figure 5.4.1: Single-particle potential for the purely displacive (a), and purely order-disorder (b) models of structural phase transitions. In (a) the dashed line depicts the harmonic part of the potential. Part (c) of the figure shows a more realistic potential for the predominantly displacive case, with a substantial anharmonic component, but the potential barrier of which at $Q = 0$ is easily overcome by thermal fluctuations.

where $a < 0$ and $b > 0$. Depending on the ratio a/b the situation may be either that in Fig. 5.4.1(b), or that in Fig. 5.4.1(c).

The potential-energy difference between the central maximum and a neighbouring minimum can be shown to be

$$\Delta E = \frac{a^2}{4b} \tag{5.4.24}$$

If $\Delta E >> k_B T$, thermal fluctuations cannot overcome substantially the central barrier, and we have an order-disorder configuration.

If $\Delta E << k_B T$, the conditions are more displacive-like.

In several perovskite structures, the soft mode changes its character from displacive to order-disorder with rising temperature (see Godefroy & Jannot 1992).

5.4.4 Overdamped and Underdamped Soft Modes

Consider a displacive-type structural phase transition. It was for such transitions that the soft-mode theory of Cochran and Anderson was originally formulated.

In Eq. 5.4.23, the first term on the right-hand side represents the harmonic part of the single-particle potential, and the second term approximately represents the anharmonicity. If only the first term were present, there would be no mutual interaction among the normal modes of vibration, and the crystal would not display even the property of thermal expansion or contraction.[2] Some anharmonicity is always present.

The anharmonic part of the interaction becomes particularly important when the amplitude of vibration of a mode is large. And this is what happens to a soft mode in the vicinity of the structural phase transition. Thus, near the transition the soft mode interacts quite strongly with other ('hard') modes of vibration. Interaction among the modes means an exchange of energy. Usually it is the soft mode which loses a large fraction of its energy to the other modes, particularly because of its weakened force constant. This results in an additional damping of the soft mode, over and above the usual damping mechanisms for all the modes. Soft modes are therefore mostly *overdamped.*

The opposite process of *underdamping* of the soft mode can also occur if there is a predominant mechanism of intermode interaction whereby energy is transferred *to* the soft mode, rather than *from* it.

5.4.5 Hard Modes and Saturation Temperature for the Order Parameter

Apart from causing an overdamping or underdamping of the soft phonon mode, hard modes make their presence felt in another significant way when the temperature of the crystal is decreased to sufficiently low values.

At sufficiently low temperatures the stability of a crystalline phase is influenced, amongst other things, by the third law of thermodynamics. The third law states that the configurational entropy must decrease with temperature, eventually becoming zero at $T = 0K$. This means that the change of entropy with temperature is zero at the absolute zero of temperature. As a result of this, the order parameter of a structural phase transition does not increase indefinitely on decrease of temperature below T_c, but rather approaches a saturation value (Salje, Wruck & Thomas 1991a; Salje, Wruck & Marais 1991b). The term *quantum saturation* is also used in this context.

This saturation of the order parameter is with respect to temperature,

[2] A soft mode assumed to be devoid of interaction with other modes is referred to as a *bare soft mode.*

and not other control parameters like composition, pressure, or other fields. Further, the phase transition temperature, T_c, is itself a function of these control parameters. Since the saturation effects occur at sufficiently low temperatures, we have to consider situations wherein the control parameters or fields are so chosen as to reduce the value of T_c. We give here some basic information from the work of Hayward & Salje (1998).

A qualitative understanding of quantum saturation can be obtained by invoking the Clausius-Clapeyron equation:

$$\frac{dp}{dT} = \frac{\Delta S}{\Delta V} \tag{5.4.25}$$

In the 'classical regime', we often have $\Delta S \sim \Delta V$, making T_c vary linearly with pressure p. But when $\Delta S \rightarrow 0$ (as demanded by the third law of thermodynamics), the linearity is broken, and quantum saturation sets in.

A more quantitative treatment is that based on the Landau theory (Salje et al. 1991a). One begins with the basic Landau expansion, namely Eq. 5.3.26, which, however, needs to be modified in view of the nonlinearity resulting from the third law of thermodynamics. Salje et al. (1991a) have argued that it is reasonably adequate to modify only the quadratic term. They write:

$$g = \frac{a'T_s}{2}\left[\coth\left(\frac{T_s}{T}\right) - \coth\left(\frac{T_s}{T_c}\right)\right]\eta^2 + \frac{b}{4}\eta^4 + \frac{c}{4}\eta^4 + \frac{c}{6}\eta^6 + \ldots \tag{5.4.26}$$

Here T_s is the *saturation temperature*: Typically, the behaviour of the system is classical for temperatures $T > 3T_s/2$, and the order parameter η is totally saturated (i.e. it is independent of temperature) for $T < T_s/2$.

T_s is also found to have a correlation with the Einstein temperature T_E (Salje, Wruck & Marais 1991b):

$$T_s \simeq T_E/2 \tag{5.4.27}$$

To determine how pressure p or composition x can push T_c towards T_s, one has to model their effect on the coefficients in Eq. 5.4.26. This is easily done in the Landau theory by introducing suitable coupling terms in this equation. The effect of pressure can be modelled (in the harmonic approximation) by including a term proportional to $p\eta^2$ in the Landau expansion. Similarly, the effect of composition variation can be incorporated by introducing a term proportional to $x\eta^2$, and that of uniaxial stress σ by a term proportional to $\sigma\eta^2$. Using the symbol x to denote any of these control parameters (and not just composition), Eq. 5.4.26 gets extended to:

$$g = \frac{a'T_s}{2}\left[\coth\left(\frac{T_s}{T}\right) - \coth\left(\frac{T_s}{T_c}\right)\right]\eta^2 + \frac{b}{4}\eta^4 + \frac{c}{4}\eta^4 + \frac{c}{6}\eta^6 + \frac{a'T_sK}{2}x\eta^2 \tag{5.4.28}$$

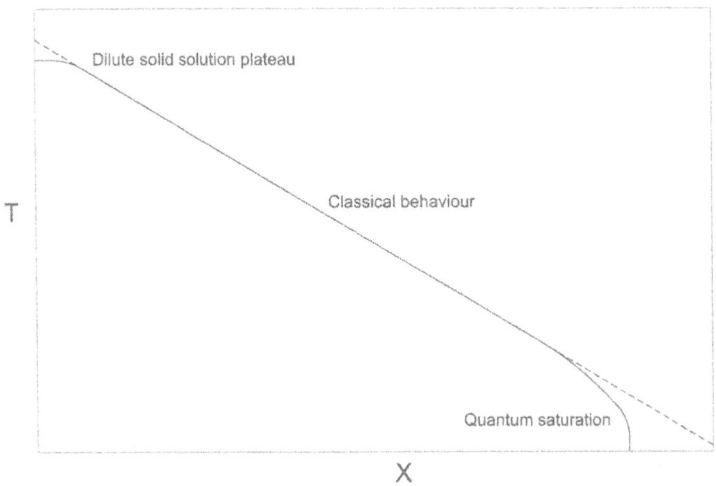

Figure 5.4.2: T-x phase diagram depicting the occurrence of quantum saturation. It also shows the *plateau effect* expected at very dilute concentrations of the solute in the solid solution. The plateau effect is discussed in §5.9. [After Hayward & Salje (1998).]

Here K is a coupling constant.

In keeping with the usual formulation of the Landau theory, one can assume that the critical temperature dependence of the order parameter is carried by the η^2 term only. The new phase transition temperature, T_c^*, is therefore determined by equating the prefactor of η^2 to zero:

$$\coth\left(\frac{T_s}{T}\right) - \coth\left(\frac{T_s}{T_c}\right) + Kx = 0, \tag{5.4.29}$$

or,

$$T_c^*(p, x) = \frac{T_s}{\coth^{-1}(\coth(T_s/T_c) - Kx)} \tag{5.4.30}$$

A composition-temperature phase diagram based on this equation is shown schematically in Fig. 5.4.2. It can be seen that it explains the quantum saturation of the order parameter.

A lattice-dynamical (soft-mode) interpretation of this saturation behaviour has been given by Dove, Giddy & Heine (1992). The total energy of the crystal can be viewed as a sum of the energies distributed among the various phonon modes (both soft and hard). If there are a large number of hard modes, they would saturate individually at rather high temperatures. And if they are coupled substantially to the order parameter (the soft mode), T_s would have a high value. If, on the other hand, the hard modes do not have a strong influence on the mechanism of the phase transition, the saturation of the order parameter would be determined mainly by the frequency of the softening mode. Since this frequency has a rather low value near T_c, a small T_s may ensue.

Hayward & Salje (1998) have listed the value of T_s for a large number of crystals. It is 0 K for SbSI, and as high as 334 K for quartz.

We return to this topic when we discuss quantum ferroelectrics in §10.5.

SUGGESTED READING

S. Aubry & R. Pick (1971). Soft modes in displacive transitions. *J de Physique*, **32**, 657.

B. B. Lavrencic & T. S. Shigenari (1973). Simple group-theoretical method to determine the symmetry of soft modes in general second order phase transitions. *Solid State Comm.*, **13**, 1329.

K. A. Muller (1981). In K. A. Muller & H. Thomas (Eds.), *Phase Transitions - I.* Springer-Verlag, Berlin.

M. T. Dove, A. P. Giddy & V. Heine (1992). On the application of mean-field and Landau theory to displacive phase transitions. *Ferroelectrics*, **136**, 33.

S. A. Hayward & E. K. H. Salje (1998). Low-temperature phase diagrams: nonlinearities due to quantum mechanical saturation of order parameters. *J. Phys.: Condens. Matter*, **10**, 1421.

K. Parlinsky, Z. Q. Li & Y. Kawazoe (1999). How to simulate a structural phase transition by the first-principles method ? *Phase Transitions*, **67**, 681.

5.5 CRITICAL-POINT PHENOMENA

The Landau theory and the soft-mode theory are essentially *mean-field theories*: The actual microscopic interactions at any point are replaced by an interaction between a test unit and the mean field generated by all the other units constituting the system. These theories thus ignore the fluctuations of the mean field, and thence of the order parameter. In this section we consider phenomena in the vicinity of the critical temperature, where thermodynamic fluctuations cannot be ignored in general. Phenomena associated with these *critical fluctuations* are referred to as critical-point phenomena, or critical phenomena.

The reader is advised to read Appendix E, which summarizes some of the principal concepts and results of thermodynamics and statistical mechanics, before reading this section.

5.5.1 Critical Fluctuations

As demonstrated in Appendix E (cf. Eq. E.2.45, and the discussion following it), when a large number of particles interact, their macroscopic thermodynamic properties are usually very close to the mean value or the ensemble average, and fluctuations from the mean value are normally not significant. This statement can become inapplicable if any of the prerequisites for its validity is violated. For example, the root-mean-square deviation from the mean value would be significantly large if the actual or effective number of particles involved in the microscopic interactions is not large. This can happen even in a physically large system if the range of the microscopic interaction is not large. It can also happen if the effective or assumed (modelled) dimensionality of the system is not sufficiently large. An extreme example is that of a 1-dimensional system, in which fluctuations destroy long-range order because the interaction between one part of the chain and another can only be through successive units of the chain and a fluctuation can obliterate a long-range ordering tendency or mean-field behaviour.

This connectivity factor becomes less important for 2-dimensional systems, wherein many interaction pathways are possible, making it more likely for the system to stay close to a mean-value configuration.

The number of connectivity pathways increases dramatically as we go to still higher dimensions, so much so that above a *critical dimension*, or *marginal dimension*, d_u (usually $d_u = 4$), fluctuations become unimportant and a mean-field theory like the Landau theory provides an analytically correct description of continuous phase transitions.

Thus, for low-dimensionality systems, or for systems governed by short-range interactions, or for both, the Landau theory fails to provide a correct description of critical phenomena for temperatures too close to the critical temperature. How close is too close? A self-consistency criterion for answering this question was formulated by Ginzburg (1961). We describe here the essence of the Landau-Ginzburg theory which incorporates partially the effect of fluctuations also, and then state the Ginzburg criterion.

5.5.2 Landau-Ginzburg Theory

The Landau theory has two main flaws. It neglects fluctuations of the order parameter, and it assumes that the free energy is an analytic function of the order parameter at the critical point. The former assumption is removed, to a good approximation, in the Landau-Ginzburg formulation of the theory. In it the spatial nonuniformity of the order parameter is not ignored, although it is assumed that the fluctuations do not occur on too fine a scale. The brief description given here follows closely the work of Plischke & Bergersen (1984).

As in §5.3.12, we assume for simplicity that the order parameter has only one component. We denote it by $\eta(\mathbf{r})$ to express its possible spatial inhomogeneity. And $X(\mathbf{r})$ is the field conjugate to it.

In view of the spatial dependence of the order parameter, $\eta(\mathbf{r})$ now denotes its *local* density, and we can write (cf. Eq. E.2.46)

$$M = < \int \eta(\mathbf{r})\, d\mathbf{r} > \tag{5.5.1}$$

For writing the free-energy expansion we work with the Helmholtz free energy, A, rather than the Gibbs free energy G (cf. Eq. 5.3.9). When the order parameter is spatially homogeneous, the two are related by

$$A = G + XM \tag{5.5.2}$$

In view of Eq. E.1.13,

$$X = \left.\frac{\partial A}{\partial M}\right|_T \tag{5.5.3}$$

The Landau-Ginzburg theory (Ginzburg 1961) assumes that the free energy can be expressed as the following volume integral (cf. Eq. 5.3.9 for comparison):

$$A = \int d(\mathbf{r})[A_0 + \frac{a}{2}\eta^2(\mathbf{r}) + \frac{b}{4}\eta^4 + \frac{c}{6}\eta^6 + \cdots + \frac{f}{2}\{\nabla\eta(\mathbf{r})\}^2] \tag{5.5.4}$$

Here the last term accounts for the spatial variation of the order parameter. The coefficient f is taken as positive, to reflect the fact that the free energy is higher when the order parameter is not the same everywhere.

Since the order parameter and the force conjugate to it are not constant over space, Eq. 5.5.3 must be replaced by the following functional derivative:

$$X(\mathbf{r}) = \frac{\delta A}{\delta\eta(\mathbf{r})} \tag{5.5.5}$$

From Eq. 5.5.4,

$$\delta A = \int d(\mathbf{r})\,[\delta\eta(\mathbf{r})\,\{a\eta(\mathbf{r}) + b\eta^3(\mathbf{r}) + c\eta^5(\mathbf{r}) + \cdots\} + f\nabla\delta\eta(\mathbf{r})\cdot\nabla\eta(bfr)] \tag{5.5.6}$$

On carrying out an integration by parts, and stipulating that $\delta\eta(\mathbf{r}) = 0$ at the surface of the sample, we obtain

$$X(\mathbf{r}) = a\eta(\mathbf{r}) + b\eta^3(\mathbf{r}) + c\eta^5(\mathbf{r}) + \cdots - f\nabla^2\eta(\mathbf{r}) \tag{5.5.7}$$

To bring out the role of spatial fluctuations of the order parameter we introduce a perturbation

$$X(\mathbf{r}) = X_0\,\delta(\mathbf{r}) \tag{5.5.8}$$

at a point $\mathbf{r}$ in the sample, and compute its effect throughout the sample under the purview of the *linear response theory* (cf. Appendix E.3). We write

$$\eta(\mathbf{r}) = \eta_0 + \phi(\mathbf{r}), \tag{5.5.9}$$

and ignore all terms that are not linear in ϕ. In particular, for use in Eq. 5.5.7, we write

$$\eta^3(\mathbf{r}) = \eta_0^3 + 3\eta_0^2\phi(\mathbf{r}) \tag{5.5.10}$$

Then, from Eq. 5.5.7,

$$\nabla^2\phi(\mathbf{r}) - \frac{a}{f}\phi(\mathbf{r}) - 3\eta_0^2\frac{b}{f}\phi(\mathbf{r}) - \frac{a}{f}\eta_0 - \frac{b}{f}\eta_0^3 = -\frac{X_0}{f}\delta(\mathbf{r}) \tag{5.5.11}$$

For $T > T_c$, $\eta_0 = 0$. And for $T < T_c$ we can obtain the value of η_0 for a continuous phase transition by putting $X(\mathbf{r}) = 0$, $c = 0$ (cf. §5.3.12 where continuous phase transitions are discussed), and $\nabla\eta(\mathbf{r}) = 0$ in Eq. 5.5.7. The result is

$$\eta_0^2 = -\frac{a}{b}, \quad T < T_c, \tag{5.5.12}$$

in agreement with Eq. 5.3.22. Eq. 5.5.11 therefore gives

$$\nabla^2\phi - \frac{a}{f}\phi = -\frac{X_0}{f}\delta(\mathbf{r}), \quad T > T_c, \tag{5.5.13}$$

$$\nabla^2\phi + 2\frac{a}{f}\phi = -\frac{X_0}{f}\delta(\mathbf{r}), \quad T < T_c \tag{5.5.14}$$

A solution of these equations after a change to spherical coordinates yields

$$\phi = \frac{X_0}{4\pi f}\frac{e^{-r/\xi}}{r}, \tag{5.5.15}$$

where

$$\xi = (f/a)^{1/2}, \quad T > T_c, \tag{5.5.16}$$

$$\xi = (-f/(2a))^{1/2}, \quad T < T_c \tag{5.5.17}$$

ξ is called the *correlation length.* Since $a = a'(T - T_c)$ (cf. Eq. 5.3.18),

$$\xi \sim |T - T_c|^{-1/2} \tag{5.5.18}$$

The correlation length thus diverges as the critical temperature is approached from both the higher and the lower sides. And we are led to another critical exponent, ν:

$$\xi \sim |T - T_c|^{-\nu} \tag{5.5.19}$$

$\nu = \frac{1}{2}$ according to the Landau-Ginzburg theory, but its actual value is different if this theory is not applicable.

We next examine the behaviour predicted for the generalized susceptibility by this theory. The order parameter (an extensive quantity) is defined by the ensemble average expressed by Eq. 5.5.1. In the canonical ensemble this average is defined by Eq. E.2.37. Therefore,

$$\frac{M}{V} = \frac{1}{V}\int d\mathbf{r}\frac{Tr[\eta(\mathbf{r})e^{-\beta\mathcal{H}}]}{Tr\, e^{-\beta\mathcal{H}}} \tag{5.5.20}$$

We assume that the Hamiltonian $\mathcal{H}$ now includes the term (Eq. 5.5.8) corresponding to the perturbing field, so that it has the form

$$\mathcal{H} = \mathcal{H}_0 - \int \eta(\mathbf{r})\, X(\mathbf{r})\, d\mathbf{r} \tag{5.5.21}$$

The canonical-ensemble average of the order parameter can then be written in accordance with Eq. E.2.37 (assuming that the system is invariant under a translation):

$$< \eta(\mathbf{r}) > = \frac{Tr[\eta(\mathbf{r})\, e^{-\beta\mathcal{H}}]}{Tr\, e^{-\beta\mathcal{H}}} \tag{5.5.22}$$

We get from Eqs. 5.5.22, 5.5.9, 5.5.21, and E.2.47,

$$\frac{\delta < \eta(\mathbf{r}) >}{\delta X(0)} = \frac{\phi(\mathbf{r})}{X_0} = \beta(< \eta(\mathbf{r})\eta(0) > - < \eta(\mathbf{r}) >< \eta(0) >) = \beta\Gamma(\mathbf{r}) \tag{5.5.23}$$

Thus

$$\phi(\mathbf{r}) \sim \Gamma(\mathbf{r}) \tag{5.5.24}$$

Also, on integrating Eq. 5.5.23, the generalized susceptibility is determined as

$$\chi = \beta \int \Gamma(\mathbf{r})\, d\mathbf{r} \tag{5.5.25}$$

This equation is an example of the *fluctuation dissipation theorem* (§E.3.3). It expresses the fact that the generalized susceptibility, i.e. the response to a perturbation, is determined by the order-parameter correlations existing in the system at equilibrium, i.e. when the perturbation is *absent.*

The correlation length ξ is related to $\Gamma(\mathbf{r})$, the Fourier transform of which is given by Eq. E.2.54:

$$\Gamma(\mathbf{k}) = <|\eta(\mathbf{k})|^2> \tag{5.5.26}$$

Following Huang (1987), we determine here the right-hand side of this equation in an approximate manner (assuming that the fluctuations are essentially of an isotropic nature) to get a feel for the energy apportioned to the kth Fourier component. Eq. 5.5.4 can be written to lowest order as

$$A = \int d\mathbf{r}\,[\frac{a}{2}\eta^2(\mathbf{r}) + \frac{f}{2}(\nabla\eta(\mathbf{r}))^2] \tag{5.5.27}$$

This integration can be written as an equivalent integration in Fourier space:

$$A = \int \frac{d\mathbf{k}}{2(2\pi)^3}(a + fk^2)\,|\eta(\mathbf{k})|^2 \tag{5.5.28}$$

Thus the free energy assignable to the kth mode is

$$A(\mathbf{k}) = \frac{1}{2}(a + fk^2)\,|\eta(\mathbf{k})|^2 \tag{5.5.29}$$

Equating this to k_BT ("equipartition of energy"), and referring to Eq. 5.5.26,

$$<|\eta(\mathbf{k})|^2> = \Gamma(\mathbf{k}) = \frac{2k_BT}{a + fk^2} \tag{5.5.30}$$

This is the so-called *Ornstein-Zernike form* of $\Gamma(\mathbf{k})$.

Taking the inverse Fourier transform of the two sides of Eq. 5.5.30 gives

$$\Gamma(\mathbf{r}) = \frac{e^{-r/\xi}}{r} \tag{5.5.31}$$

As $T \to T_c$, the correlation length becomes infinitely large (Eq. 5.5.18), and $\Gamma(\mathbf{r})$ varies as $1/r$, i.e. it becomes independent of any other characteristic length of the system. Also, it is the only characteristic length the system has for these conditions.

We have considered so far only a 3-dimensional system. The results obtained can be generalized to an arbitrary spatial dimension d. It can be shown that, instead of Eq. 5.5.15, we have

$$\phi \sim r^{-(d-2)} \quad \text{for } r << \xi, \tag{5.5.32}$$

and

$$\phi \sim e^{-r/\xi} \quad \text{for } r >> \xi, \tag{5.5.33}$$

so that, in general (and also incorporating Eq. 5.5.24),

$$\phi(\mathbf{r}) \sim \Gamma(\mathbf{r}) \sim \frac{e^{-r/\xi}}{r^{d-2}} \tag{5.5.34}$$

5.5.3 Ginzburg Criterion

Ginzburg (1961) formulated a self-consistency criterion for the Landau theory, which states that the neglect of critical fluctuations of the order parameter by the Landau theory is justifiable at those values of temperature, pressure etc. (with respect to the critical point) for which the fluctuations, averaged in all directions over distances of the order of the correlation length ξ, are small compared to the average value of the order parameter itself (Bausch 1972):

$$\int_{\Omega_\xi} d\Omega \, [< \eta(\mathbf{r})\eta(0) > - < \eta(\mathbf{r}) >< \eta(0) >] \; << \; \int_{\Omega_\xi} d\Omega \, \eta_0^2 \tag{5.5.35}$$

where η_0 is the uniform part of the order parameter (cf. Eq. 5.5.9). The integration is carried out over that part of configuration space for which the correlations are important.

Eq. 5.5.35 can be rewritten in terms of the order-parameter autocorrelation function (cf. Eq. E.2.48) as

$$\int_{\Omega_\xi} d\Omega \, \Gamma(\mathbf{r}) \; << \; C\xi^d \eta_0^2, \tag{5.5.36}$$

where we have assumed the critical fluctuations to be isotropic, and $C\xi^d$ is the volume of a hypersphere of radius ξ.

We shall be using this criterion presently for computing the so-called upper critical dimensionality.

5.5.4 Critical Exponents

The critical exponents are a measure of the singularities of the various thermodynamic quantities at and near the critical temperature, T_c.

Many of the critical exponents in use express the nature of the temperature dependence of the thermodynamic quantities (in the form of a power law). For this purpose, and also for highlighting the "universal" nature of critical phenomena (see below), one usually works with the *reduced temperature*, a dimensionless parameter:

$$t = \frac{T - T_c}{T_c} = \frac{T}{T_c} - 1 \tag{5.5.37}$$

A critical exponent is formally defined as follows. Let us consider a thermodynamic function $f(t)$ near $t = 0$, and assume that it is positive and continuous for sufficiently small values of t. We are interested in its critical-point behaviour. It is found experimentally that $f(t)$ can be expressed as

$$f(t) = At^{\lambda}(1 + Bt^{y} + \cdots), \quad y > 0 \tag{5.5.38}$$

It is also found that for sufficiently small values of t the following equation is a very good approximation:

$$f(t) = At^{y} \tag{5.5.39}$$

Therefore a plot of $\ln f$ against t will have a straight-line portion near $t = 0$, from which the factor $A\lambda$ can be determined. We can identify λ as a critical exponent, formally defined as the following limit (Stanley 1971):

$$\lambda \equiv \lim_{t \to 0} \frac{\ln f(t)}{\ln t}, \tag{5.5.40}$$

assuming that such a limit exists.

As a short-hand notation one often defines λ as a critical exponent such that

$$f(t) \sim t^{\lambda}, \tag{5.5.41}$$

although what is really involved is a relationship like Eq. 5.5.38.

We describe here the main critical exponents relevant for static phenomena near *ferromagnetic* critical points. The order parameter is the spontaneous magnetization $< m >$, and h is the external (magnetic) field conjugate to it.

The critical exponent β describes the temperature variation of the order parameter (for $h = 0$) in the vicinity of the critical temperature:

$$< m > \sim (-t)^{\beta}, \quad t < 0 \tag{5.5.42}$$

A measure of the breakdown of the Landau theory is the fact that whereas according to it $\beta = 0.5$ for all continuous phase transitions (cf. Eq. 5.3.27), some experimental values are: 0.34 ± 0.02 for Fe, 0.33 ± 0.03 for Ni, and 0.354 ± 0.005 for $YFeO_3$ (Kadanoff et al. 1967).

Another critical exponent associated with the order parameter is δ, which describes its field dependence. For very small h one derives (at the critical point, cf. Eq. 5.3.33):

$$< m > \sim h^{1/\delta}, \quad t = 0 \tag{5.5.43}$$

Observed values are: 4.2 ± 0.1 for Ni, and 4.0 ± 0.1 for Gd (Kadanoff et al. 1967), whereas the mean-field value is 3 (cf. Eq. 5.3.33).

The third important critical exponent, γ, describes the temperature dependence of the susceptibility function, χ. As T_c is approached from either side for a continuous transition, susceptibility is seen to diverge (cf. Eqs. 5.3.19 and 5.3.28):

$$\chi \sim t^{-\gamma}, \quad t > 0, \tag{5.5.44}$$

$$\chi \sim (-t)^{-\gamma'}, \quad t < 0 \tag{5.5.45}$$

The proportionality constants for these two equations are not the same (cf. Eq. 5.3.30). According to the Landau theory, $\gamma = \gamma' = 1$. Some experimental values are as follows: $\gamma = 1.333 \pm 0.015$ for Fe; $\gamma = 1.32 \pm 0.02$ for Ni; $\gamma = 1.33 \pm 0.04$ and $\gamma' = 0.7 \pm 0.1$ for $YFeO_3$; and $\gamma = 1.33$ for Gd (Kadanoff et al. 1967).

The fourth critical exponent we consider is α, which is an indicator of the temperature dependence of specific heat, C. One observes the following singular behaviour:

$$C \sim t^{-\alpha}, \quad t > 0, \tag{5.5.46}$$

$$C \sim t^{-\alpha'}, \quad t < 0 \tag{5.5.47}$$

The proportionality constants in the two cases are different. The Landau theory predicts $\alpha = \alpha' = 0$. Some experimental values are: $\alpha = \alpha' = -0.12 \pm 0.01$ for Fe (Lederman et al. 1974); $\alpha = \alpha' = -0.10 \pm 0.03$ for Ni (Kadanoff et al. 1967); and $\alpha = \alpha' = -0.09 \pm 0.01$ for EuO (Lederman et al. 1974).

The fifth important critical exponent is connected with the Fourier transform of the autocorrelation function of the order parameter (cf. Eq. E.2.54). It is instructive to see how this correlation function is obtained experimentally.

Let $m(\mathbf{r})$ denote the local density of the order parameter at a point $\mathbf{r}$ in the specimen, with M as its ensemble average (cf. Eq. E.2.46). This density distribution can be probed by a scattering experiment, e.g. by neutron scattering. Let $\mathbf{k}_i$ and $\mathbf{k}_f$ denote the wavevectors of a neutron beam before and after the scattering, with $\mathbf{k} = \mathbf{k}_f - \mathbf{k}_i$ as the momentum transferred to the specimen. In the Born approximation the scattering cross section, $\Gamma(\mathbf{k})$, is given by the square of the matrix element between the initial and the final states:

$$\Gamma(\mathbf{k}) = \langle |m(\mathbf{k})|^2 \rangle \sim \langle \left| \int d\mathbf{r}\, e^{-i\mathbf{k}\cdot\mathbf{r}}\, m(\mathbf{r}) \right|^2 \rangle \tag{5.5.48}$$

It is found experimentally that the forward-scattering cross-section ($\mathbf{k} \to 0$) diverges as $T \to T_c$. For very small values of k one observes that

$$\Gamma(k) \sim \frac{1}{k^{2-\eta}}, \quad t = 0 \tag{5.5.49}$$

The Landau theory predicts that $\eta = 0$ for all continuous transitions. Experimentally, $\eta = 0.07 \pm 0.07$ for Fe (Kadanoff et al. 1967).

We mention lastly the sixth critical exponent, ν (cf. Eq. 5.5.19). It corresponds to the divergence of the correlation length of the order parameter as the critical temperature is approached:

$$\xi \sim t^{-\nu}, \quad t > 0, \tag{5.5.50}$$

$$\xi \sim (-t)^{-\nu'}, \quad t < 0 \tag{5.5.51}$$

According to the Landau theory, $\nu = \nu' = 0.5$ (cf. Eq. 5.5.18). Experimentally observed values range between 0.59 and 1.02 (cf. Chaikin & Lubensky 1995).

5.5.5 Upper and Lower Marginal Dimensionality

The Ginzburg criterion (Eq. 5.5.36) provides a method for calculating the dimensionality of space, d_u, at or above which the Landau theory is adequate for describing critical phenomena. We write, for the right-hand side of Eq. 5.5.36,

$$\eta_0^2 \simeq |T - T_c|^{2\beta} \tag{5.5.52}$$

And for the left-hand side we substitute from Eq. 5.5.34. On carrying out the integration in spherical coordinates we obtain

$$Cd \int_0^{\xi} d\mathbf{r}\, r^{d-1}\, e^{-r/\xi-(d-2)} << C\xi^d\, |T - T_c|^{2\beta} \tag{5.5.53}$$

After some further algebra, and making use of Eq. 5.5.50, we get the following condition for the applicability of the Landau theory (Plischke & Bergersen 1994):

$$d \geq 2 + \frac{2\beta}{\nu} \tag{5.5.54}$$

According to the Landau theory, $\beta = \frac{1}{2}$ and $\nu = \frac{1}{2}$ for a continuous transition, so that, for the equality sign in Eq. 5.5.54,

$$d_u = 4 \tag{5.5.55}$$

For a tricritical point, on the other hand, $\beta = \frac{1}{4}$ (Eq. 5.3.37) and $\nu = \frac{1}{2}$, giving

$$d_t = 3 \tag{5.5.56}$$

d_u and d_t are called *upper marginal dimensionalities*, or *upper critical dimensionalities*. For dimensions greater than these, critical fluctuations are unimportant and the Landau theory provides a correct description of critical phenomena. When $d = d_u$ in Eq. 5.5.55 or $d = d_t$ in Eq. 5.5.56, only marginal corrections are required to be made to the critical exponents predicted by the Landau theory.

For dimensions less than the upper marginal dimensionality the Landau theory is inadequate. In fact, one can also define a *lower* marginal dimensionality, d_l. For $d < d_l$ the critical fluctuations or other factors are so important that the phase transition does not occur at all.

5.5.6 Models of Phase Transitions

The Landau theory, by making certain sweeping assumptions, attempts to deal phenomenologically with macroscopic parameters relevant to a continuous phase transition. With its range of validity defined by the Ginzburg criterion, the theory is quite simple, and its simplicity gives it power to deal successfully with several ferroic and other phase transitions, so much so that for temperatures not too close to the critical temperature and/or for phase transitions mediated by long-range interactions, there is hardly a need to invoke other theories for dealing with properties of ferroic materials.

For dealing with situations where the Landau theory is not applicable, and also for assigning meaning at a microscopic level to situations where the Landau theory *is* applicable, it is found useful to explain phase transitions in terms of one or the other of a variety of *models* of phase transitions. We describe the main ones here, albeit very briefly. Historically the subject of critical phenomena developed with specific reference to ferromagnetic transitions, with the net electron spin playing the role of the order parameter. Generalization to other situations has been carried out in several cases.

At the most fundamental level it should be possible to explain phase transitions in crystals in terms of the Coulomb interaction between electrons and nuclei. This, however, is too general a starting point, which does not make use of the fact that we are dealing with a *crystal*, with a well-defined lattice and band structure etc.

Therefore, at the next level of specialization we take the basic features of the crystal as known beforehand. However, since critical phenomena involve collective behaviour of electrons over a very large number of unit cells, it is enough to concentrate only on the unpaired electron spins, and ignore the effect of band structure etc., except in so far as it affects the interaction among electron spins.

The complete crystal structure can be built by assigning the contents of a primitive unit cell to each lattice point, and, for the purpose of dealing with critical phenomena a net electron spin can be assigned to each primitive unit cell. Different models exist for describing the spin-spin interaction. These models are quite crude, though also very instructive, and involve parameters which are adjusted to simulate the actual interactions as closely as possible.

Ising Model

In the Ising (1925) model the spin associated with each lattice site (or with each primitive unit cell) can take only two possible values: +1 and -1, and the interaction between spins is assumed to be the nearest-neighbour exchange interaction. The model thus has Z_2 symmetry (cf. §B.1) in the paramagnetic phase.

Let σ_l denote the spin associated with the lattice site l. Then the Ising Hamiltonian is

$$\mathcal{H}_{Ising} = -J \sum_{<l,l'>} \sigma_l \sigma_{l'}, \tag{5.5.57}$$

the summation being only over nearest-neighbour values of l and l'. The spins are purely classical in this model.

If $d < 2$, no phase transition is predicted by the Ising model even at $T = 0$. Thus the lower marginal dimensionality d_l is equal to 1 in this model.

Z_N Models

The Z_2 symmetry of the Ising model is generalized to the Z_N symmetry of the so-called *clock models* or Z_N *models*. Here the spin variable s_l at each lattice site is constrained to point along any of N equally spaced directions on a unit circle:

$$s_l = \left(\frac{\cos 2\pi n_l}{N}, \frac{\sin 2\pi n_l}{N} \right), \quad n_l = 0, 1, \ldots N-1 \tag{5.5.58}$$

The Hamiltonian for the clock models is

$$\mathcal{H}_{clock} = -J \sum_{<l,l'>} \mathbf{s}_l \cdot \mathbf{s}_l' = -J \sum_{<l,l'>} \cos[2\pi(n_l - n_l')/N] \tag{5.5.59}$$

Potts Model The N-states Potts model (Potts 1952; Wu 1982) has Z_N symmetry, like a clock model. It has a spin variable $\sigma(l)$ that can take N discrete values. The Hamiltonian is

$$\mathcal{H}_{Potts} = -J \sum_{<l,l'>} [N\delta_{\sigma_l,\sigma_l'} - 1] \tag{5.5.60}$$

Thus the energy associated with nearest-neighbour interactions is different depending on whether the neighbours are in the same state or different states.

The Ising model is a two-state Potts model.

Heisenberg Model

Heisenberg ferromagnets or antiferromagnets are those in which the crystal fields are not sufficiently strong to align the spins along crystallographic axes. The Heisenberg model has O_3 symmetry (§B.4), so that the spins are free to point along any direction in 3-dimensional space. The Hamiltonian is

$$\mathcal{H}_{Heisenberg} = -J \sum_{<l,l'>} \mathbf{s}_l \cdot \mathbf{s}_l' \tag{5.5.61}$$

For this model $d_l = 2$.

XY Model

This model is associated with systems which have O_2 symmetry, or XY symmetry in the paramagnetic phase. An example is that of an easy-plane ferromagnet, in which the spins are constrained by crystal fields to lie in the XY-plane. Its Hamiltonian can be expressed in terms of a local-angle variable by taking $s_l = (\cos\theta_l, \sin\theta_l)$:

$$\mathcal{H}_{XY} = -J \sum_{<l,l'>} \cos(\theta_l - \theta_l') \tag{5.5.62}$$

O$_N$ Models

The Heisenberg model is an O_3 model. And the O_N model with $N = \infty$ can be solved exactly (see Chaikin & Lubensky (1995) for details).

The model Hamiltonians defined here are *unit-cell Hamiltonians* or *site Hamiltonians*. We shall see presently that, because of the large correlation lengths near the critical point, we have to resort to "coarse-graining" and define blocks of spins extending over several unit cells, as also the corresponding *block Hamiltonians*.

5.5.7 Universality Classes and Scaling

The Landau theory replaces the actual local configurations of the order parameter by their mean value, and thus ignores fluctuations from the mean value. Any generalization that incorporates these spatial fluctuations is a *field theory*. In such a theory the order parameter is defined as an integral over all points in space.

The need for a field theory arises because the observed critical exponents close to the critical point do not agree with those predicted by mean-field theories like the Landau theory, and also because one must explain the existence of universality classes and scaling.

Universality Classes

Measurements of critical exponents for a wide variety of systems lead to a remarkable observation: The critical exponents do not depend significantly on the detailed nature or strength of the interaction responsible for the phase transition, or on the value of the critical temperature. Their values are determined largely by the following three factors:

- The effective spatial dimension, d, of the system.
- The symmetry and dimensionality (number of components) of the order parameter.
- The symmetry and range of the interaction.

All continuous phase transitions for which the above three factors are the same have almost the same or "universal" values of critical exponents; they constitute a universality class.

For example, for a large class of systems with $d = 3$ we have: $\alpha \simeq 0$, $\beta \simeq \frac{1}{3}$, $\gamma \simeq \frac{2}{3}$, and $\nu \simeq \frac{2}{3}$. And these values are very different from those for $d = 1$ or $d = 2$, etc.

Similarly, many of the transitions describable by the Ising model (for which the order parameter corresponds to Z_2 symmetry) have nearly the same critical exponents, examples being ferromagnetic transitions in uniaxial magnetic systems, liquid-gas transitions, and several order-disorder transitions.

Scaling

We have seen above that phase transitions in the same universality class have practically the same critical exponents. That is not all. It is also found that not all the critical exponents are independent of one another. If d is the effective dimension of a system, the following *scaling relations* are found to hold to a good approximation:

$$\gamma = \nu(2-\eta) \tag{5.5.63}$$

$$\alpha + 2\beta + \gamma = 2 \tag{5.5.64}$$

$$\gamma = \beta(\delta - 1) \tag{5.5.65}$$

$$\nu d = 2 - \alpha \tag{5.5.66}$$

These relations can be derived under the so-called *scaling hypothesis*, according to which the correlation length, ξ, is the only characteristic length of a system in the vicinity of the critical point. The validity of such a hypothesis is indicated by Eq. 5.5.19, according to which, as the critical point is approached, the correlation length becomes arbitrarily large, irrespective of other characteristic lengths of the system.

We see from Eq. 5.5.34 that, as $\xi \to \infty$, the order-parameter autocorrelation function becomes a *homogeneous function* of distance (cf. Eq. E.1.22) in that

$$\Gamma(\mathbf{r}) = b^{-(d-2)}\,\Gamma(\mathbf{r}/b) \tag{5.5.67}$$

If the only characteristic length on which $\Gamma(\mathbf{r})$ depends is ξ, and the critical exponent associated with ξ is ν (Eq. 5.5.50), then the critical exponent δ associated with $\Gamma(\mathbf{r})$ (Eq. 5.5.49) cannot be independent of ν.

And since the susceptibility χ is related to $\Gamma(\mathbf{r})$ through Eq. 5.5.25, the critical exponent γ associated with the former cannot be independent of ν and η (Plischke & Bergersen 1994). And so on.

5.5.8 Kadanoff Construction

Near the critical point the only important characteristic length of a system is the correlation length (this is the scaling hypothesis), and this length becomes larger and larger as the critical point is approached. Therefore, as was first suggested by Kadanoff (1966), we can abandon the site Hamiltonians described in §5.5.6 in favour of block Hamiltonians. The latter are obtained by averaging over blocks of unit cells, or blocks of "spins". This amounts to recognizing the fact that, near the critical point, the spin-spin correlations are in terms of large blocks (or "patches").

Let us consider a crystal with a cubic lattice (of lattice constant a) in d-dimensional space. We first take a cube of volume L^d in this crystal. This can be conveniently referred as an *L-system* (Toda, Kubo & Saito 1992).

The unit-cell spin, or site spin, $s(\mathbf{r})$ can be expressed as a Fourier series:

$$s(\mathbf{r}) = L^{-d/2} \sum_{k<\Lambda} \phi_k \, e^{i\mathbf{k}\cdot\mathbf{r}}, \tag{5.5.68}$$

where $\Lambda = 2\pi/a$, and Fourier summation is not carried out for $k > \Lambda$ because we have taken the unit cell as the smallest unit for defining the spin density.

The number of Fourier components for this system is $(L/a)^d$, because it is determined for the L-system (under the periodic boundary condition) by

$$k_i = \frac{2\pi}{L} n; \quad n = 1, 2, \ldots n_0; \quad i = 1, 2, \ldots d \tag{5.5.69}$$

Here $n_0 = L/a$. The condition $n < n_0$ in this equation is equivalent to the condition $k < \Lambda$ in Eq. 5.5.68.

Next we consider a cube of side Lb ($b > 1$). This *Lb-system* will have b^d times more particles than the L-system. The upper limit (n_0) in Eq. 5.5.69 is now Lb/a.

In the *Kadanoff construction* one chooses a block of unit cells (numbering b^d in all) as the smallest unit for defining the inter-unit spin values. In other words, the lattice constant is now ba, rather than a. Correspondingly the cutoff wavevector in Eq. 5.5.68 is $2\pi/(ba)$, rather than $2\pi/a$. Therefore, for the Lb-system,

$$k < \Lambda/b, \tag{5.5.70}$$

or, equivalently,

$$n = n_0/b = L/(ba) \tag{5.5.71}$$

To achieve this removal or suppression of the *irrelevant* degrees of freedom, one integrates out the Fourier components ϕ_k for $\Lambda/b < k < \Lambda$. The Hamiltonian $\mathcal{H}_{Lb}$ for the Lb-system is therefore determined by

$$e^{-\beta\mathcal{H}_{Lb}} = \int\cdots\int e^{-\beta\mathcal{H}_L} \prod_{\Lambda/b<k<\Lambda} d\,s_k \tag{5.5.72}$$

The Kadanoff construction amounts to replacing the spins s_{ci} in the various cells i inside a block of side ba by their average:

$$s_i = b^{-d}\sum_{cell} s_{ci}, \tag{5.5.73}$$

and writing

$$e^{-\beta\mathcal{H}_{Lb}} = \sum_{\{s_{ci}\}}\cdots\sum e^{-\beta\mathcal{H}_L}\cdot\delta(s_i - b^{-d}\sum s_{ci}) \tag{5.5.74}$$

Here the delta-function is n-dimensional if the spins are n-dimensional.

Eqs. 5.5.73 and 74 carry out a transformation which can be expressed as

$$\mathcal{H}_{Lb} = K_b\mathcal{H}_L \tag{5.5.75}$$

It is called the *Kadanoff transformation*.

As a result of this transformation the effective number of spins in the system is reduced by a factor b^{-d}:

$$N' = b^{-d}N \tag{5.5.76}$$

The *coarse-graining* resulting from the Kadanoff transformation can be applied repeatedly to obtain Hamiltonians for larger and larger blocks, till the correlation length is reached, and all the irrelevant variables have been integrated out.

Kadanoff argued that, so long as the size of the block is smaller than the correlation length, the block Hamiltonian is as good as the site Hamiltonian for describing the system in the vicinity of the critical point, and the two descriptions should have the same partition function. Therefore, in view of Eq. 5.5.76, the Gibbs free-energy densities before and after the Kadanoff transformation should be related by

$$g(h,t) = b^{-d}\, g(h',t') \tag{5.5.77}$$

We can assume that

$$t' = b^{A}t, \quad h' = b^{B}h, \tag{5.5.78}$$

so that

$$g(t,h) = b^{-d}\, g(b^{A}t, b^{B}h) \tag{5.5.79}$$

This equation expresses the homogeneous dependence of the free-energy density on distance. A similar expression can be obtained for the order-parameter autocorrelation function (cf. Eq. 5.5.67).

5.5.9 Renormalization-Group Theory

Since coarse-graining to the extent of the correlation length provides a description equivalent to that by the site-spin Hamiltonian, one can shrink the length L' of the block, at any stage of the iteration, back to the length L before the application of the Kadanoff construction: $L' = L/b$, or

$$L \to bL' \tag{5.5.80}$$

As a result of this rescaling, the spin-density changes to

$$s = \lambda_b s' \tag{5.5.81}$$

Kadanoff coarse-graining followed by rescaling, as above, is called a *renormalization-group (RG) transformation*, and is denoted by R_b. Its application changes the Hamiltonian as follows:

$$\mathcal{H}_b = R_b \mathcal{H} \tag{5.5.82}$$

The RG transformation is usually nonlinear, but the following relation holds:

$$R_{bb'} = R_b R_{b'} \tag{5.5.83}$$

The transformations R_b form a semigroup. It is not a full group because the inverse elements are missing.

Fixed Points

Eq. 5.5.83 indicates that application of a large number of successive RG transformations amounts to taking $b \to \infty$. If the Hamiltonian tends to an invariant value $\mathcal{H}^*$ as a result of this, we can write

$$\lim_{b \to \infty} R_b \mathcal{H} = \mathcal{H}^* \qquad (5.5.84)$$

Such a Hamiltonian is called a *fixed-point Hamiltonian.*

At a fixed point the system is invariant to a change of length scale, implying a correlation length that is either 0 or ∞. And the latter corresponds to a critical point.

The trajectory adopted by the Hamiltonian in reaching the fixed point corresponding to a phase transition carries information about critical exponents etc. The critical exponents are the same for all systems in the universality class defined by the critical surface containing the fixed point (cf. Huang 1987).

We stated in §5.5.2 that the Landau theory has two main flaws, namely neglect of fluctuations of the order parameter, and its assumption that the free energy is an analytical function of the order parameter at the critical point. The modification of the Landau potential by Ginzburg (§5.5.2) takes care of the first flaw, and Wilson's RG theory takes care of the second. The resulting Hamiltonian is commonly referred to as the *Landau-Ginzburg-Wilson (LGW) Hamiltonian.*

Stable fixed points for all Hamiltonians for phase transitions involving 4-, 6- and 8-component order parameters were investigated by Hatch and coworkers for all the crystallographic space groups (Kim, Stokes & Hatch 1986; Hatch, Kim, Stokes & Felix 1986; Stokes, Kim & Hatch 1987). All IRs at points of special symmetry, and all LGW Hamiltonians inducing transitions to commensurate phases were considered. In the book by Stokes & Hatch (1988), those Hamiltonians which have a stable fixed point, and can therefore lead to continuous phase transitions according to the RG theory, are marked by a double asterisk in Column 7 of Table I.

SUGGESTED READING

M. E. Fisher (1974). The renormalization group in the theory of critical behaviour. *Rev. Mod. Phys.*, **46**, 597.

S. K. Ma (1976). *Modern Theory of Critical Phenomena.* Benjamin / Cummings, Reading, Massachusetts.

J. Als-Nielsen & R. J. Birgeneau (1977). Mean field theory, the Ginzburg criterion, and marginal dimensionality of phase transitions. *Amer. J. Phys.*, **45**, 554.

P. Pfeuty & G. Toulouse (1977). *Introduction to the Renormalization Group and to Critical Phenomena.* Wiley, New York.

K. Huang (1987). *Statistical Mechanics*, second edition. Wiley, New York.

M. Toda, R. Kubo & N. Saito (1992). *Statistical Physics I: Equilibrium Statistical Mechanics*, second edition. Springer-Verlag, Berlin.

M. Plischke & B. Bergersen (1994). *Equilibrium Statistical Mechanics*, second edition. World Scientific, Singapore.

P. M. Chaikin & T. C. Lubensky (1995). *Principles of Condensed Matter Physics.* Cambridge University Press, Cambridge.

M. E. Fisher (1998). Renormalization group theory: Its basis and formulation in statistical physics. *Rev. Mod. Phys.*, **70**, 653.

5.6 SPONTANEOUS BREAKING OF SYMMETRY

The more theoretical physicists penetrate the ultimate secrets of the microscopic nature of the universe, the more the grand design seems to be ultimate symmetry and ultimate simplicity. But all of the interesting parts of the universe, at least to us, are, like the earth itself as well as our own bodies, markedly complex and markedly unsymmetric. In the most elementary sense, then, we are surrounded by "broken symmetry", the result undoubtedly of some sequence of catastrophes.

P. W. Anderson

The laws of motion do not prefer one direction in space over any other, but perch a ball symmetrically on the apex of a cone, and it will surely fall in one direction or the other. All the directions are equally probable, none has any special significance: but this symmetry will be hidden by the particular motion that results in any outcome governed by the law.

J. D. Barrow (1988)

The symmetry principle (Eq. C.1.6) states that the effect cannot be of a lower symmetry than the cause. It might appear that there is a violation of this principle when, on cooling or heating, a crystal makes a transition to a phase of lower symmetry. The "cause" (temperature) in this case is apparently a scalar, and therefore isotropic, and the effect (lowering of crystal symmetry) appears to be contrary to what we would expect from the symmetry principle. In reality there is no violation of the principle, and what we have here is an example of "spontaneous" breaking of symmetry (see, for example, Boccara 1981).

The concept of temperature is of a macroscopic and statistical nature. At a microscopic level a system undergoes thermal fluctuations (§5.5 and §E.2). These fluctuations increase in magnitude when the temperature of the continuous phase transition is approached. Each such fluctuation is a symmetry-breaking spontaneous perturbation, which is not at all isotropic, unlike the overall temperature itself. A symmetry-lowering phase transition occurs on varying the temperature if the system is *unstable*, rather than stable, under such spontaneous perturbations. Thus, even though temperature as a scalar and isotropic influence is present all through, it is not the only underlying cause. The overall cause incorporates the perturbing fluctuations also, and when these are taken into account, the symmetry principle is salvaged, as always.

5.6.1 Continuous Broken Symmetries; Goldstone Modes

Physical laws might have more symmetry than physical states.

L. Michel (1981)

The spontaneous breaking of symmetry, or the spontaneous emergence of an order parameter which breaks the symmetry of the parent configuration, is a central feature of most of the phase transitions. The symmetry broken may be either *continuous* or *discrete*. The notion of broken symmetries and of symmetry-restoring fluctuations called Goldstone modes, was first put forward for the case of continuous broken symmetries (Goldstone 1961).

To understand the basic ideas involved, consider a system with a Lagrangian density $\mathcal{L}$. Let G_0 be the (continuous) symmetry group of the system, so that $\mathcal{L}$ is invariant under its operations. The system may have either a single nondegenerate ground state, or more than one degenerate ground states. In the former case the single-valued ground state is invariant under the operations of G_0, and has the same symmetry as the Lagrangian density. For example, the s-state of a free atom possesses the full spherical symmetry of the potential.

However, if the system possesses degenerate ground states, then the possibility becomes available that some operation(s) of G_0 will transform one such state to *another*, rather than leaving it invariant. We then say that the symmetry of the Lagrangian density is spontaneously broken in this state.

Goldstone's theorem (in the nonrelativistic limit) states that when a system has spontaneously broken symmetry of the type described above, long-wavelength ($|\mathbf{k}| \to 0$) excitations exist, with a mode of frequency $\omega(\mathbf{k})$ which tends to zero in the limit $k \to 0$. This zero-frequency mode (the *Goldstone mode*) constitutes the limit of a continuous spectrum $\omega(\mathbf{k})$ of modes, the energy of which increases with increasing k .

Such considerations have played, and continue to play, a crucial role in the development of high-energy physics (see e.g. Mannheim (1986) and Gross (1995) for reviews).

Application of similar ideas to condensed-matter physics can be traced back to the work of Landau (see Anderson 1981; Chaikin & Lubensky 1995). The group G_0 introduced above now describes the symmetry of a medium that is anisotropic, homogeneous, time-symmetric, and, in certain cases, invariant under spin-rotations etc. As pointed out by Landau, the symmetry of a particular ground state of the condensed-matter system is often lower than G_0. An example is the symmetry of the ground state of a crystal of ice, compared to that of the liquid water from which it formed on freezing. The loss of translational and rotational symmetry of the fluid leads to the occurrence of $\omega(k \to 0) \to 0$ acoustic phonons.

Apart from the existence of such long-wavelength low-frequency dynamical modes, there are two other consequences of the spontaneous breaking of a continuous symmetry. One is the existence of a *rigidity* (e.g. the elastic modulus), and the other is the existence of *topological defects* (dislocations) in the broken-symmetry phase (see Chaikin & Lubensky 1995).

Landau (1937a, b, c, d) paid special attention to those situations in which the symmetry G of the ground state is a proper subgroup of G_0, and made the following very important observation:

Symmetry cannot change continuously.

Anderson (1981) calls this the *First Theorem of condensed matter physics.*

Landau also introduced in a general way the notion of the order parameter, a new thermodynamic variable the emergence of which at a phase transition heralds the broken-symmetry situation.

The magnitude of the order parameter is a measure of not only the *degree of broken symmetry*, but usually also of the degree of *ordering* of the system.

We consider now the example of a Heissenberg ferromagnet to illustrate some of the key features of such a broken-symmetry situation. Instead of treating the ferromagnetic phase as arising through a phase transition from a paramagnetic phase, we take the initial symmetry to be the continuous symmetry of the molten state of the material.

The ferromagnetic phase is governed by the Hamiltonian (cf. Eq. 5.5.61)

$$\mathcal{H} = - \sum_{<l.l'>} J_{ll'} \, \mathbf{s}_l \cdot \mathbf{s}_{l'} \tag{5.6.1}$$

Here $\mathbf{s}_l$ and $\mathbf{s}_{l'}$ are spin operators on sites l and l', and $J_{ll'}$ denotes the exchange interaction between spins on these two sites.

In the ferromagnetic phase, stable below a temperature T_c, a spontaneous magnetization $\mathbf{M}$ appears which has a certain specific direction locally, and there is thus a breaking of the initial isotropic symmetry G_0. However, all possible orientations of $\mathbf{M}$ correspond to the same energy, and a normal mode *costing zero energy to excite* can therefore exist, which results in the precession of the magnetization as a whole. This is the spin-wave mode with $\omega(k=0)=0$.

There exists thus a symmetry-restoring mode (corresponding to the precession of the magnetization as a whole) which sends the system through the entire set of degenerate states, and which costs zero energy to excite.

We note that this becomes possible because the broken symmetry is described by a *continuous group*, so that there are an infinite number of minima (ground states) for the free-energy density g. The separations of these minima in configuration space are arbitrarily small because G_0 is a continuous group, and the symmetry operators of G_0 lost on the formation of the ferromagnetic phase are the operators that map one such ground state to another.

Now the general condition for a minimum of to occur is (cf §5.3.12)

$$\frac{\partial^2 g}{\partial \eta^2} \geq 0 \tag{5.6.2}$$

Here η is the order parameter of the phase transition. Since the various minima are infinitesimally close to one another, we must have, for $T < T_c$:

$$\frac{\partial^2 g}{\partial \eta^2} = 0 \tag{5.6.3}$$

Fig. 5.6.1 depicts the difference between the situations described by $\frac{\partial^2 g}{\partial \eta^2} > 0$ and $\frac{\partial^2 g}{\partial \eta^2} = 0$. In both cases $\frac{\partial g}{\partial \eta} = 0$ at $\eta = \eta_0$, but in the latter case even the rate of variation of $\frac{\partial g}{\partial \eta}$ is zero at $\eta = \eta_0$; $g(\eta)$ has a *flat bottom* in the latter case.

Since the frequency ω of a mode is determined by the relationship

$$\omega^2 \sim \frac{\partial^2 g}{\partial \eta^2}, \tag{5.6.4}$$

the frequency of the symmetry-restoring fluctuation is thus also zero for all $T < T_c$.

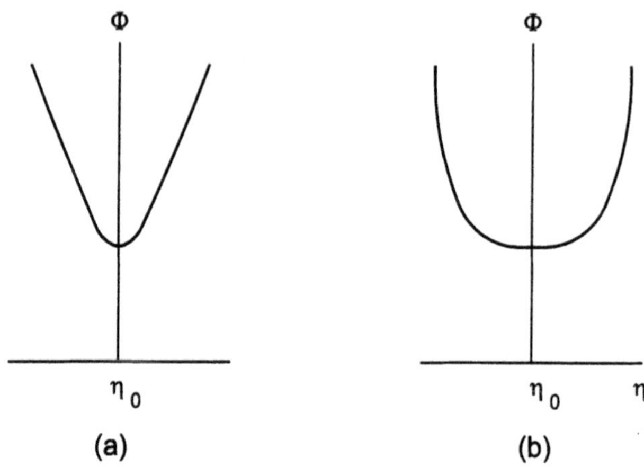

Figure 5.6.1: Dependence of free-energy density g on order parameter η. In both (a) and (b) a minimum occurs at $\eta = \eta_0$. In (b), however, even the second derivative of $g(\eta)$ is zero at $\eta = \eta_0$.

5.6.2 Discrete Broken Symmetries

In the case of a transition from one phase of a crystal to another, the initial or prototype symmetry is described by a *discrete* symmetry group. The extension of the broken-symmetry ideas to such cases was carried out by Blinc & Zeks (1974).

The fact that the initial symmetry group G_0 is a discrete group makes the following qualitative difference to the situation: There is now only a *finite* number of minima of the function $g(\eta)$. These minima, as before, can be mapped onto one another by those symmetry operators of G_0 which are lost on transition to the ferromagnetic or some other daughter phase. As we approach T_c from below, the magnitude of the order parameter decreases, and correspondingly the discrete minima come closer. If we are dealing with a *continuous* phase transition, $\omega \to 0$ continuously as $T \to T_c$ from below, and the minima of $g(\eta)$ merge into a single minimum at $T = T_c$.

It is important to note that we once again have the situation where $\partial^2 g/\partial\eta^2 = 0$, albeit only at $T = T_c$ (and not for all $T < T_c$, as is the case for continuous broken-symmetry). Thus:

> *For a continuous phase transition in a crystal, occurring at a temperature T_c, there is no difference at $T = T_c$ between the breaking of a continuous symmetry and a discrete symmetry.*

For $T < T_c$, however, the difference is of a fundamental nature. For the case of continuous broken symmetry, $\omega(\mathbf{k})$ is still zero at $k = 0$ (Fig.

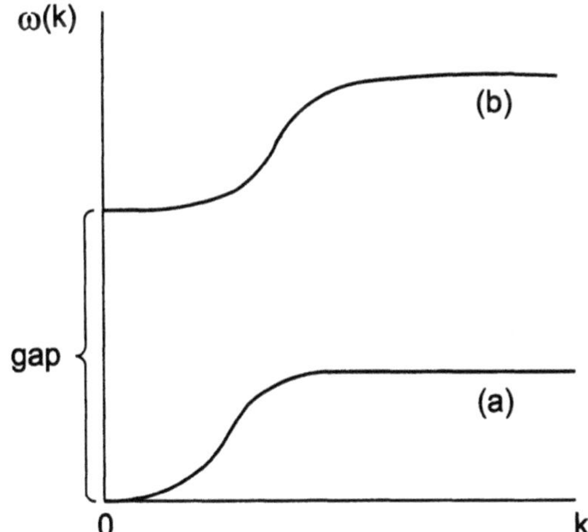

Figure 5.6.2: Dispersion curves for symmetry-restoring normal modes for $T < T_c$. Curve (a) is for the case of continuous broken symmetry, and curve (b) for a discrete broken symmetry. [After Blinc & Zeks (1974).]

5.6.2a). By contrast, for the case of discrete broken symmetry the frequency of the symmetry-restoring mode is no longer zero at $k = 0$ (Fig. 5.6.2b). This frequency does, of course, go to zero as $T \to T_c$ for a continuous phase transition. To summarize the above reasoning:

> *For a continuous phase transition, the existence of a mode of nonzero frequency, which becomes zero as $T \to T_c$, is a consequence of the breaking of a discrete symmetry* (Blinc & Zeks 1974).

The symmetry-breaking mode of the high-temperature phase becomes the symmetry-restoring mode of the low-temperature phase, provided the soft mode above T_c is nondegenerate. If it is degenerate, it splits into a number of modes when the symmetry is lowered at the phase transition, and the symmetry-restoring mode is then a linear combination of all these modes.

When a discrete symmetry is broken, there are only a finite number of ground states in the lower-symmetry phase, and there are, in general, no low-frequency excitations taking the system from one such state to another. Domain walls can be regarded as the elementary excitations separating the distinct ground states. There are no low-frequency hydrodynamic modes or low-energy excitations characterized by a rigidity, as there are for cases of breaking of continuous symmetries (see Chaikin & Lubensky 1995).

SUGGESTED READING

R. Blinc & B. Zeks (1974). *Soft Modes in Ferroelectrics and Antiferroelectrics.* North-Holland, Amsterdam.

P. W. Anderson (1981). Some general thoughts about broken symmetry. In N. Boccara (Ed.), *Symmetries and Broken Symmetries in Condensed Matter Physics.* IDSET, Paris.

P. W. Anderson (1984). *Basic Notions of Condensed Matter Physics.* Addison-Wesley, California. Chapter 2.

W. Ludwig & C. Falter (1988). *Symmetries in Physics.* Springer-Verlag, Berlin.

P. M. Chaikin & T. C. Lubensky (1995). *Principles of Condensed Matter Physics.* Cambridge University Press, Cambridge.

5.7 DISCONTINUOUS PHASE TRANSITIONS

In the context of the Landau theory of phase transitions, a continuous transition is one for which the order parameter rises continuously from the value zero as the system is taken away from the critical point into the ordered phase. Transitions for which the order parameter acquires a nonzero value in a discontinuous manner are called discontinuous phase transitions.

Two fundamentally different types of discontinuous transitions can be identified: those for which Guymont's nondisruption condition is obeyed (§5.1.1), and those for which it is not. For the former category, i.e. for NDPTs, the formalism of the Landau theory is applicable. For the latter category, which includes the so-called reconstructive phase transitions, as also most of the martensitic phase transitions, an order parameter cannot be defined in terms of the conventional Landau theory.

5.7.1 Nondisruptive Discontinuous Transitions

The classical Landau theory is applicable to NDPTs. If any of the conditions described in §5.3 is violated, it can only be a discontinuous or a first-order transition. We describe here some thermodynamic aspects of such transitions.

We start with Eq. 5.3.13, which can be rewritten as

$$[a'(T - T_0) + b\eta^2 + c\eta^4]\eta \; = \; 0 \tag{5.7.1}$$

Here T_0 is the stability limit of the higher-symmetry or disordered phase. And the coefficients a' and c are positive.

We have already considered in §5.3 the cases when $b > 0$ (continuous transitions), and $b = 0$ (tricritical points). We now take up the case when $b < 0$.

Apart from the solution $\eta = 0$ (which corresponds to the disordered phase), Eq. 5.7.1 has two real solutions:

$$\eta \; = \; \pm \left[-\frac{b}{2c} \left(1 + \sqrt{1 - \frac{4a'c}{b^2}(T - T_0)} \right) \right]^{\frac{1}{2}}, \tag{5.7.2}$$

and

$$\eta \; = \; \pm \left[-\frac{b}{2c} \left(1 - \sqrt{1 - \frac{4a'c}{b^2}(T - T_0)} \right) \right]^{\frac{1}{2}} \tag{5.7.3}$$

The solution given by Eq. 5.7.3 can be discarded. It corresponds to a free-energy *maximum*, rather than a minimum.

We next determine the stability limit of the ordered phase. It is determined by the condition expressed by Eq. 5.3.11. We rewrite it here, after assuming that $\beta = 0$, although the Landau condition (§5.3.12) does not require this for a discontinuous transition:

$$\chi_T^{-1} \; = \; \frac{\partial^2 g}{\partial \eta^2} \; = \; a + 3b\eta^2 + 5c\eta^4 \; > 0 \tag{5.7.4}$$

Note that, unlike for continuous phase transitions, we have not assumed here that $c = 0$. This is because $a = 0$ at the transition point, and $b < 0$; therefore Eq. 5.7.4 can be satisfied only if $c > 0$.

Substituting Eq. 5.7.2 into 5.7.4 we obtain

$$\chi_T^{-1} \; = \; \frac{b^2}{c} \sqrt{1 - \frac{4a'c}{b^2}(T - T_0)} \left[\sqrt{1 - \frac{4a'c}{b^2}(T - T_0)} \; + 1 \right] \tag{5.7.5}$$

The stability limit of the ordered phase is the temperature (T_0^-) above which χ_T^{-1} is no longer positive, i.e. $\chi_T^{-1}\,(T = T_0^-) = 0$. Eq. 5.7.5 gives

$$T_0^- \; = \; T_0 + \frac{b^2}{4a'c} \tag{5.7.6}$$

Thus T_0^- does not coincide with the stability limit T_0 of the disordered phase, and is higher than it. [We have assumed implicitly that the transition to the ordered phase occurs on cooling, rather than on heating.] Thus:

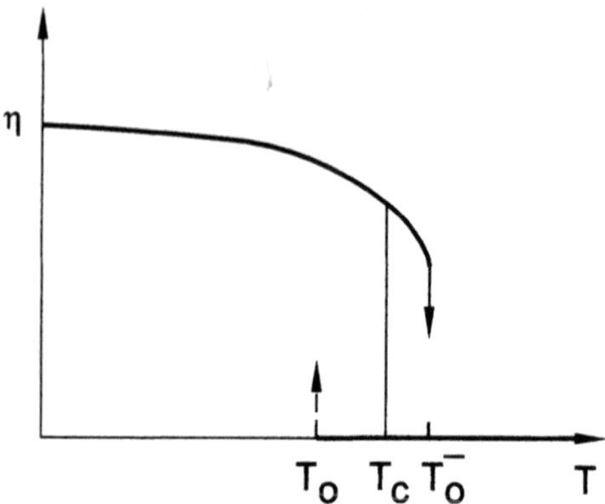

Figure 5.7.1: Temperature dependence of the order parameter for a discontinuous NDPT. See text for details.

For a discontinuous NDPT, the stability limits of the disordered phase and the ordered phase do not coincide.

If the two stability limits do not coincide, what is the phase transition temperature? It is the temperature T_c at which the free-energy curves of the two phases cross (cf. Fig. 1.1.1). It can be shown that (cf. Blinc & Zeks 1974):

$$T_c = T_0 + \frac{3}{4}\frac{b^2}{4a'c} \tag{5.7.7}$$

As we cool the crystal from the disordered phase, we have $\eta = 0$ up to $T = T_c$ because T_c is greater than the stability limit T_0 of the disordered phase. At T_c, η (discontinuously) rises to a nonzero value because the ordered phase has a lower free energy for $T < T_c$, and the ordered phase is stable for temperatures less than T_0^-, and $T_c < T_0^-$. Fig. 5.7.1 illustrates the relative values of the various temperatures involved.

Between the temperatures T_0 and T_0^- the two phases coexist. Coexistence of phases in a certain temperature range is a typical feature of discontinuous NDPTs[3].

5.7.2 Disruptive Discontinuous Transitions

We define disruptive phase transitions (DPTs) as those which violate Guymont's nondisruption condition (§5.1.1). They are necessarily of first order.

[3]It is important to point out here that, although there is a range of temperatures over which the two phases coexist, there is still a *sharp* transition at a temperature T_c. By contrast, in certain systems a *diffuse* transition occurs for which there is a *range* of T_c values. We consider ferroelectric transitions of this type in §10.3.

They include all reconstructive transitions. Most of the martensitic transitions are also of a reconstructive nature, and are therefore DPTs (cf. Fig. 5.2.2).

The classical Landau theory cannot be applied to DPTs. This is because they not only do not display a group-subgroup relationship between the phases involved, they do not permit the postulation of a prototype symmetry in the strict sense in which we have defined this concept in §5.1.

Are such transitions amenable to symmetry analysis at all? This question can be answered in the affirmative, particularly in view of the work of Toledano, Dmitriev and colleagues over the last decade (for some of the more recent accounts of this work, see Toledano & Dmitriev (1993, 1995, 1997) and Dmitriev & Toledano (1994)). They introduce a nonlinear (transcendental) "order parameter" in terms of stationary *density waves* which depend on the variational parameter associated with the mechanism of the transition. Emergence of such an order parameter results in a drastic upheaval of the crystal structure, and for certain specific values of the atomic displacements new crystallographic symmetry elements can arise. [Incidentally, this is a real-life example of the phenomenon of symmetrization depicted in Fig. C.2.1.] Needless to say, change of coordination numbers of atoms is the rule, rather than the exception.

An early attempt to rationalize symmetrization in a reconstructive phase transition was made by Cahn (1977) for explaining the fcc-bcc transition in the element Fe (also see Wadhawan (1985)). Cahn explained the origin of new symmetry elements in Fe crystals in terms of the dominant lattice strain characteristic of a martensitic transition (cf. Chapter 11). Emergence of the martensitic phase in the matrix of the parent phase leads to a large strain field. In accordance with the Curie principle, the net symmetry of the martensitic phase is the intersection of the symmetry of the unstrained martensitic phase and the symmetry of the strain field. The latter is either D_{2h} or $D_{\infty h}$. Therefore, normally the net symmetry of the strained phase cannot be higher than D_{2h} or $D_{\infty h}$, of which it is a subgroup. This can be called the process of *subgroup formation*, and is the same as the process of dissymmetrization described in Appendix C.

According to Cahn (1977), the opposite process of *supergroup formation*, or symmetrization, occurs in the case of Fe as follows: The atom-by-atom correspondence between the two phases of Fe is given the usual explanation for martensitic transitions, namely in terms of a unit-cell transformation, a homogeneous strain (the "Bain strain"), and "shuffles" (i.e. a periodic inhomogeneous strain). To understand the occurrence of the 4-fold axes of the bcc phase, the Bain strain is assumed to have symmetry $D_{\infty h}$ (rather than D_{2h}), with its unique axis parallel to a 4-fold axis of the fcc phase. Under the action of this lattice strain only this 4-fold axis survives. The other two 4-fold axes of the bcc phase cannot arise unless some additional mechanism is invoked. In the absence of such a mechanism, the resultant symmetry should be D_{4h} (body-centered tetragonal, bct), and not bcc. The bcc structure can arise as follows: The c/a ratio of the bct structure is also a function of the additional strain experienced by the growing embryo because of the crystalline environment. For a certain special configuration of the strain the c/a ratio can become unity and the four 3-fold axes necessary for the bcc structure can appear.

A similar analysis of the sequence of transitions in Fe has also been carried out by Toledano & Dmitriev (1993) in terms of the periodic character of the strain, which can be described in terms of the angle between the diagonals in the (110) cubic plane. Symmetry enhancement for special values of this angle is rather like that depicted in Fig. C.3.1 for a geometrical example. The "order parameter" in Toledano & Dmitriev's formalism is a sine function of this angle.

SUGGESTED READING

R. Blinc & B. Zeks (1974). *Soft Modes in Ferroelectrics and Antiferroelectrics.* North-Holland, Amsterdam.

K. Aizu (1991). On the variety of drastic transitions between solid phases. *Phase Transitions*, **34**, 19.

V. Dmitriev & P. Toledano (1994). Crystal geometry and phenomenological models of reconstructive phase transitions. *Phase Transitions*, **49**, 57.

P. Toledano & V. Dmitriev (1995). Theory of reconstructive phase transitions in crystals of the elements and related structures. *Key Engg. Materials*, **101-102**, 311.

5.8 TRANSITIONS TO AN INCOMMENSURATE PHASE

Incommensurate phases of crystals were described in §2.4 and 4.1.5. If the Lifshitz condition (§5.3.9) is not obeyed, we have a phase transition to an incommensurate phase.

Usually an incommensurate phase (I-phase) is straddled by a parent phase (P-phase) on the higher-temperature side and a commensurate (C) phase, or a *lock-in phase*, on the lower-temperature side. In other words, on cooling, the crystal undergoes the following sequence of transitions:

$$P \xrightarrow{T_i} I \xrightarrow{T_c} C \tag{5.8.1}$$

Here T_i is the temperature of transition to the I-phase, and T_c is the temperature at which the crystal locks-in to the C-phase.

The P-phase can be regarded as the prototypic phase for the lower-symmetry phase C commensurate to it, and a standard Landau-theory analysis can be carried out for the hypothetical P-C phase transition in terms of an order parameter labelled by a specific wavevector index $\mathbf{k}$. By contrast, the wavevector associated with the I-phase is not unique; it is strongly temperature dependent. The I-phase is therefore described by a continuous set of order parameters, which can often be interpreted in terms of a spatial modulation of the order parameter associated with the hypothetical P-C transition.

The spatial modulation of the order parameter can be modelled by including gradient invariants of the order parameter in the Landau expansion. According to the Lifshitz condition of the Landau theory (§5.3.9), such antisymmetric invariants (invariant with respect to the operations of the symmetry group of the P-phase) must be identically equal to zero if a phase transition is to be continuous and commensurate (cf. Kopsky & Sannikov 1977).

If the Lifshitz condition is not obeyed, a continuous phase transition associated with a given IR does not occur. What occurs is an incommensurate transition associated with some other IR.

The temperatures T_i and T_c in Eq. 5.8.1 usually differ by only a few degrees. Why do some crystals pass through an I-phase, rather than making a transition directly from the P-phase to the C-phase ? The answer lies in the *coupling* between *competing transition parameters.* This is particularly easy to visualize in terms of the soft-mode theory of phase transitions (§5.4). The various normal modes of vibration couple with one another to small or large extents. For example, a so-called improper ferroelectric transition occurs, not because the order parameter has the symmetry of a polarization component P, but because the order parameter, say (η, ξ), couples with P, and an invariant term of the type $\eta\xi P$ is allowed by symmetry in the Landau expansion. Similarly, the emergence of an internal macroscopic electric field associated with spontaneous polarization P may have the effect of splitting a phase transition, which would have otherwise occurred at a single temperature, into two neighbouring transitions because of the removal of degeneracy (Dvorak et al. 1975). I-phases may arise for similar reasons, except that the coupling terms are not of the type $\eta\xi P$, but rather involve gradients (cf. Eq. 5.8.2 below).

Incommensurate transitions can be classified in terms of the forms of the gradient invariants that can occur in the Landau expansion (Sannikov 1993). The inclusion of such gradient invariants in the Landau-Ginzburg expansion amounts to allowing the order parameter to be spatially inhomogeneous. This approach was adopted for magnetic helicoidal structures by

Dzyaloshinskii (1964a, b, c), and for ferroelectrics by Levanyuk & Sannikov (1976).

The most important gradient invariant is the *Lifshitz invariant* (or *L-invariant*):

$$\sigma\left(\xi\frac{\partial\eta}{\partial x}-\eta\frac{\partial\xi}{\partial x}\right) \tag{5.8.2}$$

It corresponds to a 2-component complex order parameter (η,ξ), transforming according to a 2-dimensional IR of the P-phase. An I-phase arising from an L-invariant is described as *Type I*. For it, η and ξ belong to the *same* IR, and the gradient invariant is *linear* in the spatial derivatives.

Gradient invariants are, however, possible for which η and ξ belong to *different* IRs, and only one of them (say η) is the order parameter. These have been called *Lifshitz-type invariants*, or *LT-invariants*, and the I-phase resulting from their existence is called *Type II*.

Other types of I-phases have been described by Sannikov (1993).

The presence of an L-invariant or an LT-invariant term in the Landau-Ginzburg expansion implies that the free-energy minimum at the transition point occurs, not for transition to a commensurate phase, but to a phase (the I-phase) with inhomogeneous displacements. By 'inhomogeneous displacements' we mean that no two unit cells of the P-phase undergo the same atomic displacements.

So much for the P-I transition. To account for the I-C transition (Eq. 5.8.1), we must include at least one more term which reflects the variation of the wavevector. It has the general form $(\eta^n \pm \xi^n)$, with n determining the translational symmetry of the C-phase. It can be rewritten as $\rho^n \cos n\phi$, and is called the *anisotropic term* because of its ϕ-dependence.

Near T_i (the temperature of the P-I transition) the amplitude ρ of the anisotropic term is small, and the solution minimizing the Landau-Ginzburg potential is sinusoidal.

As the temperature is decreased further, i.e. towards T_c, there is a competition between the L-invariant term and the anisotropic term, resulting in a change of the wavevector, and of the form of the modulation. Near T_c the modulation corresponds to wide domain-like regions of the C-phase, separated by narrow, shrinking, regularly-spaced, and transient "domain walls" (called *discommensurations*). Finally, below T_c the structure acquires the normal domain structure determined by the lost symmetry operators of the P-phase.

SUGGESTED READING

V. Kopsky & D. G. Sannikov (1977). Gradient invariants and incommensurate phase transitions. *J. Phys. C: Solid State Physics*, **10**, 4347.

Yu. A. Izyumov & V. N. Syromyatnikov (1984). *Phase Transitions and Crystal Symmetry*. Kluwer, Dordrecht.

R. Blinc & A. P. Levanyuk (Eds.) (1986). *Incommensurate Phases in Dielectrics*. North-Holland, Amsterdam.

G. Dolino (1988). Incommensurate phase transitions in quartz and berlinite. In S. Ghose, J. M. D. Coey & E. K. H. Salje (Eds.), *Structural and Magnetic Phase Transitions in Minerals*. Springer-Verlag, Berlin.

J. Kobayashi (1992). Incommensurate crystals and optical activity. *Condensed Matter News*, **1** (6), 17.

D. G. Sannikov (1993). New types of incommensurate phase transitions. *Crystallogr. Rep.*, **38** (4), 577.

P. Saint-Gregoire (1995). Ferroelastic incommensurate phases. *Key Engg. Materials*, **101-102**, 237.

5.9 INFLUENCE OF IMPURITIES ON STRUCTURAL PHASE TRANSITIONS

Real crystals are never perfect. The defects present may be of physical or chemical nature. We briefly review here the effect of low-concentration (less than 1 mole percent) impurities or dopants on the nature of phase transitions. Only structural phase transitions are considered here. The corresponding question in magnetic systems (spin glasses etc.) is discussed in Chapter 9.

The influence of impurities on a structural phase transition depends on a large number of factors, and the overall situation can be too complicated and diverse for a general theoretical analysis. Nevertheless some common trends can be seen for relatively simple, idealised, situations.

The classic paper by Halperin & Varma (1976) can be taken as marking the beginning of the present sophisticated theoretical analyses of the effect of defects on structural phase transitions. In general terms, the location of the defect in the crystal determines the nature of its coupling with the order parameter of the transition. The defect site may be interstitial or substitutional. And the defect may be either able to 'relax', or it may be a 'frozen' defect. In the former case it may order itself in a way that tends to match with the orientation of the order parameter, causing a 'stabilization' of the daughter phase, i.e. increasing the value of T_c. The opposite may often occur in the case of frozen defects, although the net outcome is a function of many parameters like: the time scale of the defect dynamics; whether or not the defect breaks the symmetry of the parent phase; the relative size of the defect; the average distance between defects; the nature of their interactions; etc. (Levanyuk & Sigov 1988).

Taking the results of conventional Landau theory as benchmarks, the influence of impurities and other defects in concentrations as small as $\sim 10^{18}$ defects/cm^3 may be more important than that of critical fluctuations, so far as the various anomalies related to the structural phase transition are concerned (Lebdev, Levanyuk & Sigov 1984).

Fig. 5.4.2 provides an example of this. From Vegard's law one would expect that, at least for very low values of impurity or dopant concentration x, T_c would vary linearly with x. What we see instead is a saturation effect or *plateau effect* (Salje, Bismayer, Wruck & Hensler 1991c; Salje 1993a, 1995b; Redfern & Schofield 1996; Hayward & Salje 1996).

According to Salje (1995b) the plateau effect, which occurs typically for $10^{-4} < x < 10^{-2}$, can be expected for practically all solid solutions. Its main basis is that, for very low values of x, the disturbance caused to the crystal structure may have a large *local* component (around the solute atom), rather than a strong long-ranged component encompassing the entire crystal. The result is the occurrence of short-ranged random fields, with competing interactions, a situation amenable, for example, to random-field Ising modeling (Imry & Ma 1975).

SUGGESTED READING

B. I. Halperin & C. M. Varma (1976). Defects and the central peak near structural phase transitions. *Phys. Rev. B*, **14**, 4030.

A. P. Levanyuk & A. S. Sigov (1988). *Defects and Structural Phase Transitions.* Gordon & Breach, New York.

E. K. Salje, U. Bismayer, B. Wruck & J. Hensler (1991c). Influence of lattice imperfections on the transition temperatures of structural phase transitions: The plateau effect. *Phase Transitions*, **35**, 61.

E. K. H. Salje (1995b). Chemical mixing and structural phase transitions: the plateau effect and oscillatory zoning near surfaces and interfaces. *Eur. J. Mineral.*, **7**, 791.

B. Hilczer (1995). Influence of lattice defects on the properties of ferroelectrics. *Key. Engg. Materials*, **101-102**, 95.

J. A. Krumhansl (1998). Defect induced behaviour in transforming materials. *Phase Transitions*, **65**, 109.

Chapter 6

CLASSIFICATION OF FERROIC MATERIALS. FERROGYROTROPY

In this chapter we describe a macroscopic (or thermodynamic) classification of ferroic materials, based mainly on the work of Aizu (1970a, 1970b, 1972a, 1973a). The concept of prototype symmetry (§5.1) plays a central role in this scheme.

The notions of optical and acoustical ferrogyrotropy were not included by Aizu in his formalism for ferroic materials. They were defined *formally* by Wadhawan (1979, 1982), and have their practical utility. Their recognition and formulation as distinct physical properties enables us to evolve a complete picture of ferroic materials. We discuss ferrogyrotropy in a separate section in this chapter.

6.1 FERROIC SPECIES

A ferroic phase of a crystal arises as a result of an actual or notional lowering of the point-group symmetry of the prototype. Since a point group can, in general, have more than one nontrivial or proper subgroups, a number of distinct group-subgroup pairs are possible for each prototype point group. Each such pair can describe a possible ferroic species (or *Aizu species*, as we call it sometimes). Aizu (1970a) derived 773 possible species of ferroic phases of crystals (212 of which are nonmagnetic) by making the following three stipulations:

- *Every time-symmetric point group can become the prototype point group in some species of ferroic crystals.*

- *When a prototype point group is specified, every proper subgroup of it can become the ferroic point group in some species with this prototype point group.*

- *When a prototypic and a ferroic point group are specified, all different ways in which the elements of the ferroic point group correspond to the elements of the prototype point group give so many possible species.*

A time-symmetric point group is one which includes time inversion as one of its elements.

Aizu's (1970a) derivation of the 773 ferroic species involves a straightforward enumeration of all the distinct subgroups of the various prototype point symmetries. A more compact and instructive approach to this problem can be through the use of irreducible property tensors (Jerphagnon, Chemla & Bonneville 1978; Paquet & Jerphagnon 1980; Kopsky 1979a,b).

6.1.1 Aizu Symbol for Ferroic Species

According to Aizu (1970a), ferroic crystals are said to belong to the same *species* if they have the same point-group symmetry, the same prototype symmetry, and the same correspondence between the symmetry elements of the prototype and ferroic point groups. He therefore introduced a compact symbol for each ferroic species, in which the letter F (standing for "ferroic") is put in the middle, the point group of the prototype is written at the left of F, and the point group of the ferroic phase is written to its right.

An example of the Aizu symbol is $\bar{4}2m1'Fm'm'2$. The symbol to the left of F has $1'$ included in it to denote the fact that the prototype is time-symmetric. The prototype is *always* time-symmetric, and it is customary to drop $1'$ from the full Aizu symbol when dealing with nonmagnetic ferroics.

In the above example there is no ambiguity about the fact that the m' planes of the ferroic phase can only correspond to the m' planes of the prototype, and the 2-fold axis of the ferroic phase can only originate from the $\bar{4}$-axis of the prototype.

An ambiguity *can*, however, arise in certain cases. An example of this is the symbol $\bar{4}2m1'F2'$. The $2'$-axis of the ferroic phase may correspond either to the principal axis (the $\bar{4}$-axis) of the prototype, or it may correspond to one of the $2'$-axes perpendicular to the principal axis. One makes this distinction by writing the Aizu symbol for the former case as $\bar{4}2m1'F2'(p)$, and for the latter case as $\bar{4}2m1'F2'(s)$, where p stands for "principal" and s for "side".

As emphasized by Aizu (1970a), these two situations provide two *distinct* ferroic species.

6.1.2 Orientation States

The change of directional or orientational symmetry at a ferroic phase transition implies that, in the ferroic phase in question, the crystal must have at least two equivalent states which have the same atomic structure, and differ only in their orientation (Fig. 1.1.2) (and possibly chirality) . Aizu (1962, 1969a) called them orientation states.

Since the prototype point-group symmetry P_0 is, by definition, a proper supergroup of the point group P of the ferroic phase, the validity of the following theorem (Aizu 1970a; Janovec 1972) is almost self-evident:

The number (n_d) of possible orientation states of a ferroic crystal equals the order of the prototype point group divided by the order of the ferroic point group:

$$n_d = \frac{|P_0|}{|P|} \tag{6.1.1}$$

6.1.3 F-Operations

In Eq. 6.1.1, $n_d \geq 2$ always. Let us first consider the case when $n_d = 2$. This means that a point operation p_2 exists ($p_2 \notin P$, $p_2 \in P_0$) such that

$$P_0 = P + p_2 P \tag{6.1.2}$$

In this case there are two orientation states of the ferroic crystal. Let us call them S_1 and S_2. We can take either of them (say S_1) as the "initial" state. The other state (S_2) is then obtained (under the PCA, cf. §5.1.2) by applying the operator p_2 on S_1. This is an example of an *F-operation*, with F standing for "ferroic" (Aizu 1970a).

In the general case, when $n_d \geq 2$, Eq. 6.1.2 becomes

$$P_0 = p_1 P + p_2 P + \cdots + p_{n_d} P \quad (p_1 = 1) \tag{6.1.3}$$

In this coset decomposition, $p_1, p_2, \cdots p_{n_d}$ ($\notin P$, except for p_1) constitute a *representative set of F-operations.* The choice of the members of this set is not unique.

SUGGESTED READING

K. Aizu (1970a). Possible species of ferromagnetic, ferroelectric, and ferroelastic crystals. *Phys. Rev. B*, **2**, 754.

6.2 MACROSCOPIC CLASSIFICATION OF FERROIC MATERIALS

Tensor properties like polarization, magnetization, strain, and compliance have translational invariance. Therefore it is not necessary to invoke the full space-group symmetry of the crystal for describing such properties, and it is sufficient to work at the point-group level. Since, by definition, a transition to a ferroic phase always involves a lowering of the prototype point symmetry, ferroic phases can be adequately classified in terms of their tensor properties.

6.2.1 Thermodynamic Considerations

Let us consider a ferroic crystal under the influence of an external electric field (E_i), a magnetic field (H_i), and a uniaxial stress (σ_{ij}) $(i, j = 1, 2, 3)$. Its generalized Gibbs free-energy density can be written as follows (Cady 1946; Nye 1957; Newnham 1974):

$$g = U - TS - E_i D_i - H_i B_i - \sigma_{ij} e_{ij} \tag{6.2.1}$$

Here S denotes entropy, D_i the electric displacement, B_i the magnetic induction, and e_{ij} the strain.

From the first law of thermodynamics, if a small amount of heat, dQ, flows into a unit volume of the system, and a small amount of work, dW, is done on it by the external forces, then the increase in its internal energy U is a perfect differential, given by

$$dU = dW + TdS, \tag{6.2.2}$$

where we have replaced dQ by TdS, in accordance with the second law of thermodynamics for a system undergoing a reversible change.

The work done per unit volume can be written as

$$dW = E_i dD_i + H_i dB_i + \sigma_{ij} de_{ij} \tag{6.2.3}$$

Differentiating Eq. 6.2.1, and substituting from Eqs. 6.2.2 and 6.2.3, we get

$$dg = -SdT - D_i dE_i - B_i dH_i - e_{ij} d\sigma_{ij} \tag{6.2.4}$$

Since we have taken T, (E_i), (H_i) and (σ_{ij}) as the independent variables on which g depends, we can write

$$dg = \left(\frac{\partial g}{\partial T}\right)_{E,H,\sigma} dT + \left(\frac{\partial g}{\partial E_i}\right)_{H,\sigma,T} dE_i + \left(\frac{\partial g}{\partial H_i}\right)_{E,\sigma,T} dH_i + \left(\frac{\partial g}{\partial \sigma_{ij}}\right)_{E,H,T} d\sigma_{ij} \tag{6.2.5}$$

Comparing Eqs. 6.2.4 and 6.2.5, we get

$$S = -\left(\frac{\partial g}{\partial T}\right)_{E,H,\sigma} \tag{6.2.6}$$

$$D_i = -\left(\frac{\partial g}{\partial E_i}\right)_{H,\sigma,T} \tag{6.2.7}$$

$$B_i = -\left(\frac{\partial g}{\partial Hi}\right)_{E,\sigma,T} \tag{6.2.8}$$

$$e_{ij} = -\left(\frac{\partial g}{\partial \sigma_{ij}}\right)_{E,H,T} \tag{6.2.9}$$

Eqs. 6.2.6 to 6.2.9 provide precise definitions of the dependent variables S, (D_i), (B_i) and (e_{ij}) when interplay between the various thermodynamic quantities has to be recognized explicitly.

We now write the electric displacement D_i as a sum of a *spontaneous* part $D_{(s)i}$, if any, and the *induced* parts that may arise from the presence of external fields:

$$D_i = D_{(s)i} + \epsilon_{ij}E_j + \alpha_{ij}H_j + d_{ijk}\sigma_{jk} + \cdots \tag{6.2.10}$$

Here ϵ_{ij}, α_{ij} and d_{ijk} are elements of, respectively, the dielectric permittivity tensor, the magnetoelectric tensor, and the piezoelectric tensor. We ignore higher-order contributions to the induced electric displacement vector, involving higher powers of E, H and σ and the cross terms between them.

Contributions to the total magnetic induction can be described similarly:

$$B_i = B_{(s)i} + \mu_{ij}H_j + \alpha_{ij}E_j + Q_{ijk}\sigma_{jk} + \cdots \tag{6.2.11}$$

Here μ_{ij} and Q_{ijk} are, respectively, the magnetic permeability and piezomagnetic coefficients.

The contributions to the strain term in Eq. 6.2.4 are:

$$e_{ij} = e_{(s)ij} + s_{ijkl}\sigma_{kl} + d_{kij}E_k + Q_{kij}H_k + \cdots \tag{6.2.12}$$

Here s_{ijkl} are coefficients of the elastic-compliance tensor.

It should be noted that we have used the same tensor coefficients α_{ij} in Eqs. 6.2.10 and 6.2.11, the same d_{ijk} in Eqs. 6.2.10 and 6.2.12, and the same Q_{ijk} in Eqs. 6.2.11 and 6.2.12. The justification for α_{ij}, for example, can be obtained readily as follows. We differentiate Eq. 6.2.7 with respect to H_j:

$$-\left(\frac{\partial^2 g}{\partial H_j\,\partial E_i}\right)_{\sigma,T} = \left(\frac{\partial D_i}{\partial H_j}\right)_{\sigma,T} \tag{6.2.13}$$

Next, we change the index i to j in Eq. 6.2.8, and differentiate it with respect to E_i:

$$-\left(\frac{\partial^2 g}{\partial E_i\,\partial H_j}\right)_{\sigma,T} = \left(\frac{\partial B_j}{\partial E_i}\right)_{\sigma,T} \tag{6.2.14}$$

Since the left-hand side in Eqs. 6.2.13 and 6.2.14 is the same, we get

$$\left(\frac{\partial D_i}{\partial H_j}\right)_{\sigma,T} = \left(\frac{\partial B_j}{\partial E_i}\right)_{\sigma,T} \equiv \alpha_{ij}^T, \tag{6.2.15}$$

which means that the coefficients of the magnetoelectric tensor are numerically the same as those of the converse magnetoelectric tensor.

Similarly, the coefficients of the converse piezoelectric-effect tensor can be shown to be numerically equal to those of the direct-effect tensor:

$$-\left(\frac{\partial^2 g}{\partial \sigma_{ij}\,\partial E_k}\right)_{T,H} = \left(\frac{\partial e_{ij}}{\partial E_k}\right)_{H,T} = \left(\frac{\partial D_k}{\partial \sigma_{ij}}\right)_{H,T} \equiv d_{kij}^T \tag{6.2.16}$$

Lastly, for the piezomagnetic tensor we have the following equivalence relationship:

$$-\left(\frac{\partial^2 g}{\partial \sigma_{ij}\,\partial H_k}\right)_{T,E} = \left(\frac{\partial e_{ij}}{\partial H_k}\right)_{E,T} = \left(\frac{\partial B_k}{\partial \sigma_{ij}}\right)_{E,T} \equiv Q_{kij}^T \tag{6.2.17}$$

6.2.2 Tensor Classification of Ferroics

Let us consider any pair (S_1, S_2) of the orientation states of a ferroic crystal. For S_1 we can write the free-energy density, dg_1, as given by Eq. 6.2.4, with Eqs. 6.2.10-12 providing the detailed contributions to D_i, B_i and e_{ij}. The expression for dg_1 can be integrated to obtain an expression for the free-energy density g_1 for the orientation state S_1. A similar expression can be obtained for g_2 for state S_2.

The difference Δg $(= g_2 - g_1)$ has the following form:

$$\begin{aligned} -\Delta g \;=\; & \Delta P_{(s)i}E_i + \Delta M_{(s)i}H_i + \Delta e_{(s)ij}\sigma_{ij} + \\ & \frac{1}{2}\Delta\epsilon_{ij}E_iE_j + \frac{1}{2}\Delta\mu_{ij}H_iH_j + \frac{1}{2}\Delta s_{ijkl}\sigma_{ij}\sigma_{kl} + \\ & \Delta\alpha_{ij}E_iH_j + \Delta d_{ijk}E_i\sigma_{jk} + \Delta Q_{ijk}H_i\sigma_{jk} + \\ & \frac{1}{6}\Delta\theta_{ijk}E_iE_jE_k + \cdots \end{aligned} \tag{6.2.18}$$

In writing this equation we have made use of the fact that $D^{(1)}_{(s)i} = \epsilon_0 E_i + P^{(1)}_{(s)i}$ for orientation state S_1, and $D^{(2)}_{(s)i} = \epsilon_0 E_i + P^{(2)}_{(s)i}$ for orientation state S_2, so that $D^{(2)}_{(s)i} - D^{(1)}_{(s)i} = P^{(2)}_{(s)i} - P^{(1)}_{(s)i} = \Delta P_{(s)i}$. Similarly, $B_i = \mu_i H_i + M_i$ by definition, and $\Delta B_{(s)i} = \Delta M_{(s)i}$. $\Delta P_{(s)i}$ thus denotes the *difference* (with proper sign) between the values of the ith component of spontaneous polarization between orientation states S_2 and S_1. Similarly for the other difference terms in Eq. 6.2.18. The θ_{ijk} in this equation are components of the third-rank electric permittivity tensor.

Eq. 6.2.18 is of central importance in the theory of ferroic materials. It provides the basis for the classification of these materials in terms of their tensor properties. We describe this classification now.

Full and Partial Primary Ferroics

If for at least one pair of orientation states, we have $\Delta P_{(s)i} \neq 0$ (for one or more values of i), the crystal is said to be a *ferroelectric.* [One can make a further distinction between actual and potential ferroelectrics, depending on whether or not it is possible to switch (reverse, or reorient by discrete amounts) the spontaneous polarization by applying an electric field lower than the electrical breakdown limit under "reasonable" experimental conditions. However, our definition of prototype symmetry and ferroic phase transitions in terms of the nondisruption condition is so strict that it is unlikely that the need for a distinction between actual and potential ferroics will ever arise.]

If $\Delta P_{(s)i}$ is nonzero for all pairs of orientation states, the crystal is said to be a in a *full* ferroelectric phase. If for at least one pair of states (but not all), $\Delta P_{(s)i} = 0,\ i = 1, 2, 3$, the crystal is said to be in a *partial* ferroelectric phase (Aizu 1970a, 1973a, Litvin 1984).

Full and partial *ferromagnetics* can be defined similarly by reference to the $\Delta M_{(s)i}$ term in Eq. 6.2.18.

If there are at least two orientation states which differ in at least one component of the spontaneous strain tensor, i.e. $\Delta e_{(s)ij} \neq 0$, the crystal is said to be in a *ferroelastic* phase. Like the distinction between real and potential ferroelectrics, we could impose the additional requirement of switchability of spontaneous strain for calling a crystal a ferroelastic. However, it is not very likely that a ferroelastic phase which arises in conformity with the nondisruption condition would resist ferroelastic switching, except perhaps when it is at a temperature far removed from the temperature of the ferroelastic phase transition.

If no two orientation states of the ferroelastic have all their corresponding nonzero spontaneous strain components identical, it is called a *full ferroelastic*; otherwise it is a *partial ferroelastic.*

Ferroelastics, ferromagnetics, and ferroelastics are called *primary ferroics.* For them, Δg is a linear function of either E, or H, or σ.

Secondary Ferroics

There can be six types of secondary ferroics:

Ferrobielectrics ($\Delta g \sim E^2$);

Ferrobimagnetics ($\Delta g \sim H^2$);

Ferrobielastics ($\Delta g \sim \sigma^2$);

Ferromagnetoelectrics ($\Delta g \sim EH$);

Ferroelastoelectrics ($\Delta g \sim \sigma E$);

Ferromagnetoelastics ($\Delta g \sim \sigma H$).

For secondary ferroics, Δg varies either as the square of the external field, or as one of their pairwise products.

In a *ferrobielastic* there is at least one pair of orientation states for which $\Delta s_{ijkl} \neq 0$ (cf. Eq. 6.2.18). A ferrobielastic is said to be only potentially so until a change of one orientation state to another ("switching") has been demonstrated experimentally by applying a suitable uniaxial stress to the crystal (at a suitable temperature) without causing its rupture.

Full and partial ferrobielastics are defined in a way similar to that for primary ferroics.

Ferrobielectrics and *ferrobimagnetics* are defined in terms of $\Delta\epsilon_{ij}$ and $\Delta\mu_{ij}$ respectively. Both (ϵ_{ij}) and (μ_{ij}) are polar second-rank tensors, and so is the spontaneous strain tensor (e_{ij}). Therefore, ferroelastic state shifts are always conterminous with ferrobielectric and ferrobimagnetic state shifts, although the former are brought about a uniaxial stress and the latter by E^2 or H^2.

If for at least two of the orientation states of the crystal, one or more of the magnetoelectric tensor components, α_{ij}, have different values, and if switching between the two states can be effected by a suitable combination of E_i and H_j, the crystal is said to be in a *ferromagnetoelectric* phase.

We note that even when $\alpha_{ij} \neq 0$ in Eq. 6.2.18 for a pair of orientation states, its contribution to Δg would be zero if either E_i or H_j is zero. The electric and the magnetic fields must be applied *simultaneously* for ferromagnetoelectric state shifts to occur.

We note further that a ferromagnetoelectric phase of a crystal need not also be a ferromagnetic-ferroelectric phase at the same time. The latter corresponds to the simultaneous presence of terms $\Delta M_{(s)ij}$ and $\Delta P_{(s)ij}$ in Eq. 6.2.18, and this may happen with or without the presence of the $\Delta\alpha_{ij}E_iH_j$ term governing ferromagnetoelectric behaviour.

Ferroelastoelectrics are defined in terms of Δd_{ijk}, and *ferromagnetoelastics* in terms of ΔQ_{ijk}.

Tertiary Ferroics

For tertiary ferroics,

$$\Delta g \sim E^3, H^3, \sigma^3, E^2\sigma, E^2h, E\sigma^2, EH^2, E\sigma H, \cdots \qquad (6.2.19)$$

For example, *ferrotrielastics* would be ferroics in which state shifts involve changes of free energy (or overcoming of enthalpy barriers) that are proportional to $\sigma_{ij}\sigma_{kl}\sigma_{mn}$.

The presence of lower-order ferroic behaviour can mask a higher-order effect. For example, it would be difficult to observe the ferrobielastic effect in a crystal that is also a ferroelastic with reference to the same pair of orientation states.

Demonstration of ferrotrielasticity may ordinarily require the absence of both ferroelasticity and ferrobielasticity. In any case, this question has to be addressed with reference to a specific pair of contiguous orientation states.

Stokes & Hatch (1988) have indicated, in Table I of their book, the primary or higher-order character of phase transitions arising due to points of special symmetry in the Brillouin zone (cf. §4.1.4). A more detailed description of higher-order ferroic properties, including a determination of their full or partial character etc., can be obtained easily from the computer code ISOTROPY (Stokes & Hatch 1998).

Order of a Ferroic State Shift

Aizu (1972a) has given definitions of orders of various types of ferroic state shifts.

If a shift (switching) from one orientation state to another involves a full primary ferroic property, it is called a *first-order state shift.* The change of free energy for such a state shift is proportional to the first power of the driving field.

Consider a full ferroelastic. In it, all mechanical state shifts will involve enthalpy barriers proportional to σ_{ij}.

Next consider a partial ferroelastic. Let us assume that the only external field present is a uniaxial stress; i.e. $\sigma \neq 0$, $E = H = 0$ in Eq. 6.2.18. In this case at least one, but not all, state shifts will be of *mechanical first order state shifts*. Other state shifts, not of this type, will be of second or higher orders, i.e. induced by σ^2, σ^3 terms, etc.

We next consider a situation where $\sigma \neq 0$, $E \neq 0$, and $H = 0$. For a nonmagnetic partial primary ferroic, some but not all its electrical and/or mechanical state shifts will be first-order. Aizu (1972a) introduced a so-called *electromechanical order of a state shift* in this context. If for a state shift, at least one $\Delta P_{(s)i}$ and $\Delta e_{(s)ij}$ is nonzero, its electromechanical order is unity. If $\Delta P_{(s)ij}$ and $\Delta e_{(s)ij}$ are all zero and at least one of $\Delta \epsilon_{ij}$, Δd_{ijk} and Δs_{ijkl} is nonzero, then it is an electromechnically second order state shift. Similarly for still higher orders.

The above definitions can be extended to cases where H also is nonzero.

For first-order state shifts, if the applied driving forces alternate in sign too rapidly, they become ineffective due to "inertial clamping" (Aizu 1973a).

This is generally not the case for second-order state shifts because they either depend on the square of the applied field (in which case they are independent of sign), or involve cross-terms like $E\sigma$, $H\sigma$ or EH. When cross-terms are involved, it is necessary to apply the two fields in a synchronous manner if their products are not to average out to zero. For the EH combination the synchronicity requirement is automatically satisfied when an electromagnetic field is used.

Illustrative Examples

We now discuss illustrations of some of the concepts described in this section.

Consider a crystal belonging to the ferroic species $mm2F2$. The groups $mm2$ and 2 are of orders 4 and 2 respectively, so that two orientation states are possible in the ferroic phase. Let us call these states S_1 and S_2 (Fig. 6.2.1(i)).

We work under the parent-clamping approximation (PCA) in this discussion (cf. §5.1.2), and choose a cartesian coordinate system fixed in the prototype such that the z-axis is along the diad of $mm2$, and the x- and y-axes are perpendicular to the two mirror planes.

The point group 2 of the ferroic phase is a polar group, so that a spontaneous polarization $\mathbf{P}_{(s)}$ can exist. The ferroic phase is thus at least pyroelectric. Is it also ferroelectric ? To answer this question, we first note that, in view of the point-group symmetry 2, the spontaneous polarization can only have the components (0, 0, $P_{(s)3}$), i.e. $P_{(s)1} = P_{(s)2} = 0$.

All the orientational states are equivalent, and any of them can be chosen as the "initial" state for determining the components of the various property tensors for each of the states. It is customary to choose S_1 as the initial state.

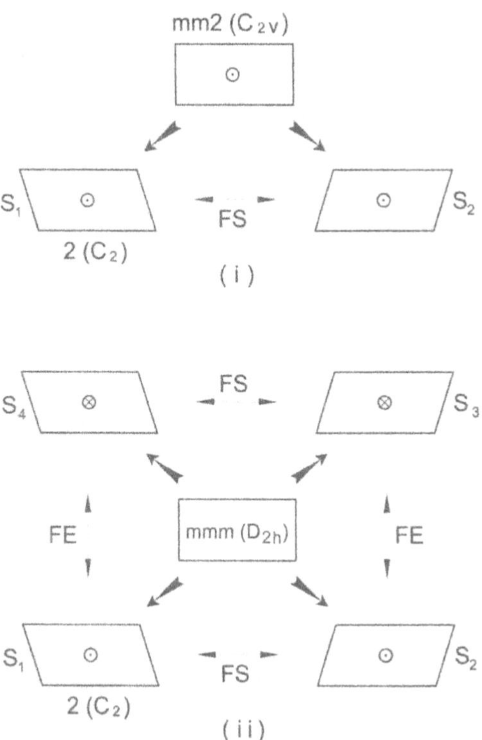

Figure 6.2.1: Ferroic state shifts possible in a monoclinic crystal of symmetry $2(C_2)$. In (i) the ferroic phase is taken as derived from a prototype symmetry $mm2(C_{2v})$. Only ferroelastic state shifts are possible in this case between the orientation states S_1 and S_2. (FS denotes " ferroelastic"). In (ii) the same monoclinic symmetry is taken as derived from a higher symmetry, namely $mmm(D_{2h})$. The possible number of orientation states is now four (S_1, S_2, S_3, S_4). Purely ferroelectric switching can now occur between S_1 and S_4, and between S_2 and S_3 (FE denotes "ferroelectric"). Similarly, purely ferroelastic switching can occur between S_1 and S_2, and between S_3 and S_4. State shifts between S_1 and S_3, and between S_2 and S_4, are simultaneously ferroelectric and ferroelastic, *even when only one kind of field (electric or mechanical) is applied.*

If (0, 0, $P_{(s)3}$) are the components of the spontaneous polarization for S_1, what are its components for S_2 ? These are determined by applying a representative set of F-operations on the matrix for the relevant property tensor. For the present example, a reflection in the yz-plane is the only distinct F-operation. It takes a material point (x, y, z) in S_1 to $(\bar{x}, y, z)$ in S_2. Under its action the spontaneous polarization tensor with components $(0, 0, P_{(s)3})$ is not changed at all. This means that $\Delta P_{(s)i} = 0$ for state shifts in crystals belonging to the ferroic species $mm2F2$, and such crystals are therefore not ferroelectric. They are only pyroelectric, as no electrical state shifts of first order are possible.

We next examine the behaviour of the polar second-rank property (ϵ_{ij}), the dielectric permittivity. Its symmetry-adapted form for S_1 is

$$\begin{bmatrix} \epsilon_{11} & \epsilon_{12} & 0 \\ \epsilon_{12} & \epsilon_{22} & 0 \\ 0 & 0 & \epsilon_{33} \end{bmatrix} \tag{6.2.20}$$

On applying the only distinct F-operation available for this species, namely a reflection across the yz-plane, we obtain the form of the dielectric-permittivity tensor for the orientation state S_2[1]:

$$\begin{bmatrix} \epsilon_{11} & -\epsilon_{12} & 0 \\ -\epsilon_{12} & \epsilon_{22} & 0 \\ 0 & 0 & \epsilon_{33} \end{bmatrix} \tag{6.2.21}$$

Therefore, in going from S_1 to S_2, we have $\Delta\epsilon_{12} = -2\epsilon_{12} \neq 0$. All other $\Delta\epsilon_{ij}$s are zero. Since at least one second-order electrical state shift is possible, the state shift is ferrobielectric. Such a state shift can, in principle, be effected by an electric field, **E**, of appropriate magnitude and direction.

We now focus our attention on the possibility of ferroelastic behaviour in this species.

The spontaneous strain tensor, $\mathbf{e}_{(s)}$, has the same transformation properties as the tensor $\boldsymbol{\epsilon}$, both being second-rank, polar, symmetric tensors. It follows that $\Delta e_{(s)12} = -2e_{(s)12} \neq 0$, and all other $\Delta e_{(s)ij}$ are zero. Ferroelastic behaviour is therefore allowed by symmetry for this species, and the state shifts are mechanically first order.

Similar arguments also hold for the magnetic permeability tensor. Thus ferroelastic, ferrobielectric, and ferromagnetic state shifts always occur concomitantly, and they are allowed by symmetry to occur for the species $mm2F2$.

[1]To understand how this form arises, we begin by noting the fact that dielectric permittivity is represented by a (polar) tensor of rank 2, and therefore its components must transform like the products of two coordinates (cf. Fumi's method of direct inspection in §3.2.3). Under the considered F-operation, $(x, y, z) \rightarrow (\bar{x}, y, z)$. Therefore: $x^2 \rightarrow x^2$, $y^2 \rightarrow y^2$, $z^2 \rightarrow z^2$, $xy \rightarrow -xy$, $yz \rightarrow yz$, and $zx \rightarrow -zx$. Therefore, for Eq. 6.2.20 the only change is that $\epsilon_{12} = \epsilon_{21} \rightarrow -\epsilon_{12}$.

Fig. 6.2(ii) depicts another situation in which the point-group symmetry of the ferroic phase is still 2, but the prototype happens to have the symmetry mmm (D_{2h}), instead of $mm2$ for Fig. 6.2(i). Since mmm is a group of order 8, the possible number of distinct orientation states is now 8/2 or 4. We call them S_1, S_2, S_3, S_4. There is a one-to-one correspondence between them and the cosets in the following coset decomposition:

$$(mmm) \;=\; (2_z) \;+\; m_x\,(2_z) \;+\; i\,(2_z) \;+\; 2_x\,(2_z) \tag{6.2.22}$$

The F-operation m_x is the same as in the previous example and takes S_1 to S_2, resulting in a ferroelastic state shift.

The inversion operation i in Eq. 6.2.22 takes S_1 to S_3. This state shift is simultaneously ferroelastic and ferroelectric.

The remaining F-operation, 2_x, takes S_1 to S_4 in a purely ferroelectric state shift.

Thus a phase belonging to the $mmmF2$ ferroic species is not only pyroelectric, but also ferroelectric. (It is also ferroelastic, ferrobielectric and ferrobimagnetic.) Some state shifts are purely ferroelectric ($S_1 \leftrightarrow S_4$; $S_2 \leftrightarrow S_3$), some are purely ferroelastic ($S_1 \leftrightarrow S_2$; $S_3 \leftrightarrow S_4$); some are both ($S_1 \leftrightarrow S_3$; $S_2 \leftrightarrow S_4$).

The state shifts $S_1 \leftrightarrow S_3$ and $S_2 \leftrightarrow S_4$ involve an inversion operation. These were defined by Aizu (1972a) as *mechanically infiniteth order*. They are also always electrically odd order. In $mmmF2$ they are electrically first order.

In the case of the species $mmmF222$ (Aizu 1972a), the state shifts between the two possible orientation states are electrically third order, mechanically infiniteth order, and electromechanically second order. Since both mmm and 222 belong to the same crystal family (orthorhombic), this is a nonferroelastic species. The F-operation here is an inversion operation:

$$(mmm) \;=\; (222) \;+\; i\,(222) \tag{6.2.23}$$

This F-operation, though not affecting the strain tensor, changes the handedness or chirality of the crystal structure. The two orientation states therefore differ in the optical gyration tensor, which governs optical activity. The ferroic species $m\bar{3}mF432$ is another example of this type. Crystals belonging to such species were called *ferrogyrotropic* by Wadhawan (1979, 1982). Aizu did not recognize this as a separate, distinct, ferroic property. We go into the reasons of this in the next section.

SUGGESTED READING

W. G. Cady (1946). *Piezoelectricity*. McGraw-Hill, New York.

J. F. Nye (1957). *Physical Properties of Crystals*. Clarendon Press, Oxford.

K. Aizu (1972a). Electrical, mechanical and electromechanical orders of state shifts in nonmagnetic ferroic crystals. *J. Phys. Soc. Japan*, **32**, 1287.

K. Aizu (1973). Second-order ferroic state shifts. *J. Phys. Soc. Japan*, **34**, 121.

R. E. Newnham (1974). Domains in minerals. *American Mineralogist*, **59**, 906.

J. Jerphagnon, D. Chemla & R. Bonneville (1978). The description of the physical properties of condensed matter using irreducible tensors. *Adv. Phys.*, **27**, 609.

V. Kopsky (1979a, b). Tensorial covariants of the 32 crystal point groups. *Acta Cryst.*, **A35**, 83 and 95.

6.3 FERROGYROTROPY

Several optically active ferroic crystals are known to exhibit a reversal of the sign of spontaneous optical rotatory power in certain directions when a suitable driving field is applied. This is the property of optical ferrogyrotropy (Wadhawan 1979, 1982).

At certain phase transitions involving the loss of a symmetry operator, an axial second-rank *i*-tensor develops non-zero components. The optical gyration tensor is such a tensor. Such transitions are therefore described as *gyrotropic phase transitions* (Konak, Kopsky & Smutny 1987; Wadhawan 1979).

Apart from the optical gyration tensor, another axial second-rank tensor is the linear magnetoelectric susceptibility tensor (Freeman & Schmid 1975). However, unlike the optical gyration tensor, it is not an *i*-tensor, but a *c*-tensor. It is thus identically equal to zero in nonmagnetic crystals. Therefore, as suggested by Wadhawan (1979), optical ferrogyrotropy should be defined only with reference to the optical gyration tensor, even for crystals which can exhibit the linear magnetoelectric effect.

It has been generally believed till recently that linear birefringence, if present for light propagating in a given direction, normally makes it difficult to measure the considerably smaller effect of circular birefringence (optical activity) in the same direction. The development of an apparatus called the *High Accuracy Universal Polarimeter* (HAUP) by Kobayashi & Uesu (1983) has overcome this difficulty. With the HAUP one can measure accurately *all* the coefficients of optical activity, linear birefringence, and rotation of the indicatrix of any crystal, even for a crystal belonging to low-symmetry systems like monoclinic and triclinic. This development has made it possible to make accurate measurements on optical ferrogyrotropy.

6.3.1 The Optical Gyration Tensor

Optical activity arises from the spatial dispersion of the dielectric permittivity tensor (see, e.g., Ramachandran & Ramaseshan 1961; Jerphagnon & Chemla 1976; Glazer 1988). The spatial dispersion of the permittivity tensor implies that the crystal structure is such that the polarization **P** at a point in the crystal depends not only on the electric field **E** at that point, but also on the field in the neighbourhood of that point (Devarajan & Glazer 1986). For such a crystal the dielectric permittivity exhibits not only the usual dispersion with respect to frequency, but also with respect to the wavevector **k** of the electromagnetic wave. One can thus write the following Taylor expansion in powers of **k** (Juretschke 1974; Portigal & Burstein 1968):

$$\epsilon_{ij}(\omega, \mathbf{k}) = \epsilon_{ij}(\omega) + \eta_{ijl}(\omega)k_l + h_{ijlm}k_l k_m + \cdots \tag{6.3.1}$$

Optical activity (optical gyration) arises from the first-order terms $\eta_{ijl}k_l$.

The electrical field energy stored in a medium is given by

$$W_e = \frac{1}{2}\mathbf{E}\cdot\mathbf{D} \tag{6.3.2}$$

If we assume that the medium is nonattenuating, the requirement of conservation of stored field energy can be written as follows (Landau & Lifshitz 1958):

$$W_e = W_e^{\dagger} \tag{6.3.3}$$

This provides the following permutation symmetry for the permittivity tensor:

$$\epsilon_{ij}(\omega, \mathbf{k}) = \epsilon_{ji}^{*}(\omega, \mathbf{k}) \tag{6.3.4}$$

The stored free energy must also have time-reversal symmetry (Portigal & Burstein 1968; Juretschke 1974):

$$W_e(t) = W_e(-t), \tag{6.3.5}$$

which leads to

$$\epsilon_{ij}(\omega, \mathbf{k}) = \epsilon_{ji}(\omega, -\mathbf{k}) \tag{6.3.6}$$

Substitution of Eqs. 6.3.4 and 6.3.6 in Eq. 6.3.1 reveals the following symmetry of the optical gyration tensor:

$$\eta_{ijl}^{*}(\omega) = -\eta_{ijl}(\omega) \tag{6.3.7}$$

This means that each component of this tensor is pure imaginary, so that we can introduce a real tensor $(\gamma_{ijl}(\omega))$ as follows:

$$\eta_{ijl}(\omega) \equiv i\,\gamma_{ijl}(\omega) \tag{6.3.8}$$

We can now rewrite Eq. 6.3.1 as follows:

$$\epsilon_{ij}(\omega,\,\mathbf{k}) = \epsilon_{ij}(\omega) + i\,\gamma_{ijl}(\omega)\,k_l + h_{ijlm}\,k_l\,k_m + \cdots \tag{6.3.9}$$

To get a further feel for the fact that the second term on the right-hand side in this equation indeed corresponds to *spatial* dispersion, we shall now arrive at it through a somewhat different route by using the direct (literal) interpretation of spatial dispersion, namely the rate of variation of the electric field with space coordinates:

$$D_i = \epsilon_{ij}\,E_j + \gamma_{ijl}\,\frac{\partial E_j}{\partial x_l} \tag{6.3.10}$$

Assuming the propagation of a plane wave, we have

$$\frac{\partial E_j}{\partial x_l} = i\,E_j\,k_l, \tag{6.3.11}$$

which means

$$D_i = (\epsilon_{ij} + i\,\gamma_{ijl}\,k_l)\,E_j, \tag{6.3.12}$$

or

$$D_i = \epsilon_{ij}\,(\omega,\,\mathbf{k})\,E_j, \tag{6.3.13}$$

where

$$\epsilon_{ij}(\omega,\,\mathbf{k}) = \epsilon_{ij}(\omega) + i\,\gamma_{ijl}(\omega)\,k_l, \tag{6.3.14}$$

which is the same as Eq. 6.3.9.

Going back to the question of the intrinsic symmetry of the gyration tensor, it follows from Eqs. 6.3.4, 6.3.6 and 6.3.9 that

$$\gamma_{jil} = -\,\gamma_{ijl} \tag{6.3.15}$$

Because of the antisymmetry of this tensor with respect to the first two indices, we can substitute it with a lower-rank (second-rank) tensor (g_{ml}) such that

$$\gamma_{ijl} = u_{ijm}\,g_{ml} \tag{6.3.16}$$

Here (u_{ijm}) is the Levi-Civita tensor or the alternating tensor (see Juretschke 1974). It is a fully antisymmetric unit tensor.

Let (l_i) denote a unit vector along the direction of propagation of the electromagnetic beam. The optical rotatory power, ρ, of the medium along this direction can be shown to be proportional to the pseudoscalar G, called the *gyration*, and defined by (see Nye 1957):

$$G = g_{mk}\, l_m\, l_k \tag{6.3.17}$$

Since G is a pseudotensor, (g_{mk}) must be a pseudotensor or axial tensor.

We can finally write Eq. 6.3.9 as follows, retaining only the (first-order) terms responsible for optical activity:

$$\epsilon_{ij}(\omega, \mathbf{k}) = \epsilon_{ij}(\omega) + i\, u_{ijm}\, G_m, \tag{6.3.18}$$

where

$$G_m = g_{mk}\, l_k \tag{6.3.19}$$

Both the tensors (γ_{ijl}) and (g_{ml}) in Eq. 6.3.16 have the same maximum number (9) of independent components, as they should. The natural optical rotatory power of crystals is described by the *symmetric* part of the tensor (g_{ml}), which can have only 6 independent components (see Nye 1957; Konak et al. 1978). The antisymmetric part of the tensor describes the polar or "weak" optical activity (Zheludev 1978; Sirotin & Shaskolskaya 1982), and accounts for the remaining three components.

The Jahn symbol for the tensor governing natural optical activity is $\epsilon[V^2]$.

6.3.2 The Hermann Theorem and Optical Gyration

We have seen above that natural optical gyration can be described either in terms of the third-rank tensor (γ_{ijl}) or in terms of the second-rank tensor (g_{ml}) (cf. Eq. 6.3.16). Is the optical gyration tensor a second-rank tensor $(r = 2)$ or a third-rank tensor $(r = 3)$? This question has to be faced when we want to make use of the Hermann theorem of crystal physics (§3.3). It turns out on detailed analysis that we have to take $r = 2$ to draw correct conclusions from the Hermann theorem.

Consider the question of optical activity in crystal classes $3m$, $4mm$ and $6mm$. For these classes, $N \geq 3$ (cf. the statement of the Hermann theorem in §3.3). The theorem tells us that for tensors with $r = 2$ these classes have the same symmetry as the point group ∞m. And for this point group all components of the tensor $\mathbf{g}$ are zero (see, e.g., Sirotin & Shaskolskaya 1982). Optical activity should thus be absent in the crystal classes $3m$, $4mm$ and $6mm$. This conclusion is in agreement with the results of detailed analysis carried out by applying the Neumann theorem. We could not have been able to apply the Hermann theorem if we had taken $r = 3$ for this problem.

For similar reasons, natural optical activity should be absent in the crystal classes $\bar{6}$, $\bar{6}m2$ and $\bar{4}3m$ (Sirotin & Shaskolskaya 1982). Here also, taking $r = 2$ (rather than 3), we can replace the 3-fold axes in these point groups by ∞-fold axes. The resulting point groups are ∞/m, ∞/mm and $\infty\infty m$, which are all centrosymmetric, and therefore do not allow the occurrence of optical gyration.[2]

Assumption of the lower of the two possible ranks for the optical gyration tensor always leads to correct results when the Hermann theorem is applied.

6.3.3 Optical Ferrogyrotropy as an Implicit Form of Ferroicity

We combine Eqs. 6.3.13 and 6.3.18 to obtain

$$D_i \;=\; \epsilon_{ij}(\omega)\, E_j \;+\; i\,(\mathbf{E} \times \mathbf{G})_i \tag{6.3.20}$$

When this is substituted in Eq. 6.3.2, an interesting result is obtained (Juretschke 1974):

$$W_e \;=\; \frac{1}{2}\, E_i\, \epsilon_{ij}\, E_j \;+\; 0 \tag{6.3.21}$$

Thus the first-order spatial dispersion term contributes nothing to the stored field energy. This has an important implication in the context of ferroic state shifts.

Suppose two orientation states differ in one or more components of the optical gyration tensor. Application of an external field can develop an enthalpy difference between the two states. However, the above result (Eq. 6.3.21) means that the contribution to this enthalpy difference from the spatial-dispersion term is zero. The driving force for switching one domain state to the other must therefore be provided by some other term occurring in Eq. 6.2.18. This fact, first pointed out by Wadhawan (1979), can be stated as follows:

> *Ferrogyrotropic state shifts cannot be mediated by the gyrotropic tensor. They can only be effected through an accompanying, explicitly ferroic, property tensor.*

Each term in Eq. 6.2.18 represents an explicitly ferroic property. There is no term possible in it for exclusively ferrogyrotropic state shifts because ferrogyrotropy is only an implicit form of ferroicity.

Nevertheless, ferrogyrotropic state shifts do occur in many ferroics. We list several examples here, grouping them under the accompanying explicitly ferroic property.

---

[2]What the Hermann theorem is saying here, in effect, is that, even though the crystal classes $\bar{6}$, $\bar{6}m2$ and $\bar{4}3m$ are not centrosymmetric, they are *effectively* centrosymmetric so far as optical activity is concerned. Such a result, though quite striking, is not unusual in crystal physics. A similar, albeit trivial, example of this type is that of density: it is a centrosymmetric property, even in noncentrosymmetric crystals.

Ferroelectrics. TGS (Shuvalov, Aleksandrov & Zheludev 1959; Hermelbracht & Unruh 1970; Kobayashi, Uesu & Takehara 1983); $LiH_3(SeO_3)_2$ (Futama & Pepinsky 1962a,b); $Pb_5Ge_3O_{11}$ (Iwasaki & Sugi 1971); and $NaNO_2$ (Chern & Phillips 1972). General papers dealing with changes in optical activity resulting from polarization reversal include those by Pepinsky (1962, 1963), Shuvalov & Ivanov (1964), and Aizu (1964a, b, c).

We consider the case of $NaNO_2$ to illustrate some features of the coupling between ferroelectric and ferrogyrotropic state shifts (Wadhawan 1982).

$NaNO_2$ belongs to the ferroic species $mmmFmm2$. As both the prototype and the ferroic phase belong to the same crystal family (orthorhombic), the ferroic phase is not ferroelastic.

The possible number of orientation states is 2, and the ferroic phase is polar. This means that ferroelectric switching is permitted by symmetry considerations, the two orientation states differing in the sign of the spontaneous polarization component P_3.

The two orientation states also differ in the sign of the gyration-tensor component g_{12} (Nye 1957), or, through a similarity transformation, in the signs of the components g_{11} and g_{22}. The change of sign is brought about by the F-operation which is either inversion or a reflection across the plane $z = 0$.

The gyration surface for a crystal of point-symmetry $mm2$ is shown in Fig. 6.3.1 (Shubnikov 1960). It is a polar plot of the pseudoscalar G (optical rotatory power), defined by Eq. 6.3.17 in the coordinate system defined by the principal axes:

$$G = -g_{11} l_1^2 + g_{22} l_2^2 \qquad (6.3.22)$$

The gyration surface is shown as white for directions along which G has a positive sign, and black when it has a negative sign.

The state shifts for the species $mmmFmm2$ are first-order electric, ∞th-order mechanical, and first-order electromechanical (Aizu 1972a). The ferroelectric state shifts are responsible for the concomitant ferrogyrotropic state shifts depicted in Fig. 6.3.1.

Ferroelastics. Dicalcium strontium propionate (DSP) (Sawada, Ishibashi & Takagi 1977; Wadhawan 1979). This crystal is also a ferroelectric below 281.7 K.

Ferromagnetics. No experimental investigation appears to have been reported, but a number of ferromagnetic crystals have been predicted to exhibit ferrogyrotropic state shifts (Wadhawan 1979): Ni – Cl boracite ($Ni_3B_7O_{13}$); Ni – I boracite; and $BaMnO_4$.

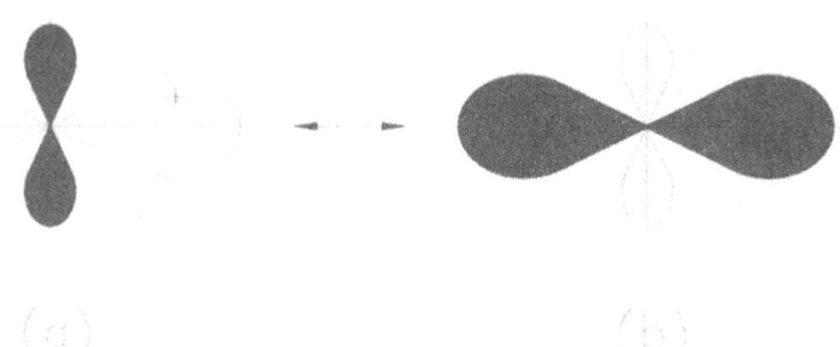

Figure 6.3.1: The optical gyration surface, and its ferroic switching, for a crystal belonging to the species $mmmFmm2$. The ferrogyrotropic state shift from (a) to (b), or vice versa, corresponds to a reflection operation in the plane of the diagram. According to Eq. 6.3.22, such an F-operation changes the sign of the gyration tensor. This is represented by colour reversal (black to white, and white to black) in going from (a) to (b) and vice versa.

Secondary and Higher-Order Ferroics. Gyrotropic phase transitions accompanied by the onset of neither ferroelectricity nor ferroelasticity have been described by Konak et al. (1978) as *pure gyrotropic phase transitions.* The ferroic phase in such cases can only undergo second or higher-orders of electric, mechanical, or electromechanical state shifts. Several crystals which undergo such transitions have been identified (Cross & Newnham 1974; Konak et al. 1978): ZrOS; CO; N_2O; Na_2ThF_6; AlF_3; $Cs_3As_2Cl_9$; TeO_2; $RbBeF_3$; and HgClBr. All of them, except ZrOS, are ferroelastoelectrics. ZrOS belongs to the ferroic species $\bar{4}3mF23$, for which the electric order of state shifts is 5 or higher, mechanical order is 3 or higher, and the electromechanical order is 3 or higher.

Other ferroelastoelectrics that exhibit ferrogyrotropic state shifts are $CuCsCl_3$ (Hirotsu 1975; Sano, Ito & Nagata 1986); and $(C_5H_{11}NH_3)_2ZnCl_4$ (Cuevas et al. 1984).

6.3.4 Optical Ferrogyrotropy vs. Ferroelasticity

The optical gyration tensor (g_{ij}) is a second-rank axial i-tensor. Under an F-operation represented by a matrix $\mathbf{A}$ its components transform as follows:

$$g'_{ij} = (\det \mathbf{A})\, a_{ik}\, a_{jl}\, g_{kl} \tag{6.3.23}$$

Here the a's are elements of the matrix $\mathbf{A}$, and det $\mathbf{A}$ = +1 if the F-operation is of the proper type, and det $\mathbf{A} = -1$ if the F-operation is an improper operation like reflection or inversion.

The spontaneous strain tensor, which governs ferroelastic state shifts, transforms as follows:

$$e'_{ij} = a_{ik}\, a_{jl}\, e_{kl} \qquad (6.3.24)$$

Unlike the gyration tensor, it is invariant under an inversion operation.

It follows from Eqs. 6.3.23 and 6.2.24 that for proper F-operations, i.e. those involving only pure rotations (no reflection or inversion), ferrogyrotropic state shifts are not distinct from ferroelastic state shifts. In view of the fact that ferroelasticity is an explicitly ferroic property, whereas ferrogyrotropy can occur only as an adjunct to some other explicitly ferroic property, it is appropriate not to regard such state shifts as ferrogyrotropic, and regard them only as ferroelastic.

For F-operations that are improper coordinate transformations, ferrogyrotropic state shifts are indeed distinct from ferroelastic state shifts.

The case of DSP provides an illustration of this (Wadhawan 1979; Glazer, Stadnicka & Singh 1981). It belongs to the nonmagnetic ferroic species $m\bar{3}mF422$ at room temperature. We can write the following direct product:

$$(m\bar{3}m) = (422) \times (3) \times (\bar{1}) \qquad (6.3.25)$$

The point groups $(m\bar{3}m)$ and (422) are of orders 48 and 8, so that 6 orientation states are possible. Since the spontaneous strain tensor is a symmetric second-rank polar tensor, it is invariant under space-inversion. It follows from the presence of the group $(\bar{1})$ on the right-hand side of Eq. 6.3.25 that the number of ferroelastically distinct orientation states is only half of the maximum number (6) of orientation states possible.

This crystalline phase is thus a *partial* ferroelastic phase. Let us call the ferroelastically distinct orientation states $S_1(e)$, $S_2(e)$, $S_3(e)$. It is clear from Eq. 6.3.25 that they correspond to the three elements of the group (3), namely 1, 3, and 3^2.

For the gyration tensor, on the other hand, all the six states are distinct. Three of these, say $S_1(g)$, $S_2(g)$, $S_3(g)$, correspond to the F-operations 1, 3, 3^2, and are thus identical to $S_1(e)$, $S_2(e)$, $S_3(e)$. These can be described in terms of tensor components as follows (Nye 1957):

$$S_1(g) = \text{diag}\,(g_{11}\; g_{11}\; g_{33}) \qquad (6.3.26)$$

$$S_2(g) = \text{diag}\,(g_{33}\; g_{11}\; g_{11}) \qquad (6.3.27)$$

$$S_3(g) = \text{diag}\,(g_{11}\; g_{33}\; g_{11}) \qquad (6.3.28)$$

Here diag$(g_{11}\; g_{11}\; g_{33})$ denotes a diagonal matrix with g_{11}, g_{11}, g_{33} as the diagonal components.

The remaining three ferrogyrotropic orientation states are obtained by applying the inversion operation to S_1, S_2, S_3:

$$S_4(g) = \text{diag}\,(-g_{11}\; -g_{11}\; -g_{33}) \qquad (6.3.29)$$

$$S_5(g) = \text{diag}\,(-g_{33} \;\; -g_{11} \;\; -g_{11}) \tag{6.3.30}$$

$$S_6(g) = \text{diag}\,(-g_{11} \;\; -g_{33} \;\; -g_{11}) \tag{6.3.31}$$

State shifts *within* the set S_1, S_2, S_3, or within the set S_4, S_5, S_6, are only ferroelastic, and are not ferrogyrotropic. State shifts from one set to the other are ferrogyrotropic; they are also ferroelastic if, say, S_1 goes to S_5 or S_6. But S_1 going to S_4 does not constitute a ferroelastic state shift. All these state shifts can be effected by applying an appropriate uniaxial stress.

State shifts in which no gyration component changes sign (like $S_1(g) \leftrightarrow S_2(g)$, or $S_4(g) \leftrightarrow S_6(g)$) are indistinguishable from the corresponding ferroelastic state shifts, and should not be referred to as ferrogyrotropic state shifts.

In the light of the above discussion, we can now formulate the following definition of a ferrogyrotropic phase of a crystal (Wadhawan 1979, 1982):

> *An optical ferrogyrotropic crystal is a ferroic (primary or of higher order), at least two orientation states of which have optical gyration tensors differing in the signs of one or more of their corresponding components.*

If the optical gyration tensors differ only in the magnitudes (and not the signs) of one or more of their corresponding components, the state shifts are *not* ferrogyrotropic; they are *ferroelastic* state shifts.

6.3.5 Partial Ferrogyrotropics

The example of DSP considered above is very suitable for introducing the notion of partial ferrogyrotropy.

On cooling, DSP undergoes a phase transition to a ferroelectric phase at 281.7 K, and its point-symmetry lowers from 422 to 4 (Kobayashi & Yamada 1962). This phase thus belongs to the ferroic species $m\bar{3}mF4$.

For this species, 48/4 or 12 orientation states are possible. It is thus a partial ferroelectric partial ferroelastic phase. It is also a *partial ferrogyrotropic* phase because the number of distinct gyration-tensor states is still 6 (and not 12), the same as in the phase above 281.7 K, which belongs to the ferroic species $m\bar{3}mF422$. This follows from the decomposition of the prototype point group $m\bar{3}m$:

$$(m\bar{3}m) = (4) \times (3) \times (2/m) \tag{6.3.32}$$

Kobayashi & Yamada (1962) reported the occurrence of occasional "spontaneous" racemization in this phase, with the tetragonal c-axes in the *dextro* and *laevo* domains remaining parallel. This can be understood in terms of ferrogyrotropic $d \leftrightarrow l$ state shifts, brought about by the F-operation m in $2/m$ in Eq. 6.3.32. The arrangement of symmetry elements implies that the number of distinct ferrogyrotropic states in only 6.

The concept of partial ferrogyrotropy described here runs parallel to that of partial ferroelasticity and partial ferroelectricity defined by Aizu (1970a).

6.3.6 The Acoustical Gyration Tensor

Acoustical activity (or acoustogyrotropy) is the mechanical analogue of optical activity. But whereas optical activity is governed by a third-rank pseudotensor, acoustical activity is governed by a fifth-rank pseudotensor. The higher rank of the latter results in a greater diversity and complexity of behaviour. In particular, all the 21 noncentrosymmetric crystal classes are acoustically active, whereas optical activity can occur in only 15 of them.

One of the manifestations of acoustical activity is that in a crystal that possesses this property, the plane of polarization of a transverse acoustic wave traveling along the acoustic axis (the mechanical analogue of an optic axis) gets progressively rotated as the wave passes through the crystal. If the propagation direction is different from an acoustic axis, the transverse and the longitudinal components generally get coupled in a complex way. In the optical case, by contrast, the longitudinal component is missing, and the effect of optical activity (circular birefringence) is simply superimposed on that of linear birefringence. However, in spite of the complexity of behaviour, use of acoustical activity as a diagnostic tool offers some distinct and unique advantages over optical activity, and interest in its study and use continues to grow (Pine 1970; Belyi 1982; Bialas & Schauer 1982).

Just as optical activity arises from first-order spatial dispersion of the dielectric permittivity tensor, acoustical activity is taken as arising from first-order spatial dispersion of the elastic-stiffness tensor. The meaning of spatial dispersion is that the interaction between stress (σ_{ij}) and strain (e_{kl}) is not only through the elastic-stiffness tensor (c_{ijkl}), but also through other terms involving nonlocal interactions:

$$\sigma_{ij} = c_{ijkl}\, e_{kl} + d_{ijklm} \frac{\partial e_{kl}}{\partial x_m} \tag{6.3.33}$$

The tensor (d_{ijklm}) is the acoustical gyration tensor.

Since (σ_{ij}) and (e_{kl}) are generally assumed to be symmetric tensors, the acoustical gyration tensor has the following intrinsic symmetry:

$$d_{ijklm} = d_{jiklm} = d_{ijlkm} \tag{6.3.34}$$

As in the optical case, the spatial-dispersion effects are taken into account by allowing for a wavevector dependence of the elastic-stiffness tensor, apart from its usual frequency dependence (Portigal & Burstein 1968):

$$c_{ijkl}(\omega, \mathbf{k}) = c_{ijkl}(\omega) + i\, d_{ijklm}(\omega)\, k_m + \cdots \tag{6.3.35}$$

Assuming time-reversal symmetry and causality, the tensor **d** can be shown to have the following antisymmetry (Portigal & Burstein 1968):

$$d_{ijklm}(\omega) = -\, d_{klijm}(\omega) \tag{6.3.36}$$

A similar antisymmetry of the third-rank optical gyration tensor enabled us to introduce an equivalent tensor of lower (second) rank. In the acoustical case also, a lower (fourth) rank tensor has been introduced by Bhagwat, Subramanian & Wadhawan (1983), which is completely equivalent to the fifth-rank tensor **d**.

The new tensor, **G**, is defined through the following equation:

$$G_{qkmn} = \frac{1}{2} u_{ilq}\, d_{iklmn} \tag{6.3.37}$$

Because of the fact that $u_{liq} = -u_{ilq}$, and because of Eq. 6.3.36, the following intrinsic symmetry of **G** follows from Eq. 6.3.37:

$$G_{qkmn} = G_{qmkn} \tag{6.3.38}$$

Another set of constraining equations can be derived by putting $q = k$ in Eq. 6.3.37, summing over k, and using Eqs. 6.3.34 and 6.3.36. We get

$$\sum_k G_{kkmn} = 0 \quad \text{for each } m, n = 1, 2, 3 \tag{6.3.39}$$

Because of Eq. 6.3.38, the tensor **G** has only 54 nonzero components, and these are subject to 9 constraining equations (Eq. 6.3.39). The net number of independent components is therefore 45. This number is the same as that for the tensor **d** (which has the intrinsic symmetry embodied in Eqs. 6.3.34 and 6.3.36).

By employing some tensor algebra described by Bhagwat et al. (1983), Eq. 6.3.37 can be inverted to yield

$$d_{iklmn} = u_{ilq}\, G_{qkmn} + u_{kmq}\, G_{qiln} \tag{6.3.40}$$

It can be readily verified that if we take Eq. 6.3.40 as the *defining* equation for **d** in terms of a tensor **G** satisfying Eqs. 6.3.38 and 6.3.39, then **d** satisfies Eqs. 6.3.34 and 6.3.36.

The tensors **d** and **G** are thus completely equivalent, and acoustical activity can be described equally well by either of them. Dealing with **G** offers the advantage that it is of a lower rank than **d**, and thus somewhat more convenient to use.

More importantly, taking $r = 4$ (rather than 5) for the acoustical activity tensor leads to correct predictions when the Hermann theorem is applied, just like the case of natural optical activity.

The 45 nonvanishing independent components of the fifth-rank tensor **d** have been identified by Kumaraswamy & Krishnamurthy (1980). The corresponding work for the fourth-rank tensor **G** has been carried out by Bhagwat et al. (1986).

Srinivasan (1988) has derived the form of the "irreducible" spectrum of a general fifth-rank tensor, using the formalism developed by Jerphagnon, Chemla & Bonneville (1978). The intrinsic symmetry of the tensor **d** is then brought in to determine the specific forms of the irreducible parts describing acoustical activity in various crystal classes.

6.3.7 Ferroacoustogyrotropy

The notion of acoustical ferrogyrotropy (or ferroacoustogyrotropy) was introduced by Wadhawan (1982) by analogy with that of optical ferrogyrotropy. A ferroacoustogyrotropic crystal is an acoustically active ferroic crystal in which at least one direction exists such that the acoustical rotatory power along this direction undergoes either a switching of sign, or a finite discrete jump in magnitude, or both, when a suitable driving force is applied.

Switching of sign, with no change of magnitude of acoustical rotatory power, has been described as *pure ferroacoustogyrotropy* by Bhagwat, Wadhawan & Subramanian (1986).

Waterman (1959) examined the existence of *pure mode axes* in crystals. If a certain direction is to be a pure mode axis for the propagation of acoustic waves, the longitudinal polarization component must be independent of the transverse polarization component, and vice versa. This requires that the crystal symmetry be such that certain components of the elastic-stiffness tensor and the acoustic gyration tensor are zero. In the absence of spatial dispersion the following two conditions are sufficient for a propagation direction x_3 in a crystal to be a pure mode axis:

- x_3 should be a proper axis of 2-fold or higher rotational symmetry; or
- x_3 should be normal to a reflection plane, or normal to a proper axis of 6-fold symmetry.

An analysis of how these conditions get modified in the presence of spatial dispersion has been given by Bhagwat et al. (1986). They have shown that pure mode axes continue to be pure mode axes even in the presence of spatial dispersion if they are along axes of 2-fold or higher rotational symmetry. However, the same may not be the case if the pure mode axis in the absence of spatial dispersion is perpendicular to a mirror plane of symmetry, or perpendicular to a 6-fold axis.

An *acoustic axis* is a *degenerate* pure mode axis (Waterman 1959; Portigal & Burstein 1968). Pure transverse acoustic waves traveling along an acoustic axis are not only completely decoupled from the pure longitudinal mode, but also have the same phase velocity in the absence of spatial dispersion.

The term *pure acoustical activity* has been introduced by Bhagwat et al. (1986) for describing acoustical activity along an acoustic axis. These authors introduce a coordinate transformation, so that the x_3 axis is along the direction of propagation of the acoustic wave. The following expression is derived for the rotation of the plane of polarization of a plane-polarized transverse acoustic wave on traversing a length l along the acoustic axis:

$$\phi = \omega^2 \, l \, \rho \, G_{3333} \, / \, 2c_{2323}^2 \tag{6.3.41}$$

Thus $\phi = 0$ if $G_{3333} = 0$. And this happens for the following eight crystal classes:

$\bar{4}$, $4mm$, $\bar{4}2m$, $3m$, $\bar{6}$, $6mm$, $\bar{6}m2$, $\bar{4}3m$.

In addition, there are 11 centrosymmetric point groups, for which $G_{3333} = 0$. Let us call this set of 19 point groups *Set A*. Crystals belonging to any class in Set A cannot show pure acoustical activity.

It is also found from the list of nonzero independent components of **G** tabulated by Bhagwat et al. (1986) that $G_{3333} \neq 0$ for the following eight crystal classes:

4, 422, 3, 32, 6, 622, 23, 432.

Let us call this *Set B*.

Coming back now to the question of pure ferroacoustogyrotropy, a pure ferroacoustogyrotropic state shift is one which involves only a change of sign of ϕ (Eq. 6.3.41), without any change of its magnitude. For this, ϕ should be identically equal to zero in the prototypic phase, and nonzero in the ferroic phase. This would be so if the prototypic point group is from among the members of Set A, and the ferroic point group belongs to Set B and is a proper subgroup of the prototype. There are 29 ferroic species of this type (Bhagwat et al. 1986):

$m\bar{3}mF4$, $m\bar{3}mF422$, $m\bar{3}mF3$, $m\bar{3}mF32$, $m\bar{3}mF23$, $m\bar{3}mF432$,

$m3F3$, $m3F23$, $\bar{4}3mF3$, $\bar{4}3mF23$,

$6/mmmF3$, $6/mmmF32$, $6/mmmF6$, $6/mmmF622$,

$6/mF3$, $6/mF6$, $\bar{6}m2F3$, $\bar{6}m2F32$, $\bar{6}F3$, $6mmF3$, $6mmF6$,

$\bar{3}mF3$, $\bar{3}mF32$, $3mF3$, $\bar{3}F3$,

$4/mmmF4$, $4/mmmF422$, $4/mF4$, $4mmF4$

Examples of crystals belonging to one or the other of these species include DSP, $KAlO_2$, $CsCuCl_3$, TeO_2, ZrOS and $Pb_5Ge_3O_{11}$ (see Bhagwat et al. (1986) for more details).

α-quartz is not mentioned in this list because, although it is ferroacoustogyrotropic, it is not *pure*-ferroacoustogyrotropic (Bhagwat et al. 1983).

6.3.8 Acoustical Ferrogyrotropy as an Implicit Form of Ferroicity

Changes of stored enthalpy during *optical* ferrogyrotropic state shifts get their entire contribution from terms other than the one responsible for optical activity (cf. Eq. 6.3.21). A similar statement applies to acoustical ferrogyrotropic state shifts. This can be seen by computing the contribution, say V, from the acoustical activity term in Eq. 6.3.35 to the potential energy of deformation per unit volume (Bhagwat, Subramanian & Wadhawan 1983):

$$V = \frac{1}{2} d_{ijklm} \, k_m \, e_{ij} \, e_{kl} \tag{6.3.42}$$

This can be written in terms of the fourth-rank tensor **G** by using Eq. 6.3.40:

$$V = \frac{1}{2} \left[u_{ikq} G_{qjlm} + u_{jlq} G_{qikm} \right] k_m \, e_{ij} \, e_{kl} \tag{6.3.43}$$

All the indices in this equation are repeated indices, and therefore a summation over all of them is carried out. It makes no difference as to what symbol is used for an index over which summation has to be carried out. We interchange the indices i and k, and also j and l, to get

$$V = \frac{1}{2} \left[-u_{ikq} G_{qljm} - u_{jlq} G_{qkim} \right] k_m \, e_{kl} \, e_{ij} \tag{6.3.44}$$

The tensor **G** satisfies Eq. 6.3.38. Therefore we can rewrite Eq. 6.3.44 as

$$V = -\frac{1}{2} \left[u_{ikq} G_{qjlm} + u_{jlq} G_{qikm} \right] k_m \, e_{kl} \, e_{ij} \tag{6.3.45}$$

Comparison of Eqs. 6.3.43 and 6.3.45 leads to the result $V = -V$, or

$$V = 0 \tag{6.3.46}$$

Thus, like the optical case, acoustical ferrogyrotropy is also an implicit form of ferroicity. It can therefore be exhibited by only those crystals which are explicitly ferroic with respect to some other tensor property. For example, such state shifts in α-quartz are mediated either by ferrobielasticity or by ferroelastoelectricity.

SUGGESTED READING

D. L. Portigal & E. Burstein (1968). Acoustical activity and other first-order spatial dispersion effects in crystals. *Phys. Rev.*, **170**, 673.

J. Jerphagnon & D. S. Chemla (1976). Optical activity of crystals. *J. Chem Phys.*, **65**, 1522.

C. Konak, V. Kopsky & F. Smutny (1978). Gyrotropic phase transitions. *J. Phys. C: Solid State Phys.*, **11**, 2493.

I. S. Zheludev (1978). The optical activity of crystals induced by an electric field (electrogyration). *Ferroelectrics*, **20**, 51.

V. K. Wadhawan (1979). Gyrotropy: An implicit form of ferroicity. *Acta Cryst.*, **A35**, 629. The term 'gyrotropy' used in this paper was later changed to 'ferrogyrotropy' to emphasize the switchability of the gyrotropy property (see Wadhawan 1982).

Yu. I. Sirotin & M. P. Shaskolskaya (1982). *Fundamentals of Crystal Physics.* Mir Publishers, Moscow.

A. Gomez Cuevas, J. M. Perez Mato, M. J. Tello, G. Madariaga, J. Fernandez, Lopez Echarri, F. J. Zuniga & G. Chapuis (1984). *Phys. Rev. B*, **29**, 2655.

A. M. Glazer & K. Stadnicka (1986). On the origin of optical activity in crystal structures. *J. Appl. Cryst.*, **19**, 108.

V. Devarajan & A. M. Glazer (1986). Theory and computation of optical rotatory power in inorganic crystals. *Acta Cryst.*, **A42**, 560.

K. V. Bhagwat, V. K. Wadhawan & R. Subramanian (1986). A new fourth-rank tensor for describing the acoustical activity of crystals. *J. Phys. C: Solid State Phys.*, **19**, 345.

T. P. Srinivasan (1988). A description of acoustical activity using irreducible tensors. *J. Phys. C: Solid State Phys.*, **21**, 4207.

J. Kobayashi, T. Asahi, S. Takahashi & A. M. Glazer (1988). Evaluation of the systematic errors of polarimetric measurements: Application to measurements of the gyration tensor of α-quartz by the HAUP. *J. Appl. Cryst.*, **21**, 479.

Z. Zikmund & J. Fousek (1989). Stabilization of polarization vector in uniaxial ferroelectrics by means of chiral dopants. Macroscopic symmetry conditions. *Phys. Stat. Solidi* (a), **112**, 625.

J. Kobayashi (1991). Applications of optical activity to studies of phase transitions. *Phase Transitions*, **36**, 95.

Chapter 7

DOMAINS

The orientational, chiral, and translational variants present in a given specimen of a ferroic material, alongwith the interfaces separating the variants, constitute its *domain structure*. We arbitrarily divide the discussion of domain structure into two parts, and put off a detailed discussion of the interfaces to Chapter 8. In this chapter we focus attention predominantly on the characteristics of the variants or domain states themselves.

Further, Chapters 7 and 8 deal mainly with the broad, common, features of domain structure of ferroics. Properties specific to the domain structure of ferromagnetics are described in Chapter 9. Similarly, Chapters 10, 11 and 12 deal, among other things, with features of domain structure peculiar to ferroelectrics, ferroelastics, and secondary and higher-order ferroics respectively.

The various domain states of a ferroic have the same crystal structure, but differ in their mutual orientation, chirality, and/or location. Therefore, when referred to a common coordinate system, they may possess different tensor coefficients. Analysis of the relationship among these tensor components enables us to devise experiments for observing and investigating the various domains. It also enables us to compute average tensor properties of polydomain materials.

The tensor classification of ferroic materials, discussed in Chapter 6, is also based on the underlying presence of domain structure. An advantage of this classification is that it provides a direct correspondence between domain structure and free energy, from which one can determine the optimum configuration of external fields needed for manipulating the domain structure to advantage.

7.1 SOME SYMMETRY ASPECTS OF DOMAIN STRUCTURE

A symmetry analysis of domain structure helps establish the mutual relationships and regularities, and makes it easier to understand the observed domain structure.

7.1.1 Derivative Structures and Domain States

Derivative structures arise as a result of loss of symmetry operators. It is generally very useful to interpret a given derivative structure as having been derived from an adequately selected prototype symmetry (§5.1).

A domain state is a homogeneous portion of a given derivative structure, which may normally coexist, either with a different phase of the crystal, or with other domain states having the same crystal structure but different orientations, chiralities, and/or locations. *Variants* is another term for domain states (van Tendeloo & Amelinckx 1974).

7.1.2 Domain Pairs

An unordered pair of domain states, D_i, D_k, considered irrespectively of their coexistence, constitutes a domain pair.

Domain pairs have an algebraic aspect, and a geometric aspect. Algebraically, they constitute an *unordered set* (Janovec 1972; Zikmund 1984; Janovec, Richterova & Litvin 1992):

$$\{D_i, D_k\} = \{D_k, D_i\} \tag{7.1.1}$$

Geometrically, they amount to a *superposition* of domain states D_i and D_k. Therefore the symmetry group, J_{ik}, of a domain pair has two types of contributions:

- the symmetry group H_i of D_i; and
- an operator j_{ik} of the prototype group G, which transforms D_i to D_k:

$$J_{ik} = H_i + j_{ik} H_k \tag{7.1.2}$$

Here

$$H_i = \{g \in G \,|\, gD_i = D_i\} \tag{7.1.3}$$

Domain pairs are similar in concept to "dichromatic complexes" (§7.3.4). The latter concept (Pond & Vlachavas 1983) has a more general range of applicability, and applies, for example, even to grain boundaries in ceramics.

Transposable Domain Pairs

If there exists an operation $j_{ik} \in G$ such that $j_{ik}D_i = D_k$ and $j_{ik}D_k = D_i$, then $\{D_i,\ D_k\}$ is a transposable or ambivalent domain pair.

Nonferroelastic and Ferroelastic Domain Pairs

Let $\mathbf{e}^{(i)}$ and $\mathbf{e}^{(k)}$ be the spontaneous deformations, or spontaneous strains, of the domain states D_i and D_k, referred to a common coordinate system. The two domain states constitute a nonferroelastic domain pair if $\mathbf{e}^{(i)} = \mathbf{e}^{(k)}$, and a ferroelastic domain pair if $\mathbf{e}^{(i)} \neq \mathbf{e}^{(k)}$.

7.1.3 Single-Domain States

A phase transition from a prototype structure results in the formation of a heterogeneous aggregate of homogeneous regions (domains). The nature of the aggregate (the polydomain specimen), apart from depending on the space-group symmetries before and after the phase transition, can also depend on the presence or absence of specific domain pairs. If none of the domain pairs present is a ferroelastic domain pair, the mutual orientations of the domains present are not affected by the presence or absence of any of them. In such a case each domain state is identical to what it would be if it were the only domain present (ignoring any surface effects). Such domain states are referred to as single-domain states.

7.1.4 Disorientations

Because of the spontaneous lattice distortion involved in a ferroelastic phase transition, members of a ferroelastic domain pair undergo (usually small) rotations to make physical contact at domain boundaries and to minimize the overall strain energy of the specimen. These rotations are referred to as disorientations. Their magnitudes and directions are generally a complicated function of sample history.

Single-domain states correspond to configurations for which all the disorientations are either actually zero, or are treated as zero (under the PCA) for the sake of simplicity of analysis.

7.1.5 Antiphase Domains

Antiphase domains, or out-of-phase domains, have parallel axis systems (under the PCA) (Wondratschek & Jeitschko 1976). They arise when there is a lowering of the translational symmetry of a crystal at a phase transition. We also use the term *translation twins*, or *T-twins*, for them (cf. §7.4).

7.1.6 Orientational Twins

We define orientational twins as consisting of two contiguous domains related by a constant and recurring orientation-changing Seitz operator (involving a rotation, or inversion, or both). Their axis systems are at least partially nonparallel or antiparallel (even under the PCA).

7.1.7 Rotational Domains

We use the term rotational domains for contiguous or noncontiguous domain pairs related by a Seitz operator with a rotational component. The rotation involved may be proper or improper.

Contiguous rotational domains constitute an orientational twin (provided the orientational relationship is of a constant and recurring nature).

The term 'domain' is normally used only in the context of transformation twinning. Therefore, not all orientational twins are contiguous rotational domains. For example, nonferroelastic mechanical twins like those occurring in a cubic or hexagonal phase of a crystal are not rotational domains, but they are orientational twins.

7.1.8 Domain Structure and the Curie Principle

> *The whole process of a phase transition obeys the astonishing law discovered by philosophers* - the symmetry compensation law*: If symmetry is reduced at one structural level, it arises and is preserved at another !*
>
> A. V. Shubnikov & V. A. Koptsik (1974)

When the symmetry of a crystal changes from G to H on cooling through the transition temperature T_c, the phenomenon can be formally regarded as occurring under a *scalar* influence, namely temperature (Zheludev 1971). Assuming that this scalar has the symmetry of the continuous group $O(3)$ (cf. §B.4), the Curie principle (§C.1) tells us that the net symmetry below T_c should

$$G_d = G \cap O(3) = G \tag{7.1.4}$$

Thus, below T_c, the average symmetry *as a whole* should still be G. However, we know that the microscopic symmetry below T_c is H, and $H \subset G$. The situation can be reconciled only if the crystal splits into domains below T_c. Assuming that all domains are able to develop to the same extent, the average symmetry can then be G, even though the symmetry group of each domain is H (or a group conjugate to it through elements of G).

It also follows that, since each domain has symmetry H, symmetry operators which can map one single-domain state to another (under the

PCA) are those which are present in G but not in H, i.e. the symmetry operators *lost* at the phase transition. We are thus led to the following result (Aizu 1970a; Janovec 1972, 1976):

> *For a crystal that undergoes a phase transition with a space-group symmetry reduction from G to H, whereas H determines the symmetry of the order parameter (or vice versa) (cf. §5.3), it is the symmetry operations* lost *in going from G to H which determine the domain structure in the phase with symmetry H.*

We have already come across an elementary example of this in the description of Fig. 1.1.2. As another example we consider the cubic-to-tetragonal phase transition in $BaTiO_3$. The spontaneous polarization vector $\mathbf{P}_s$ in the tetragonal phase can point along any of the six directions $+x$, $+y$, $+z$, $-x$, $-y$, or $-z$ in the Cartesian frame of reference of the prototypic (in this case the cubic) phase. It is readily verified that these six possible configurations of $\mathbf{P}_s$ (and therefore of the corresponding ferroelectric domains) can be transformed among one another by only six distinct symmetry operations which belong to the group $Pm\bar{3}m$ of the prototypic phase, but not to the group $P4mm$ of the tetragonal ferroelectric phase.

7.1.9 Symmetry of Single-Domain States

The symmetry group H of a single-domain state is determined by the group G of the prototype phase and the group Γ of the order parameter, in accordance with the Curie principle:

$$H = G \cap \Gamma \tag{7.1.5}$$

Since a change of space-group symmetry is involved at the phase transition, it is better to choose a frame of reference in the prototype phase, and refer all transformations in any of the phases to this common coordinate system.

The order-parameter vector can have more than one orientations in representation space (§5.3.3). Γ in Eq. 7.1.5 is the symmetry group of the order parameter for a definite orientation of it.

The intersection group H, as determined by Eq. 7.1.5, is the highest common subgroup of G and Γ. It is thus a *maximal* subgroup of G which leaves the order parameter invariant, and Eq. 7.1.5 is in conformity with the maximality conjecture (§5.3.10).

The creation of domain walls costs energy. The maximality conjecture implies that the number of domain types, and therefore the number of domain walls, is the minimum necessary.

We consider phase transitions in $BaTiO_3$ for illustrating the applicability of Eq. 7.1.5.

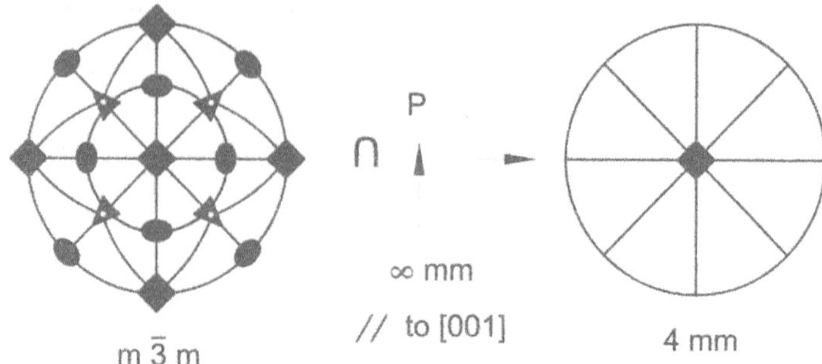

Figure 7.1.1: A Curie-principle interpretation of the phase transition in $BaTiO_3$ from the cubic (prototypic) phase of symmetry $m\bar{3}m$ to the tetragonal phase of symmetry $4mm$. Spontaneous polarization **P** (having the symmetry ∞mm ($C_{\infty v}$) of a cone or a single-headed arrow) is the order parameter of this transition. [After Perelomova & Tagieva 1983.]

Fig. 7.1.1 provides a (point-group level) Curie-principle interpretation of the cubic-to-tetragonal phase transition in this crystal.

Since the tetragonal phase, as well as the other two phases occurring at lower temperatures (Eq. 5.1.2), are ferroelectric phases, they are all described by polar symmetry groups. The Curie principle, alongwith the maximality conjecture, enables us to make the following statement (Ascher 1966): *The symmetry group of a ferroelectric phase of* $BaTiO_3$ *is a maximal polar subgroup of the prototype group.*

The prototype space group O_h^1 ($Pm\bar{3}m$) has only three maximal polar subgroups, namely C_{4v}^1 ($P4mm$), C_{2v}^{14} ($Amm2$), and C_{3v}^5 ($R3c$). The three observed ferroelectric phases of $BaTiO_3$ have exactly these symmetries (Eq. 5.1.2).

The tetragonal phase is obtained when the order parameter (spontaneous polarization) is directed along a $< 100 >$ direction of the prototype (Fig. 7.1.1). The orthorhombic phase of symmetry C_{2v}^{14} is obtained when the polarization vector points along a $< 110 >$ direction. And the trigonal phase of symmetry C_{3v}^5 is obtained as an intersection symmetry when the polarization vector points along any of the $< 111 >$ directions.

The tetragonal space group C_{4v}^1 can be obtained as a maximal polar subgroup of O_h^1 in three ways, corresponding to the three equivalent polar directions [100], [010] and [001]. Following Ascher (1966), we denote the corresponding point groups as C_{4v}^x, C_{4v}^y and C_{4v}^z, where the superscript denotes the polar axis. [Actually, the possible number of equivalent orientation states is 6, and not 3, because, e.g. [100] and [$\bar{1}$00] are distinct polar directions; however, the point-group symbol for both is just C_{4v} or C_{4v}^x.]

We now consider the phase transition to the orthorhombic phase. The point group O_h has 12 subgroups C_{2v}. Six of these have an intermediate supergroup, namely C_{4v} (and therefore are not maximal subgroups), and the other six are direct or maximal subgroups of O_h.

We first consider the former. Fig. 7.1.2 describes how the point-group symmetry C_{4v}^z can be lowered to orthorhombic symmetry C_{2v} in two possible ways. The resultant subgroups are denoted as $C_{2v}^{[z]xy}$ and $C_{2v}^{[z]ef}$, where $e = [110]$ and $f = [1\bar{1}0]$. In both cases the spontaneous polarization is directed along the [001] direction or the z-axis, although the mirror-planes of symmetry are oriented differently. This amounts to *nonuniqueness* because two different types of domains are possible, both having the same direction of the spontaneous polarization. Such configurations have not been found to exist. Similarly, C_{4v}^x and C_{4v}^y each can give rise to two nonmax-

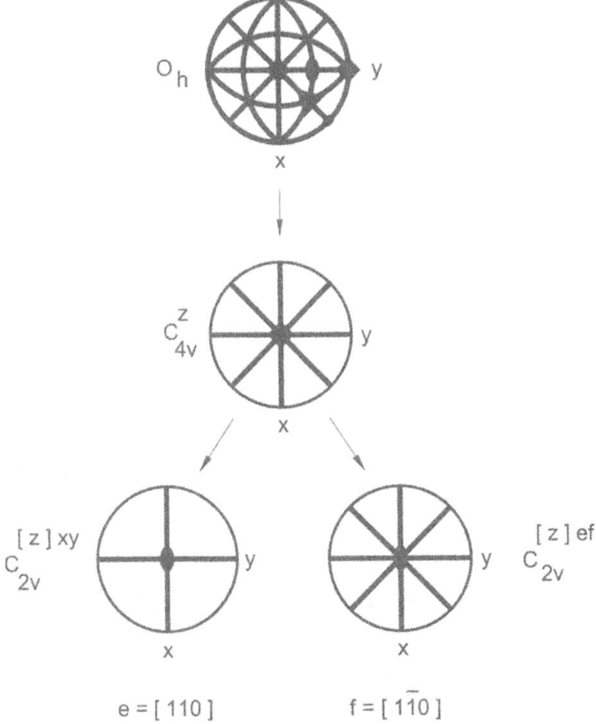

Figure 7.1.2: Derivation of polar, orthorhombic, nonmaximal subgroups of the point group O_h via one of the intermediate groups C_{4v}

imal, polar, orthorhombic subgroups of O_h. These also have never been observed.

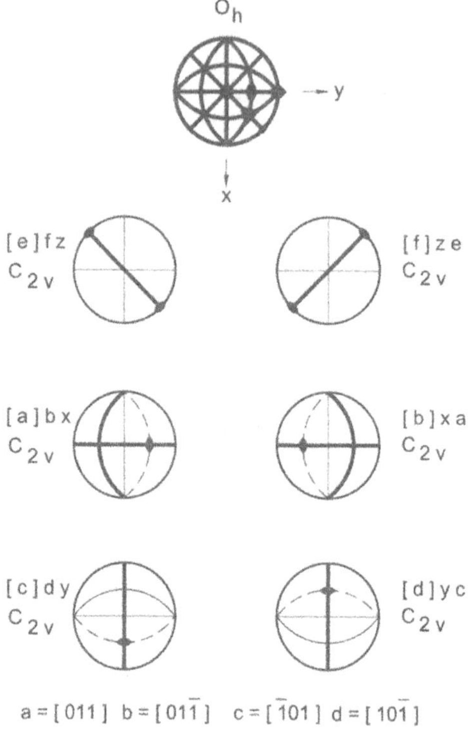

Figure 7.1.3: Derivation of the six maximal, polar, orthorhombic subgroups of the prototype group O_h. [After Ascher (1966).]

Experimentally, one observes only the six *maximal* polar orthorhombic subgroups of O_h depicted in Fig. 7.1.3. Each of these corresponds to a *unique* direction of the spontaneous-polarization vector.

7.1.10 Enumeration of Single-Domain States

The space group G of the prototype phase and the space group H of the distorted phase determine the possible number, n, of single-domain states (Aizu 1970a, 1974; Janovec 1972, 1976):

$$n = (|G_p| : |H_p|)\,(Z_H : Z_G) \tag{7.1.6}$$

Here $|G_p|$ is the order of the point group G_p underlying the space group G, and $|H_p|$ is the same for the space group H. The ratio $Z_H : Z_G$ is the number of times the primitive unit cell of the distorted phase is larger than that of the prototype. Z_H and Z_G can also be identified with the number of formula units of the crystal in the primitive unit cells of the distorted and prototype phases respectively.[1]

The single-domain states can be numbered as $D_1, D_2, \cdots D_n$, with n given by Eq. 7.1.6.

All the domain states have the same symmetry described by the group H. However, because they differ in orientation, chirality, and/or relative location, the positions and orientations of the symmetry elements in each domain are different. We can specify this by saying that the symmetry groups of $D_1, D_2, \cdots D_n$ are $H_1, H_2, \cdots H_n$, all referred to a common frame of coordinates fixed in the prototype.

Let us consider the group H_1 for the domain state D_1. All symmetry operators of H_1 will transform D_1 back into itself. Symmetry operators belonging to G but not belonging to H_1 will transform D_1 into other domain states $D_2, D_3, \cdots D_n$. It can be shown readily (Aizu 1970a; Janovec 1972) that all operations of G that transform D_1 to a particular state D_j are given by the left coset $g_j H_1$:

$$D_j = (g_j H_1) D_1, \quad g_j \in G, \ j = 1, 2, \ldots n \tag{7.1.7}$$

The choice of operators g_j is not unique. However, there can be only n, and exactly n, of them. This statement is made under the parent-clamping approximation (cf. §5.1.2).

Since H_1 is a proper subgroup of G, the following coset decomposition can be written:

$$G = g_1 H_1 + g_2 H_1 + \cdots g_n H_1 \tag{7.1.8}$$

There is a one-to-one correspondence between the domain states D_1, $D_2, \cdots D_n$ and the left cosets in Eq. 7.1.8. This also means that we can identify g_1 with the identity operator $\{\mathbf{E}\,|\,0\}$.

As mentioned earlier, the domain states $D_1, D_2, \cdots D_n$ have identical inherent symmetry. The set $\{D_1, D_2, \cdots D_n\}$ forms an *orbit* of D_1 in G (§2.2.6).

With reference to our discussion in §5.3.3 we can identify the group of all elements $g \in G$ that leave the domain state D_1, and therefore the order-parameter vector for that state, invariant as the *isotropy group*, also called the *stabilizer* (§2.2.6) of D_1 in G.

We note that H_1 can be a subgroup of G only if the translation subgroup of H_1 is also a subgroup of G. Unless stated otherwise, we assume that

[1] A phase transition from one ferroic phase to another ferroic phase is conceivable such that the daughter phase has a *smaller* Z-number in Eq. 7.1.6 than the starting phase, making the second factor in this equation *less than unity*. In such a situation, n in this eqation is *not* a product of the two factors. The important rule to follow is: single-domain states are determined by *lost* symmetry operators (including translational symmetry operators), and not by gained symmetry operators (cf. Guymont 1981).

this is indeed the case. We also assume that the nondisruption condition is satisfied, and that the parent-clamping approximation can be made (§5.1.2).

We now describe an alternative procedure, due to Guymont (1981), for the symmetry analysis of domain structures. In this there is no need to make a reference to the prototype symmetry. We simply assume that the phase transition $G \leftrightarrow H$ is such that the nondisruption condition is obeyed. We do not assume that G is the prototype symmetry for H; in the most general case $H \not\subset G$ and $G \not\subset H$.

The first step in this procedure is to choose a sufficiently large and common unit cell, such that the symmetry elements of both the phases can be described with reference to it. The fact that the nondisruption condition is satisfied ensures that this is possible.

The second step is to write down the symmetry operators of the groups G and H with respect to this common unit cell.

The third step is to identify the intersection group I:

$$I = G \cap H \tag{7.1.9}$$

In going from the initial phase (either G or H) to the final phase, only the *lost* operators are relevant for enumerating the domain states or variants. We do not take any special note of *gained* operators.

For the phase transition $G \rightarrow H$, the lost symmetry operators are members of the set $(G - I)$. And for the transition $H \rightarrow G$, the lost operators are members of the set $(H - I)$.

Suppose g is a lost operator for the transition $G \rightarrow H$. Then the coset gI contains only lost operators. If $|G| : |I| = 2$, then gI contains *all* the lost operators.

If $|G| : |I| > 2$, the lost operators are distributed among the cosets gI, $g'I$, $g''I$, $\cdots$, where g, g', g'' $\cdots$ are symmetry operators present in G but not in H.

A domain boundary is fully characterized by the corresponding lost operator.

We illustrate this procedure with the example of the phase transition $Pnam \leftrightarrow P112_1/m$ in $KClO_3$, a ferroelastic crystal (Wadhawan 1980; Guymont 1981).

First a common unit cell has to be chosen. As shown in Fig. 7.1.4(a), a C-centered orthorhombic cell serves the purpose. This cell represents the following lattice correspondence between the two phases:

$$\mathbf{a}_m = \mathbf{a}_o; \ \ \mathbf{b}_m = \frac{1}{2}(-\mathbf{a}_o + \mathbf{b}_o; \ \ \mathbf{c}_m = \mathbf{c}_o) \tag{7.1.10}$$

The chosen unit cell has twice the volume of the primitive unit cell of the monoclinic phase (under the PCA). Therefore the number of symmetry operators with reference to this (common) unit cell is double of what it is

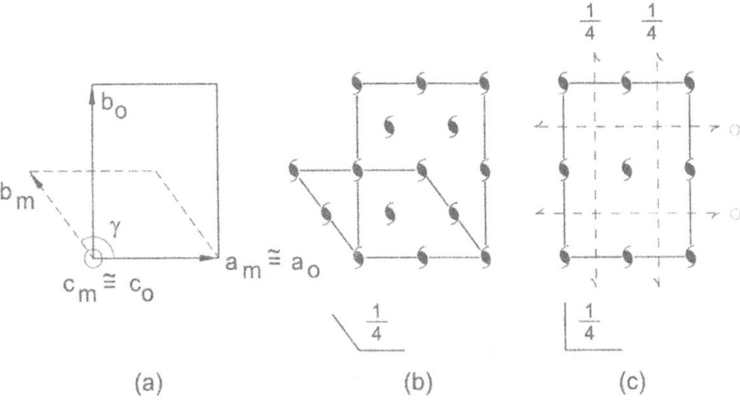

Figure 7.1.4: (a) Choice of a common unit cell for the orthorhombic and monoclinic phases of $KClO_3$. The common cell is the C-centered orthorhombic cell defined by a_o, b_o, c_o. (b) Symmetry elements of the monoclinic phase. (c) Symmetry elements of the orthorhombic phase. [After Guymont (1981).]

for the conventional cell:

$$\begin{aligned} H &= P2_1/m = P112_1/m = \\ &[\{1|000\},\ \{\bar{1}|000\},\ \{2_{1[001]}|00\tfrac{1}{2}\},\ \{m_{[001]}|00\tfrac{1}{2}\}] + \\ &[\{1|\tfrac{1}{2}\tfrac{1}{2}0\},\ \{\bar{1}|\tfrac{1}{2}\tfrac{1}{2}0\},\ \{2_{1[001]}|\tfrac{1}{2}\tfrac{1}{2}\tfrac{1}{2}\}, \{n_{[001]}|\tfrac{1}{2}\tfrac{1}{2}\tfrac{1}{2}\}] \end{aligned} \quad (7.1.11)$$

Similarly, for the orthorhombic phase,

$$\begin{aligned} G &= Pnam = \\ &[\{1|000\},\ \{\bar{1}|000\},\ \{2_{1[001]}|00\tfrac{1}{2}\},\ \{m_{[001]}|00\tfrac{1}{2}\}] + \\ &[\{2_{1[100]}|\tfrac{1}{2}\tfrac{1}{2}\tfrac{1}{2}\},\ \{2_{1[010]}|\tfrac{1}{2}\tfrac{1}{2}0\},\ \{n_{[100]}|\tfrac{1}{2}\tfrac{1}{2}\tfrac{1}{2}\}, \{a_{[010]}|\tfrac{1}{2}\tfrac{1}{2}0\}] \end{aligned} \quad (7.1.12)$$

It is now straightforward to pick out the common operators between Eqs. 7.1.11 and 12 to identify the intersection group:

$$I = G \cap H = \{1|000\},\ \{\bar{1}|000\},\ \{2_{1[001]}|00\tfrac{1}{2}\},\ \{m_{[001]}|00\tfrac{1}{2}\} \quad (7.1.13)$$

For the transition $Pnam \to P2_1/m$, the lost operators are those within the second square bracket in Eq. 7.1.12. They all involve rotational operators, so that orientational twinning is expected in the monoclinic phase.

The point group of the orthorhombic phase is mmm, and that of the monoclinic phase is $2/m$. These are of orders 8 and 4 respectively, so that two orientation states are expected in the monoclinic phase. Therefore the four lost operators in Eq. 7.1.12 can give only two (and not four) distinct orientation states. We can take $\{a_{[010]}|\frac{1}{2}\frac{1}{2}0\}$ as a representative variant-producing operator. The two variants are then described by the cosets $P2_1/m$ and $\{a_{[010]}|\frac{1}{2}\frac{1}{2}0\}\,P2_1/m$.

On a macroscopic scale the second variant corresponds to a reflection across the plane $y = 0$ (under the PCA). This plane thus defines one type of domain wall. Another domain-wall type, perpendicular to the first, and defined by the equation $x = 0$, can also be expected (Wadhawan 1980). This corresponds to the lost operator $\{n_{[100]}|\frac{1}{2}\frac{1}{2}\frac{1}{2}\}$ in Eq. 7.1.12.

We next consider the reverse transition $P2_1/m \rightarrow Pnam$. Here there is no loss of point-symmetry operators in going from the initial phase to the final phase. No orientational variants are therefore expected in the final (orthorhombic) phase. Translational symmetry of the final phase is, however, lower (by a factor of 2) compared to the initial phase, so that antiphase domains (or T-twins, cf. §7.4.4) can be expected in the final phase. The relevant lost operator can be taken as any of the four inside the second square bracket in Eq. 7.1.11. We can take, for example, $\{1|\frac{1}{2}\frac{1}{2}0\}$ as a representative operator describing the formation of T-twins in the orthorhombic phase.

7.1.11 Symmetry-Labeling of Domain States and Domain Walls

Domain States

The symmetry operations lost in going from the prototype symmetry to the symmetry of a ferroic phase may be both orientational and translational. Ignoring disorientations (§7.1.4) if any, i.e. working under the parent-clamping approximation, these lost operations are precisely those which can map one domain state to another. Domain states which can be mapped onto one another by merely a translational operator lost at the ferroic transition are said to belong to the same *orientation state*. Only orientational operators lost at the transition can map a domain state belonging to one orientation state to a domain state belonging to another orientation state.

These symmetry-related facts enable us to label domain states in a unique and complete manner (Janovec & Dvorak 1986).

Let p be the number of orientation states, and t the number of translational states possible in each orientational state. With reference to Eq. 7.1.6, $n = p\,.\,t$, $p = |G_p| : |H_p|$, and $t = Z_H : Z_G$. Any domain state can then be given a label A_a, where $A = 1, 2, \cdots p$, and $a = 1, 2, \cdots t$.

For example, if for a given ferroic phase, $p = 2$ and $t = 4$, then it has the following 8 possible domain states: 1_1, 1_2, 1_3, 1_4, 2_1, 2_2, 2_3, 2_4.

Domain Walls

Two contiguous states meet at an interface which is usually planar, and is called a domain wall.

Clearly, a domain wall is characterized by its orientation and by the two domains it separates. Therefore it can be specified by a symbol $A_a/\mathbf{n}/B_b$, or simply A_a/B_b (Janovec & Dvorak 1986). Here $\mathbf{n}$ denotes a unit vector normal to the domain wall, A_a is the domain at the negative end of this vector, and B_b the domain at its positive end.

The symmetry of the slice of the crystal structure which constitutes a domain wall is determined not only by $\mathbf{n}$, but also by the *position* of the central plane of the wall with respect to the intersecting lattices of the domains separated by it. This symmetry is diperiodic, and is defined by a layer group (Janovec 1981; Pond & Vlachavas 1983; Zikmund 1984).

Prominent Orientation of a Wall

A domain wall is said to have a *symmetrically prominent orientation* if a small change of this orientation results in a lowering of its symmetry.

One can associate an energy σ per unit area of a domain wall. This energy passes through an extremum at the prominent orientation of the wall.

SUGGESTED READING

K. Aizu (1970). Possible species of ferromagnetic, ferroelectric, and ferroelastic crystals. *Phys. Rev. B*, **2**, 754.

V. Janovec (1972). Group analysis of domains and domain pairs. *Czech. J. Phys.*, **B22**, 974.

M. Guymont, D. Gratias, R. Portier & M. Fayard (1976). Space-group-theoretical determination of translation, twin, and translation-twin boundaries in cell-preserving phase transitions. *Phys. Stat. Sol.* (a), **38**, 629.

M. Guymont (1981). Symmetry analysis of structural transitions between phases not necessarily group-subgroup related. Domain structures. *Phys. Rev. B*, **24**, 2647.

V. Kopsky (1982). *Group Lattices, Subduction of Bases, and Fine Domain*

Structures for Magnetic Crystal Point Groups. Academia, Prague.

L. A. Shuvalov, E. F. Dudnik & S. V. Wagin (1985). Domain structure geometry of real ferroelastics. *Ferroelectrics*, **65**, 143.

V. K. Wadhawan & M. S. Somayazulu (1986). Symmetry analysis of the atomic mechanism of ferroelastic switching and mechanical twinning: Application to thallous nitrate. *Phase Transitions*, **7**, 59.

V. Janovec, W. Schranz, H. Warhanek & Z. Zikmund (1989). Symmetry analysis of domain structure in KSCN crystals. *Ferroelectrics*, **98**, 171.

D. M. Hatch, R. A. Hatt & H. T. Stokes (1997). Systematic approach to the study of domain twins in ferroic phase transitions. *Ferroelectrics*, **191**, 29. This paper, and the relevant references therein, are part of a very comprehensive computerized approach to the Landau-Ginzburg theory of phase transitions and domain structure. The computer code ISOTROPY developed by Stokes & Hatch (1998) is based on space-group representations, and is being expanded continually. It was used ealier for generating and tabulating the subgroups of the 230 crystallographic space groups (Stokes & Hatch 1988). It has been extended to generate comprehensively the expected characteristics of the domain structure of a ferroic phase (domains and their orientations; tensorial properties; gradient invariants; etc.). Among its many advantages is the fact that it is free from human error. In fact, Stokes & Hatch (1988, 1998) identified a surprisingly large number of errors in the calculations done manually by earlier workers.

R. A. Hatt & D. M. Hatch (1999). Order-parameter profiles in ferroic phase transitions. *Ferroelectrics*, **226**, 61.

7.2 TWINNING

Twinning in crystals is a more general phenomenon than the occurrence of domain structure. Domain structure is "transformation twinning" (Wadhawan & Boulesteix 1992), in the sense that lowering of the prototype symmetry at a phase transition (or phase transformation) results in the occurrence of two or more equivalent domain states or twins. But twinning can occur even when there is no relevance of a phase transition; e.g. "mechanical twinning". Another type of twinning with features quite different from those of transformation twinning is "growth twinning". The Brazil twins of quartz are growth twins, whereas its Dauphiné twins are transformation twins.

Although our main interest in this book is in transformation twinning, we take a more general approach in this chapter, so that the reader can view transformation twinning in a proper perspective.

7.2.1 Definition of Twinning

A rigorous and unique definition of twinning is not easy to give. We could adopt, for example, the following definition given by Cahn (1954) on the basis of the work of Friedel (1926):

Definition 1. *A twin is a polycrystalline edifice, built up of two or more homogeneous portions of the same crystal species in juxtaposition, and oriented with respect to each other according to well-defined laws.*[2]

The twin laws do not comprise any symmetry operators of the crystal species. If they did, it would not be possible to define a twin boundary, and a single, untwinned, crystal would ensue.

The "well-defined" nature of the twin law implies a constant and recurring orientational relationship between the component crystals of the twin. The emphasis on the constancy of orientational relationship is necessary to distinguish between *twin walls* and, for example, *grain boundaries*. However, as discussed by Wadhawan & Boulesteix (1992) by considering the example of (Y-Ba-Cu-O), a whole gradation of situations are possible between the so-called "true twins" defined by Friedel (1926) and a grain-boundary type configuration. The complications arise because of local stresses and/or stoichiometry variations. Taking resort to the PCA (§5.1.2) is not an entirely satisfactory solution to the problem of defining twinning rigorously in terms of a well-defined twin law.

The term *bicrystal* is convenient to use when one does not, or cannot, make a distinction between a twin boundary (in terms of a well-defined twin law) and a grain boundary.

It has been pointed out by Bendersky, Cahn & Gratias (1989) that G. Friedel had initially proposed the following definition of twinning:

Definition 2. *A twin is a homogeneous crystalline aggregate in which one observes so large a number of identical mutual orientations between crystals that one can rule out a random cause.*

In this definition the emphasis is on the orientation of variants, and there is no mention of the crystal lattice which is merely a geometrical construct.

[2]Continuity of the strength of bonding across the twin interface can be introduced as an additional requirement for defining a twin.

Consider an aggregate of N variants of a crystal. If their mutual orientations are totally random, $3N$ linearly independent reciprocal-lattice vectors would be required for indexing the diffraction pattern of the aggregate. Any nonrandom relationship between even one pair of variants would make it possible to index the diffraction pattern with less than $3N$ reciprocal-lattice vectors. Bendersky et al. (1989), generalizing on the work of Gratias & Thalal (1988), have therefore proposed the following definition of twinning, which also conforms in spirit with the initial definition given by Friedel (Definition 2 above):

Definition 3. *If the diffraction pattern of a polycrystalline edifice of N orientational variants with a fixed, not random, orientational relationship between pairs can be indexed with less than 3N reciprocal-lattice vectors, twinning (or "hypertwinning") can be said to exist.*

7.2.2 Transformation Twins

Various terminologies are in use in the literature for the description of twinning in crystals (for reviews, see Cahn 1954; Klassen-Neklyudova 1964; Wadhawan 1987b; Wadhawan & Boulesteix 1992). It is necessary for the student of this subject to be familiar with the main descriptions and classification schemes, and the jargon employed.

Twinning caused by, or attributable to, phase transitions is called transformation twinning. The term "domain structure" normally refers to this type of twinning.

The concept of prototype symmetry (§5.1) is very important for explaining systematically the occurrence of transformation twins in a crystal. And the prototype structure can sometimes be a hypothetical one (Dvorak 1978). This is best exemplified by the case of twinning observed in $(NH_4)_2SO_4$ (Cahn 1954; Makita, Sawada & Takagi 1976; Sawada, Makita & Takagi 1976; Izyumov & Syromyatnikov 1990). On cooling, this crystal undergoes a phase transition at about -50^oC from a paraelectric phase of orthogonal symmetry D_{2h}^{16} to a ferroelectric (or rather ferrielectric) phase of monoclinic symmetry C_{2v}^9. The phase stable at room temperature is the orthorhombic phase, and no further phase transitions are observed on raising the temperature, before decomposition occurs. However, the room-temperature orthorhombic phase exhibits ferroelastic switching, and has a domain structure such that, normal to the a_o-axis, the orientations of domains differ by an angle very close to 60°. The lattice parameters b_o and c_o are such that $b_o/c_o = 1.774$, which is very close to the value $\sqrt{3}$ expected for the unit cell of a lattice with hexagonal symmetry. One therefore describes this orthorhombic lattice as pseudo-hexagonal. From an analysis of the domain structure and the observed metrics it is possible to deduce the hypothetical hexagonal prototype symmetry (D_{6h}^4). The transformation twinning observed at room temperature is then explained in a very satisfactory manner in terms of a hypothetical phase transition: $D_{6h}^4 \rightarrow D_{2h}^{16}$.

7.2.3 Growth Twins

Crystal growth from the liquid phase or the vapour phase is also a process of phase transformation, and in that very broad sense growth twins could also be regarded as transformation twins. However, certain bonds (particularly covalent bonds), once formed in the crystal, cannot be broken easily during a solid-to-solid phase transformation. By contrast, even such bonds *can* be altered at the stage of formation of the crystal from the liquid or the vapour phase (Cahn 1954), particularly at the *nucleation* stage (§2.1.1) of the overall crystal-growth process. Thus it makes sense to treat growth twins as distinct from transformation twins.

When the atomic bonding allows it, even growth twins can exhibit ferroic switching. The ferroelastic behaviour of $(NH_4)_2SO_4$ described above is an example of this.

By contrast, the Brazil twins of quartz are more typical of what we normally observe in the case of growth twins. It is almost impossible to detwin a crystal of quartz having Brazil twins. Extreme measures like high temperatures and large mechanical shearing stresses are often needed for detwinning growth twins, if at all such procedures are successful.

Cahn (1954) has described several varieties of growth twins: annealing twins, repeated twins, mimetic twins, contact twins, lamellar twins, and polysynthetic twins.

7.2.4 Mechanical Twins

For certain crystals it is possible to cause a cooperative movement of atoms by applying mechanical stress of appropriate magnitude and direction such that the atom movements result in a new crystal of different orientation but identical structure. Thus a portion of the parent crystal is changed into its twin, and the two together constitute a mechanical twin. The changed part of the crystal undergoes a macroscopic change of shape called a *simple shear*: A lattice of points is said to undergo a simple shear if each lattice point moves parallel to a certain direction (called the *shear direction* η_1, see Fig. 7.2.1) by a distance proportional to the distance of the point from a particular plane called the *habit plane*, or the *twin plane* K_1.

The habit plane is the interface between the parent crystal and its twin produced by the shear deformation. It thus remains invariant during the gliding of the lattice planes that results finally in the formation of the

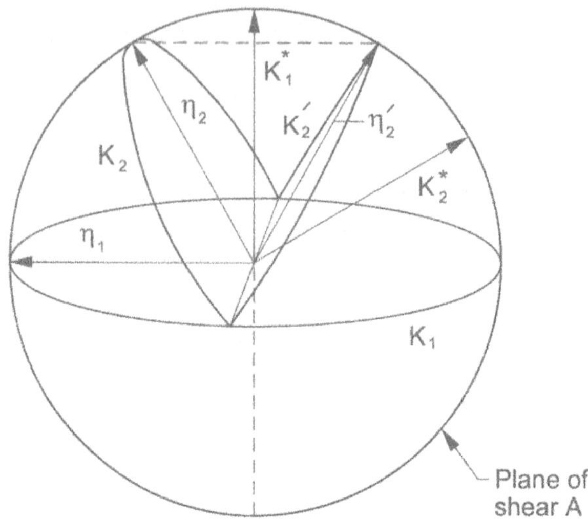

Figure 7.2.1: Description of the symmetry elements in mechanical twinning. K_1 is the habit plane, or twin plane, K_2 the second undeformed plane, and η_1 the shear direction. The shear plane is perpendicular to the plane K_1 and contains the direction η_1. η_2 is the intersection of the shear plane with K_2. [After Hall (1954) and Stark (1988).]

mechanical twin.

We can imagine a reference sphere in the parent crystal, divided into two equal parts by the twin plane K_1. After mechanical twinning the hemisphere belonging to the parent crystal remains undeformed, and the other hemisphere deforms to a spheroid. In this spheroid one can identify a plane (K_2) which is also an invariant plane like K_1.

The *shear plane*, A, is perpendicular to the twin plane K_1 and contains the shear direction $\boldsymbol{\eta}_1$.

The intersection of the plane of shear with the invariant plane K_2 defines a direction, $\boldsymbol{\eta}_2$, which is an invariant direction.

The shear plane has the property that it contains the normals $\mathbf{K}_1^*$ and $\mathbf{K}_2^*$ to the invariant planes K_1 and K_2.

With reference to the Bravais lattice of the parent crystal, the invariant planes K_1 and K_2 may or may not have rational Miller indices. Similarly, the invariant directions $\boldsymbol{\eta}_1$ and $\boldsymbol{\eta}_2$ may or may not be rational directions with respect to the crystal lattice.

Type 1 mechanical twins are defined as those for which K_1 and $\boldsymbol{\eta}_2$ and the plane of shear, A, have rational indices.

Type 2 mechanical twins are characterized by a rational plane K_2 and a rational direction η_1.

For a *compound twin* (Cahn 1954), K_1, K_2, η_1, η_2 all have rational indices.

Type 1 twins are also called *mirror twins* (Cahn 1954) because for them the twin plane K_1 is a mirror plane of symmetry. As emphasized by Cahn (1954), the mirror operation defining the twin law must be parallel to a lattice plane in both components of the twin.

Type 2 twins are also called *rotational twins* (Cahn 1954), with η_1 acting as the direction of 2-fold rotation axis providing the twin law. This 2-fold axis must be parallel to a lattice row common to both components of the twin.

Following Stark (1988), we can express the twinning operator for Type 1 and Type 2 twins, respectively, as follows:

$$\alpha_1 = \{m_{K_1} \,|\, \mathbf{t}_1\} \tag{7.2.1}$$

$$\alpha_2 = \{2_{\eta_1} \,|\, \mathbf{t}_2\} \tag{7.2.2}$$

For compound twins, Stark (1988) has provided detailed verification for the validity of the following statement:

> *The only way in which a compound twin is possible in a crystal is when the mirror, m_A, associated with the plane of shear (A) is a member of the space group of the parent crystal:*

$$\{m_A \,|\, \tau\} \in S_1 \tag{7.2.3}$$

Here S_1 is the space group of the parent crystal. Thus it is not sufficient for a compound twin that the shear plane be a mirror plane in a macroscopic (point group) sense. It must also conform to the *space group* symmetry of the parent crystal.

If the parent crystal structure is such that a mirror plane of symmetry and a 2-fold axis perpendicular to the mirror plane exist simultaneously, then the rotation twin and the mirror twin for such a case are said to be *reciprocal* to each other. Their shear elements are determined at the same time: K_1 for one corresponds to K_2 for the other, and vice versa; similarly, η_1 and η_2 have mutually reciprocal roles.

7.2.5 Friedel's Four Twin Types

G. Friedel (1926) specified four categories of twinning in crystals (see Cahn 1954). For describing these it is necessary to define the concepts of *twin index* Σ, and *twin obliquity* ω .

Σ is the inverse of the fraction of lattice sites common to the two components of the twin.

And twin obliquity ω is a measure of the disorientation of one component with respect to the other.

The primary subdivision of twinning in Friedel's scheme is in terms of whether $\Sigma = 1$, or $\Sigma > 1$. And for each case one has to consider whether $\omega = 0$, or $\omega \neq 0$. Four types of twinning are thus recognized.

Twinning by merohedry ($\Sigma = 1$, $\omega = 0$). This concept of twinning (attributed to the work of Bravais) corresponds to situations where the crystal structure has a lower symmetry than the Bravais lattice on which it is based. Because of the lower symmetry of the crystal structure, the morphology of the individual crystal displays a smaller number of faces than would be possible for the full symmetry of the lattice on which it is based. The additional symmetry elements present in the lattice provide the twin laws.

An example is that of pyrite, having point-group symmetry $m3$, whereas the symmetry of the underlying lattice is $m\bar{3}m$. The additional plane of symmetry, $m_{[110]}$, present in the lattice provides the twinning operator.

Another well-known example is that of Dauphiné twinning in quartz.

Twinning by pseudomerohedry ($\Sigma = 1$, $\omega \neq 0$). This type of twinning was first recognized by Mallard. Here we can have either a twin plane or a twin axis such that they are, respectively, parallel to a lattice plane or lattice row which are *almost* a symmetry plane or a symmetry axis of the crystal structure.

Twinning by reticular merohedry ($\Sigma > 1$, $\omega = 0$). Here, although the lattices of the component individuals are not parallel, a larger unit cell can be identified for the twin which continues without disturbance across the twin boundary. In this context one speaks of a *lattice of coincidence sites*, or a *coincidence lattice*, with Σ providing a measure of the fraction of the lattice sites that are coincident for the entire twin. This type of twinning is common in crystals belonging to the cubic system. Growth twins in galena, fluorite, diamond, silicon and germanium are examples (Chen et al. 1992).

Twinning by reticular pseudomerohedry ($\Sigma > 1$, $\omega \neq 0$). Twinning in aragonite ($CaCO_3$) provides an example of this type. The symmetry of the crystal is orthorhombic pseudohexagonal, and the planes of pseudosymmetry act as twin planes. Here, rather than the crystal lattice itself, a small *multiple* of the lattice coincides *approximately* with the corresponding multiple of the lattice of the other component.

7.2.6 Manifestation of Twin Type in the Diffraction Pattern

Before the advent of X-ray crystallography, optical and contact goniometry were the main tools for studying twinning in crystals. Friedel's rules, described above, were largely based on such information. Donnay & Donnay (1974) proposed a classification of twinned crystals which is based on distinctive features of the diffraction pattern of the twinned crystal. The obvious characteristic to look for in the diffraction pattern is whether or not there is a splitting of the diffraction spots. This depends on whether or not the twin obliquity, ω, is zero or nonzero.

If $\omega = 0$, we have *twinning by twin-lattice symmetry* (TLS).

If $\omega \neq 0$, we have *twinning by twin-lattice quasi symmetry* (TLQS).

For each of these broad categories, further subdivision is made in terms of the twin index Σ.

7.2.7 Hypertwins

Rapidly solidified aggregates of Al – Mn – Fe – Si display twinning such that the icosahedral motifs in all the twin components are parallel (Bendersky, Cahn & Gratias 1989). The point-group symmetry of the individual components is $m3$, and any two such components are related by a rotation of 72° (or its multiples) about an irrational direction $(1, \tau, 0)$, where τ is the Golden mean (= 1.61803..). Thus, although each aggregate comprises hundreds of crystals, only five orientations of the variants occur. Further rotations about any of the six $< 1, \tau, 0 >$ axes do not produce any additional orientations for the variants.

We have seen above that merohedral twinning occurs when the lattice has a higher symmetry than the crystal structure. The additional symmetry operators of the lattice flip the crystal structure across the twin interface without disturbing the lattice orientation and continuity. In the case of the Al – Mn – Fe – Si alloy, it is the *lattice* which gets flipped around across the twin interfaces, and the orientation of the icosahedral motif remains unchanged.

Bendersky et al. (1989) (also see Thalal & Gratias 1988) have given a general prescription for describing an aggregate of N individuals of a crystal. We know that the lattice vectors of each component crystal provide the geometrical basis in the actual 3-dimensional space. Simplification occurs when the six basis vectors corresponding to any particular bicrystal are taken as defining a 6-dimensional hyperspace, in which each component crystal is represented as a 3-dimensional facet. There can be a maximum of $3N$ independent lattice vectors for the entire aggregate, the maximum corresponding to the situation in which the orientations of all the compo-

Table 7.2.1: Examples of hypertwinning. [After Bendersky et al. (1989).]

Type of boundary	N	Rank r	Reduction $(3N - r)$
General grain	2	6	0
Special grain (3-D coincidence lattice)	2	3	3
Merohedral twin	2	3	3
Twin aggregate	N	3	$3N - 3$
Mechanical twin (2-D coincidence lattice)	2	4	2
Transformation Twinning			
$6/mmm \rightarrow mmm$	3	3	6
$8/mmm \rightarrow 4/mmm$	2	5	1
$10/mmm \rightarrow mmm$	5	5	10
$12/mmm \rightarrow 6/mmm$	2	5	1
$m35 \rightarrow \bar{3}m$	10	6	24
$m35 \rightarrow m3$	5	6	9

nents are entirely random. For a nonrandom configuration the rank r of the matrix describing the lattice relationships will be less than $3N$. The condition

$$3 < r < 3N \tag{7.2.4}$$

thus defines a *hypertwin*. Table 7.2.1 gives a number of examples of the reduction $(3N - r)$ of the rank r.

In this table, the transformation $m35 \rightarrow m3$ describes the situation for Al − Mn − Fe − Si. For any bicrystal of the aggregate of this material the interface contains a 3-fold axis (corresponding to a 1-dimensional coincidence lattice). Therefore only five lattice vectors (instead of six) are independent. If we attach another component to this bicrystal, only one new basis vector is added because two of the 3-fold axes of the third crystal are already specified, and the third crystal has therefore only one degree of freedom. Addition of a fourth and a fifth variant adds no new degrees of freedom because their orientations are fully specified by the 3-fold axes they have in common with the first three variants. Therefore r= 6, and $3N - r = 9$. The diffraction pattern of this 5-component hypertwin can thus be indexed, not by 15, but only 6 reciprocal-lattice vectors.

7.2.8 Hermann's Space-Group Decomposition Theorem

For a symmetry analysis of the observed or expected domain structure of a given crystal, once the prototype symmetry has been selected with due care and certitude, the next question is that of the various modes available for the lowering of this symmetry. Hermann's space-group decomposition theorem has a valuable systematizing influence in this respect (Hermann 1929; Wondratschek & Jeitschko 1976; Deonarine & Birman 1983). According to this theorem:

Every subgroup H of a space group G is a class-equivalent subgroup of a translation-equivalent subgroup Z of the parent group G:

$$H \subseteq Z \subseteq G \tag{7.2.5}$$

In other words, there always exists a space group Z such that G and Z have the same translational symmetry, and Z and H have the same point-group symmetry.

There are thus three ways in which the prototype space-group symmetry G can be lowered:

$$H = Z \subset G \tag{7.2.6}$$

$$H \subset Z = G \tag{7.2.7}$$

$$H \subset Z \subset G \tag{7.2.8}$$

Eqs. 7.2.6 and 7.2.8 correspond to ferroic phase transitions, because the point-group symmetry of H is lower than that of G.

If Eq. 7.2.7 applies we have a nonferroic phase transition, resulting in antiphase domain boundaries only.

Eq. 7.2.6 would give only rotational domains (with no translation domains present), whereas both rotational domains and antiphase domains would be present in a crystal that undergoes a phase transition that conforms to Eq. 7.2.8.

Fig. 7.2.2 provides a schematic illustration of the three types of situations described by Eqs. 7.2.6-8.

An example of the occurrence of only rotational domains is provided by the domain structure of α-quartz, resulting from the $\beta \rightarrow \alpha$ phase transition, when the symmetry changes from $P6_222\,(D_6^4)$ to $P3_221\,(D_3^6)$.

An example of the occurrence of purely translational domains (Eq. 7.2.7) is that of the ordered phase of the alloy Cu_3Au. In this case an order-disorder, nonferroic, phase transition occurs, with symmetry changing from $Fm\bar{3}m\,(O_h^5)$ for the disordered phase to $Pm\bar{3}m\,(O_h^1)$ for the ordered phase.

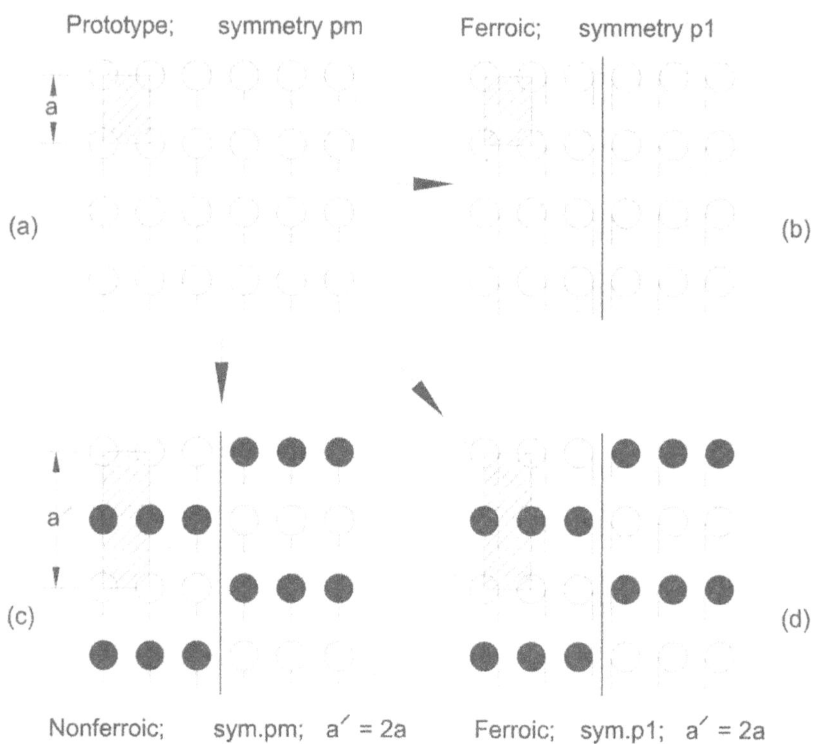

Figure 7.2.2: Three possible ways in which a prototype symmetry can be lowered at a phase transition. Part (a) shows a prototypic structure, having plane-group symmetry *pm*. If the point-group symmetry continues to be m after the phase transition, and there is a lowering of only the translational symmetry, ($a' = 2a$), we have a nonferroic phase transition ($(a) \rightarrow (c)$). Thick vertical lines represent domain walls. The domain wall in (c) is purely of the antiphase type. If the point-group symmetry is lowered (from m to 1 in (b) and (d)), we have a ferroic phase transition. This transition occurs without a lowering of translational periodicity in (b). In (d) there is a lowering of both translational and point symmetry. Both rotational domains and antiphase domains occur in (d), whereas only rotational domains occur in (b). [After Wondratschek & Jeitschko (1976).]

$\beta - Gd_2MoO_4$ (GMO), having space-group symmetry $Pba2\,(C_{2v}^8)$, provides an example of domain structure in which both rotational domains and antiphase domains occur together. The prototype symmetry in this case is $P\bar{4}2_1m\,(D_{2d}^3)$, and the unit cell of the lower-symmetry phase has a volume twice that of the prototype. The intermediate space group Z of Eq. 7.2.8 can be identified as $Cmm2$ (Dvorak 1971; Levanyuk & Sannikov 1974; Wondratschek & Jeitschko 1976; Janovec 1976).

SUGGESTED READING

R. W. Cahn (1954). Twinned crystals. *Adv. Phys.*, **3**, 363.

V. Janovec (1976). A symmetry approach to domain structures. *Ferroelectrics*, **12**, 43.

V. K. Wadhawan & C. Boulesteix (1992). Transformation twinning and related phenomena. *Key Engg. Materials*, **68**, 43.

7.3 BICRYSTALLOGRAPHY

Any two crystals sharing an interface constitute a bicrystal. For mathematical convenience the interface is usually taken as planar, and a bicrystal is defined as *two semi-infinite crystals sharing a planar interface.*

Two basic types of bicrystals are possible: those having two different crystalline species across the interface, and those having the same crystal (in two different orientations, chiralities and/or locations) across the interface (Fig. 7.3.1). In the first case the interface is a *heterophase interface* or an *interphase boundary.* In the second case we have a *homophase interface.*

Interphase boundaries can arise during phase transitions, or in epitaxial crystal growth. We shall also come across them when we discuss morphology of the ferroic phase, in the next chapter.

Homophase boundaries are encountered as domain walls in domain twins, interfaces in stacking faults, and as grain boundaries in general. Low-angle and high-angle grain boundaries, particularly the latter, continue to be the subjects of substantial interest (Fischmeister 1985).

The subject of the domain structure of ferroic materials overlaps considerably with bicrystallography, although the latter does not depend on the notion of prototype symmetry as its central premise. Bicrystallography has a rather large range of applicability in physics and materials science. For example, it is relevant to crystal growth and characterization. Even a carefully grown crystal may contain growth-sector boundaries, growth bands, misfit dislocations, etc. (see, for example, Bhat 1985). The formalism and the language of bicrystallography can be used for a detailed

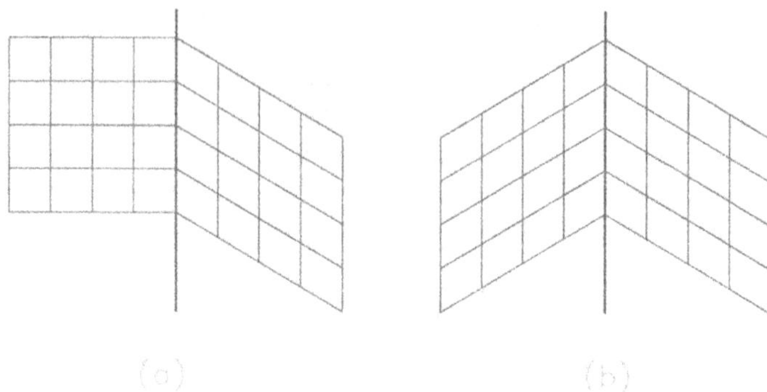

Figure 7.3.1: Schematic depiction of the two basic types of bicrystals. In (a) the two crystals across the interface are different. In (b) we have the same crystal on the two sides, but the two individuals may have different orientations, chiralities, and/or mutual disposition.

description of an as-grown crystal. Similarly, information about the symmetry group of a bicrystal (see below) can enable us to predict and classify comprehensively the types of extended defects that can exist in the interface (Pond & Vlachavas 1983; Kalonji 1985). Phase transitions occurring at interfaces, and the resultant domain structure at these interfaces, can be described quite adequately in bicrystallographic terminology. Constraints on the vibrational modes of the interface can be derived from a knowledge of the space-group symmetry of the interface. This information can be used for understanding not only interfacial phase transitions, but also diffusion and other migration mechanisms at the interface.

Looking at history that is not too distant, Bollmann's (1970) book on crystal defects and crystalline interfaces marked the beginning of many of the current geometrical ideas of bicrystallography. His latter book (Bollmann 1982) further consolidated this work. However, his analysis applied only to symmorphic structures with one atom per primitive unit cell. Pond & Bollmann (1979) defined the notion of the dichromatic pattern (described below). This was extended later to the concept of the dichromatic complex by Pond & Vlachavas (1983). This notion overlaps substantially with that of domain pairs defined by Janovec (1972).

Cahn & Kalonji (1981), Fayard, Portier & Gratias (1981), Gratias & Portier (1982), Kalonji & Cahn (1982), and Kalonji (1985) developed alternative formalisms for bicrystallography, with special attention to interfacial properties of bicrystals.

The English edition of Shubnikov & Koptsik's book on symmetry in science and art appeared in 1974. This book forms the basis of the bicrystallographic formalism developed by Pond & Vlachavas (1983), which we sketch very briefly in this section. We also draw from the work of Gratias and coworkers, as well as Kalonji and Cahn.

7.3.1 General Methodology

Intersection Group

The symmetry group of a bicrystal that has two different crystals across the interface comprises symmetry elements (if any) common to the component crystals. It is thus the intersection group (I) of the groups G_1 and G_2 describing the symmetry of the component crystals, in accordance with the Curie principle (§C.1):

$$I = G_1 \cap G_2 \tag{7.3.1}$$

The symmetry of such a bicrystal (Fig. 7.3.1(a)) is always lower than that of the components, and exemplifies the process of *dissymmetrization* (§C.1).

Operations of I simultaneously map each component crystal back onto itself.

Intersection groups are of central importance to bicrystallography, just as crystallographic point groups are to monocrystallography (Cahn & Kalonji 1981). We shall have occasion to refer to them frequently.

Symmetrizer

When the two parts of a bicrystal are comprised of the *same* crystalline species, the possibility of an additional type of symmetry operation arises, which can map each component crystal to *the other*. The overall symmetry, say G_s, of such a bicrystal can thus be higher than the intersection-group symmetry I. This is a case of *symmetrization*, occurring in accordance with the Curie-Shubnikov principle (§C.2). One obtains G_s from I as an extended group (cf. Eq. C.2.2):

$$G_s = I \cup M', \tag{7.3.2}$$

where M' is an appropriate symmetrizer (§C.2).

The mutual symmetry relationship between Crystal 1 and Crystal 2 comprising such a bicrystal can be specified by the Seitz operator $\{\mathbf{R}|\mathbf{T}_{frac}\}$, where $\mathbf{R}$ is a point-group operator that maps the lattice of Crystal 1 to the lattice of Crystal 2, and $\mathbf{T}_{frac}$ is an additional rigid-body translation that might be needed for superimposing the corresponding ("homologous") atomic sites.

Any point $\mathbf{r}_1$ in Crystal 1 is related to the corresponding point $\mathbf{r}_2$ in Crystal 2 as follows:

$$\mathbf{r}_2 = \{\mathbf{R}|\mathbf{T}_{frac}\}\mathbf{r}_1 = \mathbf{R}\mathbf{r}_1 + \mathbf{T}_{frac} \tag{7.3.3}$$

Therefore the space group G_2 of Crystal 2 is related to space group G_1 of Crystal 1 as follows (the two groups are isomorphic):

$$G_2 = \{\mathbf{R}|\mathbf{T}_{frac}\}\, G_1\, \{\mathbf{R}|\mathbf{T}_{frac}\}^{-1} \tag{7.3.4}$$

The intersection group I (Eq. 7.3.1) is therefore the following:

$$I = G_1 \cap \{\mathbf{R}|\mathbf{T}_{frac}\}\, G_1\, \{\mathbf{R}|\mathbf{T}_{frac}\}^{-1} \tag{7.3.5}$$

The groups G_1, G_2 and I are ordinary or classical space groups (Fedorov groups). The elements of the symmetrizer M in Eq. 7.3.2 are also ordinary coordinate transformations, but with the distinctive feature that they interchange the identities of Crystal 1 and Crystal 2. It is found useful to keep track of those symmetry operators of the bicrystal which map Crystal 1 onto Crystal 2, and Crystal 2 onto Crystal 1. This is easily done by arbitrarily calling Crystal 1 "white" and Crystal 2 "black", and treating the operations involving the symmetrizer M' as antisymmetry operations, or colour-reversal operations (§2.2.18).

With this proviso we can write the following defining equation for the symmetrizer:

$$M' = \{\mathbf{R}'|\mathbf{T}'_{frac}\} \times I, \tag{7.3.6}$$

where we write $\mathbf{R}'$ and $\mathbf{T}'_{frac}$ instead of $\mathbf{R}$ and $\mathbf{T}_{frac}$ to represent the fact that they are colour-reversal operators.

The group G_s is thus made into a Shubnikov group, although it is possible to design an alternative formalism wherein one deals with Fedorov groups only.

In the case of a bicrystal comprising two *different* crystals, M' is an empty set, and $G_s = I$.

The Maximum Symmetry Group

The groups G_1 and G_2 can be, in general, nonsymmorphic (§2.2.17), and nonholohedral (§2.2.16). In the approach of Pond & Vlachavas (1983), the first stage in the symmetry analysis of a bicrystal is the determination of the maximum symmetry group for the problem at hand. For this, one has to first obtain from G_1 and G_2 the corresponding symmorphic and holohedral space groups. These latter are nothing but the symmetry groups of the underlying lattices of Crystal 1 and Crystal 2. These are denoted by symbols Φ_1^* and Φ_2^*, where the use of the letter Φ signifies the fact that these are Fedorov groups (and not Shubnikov groups; see below). For dealing with magnetic crystals the Fedorov groups must be replaced by Shubnikov groups. The superscript '*' is used consistently for symmetry groups of *lattices*, rather than of crystal structures.

The following intersection group is constructed next:

$$\Phi(p) = \Phi_1^* \cap \Phi_2^* \tag{7.3.7}$$

The maximum symmetry group is then obtained as the following extended group:

$$Ш^*(p) = \Phi(p) \cup M'(p), \tag{7.3.8}$$

where

$$M'(p) = \Phi(p) \times \{\mathbf{R}'|\mathbf{T}'_{frac}\} \tag{7.3.9}$$

The maximum symmetry group $Ш^*(p)$ describes the symmetry of the *dichromatic pattern* (DCP). The DCP is the *lattice* defined by the superposition of the white lattice and the black lattice.

The idea behind establishing the *maximum*-symmetry group (or the universal group) is that the various stages of symmetry-descent from this group (due to various internal causes to be described below) provide *exhaustively* the complete set of variants or domain states.

Symmetry Descent

Having determined the symmetry group of the DCP, the next step is to determine the symmetry group of the *dichromatic complex* (DCC). The DCC is the superposition of the white and the black *lattice complexes*. [A lattice complex, we may recall (§2.2.17), is a set of points obtained by applying on any point in a crystal all the symmetry operations of the space group of the crystal.] We first construct the following intersection group:

$$\Phi(c) = \Phi_1 \cap \Phi_2, \tag{7.3.10}$$

where Φ_1 and Φ_2 are the space groups of the white and the black crystal structures.

The Fedorov group $\Phi(c)$ defined by Eq. 7.3.10 does not take cognisance of any colour-reversal symmetry that may be present. A suitably extended group, $Ш(c)$, is therefore constructed from $\Phi(c)$ (Pond & Vlachavas 1983).

The group $Ш(c)$ so constructed may or may not be holosymmetric, i.e. its symmetry may be equal to, or may be less than, that of the underlying lattice. In the latter case we determine, from the knowledge of $Ш(c)$, the holosymmetric group $Ш^*(c)$ for the DCC.

So far we have dealt only with the interpenetrating white and black lattice complexes. The next step is to introduce the interface. This is done in two steps. First an *unrelaxed* or *idealized* interface is defined, and then atoms near the interface are allowed to relax.

Fig. 7.3.2 shows the stepwise construction of the idealized homophase interface. We begin with two superimposed crystals, as in part (a) of the figure. A rotation or reflection ($\mathbf{R}$) is carried out next, as in (b). Part (c)

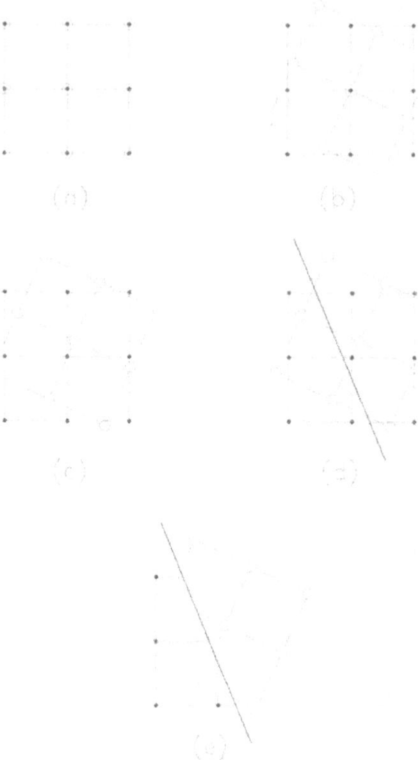

Figure 7.3.2: Idealized construction of a homophase interface in a bicrystal. See text for description. [After Kalonji (1985).]

shows the operation of the translation $\mathbf{T}_{frac}$ (cf. Eq. 7.3.6). An unrelaxed interface is brought in next, as shown in (d). Finally black atoms are wiped out on one side of the interface, and white atoms on the other, as in (e).

Operations of the space group III(c) that leave the interface invariant define the bicrystal group III(b). Knowing III(b), one can construct the holohedral group III*(b), in case the former is not already so.

The arbitrary introduction of the interface into the DCC can result in a thermodynamically metastable configuration. The structure near the interface therefore relaxes to a stable state, with a concomitant lowering of the symmetry III*(b).

Fig. 7.3.3 summarizes the entire procedure, involving the construction of the maximum symmetry group III*(p), and the various stages of the descent of this symmetry. Symmetry operators lost at each stage result in variants or domain states. Extension of such a formalism to *tricrystals* has

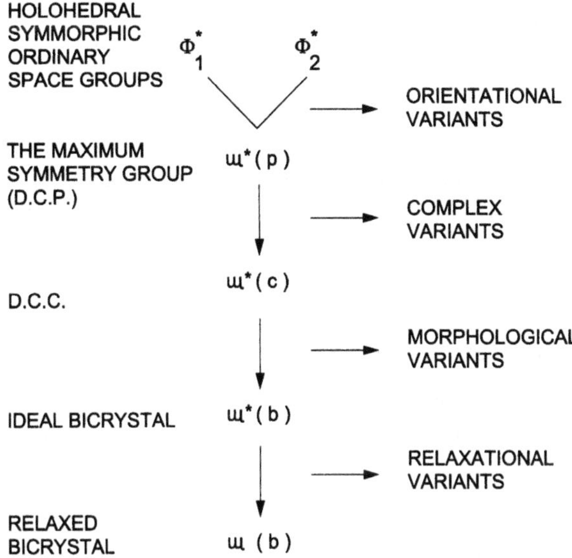

Figure 7.3.3: The various types of variants possible in a bicrystal, as determined by successive loss of symmetry operators. [After Pond & Vlachavas 1983.]

been carried out by Dimitrakopulos & Karakostas (1996).

7.3.2 Dichromatic Pattern

The concept of the dichromatic pattern (DCP) is fundamental to the symmetry description of bicrystals. Eq. 7.3.8 defines the symmetry group of the DCP, with Eq. 7.3.9 defining the symmetrizer $M'(p)$.

Pond & Vlachavas (1983) have derived the 12 possible symmetry groups $\{1, M'(p)\}$ for bicrystals. A typical example is the group $\{1, 4'(\text{mod2})\}$. Here the second operation can be expressed in the *axis-angle notation* $\{[uvw]\,\theta\}'$ as $\{[001]\,90^o\}'$. Its meaning is that there is a 90° rotation about the [001] axis, followed by colour reversal. Obviously, if this operation is carried out twice, there is no change of colour, and such a double operation cannot qualify as a symmetrizing factor. Hence the need to put 'mod 2' (modulo 2) in $4'$ (mod 2). Similarly, the operator $12'$ (mod 6) implies that alternate operations of the 12-fold colour-reversing rotation axis are excluded.

Let T be the translation group underlying the symmetry group G_1 of Crystal 1 in Eq. 7.3.5. Then, from this equation, the translational symmetry group of the DCP can be written as

$$T(p) = T \cap \{\mathbf{R}|\mathbf{T}_{frac}\} T \{\mathbf{R}|\mathbf{T}_{frac}\}^{-1} \tag{7.3.11}$$

The translation group $T(p)$ is an infinite set in 1, 2 or 3 dimensions, or it is an empty set (corresponding to no periodicity at all). Patterns with no translational symmetry are described by point groups. Patterns with 1-dimensional translational symmetry are described by rod groups, and those with 2-dimensional periodicity by layer groups (§2.2.20; also see Shubnikov & Koptsik 1974).

7.3.3 Coincidence Lattice

A 3-dimensional $T(p)$ implies the existence of a coincidence lattice.

Vectors of the coincidence lattice are solutions of the equations

$$\mathbf{t}2j = \mathbf{R}' \mathbf{t}_{1i}, \tag{7.3.12}$$

where $\mathbf{t}_{1i}$ and $\mathbf{t}_{2j}$ are vectors spanning the lattices T_1 of Crystal 1 and T_2 of Crystal 2:

$$T_1 = \{\mathbf{E}|\mathbf{t}_{1i}\}, \tag{7.3.13}$$

$$T_2 = \{\mathbf{E}|\mathbf{t}_{2i}\} \tag{7.3.14}$$

The vector $\mathbf{T}_{frac}$ in Eqs. 7.3.3 and 7.3.11 denotes any displacement of the black lattice (or Crystal 2) with respect to the white lattice (Crystal 1), away from a reference pattern in which at least one black lattice site is coincident with a white lattice site.

$T(p)$ defined by Eq. 7.3.11 is independent of $\mathbf{T}_{frac}$. To prove the validity of this statement, we note that, for a coincidence lattice to exist, we must have

$$\{\mathbf{E}|\mathbf{t}_{1i}\} = \{\mathbf{R}|\mathbf{T}_{frac}\}^{-1} \{\mathbf{E}|\mathbf{t}_{2j}\} \{\mathbf{R}|\mathbf{T}_{frac}\} \tag{7.3.15}$$

On using Eqs. 2.2.13 and 2.2.14 we get

$$\{\mathbf{E}|\mathbf{t}_{1i}\} = \{\mathbf{R}^{-1}|-\mathbf{R}^{-1}\mathbf{T}_{frac}\} + \{\mathbf{E}|\mathbf{R}^{-1}\mathbf{T}_{frac} + \mathbf{R}^{-1}\mathbf{t}_{2j} - \mathbf{R}^{-1}\mathbf{T}_{frac}\}$$

$$= \{\mathbf{E}|\mathbf{R}^{-1}\mathbf{t}_{2j}\} \tag{7.3.16}$$

The terms involving the displacement vector $\mathbf{T}_{frac}$ thus cancel out. For this reason, Gratias & Portier (1982) and Kalonji (1985) have advocated the use of the term 'coincidence lattice', rather than 'coincidence *site* lattice'. The coincidence lattice is the same, no matter which site in it is chosen for coincidence; the translational symmetry $T(p)$ is independent of $\mathbf{T}_{frac}$.

We consider coincidence lattices again in §7.4.

7.3.4 Dichromatic Complex

Eq. 7.3.10 defines the Fedorov-group symmetry, $\Phi(c)$, of the dichromatic complex (DCC). Its translational symmetry is the same as that of the DCP, and is therefore given by the group $T(p)$ (Eq. 7.3.11).

From $\Phi(c)$ one constructs the Shubnikov group Ш(c) of the DCC, so as to take cognizance of any antisymmetry operations (Pond & Vlachavas 1983).

The group Ш(c) may or may not be the same as the corresponding holosymmetric symmorphic group Ш$^*(c)$.

Unlike the DCC, the symmetry group Ш$^*(p)$ of the DCP is constructed from two space groups that are both holohedral and symmorphic. The symmorphic nature of the two constituent groups ensures that, for each of them, at least one point exists at which all the point symmetry elements intersect (§2.2.17). The coincidence of one such point in Crystal 1 with a similar point in Crystal 2 provides a natural *reference pattern*, corresponding to $\mathbf{T}_{frac} = 0$ in Eq. 7.3.3.

In nonsymmorphic lattice complexes the symmetry elements do not all intersect at a single point. In the work of Pond & Vlachavas (1983), a particular nonsymmorphic complex is taken as based on the symmorphic complex to which it is isomorphic, so as to define the reference structure corresponding to $\mathbf{T}_{frac} = 0$.

For nonsymmorphic bicrystals a whole set of equivalent DCCs can occur. These are given by the coset decomposition of Ш$^*(c)$ with respect to Ш(c).

7.3.5 Unrelaxed or Ideal Bicrystal

Introduction of an interface into the interpenetrating white and black lattice complexes, followed by wiping out of the black lattice complex on one side of the interface and the white on the other, gives us a bicrystal (unrelaxed or ideal) of translational symmetry $T(b)$:

$$T(b) \;=\; T(p) \cap G_{interface} \tag{7.3.17}$$

The possible bicrystal groups can be classified in terms of the dimensionality of the group $T(b)$: If this group has 2-dimensional periodicity, the symmetry group of the bicrystal is a *two-sided layer group*. If it is 1-dimensional, the symmetry is that of a *two-sided band* in the terminology of Shubnikov & Koptsik (1974). If $T(b)$ has no translational symmetry, one speaks of a *two-sided rosette symmetry*.

The following symmetry operations of the DCP can survive the introduction of an interface (Kalonji 1985): classical (i.e. non-colour-reversing) rotation axes and mirrors perpendicular to the interface; classical translations lying in the (planar) interface; antimirrors with any glide component parallel to the interface; antidiads of any orientation parallel to the interface; and antiinversion centres lying on the interface.

7.3.6 Relaxed Bicrystal

The holosymmetric ideal bicrystal of symmetry III*(*b*) described above may be only a metastable structure, which tends to relax to a thermodynamically more stable configuration, with a concomitant reduction of symmetry from III*(*b*) to III(*b*). Pond & Vlachavas (1983) have considered the following routes of relaxation:

- Rigid-body translation;
- Movement of the interface to a more stable position;
- Local atomic relaxation; and
- Insertion or removal of material at the interface.

7.3.7 The Six Bicrystal Systems

The possible bicrystal space groups have been derived and tabulated by Pond & Vlachavas (1983). For a more fundamental discussion of such groups the book by Shubnikov & Koptsik (1974) should be consulted.

Bicrystals belong to one or the other of six bicrystal systems. A bicrystal belonging to a particular system must possess the minimum essential point symmetry compatible with that system. The cubic system is not allowed. This is because a bicrystal must always have a *unique plane* (namely the interface), whereas the cubic system cannot allow a unique plane. The six bicrystal systems are: monadic (triclinic), diagonal (monoclinic), orthodiagonal, tetragonal, trigonal, and hexagonal.

Rosettes (an empty set $T(b)$) and layers (a 2-dimensional $T(b)$) can be associated with any of the six bicrystal systems, whereas bands (1-dimensional $T(b)$) can be compatible with only monadic, diagonal and orthodiagonal systems. This brings up an interesting difference between monocrystals and bicrystals. Whereas for the former it is possible to assign a crystal system unambiguously, for the latter the translational symmetry must also be specified simultaneously to avoid ambiguity. For example, the *diagonal band* bicrystal system is different from the *diagonal layer* system; the former is 1-dimensional, and the latter 2-dimensional, although both are monoclinic.

7.3.8 Bicrystallographic Variants

Spontaneous loss of a symmetry operator results in the occurrence of two or more equivalent states or variants. Fig. 7.3.3 lists the various types of variants possible in a bicrystal. Variants for each symmetry descent can be enumerated by writing a coset decomposition of the higher symmetry group with respect to the lower.

Orientational Variants

These variants can be mapped on to one another by those symmetry operators of Φ_1^* and Φ_2^* which are not present among the ordinary (classical) operators of $\Phi(p)$.

Complex Variants

The symmetry reduction, if any, in going from $\mathrm{III}^*(p)$ to $\mathrm{III}^*(c)$ leads to the formation of complex variants. Although $\mathrm{III}^*(c)$ is, by definition, a holosymmetric group, complex variants can arise if either the white or the black crystal is nonholosymmetric. They can also arise if one or both the crystals are nonsymmorphic.

Morphological Variants

These arise due to dissymmetrization from $\mathrm{III}^*(c)$ to $\mathrm{III}^*(b)$ (Fig. 7.3.3). They are called morphological variants because they either correspond to different orientations, or to different locations, of the interface. The various equivalent configurations of the interface have a one-to-one correspondence with the cosets in the coset decomposition of $\mathrm{III}^*(c)$ with respect to $\mathrm{III}^*(b)$. *Grain-boundary facetting and precipitate morphology are directly determined by the number and mutual disposition of morphological variants.*

Relaxational Variants

Symmetry reduction from $\mathrm{III}^*(b)$ to $\mathrm{III}(b)$ leads to relaxational variants. Since the relaxational processes occur around the interface only, these variants occur only on the (2-dimensional) interface, and are separated by 1-dimensional or line defects.

An Illustrative Example

We illustrate some of the concepts described above by considering the example of epitaxial growth of thin films of CdS on an NaCl substrate (Holt & Wilkox 1971; Multi & Holt 1972; Pond & Vlachavas 1983).

Since the interface is between two different crystals (interphase boundary), no antisymmetry operations are involved in this example.

CdS has the sphalerite structure. Let us first suppose that a thin layer of CdS in the (001) plane is deposited on the (001) plane of the NaCl substrate. It is observed in practice that the [110] direction of CdS is parallel to the [110] direction of the NaCl substrate.

For the substrate,

$$\Phi^*(\lambda) = Fm\bar{3}m, \tag{7.3.18}$$

and

$$\Phi(\lambda) = F\bar{4}3m, \tag{7.3.19}$$

and for the CdS deposit,

$$\Phi^*(\mu) = \Phi(\mu) = Fm\bar{3}m \tag{7.3.20}$$

λ and μ refer to the white and the black members of the bicrystal.

The translation group, $T(p)$, of the DCP for this problem is an empty set. Therefore the maximum symmetry group, $\text{III}^*(p)$, is the same as the underlying point group $G(p)$, namely $m\bar{3}m$. For the same reason, the group of the holosymmetric DCC can be identified with its underlying point group: $\text{III}^*(c) = G(c) = \bar{4}3m$. And the group of the bicrystal is $\text{III}^*(b) = G(b) = 2mm$. Thus the bicrystal belongs to the orthodiagonal rosette system.

There are no orientational variants in this case because $\text{III}^*(p) = \Phi^*(\lambda) = \Phi^*(\mu)$.

There can be two complex variants because

$$(m\bar{3}m) = (\bar{4}3m) \cup (\bar{1}) \tag{7.3.21}$$

One complex variant is related to the other through an inversion operation.

Similarly, the possible morphological variants are given by the following coset decomposition:

$$(\bar{4}3m) = (2mm) \cup (1,\, 2_x,\, m_{(101)},\, m_{(\bar{1}01)},\, m_{(01\bar{1})},\, m_{(011)}) \tag{7.3.22}$$

These can be visualized as the six {001} interfaces bounding a CdS crystal in the shape of a cube embedded in the NaCl matrix.

As a variation of this example, let us now suppose that the NaCl crystal chosen for depositing the thin film is parallel to the plane (110), instead of (001). The CdS film in this case is found to be oriented such that its (110) plane is parallel to the (110) plane of the substrate. The DCP and the DCC symmetries are the same as before. But the bicrystal now has the symmetry of a diagonal rosette:

$$G(b) = (m) = (1,\, m_{(\bar{1}10)}) \tag{7.3.23}$$

Coset decomposition of $G(c)$ with respect to $G(b)$ now gives 12 morphological variants. These correspond to dodecahedral facetting, compared to the cubic facetting described above.

SUGGESTED READING

A. V. Shubnikov & V. A. Koptsik (1974). *Symmetry in Science and Art.* Plenum Press, New York.

R. C. Pond & D. S. Vlachavas (1983). Bicrystallography. *Proc. R. Soc. London*, **A386**, 95.

H. F. Fischmeister (1985). Structure and properties of high-angle grain boundaries. *J. de Physique*, Colloque C4, **46**, 3.

A. P. Sutton & R. W. Balluffi (1996). *Interfaces in Crystalline Materials.* Clarendon Press, Oxford.

7.4 A TENSOR CLASSIFICATION OF TWINNING

Several classification schemes for twinning in crystals are in use. The simplest way of distinguishing between various types of twins is in terms of physical appearance: contact twins, lamellar twins, polysynthetic twins, mimetic twins, penetration twins, etc. (see Cahn 1954; Klassen-Neklyudova 1964; Milovsky & Kononov 1985).

Another classification is in terms of the origin of twins: transformation twins, mechanical twins, growth twins. These were considered in §7.2.

The pioneering and systematic work of G. Friedel (1926) resulted in the recognition of four twin types (§7.2): twinning by merohedry ($\Sigma = 1$, $\omega = 0$); twinning by pseudomerohedry ($\Sigma = 1, \omega \neq 0$); twinning by reticular merohedry ($\Sigma > 1$, $\omega = 0$); and twinning by reticular pseudomerohedry ($\Sigma > 1, \omega \neq 0$).

Friedel's classification was based on data obtained by optical and contact goniometry. After the advent of X-ray crystallography, another classification scheme was proposed by Donnay & Donnay (1974), cf. §7.2. It differed from Friedel's classification in that it interchanged the importance attached to the parameters Σ and ω. The primary subdivision of twinning in this scheme was done in terms of ω (rather than Σ, as in Friedel's scheme). If $\omega = 0$, we have twinning by twin-lattice symmetry (TLS), and if $\omega \neq 0$ we have twinning by twin-lattice quasisymmetry (TLQS). Further subdivision was done for each of these main categories in terms of the twin index Σ.

All these classification schemes, though serving their intended purposes very well, do not pack a large amount of crystallophysical information, and are therefore not very discriminative. For example, both Dauphiné and

Brazil twins of quartz come under the TLS category of Donnay & Donnay (1974), in spite of the fact that the response of these two types of twins to mechanical stress is very different. Dauphiné twins are transformation twins, and can be readily detwinned (Chapter 12). Brazil twins, on the other hand, are growth twins, and it is almost impossible to detwin them because of the nature of the bonding frozen into the interface at the nucleation/growth stage of the crystal.

Similarly, twinning (domain structure) resulting from a nondisruptive phase transition has several features quite distinct from those of twinning due to a reconstructive phase transition (§5.7.2). The very process of reconstruction in the latter case makes movement of twin walls under the action of applied forces very difficult, if not impossible. The old classification schemes for twinning have no provision for making use of such information.

Twinning in NH_4Cl is another example. Here the twin individuals differ in the sign of the piezoelectric coefficient d_{123}. However, neither the description "twinning by merohedry" (Friedel), nor "twinning by TLS" (Donnay & Donnay) conveys any information about this fact.

To absorb such additional information, one of the things one must do is to invoke the full space-group symmetry of the crystal. Twinning is determined by crystal structure, and crystal structure is properly described by the space group to which a crystal belongs.

A classification scheme for twinning based on space-group considerations has to deal with relationships between *two* space groups: the actual space-group type of the individual components comprising the twin, and another space group (not necessarily a supergroup of the first space group), which has at least one additional symmetry operator among its elements. This additional operator maps one twin individual to the other. Quite often, it is desirable, even indispensable, to choose the space group containing the twin-mapping operator as a supergroup of the other space group. One can then bring in the important notion of prototype symmetry, defined by us rigorously in §5.1 in terms of Guymont's nondisruption condition.

We introduce here a classification scheme for twinning that is centered around the important notion of prototype symmetry, and is designed to absorb information about the tensor distinction of domains and twin components (Wadhawan 1997).

We begin by dividing all twins into two main categories: those which differ in at least one tensor coefficient, and those for which all macroscopic tensor coefficients are the same (with reference to a common frame of reference) (Fig. 7.4.1). We call the latter *translation twins* or *T-twins.*

Twins which differ in at least one macroscopic tensor property (i.e. rotational twins, cf. §7.1.1) can be of two types: those for which a prototype symmetry is definable (we call them *Aizu twins*), and those for which the prototype is not definable (we call them *Bollmann twins* or *B-twins*).

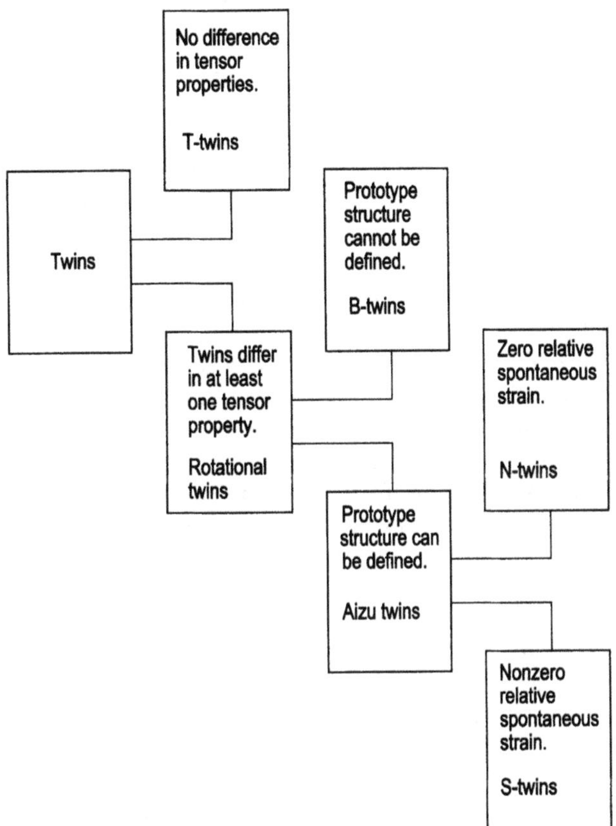

Figure 7.4.1: Tensor classification of twinning.

[Bollmann's (1970) book on crystalline interfaces marked the beginning of many of the current geometric ideas in bicrystallography. We propose to associate non-Aizu rotational twins with the name of Bollmann.]

For Aizu twins, within the parent-clamping approximation (§5.1.2), all the twin mapping operators are from among the n domain mapping operators g_j occurring in Eq. 7.1.8. Since we have already recognized T-twins as a separate category, all Aizu twins are necessarily *ferroic twins*, i.e., for them the mapping operator necessarily has a rotational component (in addition to the fact that it belongs to the prototype group); it may also have a fractional-translation component sometimes, arising from a screw-axis or glide-plane operation (Guymont, Gratias, Portier & Fayard 1976).

Aizu twins can be of two inherently different types: *ferroelastic twins* (or *S-twins*), and *nonferroelastic-ferroic twins* (or *N-twins*).

An S-twin can be considered as arising from a ferroelastic phase transition. Proper ferroelastic transitions are well described by mean-field theories like the Landau theory. This is because of the long-ranged nature of the elastic interaction. The same cannot be said, in general, about nonferroelastic-ferroic transitions, from which N-twins arise.

All twins can thus be divided into four fundamentally different classes: S-twins, N-twins, B-twins, and T-twins (Fig. 7.4.1). We shall refer to it as the *SNBT classification.*

This classification is based on concepts drawn from the theory of transformation twinning. But what is its validity for mechanical twins (§7.2.4), and for growth twins (§7.2.3)?

Two types of mechanical twins can be distinguished. Those corresponding to ferroelastic domain pairs, and the rest. The former are just S-twins, and the latter should be classifiable as B-twins. The validity of the second part of this statement follows from the description of mechanical twins in §7.2.4. The twin operator for Type 1 mechanical twins, Type 2 mechanical twins, and compound twins is a Seitz operator defined respectively by Eqs. 7.2.1, 7.2.2 and 7.2.3. If this operator can be identified with one of the g_js in Eq. 7.1.8, we have an S-twin; otherwise a B-twin. But in either case the SNBT classification scheme covers mechanical twins adequately.

We consider growth twins next. Twinning can occur both during the nucleation stage and the growth stage of a crystal (§2.1). The cluster-to-crystal transition (§2.1.2) necessarily involves a change of symmetry (Multani & Wadhawan 1990; Haberland 1994). For every symmetry operator lost at such a transition, equivalent configurations (twins) can appear in the microcrystal, and the possible twin states can be enumerated by using Eq. 7.1.8.

The symmetry of the clusters need not be from among the 32 crystallographic point groups. Icosahedral symmetry is favoured quite frequently. Gold clusters, for example, have icosahedral symmetry in the 4-15 nm regime. Since icosahedral symmetry is not compatible with translational periodicity of a bulk crystal in 3-dimensional space, the cluster-to-crystal transition in such a case is not at all a nondisruptive phase transition. It may also involve multiple twinning as an adjustment mechanism (§2.1.2).

In the ultimate analysis, peculiarities of the atomic structure determine the laws of twinning at the nucleation or growth stages. For example, the energy of formation of a faulted 2-dimensional embryo having the configuration of a rotational twin is very low for Si and Ge crystals (Tiller 1991a). Twinning at even the nucleation stage is therefore quite likely to occur for them.

Twinning can occur not only during the nucleation stage, but also during the growth stage of a crystal. The likelihood of a particular habit face becoming a twin plane is high if it has a high density of atoms and if a large fraction of the atomic sites is common to the individuals comprising the twin. For example, in crystals with fcc symmetry the twin plane is parallel to the octahedral face, which has the maximum density of atoms. This is typical of the spinel law of twinning. Similarly, in aragonite twins a fraction of the structure of $CaCO_3$ has a common orientation in the two components (Milovsky & Kononov 1985).

In growth twins, a twin-operation configuration can provide additional reentrant corners or junctions where the growth units can bind more strongly, resulting in enhanced growth rates compared to surface or terrace sites. However, since dislocations are also normally present in real specimens, the generation rate of layers of the growing crystal is determined by three primary competing mechanisms: (i) 2-dimensional or "pill-box" nucleation; (ii) screw dislocations; and (iii) twin-plane reentrant corners (Sunagawa 1987). In both (i) and (iii), layers are initiated by nucleation of 2-dimensional pill-boxes, but in (iii) only a partial pill-box, with a lower formation energy, is needed (Tiller 1991a).

Mechanical twinning can also occur during the growth of a crystal due to internal mechanical and thermal stresses (Tiller 1991b).

In the laboratory large crystals are often grown by starting with a seed crystal, which grows in size on being surrounded by the nutrient fluid under appropriate conditions. Twinning may exist beforehand in the seed selected. For example, the microtwins observed in hydrothermally grown quartz mainly originate at the surface of the seed, having been produced by the stress at high temperatures generated by the sawing procedures used for obtaining the seed from a larger crystal. The mechanical twins so produced on the surface of the seed crystal are not always dissolved away fully by the etching practices adopted (Kotru & Raina 1982; Kotru, Kachroo & Raina 1985).

To sum up, twinning can occur both during the nucleation stage and the growth stage of a crystal obtained from a liquid or a vapour state. The formation of variants at the cluster-to-crystal transition is not a thoroughly investigated subject yet. However, in view of the noncrystallographic symmetry often adopted by clusters, such growth twins are not very likely to be S-twins or N-twins. After the nucleation has occurred and growth is progressing, twinning can also arise as a result of "probability accidents" (Tiller 1991b), especially when such "accidents" can lead to increased growth rates because of the peculiarities of the atomic structure at certain habit faces. And the reentrant sites appearing as a result of such twinning operations can serve as ledges where attachment of growth units can occur at a faster rate than on the surface or terrace sites of the growing crystal.

We conclude that growth twins can be expected to be B-twins, in general (although exceptions are possible). The twinning operations involved in them are normally not traceable to a prototype symmetry group in the spirit of Eq. 7.1.8. We have already mentioned the case of Brazil twins in quartz. Another example is that of twins in III-V compound semiconductors, which have been identified by Chen et al. (1992) to be 60°-rotation twins, a nonferroic operation in the present context.

Before we discuss the characteristics of each of the four types of twins identified in the SNBT scheme, it remains to deal with one more aspect of twinning, namely that related to coincidence lattices.

The notion of the dichromatic pattern (DCP) was explained in §7.3.2. The translation group of the DCP is the translation group (T_I) underlying the intersection group I (Eq. 7.3.5). If T is the translation group underlying the group G of Component 1 of the twin, Eq. 7.3.5 gives

$$T_I = T \cap [\{\mathbf{R}|\mathbf{T}_{frac}\} T \{\mathbf{R}|\mathbf{T}_{frac}\}^{-1}] \tag{7.4.1}$$

A 3-dimensional T_I corresponds to the existence of a coincidence lattice, running right across the interface in the twin.

Although the group T_I comprises only lattice translations, the presence of the Seitz operator $\{\mathbf{R}|\mathbf{T}_{frac}\}$ (and its inverse) in Eq. 7.4.1 makes it possible to identify point-group operations which achieve the same invariance of the interpenetrating lattices of the bicrystal as that achieved by pure lattice translations of T_I. This is represented conveniently in terms of the *axis-angle pair* $\{[uvw]\,\theta\}$, where θ denotes the misorientation of Crystal 2 with respect to Crystal 1 through a rotation about an axis $[uvw]$ defined in the coordinate system of Crystal 1 (Pumphrey & Bowkett 1971).

Certain axis-angle pairs can result in twinning configurations (Woirgard & de Fouquet 1972; Ranganathan & Roy 1992). Coincidence lattices with twin index Σ can be generated from the Ranganathan equation (Ranganathan 1966):

$$\Sigma = x^2 + N y^2, \tag{7.4.2}$$

where x and y are integers, and $N = u^2 + v^2 + w^2$. If Eq. 7.4.2 gives an even value for Σ, it is repeatedly divided by 2 until an odd number is obtained. The rotation angle θ which results in a twin configuration is given by (Woirgard & de Fouquet 1972; Pumphrey & Bowkett 1972):

$$\theta = 2 \tan^{-1}(y/x)\sqrt{N} \tag{7.4.3}$$

Coincidence lattices can occur for both Aizu twins and Bollmann twins. The presence or absence of a coincidence lattice makes no difference to our tensor classification of twinning. If a coincidence lattice is present in a twin (whether of type S, N, B or T), this is an additional piece of information which can have a bearing on, for example, the energy of formation of the composition plane: the smaller the value of Σ, the more stable may be the composition plane (Gratias et al. 1979).

A crystal structure can be considered as consisting of a number of sublattice structures corresponding to the various sets of Wyckoff positions. Often, even a subset of one type of Wyckoff positions can constitute a sublattice (with an underlying translation group) (Gratias et al. 1979; Doni et al. 1985). A sublattice L is called a *total sublattice* by Gratias et al. (1979) if it consists of one or more complete sets of Wyckoff positions. If at least one set of Wyckoff positions is included only partially in the sublattice, it is a *partial sublattice.* Any number of sublattices can be constructed from a total or partial sublattice.

Several important features of total sublattices have been discussed by Gratias et al. (1979). The space group of a crystal is always a subgroup of the space group of any of its total sublattices. Further, if a total sublattice consists of only one set of Wyckoff positions, the twin mapping operation (if present) is always "translation reducible" (comprises only a point operation).

Available information about the total or partial nature of a sublattice in a given twinned crystal can be readily incorporated in the SNBT classification scheme by appending a subscript t (for total) or p (for partial) to the symbols S, N or B. For T-twins the twinning operation is only a rigid-body translation of one component with respect to the other, and a coincidence lattice is independent of such an operation (cf. §7.3.2).

The (110) mirror twins of pyrite (FeS_2) are an example of twins with a total underlying sublattice (Gratias et al. 1979). They are also B-twins. One can therefore represent them by the twin symbol B_t (cf. §7.4.5 below).

The (111) mirror twins commonly found in fcc metals are B-twins having a partial underlying coincidence lattice (twin symbol B_p).

7.4.1 S-TWINS

The distinguishing feature of S-twins is that they differ in at least one component of all second-rank polar tensor properties. Their presence is readily revealed under the polarizing microscope because of the relative disorientation of the optical indicatrices, or in a diffraction experiment because of the difference in the orientations of the respective crystallographic axes. Examples of purely ferroelastic S-twins are: $BaCl_2.2H_2O$, $BiVO_4$, and $Pb_3(PO_4)_2$.

Table 7.4.1 provides a comparison of the attributes of the four basic types of twinning. S-twins may differ not only in second-rank polar tensor properties, but also in other macroscopic properties like spontaneous polarization, spontaneous magnetization, spontaneous optical activity, compliance-tensor coefficients, etc. For example, ferroelectric S-twins occur in $BaTiO_3$, ferromagnetic S-twins in Mn_3O_4, and ferroelectric-ferromagnetic S-twins in Ni – I boracite. Some of the twins occurring in

Table 7.4.1: Comparison of attributes of the four basic types of twinning.

Attribute → Twin type ↓	Is a prototype structure conceivable?	Twins differ in which macroscopic tensor property?	Generalized susceptibilities near T_c	Is detwinning possible?	Origin of the twin
S	Yes	Second-rank polar. May differ in other properties also	At least one of them becomes arbitrarily large	Yes. Theory of ferroic transitions provides a systematic approach	Phase transition from the prototypic phase
N	Yes	Other than second-rank polar	At least one of them becomes arbitrarily large	Yes. Theory of ferroic transitions provides a systematic approach	Phase transition from the prototypic phase
B	No	No restriction on rank, but such twins are not derivative structures	?	Very difficult. No general theoretical approach exists	Growth twin; non-ferroelastic mechanical twin; disruptive phase transition
T	'Yes' in some cases, 'no' in others	None	There is no relevant generalized susceptibility	No	Transformation twin, or some types of stacking fault

dicalcium strontium propionate (DSP) are ferrogyrotropic S-twins (Glazer, Stadnicka & Singh 1981).

The spontaneous distortion of the prototype lattice, responsible for the occurrence of S-twins, necessitates the occurrence of disorientations (§7.1.4). Another consequence of this distortion is the formation of a variety of "tweed structures" (Salje 1990, 1994). Also, the critical fluctuations near the Curie temperature are influenced strongly by this distortion. The emerging phase tends to suppress the critical fluctuations because of the mismatch between the lattices of the old phase and the new phase. This has a bearing, not only on the critical fluctuations, but also on response functions of the crystal in the vicinity of the phase transition (Friedel 1981; Wadhawan 1985; Salje 1995).

7.4.2 N-Twins

N-twins differ in at least one macroscopic tensor property other than a second-rank polar tensor property, and, in addition, their twinning pattern is describable in terms of a prototype space group. We mention here some examples of N-twins: ferroelectric-ferrobielastic twins in $Pb_5Ge_3O_{11}$, TGS, SbSI, $LiNbO_3$ and $BaTiO_3$ (180° twins); ferrobielastic-ferroelastoelectric twins (Dauphiné twins) in α-quartz; ferromagnetoelastic twins in CoF_2 and $FeCO_3$; ferromagnetoelectric twins in Cr_2O_3; and several types of inversion twins.

Twins related purely by an inversion operation can be either N-twins or B-twins. They cannot be S-twins; this is because second-rank polar tensors are invariant under an inversion operation.

7.4.3 B-Twins

Unlike for S-twins and N-twins, a meaningful prototype structure is not conceivable for B-twins. Their main subclasses are as follows:

(i) Twinning occurring in a prototypic phase, rather than in a ferroic phase. For example, mechanical twinning in crystals with a cubic or a hexagonal point symmetry. A case in point is that of twinning induced in Mg crystals across $(10\bar{1}2)$ planes by a shear force along $[10\bar{1}\bar{1}]$ (Laves 1975).

(ii) Most growth twins are likely to be B-twins. An example is that of 60° rotation-twins in crystals of GaAs, GaP and InAs (Chen et al. 1992). Another example is that of Brazil twins in α-quartz and β-quartz.

(iii) Twinning configurations arising from special coincidence-lattice situations (cf. Eqs. 7.4.2 and 7.4.3).

(iv) Twins resulting from any phase transition in which the nondisruption condition is violated, and in which there is a loss of at least one rotational-symmetry operator.

B-twins differ from S-twins in that the latter are transformation twins which disappear on heating to the prototypic phase. That is, the atomic displacements involved in S-twinning are thermodynamically reversible; this is not the case for B-twins.

A symmetry analysis of B-twins can be carried out in terms of the intersection-group approach (cf. §7.3.1). The intersection group is determined by Eq. 7.3.5, which we rewrite here after replacing the symbol G_1 for the space group of Component 1 of the twin by the symbol G:

$$I = G \cap \{\mathbf{R}|\mathbf{T}_{frac}\} G \{\mathbf{R}|\mathbf{T}_{frac}\}^{-1} \tag{7.4.4}$$

Seitz operators g_j present in G but absent in I define the possible variants of the B-twin. The variants can be identified with the cosets in the following coset decomposition:

$$G = I + g_2 I + \cdots + g_j I + \cdots g_m I \tag{7.4.5}$$

Here m is the ratio of the orders of the groups G and I.

If the B-twins have resulted from a disruptive phase transition, the intersection group is the intersection of two distinct space-group types, and is given by Eq. 7.3.1. For the transition $G_1 \rightarrow G_2$ the B-twin states in the phase with symmetry G_2 have a one-to-one correspondence with the cosets in the following equation:

$$G_1 = I + g_{12} I + \cdots + g_{1j} I + \cdots g_{1k} I \tag{7.4.6}$$

Here k is the index of I in G_1.

Similarly, the reverse transition $G_2 \rightarrow G_1$ can result in variants identifiable with the cosets in the following:

$$G_2 = I + g_{22} I + \cdots + g_{2j} I + \cdots g_{2l} I \tag{7.4.7}$$

Here l is the index of I in G_2.

7.4.4 T-Twins

T-twins do not differ in any macroscopic tensor property at all. Their detection requires the use of techniques like HRTEM, etching, and X-ray diffraction topography with a superlattice reflection.

A familiar example of T-twins is the twinning observed across antiphase boundaries in the alloy Cu_3Au below 667 K. The lowering of symmetry from $Fm\bar{3}m\,(O_h^5)$ to $Pm\bar{3}m\,(O_h^1)$ at the disorder-order transition leads to the occurrence of four possible translation states. With reference to Eq. 7.1.8 these correspond to

$$g_1 = \{1|000\},\ g_2 = \{1|\frac{1}{2}\frac{1}{2}0\},\ g_3 = \{1|\frac{1}{2}0\frac{1}{2}\},\ g_4 = \{1|0\frac{1}{2}\frac{1}{2}\} \qquad (7.4.8)$$

A surprisingly large number (about 50) of crystals (mainly organic compounds and metallic alloys) in which nonferroic phase transitions occur, leading to the formation of T-twins, have been listed by Toledano & Toledano (1982). These authors also establish the following theorem in this context:

> *A nonferroic phase transition can only be induced by an IR whose small representation τ_n is 1-dimensional (real or complex).*

7.4.5 A Symbol for Twinning

> *A good notation has a subtlety and suggestiveness which at times make it seem like a live teacher.*
>
> Bertrand Russell

We introduce a compact and informative symbol for twinning. It consists of one of the four letters S, N, B or T corresponding to the four types of twinning, followed by one or more lower-case letters in brackets which represent the tensor properties in which the twins differ (Wadhawan 1997).

Consider the case of twinning in α-quartz. This crystal is ferrobielastic, as well as ferroelastoelectric. Our twin symbol for its Dauphiné twins is therefore $N(d, s)$, where d denotes the fact that the two components of the twin differ in at least one piezoelectric coefficient, and s represents their difference with respect to the compliance tensor. This type of twinning disappears when the crystal makes a transition to the higher-symmetry β-phase on heating.

By contrast, the Brazil twins of quartz are growth twins, with a mirror operation parallel to the optic axis as the twinning operation. This type of twinning does not disappear on transition to the β-phase. In any case, mirror symmetry is not present in the space-group symmetry $P6_222$ (or $P6_422$) of β-quartz, and cannot be a ferroic mapping operation g_j of Eq. 7.1.8. These are thus B-twins, and not N-twins, and their twin symbol is B(g), where g denotes the difference of the twin components with respect to the optical gyration tensor. Depending on the context, as well as the availablity of information, this symbol can be expanded to include other tensor properties in which the twins differ (e.g. the acoustical gyration tensor).

Table 7.4.2 shows several examples of the twin symbol in use. It also serves to demonstrate the higher information content of the tensor classification scheme described here. In this table, when the twins differ in spontaneous polarization, the letter p appears in the twin symbol. Similarly, e denotes that the twins differ in spontaneous strain, m stands for spontaneous magnetization, g for a coefficient of the optical-gyration tensor, d for a piezoelectric coefficient, s for compliance, q for magnetoelastic, and α for magnetoelectric coefficient(s). The subscripts t and p denote the existence of total and partial coincidence sublattices, respectively, across the twin interface.

The SNBT classification scheme for twinning in terms of tensor properties makes it possible to make practical use of the results of group-theoretical analyses of tensor distinction of domains resulting from ferroic phase transitions (Janovec, Richterova & Litvin 1993; Litvin, Litvin & Janovec 1995).

SUGGESTED READING

M. Guymont (1981). Symmetry analysis of structural phase transitions between phases not necessarily group-subgroup related. *Phys. Rev. B*, **24**, 2647.

P. Toledano & J.-C. Toledano (1982). Nonferroic phase transitions. *Phys. Rev. B*, **25**, 1946.

V. Janovec, L. Richterova & D. B. Litvin (1993). Nonferroelastic twin laws and distinction of domains in nonferroelastic phases. *Ferroelectrics*, **140**, 95.

D. B. Litvin, V. Janovec & S. Y. Litvin (1994). Nonferroelastic magnetoelectric twin laws. *Ferroelectrics*, **162**, 275.

V. K. Wadhawan (1997). A tensor classification of twinning in crystals. *Acta Cryst.*, **A53**, 546.

7.5 THE GROUP-TREE FORMALISM

Consider a crystal the symmetry of which decreases from G_0 to G_1 on cooling through a phase transition temperature T_c. The lower-symmetry phase has to arise and grow in the anisotropic crystalline environment of the parent phase. Therefore its actual symmetry is different, usually lower, than the inherent symmetry G_1 it would have in an isotropic environment

Table 7.4.2: Comparison of the SNBT classification scheme for twinning with two other schemes. See text for details.

Twin	Present scheme (twin type)	Friedel's scheme (twinning by)	Donnay & Donnay's scheme (twinning by)	References
α-quartz (Dauphiné)	N(d,s)	Merohedry	TLS	1,2
$Pb_5Ge_3O_{11}$	N(p,s,g)	Merohedry	TLS	3,4
$BaTiO_3$ (180° twin)	N(p,g)	Merohedry	TLS	1,4
NH_4Cl	N(d)	Merohedry	TLS	1,4
$CuCsCL_3$	N(d,g)	Merohedry	TLS	1,4,5
CoF_2, $FeCO_3$	N(q)	Merohedry	TLS	1,6
Cr_2O_3	N(α)	Merohedry	TLS	1,6
α-quartz (Brazil)	B(g)	Merohedry	TLS	2
Pyrite [(110) mirror twin]	B_t	Merohedry	TLS	7,8
F.c.c. metals [(111) mirror twin]	B_p	Reticular merohedry	TLS	7
Mg	B	?	TLQS	1
Cu_3Au	T	?	TLS	9
$BaTiO_3$ (90° twin)	S(p,e)	Pseudo-merohedry	TLQS	1
Fe_3O_4	S(m,e)	Pseudo-merohedry	TLQS	10
Ni-I boracite	S(p,m,e)	Pseudo-merohedry	TLQS	10
Aragonite	S(e)	Reticular pseudo-merohedry	TLQS	8

1. Wadhawan (1982); 2. Donnay & Donnay (1974); 3. Toledano & Toledano (1976); 4. Aizu (1972); 5. Wadhawan (1979); 6. Newnham & Cross (1974); 7. Gratias et al. (1979); 8. Cahn (1954); 9. Portier & Gratias (1982); 10. Aizu (1970a).

(e.g. on growth from a fluid phase). Sometimes the dissymmetrizing influence is external (e.g. an applied field), rather than, or in addition to, internal. Irrespective of the nature of the influence, the net symmetry of any crystal (e.g. a ferroic phase coexisting in an anisotropic environment with its parent phase) can be determined by invoking the Curie principle (§C.1). Such considerations have been developed by Portier & Gratias (1982) for determining the enhanced number of variants possible for the daughter phase because of the additional loss of symmetry elements caused by the anisotropic influences. Their procedure has come to be known as the group-tree formalism (cf. Fig. 7.5.1 below, which has the appearance of a tree).

For describing this procedure we begin by noting that the internal or external influence is present, not only for the daughter phase, but also for the parent phase. Let (g) be the symmetry group of the environment or "solicitation". For example, (g) may be the symmetry of just a scalar field like temperature if the system is simply cooled or heated through the phase transition and it is assumed that no other influences are present. Or (g) may be either the group mmm or ∞/mm if some internal or external strain field is present.

We first construct the following intersection group:

$$I_0 = G_0 \cap G_1 \qquad (7.5.1)$$

Portier & Gratias (1982) call I_0 *the group of isoprobability of nucleation*. This is because if a particular nucleus of the daughter phase can be formed, so also can, with equal probability, all other nuclei obtainable from it by application of the operations of I_0. When (g) has the symmetry of a scalar, I_0 is the same as G_0.

Effectively, the final phase of symmetry G_1 arises from the initial phase of symmetry I_0, rather than from G_0.

To find out how many variants of the final phase can form, we have to identify symmetry elements present in I_0 but absent in G_1. For this we construct another intersection group N_{01}:

$$N_{01} = I_0 \cap G_1 \qquad (7.5.2)$$

The subscripts 0 and 1 refer to the initial and final phases.

The group N_{01} comprises elements common to both I_0 and G_1, and a representative set of n_{01} symmetry elements present in I_0 but absent in G_1 will provide, through their action on G_1, all the distinct variants of the final phase, with n_{01} given by

$$n_{01} = |I_0| \,/\, |G_1| \qquad (7.5.3)$$

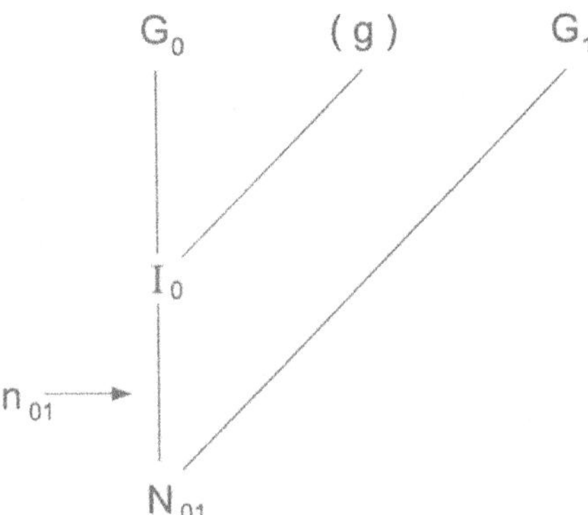

Figure 7.5.1: Depiction of the basic group-tree formalism. [After Portier & Gratias 1982.]

The group I_0 can be split into n_{01} left cosets with respect to the intersection group N_{01}, and there is a one-to-one correspondence between the left cosets and the variants.

We depict in Fig. 7.5.1 the complete group-tree for the symmetry descent from G_0 to G_1. We shall have occasion to use this formalism later in the book (e.g. when discussing martensitic phase transitions in Chapter 11). However, it is necessary to sound here a word of caution regarding its indiscriminate use based on symmetry arguments alone, as pointed out by Wadhawan (1988) by considering the example of the possible number of variants in the ferroelastic superconductor Y – Ba – Cu – O (also see Wadhawan & Boulesteix (1992)).

This crystal has a tetragonal phase of symmetry $G_0 = 4_z/m_z m_x m_{xy}$ at high temperatures. On cooling from this phase, an apparently continuous phase transition occurs to a ferroelastic phase of symmetry $G_1 = m_x m_y m_z$. We can assume, for the sake of simplicity of argument, that the tetragonal phase is prototypic for this crystal (see Wadhawan & Glazer 1989; Tiwari & Wadhawan 1991).

As the crystal is cooled from above the transition temperature, its spontaneous strain increases gradually from the value zero at the Curie temperature T_c. The point groups $4/mmm$ and mmm are of orders 16 and 8 respectively, so that 16/8 or two ferroelastic orientation states (differing in the interchange of m_x and m_y) are possible. However, because of the nonzero spontaneous strain, a rotation or disorientation (of 0.45^o at room temperature) must occur about the z-axis if the domains are to make contact with one another without developing cracks at the domain boundaries (Wadhawan 1988). Therefore the true point-group symmetry of a domain has to be taken as $M_x M_y m_z$, where M_x and M_y denote mirror planes not strictly parallel to m_x and m_y.

The group tree for this system can therefore be drawn as in Fig. 7.5.2.

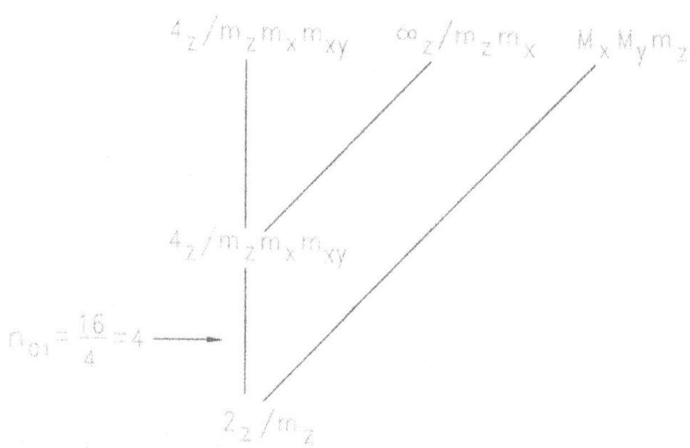

Figure 7.5.2: Group tree for the ferroelastic superconductor Y–Ba–Cu–O.

In this group tree the solicitation (g) is shown as having the symmetry of a cylinder ($\infty_z/m_z m_x$), which is taken as representing the symmetry of the strain field in which the ferroelastic domains exist. We see that $I_0 = 4_z/m_z m_x m_{xy}$, so that

$$N_{01} = 4_z/m_z m_x m_{xy} \cap M_x M_y m_z = 2_z/m_z \tag{7.5.4}$$

The number of disorientation states predicted by this simplified analysis is the index of the group $4_z/m_z m_x m_{xy}$ in the group $2_z/m_z$, namely 4. However, from a more detailed analysis which takes due note of the physics of the problem, the possible number of disorientation states has been shown to be a function of sample history, and in any case much larger than 4 (Wadhawan 1988). The general result is that suborientation states can occur, in principle, at any of the following orientations:

$$[(1 \pm 2n)\,\theta] \text{ or } [(1 \pm 2n)\theta + 90^o] \;\; [\text{mod}180^o],$$

with $\theta = 0.45^o$ and $n = 0, 1, 2, \ldots$

The reason for the failure of the group-tree formalism is that the symmetry arguments given above cannot make a distinction between *one* disorientation operation (a rotation of 0.45°) and *multiple* disorientations which can occur in a real specimen. We get 'M_x not parallel to m_x' irrespective of the number of times the basic disorientation (of 0.45^o) is repeated. Only physical arguments can make this distinction.

Symmetry arguments alone, though elegant and systematizing, cannot be trusted always to produce the right numbers.

SUGGESTED READING

R. Portier & D. Gratias (1982). Symmetry and phase transformation. *J. de Physique*, Colloque C4, **43**, 17.

Chapter 8

DOMAIN WALLS

In Chapter 7 we focussed attention on domains themselves, and not so much on the interfaces that separate one domain from another. We take a closer look at interfaces in this chapter. As in Chapter 7, we adopt a somewhat general approach, and discuss not only domain walls in ferroics but also interfaces in nonferroic materials, as well as interfaces between non-identical structures. This would help us in viewing phase boundaries and domain boundaries in ferroics in the larger perspective of heterophase and homophase interfaces in materials.

8.1 ORIENTATIONAL DEPENDENCE OF PROPERTIES OF INTERFACES

8.1.1 Morphology of Crystals Grown from Crystalline Matrices

Group of the Wulff Plot

A crystalline phase of a material can grow either from a fluid phase (vapour, melt, solution), or from another solid phase. In the former case the morphology that the crystal adopts is in conformity with the point-group symmetry underlying the space-group of the crystal. In fact, not only shape, but also all other macroscopic properties of the crystal have directional symmetry that is equal to, or higher than, the point-group symmetry, in accordance with the Neumann theorem.

What is the symmetry of the shape of a crystal that grows, as a result of a phase transition, inside a crystalline parent phase? Which point-group does this morphology conform to? The answer is provided by the simple notion of *intersection-group symmetry*, which we encounter repeatedly in our discussion of domain boundaries and phase boundaries in this and the previous chapter (see Eq. 7.3.1, for example) (Cahn & Kalonji 1981; Portier & Gratias 1982; Kalonji & Cahn 1982; Kalonji 1985).

Let the old or parent phase be called Crystal 1, and the new or daughter phase Crystal 2. We are interested in knowing the directional symmetry of the interfaces between the two phases, or the growth morphology of Crystal 2 that has grown inside Crystal 1.

Let S_1 and S_2^0 denote the symmetry groups of Crystal 1 and Crystal 2. The symmetry operators of S_1 are defined with reference to the coordinate axes oriented and located with respect to atomic positions in Crystal 1. Similarly the symmetry elements of S_2^0 are defined with respect to Crystal 2. To determine the intersection group for the bicrystal, we must first obtain the symmetry group (say S_2) for Crystal 2 in the frame of reference of Crystal 1, so that their elements are defined in a common frame of reference.

Let $\{\mathbf{R}|\mathbf{T}_{frac}\}$ be the Seitz operator which, when applied to Crystal 1 (the coordinate system of which we have chosen as a common frame of reference for defining all symmetry elements and operators, including $\mathbf{R}$ and $\mathbf{T}_{frac}$), maps it to Crystal 2. The symmetry group (S_2) of Crystal 2 in the common frame of reference is then given by

$$S_2 = \{\mathbf{R}|\mathbf{T}_{frac}\} \, S_2^0 \, \{\mathbf{R}|\mathbf{T}_{frac}\}^{-1} \tag{8.1.1}$$

We next single out those symmetry operators (if any) which map Crystal 1 onto itself and, simultaneously, Crystal 2 onto itself. These operators define the intersection group I:

$$I = S_1 \cap S_2, \tag{8.1.2}$$

with S_2 given by Eq. 8.1.1, so that

$$I = S_1 \cap \{\mathbf{R}|\mathbf{T}_{frac}\} \, S_2^0 \, \{\mathbf{R}|\mathbf{T}_{frac}\}^{-1} \tag{8.1.3}$$

Let a unit vector $\mathbf{n}$ define the orientation of the (planar) interface between Crystal 1 and Crystal 2. We define $\mathbf{n}$ in the common frame of reference chosen above, which is also the frame of reference inherent to Crystal 1.

Operators of the group I leave both Crystal 1 and Crystal 2 invariant, but can flip around and translate the interface defined initially by its normal $\mathbf{n}$. The set of planes defined by all the distinct directions of the plane-normals obtained by the operations of I defines a closed or open simple form (§2.2.16). Cahn & Kalonji (1981) call I *the group of the Wulff plot.*

The use of the name Wulff refers to the well-known *Wulff theorem* (sometimes called the *Gibbs-Curie-Wulff theorem* in the theory of crystal morphology (see Chernov 1984; Pimpinelli & Villain 1998)).

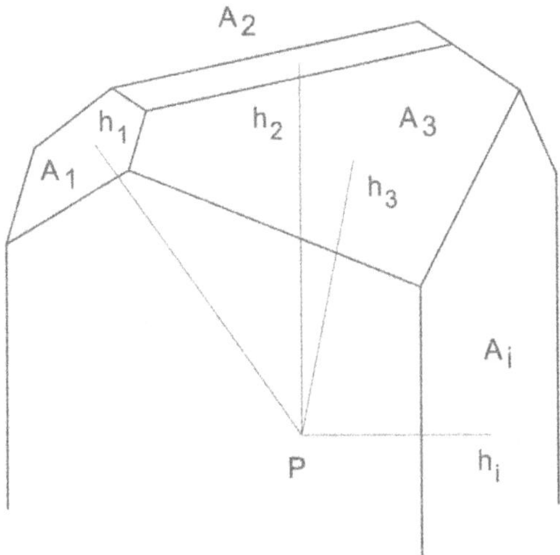

Figure 8.1.1: The Wulff construction.

The Wulff theorem determines the growth morphology of a crystal which is at equilibrium with its fluid (isotropic) surroundings. Imagine a crystal of volume V, having facets of area A_i and surface-free-energy densities σ_i. At thermodynamic equilibrium, we must have

$$\delta \sum_i \sigma_i A_i = 0, \tag{8.1.4}$$

subject to the constraint $\delta V = 0$. Solution of these simultaneous equations leads to the following result:

$$\sigma_1 : \sigma_2 : \ldots = h_1 : h_2 : \ldots \tag{8.1.5}$$

Here the h_i are distances from a point P inside the crystal (Fig. 8.1.1), and are defined by the following statement which embodies the *Wulff construction* or the Wulff theorem:

> *When a crystal is in its equilibrium shape in an isotropic environment, there exists a point inside it whose perpendicular distances (h_i) from all faces of the crystal are proportional to their specific surface free energies.*

The directional symmetry of the Wulff construction is equal to, or higher than, the point-group symmetry of the crystal, in conformity with the Neumann theorem.

The group of the Wulff plot consists of symmetry operators which generate from a given interface or surface the entire set of surfaces physically equivalent to the initial surface.

The operations of I are symmetry operations for the simple form defined by the Wulff plot or the Wulff construction. The group of the Wulff plot therefore represents the symmetry of all interfacial properties.

It is often justifiable to ignore translational operations, and work only with the point groups underlying the space groups S_1 and S_2.

We see that the group of the Wulff plot has a direct bearing on the morphology of the simple forms which appear when a new crystalline phase emerges in the crystalline environment of the parent crystal. This group thus plays a role similar to that played by the full point group of a crystal that grows in an isotropic environment, i.e. from a fluid phase.

8.1.2 Homophase Interfaces

For heterophase interfaces discussed above, because of the different crystal lattices across the interface, very little symmetry may survive when the intersection group I (Eq. 8.1.3) is constructed. This symmetry can be substantially higher when we have the same crystalline material on the two sides of the interface, i.e. when the interface is a homophase interface. Domain walls and grain boundaries are examples of this type of an interface. The enhancement of symmetry results from the fact that now an additional set of symmetry operations (equal in number to the order of the intersection group I) can arise which map Crystal 1 to the *other* member of the bicrystal, namely Crystal 2, and Crystal 2 to crystal 1. These were called the antisymmetry or colour-reversal operations in §7.3.1. We examine here the effect of this additional symmetry on the orientational dependence of the properties of the homophase interface.

Let us arbitrarily call Crystal 1 white, and Crystal 2 black. This makes it convenient for us to describe the symmetry of the homophase interface in terms of a Shubnikov group III. The group I is a classical (or Fedorov) subgroup of III.

If we operate on the bicrystal by any operator of I, each member crystal maps back onto itself, and the interface moves to an equivalent orientation and position.

The colour-reversing operations are all contained in the set (III $- I$), with the group III defined by Eqs. 7.3.2 and 7.3.6. Any operation from this set, followed by an inversion of the planar interface, will map Crystal 1 to Crystal 2, Crystal 2 to Crystal 1, and the interface to a new, equivalent, orientation and position.

The group of the Wulff plot for a bicrystal with a homophase interface is thus the following:

$$W = I \cup (\text{III} - I)\,\bar{1}, \tag{8.1.6}$$

where the inversion operator is required to be introduced if we follow the convention that the normal to the interface be taken as pointing from the white member to the black member. [Any colour-reversal operation on the bicrystal will invert the sense of this normal.]

We combine Eqs. 7.3.2 and 7.3.6 to write the following coset decomposition:

$$\text{III} \equiv I + (\text{III} - I) = I + \{\mathbf{R}'|\mathbf{T}'_{frac}\}I \qquad (8.1.7)$$

Thus there are only two cosets in the decomposition of III with respect to its subgroup I.

For further analysis, we first recall here that any two cosets have either all elements in common, or none at all (§B.1).

Case 1: $\bar{1} \in (\text{III} - I)$. Since the inversion operator is in the coset $(\text{III} - I)$, it cannot be in the other coset in Eq. 8.1.7, namely the group I. Also, if $(\text{III} - I)$ contains the inversion operator, then $(\text{III} - I)\bar{1}$ must contain the identity element E (because $\bar{1}\bar{1} = E$). But the identity element is also present in coset (I), the latter being a group. Since the two cosets cannot share an element, unless they are identical, it follows that, for Case 1, the group of the Wulff plot is

$$W_1 = (I) \cup (I) = I \qquad (8.1.8)$$

Thus, although we have a bicrystal with a homophase interface, the possible morphological symmetry of the interface is still only I in this case, in spite of the fact that additional symmetry exists in the form of interchangeability of the member crystals.

Case 2: $\bar{1} \in (I)$. Here, since the inversion operator cannot be in the coset $(\text{III} - I)$, the identity element E cannot be in the set $(\text{III} - I)\bar{1}$, which is therefore not a group; it is thus distinct from the coset (I) which is a group. Eq. 8.1.4 therefore reduces to

$$W_2 = \text{III} \qquad (8.1.9)$$

Case 3: $\bar{1} \notin \text{III}$. Here inversion symmetry is absent from the intersection group, as well as from the colour-reversing operations. The morphological group is therefore that given by Eq. 8.1.6:

$$W_3 = (I) \cup (\text{III} - I)\bar{1} \qquad (8.1.10)$$

We shall have occasion to refer to the Wulff plot when we consider martensitic transformations (Chapter 11).

It is important to mention here that the symmetry groups of Wulff plot derived so far are correct only for an isotropic ambient for the specimen as a whole. If an additional anisotropic force field, e.g. a uniaxial stress (say of symmetry F) is present, another intersection group has to be constructed, in accordance with the Curie principle, to determine the net symmetry of the Wulff plot:

$$W' = W \cap F, \tag{8.1.11}$$

with W given by Eqs. 8.1.8, 8.1.9, or 8.1.10.

8.1.3 Symmetry-Dictated Extrema

The use of variational principles in physics is well known. For example, equilibrium states of a system can be determined by first defining its free energy, $\Phi(X)$, in terms of all the relevant parameters, and then determining those values of the parameters which correspond to minima of $\Phi(X)$ in the configuration space defined by the parameters X. Often an analytical solution of such a problem can be an extremely complex exercise, and one is then interested in knowing only the *minimal* symmetry possessed by $\Phi(X)$. The idea behind such an approach is the following (Cahn & Kalonji 1981; Kalonji 1982, 1985; Gratias & Portier 1982; Portier & Gratias 1982):

Suppose the scalar $\Phi(X)$ is invariant under operations of a group G operating on the configuration space defined by the parameters X. G is also the symmetry group of the constant-Φ surface defined in the configuration space. Certain *special points* on this surface are *symmetry-dictated extrema.* This comes about because of the following theorem (for the proof of which the paper by Portier & Gratias (1982) may be consulted):

> *The little group of the gradient of a G-invariant scalar function at a point X is the same as the little group of the point X.*

Symmetry dictated extrema, or special points, are those points X in configuration space which do not remain invariant under operations of G. At such points the gradient of a scalar function like $\Phi(X)$ must vanish, implying that $\Phi(X)$ has a maximum, or a minimum, or a saddle point. Further, if an extremum is found *experimentally*, it must be a minimum.

Cahn & Kalonji (1981) have analysed the occurrence of symmetry dictated extrema in bicrystals as a function of their relative orientation (specified by $\mathbf{R}$ in the Seitz operator $\{\mathbf{R}|\mathbf{T}_{frac}\}$). The dependence of symmetry-dictated extrema on rigid-body translations $\mathbf{T}_{frac}$ (e.g. when stacking faults occur) has been examined by Gratias & Portier (1982). Several other applications of this approach for interfaces in bicrystals have been mentioned by Kalonji (1985).

Symmetry dictated extrema are an important concept for analysing precipitate morphology, facetting, eutectic solidification, and martensite morphologies.

SUGGESTED READING

J. W. Cahn & G. Kalonji (1981). In *Proc. Int. Conf. on Solid-Solid Transformations*, Carnegie-Melon University, Pittsburgh, Aug. 10-14, 1981. Published by the Metals Society of the AIME, USA.

R. Portier & D. Gratias (1982). Symmetry and phase transformation. *J. de Physique*, Colloque C4, **43**, 17.

G. Kalonji (1985). A roadmap for the use of interfacial symmetry groups. *J. de Physique*, Colloque C4, **46**, 249.

V. Janovec, W. Schranz, H. Warhanek & Z. Zikmund (1989). Symmetry analysis of domain structure in KSCN crystals. *Ferroelectrics*, **98**, 171.

R. C. Pond & P. E. Dibley (1990). On the morphology of included crystals: bicrystalline kaleidoscopes. *Colloque de Physique*, **51**, C1.

I. Lyuksyutov, A. G. Naumovets & V. Pokrovsky (1992). *Two-Dimensional Crystals*. Academic Press, New York.

D. M. Hatch, P. Hu, A. Saxena & G. R. Barsch (1996). Systematic technique for the study of interphase boundaries in structural phase transitions. *Phys. Rev. Lett.*, **76**, 1288.

D. M. Hatch & W. Cao (1999). Determination of domain and domain wall formation at ferroic transitions. *Ferroelectrics*, **222**, 1.

8.2 STRUCTURAL EXTENDED DEFECTS

Domain walls and other interfaces in a crystalline specimen can be viewed as defects, in the sense that the atomic structure in these regions is not the same as the regular structure in regions far away from them. It is instructive to view them from the perspective of an overall classification of defects in crystals.

Defects in crystals can be divided into four broad categories: point defects, line defects, defect clusters, and extended defects (see Alario-Franco (1987) for a review).

Extended defects are defined as those having more than one dimension (Alario-Franco 1987). There are two main types: structural extended defects (SEDs), and compositional extended defects (CEDs).

SEDs are also called *conservative* or *physical* defects. By contrast, CEDs are *nonconservative* (Synkers, Delavignette & Amelinckx 1971) or *chemical* defects because they produce (or accommodate) a variation of composition where they occur. We discuss CEDs in a separate section (§8.3).

The main types of SEDs are: antiphase boundaries, stacking faults, twin walls, and grain boundaries. We have already dealt with their symmetry aspects in §7.3 and 7.4. Some additional structural aspects are described here from a different vantage point.

8.2.1 Aristotype and Hettotype Structures

Certain SEDs can be regarded as special or limiting cases of CEDs, and it is useful to describe such SEDs in the terminology of CEDs.

Very often the description of CEDs can be systematized by taking recourse to the notions of structural families, aristotypes, and hettotypes.

> *Structures may be said to belong to the same structural family if there is a one-to-one correspondence between all their atoms and between all their interatomic bonds* (Megaw 1973).

For example, α-quartz and β-quartz belong to the same structural family, even though their crystal structures are different. Similarly, diamond and zinc blende belong to the same structural family. But zinc blende is not in the same structural family as wurtzite which has a topologically different linkage pattern.

The *aristotype* is the simplest and the most symmetrical member of a structural family (Megaw 1973). It is a parent structure with perfect stoichiometry, from which a whole family of structures can be derived by a variety of "shear" displacements on appropriate planes, alongwith the requisite elimination of excess atoms from the initial structure (Wadsley 1964; Anderson & Hyde 1965, 1967).

A *hettotype* is a member of a structural family with a symmetry lower than that of the aristotype. Hettotypes are *derivative structures* of an aristotype.

Fig. 8.2.1 shows an example of an aristotype, namely the structure of ReO_3. We shall consider below, as well as in the next section, a number of hettotypes that can be derived from this aristotype.

8.2.2 Antiphase Boundaries

Antiphase boundaries (APBs) are interfaces between antiphase or out-of-step domains (or T-twins). They can arise, for example, when there is a

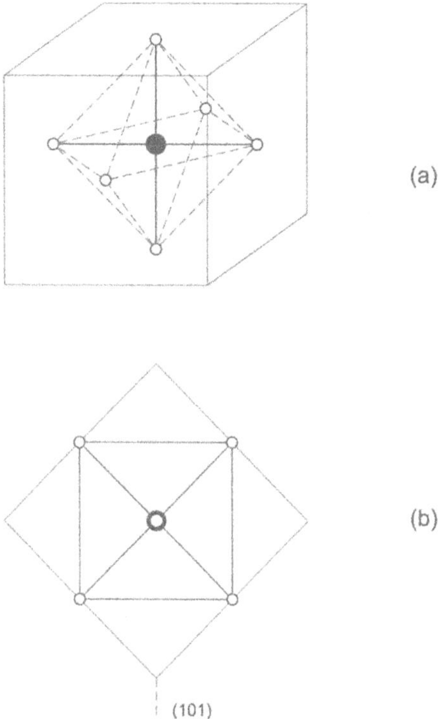

Figure 8.2.1: Unit cell of an idealized aristotype structure of ReO_3. (a) 3-dimensional view; (b) projection along [001]. This projection is used in Fig. 8.2.2 and in subsequent figures in this section.

lowering of translational symmetry (cf. Fig. 7.2.2).

Fig. 8.2.2 shows how an APB can arise from an ReO_3 type of aristotype structure. Starting from the ideal initial structure (Fig. 8.2.2a), with corner-sharing octahedra $[M - O_6]$, we imagine (101) as the slip plane (Fig. 8.2.2b). The structure to the left of this plane is kept fixed, and that to its right is given a translation $\frac{1}{2} < \bar{1}01 >$, resulting in an APB in Fig. 8.2.2c.

Before the imaginary formation of the APB, the coordination number for metal atoms is 6 [M-O_6, octahedral], and that for oxygen atoms is 2 [O-M_2, linear]. As seen from Fig. 8.2.2(c), these numbers do not change at the APB. This means that the introduction of the APB does not lead to any change in the overall composition, in keeping with its description as a *structural* extended defect, rather than a compositional extended defect.

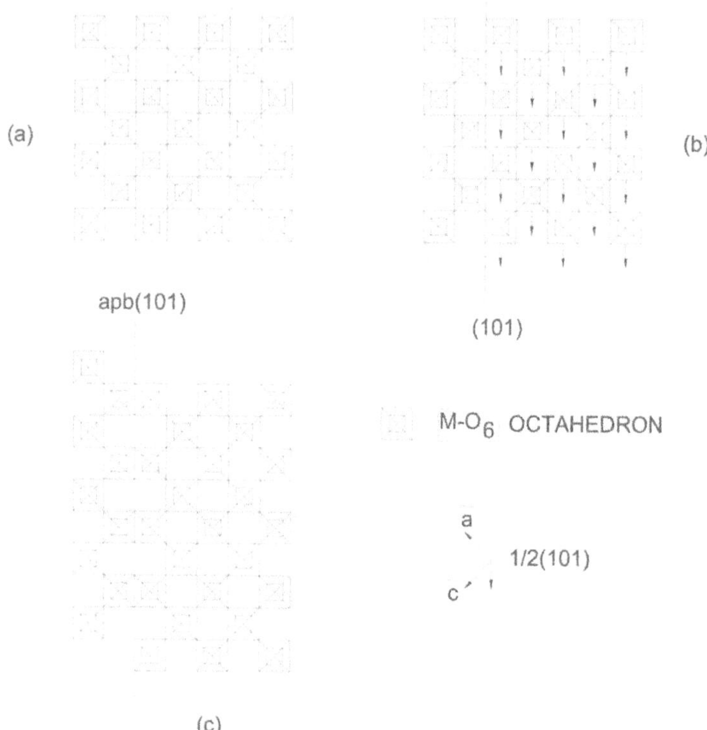

Figure 8.2.2: Construction of an antiphase boundary in the ReO_3 structure by the crystallographic operation $\frac{1}{2} < \bar{1}\,0\,1 > (101)$. [After Alario-Franco 1987.]

8.2.3 Stacking Faults

In certain close-packed structures, a sequence ..ABCABCABC.. of layers is sometimes broken by a stacking fault to become, say, ..ABCABABC.. This extended defect is again of a conservative nature, with no net change of composition around it.

8.2.4 General Twin Walls

Structural twinning (as opposed to chemical twinning described in the next section) preserves coordination numbers, on the whole, in the twin-boundary region. The reasons for this are similar to those depicted in Fig. 8.2.2 for the case of an APB.

8.2.5 Grain Boundaries

Grain boundaries are similar to twin boundaries (in terms of stoichiometry conservation), but the mapping operations for them are not of a constant or recurrent nature. The non-constancy of the mapping operation usually does not have a serious effect on the overall constancy of the composition of the material in regions near or away from the grain boundary.

SUGGESTED READING

M. A. Alario-Franco (1987). Extended defects in inorganic solids. Cryst. Latt. *Def. & Amorph. Mat.*, **14**, 357.

8.3 COMPOSITIONAL EXTENDED DEFECTS

Compositional extended defects (CEDs) are relevant in the context of variation of stoichiometry in crystals. When the deviations from perfect stoichiometry are small, the point-defect model of Schottky & Wagner (1931) is adequate. However, if the number of sites with vacancies or interstitials exceeds a fraction of a percent, the point-defect model is no longer appropriate (see, e.g., Rao 1984). Usually an ordering of such defects takes place, giving rise to a rich variety of structures, microstructures, and nanostructures.

For a large class of materials, especially oxides of mixed-valence cations, overall nonstoichiometry is widely prevalent. The cation in such (usually binary) compounds can adopt suitable oxidation states to match the local valence requirements of nonstoichiometry.

The nonstoichiometry results in vacant cation sites. These vacancies are seldom distributed randomly, except at very high temperatures and/or for very small deviations from perfect stoichiometry. The third law of thermodynamics (cf. §5.4.5) is partly responsible for this (Anderson 1973; Khachaturyan 1983). It leads to a tendency for ordering (reduction of entropy). Vacancy ordering results in regions of different compositions, separated by CEDs. Wadsley's (1964) work marked some major initial advances in the understanding of CEDs.

We now consider, very briefly, some specific CEDs, following mainly the description given by Alario-Franco (1987).

8.3.1 Crystallographic Shear Planes

To illustrate how compositional variations are produced or accommodated at CEDs, we consider again the aristotype ReO_3 depicted in Fig. 8.2.2(a)

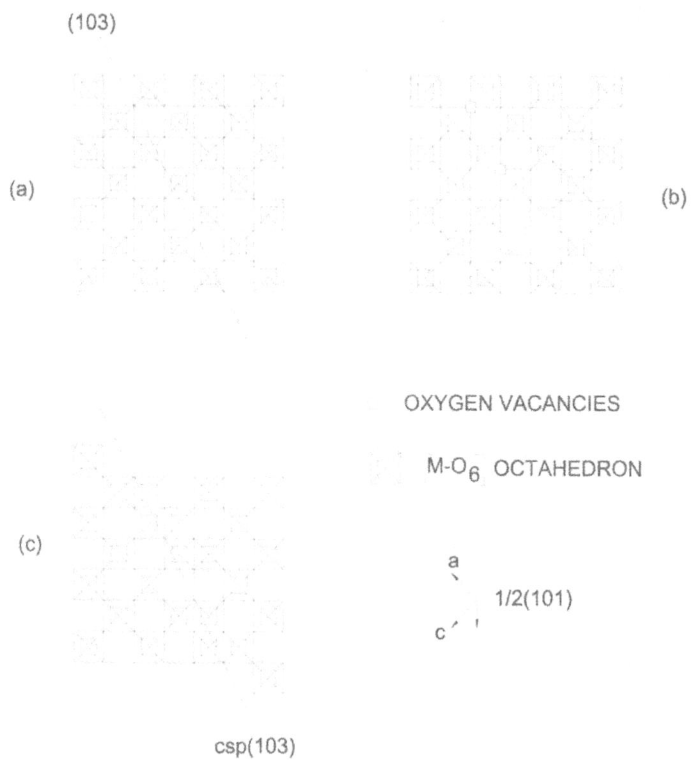

Figure 8.3.1: Formal construction of a rational (or crystallographic) shear plane through the crystallographic operation $\frac{1}{2} < \bar{1}01 >$ (103). [After Alario-Franco 1987.]

(and redrawn in Fig. 8.3.1(a)). The difference we introduce here is that the shear plane is taken as (103), rather than (101) taken for Fig. 8.2.2. The displacement vector is the same as before, namely $\frac{1}{2} < \bar{1}01 >$. In other words, the displacement vector is no longer parallel to the slip plane. This can be possible only if certain selected atoms near the slip plane are absent (which amounts to a local change of composition); otherwise some atoms will lie too close to each other. This is shown in Fig. 8.3.1(b).

Fig. 8.1.3(c) shows the final result of the shear applied on the *crystallographic slip plane* (CSP) (103). There is a faulted region, a slice of the structure parallel to the slip plane, in which the coordination numbers stand altered because the [M-O_6] octahedra share *edges* rather than corners. The coordination number of some anions has changed from 2 [M-O_2] to 3 [M-O_3], although the coordination number of cations still remains 6

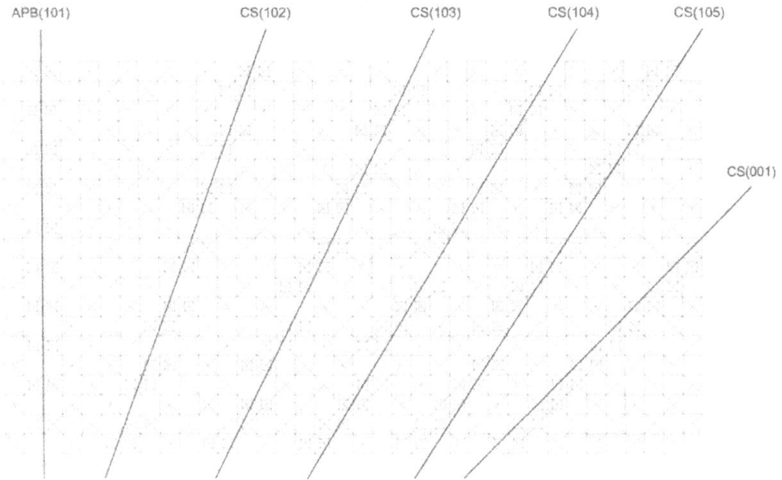

Figure 8.3.2: Variation of orientation and stoichiometry of an ReO_3 structure with index l for the crystallographic slip plane $\frac{1}{2} < \bar{1}01 > (1\,0\,l)$. [After Alario-Franco 1987.]

$[M\text{-}O_6]$. There is thus an altered stoichiometry, a feature characteristic of CEDs.

A whole set of off-stoichiometric configurations can be obtained by varying the index l of the slip plane $(10l)$; $l = 1$ for Fig. 8.2.2, and $l = 3$ for Fig. 8.3.1.

Fig. 8.3.2 shows the situation for a number of values of l, all corresponding to the same displacement vector $\frac{1}{2} < \bar{1}01 >$. The orientation, and therefore the stoichiometry, of the CSP changes with l. It can be verified that the number of octahedra sharing edges in each case is $2l$.

The structure of $WO_{2.90}$ can be viewed as resulting from WO_3 by a $\frac{1}{2} < \bar{1}01 > (103)$ crystallographic shear operation on one out of every ten oxygen-only planes (Magneli 1950).

One can obtain a homologous series of structures by varying the *spacing* between the ordered CSPs. And by changing the *orientation* of the CSPs, as in Fig. 8.3.2, a "family of families" of crystalline solids can be obtained, all having the same structural basis but different physical and chemical properties. Aperiodic configurations are also quite common in systems such as WO_3 and TiO_2.

8.3.2 Irrational Shear Planes

The existence of irrational (or noncrystallographic) shear planes was recognized by Anderson (1973) in the so-called *infinitely adaptive structures.*

Certain composition ranges, such as Ti_nO_{2n-1} for $n = 10$ to 14, exhibit such structures. For them, *every* composition orders into a *perfect* but *different* superlattice. There is neither a discrete succession of phases, nor a forming of nonstoichiometric or solid solutions with disordered defect structures. Instead, each composition exhibits a diffraction pattern that can be indexed in terms of a *single phase* with a (usually) large unit cell. The situation is analogous to that of polytypism in SiC, ZnS, CdI etc.

In such structures there exist, say, i basic subunits, each with a unique configuration derived from the aristotype with specific site occupancies, CSP orientations, and stacking sequences. Suppose these subunits constitute a set with unit cells that are $m_1, m_2, \cdots m_i$ times the unit cell of the aristotype, with $x_1, x_2, \cdots x_i$ as the corresponding compositions. Then the overall multiplicity m^*, and the composition x^* are built up from $a_1, a_2, \cdots a_i$ of these subunits (Anderson 1973):

$$m^* = a_1 m_1 + a_2 m_2 + \cdots a_i m_i, \tag{8.3.1}$$

$$x^* = a_1 x_1 + a_2 x_2 + \cdots a_i x_i \tag{8.3.2}$$

Any atomic ratio can give an adaptive structure by a choice of a sufficiently large value of m^*.

8.3.3 Chemical Twin Planes

Chemical twins differ from the usual *structural* twins in that certain atoms at the twin plane must be absent for stereochemical registry to occur. The notion of chemical twins was introduced by Anderson & Hyde (1974). The example of the NaCl structure can be used for illustrating its meaning (Baker & Hyde 1978; Alario-Franco 1987).

If we reflect the NaCl structure across the plane (113), and eliminate one member of each pair of atoms which lie too close to each other as a result of this operation, and allow the remaining atoms at the interface to relax to equilibrium positions, an acceptable trigonal configuration is obtained (Fig. 8.3.3). Such twinning at the unit-cell level can result in a variety of new structures and stoichiometries. For example, we can obtain the structure of $CaTi_2O_4$ from the NaCl structure by alternating twinning every four (113) planes. Repeating the chemical twinning every three (113) planes gives the structure of Re_3B (Alario-Franco 1987; Andersson & Hyde 1974).

The fascinating concepts developed in the field of CEDs have direct relevance to the subject of the domain structure of nonstoichiometric ferroics. Much remains to be done in this direction, although several detailed

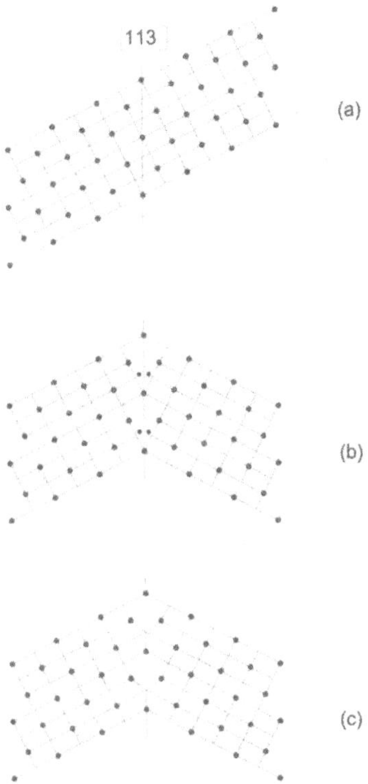

Figure 8.3.3: Illustration of chemical twinning. (a) Projection of the NaCl structure on the $(1\,\bar{1}\,0)$ plane. (b) Reflection of the structure across the plane (113). (c) Elimination of half the atoms at the interface that are unphysically close to each other as a result of the reflection across (113). [After Hyde et al. 1974.]

studies on the microstructure and nanostructure of rare-earth-oxide ferroics already exist (see Boulesteix (1983), Schweda (1992), and Ben Salem & Yangui (1995)).

SUGGESTED READING

J. S. Anderson (1973). On infinitely adaptive structures. *J. Chem. Soc. Dalton*, 1107.

C. N. R. Rao & K. J. Rao (1978). *Phase Transitions in Solids: An Approach to the Study of the Chemistry and Physics of Solids.* McGraw-Hill, New York.

M. A. Alario-Franco (1987). Extended defects in inorganic solids. *Cryst. Latt. Def. & Amorph. Mat.*, **14**, 357.

8.4 ATOMIC DISPLACEMENTS UNDERLYING THE MOVEMENT OF DOMAIN WALLS

Under the action of an appropriate driving field, a ferroic domain can be often made to grow at the expense of a neighbouring domain. When this happens, the wall separating the domains moves away from the domain growing in size, and towards the domain decreasing in size. If we know the crystal structures of the prototype and the ferroic phases, we can readily determine the net atomic displacement vectors, $\mathbf{\Delta}$, instrumental in the switching of one domain to the other.

Two basic types of information are used in a formal (group-theoretical) analysis of the problem (Wadhawan & Somayazulu 1986). One is that the symmetry group of the ferroic phase is a subgroup of the prototype space group G (Eq. 7.1.8). The other is that the atomic structure is only a slight distortion of the prototype structure, brought about by a nondisruptive phase transition. Consequently, apart from the symmetry relations which exist in the ferroic structure in keeping with its space-group symmetry H, there exist *pseudosymmetry relations* (Abrahams 1971; Abrahams & Keve 1971). The latter are really the symmetry relations of the higher-symmetry group G of the prototype, but are approximately true even for the ferroic phase because the ferroic phase is only a slight distortion of the prototype structure. Each such pseudosymmetry relation determines an atomic displacement vector $\mathbf{\Delta}$ for every atom. Atoms of the ferroic phase undergo these (net) displacements when a ferroic domain changes to another ferroic domain.

We first consider a very simple example to illustrate these ideas.

$BaCl_2.2H_2O$ is a nonferroelectric ferroelastic crystal (Wadhawan 1978a, 1982). It has been assigned the Aizu symbol $mmmF2/m$. The space-group symmetry of the room-temperature monoclinic phase is $P2_1/n(C_{2h}^5)$. Purely from an inspection of its atomic coordinates it is readily inferred that the following pseudosymmetry relation holds for it:

$$x_2,\, y_2,\, z_2 \;=\; (\tfrac{1}{2} - x_1,\, y_1,\, \tfrac{1}{2} + z_1) + \mathbf{\Delta}, \qquad (8.4.1)$$

where the magnitude of $\mathbf{\Delta}$ varies between 0.265 and 1.018 Å for the various atoms.

Eq. 8.4.1 with $\Delta = 0$ is *not* a symmetry operation of the ferroic-phase space group $P2_1/n$. The existence of such a relation is indicative of a higher, but approximate, symmetry.

The crystal structure with space-group $P2_1/n$ symmetry has four general Wyckoff positions. Therefore, we can generate three more pseudosymmetry relations similar to Eq. 8.4.1 by using the other three members of the Wyckoff set generated from the general point x_1, y_1, z_1. By putting $\Delta = 0$ in all these, we obtain a total of 8 equivalent points, four of them belonging to the group $P2_1/n$, and four additional ones. These 8 points are readily seen to comprise the general Wyckoff set for the space group $Pcnb\,(D_{2h}^{14})$. This space group has been therefore assigned as the prototype symmetry for this structure (Wadhawan 1978a).

On heating, the crystal gradually loses its water molecules, and finally decomposes. The prototypic phase is thus hypothetical in this case, and its space group has been inferred from the pseudosymmetry displayed by the atomic coordinates in the ferroelastic phase at room temperature, *with further corroboration provided by the observed domain structure.*

Two mutually perpendicular types of ferroelastic domain walls can be easily created and moved in this crystal by applying a small uniaxial stress. The domain walls are defined by the equations $x = 0$ and $z = 0$ (under the PCA).

On a macroscopic level, the point group $2/m$ of the ferroic phase has only one plane of mirror symmetry, namely $y = 0$. And the point group mmm of the prototype has three such planes: $x = 0$, $y = 0$, and $z = 0$. The mirror planes $x = 0$ and $z = 0$ are lost on going to the ferroic phase, and therefore become candidates as domain walls, as indeed observed experimentally (see Wadhawan 1982). Although Eq. 8.4.1 with $\Delta = 0$ represents the operation of a c-glide normal to the x-axis, on a macroscopic (point-group) level, it corresponds to a reflection across the plane $x = 0$. Therefore the various values of $\mathbf{\Delta}$ in this equation for various atoms give the (net) atomic displacement vectors responsible for the movement of the domain wall $x = 0$. The movement of the other domain wall, namely $z = 0$, is brought about by atomic displacement vectors $\mathbf{\Delta}'$ given by a pseudosymmetry relation corresponding to a different operation of the prototype space group $Pcnb$, namely

$$x_2,\, y_2,\, z_2 \;=\; (x_1,\, \tfrac{1}{2} + y_1,\, \bar{z}_1) + \mathbf{\Delta}', \tag{8.4.2}$$

This corresponds to the b-glide normal to the z-axis.

The procedure for calculating $\mathbf{\Delta}$ and $\mathbf{\Delta}'$ for various atoms is quite simple: We generate the atomic positions (x_2, y_2, z_2) by using Eq. 8.4.1 and 8.4.2 with $\mathbf{\Delta} = \mathbf{\Delta}' = 0$. Since the two ferroelastic orientation states are only slight distortions of the prototype, and therefore of each other, the calculated positions (x_2, y_2, z_2) will be found to be close to actual coordinates of atoms of the same species. The difference between (x_2, y_2, z_2) and the actual coordinates of the atom nearby gives the displacement vectors $\mathbf{\Delta}$ or $\mathbf{\Delta}'$. Atoms move by vectors $\mathbf{\Delta}$ (or $\mathbf{\Delta}'$) across the domain wall when the old ferroic domain becomes the new domain under the influence of the driving field. Fig. 8.4.1 shows this for the case of the movement of the domain wall $x = 0$ in $BaCl_2.2H_2O$.

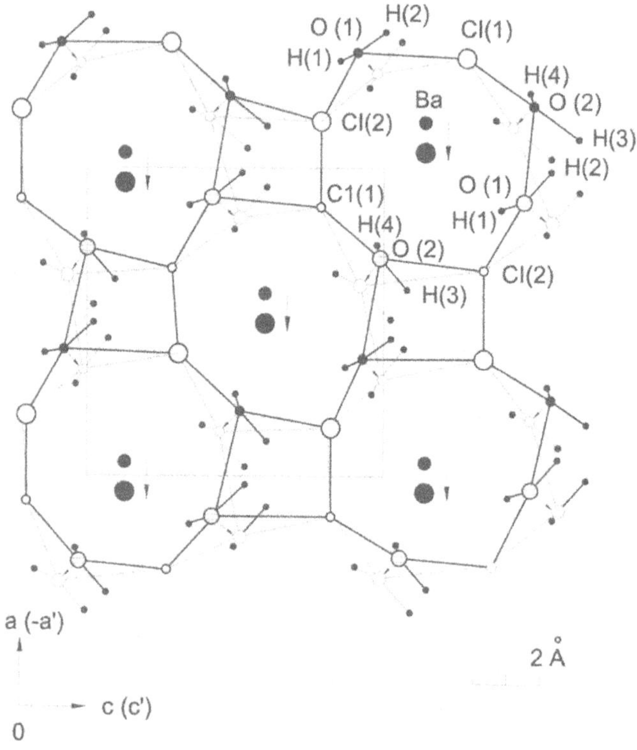

Figure 8.4.1: Atomic positions in $BaCl_2.2H_2O$ before (solid lines) and after (dashed lines) a ferroelastic state shift accompanying the movement of a domain wall perpendicular to the a-axis. The monoclinic 2-fold axis is perpendicular to the plane of the diagram. The atomic displacement vectors Δ are determined by Eq. 8.4.1 as atoms move from (x_1, y_1, z_1) to (x_2, y_2, z_2). The numbers near the atomic positions are the y-coordinates in Å units. Atoms move in the (net) directions of the arrows when the domain wall moves.

A special and very interesting situation arises sometimes, wherein there is no actual atom of the same species close to the calculated position (x_2, y_2, z_2). This was first noted by Wadhawan (1978b) in the case of

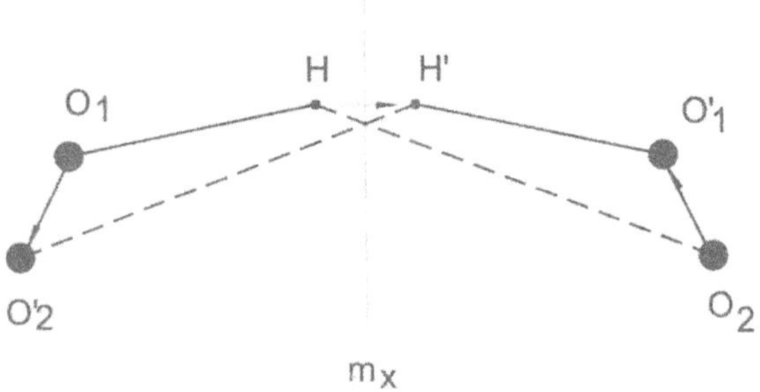

Figure 8.4.2: Effect of a pseudosymmetry operation on a hypothetical $O_1 - H \cdots O_2$ hydrogen bond in the xy-plane. O_1 is the donor and O_2 the acceptor oxygen before the pseudosymmetry operation, namely a reflection (m_x) normal to the x-axis, is applied. This operation maps the coordinates of O_1, H and O_2 to O_1', H' and O_2' respectively. The oxygen atom closest to the generated position O_1' is O_2, so O_2 moves to the position O_1' when the domain wall moves. Similarly O_1 moves to the generated position O_2'. But the situation is different for the hydrogen atom. Whereas there are two oxygen atoms, there is only one hydrogen. And whereas the oxygen atoms are quite distant from m_x, the hydrogen atom is close to it. The fact there are two oxygen atoms makes it possible for their identities to be interchanged by the pseudosymmetry operation. By contrast, since there is only one atom H, the *same* atom must move to the new position H'. As a result, the roles of the donor and acceptor oxygen atoms are interchanged. This also illustrates a very important difference between a *pseudo*symmetry operation and a real symmetry operation. The latter can never lead to an interchange of the roles of donor and acceptor atoms, whereas the former can do so.

H_3BO_3 crystals, and has since been reported for n-heptyl- and n-octylammonium dihydrogen phosphate crystals also (Fabry et al. 1997). Urea inclusion compounds may present another such possibility (Brown & Hollingsworth 1995). In all such cases the concerned atoms are hydrogen atoms, involved in hydrogen-bonding.

The stress-induced movement of the hydrogen atoms to the calculated sites (x_2, y_2, z_2) in all these cases has the effect of interchanging the roles of donor and acceptor atoms in the hydrogen bond. Fig. 8.4.2 illustrates this schematically, and its caption explains why this happens.

What happens to hydrogen atoms during ferroelastic switching in crystals like H_3BO_3 happens to the basal-plane oxygen atoms in the case of the high-T_c superconductor Y – Ba – Cu – O, which is also a ferroelastic at room temperature (Somayazulu, Rao & Wadhawan 1989). This crystal is a nonstoichiometric ferroelastic (Wadhawan 1991). The stress-induced hopping of basal-plane oxygen atoms in it is analogous to 'Snoek relaxation' (§11.1.4). We could turn the analogy around, and call the hydrogen-hopping in H_3BO_3 an example of Snoek relaxation *within* hydrogen bonds.

All in all, a reasonably accurate, though somewhat idealized, calculation of atomic displacement vectors can be carried out (under the PCA) from all possible relations of the type

$$x_2,\, y_2,\, z_2 \;=\; S\,(x_1,\, y_1,\, z_1) + \boldsymbol{\Delta}, \tag{8.4.3}$$

where $S(x_1, y_1, z_1)$ is a space-group operation of the prototype symmetry that is not an operation of the space-group symmetry of the ferroic phase in question.

Group-Theoretical Determination of Atomic Displacement Vectors

The above approach, though simple, is not very general. It may also not always ensure completeness. The most general approach is through the use of domain-structure systematics (Wadhawan & Somayazulu 1986).

One begins by writing the following left-coset decomposition (Eq. 7.1.8):

$$G \;=\; H \;+\; g_2 H \;+\; \cdots \; g_n H \tag{8.4.4}$$

Here n is the index of H in G, and the g's are a representative set of elements of G not belonging to H (Aizu 1974; Guymont 1981). Each coset corresponds to a particular domain-type or variant.

Let G_p and H_p denote the point groups underlying the space groups G and H. One can write a coset decomposition in terms of the point groups also:

$$G_p \;=\; H_p \;+\; f_2 H_p \;+\; \cdots \; f_q H_p \tag{8.4.5}$$

Here q is related to n in Eq. 8.4.4 through

$$n \;=\; q\,m, \tag{8.4.6}$$

where m is the number of times the primitive unit cell of the ferroic phase is larger than that of the prototype.

Eq. 8.4.4 provides a complete set of pseudosymmetry relations for determining the atomic displacement vectors, one for each coset in this equation. One identifies the g's with the operation S in Eq. 8.4.3 one by one, and determines the displacement vectors in each case for every atom in the asymmetric unit of the ferroic phase.

If H is not a normal subgroup of G, several conjugate subgroups are possible. A coset decomposition must be made with respect to each of these (Guymont 1978).

For writing a coset decomposition like Eq. 8.4.4, one has to deal with two space groups, namely the prototype group and the ferroic group. It is necessary to choose a common origin for this purpose, as also a common coordinate system (see Wadhawan & Somayazulu 1986 for a detailed example).

If one is dealing with a ferroelastic domain pair, disorientations of the domains introduce a complication in the choice of common coordinate axes. Various approximations have been employed for dealing with this problem (Guimaraes 1979a; David, Glazer & Hewat 1979).

Ferroelastic Switching and Acoustic Emission

We conclude this section by pointing out that pseudosymmetry relations described above give only the *net* displacement vectors $\mathbf{\Delta}$ for the atoms in a unit cell after a ferroic switching of a domain has occurred. One determines by the above procedure the *total* vector displacement each atom undergoes for the domain switching to occur. These net displacement vectors are actually made up of a number of zigzag displacement vectors:

$$\mathbf{\Delta} = \mathbf{\Delta}_1 + \mathbf{\Delta}_2 + \ldots \qquad (8.4.7)$$

This is because during ferroic switching a domain wall has to move across the crystal, and the domain wall has a symmetry group different from those of the domains separated by it. The symmetries of the sites occupied by the atoms can change several times when the domain wall is passing through their positions, especially if the domain wall has substantial effective thickness. The net atomic displacement vectors can therefore be composed of a whole series of successive components (Eq. 8.4.7). This has a bearing on, for example, the acoustic emission that accompanies ferroic switching (Mohamad, Zammit-Mangion, Lambson & Saunders 1982; Zammit-Mangion & Saunders 1984).

SUGGESTED READING

S. C. Abrahams & E. T. Keve (1971). Structural basis of ferroelectricity and ferroelasticity. *Ferroelectrics*, **2**, 129.

M. Guymont (1978). Domain structure arising from transitions between two crystals whose space groups are group-subgroup related. *Phys. Rev. B*, **18**, 5385.

M. Guymont (1981). Symmetry analysis of structural transitions between phases not necessarily group-subgroup related. Domain structures. *Phys. Rev. B*, **24**, 2647.

V. K. Wadhawan & M. S. Somayazulu (1986). Symmetry analysis of the atomic mechanism of ferroelastic switching and mechanical twinning. Application to thallous nitrate. *Phase Transitions*, **7**, 59.

8.5 DOMAIN STRUCTURE OF INCOMMENSURATE PHASES

Several ferroic materials pass through an incommensurate phase (I) before making a transition to the parent phase (P) on heating (cf. Eq. 5.8.1):

$$P \xrightarrow{T_i} I \xrightarrow{T_c} C \tag{8.5.1}$$

Notwithstanding some thermal hysteresis (Janovec, Godefroy & Godefroy 1984), the I-phase can be taken as existing between the temperatures T_i and T_c. At T_c it changes to the C-phase, which is *commensurate* to the parent phase or the P-phase.

Near the T_i-end the I-phase has a sinusoidal modulation with respect to the C-phase, and near the T_c-end one usually sees discommensurations (cf. §5.8). The latter are narrow transient domain walls which separate wide domain-like regions.

The C-phase is usually a ferroic derivative of the prototypic P-phase. This means that the C-phase can exist in two or more orientation states. In the I-phase, near its T_c-end, orientation states or domains appear which are almost commensurate regions, with discommensurations serving as domain walls.

Since several orientations are possible for the domains and the domain walls in the I-phase (near the T_c-end), the term *domain texture of the I-phase* has been used in the literature for describing the microstructure (Janovec & Dvorak 1986).

Janovec (1981, 1983) has developed a pictorial representation for domains and domain walls in general, which is particularly useful in visualizing the discommensuration texture in an incommensurate phase.

The $P - I - C$ sequence of phase transitions is typically induced by a two-component order parameter $\eta(p, q)$, and in the I-phase there occurs a modulation of this order parameter.

Following Janovec (1981), we shall consider the concrete example of the ammonium fluoberyllate (AFB) crystal (Iizumi & Gesi 1977) to illustrate the symmetry analysis of domain texture in an I-phase.

For AFB the P-phase has the symmetry *Pnam*, and the symmetry group of the (ferroelectric) C-phase is $Pn2_1a$. The primitive unit cell of the C-phase is twice as large as that of the P-phase. The ferroic species $mmmFm2m$ to which this crystal belongs has two orientation states, say

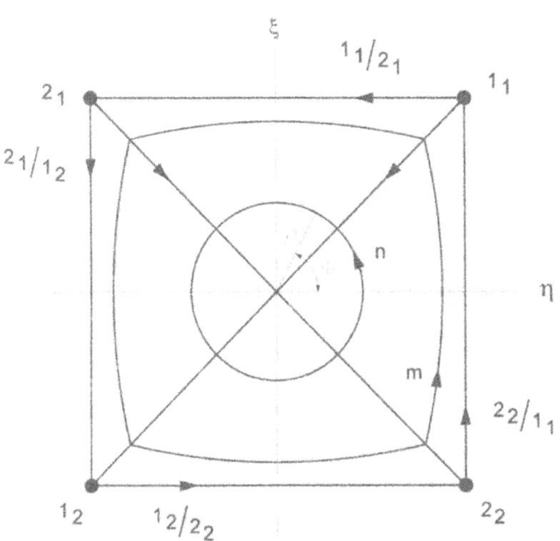

Figure 8.5.1: Order-parameter-space representation of domain states, domain walls, and incommensurate modulations in an ammonium fluoberyllate crystal. [After Janovec (1981).]

1 and 2. The cell-doubling along the x-axis on transition to the C-phase means that two translational domain-states are possible for each of the two orientation states. Therefore, in terms of the notation introduced in §7.1.11, we can represent the four domain states as 1_1, 1_2, 2_1, and 2_2. In the C-phase the states 1_1 and 1_2 have the same spontaneous-polarization vector, pointing in the a-direction, opposite to that for states 2_1 and 2_2.

The order parameter for the hypothetical $P \rightarrow C$ phase transition has two components (Ishibashi & Dvorak 1978), which we denote by η and ξ. In the pictorial representation introduced by Janovec (1981), a domain state corresponds to a point in the order-parameter space spanned by η and ξ. The four single-domain states are represented by the points marked 1_1, 1_2, 2_1, and 2_2 in Fig. 8.5.1.

A finite-thickness domain wall between any two domain states is represented by a directed line segment connecting the points which represent the domain states. Such a depiction takes note of the fact that a thick domain wall has a continuously varying structure.

Note that in Fig. 8.5.1 the straight lines joining 1_1 with 1_2, and 2_1 with 2_2 represent antiphase domain walls. In the notation of §7.1.11 we denote these walls by $1_1/1_2$ and $2_1/2_2$ respectively.

Similarly, the walls $1_1/2_1$, $2_1/1_2$, $1_2/2_2$, and $2_2/1_1$ separate (ferroelectric) orientational twins.

Let $x = 0$ represent the central plane of a domain wall. The atomic structure varies most rapidly in the vicinity of this plane, and the variation becomes small for large values of x. The following function provides a good description of the rate of change of structure with x:

$$h(x) = \left[\left(\frac{d\eta}{dx}\right)^2 + \left(\frac{d\xi}{dx}\right)^2\right]^{1/2} \tag{8.5.2}$$

This function is large for small x, and small for large x. The local symmetry of a domain wall can be described by a layer group only for positions for which h is large; otherwise the symmetry is that described by a 3-dimensional space group (Janovec 1981).

Near the T_c-end of the range of existence of the I-phase the domain walls are quite thick, and occur in sufficiently large numbers to be viewed as a lattice of domain walls of negative energy (Bruce, Cowley & Murray 1978). This is sometimes referred to as the *multidomain approximation*, or the *soliton approach*.

The sequence $P - I - C$ of phase transitions is described in terms of a complex order parameter $\rho(x)\, e^{i\phi(x)}$ ($\eta = \rho\cos\phi$, $\xi = \rho\sin\phi$ in Fig. 8.5.1). The four domain states (or discommensurations) of AFB correspond to $\phi(x) = \pi/4$, $3\pi/4$, $5\pi/4$ and $7\pi/4$. In the I-phase the domain walls are represented by a closed oriented curve, the shape and location of which changes with temperature. Two such curves, m and n, are shown in Fig. 8.5.1, the former for temperatures near T_c, and the latter for temperatures near T_i.

Loop m corresponds to the occurrence of sequence of domains 1_1 / 2_1 / 1_2 / 2_2 / 1_1 ..., with spontaneous polarization alternating in sign over successive domains.

Since the walls (discommensurations) separating these domains have a transient character, the average of any spontaneous quantity like polarization over a period of this siding-phase lattice is zero. This implies that the macroscopic or point-group symmetry of the I-phase is the same as that of the P-phase, namely mmm.

Although the average spontaneous polarization is zero, application of an electric field nevertheless has the effect of favouring the growth of domains with spontaneous polarization oriented along it. This domain growth occurs by the movement of domain walls, the energy of which varies with position. They may get pinned at *pinning sites* like defects, impurities, etc. Thus the system can exhibit hysteresis in the I-phase, even though its macroscopic spontaneous polarization is zero (Hamano et al. 1980).

When phase modulations exist along more than one directions, the permitted (i.e. energetically favourable) sequences of regularly spaced discommensurations or domain walls occur along each of them. The resulting *domain texture* of the I-phase has been analysed for several situations by

Janovec & Dvorak (1986) and Saint-Gregoire (1995).

SUGGESTED READING

V. Janovec (1981). Symmetry and structure of domain walls. *Ferroelectrics*, **35**, 105.

V. Janovec & V. Dvorak (1986). Perfect domain textures of incommensurate phases. *Ferroelectrics*, **66**, 169.

P. Saint-Gregoire (1995). Ferroelastic incommensurate phases. *Key Engg. Materials*, **101-102**, 237. This volume is also available in book form as C. Boulesteix (Ed.), *Diffusionless Phase Transitions in Oxides*. Trans Tech Publications, Zurich.

Part B

CLASSES OF FERROICS, MICROSTRUCTURE, NANOSTRUCTURE, APPLICATIONS

The pedagogical approach adopted in this book is as follows. In Part *A* the general, and mostly common, features of ferroic phase transitions and ferroic materials were introduced, with the barest minimum details. In Chapters 9 to 12 of Part *B* we concretize the treatment of the subject to, respectively, ferromagnetics, ferroelectrics, ferroelastics, and secondary and higher-order ferroics. By and large, we do not deal with the effects of particle size in these four chapters.

Size effects are very important, and are described in Chapter 13, which deals with microstructure and nanostructure.

Finally, in Chapter 14 we explain how the special properties of ferroics, including those determined by particle size and coexisting phases, can be exploited for practical applications.

Chapter 9

FERROMAGNETIC CRYSTALS

9.1 SOME MAGNETIC PROPERTIES OF ORDERED CRYSTALS

9.1.1 Magnetic Moment and Exchange Interaction

The charge-density function $\rho(\mathbf{r})$ describes the atomic structure of a crystal, and the current-density function $j(\mathbf{r})$ describes its magnetic structure. Unless cancelled by opposing influences, magnetic moments can arise in a crystal because a moving or spinning charge has a magnetic moment associated with it.

A crystal can be thought of as consisting of atomic and/or molecular ionic cores and a set of completely or partially delocalized electrons. Apart from contributions from nuclear magnetic moments, the magnetic moment of a crystal can arise from either the orbital motion or the spin of the electrons in it.

The magnetic properties of materials can be understood in terms of two basic notions: (i) One can associate a magnetic moment with the atoms or ions constituting the materials; and (ii) the interaction among these magnetic moments is predominantly of quantum-mechanical origin, including *exchange forces.* The Pauli exclusion principle forms the basis of all exchange forces between fermions.

An atom has a net magnetic moment when an inner d-shell or f-shell of electrons is not filled completely, resulting in only a partial cancellation of the spins and orbital moments of the electrons in that shell. Across the Periodic Table, the electron shells for which this may happen are: $3d$, $4d$,

$4f$, $5d$, and $5f$. The electrons in such shells are commonly referred to as *magnetic electrons*. It is assumed that the magnetic moment of an atom persists to a large extent when it is incorporated in a solid.

The usual ferromagnetic materials possess spontaneous magnetization and exhibit hysteresis in their magnetization or magnetic induction versus magnetic-field plots (*B-H curves*) (cf. Fig. 1.2.1; we shall discuss this in more detail presently). Further, above a temperature T_f, called the *ferromagnetic Curie temperature*, the spontaneous magnetization vanishes and the material becomes paramagnetic.

For temperatures substantially above T_f the magnetic susceptibility obeys the Curie-Weiss law:

$$\chi = \frac{C}{T - T_p} \tag{9.1.1}$$

C is called the Curie constant, and T_p the *paramagnetic Curie temperature* ($T_p > T_f$).

Several basic features of ferromagnetic materials were explained by Weiss (1907) in terms of the following two postulates:

- A ferromagnetic crystal comprises several *domains*, each having a spontaneous magnetization, the direction of which is different in different domains. The net magnetization of a specimen crystal is the vector sum of the magnetization vectors of the various domains, and is thus of small magnitude in the absence of a substantial applied magnetic field.

- The spontaneous magnetization of a domain arises from the existence of a *molecular field* which tends to align the atomic magnetic moments parallel to one another.

Weiss (1907) assumed that the effective molecular field acting at a point in a ferromagnetic domain can be expressed as

$$\mathbf{H}_m = \mathbf{H} + \lambda \mathbf{M}, \tag{9.1.2}$$

where $\mathbf{H}$ is the applied magnetic field, and $\mathbf{M}$ is the magnetization (magnetic moment per unit volume). λ is called the *Weiss constant*, and is a measure of the long-ranged cooperative ordering of the atomic magnetic moments.

For the paramagnetic phase (in which there is no long-ranged ordering), the extent of alignment of magnetic moments on application of the magnetic field is small, and we can assume that the magnetization can be written as a constant χ times the field. Further, the temperature dependence of χ is given by the Curie law: $\chi = C/T$. Thus, $MT = C(H + \lambda M)$, giving

$$M(T - C\lambda) = CH \tag{9.1.3}$$

Therefore,

$$\chi \equiv \frac{M}{H} = \frac{C}{T - C\lambda}, \tag{9.1.4}$$

which is the Curie-Weiss law (Eq. 9.1.1), with $C\lambda = T_p$.

A rough estimate of the magnitude of the molecular field H_m can be made by noting that the energy of an atomic magnetic moment in this field should be of the order of $k_B T_p$:

$$\mu_B H_m \simeq k_B T_p \tag{9.1.5}$$

Here μ_B is the Bohr magneton. Taking $T_p = 1000\,K$ gives $H_m \simeq 10^7$ gauss.

It follows that the molecular field is not due to simple (classical) dipole-dipole interactions; such interactions are mediated by fields of the order of 10^3 gauss only.

An attempt to explain the nature of the Weiss molecular field was first made by Heisenberg (1928) in terms of the quantum-mechanical *exchange interaction* between electrons. In Heisenberg's model, the interaction energy between the spin $\mathbf{S}_i$ of an atom and the spins $\mathbf{S}_j$ on neighbouring atoms was written as

$$U = -2J \sum_j \mathbf{S}_i \cdot \mathbf{S}_j, \tag{9.1.6}$$

where J is the *exchange integral.* It is determined by the degree of overlap of the charge distributions of the atoms involved.

Magnetization of a Virgin Ferromagnetic Specimen

Fig. 9.1.1 shows a typical magnetization curve of a virgin specimen of a ferromagnetic crystal. The shape of such a curve was first explained by Weiss (1907) by postulating the existence of domains and domain walls.

In the absence of an applied magnetic field, the domains may be oriented randomly, so that the vector sum of their individual magnetic moments is zero, or nearly so.

Application of an external field results in an increase of the net magnetization of the polydomain specimen.

It is often more convenient to deal with magnetic induction $\mathbf{B}$, rather than with magnetization $\mathbf{M}$. The two are related by

$$\mathbf{B} = \mu_0 (\mathbf{H} + \gamma_B \mathbf{M}) \tag{9.1.7}$$

Here μ_0 is the permeability of free space, and γ_B is a multiplier which takes care of the system of units used (Brown 1962). $\gamma_B = 1$ for the SI units, and $\gamma_B = 4\pi$ (and $\mu_0 = 1$) for the CGS gaussian units.

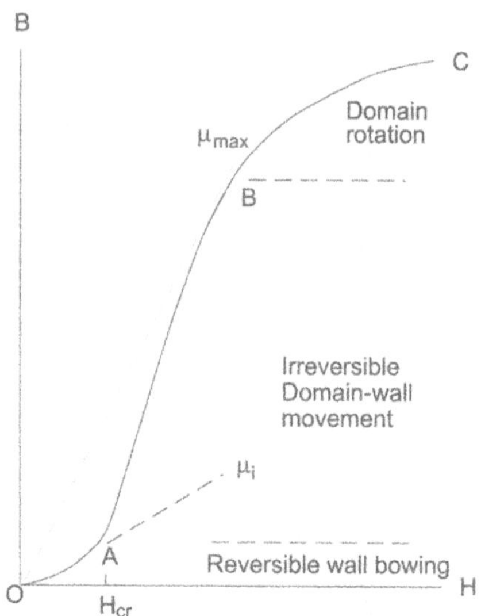

Figure 9.1.1: A representative B-H curve for a virgin specimen of a ferromagnetic crystal. Regions OA, AB and BC correspond to the different magnetization processes dominating at different values of the driving magnetic field H. [After Kittel (1949) and Valenzuela (1994).]

To the extent that Eq. 9.1.7 is a valid expression for a given crystal, and this is indeed the case for diamagnetic and paramagnetic crystals, one can also write

$$\mathbf{B} = \mu \mathbf{H}, \tag{9.1.8}$$

with

$$\mu = \mu_0 (1 + \gamma_B \chi) = \mu_0 \mu_r \tag{9.1.9}$$

μ is the magnetic permeability of the medium, and μ_r is the *relative* magnetic permeability. μ_r is a dimensionless quantity, independent of the system of units used. $\mu_r = 1$ for empty space.

Because of their domain structure, and the resultant highly nonlinear response to magnetic field, ferromagnetic materials are not well-described by Eq. 9.1.7 (**M** is not linearly proportional to **H**).

The *effective permeability* of a ferromagnetic specimen depends on its history and on the value of the field applied. It is therefore defined as a derivative:

$$\mu_{eff} = \frac{\partial B}{\partial H} \tag{9.1.10}$$

A typical ferromagnetic specimen has its share of defects like impurities, dislocations, inclusions, etc., which tend to *pin* the domain walls to their existing locations. With reference to Fig. 9.1.1, when the field applied is smaller than a critical value H_{cr} (called the *critical field*), it does not result in the movement of the domain walls because the pinning centres resist their movement. Instead, only a (reversible) bowing of the domain walls takes place. Further, because of the smallness of the field applied, the magnetization curve is a straight line for $H < H_{cr}$, the slope of this line defining the *initial permeability*.

For a chosen direction of the applied field the spins in some domains will be closer in their orientation with respect to that of the field than others. The field tends to alter the directions of all spins to make them closer to its own direction. As a consequence, the total volume of some domains will grow at the cost of others. For $H < H_{cr}$, this happens mainly by the mechanism of bowing of domain walls mentioned above. For higher values of the field the pinning centres give way, successively, resulting in a sharp upward change in the slope of the B-H curve (region AB in Fig. 9.1.1). This region is characterized by a movement of domain walls, and the process is largely irreversible because there may be hardly any reason for the domain walls to retrace their paths when the field is decreased or switched off.

The region AB is characterized by the occurrence of *Barkhausen effect* (see Jiles 1991): Every time a pinning centre gives way, allowing a domain wall to move, there is a *discontinuous* change in the flux density $\mathbf{B}$ in the specimen. If Fig. 9.1.1 were to be plotted on a sufficiently expanded scale, part AB would appear zigzag because of the Barkhausen effect, rather than continuous as shown.

In addition to the Barkhausen effect there is also an *acoustic emission* of pulses in part AB, connected with discontinuous changes in the magnetoelastic energy as the pinning centres let go the domain walls, successively (see Jiles 1991).

The point B in Fig. 9.1.1 is a point of inflexion, beyond which movement of domain walls is not the main mechanism of magnetization because practically all the domains unfavourably oriented with respect to the field direction have been already obliterated, and any further increase of magnetization can occur only by a *rotation* of the domains towards the field direction. Point B corresponds to the maximum slope of the curve, and defines the *maximum permeability*, μ_{max}.

The point C in the magnetization curve defines the *saturation magnetization*, $\mathbf{M}_s$, which is almost as large as the spontaneous magnetization in any of the individual ferromagnetic domain states.

Easy Directions of Magnetization

The energy required for the creation of a domain with spontaneous magnetization pointing along a specific direction depends on the existence of the so-called easy directions of magnetization. For example, for a crystal of Fe, which has cubic symmetry, [100] directions are easy directions. By comparison, [111] directions are hard directions. For a crystal of Co, which has hexagonal symmetry, the hexagonal axis is the direction of easy magnetization.

The existence of directions of easy magnetization is related to the existence of a *magnetocrystalline* or *anisotropy energy*, which in turn is determined to a large extent by the anisotropy of overlap of electron distributions on neighbouring ions.

The Anisotropy Energy

Spin-orbit coupling of electrons results in a non-spherical charge distribution, which influences the anisotropy energy through electrostatic fields and overlapping wavefunctions between neighbouring atoms (van Vlack 1947).

The expression for this energy naturally satisfies the symmetry requirements of the crystal. For example, for a crystal with cubic symmetry, if α_1, α_2, α_3 are the direction cosines of the magnetization vector, the anisotropy energy has the form

$$U_{aniso} = K_1 (\alpha_1^2\alpha_2^2 + \alpha_2^2\alpha_3^2 + \alpha_3^2\alpha_1^2) \qquad (9.1.11)$$

Similarly, for a crystal like Co with hexagonal symmetry, if θ is the angle between the magnetization vector and the axis of 6-fold symmetry,

$$U_{aniso} = K_1 \sin^2\theta + K_2 \sin^4\theta \qquad (9.1.12)$$

The constants K_1 and K_2 are sensitive functions of temperature.

In view of the anisotropy energy, Eq. 9.1.6 must be generalized suitably. In the most general case (e.g. in partially ordered or completely disordered systems) the crystal field may vary from point to point. There is then a *local* easy direction of magnetization, determined by the local crystal field. Assuming a cylindrical local symmetry (Kanamori 1963), the anisotropy energy at a site i can be expressed as $D_i S_{zi}^2$, where D_i is the strength of the local field, and S_{zi} is the component of the spin along the local easy direction of magnetization.

The total interaction energy in a magnetic system thus has an anisotropy contribution and an exchange contribution:

$$U = -\sum_i D_i (S_z)_i^2 - \sum_{ij} J(r_{ij})\, \mathbf{S}_i \cdot \mathbf{S}_j, \qquad (9.1.13)$$

where we have further generalized Eq.9.1.6 by not assuming J to be a universal constant.

9.1.2 Magnetic Ions in Solids

The electron has not only a charge, but also a spin or magnetic moment. In a crystal, whereas the ions are practically immobile, some of the electrons are highly mobile, and thus act as carriers of both charge and magnetic moment.

Electrons in the valence shells of neighbouring ions interact strongly, especially in metals, resulting in their delocalization and in the formation of energy bands. The relevant shells are $3d$, $4d$, $5d$, and $4f$, $5f$.

The Kondo Effect

The f-shell is more localized than the d-shell, and it is the d-shell which is more affected by the bonding between neighbouring ions.

There are various $s - d$ interaction models for the interaction between a magnetic ion and the delocalized electrons. We speak of *s-d mixing*, rather than *hybridization*, when the s-conduction levels and the local d-levels overlap in energy.

In a metal the itinerant s-electrons spend some time in the vicinity of the d-electrons of the magnetic ion. During this period their spin is polarized antiferromagnetically to that of the magnetic moment of the ion. After a temporary stay in the d-level of the magnetic ion the electrons tunnel back to delocalized states. The term *Kondo effect* is used for the localization of a cloud of antiferromagnetically polarized conduction-electron spins around an isolated magnetic ion (Bell & Caplin 1975).

The Kondo binding disappears above the *Kondo temperature* T_K. Below T_K the antiparallel spins of the electron charge cloud effectively cancel the observable moment of the magnetic ion. The so-called *Kondo regime* is an idealization, which can be realized or approached in very dilute magnetic alloys at very low temperatures.

The Kondo effect hinders interactions between the spins on the magnetic ions.

Itinerant-Electron Magnetism

When the percentage of magnetic ions is very high, there is sufficient interaction between neighbouring magnetic ions to make the magnetic electrons itinerant, occupying energy states in narrow bands formed by the interaction between the neighbouring ions. One then speaks of itinerant-electron magnetism.

This type of magnetic ordering is the result of a trade-off between exchange energy and kinetic energy. The exchange effects are a consequence of the Pauli exclusion principle: electrons with parallel spins avoid one another. This results in the creation of local *exchange holes* or *Fermi holes*: each electron is surrounded by a void due to a local depletion of electrons of spins parallel to its spin. An exchange hole around an electron has the effect that the electron experiences a less repulsive or more attractive Coulomb potential in the presence of parallel-spin electrons.

In addition to the exchange hole, there is also a *correlation hole* around an electron, resulting from the ordinary Coulomb repulsion between electrons of any spin.

The decrease in the potential energy due to the existence of the exchange hole and the correlation hole must be compensated by a corresponding increase in the kinetic energy. There is thus a competition between exchange and kinetic energies.

The magnetism of ions in such systems is due to the exchange hole. Although the magnetic electrons are itinerant to a substantial extent, there is also an interchange of the electron population between the delocalized band-electrons and the localized shell electrons, and, statistically speaking, some of the electrons may stay long enough in the atom shell to align the other spins on the atom, giving it a net local magnetic moment (the itinerant-electron magnetic moment).

Dependence of Ionic Magnetic Moment on Environment

The magnetic moment of an ionic species in a solid is not a unique quantity; it depends on the environment of the ion. And the environmental conditions depend, for example, on the degree of dilution of the magnetic ions in the nonmagnetic host crystal (e.g. Mn ions in a Cu host crystal). The main reason for the dependence of the magnetic moment of ions on the separation and geometrical arrangement of their ligands is the anisotropic nature of the charge distribution of the unfilled d-shell or f-shell. Although this variation occurs even in normal crystalline materials, it is particularly striking in disordered solids (cf. §9.2), especially for d-shell ions.

9.1.3 Coupling Between Magnetic Moments

In ferromagnetic crystals, the ferromagnetic ordering arises due to a cooperative exchange coupling among the magnetic moments. There are two types of exchange interaction: direct and indirect.

Direct Exchange

Direct-exchange coupling between magnetic moments occurs when they are close enough to have overlapping wave functions. This strong coupling decreases rapidly with distance between the interacting ions.

In a simple situation it may be *isotropic*, depending only on the distance r_{ij} between the two atoms or ions. If $\mathbf{S}_i$ and $\mathbf{S}_j$ are the spins of these atoms, the exchange energy has the form

$$\mathcal{H} = -\sum_{ij} J(r_{ij})\, \mathbf{S}_i \cdot \mathbf{S}_j \tag{9.1.14}$$

The exchange parameter J may be positive or negative, depending on the competition between the Coulomb energy and the kinetic energy.

Indirect Exchange

The indirect-exchange coupling can occur between ions that are not in nearest-neighbour configurations. It naturally requires an intermediary, which may be itinerant electrons (as in metals), or nonmagnetic ions (as in insulators). It is known as *RKKY coupling* in the former case, and *superexchange* in the latter.

The RKKY Interaction

In the Ruderman-Kittel-Kasuya-Yosida interaction (Ruderman & Kittel 1954; Kasuya 1956; Yosida 1957), the exchange parameter J (cf. Eq. 9.1.14) oscillates in sign as the distance between the ions changes. Itinerant electrons are the intermediaries for this interaction. To understand why the RKKY interaction has an oscillatory character even for an ordered magnetic crystal, we recall that whereas a periodic or oscillatory function can be expressed as a Fourier series, a nonoscillatory function must be written as a Fourier integral, implying that the number of Fourier components for the latter case is infinite. If we are compelled to express a nonoscillatory function as an incomplete set of Fourier components, the representation is at best a poor approximation, resulting in an artificially oscillatory behaviour of the nonoscillatory function. Something similar happens in the case of an itinerant-electron system. Just as these electrons tend to screen out the charge on a positive ion, they also tend to screen out its magnetic moment by preferentially adopting antiparallel spin configurations in its vicinity. But a complete screening of the magnetic moment of the ion will require the presence of electrons of all possible wavenumbers, whereas in reality the maximum wavenumber available is only $2k_F$, where k_F is the wavenumber of an electron on the Fermi surface. As a result, the magnetic ion induces an oscillatory spin polarization in the mobile electrons in its neighbourhood. The effect of this oscillating spin polarization is felt by the magnetic moments of other ions even at substantial distances from a given ion, resulting in an indirect exchange coupling that oscillates in sign. This is a unique feature of the RKKY interaction.

Superexchange

Indirect exchange coupling of non-nearest-neighbour magnetic moments via a nonmagnetic ligand, rather than via conduction electrons, is referred to as superexchange (White & Geballe 1979). A typical situation is that of a pair of magnetic cations separated by, or coupled by, a diamagnetic anion. The $R^{3+} - Fe^{3+}$ coupling in a garnet crystal is an example of this, with R^{3+} denoting a rare-earth cation. The coupling in this case is a hybridization of the orbitals of Fe^{3+} and R^{3+} with that of the intervening O^{2-} anion. Because of the superexchange coupling via the O^{2-} anion, any reorientation of the magnetic moment on the ferric ion changes the degree of the overlap of the charge clouds of the three ions, leading to a strongly anisotropic exchange energy.

9.1.4 Diamagnetism and Paramagnetism

Diamagnetism is present in all materials, and is a manifestation of Lenz's law. All materials get a small, negative, temperature-independent contribution to their overall magnetic susceptibility, this being the result of shielding currents induced in the filled shells of an ion by the applied magnetic field (Fig. 9.1.2(a)).

If at least some of the constituents of a crystal have spontaneous magnetic moments, and if there is very little exchange coupling between them, the resulting (paramagnetic) magnetic susceptibility is positive and follows the Curie law ($\chi = C/T$). If some exchange coupling is present, the temperature dependence of the susceptibility is described by the Curie-Weiss law: ($\chi = C/(T - \theta)$), where θ is a measure of the cooperative coupling of the magnetic moments (Fig. 9.1.2(b)).

9.1.5 Ferromagnetism, Antiferromagnetism, and Ferrimagnetism

Ferromagnetism is a manifestation of long ranged cooperative coupling among the constituent magnetic moments in a crystal. A ferromagnetic crystal possesses a spontaneous magnetic moment M on a macroscopic scale, although the magnetic moments of the various domains, pointing in

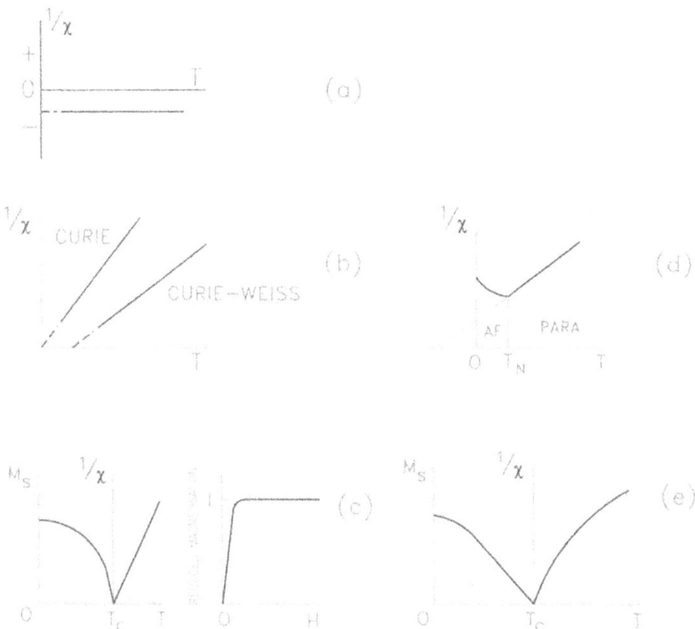

Figure 9.1.2: The five 'classical' types of magnetism in crystals. (a) Diamagnetism; (b) paramagnetism; (c) ferromagnetism; (d) antiferromagnetism; and (e) ferrimagnetism. χ is the magnetic susceptibility, T the temperature, and M_s the saturation value of the magnetization M (also equal to the magnetization in a single domain. The reduced magnetization is M/M_s. T_c is the Curie temperature, and T_N the Néel temperature. [After Hurd (1982).]

different directions, may lead to a zero or nearly zero vector sum. Application of even a moderate magnetic field is then sufficient to make the net magnetization rise to the final, saturation value M_s, which changes little on further increase of the applied field as it is already close to the single-domain-state value of the spontaneous magnetization (Fig. 9.1.2(c)). Above a certain temperature T_c the thermal effects overcome the ordering tendency of the crystal, and it becomes a paramagnetic crystal, with the susceptibility obeying the Curie-Weiss law.

The long-ranged cooperative interaction leading to ferromagnetic ordering may be either through nearest-neighbour exchange, or through itinerant electrons. Fe, Co and Ni are examples of itinerant-electron ferromagnetic crystals.

Like ferromagnetism, antiferromagnetism is also a manifestation of long-ranged cooperative ordering of magnetic moments, except that neighbouring magnetic moments adopt an antiparallel (rather than parallel) configuration. The overall spontaneous magnetization is therefore zero.

Above a certain temperature called the *Néel temperature*, T_N, an antiferromagnet behaves like a paramagnet, with its susceptibility exhibiting Curie-Weiss behaviour (Fig. 9.1.2(d)). As the temperature is lowered below T_N, the susceptibility falls, reaching its lowest value at $T = 0$, at which temperature all the magnetic moments are aligned in a perfect antiparallel configuration.

When the long-ranged antiparallel ordering of spins involves unequal spins, we speak of ferrimagnetic ordering (Fig. 9.1.2(e)), rather than antiferromagnetic ordering. The most familiar ferrimagnetic crystal is magnetite ($FeO.Fe_2O_3$). The antiparallel ordering in it is between the unequal spins of the Fe^{2+} and Fe^{3+} ions. These two types of ions occupy different Wyckoff sites in the crystal (say sites A and B). Ions on Site A are ferromagnetically aligned among themselves (below T_c), and so are those on Site B. But the coupling between ions on Site A and those on Site B is of the antiferromagnetic type. Since the moments on sites A and B have different magnitudes, a net spontaneous magnetization results.

Above T_c the magnetic susceptibility of a ferrimagnet follows the Curie-Weiss law only approximately, particularly for temperatures not far above T_c (Fig. 9.1.2(e)).

9.1.6 Molecular Ferromagnets

Apart from the conventional atom-based ferromagnets involving d-shell or f-shell atoms, molecular ferromagnets based on organic compounds have also been synthesized and are, in fact, an area of intense current research.

A charge-transfer salt, namely decamethylferrocenium tetracyanoethenide, $[Fe^{III}(C_5Me_5)_2]^+[TCNE]^-$, was the first organic ferromagnet to be synthesized (Chittipedi et al. (1987), Miller et al. (1988)).

Other approaches for synthesizing molecular ferromagnets have been those employing bimetallic ferrimagnetic chains, and the metal-radical approach (see Chavan, Yakhmi & Gopalakrishnan (1995) for a review).

9.1.7 Metamagnetism and Incipient Ferromagnetism

In the usual ferromagnets and antiferromagnets the exchange fields and the anisotropic crystal fields are very strong. In the so-called metamagnets and incipient ferromagnets the exchange and anisotropy effects are generally rather weak.

A *metamagnet* is basically an antiferromagnet, except that the magnetocrystalline anisotropy forces in it are only moderately strong (Stryjewski

& Giordano 1977). As a result, an external magnetic field can easily disturb its antiferromagnetic ordering below the Néel temperature, causing a transition from a phase of low magnetization (of the component antiparallel configurations) to a phase of high magnetization. The magnetic susceptibility, however, continues to be low before and after the field-induced phase transition.

Metamagnetic behaviour can result not only from the overcoming of the anisotropic internal fields by an external field, but also by the effect of (rather large) external fields on itinerant electrons (Wohlfarth 1980). Systems in which this occurs are typically 3*d*-transition metal compounds (YCO_2, $TiBe_2$, $FePt_3$).

Apart from metamagnetism, another variation on the normal ferromagnetic behaviour occurs in what are called *incipient ferromagnets*. An incipient ferromagnet, or an *exchange-enhanced metal*, is a metal in which the effect of itinerant electrons on magnetic ordering is not strong, but it is nevertheless nonzero. Therefore, at sufficiently low temperatures, this ordering tendency manifests itself over small regions of the metal, leading to a localized alignment of magnetic moments for substantial periods of time. Such regions of magnetic order are known as *paramagnons* or *localized spin fluctuations*. Palladium and platinum are examples of metals in which this occurs (Hurd 1982).

9.1.8 Helimagnetism

$MnAu_2$ is an example of a helimagnetic crystal. It has a body-centered tetragonal structure. The magnetic moments on the Mn ions point along directions normal to the *c*-axis, and the direction rotates by about 50° from plane to plane normal to the *c*-axis, thus tracing out a spiral or a helix. Above a certain disordering temperature, this ordered orientation of spins is destroyed and the crystal enters a paramagnetic phase.

Some rare-earth metals exhibit a different form of helimagnetism, wherein the spins rotate along the surface of a cone, rather than in a plane.

Weak Ferromagnetism

Weak ferromagnetism (also known as *canted ferromagnetism*) involves two or more antiferromagnetic configurations canted at an angle, rather than being collinear, so that there is a net spontaneous magnetization (Moriya 1963).

Two factors may contribute singly or jointly to the occurrence of weak ferromagnetism. One is the difference between the local anisotropies for the two sublattices. The other is the so-called *Dzyaloshinsky-Moriya (DM) interaction* (Moriya 1963). The DM interaction results from the asymmetric upsetting of the indirect coupling between cations by spin-orbit interactions.

Examples of weak ferromagnets include NiF_2, $\beta - MnS$, $\alpha - Fe_2O_3$, MnSi, CrF_3, and $CoCO_3$ (Hurd 1982). Since weak ferromagnetism is a result of a delicate trade-off between opposing forces, many of these crystals are also metamagnets.

SUGGESTED READING

C. Kittel (1949). Physical theory of ferromagnetic domains. *Rev. Mod. Phys.*, **21**, 541.

C. M. Hurd (1982). Varieties of magnetic order in solids. *Contemp. Phys.*, **23**, 469. An excellent overview of the subject, for the beginner.

D. Jiles (1991). *Introduction to Magnetism and Magnetic Materials.* Chapman & Hall, London.

S. A. Chavan, J. V. Yakhmi & I. K. Gopalakrishnan (1995). Molecular ferromagnets - a review. *Mater. Sci. & Engg.*, **C3**, 175.

A. Aharony (1996). *Introduction to the Theory of Ferromagnetism.* Clarendon Press, Oxford.

P. M. Lahti (Ed.) (1999). *Magnetic Properties of Organic Materials.* Marcel Dekker, New York.

9.2 SPIN GLASSES AND CLUSTER GLASSES

The history of spin glass may be the best example I know of the dictum that a real scientific mystery is worth pursuing to the ends of the Earth for its own sake, independently of any obvious practical importance or intellectual glamour.

P. W. Anderson

So far in this chapter we have considered only the more conventional types of magnetism. A fascinating new world of condensed-matter physics comes into view when we introduce a certain amount of random magnetic (or spin) disorder into an otherwise nonmagnetic crystal.

Glasses are the very antithesis of perfect single crystals, in that they lack the order responsible for the translational and certain other symmetries

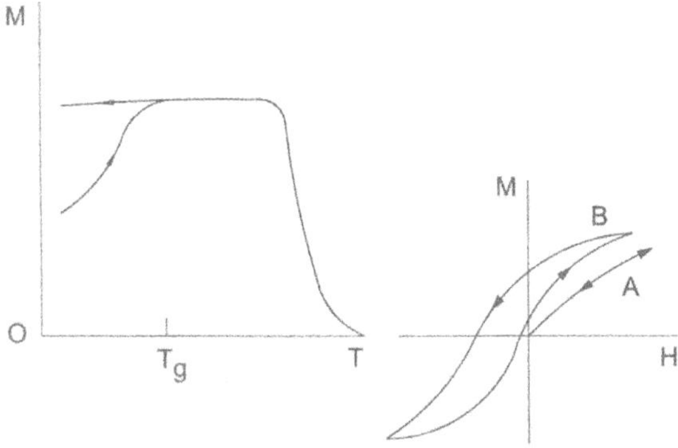

Figure 9.2.1: Typical $M(T)$ and $M(H)$ curves for a spin- or cluster-glass crystal. Curve A is for a crystal cooled without the influence of an external magnetic field, and Curve B is for a field-cooled specimen. [After Hurd (1982).]

possessed by crystals. The term *spin glasses* is used for crystals in which a small fraction of lattice sites are occupied randomly by magnetic ions, and the magnetic ions are oriented randomly in the absence of an external magnetic field. A classic example of this is a crystal of Cu, alloyed with a small amount of Mn. Cu is nonmagnetic, whereas Mn carries a spin of 5/2. In this magnetic alloy the magnetic and nonmagnetic ions are distributed quite randomly. There is thus *positional disorder*, in addition to other kinds of disorder.

Spin glasses generally have small linear magnetic susceptibilities (comparable to those of paramagnetic crystals). But they also exhibit, at low temperatures, features like hysteresis and remanence which are characteristic of ferromagnets. In fact, in a certain sense they are even more hysteretic than conventional ferromagnetics in that a field-cooled specimen of a spin glass exhibits a hysteresis curve that is *shifted* along the H-axis (see Fig. 9.2.1, and §9.2.10), indicating that the specimen has a memory of the direction in which the field was applied while cooling it.

The essence of a spin glass is the presence of an exchange interaction that is *random* in sign and magnitude. Further, the net interaction is the result of *competing* interactions, resulting in "frustration" (see below).

Investigations on spin glasses have led to the introduction of concepts which have far-reaching consequences, not only for condensed-matter physics, but also for neural networks, protein folding, etc. (Theumann & Koberle 1990; Stein 1992).

The behaviour of the host crystal, which is nonmagnetic to start with, and in which a certain fraction of lattice sites are occupied randomly by magnetic dopants or impurities, depends on the concentration of these dopants, as also on several other factors. We get giant-moment and spin-glass characteristics at low concentrations, cluster-glass characteristics at intermediate concentrations, and percolation and long-range magnetic order at still higher concentrations.

9.2.1 Giant-Moment Ferromagnetism

This is a phenomenon involving long-ranged but nonuniform alignment of spins parallel to one another, mediated by conduction electrons. A typical example is that of the dilute transition-metal alloy Pd-(0.1 at % Fe) (see Mydosh & Nieuwenhuys 1980). The dopant Fe has a magnetic moment of about 4 Bohr magnetons. However, the host lattice of Pd has a highly polarizable population of itinerant electrons. The magnetic dopant therefore induces indirectly (via itinerant electrons) a large magnetic moment in the host atoms, resulting in a "giant" total (dopant plus host) magnetic moment of about 12 Bohr magnetons. The induced polarization extends to about 10 Å around the Fe impurity, affecting about 200 Pd atoms.

9.2.2 Characteristics of Spin Glasses

Like giant-moment ferromagnetism, spin-glass behaviour is also mediated by itinerant electrons.

A typical spin glass exhibits the following properties:

(a) There is a sharp cusp in the *low-field* low-frequency a.c. susceptibility $\chi(T)$ at a certain temperature T_{sg}, called the *spin-glass transition temperature* (Fig. 9.2.2). This is a *defining* feature of a spin glass.

(b) A magnetic field as small as 50 gauss is able to flatten the cusp (Fig. 9.2.2)(Canella, Mydosh & Budnick 1971; also see Weissman 1998). This again is a very characteristic feature of a spin glass.

(c) In the absence of an external magnetic field, no sharp anomaly is observed at T_{sg} in the temperature dependence of the specific heat.

(d) The magnetic susceptibility at T_{sg} depends on experimental conditions and on sample history. The susceptibility measured in a zero-field-cooled sample is lower than that in a field-cooled sample.

(e) The remanent magnetization below T_{sg} decays very slowly with time.

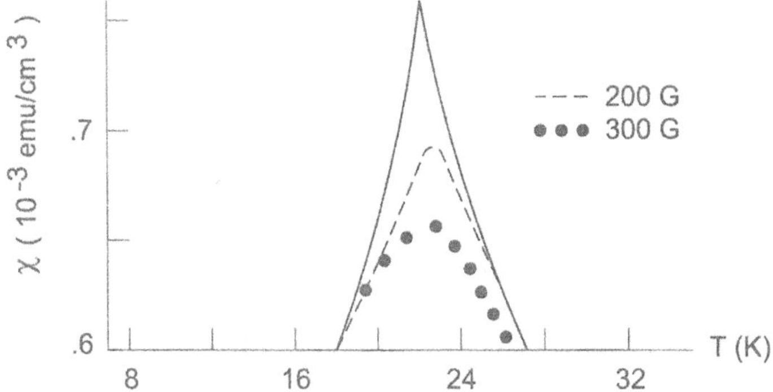

Figure 9.2.2: Temperature variation of low-frequency susceptibility of the alloy $Au_{0.95}Fe_{0.05}$. The sharp cusp at the spin-glass transition temperature T_{sg} becomes flattened even when a biasing magnetic field as small as 50-100 G is applied. [After Canella & Mydosh (1972).]

(f) Spin glasses display hysteresis below T_{sg}, but the M-H curve is shifted laterally from the origin.

(g) No long-ranged order is generally observed below T_{sg} in the neutron-diffraction pattern of spin glasses. There are no Bragg peaks attributable to the magnetic contribution to the diffraction pattern. There is, however, a splitting of some Mossbauer lines near T_{sg}, indicative of a quasi-static internal field below T_{sg} (see Fischer (1983) for a review).

(h) Even for temperatures far above T_{sg}, a deviation from Curie-Weiss behaviour begins to occur on cooling.

(i) The range of dopant concentrations over which spin-glass behaviour is observed is quite restricted. At large dopant concentrations, either ferromagnetic or antiferromagnetic ordering occurs.

(j) There is a very large range of relaxation times, extending from 10^{-13} sec to several hours. Postulation of the existence of *clusters of spins* of various sizes, rather than the existence of only isolated spins, can explain this (Binder 1979). Reorientation of clusters of spins is necessarily a slower process compared to flipping of single spins.

The structure of spin glasses has two key features. One is the presence of *quenched disorder*, and the other is the presence of *competing interactions* (e.g. ferromagnetic and antiferromagnetic) between the spins, resulting in a *frustration configuration* (cf. §9.2.6 below).

By quenched disorder we mean that the disorder is of a nonthermal nature and comprises a frozen random occupation and/or random orientation of the spins.

We give here a very elementary and introductory account of concepts and models, many of them from other fields, that have been used for trying to explain the properties of spin glasses.

9.2.3 The Glassy Phase and the Glass Transition

There are several definitions of glass. In general terms, any noncrystalline solid is a glass.

An empirical definition of glass, due to Vogel (1921) and Fulcher (1925), assumes that a glass is one which obeys the following equation (now called the *Vogel-Fulcher equation*) for the temperature dependence of relaxation time:

$$\tau = \tau_0 \, e^{T_0/(T-T_f)}, \tag{9.2.1}$$

with τ_0, T_0 and T_f as 'best-fit' parameters.

However, the Vogel-Fulcher (V-F) law is not obeyed universally, and several other empirical relationships have been proposed (see, e.g., Cheng et al. 1997). Further, the parameters τ_0 and T_0 are found to be dependent on the temperature range in which measurements are made (see Bessada et al. 1987).

A thermodynamic definition of glass, based on experimental criteria, was introduced by Suga & Seki (1974, 1981). The two experimental criteria specified are: existence of a *glass transition*, and the existence of a *residual entropy* at $T = 0K$.

We summarize here the main features of a glass transition (see Rao & Rao (1978) for more details).

A conventional or *canonical* glass is usually obtained by a rapid cooling or "quenching" of a melt. When the melt is cooled, its volume V decreases, or the density increases. As the density approaches the single-crystal value, the rate of decrease of volume becomes smaller. A temperature T_g, called the glass-transition temperature, can be usually identified such that $|\partial V \,/\, \partial T|$ for temperatures above T_g is substantially higher than for temperatures below T_g.

Other differences can also be observed above and below T_g. There is a large increase in the viscosity below T_g. The specific heat suddenly drops to a lower value on cooling to T_g (Fig. 9.2.3(b)).

The glassy state has frozen disorder. Its configurational entropy can therefore be expected to be nonzero at $T = 0K$, as stipulated in the Suga-

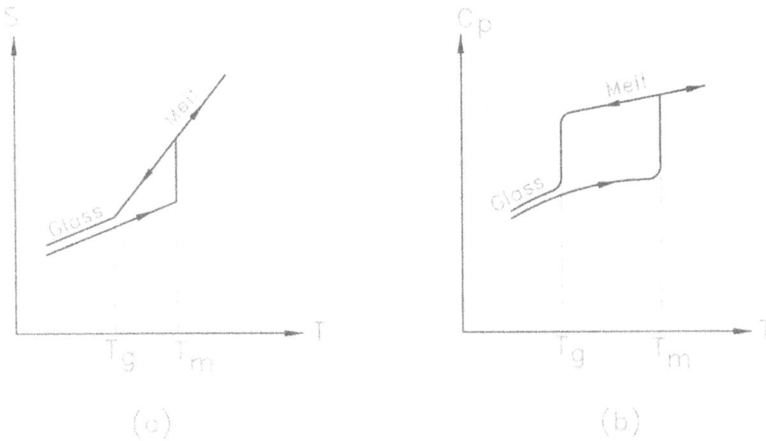

Figure 9.2.3: Temperature dependence (schematic) of the entropy (a) and specific heat (b) of a glass-forming material. T_g is the glass-transition temperature, and T_m the temperature below which there is a marked departure from melt-like behaviour. [After Rao & Rao (1978).]

Seki definition of glass, stated above.

Although there are several points of similarity between canonical glasses and spin glasses (as also orientational glasses to be described in Chapters 10 and 11), there is also a very important point of difference. Canonical glasses are obtained from a melt by quenching. Quenching is resorted to for preventing crystallization through diffusion processes. By contrast, the glass transition in a spin glass or orientational glass does not involve diffusion, and can take place at any cooling rate. Thus, although there is quenched disorder in the latter case also, sudden cooling or quenching is not necessary for effecting it.

9.2.4 Two-Level Model for Tunneling or Thermal Hopping in Glasses

At low temperatures the specific heat of a variety of glasses, including spin glasses, is observed to have a contribution that varies linearly with temperature. A two-level-tunneling (or thermal-hopping) model, based on a statistical distribution of localized tunneling levels, was advanced by Anderson, Halperin & Varma (1972) to explain this fairly universal phenomena (see Chowdhury (1986) for a more detailed discussion of this than is presented here).

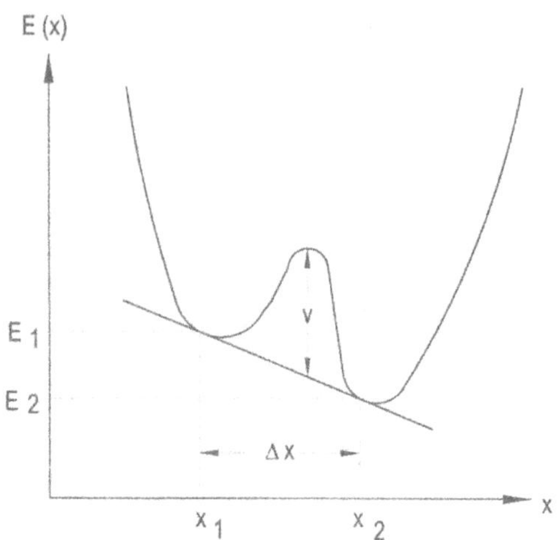

Figure 9.2.4: Representation of a two-level system, depicting the dependence of the energy E of a glassy system on a generalized coordinate x, measuring position along a line connecting two nearby local minima of E. [After Anderson, Halperin & Varma (1972).]

The main stipulation of this model is that in any disordered or glassy system there are a certain fraction of atoms or groups of atoms which can occur, with almost equal probability, in *two* equilibrium sites or orientations.

We can imagine a $3N$-dimensional configuration space for a set of N atoms constituting a glass, and take a section along some appropriate position or orientation coordinate x for a relevant atom or group of atoms. There will then be two local minima for the energy $E(x)$, with a barrier separating the two minima (Fig. 9.2.4).

Those atoms make a contribution to the linearly temperature-dependent specific heat for which the energy barrier is sufficiently small for substantial tunneling to occur between the two levels, so that thermal equilibrium is reached during the time required for a typical specific-heat experiment (from 10^{-10} sec to 10^3 sec). For the relevant atoms there is not only an upper limit for the energy barrier, but also a lower limit, which is set by the requirement that the barrier should be sufficiently large so that "resonant tunneling" between the two local minima does not occur.

For low temperatures in the range $0.1K$ to $10K$, the "window" for acceptable energy barriers is provided by those atoms for which the energies for the two local minima are accidentally degenerate to within an energy of the order of k_BT.

Because of the quenched disorder, the splitting of the two minima, Δx, is a random function. Anderson et al. (1972) assumed that the distribution function, ρ, for this is a constant. Then the energy is determined by

$$E = \int \rho(\Delta E) \left[\frac{\Delta E}{1 + \exp(\Delta E/k_B T)}\right] d(\Delta E), \qquad (9.2.2)$$

which leads to the following (linear) temperature dependence for specific heat:

$$C(T) \sim T\rho \qquad (9.2.3)$$

Several other results also follow from this model. For example, it explains the T^2-variation of the thermal conductivity of glasses at low temperatures.

9.2.5 Broken Ergodicity

In the Gibbs-Boltzmann formulation of statistical mechanics one makes the *ergodicity hypothesis* (§E.2), according to which a system in equilibrium can find itself in any of the microscopic states allowed to it, the probability of occupation of any such state being given by the Boltzmann factor $e^{-\beta\mathcal{H}}$ (Eq. E.2.19). In other words, a system is said to be ergodic if its macroscopic properties are in conformity with the corresponding ensemble averages.

In spin glasses there is no *simple* long-ranged ferromagnetic or antiferromagnetic order. The quenched or static nature of the disorder, coupled with the presence of very long relaxation times, leads to a breaking of ergodicity (Palmer 1982).

Fig. 9.2.5 shows schematically the variation of free-energy density of such a glassy system with some phase-space coordinate. It typically has a number of minima ($\phi_j^{(1)}$, $\phi_j^{(2)}$, $\cdots$), separated by large barriers. In contravention of the ergodicity hypothesis, all states of the system may not be accessible with a probability determined by the Boltzmann factor (although around a particular energy minimum they are). And yet, although there may be no apparent breaking of symmetry, and no conventional phase transition at T_{sg}, several metastable states can exist below T_{sg}.

This is inherently a very complex problem. For example, does one explain the existence of a wide range of relaxation times in terms of static correlations, or does one invoke the presence of dynamic effects ? Not all answers are known yet.

9.2.6 Frustration

A key feature of spin glasses is the presence of competing interactions, leading to frustration (Toulouse 1977). Frustration is a cause of broken

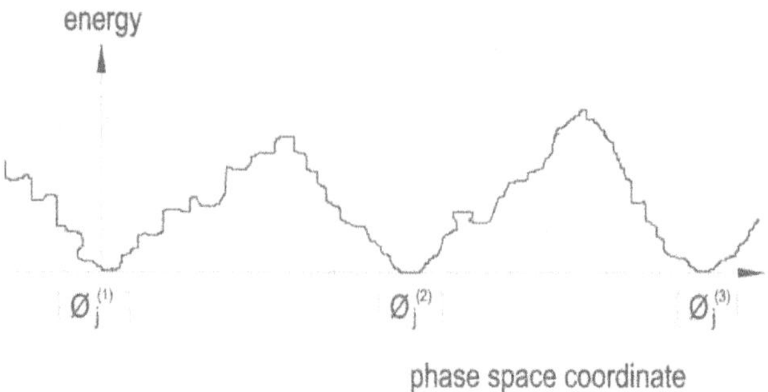

Figure 9.2.5: Schematic depiction of the dependence of the free-energy density of a glassy system at low temperature on a phase-space coordinate. [After Binder (1980).]

ergodicity in spin glasses. The presence of frustration is believed to be a necessary condition for a system to behave like a spin glass, because it leads to the occurrence of a highly degenerate ground state.

To understand how that happens, we refer to Fig. 9.2.6. Shown there is a unit of a square lattice for two different situations. Such a unit is called a *plaquette.*

At the corners of the square are spins, and the line segments joining them are labelled either '+' or '-', the former representing a ferromagnetic interaction, and the latter an antiferromagnetic interaction.

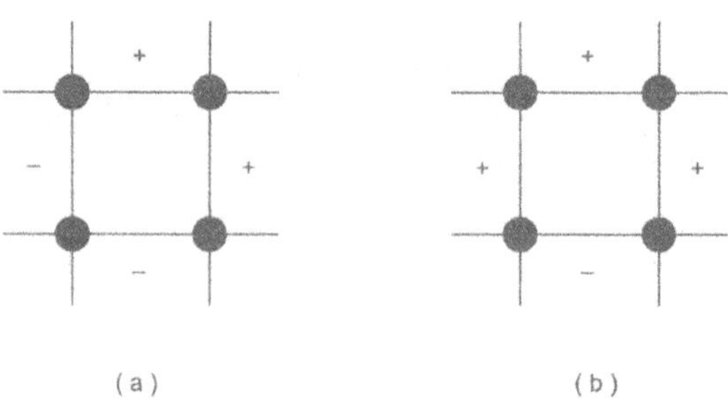

Figure 9.2.6: Unfrustrated (a) and frustrated (b) plaquettes on a square lattice.

In Fig. 9.2.6(a) the number of negative bonds is even, and it is odd in Fig. 9.2.6(b). In the former case, if we start from any spin (or corner of the square), and move around the square, assigning to each spin a value determined by that of the previous spin multiplied by the sign of the bond connecting them, we end up in a self-consistent configuration when we return to the spin we started with. This is therefore an unfrustrated plaquette.

By contrast, Fig. 9.2.6(b), which has an *odd* number of antiferromagnetic bonds on a square lattice, is an example of a frustrated plaquette. This is because when we go around it to arrive at the initial spin, the spin has to be the opposite of what we started with. There is thus a conflict (leading to frustration, just like in a sociological situation) regarding what should be the actual configuration.

On generalizing this to a real 3-dimensional structure, one can conclude that a frustrated system (e.g. a spin glass) has a large number of equally likely but distinct ground states. Thus, competing interactions lead to frustration, and thence to a *strong degeneracy* of the ground state.

9.2.7 Edwards Anderson Model and Sherrington Kirkpatrick Model

The basic model for dealing with quenched-spin systems is the Edwards-Anderson (EA) model (Edwards & Anderson 1975). It has also been extended to other glassy crystals like orientational glasses (see, e.g., Binder & Reger (1992) for a review).

The essence of the model is best introduced with the help of an analogy given by Edwards & Anderson (1975). The analogy is that of gelation in a polymer. As the temperature of a solution of very long molecules is decreased, its density increases, and the mobility of the molecules decreases. Near a particular density the mobility of the molecules drops to zero, and the molecules nearly freeze to a random-orientation static (or "quenched") configuration. This static or quenched disorder differs from disorder produced by thermal causes in that, if we view it at a later instant, it has the *same* random configuration.

According to the Edwards-Anderson (1975) theory, since the interaction between the spins in a spin glass oscillates in sign with distance, there is little ferromagnetic or antiferromagnetic ordering, but there is still *a definite ground state*, with spins pointing in a specific set of directions. It was shown by them that the existence of such a ground state is sufficient to cause the existence of a cusp in the susceptibility curve at a critical temperature, and that the cusp can be smoothed by an external magnetic field.

Let a particular spin be measured as $\mathbf{S}_i^{(1)}$ at time zero. And let $\mathbf{S}_i^{(2)}$ be the result of a measurement of the same spin at a later time t. The theory shows that the following probability is nonzero:

$$q(T) \equiv \lim_{t\to\infty} < \mathbf{S}_i^{(1)} \cdot \mathbf{S}_i^{(2)} > \neq 0, \tag{9.2.4}$$

One expects that $q = 1$ at $T = 0$. And for temperatures above the susceptibility-cusp temperature T_c , $q = 0$. The parameter q is thus the local order parameter for the spin-glass transition. It provides a measure of the local order, without our having to specify the nature of any long-range correlations in the random system of spins. It is an indicator of the long-*time* correlations, rather than long-ranged *spatial* correlations, among the spins.

The absence (or near-absence) of long-ranged spatial ordering and the presence of long-time ordering was aptly described as nonergodicity by Palmer (1982) because such a system does not 'visit' all possible states available to it in the course of time (cf. Fig. 9.2.5).

In an actual spin glass the spins randomly occupy lattice sites, and are few and far between. In the EA model a great simplification is introduced by replacing the actual system by a system in which the spins lie on the sites of a regular (i.e. translationally invariant) lattice, and it is the *interactions* between them that are random. There is thus a replacement of the *random site* problem by a *random bond* problem.

The Hamiltonian is assumed to be the random Heisenberg Hamiltonian:

$$\mathcal{H} = -1/2 \sum_{ij} J_{ij}\, \mathbf{S}_i \cdot \mathbf{S}_j \; - D \sum_i S_i^z, \tag{9.2.5}$$

where the exchange constants J_{ij} have a random distribution, specified by a density function $\rho(J_{ij})$. A symmetric Gaussian distribution function is a natural choice to start with:

$$\rho(J_{ij}) = \frac{1}{(2\pi\Delta_{ij})^{1/2}}\, e^{-J_{ij}^2/2\Delta_{ij}} \tag{9.2.6}$$

In this case the strength of the short-ranged exchange interaction depends only on the distance, r, between the spins.

Let ϵ_{ij} be a parameter which is unity or zero depending on whether or not the sites i and j are occupied by spins. In the EA model the *average* exchange parameter $\bar{J}$ is zero on any length scale, i.e.

$$\bar{J} = \sum_{i,j} J_{ij}\epsilon_{ij} = 0, \tag{9.2.7}$$

although the *variance*, $Var(J)$, is nonzero. In fact, it is shown that

$$kT_{sg} = \left(\sum_{i,j} < \frac{2}{9} J_{ij}^2 \epsilon_{ij} > \right)^{1/2} \tag{9.2.8}$$

An infinite-range version of the EA model was formulated by Sherrington & Kirkpatrick (1975), and is commonly called the *SK model.* In this model the *same* distribution $\rho(J_{ij})$ is assumed to hold for any pair of spins, *irrespective of the distance between them.* This infinite-range model is important because it carries the essence of a mean-field theory of spin glasses.

To compute the macroscopic thermodynamic properties from the EA or SK models, one must first calculate averages of extensive quantities like free energy and entropy (cf. §E.2). These can be determined from the logarithm of the partition function (Eq. E.2.35):

$$\mathcal{Z} = Tr\,(e^{-\beta\mathcal{H}}) \tag{9.2.9}$$

This function grows exponentially with the size of the system, characterized by the number N of lattice sites (each having effectively a spin, which interacts with other spins through the random-exchange integral). It also fluctuates widely because of the random nature of the interaction, and the fluctuations grow with N. For these reasons the conventional statistical-mechanical method of obtaining the ensemble average is not applicable to spin glasses.

The origin of the difficulty lies in the fact that we are dealing with a *quenched* random system: the J_{ij} in Eq. 9.2.5 are fixed for all time by the way the specimen was prepared. And yet we want to compute an ensemble average for a macroscopically large crystal (in which several different distributions for J_{ij} occur in different parts of the crystal) such that the ensemble average is sensibly representative of the crystal as a whole. We also want that our statistical averages should not diverge as $N \to \infty$.

In the EA model this objective was sought to be achieved by the so-called *replica method.*

The initial step in this method is to take note of the identity

$$\ln \mathcal{Z} \equiv \lim_{n \to 0} \frac{\mathcal{Z}^n - 1}{n}, \tag{9.2.10}$$

which means that averaging over $\ln \mathcal{Z}$ can be achieved if we can average over $\mathcal{Z}^n$ for $n = 1, 2, 3, ..$, and then extrapolate the results to the limit $n \to 0$.

The averaging over $\mathcal{Z}^n$ is made tractable by the fact that Z is defined by Eq. 9.2.9, and thus has an exponential dependence on the Hamiltonian, and thence on a *random* J_{ij} (cf. Eq. 9.2.5).

For carrying out the averaging, one considers n identical replicas of the system (labelled by an index α, $\alpha = 1, 2, ..n$. The result is

$$< \mathcal{Z}^n > = Tr_{(S^\alpha)} \int \exp(-\beta \sum_{\alpha=1}^{n} \sum_{ij} J_{ij} \mathbf{S}_i^\alpha \cdot \mathbf{S}_j^\alpha)\, \rho(J_{ij})\, d(J_{ij}) \quad (9.2.11)$$

This integration is easy to carry out if $\rho(J_{ij})$ is a Gaussian, given by Eq. 9.2.6. We get

$$< \mathcal{Z}^n > = Tr_{(S^\alpha)} \exp\left[\frac{\beta^2 J^2}{2} (\sum_{\alpha=1}^{n} \mathbf{S}_i^\alpha \cdot \mathbf{S}_j^\alpha)^2\right] \quad (9.2.12)$$

The main assumption of the replica method is that the various replicas $(\alpha, \beta, ..)$ are coupled, and that the replica-replica correlation function

$$q_{\alpha\beta} \; = < S_i^\alpha \, S_i^\beta > \quad (9.2.13)$$

behaves the same way as the EA order parameter defined by the time-correlation function $< S_i(0)\, S_i(\infty) >$. What this means is that the various replicas correspond to the same spatially frozen configuration probed at widely separated times.

Eq. 9.2.12 is difficult to evaluate, except mainly for the mean-field approximation of the SK model. One can make the further simplifying assumption that S_i^α and S_j^α are uncorrelated.

While the application of the replica method to the EA (or rather the SK) model explains qualitatively the cusp in the susceptibility curve, and also some other observations, it ends up with the so-called *negative entropy catastrophe* at low temperatures. In the SK model each of the N spins interacts through a random exchange interaction, $J_{ij}/\sqrt{N}$, with every other spin. Application of the replica method leads to a solution of the SK model in which the entropy approaches zero and then a negative value as $T \to 0$. This is not acceptable because in statistical mechanics entropy is the logarithm of an integer.

We touch on two attempted solutions of this problem. One is due to Parisi (1979, 1980a-e, 1983), and the other is the "TAP" theory (Thouless, Anderson & Palmer 1977), which avoids the use of the replica trick.

9.2.8 Breaking of Replica Permutation Symmetry

The Edwards-Anderson-Sherrington-Kirkpatrick (EA-SK) model operates in the infinite-range mean-field (Gaussian) approximation. It also uses the random Ising Hamiltonian, which is simpler than the random Heisenberg

Hamiltonian. It applies the replica method for computing the partition function for the random-interaction system that a spin-glass crystal is. The replica method introduces an order-parameter matrix $(q_{\alpha\beta})$ (Eq. 9.2.13) as a limit $(n \to 0)$ of an $n \times n$ symmetric matrix. The diagonal elements of this matrix are zero; i.e. $q_{\alpha\alpha} = 0$ for $\alpha = 1, 2, \cdots n$ (otherwise we would have $q = q_{\alpha\alpha} = 0$). The assumed permutation symmetry in replica space $(q_{\alpha\beta} = q_{\beta\alpha})$ is commonly referred to as the *replica symmetry*. There is a possibility that the *ansatz* regarding this symmetry may be the reason for unacceptable results like negative entropy mentioned above. That is, a breaking of the replica permutation symmetry may actually be taking place in a spin glass as one cools it below the spin-glass transition.

It was proved by Thouless & Almeida (1978) that this is indeed the case: Not all pairs of replicas produce the same average correlation $q_{\alpha\beta}$. They found that the reason for the failure for of the replica method was that below a certain line (now called the *AT instability line*) in the $\mathcal{H}$ versus T diagram, a replica-symmetric solution of the EA-SK model is dynamically unstable against replica-symmetry breaking. Above the AT line the EA-SK solution is stable.

Parisi (1979, 1980a-e, 1983) proceeded to achieve a breaking of the permutation symmetry of the replicas by first imposing the requirement that all quantities (e.g. magnetic moment m) involving any particular replica be replica-independent. In other words, for any integer k and any replica pair (β, γ), we must have

$$m_\beta = m_\gamma, \tag{9.2.14}$$

$$\sum_{\alpha=1}^{n} (q_{\alpha\beta})^k = \sum_{\alpha=1}^{n} (q_{\alpha\gamma})^k \tag{9.2.15}$$

Parisi's *ansatz* for the degree of resemblance among replicas (or the replica-replica correlation) can be represented in the form of an *ultrametric tree* (Fig. 9.2.7).

The circles at the top (corresponding to the maximum values of the order parameters) are the n replicas. The fact that the exponent k in Eq. 9.2.15 can take any value (up to infinity) corresponds to having an infinite number of bifurcations of the ultrametric tree. What this means is that there is no *unique* locally stable state of the spin glass. Instead, there are many such states, with varying degrees of replica-overlap or resemblance. The various solutions of the replica method can be viewed as clusters of states in the N-dimensional space of the N spins. At the highest level in the ultrametric tree, let q_1 denote the overlap or correlation between any pair of states in the same cluster. The overlap between two states or replicas belonging to two *different* clusters (say α and β in Fig. 9.2.7) is

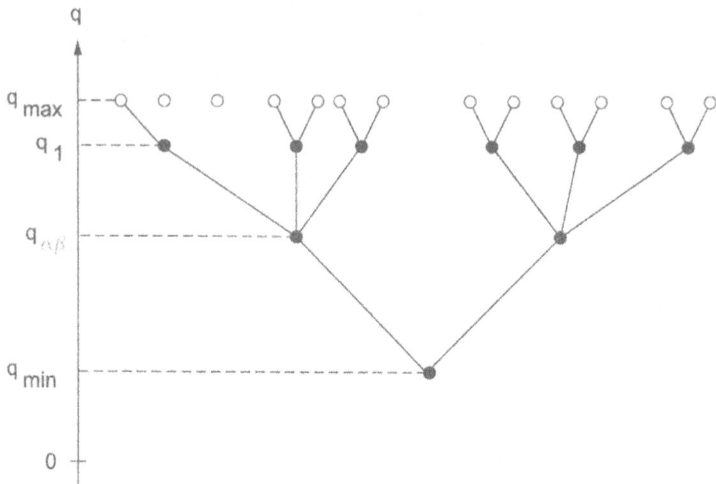

Figure 9.2.7: Parisi's ultrametric-tree scheme for breaking the permutation symmetry of replicas. The circles at the top are the n replicas. To find a particular replica-replica correlation, say $q_{\alpha\beta}$ between two replicas α and β, we trace downwards from the circles for α and β along the branches of the tree until they join.

obtained by tracing downwards from the circles for α and β along branches of the tree until they meet a node. If q_2 is the order parameter for this node, then $q_2 \leq q_1$. Similarly, the overlap between any of the 7 states in the left part of Fig. 9.2.7 with any of the 6 states on the right is $q_3 < q_2$.

This procedure works for $n > 0$ and k finite in Eq. 9.2.15. It is assumed next that the limits $n \to 0$ and $k \to \infty$ exist, corresponding to an infinite number of order parameters.

This theory provides a stable solution for the spin-glass transition, overcomes the negative-entropy problem, and, as we shall see later, has several other important fallouts.

9.2.9 Thouless-Anderson-Palmer Theory

The TAP (1977) theory of spin glasses is a mean field, infinite range interaction, theory which was formulated with the express purpose of avoiding the use of the replica method.

A mean-field treatment of a conventional ferromagnetic phase transition leads to the result that the macroscopic spontaneous magnetization M is given by the stable solution of the equation (Stanley 1971):

$$M = \tanh(\beta J M), \tag{9.2.16}$$

where $J = \sum_{j \neq i} J_{ij}$.

A generalized version of the above equation is

$$m_i = \tanh(\beta \sum_j J_{ij} m_j), \tag{9.2.17}$$

where $m_i =< S_i >$ is the local expectation value of the spin at a particular lattice site.

The TAP theory extended this approach to spin glasses by working out the site magnetizations *within* a configuration valley (cf. Fig. 9.2.5). The replica method of averaging was sought to be avoided by deferring it till the end.

The theory incorporated a local-field correction to Eq. 9.2.17 by adopting the cavity-field method of Onsager and Bethe (cf. Brout & Thomas 1967). The correction (of the order of $1/N$) accounts for the response of all the N spins affected by the fluctuations of a particular spin. Whereas $N \to \infty$ for a conventional ferromagnet, such is not the case for a local spin-glass configuration in a "rugged energy landscape".

The results of this formulation agree with those of the EA-SK model for $T > T_{sg}$. Moreover, below T_{sg} several new features emerge. In particular, it gets rid of the negative-entropy problem.

9.2.10 Cluster Glasses, Mictomagnets, Superparamagnets

In canonical spin glasses the interaction between magnetic ions is usually mediated by itinerant electrons, and is thus, by and large, an *indirect* interaction. The occurrence of direct interaction by overlap of wave functions of the $3d$ electrons belonging to nearest-neighbour (nn) or next-nearest-neighbour (nnn) magnetic ions increases as the concentration of these ions is increased. This short-ranged (usually ferromagnetic) interaction results in the formation of magnetic clusters of various sizes and wide-ranging relaxation times (Levin, Soukoulis & Grest 1979; Fischer 1983). Even phase separation (*chemical clustering*) can occur at suitably high dopant concentrations. The net result can be the occurrence of extremely large effective magnetic moments, ranging up to 20,000 Bohr magnetons (see Mydosh & Nieuwenhuys (1980) and Hurd (1982) for reviews). Such systems are referred to as cluster glasses or mictomagnets.

For a given cluster size there is a freezing temperature, or a glass-transition temperature, below which clusters of that or larger size freeze into a random-orientation configuration. A cluster glass is thus akin to a spin glass in certain respects, but its magnetic moments reside mostly in clusters (of different sizes) arising from *direct* exchange interactions.

Levin et al. (1979) have performed extensive calculations on the cluster model for spin glasses, assuming a Heisenberg Hamiltonian of the form

$$\mathcal{H} = -1/2 \sum_{\mu\nu} J_{\mu\nu}\, \mathbf{S}_\mu \cdot \mathbf{S}_\nu - 1/2 \sum_{i,j,\nu} J_{ij}^0 \mathbf{S}_{i\nu} \mathbf{S}_{j\nu} - \sum_{i,\nu} h \mathbf{S}_{i\nu}, \quad (9.2.18)$$

and a Gaussian distribution for the inter-cluster exchange integral $J_{\mu\nu}$. Here the first term takes account of inter-cluster interactions, the second that of intra-cluster (ferromagnetic) interactions, and the third recognizes the presence of a magnetic or 'ordering' field h (internal or external).

A typical ferromagnetic crystal splits into an optimum configuration of domains (and domain walls) to minimize the overall free energy. As the size of the crystal is reduced, its surface energy becomes more and more dominant compared to the bulk energy. Below a certain size of the crystal, domain walls are not favoured and the whole crystallite exists as a single domain. Such a situation is also encountered in certain small-sized magnetic particles embedded in rock materials. Here the role of the surface energy of free magnetic crystallites is replaced by that of interfacial energy between the magnetic particle and the rock material, with additional contributions to the overall energy coming from factors such as magnetostrictive strain anisotropy, as also shape anisotropy. Further, at high temperatures the magnetic particles cannot behave like normal paramagnets with thermally disordered orientations of individual atomic magnetic moments. Instead, they are constrained by the surrounding material to have all the magnetic moments point in the same easy direction of magnetization. Because of the resulting large magnetic moment (which can be several thousand Bohr magnetons), the term *superparamagnets* is used for such particles.

It may be mentioned here that, since ferromagnetic ordering is a cooperative effect, the energies involved scale with the volume of the particle. As particle size decreases, the energy barrier opposing a flipping from one easy direction of magnetization to another becomes comparable to $k_B T$, and the giant superparamagnetic moment can orient as a whole, quite easily, from one energy minimum to another.

Under the action of an external magnetic field these ferromagnetically ordered moments rotate coherently as a whole, a process opposed by the various anisotropies mentioned above. A strong enough magnetic field can even induce domain reversal (without a movement of domain walls because none exists). Therefore the B-H curve of a superparamagnet at a given temperature does not display a sizeable hysteresis (Bean & Livingston 1959; Cullity 1972).

Because of the increasing effect of thermal fluctuations at higher temperatures, the field $H(T)$ required for effecting domain reversal decreases with increasing T. In fact, the total magnetization is found to be a *universal function* of H/T.

As a superparamagnetic particle is cooled, at a certain temperature called the *blocking temperature* (Néel 1949) its magnetization freezes to a stable state. The blocking temperature varies linearly with the volume V of the particle. It is the temperature below which the relaxation time (see Eq. 9.2.20 below) becomes larger than the observation time of the experimental probe used.

There are several points of similarity between superparamagnets and cluster glasses (Wohlfarth 1977), although the two are quite different from the point of view of atomic structure. For instance, the spread of blocking temperatures in cluster glasses is related to the spread in their volumes V.

We next consider the effect of an external magnetic field on a cluster glass. Field-cooling of a cluster glass (in a relatively small field of the order of 1 tesla) through the average freezing or blocking temperature results in a configuration with a frozen preferred orientation of magnetic moments for a certain fraction of the clusters. This leads to a *shifting* of the hysteresis loop along the field axis. Let M_s denote the saturation magnetization at the temperature at which the field is switched off. The relaxation rate for the remanent magnetization M_r has typically the following form:

$$M_r = M_s\, e^{-t/\tau}, \qquad (9.2.19)$$

with τ given by

$$1/\tau = (1/\tau_0)\, e^{-KV/k_B T} \qquad (9.2.20)$$

Here K is some appropriate anisotropy constant, and τ_0 is of the order of 10^{-9} sec.

If τ_m is a typical measurement time, clusters for which the product KV in Eq. 9.2.20 is larger than a certain critical value will, for all practical purposes, act like frozen clusters because $\tau > \tau_m$ for them.

Since a whole range of cluster sizes exists, the glass transition for a cluster glass has a somewhat smeared appearance. Let $f(T)$ denote the distribution function for glass-transition temperatures or blocking temperatures. The susceptibility is then determined by the equation (Wohlfarth 1979):

$$\chi(T) = (C/T) \int f(T')\, dT', \qquad (9.2.21)$$

where C is the Curie constant.

Only those clusters contribute to the susceptibility which have not frozen at the temperature of the experiment, and which have a response time smaller than the probing time.

9.2.11 Percolation-Related Magnetic Order

If we start with a cluster glass, and increase progressively the concentration of the randomly distributed magnetic dopant, a concentration level will be reached above which practically every magnetic ion will have at least one nearest-neighbour magnetic ion. This is the so-called *percolation limit*: for concentrations higher than it an uninterrupted chain of magnetic ions extends in the crystal from end to end, resulting in long-ranged magnetic order.

Such an order is, however, different from the order in conventional ferromagnetic or antiferromagnetic crystals, in that it is highly inhomogeneous spatially. In addition to the 'central' chain(s), there may exist spins and/or clusters of spins, some of which may be coupled to the central chains (see Mydosh & Nieuwenhuys 1980). Such configurations can be rich in interesting possibilities. For example, Gabay & Toulouse (1981), while studying a system of m-component spins in the SK-model, reported a coexistence of the spin-glass ordering with a ferromagnetic ordering. Using the Sherrington-Kirkpatrick replica trick, they found that, for $m \neq 1$, there are two mixed phases, M_1 and M_2, in the region where there is a transition from spin-glass ordering to ferromagnetic ordering. The transition region is a mixed-phase region. The mixed phase M_1 is characterized by a coexistence of spontaneous magnetization and spin-glass ordering of the transverse components of the spins. The mixed phase M_2 has the same coexistence of orderings as M_1, and, in addition, has a spontaneous breaking of replica symmetry.

The term *re-entrant spin glass* is sometimes used in the literature in the context of such mixed phases.

There exist a number of experimental results, particularly for metallic systems, which can be interpreted in terms of the above model for the magnetic structure, employing the 'transverse spin-component freezing' approach (see Campbell & Senoussi (1992)).

A very similar coexistence of long-ranged magnetic order and frozen transverse-spin components had been predicted earlier by Mookerjee for a quenched random Heissenberg model with isotropic RKKY interaction (Mookerjee 1979). This alternative approach did not take recourse to the replica trick, and involved a mean-field effective-medium approximation. The free energy was minimized with respect to the local magnetization prior to configuration averaging. The various phases in this scheme were characterized by the probability distribution of the local molecular field or, equivalently, the local magnetization. Mookerjee & Roy (1983,1984) have applied this approach explicitly to the Au-Fe system.

9.2.12 Speromagnets and Sperimagnets

Speromagnets

The general term 'speromagnets' has been used in the literature (Coey 1978) for covering both spin glasses and cluster glasses (conductors as well as insulators). It applies to crystalline as well as amorphous materials.

Sperimagnets

Crystals of $FePd_{1.6}Pt_{1.4}$ are examples of sperimagnets (Coey 1978). In them the spins on only the Fe ions are predominantly frozen in random orientations. The other magnetic species (Pd, Pt) has predominantly a long-ranged order of the conventional ferromagnetic variety. The structure thus has a net spontaneous magnetization. Sperimagnets are the glassy counterpart of ferrimagnets.

9.2.13 Nonexponential Relaxation in Materials

Relaxation phenomena in spin glasses are a good example of what may be called 'dynamical heterogeneity'. Several aspects of their dynamical response to small perturbations have much in common with what is observed in many other condensed-matter systems (liquids, glasses, polymers, and certain 'mixed crystals'). When any such system is disturbed from its equilibrium state by a small perturbing field, it tends to relax back to equilibrium. And as it is doing so, its various properties display a characteristic time-evolution towards equilibrium (§E.3). Let $\Phi(t)$ denote one such function of time. The simplest possible situation one can imagine is that in which the rate of relaxation towards equilibrium is linearly related to the existing deviation of the property from its equilibrium value $\Phi(\infty)$:

$$\frac{d\Phi}{dt} = -\frac{\Phi(t) - \Phi(\infty)}{\tau} \tag{9.2.22}$$

If the 'time constant' τ is really a *constant*, and there is only one such time constant involved, an ordinary exponential relaxation can be expected:

$$\Phi(t) \sim \exp(-t/\tau) \tag{9.2.23}$$

Such a Debye-like relaxation is generally only an idealization. Real systems generally exhibit *nonexponential relaxation.*

This very old field of enquiry has been reviewed recently by Chamberlin (1998). A remarkable general observation is the high degree of *universality* exhibited by diverse systems. In fact the relaxation may be one of just two main types, namely *KWW relaxation* and *CvS relaxation.* Even the deviations from these two types show a notable degree of universality (Chamberlin 1998).

The KWW (Kohlrausch-Williams-Watts) response (Kohlrausch 1854) is of the 'stretched exponential' type:

$$\Phi(t) \sim \exp[-(t/\bar{\tau})^{\beta}] \tag{9.2.24}$$

Here $\bar{\tau}$ is a characteristic relaxation time, and β is the stretching exponent ($0 < \beta < 1$).

The (less common) CvS response (Curie-von Schweidler response) (Curie 1889; von Schweidler 1907) involves a power-law relaxation:

$$\Phi(t) \sim (t/\bar{\tau})^{-\alpha}, \quad \alpha > 0 \tag{9.2.25}$$

A general theory of these two fairly universal rates for slow response has not been available for a long time. Such a theory, with a thermodynamic backing, has been recently formulated by Chamberlin (Chamberlin & Haines 1990; Chamberlin 1993, 1994, 1996, 1998).

Chamberlin & Haines (1990) made three basic postulates in their model:

- Nonexponential relaxation is due to a *distribution* of relaxation times.
- Relaxation rates vary exponentially with *inverse* size.
- The argument in the exponent of the relaxation rates can be either positive or negative.

Thus the theory works on the concept that the primary response of the perturbed system is *dynamically* heterogeneous. It is assumed that there are *dynamically correlated domains* (DCDs) which relax 'independently'. The relaxation of the specimen as whole is parametrized in terms of the distribution (n_s) of domain sizes (s), with a size-dependent relaxation rate

$$\omega_s = 1/\tau \tag{9.2.26}$$

The net relaxation rate is then determined by the product of the probability (sn_s) that a given particle belongs to a DCD of size s and the probability ($e^{-\omega_s t}$) that this domain has not yet relaxed, summed over all sizes:

$$\Phi(t) \sim \sum_{s=1}^{\infty} [sn_s] e^{-\omega_s t} \tag{9.2.27}$$

According to Chamberlin, the observed nature of the slow dynamics in most condensed materials is well characterized by one of just two distinct size distributions.

For ergodic systems (like liquids) the Gaussian distribution applies for DCD sizes:

$$n_s \sim e^{-(s-\bar{s})^2/\sigma^2} \tag{9.2.28}$$

Here σ has the usual meaning of standard deviation, and $\bar{s}$ is an average or characteristic size.

For glasses which can be modelled as quenched systems with isotropic local order, the following Poisson-like function from percolation theory is appropriate:

$$n_s \sim s^{1/9}\, e^{-(s/\sigma)^{2/3}} \tag{9.2.29}$$

The scaling parameter σ in this equation diverges at the percolation threshold.

Thus, relaxation rates are (inversely) size dependent, and the sizes of the DCDs may have either a Gaussian or a Poisson-like distribution.

The size-dependent relaxation rate is often expressed in the literature by the Arrhenius equation:

$$\omega_s \sim e^{-\delta E_s/k_B T}, \tag{9.2.30}$$

where δE_s is an appropriate 'activation' energy, and k_B is Boltzmann constant. According to Chamberlin (1996), the more common KWW-like response corresponds to an *inverse* Arrhenius law, i.e. δE_s is *negative* for it. He writes

$$\frac{\delta E_s}{k_B T} = \frac{C\sigma}{s}, \tag{9.2.31}$$

where the correlation coefficient C is negative. The justification for this assignment of sign (and later rationalization in terms of a thermodynamic theory) is that, since the Gaussian distribution is symmetric, the observed asymmetry in the dynamic response (which is skewed towards high frequencies) must be attributed to the size-dependent energy scale postulated in Eq. 9.2.31.

For CvS-like response, which is skewed towards *lower* frequencies, C must be positive, corresponding to the usual Arrhenius law (rather than the inverse law).

A thermodynamic theory, which leads to energy scales which vary inversely with domain size, and which also yields direct or inverse Arrhenius laws in appropriate limits, has been proposed by Chamberlin (1998).

SUGGESTED READING

S. F. Edwards & P. W. Anderson (1975). Theory of spin glasses. *J. Phys. F: Metal Phys.*, **5**, 965.

J. M. D. Coey (1978). Amorphous magnetic order. *J. Appl. Phys.*, **49**(3), 1646.

J. A. Mydosh & G. J. Nieuwenhuys (1980). Dilute transition metal alloys; spin glasses. In E. P. Wohlfarth (Ed.), *Ferromagnetic Materials*, Vol. 1. North-Holland, Amsterdam.

K. Binder & A. P. Young (1986). Spin glasses: Experimental facts, theoretical concepts, and open questions. *Rev. Mod. Phys.*, **58**, 801.

D. Chowdhury (1986). *Spin Glasses and Other Frustrated Systems.* Princeton University Press, Princeton.

P. W. Anderson (1988-90).
Spin glass I: A scaling law rescued. *Phys. Today*, **41**(1), 9 (1988a).
Spin glass II: Is there a phase transition ? *Phys. Today*, **41**(3), 9 (1988b).
Spin glass III: Theory raises its head. *Phys. Today*, **41**(6), 9 (1988c).
Spin glass IV: Glimmerings of trouble. *Phys. Today*, **41**(9), 9 (1988d).
Spin glass V: Real power brought to bear. *Phys. Today*, **42**(7), 9 (1989a).
Spin glass VI: Spin glass as cornucopia. *Phys. Today*, **42**(9), 9 (1989b).
Spin glass VII: Spin glass as paradigm. *Phys. Today*, **43**(3), 9 (1990).

K. H. Fischer & J. A. Hertz (1991). *Spin Glasses.* Cambridge University Press, Cambridge.

J. Mydosh (1992). *Spin Glasses.* Taylor & Francis, London.

H. T. Diep (Ed.) (1994). *Magnetic Spin Systems with Competing Interactions (Frustrated Spin Systems).* World Scientific, Singapore.

R. V. Chamberlin (1994). Mesoscopic model for the primary response of magnetic materials. *J. Appl. Phys.*, **76**, 6401.

A. P. Young (Ed.) (1998). *Spin Glasses and Random Fields.* World Scientific, Singapore.

W. Kleemann & E. K. H. Salje (Eds.) (1998). *Non-Exponential Relaxation and Rate Behaviour. Phase Transitions*, **65**, 1-290 (special issue).

9.3 FERROMAGNETIC PHASE TRANSITIONS

The prototype of every ferroic crystal, whether magnetically ordered or not, is nonmagnetic. A phase transition from a prototype to a ferromagnetic phase involves a loss of time-reversal symmetry of the underlying point group. Such a transition, unlike many structural phase transitions, does not involve a drastic change of the atomic (structural) configuration if it is of the *proper* ferromagnetic type. Therefore it usually satisfies the rather strict nondisruption condition imposed by us in the definitions of prototype symmetry and ferroic phase transition in §5.1.

For describing magnetic structures in general, and ferromagnetic phases in particular, one has to specify two sets of parameters, namely the atomic coordinates $\mathbf{r}_i$ and the spin (or rather magnetic moment) vectors $\mathbf{S}_i$. If the spins adopt only two configurations, namely 'up' and 'down', the symmetry of the ferromagnetic phase of a crystal is given by one of the 1651 Shubnikov groups described in §2.2.18.

However, some ferromagnetic phases adopt more complicated arrangements of spins, and Shubnikov groups are not adequate for describing them. For example, $ZnCr_2Se_4$, which has the Fedorov-group symmetry O_h^7, adopts a helical configuration of spin orientations, with an angle ϕ specifying the successive spin orientations. A Shubnikov-group symmetry can allow for only two values of ϕ, namely $\phi = 0$ and $\phi = 180^o$. Here we shall briefly touch on transitions to such phases also, which require the use of colour-symmetry groups (§2.2.21) for their description and analysis.

9.3.1 Prototype Symmetry for a Ferromagnetic Transition

Aizu (1970a) carried out a symmetry analysis for determining all possible species of full or partial primary ferroics. He also identified the species in which two or all three of the properties of ferromagnetism, ferroelectricity and ferroelasticity can couple completely or incompletely with one another.

We introduced in §5.1.3 a rigorous definition of prototype symmetry. Similarly, ferroic phase transitions were defined according to the new approach in §1.1 and §5.2.2. We introduce here an even more comprehensive definition (which covers magnetic ordering explicitly):

> *A ferroic phase transition is a nondisruptive phase transition, either from the prototypic phase, or from another ferroic phase, involving a change of chemical, magnetic, or colour point-group symmetry of the crystal.*

For such a transition, a prototype symmetry (as defined by us) can be always assigned.

Aizu (1970a) stipulated that the prototype of every ferroic crystal, irrespective of whether the ferroic phase is magnetic or not, is nonmagnetic. This means that the prototype symmetry is nonmagnetic even for a ferromagnetic phase transition. The argument for this rests on the belief that if a certain domain state exists with a certain configuration of spins, then a domain state with the same atomic coordinates, but with all spins reversed in sign, is also equally likely to occur. [Such a domain pair, with time-inversion as an F-operation, is described as a *time conjugate domain pair.*]

When there is a phase transition from a nonmagnetic prototype to a ferromagnetic ferroic phase, the two types of time-conjugate domains are equally likely to appear, and there may only be a small enthalpy barrier for state shifts between them. However, it must be remembered that time-inversion is not a control parameter available to the experimenter in the laboratory. What is available in this context is magnetic field, reversing the sign of which may not always achieve the desired ferromagnetic state shift. For example, if the ferromagnetic phase has colour symmetry, instead of black-white symmetry, a properly configured magnetic field is hardly ever available to the experimenter for effecting a reversal of all the spins in the ferromagnetic domain to be switched.

Starting from each of the 32 time-symmetric crystallographic point groups, Aizu (1970a) worked out a total of 773 group-subgroup combinations (ferroic species) in accordance with the criteria described in §6.1. Out of these, 327 species are ferromagnetic.

9.3.2 Ferromagnetic Species of Crystals

There are 327 possible species of ferromagnetic crystals. Out of these, 126 are full, and 201 are partial ferromagnetic species.

Aizu's (1970a) definition for partial ferroelectric and partial ferroelastic species which are also ferromagnetic is slightly different from his definition when ferromagnetic order is absent, in that time-conjugate pairs of orientation states are treated as *single* entities for determining the full or partial status of a species with respect to ferroelectricity or ferroelasticity.

If the prototype and ferroic point groups involved are such that the ratio of their orders is q, then the possible q orientation states are divided into $q/2$ time-conjugate pairs. The species is considered as full ferroelectric if the $q/2$ pairs all have different spontaneous-polarization vectors. If any two of the pairs have the same spontaneous polarization vector, the species is regarded as partial-ferroelectric. Similarly for the full or partial character of a ferroelastic species.

The reason for this distinction compared to nonmagnetic species is that polarization and strain tensors are time-symmetric (they are i-tensors). Therefore it makes sense to treat time-conjugate domain pairs as nondistinct for determining the full or partial character of such species with respect to ferroelectric and/or ferroelastic state shifts.

It is instructive to consider here some results and examples from Aizu's (1970a) symmetry analysis of ferroic species.

(i) The spontaneous magnetization vector is invariant under space inversion; i.e. the presence or absence of the space-inversion operator in the symmetry group of the crystal makes no difference to the magnitude and sign of the spontaneous magnetization of an orientation state. Therefore, if the space-inversion operator is lost at a ferromagnetic phase transition, those domain pairs which are related by the F-operation of space inversion will be degenerate with respect to spontaneous magnetization. Therefore such a species cannot be full ferromagnetic. For example, the species $2/m1'\,F\,1$ has four pairs of time-conjugate orientation states, and is a partial ferromagnetic, full ferroelectric, and partial ferroelastic species.

(ii) If a species is full ferromagnetic and has more than two orientation states, then it is a full ferroelastic species. For example, the species $4/m1'F\bar{1}$ possesses four pairs of time-conjugate orientation states, and is full ferromagnetic, nonferroelectric, and full ferroelastic.

(iii) In any species that is simultaneously full ferromagnetic and full ferroelastic, there is a complete coupling of the spontaneous magnetization vector and the spontaneous strain tensor. Similarly for simultaneously full ferromagnetic and full ferroelectric species.

(iv) The species $41'\,F\,2'$ serves to illustrate several concepts. It has four orientation states, which can be divided into two pairs of time-conjugate states.

The ferroic phase can allow a spontaneous polarization, but it is the same for all the four states. Thus, although there is a nonzero *absolute* spontaneous polarization, the *relative* spontaneous polarization for any domain pair is zero because all domains have the same absolute spontaneous polarization (cf. §10.1.7). Therefore it is a pyroelectric but nonferroelectric species.

It is a full ferromagnetic species; the four orientation states are related by 90° rotations and/or time-inversion operations.

It is also a full ferroelastic species; the spontaneous strain tensors of the two time-conjugate pairs of states are related by a 90° rotation.

(v) Crystals of Fe have a ferromagnetic tetragonal phase at room temperature, which changes to a nonferromagnetic bcc phase at 1183 K. The bcc phase is prototypic, and Fe at room temperature belongs to the Aizu species $m\bar{3}m1'\,F\,4/mm'm'$, which is full ferromagnetic, nonferroelectric, and full ferroelastic.

(vi) Cobalt belongs to the species $6/mmm1'\ F\ 6/mm'm'$ at room temperature, which is a full ferromagnetic, nonferroelectric, nonferroelastic species.

9.3.3 Proper Ferromagnetic Transitions and Critical Phenomena

The Landau theory of phase transitions is a mean-field theory, and therefore has a greater chance of success in dealing with transitions driven by long-ranged interactions in high-dimensional space. The exchange interaction is an extremely short-ranged interaction, and if it is the primary interaction responsible for the occurrence of a ferromagnetic transition, the Landau theory cannot be expected to provide a satisfactory explanation of the observed critical exponents in the close vicinity of the critical point. For this reason, when theories of critical phenomena, including the RG theory (§5.5.9) were developed, they made repeated reference to experimental results in the vicinity of ferromagnetic phase transitions. For the same reason, when we discussed critical phenomena in a general way in §5.5, many of the results stated were for ferromagnetic transitions. Nevertheless, it is instructive to recapitulate here the basic approach specifically for ferromagnetic transitions.

In the theory of proper ferromagnetic transitions the observed critical transitions are sought to be explained by making two main assumptions:

(i) For temperatures sufficiently close to T_c the asymptotic behaviour of the thermodynamic properties always varies as a *power law* in $|t|$ (Eq. 5.5.42), where $t = (T/T_c) - 1$ (Eq. 5.5.37).

(ii) In the close vicinity of T_c, the temperature dependence of all the relevant thermodynamic properties is only through their dependence on the correlation length of the order parameter (cf. Eqs. 5.5.16 and 5.5.17).

The second assumption is equivalent to accepting the validity of the scaling hypothesis described in §5.5.7.

The critical exponents for different universality classes of ferromagnetic transitions (§5.5.7) are calculated using the RG theory (§5.5.9).

Actual values of the critical exponents for several ferromagnetic transitions were given in §5.5.4.

Aharony (1996) has drawn attention to the inherent complexity of the experimental situation pertaining to ferromagnetic phase transitions when it comes to measuring critical exponents. The quantum-mechanical exchange interaction responsible for causing ferromagnetic ordering at and below T_c is a function of temperature, and its value at T_c and at a temperature slightly away from T_c may not be the same. This can happen, for example, because interatomic distances vary with temperature, and the exchange interaction is a very sensitive function of distance. Such effects can distort the effective values of critical exponents, especially if they are determined, not by a simpler relationship like that expressed by Eq. 5.5.39, but by a more complex one like

$$f(t) = At^{\lambda}\,|\ln t|^{\mu}\,(1 + Bt^{y} + \cdots) \tag{9.3.1}$$

As $t \to 0$, a plot of $\ln f$ against t may not necessarily have a straight-line portion from which to extract a 'constant' and meaningful value of the critical exponent λ.

Unless due care is taken in the planning and execution of experiments, further complications can come from the assumption usually made in the theoretical calculation of critical exponents that the specimen under investigation is a single-domain specimen of infinite size.

9.3.4 Colour Symmetry and the Landau Potential

In §5.3 we described the extended Landau theory of continuous phase transitions. A number of criteria and conditions were discussed there for shortlisting the Landau expansions to be considered for enumerating the possible phase transitions from a given initial symmetry.

These considerations have been extended to crystallographic colour groups (§2.2.21) by Litvin et al. (1982). These authors also introduced an additional criterion, called the *kernel-core criterion*, which further limits the groups and IRs which can be associated with a phase transition.

This subject has been further discussed in considerable depth by Izyumov & Syromyatnikov (1990), who have specifically constructed the Landau potential for crystals of $FeSn_2$ and $FeGe_2$.

9.3.5 Incommensurate Ferromagnetic Transitions

It can happen that an interaction causing a magnetic ordering of a crystal results in a period that is not commensurate with that of the underlying crystal structure. The result is an incommensurate magnetic transition.

The incommensurate nature of the configuration is usually a consequence of competing interactions. Several types of interactions are possible. One such example occurs in crystals of MnSi and FeGe, which belong to the Fedorov or chemical space group T^4, with a simple-cubic Bravais lattice (see Landau, Lifshitz & Pitaevsky 1984). Here the competing interactions are the exchange interaction and the non-exchange relativistic interaction involving (symmetry-permitted) products of spatial derivatives of the spontaneous magnetization. The crystal class T allows the existence of a term of the form $\mathbf{M}\cdot\mathrm{curl}\,\mathbf{M}$.

The presence of this small extra term in the Landau-Ginzburg expression results in the occurrence of a helicoidal magnetic structure, over and above the basic ferromagnetic structure. One can identify planes in which the magnetic moments lie, such that as we go from one plane to the next, the net magnetic moment reorients gradually, like in a helix. The pitch of the helix is large compared to the lattice period normal to these planes, and is, in general, incommensurate with the crystal structure.

SUGGESTED READING

K. Aizu (1970a). Possible species of ferromagnetic, ferroelectric, and ferroelastic crystals. *Phys. Rev. B*, **2**, 754.

D. B. Litvin, J. N. Kotzev & J. L. Birman (1982). Physical applications of crystallographic color groups: Landau theory of phase transitions. *Phys. Rev. B*, **26**, 6947.

R. J. Elliot (1983). Magnetic phase transitions. In M. Ausloos & R. J. Elliot (Eds.), *Magnetic Phase Transitions.* Springer-Verlag, Berlin.

Yu. A. Izyumov & V. N. Syromyatnikov (1990). *Phase Transitions and Crystal Symmetry.* Kluwer, Dordrecht.

9.4 DOMAIN STRUCTURE OF FERROMAGNETIC CRYSTALS

The various domains existing in a specimen of a ferromagnetic crystal, together with the domain walls separating them, constitute the domain structure of the crystal.

The domain structure of a ferromagnetic crystal, like that of any other type of ferroic crystal, has a symmetry aspect and a thermodynamic aspect. The symmetry aspect is that the number and type of single-domain states or variants is determined by the symmetry operators lost in going from the prototype symmetry to the ferroic-phase symmetry. The thermodynamic (or rather the thermostatic) aspect is that the actual shapes and sizes of the domains and domain walls are determined by the criterion of minimization of the overall free energy.

9.4.1 The Various Contributions to the Internal Energy

The free energy has a contribution from the entropy term, and various contributions from the internal-energy terms. We summarize here some salient features of the latter. For more details, the excellent recent texts by Jiles (1991), Valenzuela (1994), and Aharony (1996) should be consulted.

Exchange Energy

The dominant interaction involved in ferromagnetic ordering is the (quantum mechanical) exchange interaction. It is a very strong, and very short-ranged, interaction, hardly extending beyond nearest neighbours. We have already discussed some aspects of it in §9.1.1 (cf. Eq. 9.1.6).

Anisotropy Energy

The magnetocrystalline anisotropy energy was discussed in §9.1.1, and Eq. 9.1.13 was written as an extension of the original Heisenberg exchange integral.

Magnetostriction

Magnetization, whether spontaneous or induced, results in a change of physical dimensions (strain). The change of dimensions gets contributions from various effects. That part which depends linearly on the magnetization is called the *linear piezomagnetic effect*, and the part which depends on second-degree terms is called *magnetostriction*. The latter has an isotropic (or volume) part and an anisotropic part (see, for example, du Tremolet de Lacheisserie 1993). The anisotropic part is described by a polar tensor of rank 4 through the equation

$$e_{ij} = \lambda_{ijkl} M_k M_l \qquad (9.4.1)$$

The magnetostriction tensor (λ_{ijkl}) is an i-tensor (like the electrostriction tensor), even though it connects a c-tensor (magnetization) to an i-tensor (strain). This is because it describes the quadratic part of the dependence of strain on magnetization, rather than the linear (or odd-order) part.

The anisotropic part of magnetostriction has a small contribution and a large contribution. The small part comes from the magnetic dipole-dipole interaction, and is a 'form' effect. The larger part, called *Joule magnetostriction*, stems from the much stronger short-ranged exchange interaction.

Two types of magnetostriction can be identified: spontaneous and induced, depending on whether the magnetization causing the strain is spontaneous or induced.

When spontaneous magnetization arises as a result of a proper ferromagnetic phase transition, there is an accompanying onset of spontaneous strain as a faint variable, the two being related at least by Eq. 9.4.1, if not by terms involving other powers of the order parameter.

By contrast, consider a specimen which is already in a ferromagnetic phase, and we apply a magnetic field to it. The various domains are randomly oriented to start with, and we can regard this as the zero relative-strain state so far as the *induced* magnetostriction effect is concerned. Application of the magnetic field induces a net magnetization, which results in a net induced strain, again via Eq. 9.4.1. This is the phenomenon of induced magnetostriction.

The strain produced by magnetostriction is generally a small effect (of the order of 10^{-5} or less) when compared to strain via electrostriction. For a crystal of Fe, for example, $e_{11} = 21 \times 10^{-6}$ (cf. Jiles 1991). An exception is the alloy $Tb_{0.3}Dy_{0.7}Fe_2$, also known as TERFENOL$-$D (see du Tremolet de Lacheisserie 1993; Chaudhry & Rogers 1995). For it the saturation value of the strain produced by Joule magnetostriction is of the order of 10^{-3}.

Like magnetocrystalline anisotropy, magnetostriction is also related to spin-orbit coupling. Changes in spin lead to changes in orientations of orbitals, with a concomitant change of physical dimensions of the material.

Magnetostriction results in a complex internal-stress pattern in a polydomain crystal. Landau et al. (1984) have discussed some salient aspects of how one can model this contribution to the total internal energy of a magnetic crystal.

Demagnetization-Field Energy

If there are no domain walls and the whole specimen is just a single ferromagnetic domain, there will be a large magnetic field outside the specimen, extending over a large region of space. It is called the *demagnetization field*, $\mathbf{H}_d$, because it acts in opposition to the field used for magnetizing the finite-sized specimen, i.e. for turning a whole polydomain specimen into a single domain. The energy associated with this field is $(1/8\pi)\int H_d^2\, dV$, and it depends on the magnetization effected:

$$\mathbf{H}_d = N_d \mathbf{M} \quad (9.4.2)$$

N_d is a factor determined by the shape of the specimen crystal (see Jiles 1991). The term *shape anisotropy* is used in this context. In particular, $N_d = 0$ for a very long and thin specimen.

Because of the existence of $\mathbf{H}_d$, the net field $\mathbf{H}_{in}$ inside the specimen is different from the applied field $\mathbf{H}_a$:

$$\mathbf{H}_{in} = \mathbf{H}_a - \mathbf{H}_d \quad (9.4.3)$$

Also, $\mathbf{M} = \chi \mathbf{H}_{in}$. Therefore,

$$\mathbf{H}_{in} = \mathbf{H}_a / (1 + \chi N_d) \quad (9.4.4)$$

Near T_c, $\chi \rightarrow$ inf for a proper ferromagnetic phase transition. This means that $H_{in} \rightarrow 0$ as T_c is approached, making it problematic to measure χ accurately in the close vicinity of T_c.

Demagnetization-field energy is mainly a result of long-ranged dipole-dipole type interactions. The system can minimize this energy very substantially by splitting, if possible, into differently oriented domains, separated by domain walls. This results in a great reduction of the spatial extension of the lines of force outside the crystal. However, the creation of domain walls costs energy, and the system settles for an optimum number and configuration of domain walls. An interesting configuration often encountered is that of "closure domains" (see Kittel 1949).

Application of a magnetic field disturbs the existing optimum domain structure, resulting in the creation of a demagnetization field which opposes the magnetizing field.

Prediction or rationalization of the domain structure of a ferromagnetic specimen comprising two or more domains involves free-energy minimization, taking due note of the various contributions to the total energy. This is a highly complex problem, far from understood properly at present.

9.4.2 Orientations of Walls between Ferromagnetic Domain Pairs

Walls between ferromagnetic domain pairs are usually planar. And their orientations are determined predominantly by constraints ensuring a minimization of the *magnetoelastic* energy for the domain pair in question.

9.4.3 Thickness of Walls Separating Ferromagnetic Domain Pairs

The ferromagnetic exchange interaction tends to align spins in a parallel arrangement. Imagine a ferromagnetic domain pair separated by a domain wall. How thick is this wall ?

One possibility we should examine is that of a wall thickness of the order of an inter-atomic spacing, or, in the case of a ferrimagnetic rather than a ferromagnetic crystal, a unit-cell spacing. In each of the two domains separated by the wall the spins are likely to be aligned along an easy direction, so that the situation is favourable in terms of the magnetocrystalline-anisotropy contribution to the total energy. But it is highly unfavourable so far as the much stronger exchange energy is concerned.

We next consider another possibility, in which the rotation of the spins (say by 180°) in going from one domain to the other across the wall is spread over a large number, N, of unit cells. The spins in the unit cells in the domain wall are now inclined at nonzero angles to the easy directions of magnetization, resulting in an increase in the magnetocrystalline energy. But the exchange energy decreases dramatically for large N. The system settles for a compromise value of N, which may be as high as 100-200.

These *Bloch walls* separating ferromagnetic domains in bulk crystals are very thick. This situation is very different from that in ferroelectric or ferroelastic domain twins.

9.4.4 The Ferromagnetic Hysteresis Loop

We have already discussed, in §9.1.1 using Fig. 9.1.1, the magnetization process for a virgin ferromagnetic multidomain crystal. The dashed line in Fig. 9.4.1 shows the virgin magnetization curve again.

Also shown in Fig. 9.4.1 is a small loop around the origin. It is called a *minor loop*; it depicts what happens when the gradual increasing of the driving magnetic field is stopped at a value less than what is needed to achieve a state of saturation magnetization (cf. §9.1.1). When this intermediate field is decreased to the value zero, the specimen has a reduced, but nonzero, magnetization, and therefore magnetic induction B. The induction can be brought to zero only by further decreasing the applied field H to a suitable negative value. The sign of the induction becomes negative on still further decrease of the applied field. If we stop decreasing the applied field before reaching the saturation magnetization in the opposite direction, and start increasing it towards the origin, and beyond on the positive-H side, the rest of the minor loop is traced.

By increasing the maximum magnitude of the applied field, successively larger minor loops can be traced, till the field applied is so high that the

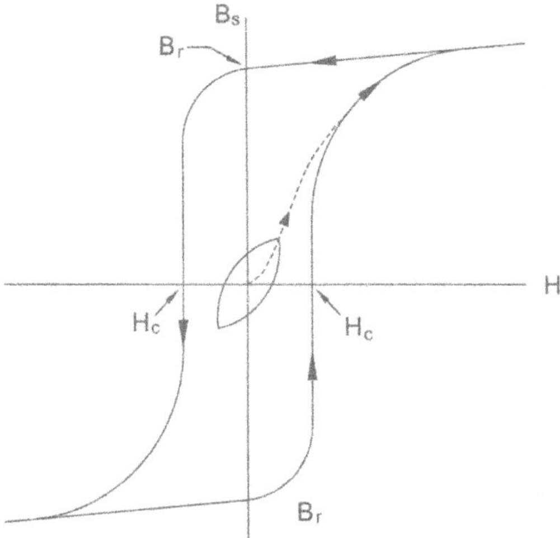

Figure 9.4.1: A typical *limiting* hysteresis loop for a ferromagnet. Also shown is the virgin hysteresis curve (dashed line), as well as a minor loop.

saturation magnetization M_s (or the corresponding saturation induction B_s) is reached. This outermost hysteresis loop is called the *limiting hysteresis loop*. Its interception on the B-axis defines the *remanent induction* B_r (and the corresponding *remanent magnetization*). B_r is the maximum attainable magnetic induction at $H = 0$.

And H_c in Fig. 9.4.1 is the *coercive field* needed to bring to zero the net induction in the specimen.

Soft and Hard Ferromagnetic Materials

A soft ferromagnetic material (single crystal or polycrystal) is one which has a high initial permeability μ_i (cf. Fig. 9.1.1), a high maximum permeability μ_{max}, and a low coercive field H_c (lower than $\sim 10A/m$).

A ferromagnetic material not meeting the above criteria is a hard ferromagnetic material.

Hardness of ferromagnetic materials is particularly important in the context of making permanent magnets, i.e. magnets which provide a fairly constant (and large) magnetic field without requiring a real-time use of electric current, and which do not deteriorate excessively or too soon with the passage of time.

A permanent magnet, by its very nature, possesses *stored energy*, which was spent on it at the time of magnetizing it. Since it is desirable to produce the highest possible magnetic field from the permanent magnet, the material used should have a high B_s. It should also be stable against high external magnetic fields, which means that it should have a high coercivity, H_c. The product $(BH)_{max}$, which has the dimensions of energy, is a good measure of the quality of a permanent magnet. This product varies directly with the area enclosed by the BH-curve. A large remanent induction, B_r, is therefore desirable for a permanent magnet.

SUGGESTED READING

A. Globus (1977). Some physical considerations about the domain wall size theory of magnetization mechanisms. *J. de Physique* (Paris), **C1-38**, C1-1.

D. Jiles (1991). *Introduction to Magnetism and Magnetic Materials.* Chapman & Hall, London.

R. Valenzuela (1994). *Magnetic Ceramics.* Cambridge University Press, Cambridge.

A. Aharony (1996). *Introduction to the Theory of Ferromagnetism.* Clarendon Press, Oxford.

9.5 DYNAMICS OF FERROMAGNETIC BEHAVIOUR

The response of a ferromagnetic specimen to an oscillating magnetic field is a complex phenomenon, involving a variety of processes, each with its own characteristic time constant. In the sub-giga-Hertz regime of frequencies, the response is determined mainly by domain-wall dynamics, especially for insulator ferromagnets.

We have seen in §9.1.1 and 9.4.4 that, except for high applied fields, bowing of the domain walls and domain-wall motion are the two main processes responsible for the magnetization behaviour, with the former as the only process at low applied fields. The following equation of motion serves to model the main features (Valenzuela 1994):

$$m\frac{d^2x}{dt^2} + \beta\frac{dx}{dt} + \alpha x = 2M_sH(t) \qquad (9.5.1)$$

The first term models the wall inertia, with x as the wall displacement coordinate, and m the effective mass. The second is a damping term, with β denoting a 'viscosity' factor. The third term stands for the restoring force, associated with pinning centres.

For a sinusoidal driving field of angular frequency ω, the real and the imaginary parts of the complex susceptibility have the following forms:

$$\chi' = \chi_0 \frac{1 - (\omega/\omega_s)^2}{(1 - \omega^2/\omega_s^2)^2 + (\omega/\omega_x)^2}, \tag{9.5.2}$$

$$\chi'' = \chi_0 \frac{\omega/\omega_x}{(1 - \omega^2/\omega_s^2)^2 + (\omega/\omega_x)^2}, \tag{9.5.3}$$

Here χ_0 is the static susceptibility. And ω_s and ω_x are the resonance and the relaxation frequencies:

$$\omega_s = (\alpha/m)^{1/2}, \tag{9.5.4}$$

$$\omega_x = \alpha/\beta \tag{9.5.5}$$

For very high frequencies, the domain walls do not respond significantly, and spin rotation within the domains is almost the only magnetization mechanism. The applied magnetic field tends to reorient the spins towards itself, and away from the easy direction of magnetization. The spins precess around the field direction (with the well-known Larmor frequency ω_L) for a certain relaxation time τ, before settling in the new orientation. By choosing $\omega = \omega_L$, resonant absorption (*ferromagnetic resonance*) can be achieved. Such experiments can provide data regarding the anisotropy fields present in the specimen.

SUGGESTED READING

R. Valenzuela (1994). *Magnetic Ceramics.* Cambridge University Press, Cambridge.

Chapter 10

FERROELECTRIC CRYSTALS

We have defined ferroic phase transitions as nondisruptive phase transitions involving a change of point-group symmetry. For transitions defined in this way a prototype symmetry can always be assigned or chosen. Ferroelectric phase transitions are a subset of ferroic transitions such that the ferroic phase belongs to one of the 10 polar classes, namely 1, 2, 3, 4, 6, m, $mm2$, $3m$, $4mm$, $6mm$.

These 10 crystallographic point groups are subgroups of the limit group ∞m (§2.2.19).

A ferroelectric material is one which can, or can be realistically conceived to, undergo one or more ferroelectric phase transitions.

We begin this chapter with a description of some basic properties of dielectric crystals.

10.1 SOME DIELECTRIC PROPERTIES OF ORDERED CRYSTALS

10.1.1 Polarization

Crystals and several other forms of matter are composed of positively and negatively charged particles (nuclei and electrons), with a tendency towards overall balancing of charge. The main identifiable units are: neutral atoms, positive ions, negative ions, molecules, and electrons. A molecule may be neutral, and yet the centres of mass of its positive and negative constituents may not coincide. If their separation is, say, l, and the charges are $\pm q$, the molecule has a dipole moment p, defined by the product $q\,l$.

Such a dipole can exist even when no field is applied. A simple example is that of the HCl molecule. Because of the different electronegativities of H and Cl, there is a shift of the electronic charge cloud towards Cl, giving rise to a *permanent* dipole moment. Molecules with a permanent dipole moment are called *polar molecules.*

For a molecule composed of identical atoms, symmetry prevents the existence of a *spontaneous* dipole moment. The same is true of an isolated atom or ion. But an *induced* dipole moment can exist under the action of an external field. The charge separation comes from the net movement of the valence electrons and the atomic cores. The induced dipole moment is given by

$$p_i = \alpha_{ij} E_j \tag{10.1.1}$$

(α_{ij}) is called the *polarizability tensor* of the atom or the molecule.

The spontaneous or induced separation between positive and negative charges in matter is called *polarization.* It is defined quantitatively in a general way as follows:

Imagine a charge $+q$ at a position $\mathbf{r}_1$, and a charge $-q$ at $\mathbf{r}_2$. The dipole moment of this pair is

$$\mathbf{p} = q\mathbf{r}_1 - q\mathbf{r}_2 = q(\mathbf{r}_1 - \mathbf{r}_2), \tag{10.1.2}$$

and is directed from the negative charge toward the positive charge.

Eq. 10.1.2 can be rewritten as

$$\mathbf{p} = q\mathbf{r}_1 + (-q)\mathbf{r}_2 \tag{10.1.3}$$

This admits of a generalization. Imagine a set of charges q_i, with position vectors $\mathbf{r}_i$. The total dipole moment of this set is defined as

$$\mathbf{p} = \sum_s q_s \mathbf{r}_s \tag{10.1.4}$$

The polarization $\mathbf{P}$ is

$$\mathbf{P} = \mathbf{p}/V, \tag{10.1.5}$$

where V is the volume occupied by the charge distribution.

$\mathbf{P}$ can also be defined as the polarization charge per unit area perpendicular to the direction of total dipole moment.

10.1.2 Pyroelectric Effect

For a crystalline phase belonging to any of the 10 polar classes, there exists a direction (the polar axis) that is not equivalent to any other direction under the operations of the point group of the crystal. The charge configuration along the two ends of this direction is therefore not the same, resulting in a net dipole moment for the unit cell, even in the absence of an external electric field. Such a dipole moment, $\mathbf{P}_s$, called the spontaneous polarization, naturally varies with temperature, giving rise to the pyroelectric effect:

$$\Delta P_{si} = p_i \, \Delta T \tag{10.1.6}$$

[The symbol p_i used for the components of the pyroelectric-effect tensor is not to be confused with the symbol used in Eqs. 10.1.2-4 for the dipole moment.]

There are two types of contributions to (ΔP_{si}) when the temperature is changed. One is due to the thermal expansion or contraction of the free or unclamped crystal, resulting in its deformation. This 'piezoelectric' contribution to the pyroelectric effect is called the *spurious* or *secondary* pyroelectric effect, with (p_i'') denoting the corresponding tensor coefficients.

The *true* or *primary* pyroelectric effect occurs even when the crystal is clamped and thus prevented from deforming, and is the result of a genuine change in the charge distribution on change of temperature. Denoting the corresponding tensor by (p_i'), the total pyroelectric effect can be written as

$$\Delta P_{si} = (p_i' + p_i'') \, \Delta T \tag{10.1.7}$$

Here we have neglected the possible dependence of ΔP_{si} on higher powers of ΔT. In this *linear* pyroelectric effect, the primary contribution is usually much *smaller* than the secondary contribution.

A pyroelectric phase is also a ferroelectric phase if it possesses at least two orientation states which differ in spontaneous polarization. In other words, there should be at least one domain pair for which the 'relative' spontaneous polarization is nonzero (cf. §10.1.7 below). In view of the definition adopted by us for prototype symmetry and ferroic phase transitions in terms of the nondisruption condition, a switching from one orientation state to another should be always possible (under the action of a suitable external field).

10.1.3 Effect of Static Electric Field

Dielectric Permittivity

Let $\mathbf{E}$ denote the intensity of the electric field. The electric displacement, or electric flux density, is defined as

$$D_i = \epsilon_0 E_i + P_i, \quad i = 1, 2, 3 \tag{10.1.8}$$

Here ϵ_0 denotes the permittivity of free space.

The polarization may have an induced part and a spontaneous part:

$$P_i = P_{Ei} + P_{si} \tag{10.1.9}$$

If the field $\mathbf{E}$ is not too strong, the following linear relationship holds for induced polarization:

$$P_{Ei} = \epsilon_0 \chi_{ij} E_j, \tag{10.1.10}$$

where (χ_{ij}) is the dielectric susceptibility tensor. Then, from Eq. 10.1.8,

$$D_i = \epsilon_0(\delta_{ij} + \chi_{ij})E_j + P_{si} \equiv \epsilon_{ij}E_j + P_{si}, \tag{10.1.11}$$

where

$$\epsilon_{ij} = \epsilon_0(\delta_{ij} + \chi_{ij}) \tag{10.1.12}$$

(ϵ_{ij}) is called the dielectric permittivity tensor of the medium.

Since ϵ_0 is a scalar constant, it is convenient to introduce a dimensionless tensor

$$\varepsilon_{ij} = \epsilon_{ij}/\epsilon_0, \tag{10.1.13}$$

called the relative-dielectric-permittivity tensor, or simply the 'dielectric-constant' tensor.

From Eqs. 10.1.10 and 10.1.12, the following relationship can be arrived at:

$$\epsilon_0(\varepsilon_{ij} - \delta_{ij}) = P_{Ei}/E_j \tag{10.1.14}$$

For an isotropic medium this reduces to

$$P_E = \epsilon_0(\varepsilon - 1)E \tag{10.1.15}$$

Depolarizing Field

Consider a finite-sized dielectric crystal (pyroelectric or nonpyroelectric), to which an external field $\mathbf{E}_a$ is applied. The applied field induces charges on the surface of the crystal. In addition, surface charges may exist because of the spontaneous polarization, if any. The field produced in the interior of the crystal by these surface charges is called the depolarizing field, $\mathbf{E}_d$. It is called a *depolarizing* field because it acts in opposition to the applied field, as well as to the spontaneous polarization if any.

The surface charge density obeys the Poisson equation:

$$\rho = \nabla \cdot \mathbf{D} \tag{10.1.16}$$

Substituting for $\mathbf{D}$ from Eq. 10.1.11,

$$\rho = \epsilon \nabla \cdot \mathbf{E} + \nabla \cdot \mathbf{P}_s \tag{10.1.17}$$

This can be rewritten as

$$\nabla \cdot \mathbf{E} = (\rho - \nabla \cdot \mathbf{P}_s)/\epsilon \tag{10.1.18}$$

The spontaneous polarization $\mathbf{P}_s$, which is nonzero for a pyroelectric crystal, has no macroscopic spatial variation inside the bulk of an (infinite) crystal; therefore $\nabla \cdot \mathbf{P}_s = 0$, leading to

$$\nabla \cdot \mathbf{E} = \rho/\epsilon \tag{10.1.19}$$

This result for the *interior* of the crystal is thus the same as for a nonpyroelectric crystal.

However, the situation becomes radically different in the vicinity of the *surface* of a pyroelectric dielectric, where $\mathbf{P}_s$ has a strong spatial variation, dropping sharply to zero at the surface. The $\nabla \cdot \mathbf{P}_s$ term then becomes a major contributor to the depolarizing field.

The energy associated with the depolarizing field is given by (Lines & Glass 1977):

$$W_E = \frac{1}{2} \int_V \mathbf{D} \cdot \mathbf{E} \, d\mathbf{r} \tag{10.1.20}$$

When a crystal is cooled from a paraelectric phase to a pyroelectric (or a ferroelectric) phase, it tends to split into a domain configuration which can annul, or at least minimize, the overall depolarizing field. In addition, the depolarization effect may also tend to be compensated by those mobile charges which can reach the surface, either by electrical conduction within the crystal, or by their trapping from the surroundings, or both.

Internal Field

Compared to magnetic susceptibility, electrical susceptibility of crystals is generally quite large ($\chi_{ij} >> 1$). Therefore the internal field, i.e. the local field ($\mathbf{E}_{loc}$) at a point inside a dielectric crystal, can be very different from the applied field ($\mathbf{E}_a$) at the same point in the absence of the crystal.

To calculate $\mathbf{E}_{loc}(\mathbf{r})$, the following procedure was adopted by Lorentz (see, e.g., Dekker (1957)). A small sphere (the *Lorentz cavity*) is imagined to be constructed around the point $\mathbf{r}$. The radius of the sphere is chosen to be the smallest possible for which the region outside it can be regarded as a polarized continuum, making a contribution $\mathbf{E}_L$ to $\mathbf{E}_{loc}(\mathbf{r})$. And the atoms inside the sphere are treated individually for their contribution ($\mathbf{E}_c$) to $\mathbf{E}_{loc}$ at the centre of the imaginary sphere.

To conduct measurements of polarization and permittivity etc., one applies a voltage V_a between two plates of a capacitor. Let $\mathbf{E}_a$ be the field generated by the applied voltage *before* the crystal specimen is inserted between the plates. On insertion of the specimen, a depolarization field $\mathbf{E}_d$ is induced on the surface of the crystal (corresponding to a surface polarization $\mathbf{P}$), and this field is neutralized by a flow of current in the capacitor circuit. There are thus four contributions to $\mathbf{E}_{loc}(\mathbf{r})$:

$$\mathbf{E}_{loc}(\mathbf{r}) = (\mathbf{E}_a + \mathbf{P}/\epsilon_0) + \mathbf{E}_d + \mathbf{E}_L + \mathbf{E}_c \tag{10.1.21}$$

The depolarization field $\mathbf{E}_d$ is defined by Eq. 10.1.17.

$\mathbf{E}_L$ is the field (the *Lorentz field*) resulting from the charges (spontaneous or induced) on the inner surface of the Lorentz cavity.

And $\mathbf{E}_c$ is the field contributed at the centre of the (spherical) Lorentz cavity by the atoms filling this cavity.

The Lorentz field is proportional to the polarization of the crystal:

$$\mathbf{E}_L = \gamma \mathbf{P}/\epsilon_0, \tag{10.1.22}$$

where γ is a constant called the *Lorentz factor.*

The value of $\mathbf{E}_c$ in Eq. 10.1.21 depends on the crystal structure, and on the position vector $\mathbf{r}$. One gets $\mathbf{E}_c = 0$ for several special situations (see Dekker (1957)).

The depolarizing field $\mathbf{E}_d$ depends on the shape of the specimen. For a flat isotropic specimen in a uniform field $\mathbf{E}_a$ perpendicular to its faces,

$$\mathbf{E}_d = -\mathbf{P}/\epsilon_0 \tag{10.1.23}$$

One can therefore write the following expression for the local field (with $\mathbf{E}_c = 0$):

$$\mathbf{E}_{loc} = \mathbf{E}_a + (\mathbf{P} - \mathbf{P} + \gamma \mathbf{P})/\epsilon_0 \tag{10.1.24}$$

For certain simple situations, $\gamma = 1/3$ (see Dekker (1957)); thence

$$\mathbf{E}_{loc} = \mathbf{E}_a + \mathbf{P} / (3\epsilon_0) \tag{10.1.25}$$

We can express this in terms of the static dielectric function by noting that $\mathbf{D} = \epsilon_0 \mathbf{E}_a + \mathbf{P}$, and also $\mathbf{D} = \epsilon \mathbf{E}_a$, so that

$$\mathbf{P} = \epsilon_0(\varepsilon - 1)\mathbf{E}_a \tag{10.1.26}$$

Then

$$\mathbf{E}_{loc} = \mathbf{E}_a + \frac{(\varepsilon - 1)\mathbf{E}_a}{3} = \frac{\mathbf{E}_a}{3}(\varepsilon - 1 + 3) = \frac{\mathbf{E}_a}{3}(\varepsilon + 2) \tag{10.1.27}$$

It is important to make a clear distinction between two types of electric field. One is the *average* field $\mathbf{E}_M$, which enters Maxwell's equations. The other is the local or internal field $\mathbf{E}_{loc}(\mathbf{r})$ experienced by a small positive test charge at a point $\mathbf{r}$ inside the crystal. $\mathbf{E}_M$ is the sum of the applied field $\mathbf{E}_a$ and the average field produced by the charges at *all* the points in the crystal. $\mathbf{E}_{loc}(\mathbf{r})$, on the other hand, is the field experienced at the point $\mathbf{r}$ due to the external field $\mathbf{E}_a$ and all the charges in the crystal, *except* the charge at $\mathbf{r}$ (if any).

10.1.4 Thermodynamics and Symmetry of Dielectric Properties

We have already discussed in §6.2.1 several thermodynamic aspects of ferroic crystals. We note, in particular, that entropy S (Eq. 6.2.6) and electric displacement $\mathbf{D}$ (Eq. 6.2.7) are *first derivatives* of the Gibbs free-energy density.

The dielectric permittivity is a *second derivative* of the free-energy density:

$$\epsilon_{ij} = \left.\frac{\partial D_i}{\partial E_j}\right|_{H,\sigma,T} = -\left.\frac{\partial^2 g}{\partial E_i\,\partial E_j}\right|_{H,\sigma,T} \tag{10.1.28}$$

Another example of a second derivative of g is the isothermal piezoelectric tensor:

$$d_{kij}^{H,T} \equiv \left.\frac{\partial D_k}{\partial \sigma_{ij}}\right|_{H,T} = -\left.\frac{\partial^2 g}{\partial \sigma_{ij}\partial E_k}\right|_{H,T} \tag{10.1.29}$$

An example of a property defined by a *third derivative* of free energy is *electrostriction*. This property, which is a measure of the quadratic dependence of strain on electric field, is present in all materials:

$$e_{ij} = M_{ijkl}\,E_k\,E_l \tag{10.1.30}$$

The electrostriction tensor is defined in differential form as follows:

$$M_{ijkl}^{H,T} = \left.\frac{\partial^2 e_{ij}}{\partial E_k \partial E_l}\right|_{H,T} = \left.\frac{\partial^3 g}{\partial E_k \partial E_l \partial \sigma_{ij}}\right|_{H,T}, \tag{10.1.31}$$

where we have made use of Eq. 6.2.9.

In contrast to electrostriction, which is present in all materials, piezoelectricity (Eq. 10.1.29), which is a measure of the *linear* dependence of strain on electric field (or of electric displacement on stress), is absent in the 11 centrosymmetric crystal classes. This is because, in Eq. 10.1.29, (D_k) (being a polar vector) vanishes in centrosymmetric crystals, and (σ_{ij}) is invariant under an inversion operation.

In addition to the 11 Laue classes, the piezoelectric tensor is also zero for the crystal class 432. It is nonzero for the remaining 20 noncentrosymmetric crystal classes, which are therefore referred to as the *20 piezoelectric classes*. They include all the 10 polar classes.

10.1.5 A Crystallophysical Perspective for Ferroelectrics

Figure 10.1.1 provides a perspective for ferroelectric phases of materials.

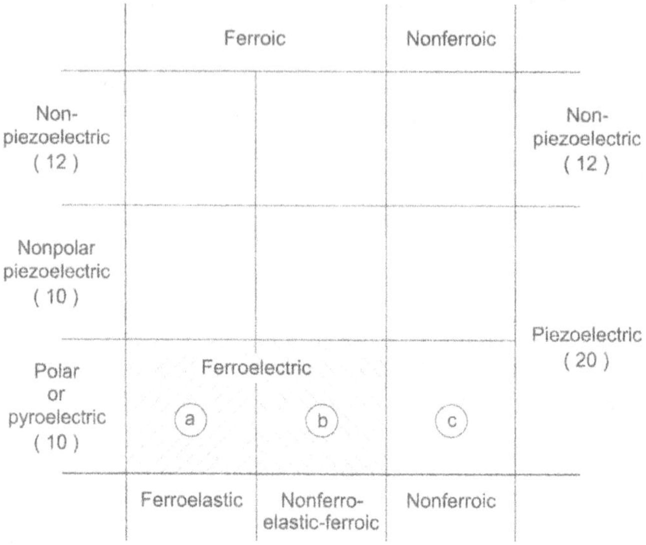

Ferroic
Nonferroic
Non-piezoelectric (12)
Non-piezoelectric (12)
Nonpolar piezoelectric (10)
Piezoelectric (20)
Polar or pyroelectric (10)
Ferroelectric
a
b
c
Ferroelastic
Nonferro-elastic-ferroic
Nonferroic

Figure 10.1.1: Ferroelectric and nonferroelectric phases of crystals. Only the hatched regions represent ferroelectric phases. See text for details.

Ferroelectric phases are a subset of ferroic phases. Ferroic phases can be of two types: nonpiezoelectric and piezoelectric. Piezoelectric phases may be either nonpolar or polar; only the polar subset can possibly be ferroelectric.

The division of ferroic phases into ferroelastics and nonferroelastics is in accordance with our crystallographic classification of phase transitions (cf. Fig. 5.2.1).

It follows from Fig. 10.1.1 that polar phases can be of three types:

(a) *Ferroelastic-ferroelectric.* An example is the room-temperature tetragonal phase of $BaTiO_3$ (Aizu species $m\bar{3}mF4mm$).

(b) *Nonferroelastic-ferroelectric.* The room-temperature phase of triglycine sulphate (TGS) is an example of this (Aizu species $2/mF2$).

(c) *Nonferroelectric-polar.* An example of this would be any crystal in a polar prototypic phase.

10.1.6 Dielectric Response and Relaxation

The reader is advised to go through §E.3 before reading this section.

Stationary Processes

Consider a dielectric crystal at equilibrium. At a time $t = t_1$ an 'up-step' electric field of moderate magnitude is switched on, and kept at a constant value indefinitely. After a sufficiently long time all transients die out and a steady state is reached, and only the so-called stationary processes survive.

Two distinct types of stationary processes can be recognized. Charges which cannot be dislodged from their sites of occupancy by the applied field occupy new sites nearby, and stay there (we ignore thermal vibrations). This is the process of *polarization* (or, more generally, *change* of polarization).

By contrast, charges which are mobile acquire a certain constant average velocity. This is the process of *conduction.*

We consider the two stationary processes in turn.

Change of Polarization

There are two basic kinds of change of polarization on application of electric field. Although they occur concomitantly, it is convenient to consider them separately.

One type is the relative separation of the positive and negative charges. This is *induced polarization.*

The other is the change of orientation (on the whole) of any existing permanent dipole moments. This is *orientational polarization.*

Induced Polarization. The external electric field can cause a relative shift of the electrons with respect to the nuclei, giving rise to *electronic-displacement polarization*, $\mathbf{P}_e$. The field can also cause a shift of the ions with respect to one another, resulting in *ion-displacement polarization*, or *atom-displacement polarization*, $\mathbf{P}_a$. Both these processes are fairly independent of temperature.

Atoms and molecules can be regarded as consisting of internal ionic cores and weakly bound valence electrons, and an applied electric field causes a shift between the two. This means that electronic-displacement polarizability is nonzero for all dielectrics. It is the smallest in the noble gases because the completely filled electron shells provide quite effective shielding of the nuclei. By contrast, an atom like sodium has a very large electronic polarizability because of its highly polarizable valence electron. Similarly, compared to single atoms or ions, the polarizabilities of molecules are larger because of the greater spatial freedom their bond electrons have for getting displaced in response to an electric field.

Crystals involving ionic bonding (e.g. NaCl) respond to an applied field through the additional mechanism of ion-displacement polarization or atomic-displacement polarization, $\mathbf{P}_a$.

Orientational Polarization. This mode of change of polarization has a strong temperature dependence. Another name for it is *dipolar polarization*, $\mathbf{P}_d$. This name is given because the two main phenomena occurring here are: reorientation of free or weakly bound permanent dipoles; and net change of polarization by the *hopping* of charge carriers. [We are considering only a static external field.]

The case of net reorientation or relaxation (*Debye relaxation*) of freely rotating polar molecules on application of an electric field was first analyzed in the classic work of Debye (1945). Assuming that the polar molecules exist in a dielectrically inert non-polar fluid, and are non-interacting (an assumption that can be substantially valid only for dilute gases), the following *Debye-Langevin equation* can be derived for the average dipole moment per unit volume under the action of an electric field (cf. Jonscher 1983):

$$\bar{P} = P(\coth a - 1/a) = PL(a) \qquad (10.1.32)$$

Here P is the permanent dipole moment of the molecule (per unit volume), and

$$a = PE/(k_B T), \qquad (10.1.33)$$

E being the electric field. $L(a)$ is called the *Langevin function.*

Under the low-field approximation ($a << 1$, or $PE << k_B T$), Eq. 10.1.32 implies a linear dependence of the average dipole moment on inverse-temperature:

$$\bar{P} = P^2 E/(3k_B T) \qquad (10.1.34)$$

Under this approximation the static (zero-frequency) dielectric susceptibility for N non-interacting dipoles is given by

$$\chi(0) = NP^2/(3\epsilon_0 k_B T) \qquad (10.1.35)$$

In our discussion of dielectric response and relaxation, we have considered, so far, systems in which the positive and negative charges are bonded quite strongly to one another, and the identities of the positive-negative units (whether atoms, or ion-pairs, or molecules) are preserved before and after the application of the electric field. Another quite common situation is encountered, particularly in dielectrics with impurities, as also in strongly disordered solids, wherein certain charged species are localized most of the time, but may occasionally make a *hopping transition* to a different localized site (or sites). This phenomenon is well-known in the field of ionic conduction. It also occurs in strongly disordered solids, for which standard band theory cannot be invoked for understanding electronic conduction;

the electrons responsible for conduction can only hop from one localized site to another (Mott & Davis 1979).

The rate at which the thermally activated hopping transitions occur is determined by the enthalpy barriers to be overcome, and by the distances between the sites involved. Application of the electric field leads to a modification of the activation energies, favouring some hopping transitions over others, resulting in a net change of polarization.

Clausius-Mosotti Equation

As discussed above, application of electric field can effect three different types of change of polarization. These must be summed to obtain the total change:

$$\mathbf{P} = \mathbf{P}_e + \mathbf{P}_a + \mathbf{P}_d \tag{10.1.36}$$

Here P_e is the electronic polarization, P_a the ionic or atomic polarization, and P_d the orientational or dipolar polarization (including any contribution from hopping of charge carriers).

Let us first consider a simple case when $P_a = P_d = 0$, so that $P = P_e$. This can happen only in elemental crystals, such as silicon. P_e is determined by the internal or local electric field E_{loc}, and the electronic polarization is:

$$\mathbf{P}_e = \epsilon_0 N \alpha_e \mathbf{E}_{loc}, \tag{10.1.37}$$

α_e being the contribution from the electronic polarizability of atoms. Substituting from Eq. 10.1.27,

$$\mathbf{P}_e = N \alpha_e \mathbf{E}_a (\varepsilon + 2)/3 \tag{10.1.38}$$

Eliminating $\mathbf{P}_e$ from Eqs. 10.1.38 and 10.1.26 we obtain the well-known Clausius-Mosotti equation:

$$(\varepsilon - 1)/(\varepsilon + 2) = N \alpha_e \tag{10.1.39}$$

In crystals such as Na^+Cl^-, which contain more than one type of ions but no permanent dipole moment, $P_e \neq 0$, $P_a \neq 0$ in Eq. 10.1.36, but $P_d = 0$. Eq. 10.1.39 then generalizes to the following approximate expression (see Dekker 1957):

$$(\varepsilon - 1)/(\varepsilon + 2) = N(\alpha_{e^+} + \alpha_{e^-} + e^2/f) \tag{10.1.40}$$

Here α_{e^+} is the polarizability of the positive ion, and α_{e^-} that of the negative ion; e is the magnitude of the charge on each ion; and f is the restoring force constant operative when the oppositely charged ions are displaced in opposite directions by the applied field.

The case when all three contributions to polarization in Eq. 10.1.36 are nonzero is quite complicated, and will not be discussed here.

Electrical Conduction

Apart from change of polarization, another stationary process brought about by a static electric field applied to a dielectric is that of conduction. For dielectric crystals the principal mechanism is that of ionic conduction, although for strong fields substantial electronic conduction also occurs.

The conductivity is defined by the relation

$$\sigma = ne\mu, \tag{10.1.41}$$

where n is the number density of charge carriers, each of charge e and average mobility μ.

The electrical conductivity of a dielectric may have a contribution both from lattice ions and impurity ions, with, say B_1 and B_2 as the corresponding activation energies. The temperature dependence of the overall process is therefore described by an equation of the type

$$\sigma = A_1 e^{-B_1/T} + A_2 e^{-B_2/T} \tag{10.1.42}$$

Time-Dependent Processes

Application of a time-varying field $\mathbf{E}(t)$ to a dielectric invokes a delayed response because of the inherent inertia of the processes involved. By contrast, the response of free space to such a field is relatively instantaneous, and therefore in phase with the applied field. Eq. 10.1.8 can therefore be generalized to

$$D(t) = \epsilon_0 E + P(t) \tag{10.1.43}$$

The *dielectric response function*, $\chi(t)$, for a delta-function type of perturbation is defined by the relation

$$P(t) = \epsilon_0 (E\Delta t)\, \chi(t), \tag{10.1.44}$$

where $(E\Delta t)$ approaches a constant value as $\Delta t \to 0$.

Under the assumption that *linear response theory* (LRT) is applicable (cf. §E.3), the time-dependence of the polarization is given by (cf. Eq. E.3.20)

$$P(t) = \epsilon_0 \int_{-\infty}^{t} dt'\, \chi(t')\, E(t-t') \tag{10.1.45}$$

The electric field, as a general function of time, can be expressed as a Fourier integral (cf. Eq. E.3.30):

$$E(t) = \frac{1}{2\pi} \int_{-\infty}^{\infty} E(\omega)\, e^{i\omega t}\, d\omega, \tag{10.1.46}$$

which can be Fourier-inverted to yield

$$E(\omega) = \int_{-\infty}^{\infty} E(t)\, e^{-i\omega t}\, dt \tag{10.1.47}$$

Similar equations can be written for $P(t)$ and $P(\omega)$:

$$P(t) = \frac{1}{2\pi}\int_{-\infty}^{\infty} P(\omega)\, e^{i\omega t}\, d\omega, \tag{10.1.48}$$

$$P(\omega) = \int_{-\infty}^{\infty} P(t)\, e^{-i\omega t}\, dt \tag{10.1.49}$$

To obtain the Fourier transform of Eq. 10.1.45, we substitute it into Eq. 10.1.49, and use Eq. 10.1.46 (cf. Eq. E.3.35):

$$P(\omega) = \epsilon_0\, \chi(\omega)\, E(\omega) \tag{10.1.50}$$

Here $\chi(\omega)$, called the frequency-dependent susceptibility, is the Fourier transform of the response function $\chi(t)$:

$$\chi(\omega) = \int_0^{\infty} \chi(t)\, e^{i\omega t}\, dt \equiv \chi'(\omega) + i\chi''(\omega) \tag{10.1.51}$$

It follows that the real and the imaginary parts of the dielectric susceptibility are defined by

$$\chi'(\omega) = \int_0^{\infty} \chi(t)\, \cos\omega t\, dt, \tag{10.1.52}$$

$$\chi''(\omega) = \int_0^{\infty} \chi(t)\, \sin\omega t\, dt \tag{10.1.53}$$

Fourier inversion of these two equations yields

$$\chi(t) = \int_0^{\infty} \chi'(\omega)\, \cos\omega t\, d\omega = \int_0^{\infty} \chi''(\omega)\sin\omega t\, d\omega \tag{10.1.54}$$

Thus a knowledge of either $\chi'(\omega)$ or $\chi''(\omega)$ is sufficient to determine $\chi(t)$. $\chi'(\omega)$ and $\chi''(\omega)$ are not independent of each other. This fact is also reflected in the well known Kramers-Kronig dispersion relations (see, for example, Bonin & Kresin (1997)):

$$\epsilon'(\omega) - \epsilon(\infty) = 2/\pi \int_0^{\infty} \frac{\epsilon''(\omega')\omega'\, d\omega'}{\omega'^2 - \omega^2}, \tag{10.1.55}$$

$$\epsilon''(\omega) = 2/\pi \int_0^{\infty} \frac{(\epsilon'(\omega') - \epsilon(\infty))\omega\, d\omega'}{\omega^2 - \omega'^2} \tag{10.1.56}$$

Dielectric Losses

The appearance of heat on application of an electric field to a dielectric constitutes a 'loss'. Two principal mechanisms of loss are: nonzero resistivity, and nonzero inertia (and viscosity) of the system in following the time-variations of the applied field.

Under the assumption of the validity of the linear response theory, it is sufficient to consider the behaviour of the system for any *one* frequency (say ω); the effect of all the other harmonics can then be compounded by linear superposition. Let the electric field, and the resulting polarization, have the following time dependence:

$$\mathbf{E} = \mathbf{E}_0 \, e^{i\omega t}, \tag{10.1.57}$$

$$\mathbf{P} = \mathbf{P}_0 \, e^{i\omega t + \psi} \tag{10.1.58}$$

In conformity with Eq. 10.1.10, the dielectric susceptibility is

$$\chi = P_0/(\epsilon_0 E_0) \, e^{i\psi} = (P_0/(\epsilon_0 E_0)) \, (\cos\psi + i \sin\psi) \tag{10.1.59}$$

A comparison with Eq. 10.1.51 yields

$$\tan\psi = \chi''/\chi' \tag{10.1.60}$$

Similarly, if the phase of the electric displacement vector is shifted with respect to the electric applied field by an angle δ, we have

$$\mathbf{D} = \mathbf{D} \, e^{i(\omega t + \delta)} \tag{10.1.61}$$

We can Fourier-invert Eq. 10.1.11, and use Eq. 10.1.51 to get

$$D(\omega) = \epsilon_0 \, (1 + \chi'(\omega) + i\chi''(\omega)) \, E(\omega) = \epsilon(\omega) \, E(\omega) \tag{10.1.62}$$

If there are several mechanisms contributing to the overall susceptibility, this equation has to be generalized to

$$D(\omega) = \epsilon_0 \, (1 + \sum_m \chi_m'(\omega) + i \sum_m \chi_m''(\omega)) \, E(\omega) = \epsilon(\omega) \, E(\omega) \tag{10.1.63}$$

The dielectric function is thus a complex quantity, with a real part and an imaginary part:

$$\epsilon(\omega) \equiv \epsilon'(\omega) + i\epsilon''(\omega), \tag{10.1.64}$$

$$\epsilon'(\omega) = \epsilon_0(1 + \sum_m \chi'(\omega)), \tag{10.1.65}$$

$$\epsilon''(\omega) = \epsilon_0 \sum_m \chi''(\omega) \tag{10.1.66}$$

With reference to Eq. 10.1.61 and 10.1.57,

$$\epsilon = (D_0/E_0)\, e^{\delta} = (D_0/E_0)(\cos\delta + i\sin\delta) \qquad (10.1.67)$$

A comparison with Eq. 10.1.64 yields

$$\tan\delta = \epsilon''(\omega)/\epsilon'(\omega) \qquad (10.1.68)$$

To demonstrate that $\tan\delta$ is a measure of dielectric loss (it is called the *loss factor*), we write the total current as a sum of the direct current and the displacement current:

$$I = \sigma E + \frac{\partial D}{\partial t} \qquad (10.1.69)$$

Fourier inversion and use of Eq. 10.1.61 gives

$$I(\omega) = \sigma\, E(\omega) + i\omega D(\omega) \qquad (10.1.70)$$

On substituting from Eq. 10.1.63 this becomes

$$I(\omega) = [\{\sigma + \epsilon_0\omega \sum_m \chi''(\omega)\} + i\omega\epsilon_0 \{1 + \sum_m \chi'_m(\omega)\}]\, E(\omega) \qquad (10.1.71)$$

Use of Eq. 10.1.65 and 10.1.66 gives

$$I(\omega) = [\{\sigma + \omega\epsilon''(\omega)\} + i\omega\epsilon'(\omega)]\, E(\omega) \qquad (10.1.72)$$

Thus the real part of the dielectric permittivity (which also incorporates the contribution from free space) is responsible for the displacement current without a power loss. And the imaginary part, alongwith the dc conductivity, is a measure of the dielectric loss.

Dielectric Relaxation

Application of electric field causes a change of polarization in a dielectric crystal. The electronic polarization, $\mathbf{P}_e$, follows the variations of the field almost instantaneously. For frequencies considerably smaller than IR frequencies, the ionic or atomic polarization $\mathbf{P}_a$ can also be taken as following the field variations almost instantaneously. The dipolar polarization, $\mathbf{P}_d$, on the other hand, can display a wide range of response times, anywhere from a few picoseconds to several days. Debye (1945) gave the basic theory of dielectric response of dipolar molecules suspended in a fluid. Although the theory was initially formulated for liquids, it provides an important reference point for discussing crystals with dipolar molecules.

Debye introduced a relaxation time, τ, such that the value of the dipolar polarization at a time t can be expressed as

$$P_d = P(\infty)\,(1 - e^{-t/\tau}) \tag{10.1.73}$$

Assuming that the applied field has the time variation given by $E = E_0 e^{i\omega t}$, Debye derived the following expression for the polarizability associated with the dipolar part of the change of polarization:

$$\alpha(\omega) = \frac{\alpha(0)}{1 - i\omega\tau} = \frac{\alpha(0)[1 + i\omega\tau]}{1 + \omega^2\tau^2} \tag{10.1.74}$$

In the presence of dielectric relaxation and loss, the dielectric permittivity is a complex quantity ($\epsilon^*(\omega) = \epsilon'(\omega) + i\epsilon''(\omega)$). The following *Debye equations* can be derived:

$$\epsilon'(\omega) = \epsilon'(\infty) + \frac{\alpha(0)\,N}{1 + \omega^2\tau^2} \tag{10.1.75}$$

$$\epsilon''(\omega) = \frac{\alpha(0)\,\omega\,\tau\,N}{1 + \omega^2\tau^2} \tag{10.1.76}$$

It follows from Eq. 10.1.76 that $\epsilon''(\omega)$ (and therefore the dielectric loss) is maximum at $\omega = 1/\tau$.

Also, for $\omega << 1/\tau$, $\epsilon'(\omega)$ becomes independent of ω, approaching its value for static fields. For such frequencies, $\epsilon''(\omega) \to 0$.

And for $\omega >> 1/\tau$, the dipoles are no longer able to follow the rapid variations of the field; thence $\epsilon''(\omega) \to 0$.

10.1.7 Absolute and Relative Spontaneous Polarization

Spontaneous polarization is polarization in the absence of an applied electric field. In the context of ferroelectric phase transitions, and to make clear the meaning of the phrase 'nonferroelectric pyroelectric', we introduce here a distinction between absolute and relative spontaneous polarization.

Absolute spontaneous polarization is simply the ordinary spontaneous polarization possessed by any pyroelectric (whether ferroelectric or not). It can be defined without any reference to the spontaneous polarization of other domain states (if any exist).

We define *relative* spontaneous polarization by analogy with relative spontaneous strain (cf. §11.1.3). For defining relative spontaneous polarization it is mandatory to make a reference to the prototype symmetry, from which a given ferroelectric phase can be taken to have arisen through loss of point-symmetry operators in a nondisruptive manner, so that there are, say, q equivalent and distinct ferroelectric orientation states. We calculate the average spontaneous polarization of these q states, and then subtract this average from the absolute spontaneous polarization of each orientation state to obtain the relative spontaneous polarization for that orientation state:

$$\mathbf{p}_{(s)av} = \sum_{i=1}^{q} \mathbf{p}^{abs}_{(s)i}/q \tag{10.1.77}$$

$$\mathbf{p}^{rel}_{(s)i} = \mathbf{p}^{abs}_{(s)i} - \mathbf{p}_{(s)av} \tag{10.1.78}$$

Unlike relative spontaneous polarization, no need was felt to introduce the concept of *relative* spontaneous magnetization in Chapter 9. This is because ferromagnetic materials always have (or can be reasonably conceived to have) a nonmagnetic (time-symmetric) prototype. Therefore, examples of pyromagnetic nonferromagnetic materials are unlikely to exist (unless we insist on the switchability condition for calling a crystal a ferromagnet).

Purely ferromagnetic transitions can be expected to be always of the nondisruptive variety. Normally this should ensure switchability between contiguous ferromagnetic domain pairs. An exception can occur in the case of a crystal with a helical (or more complex) ferromagnetic configuration. In such a system, although ferromagnetic switching may not be experimentally feasible, the crystal still has two or more orientation states, and each orientation state has a nonzero *relative* spontaneous magnetization (which we can define by replacing $\mathbf{p}$ by $\mathbf{m}$ in Eqs. 10.1.77 and 10.1.78). By contrast, a nonferroelectric pyroelectric has *zero* relative spontaneous polarization because $q = 1$ for it in Eq. 10.1.77.

SUGGESTED READING

A. J. Dekker (1957). *Solid State Physics.* Prentice Hall, New York.

C. Kittel (1966). *Solid State Physics*, 3rd edition. Wiley, New York.

A. Chelkowski (1980). *Dielectric Physics.* Elsevier, Amsterdam.

A. K. Jonscher (1983). *Dielectric Relaxation in Solids.* Chelsea Dielectrics Press, London.

10.2 STRUCTURAL CLASSIFICATION OF FERROELECTRICS

In view of the diversity of crystal structures which exhibit the property of ferroelectricity, it is helpful to classify them in terms of various criteria.

There are at least two classification schemes based on the nature of the structural phase transition giving rise to a ferroelectric phase. One is depicted in Fig. 10.1.1, according to which a ferroelectric phase may be either ferroelastic (as in $BaTiO_3$), or nonferroelastic (as in TGS).

Another classification scheme divides ferroelectrics (or rather ferroelectric phases) into proper, pseudoproper, and improper categories, depending on the nature of the order parameter driving the ferroelectric phase transition. A description of these will be given in §10.3.

Crystal structure and the nature of bonding form another useful basis for classifying ferroelectrics (Jona & Shirane 1962; Lines & Glass 1977). Classifications along these lines have been influenced by the historical development of the subject (§1.2). A particularly comprehensive *structural* classification has been given by Bunget & Popescu (1984), which we summarize here; we also update and expand it, although not exhaustively.

In this scheme, all ferroelectrics are first recognized as either hydrogen-bonded, or non-hydrogen-bonded. The former may be either structures with linear ordering of protons, or structures with ordering of radicals. The non-hydrogen-bonded ferroelectrics may be either structures with oxygen octahedra, or without oxygen octahedra. Further subcategories are recognized for most of these categories.

10.2.1 Hydrogen-Bonded Ferroelectrics

Ferroelectrics with Linear Ordering of Protons

Four subcategories are possible.

(a) **Structures with chains of ions**. The triglycine sulphate (TGS) family is an example of this type. The general formula is $(NH_2CH_2COOH)_3.H_2AB_4$, with $A = S, Se, Ba$, or B, and $B = O$ or F. The crystal structure consists of chains of dipolar ions of $NH_3^+CH_2$ and SO_4^-.

(b) **Structures with tetrahedral units**. This group includes dibasic or dihydrogenated phosphates or arsenates of alkali metals and ammonium; ammonium sulphate; and ammonium fluoberyllate ($(NH_4)_2BeF_4$). The best known member is KDP (KH_2PO_4).

(c) **Structures with octahedral units**. Some members of this group are: methyl-ammonium alum ($(CH_3NH_3)Al(SO_4)_2.12H_2O$); $KAl(SO_4)_2.12H_2O$; hexacyanides of potassium and metals of the 8th group ($K_4Fe(CN)_6.3H_2O$); and periodated compounds of ammonium ($(NH_4)_2H_3IO_6$) and silver.

(d) **Structures with pyramidal units**. Lithium hydrogen selenate, $LiH_3(SeO_3)_2$

and sodium selenate are examples of this type of ferroelectrics. The ions of $(SeO_3)^{2-}$ have a pyramidal structure.

Ferroelectrics Having Ordered Radicals

Rochelle salt is the best known example of this class. It is a tartrate of sodium and potassium, alongwith water molecules of crystallization ($NaKC_4H_4O_6.4H_2O$). Other members are: double tartrates of Li and Tl; dicalcium strontium propionate ($Ca_2Sr(C_2H_5CO_2)_6$); and tetra-methyl-ammonium bromo- or chloro mercurate ($N(CH_3)_4Hg(Br$ or $Cl)_3$).

10.2.2 Non-Hydrogen-Bonded Ferroelectrics

It is convenient to divide this class of ferroelectrics into two: those with oxygen octahedra, and those without them.

Ferroelectrics with Oxygen Octahedra

There are four groups in this category: the perovskite group, the pyrochlore group, the pseudo-ilmenite group, and the tungsten-bronze group.

(a) **Perovskite group**. Perovskite is the name of the mineral $CaTiO_3$. The general formula for this group is ABX_3, where A is a large-radius cation (Ba, Ca, Pb, Na, K) having a coordination number of 12, and B is a small-radius cation (Ti, Zr, Nb, Ta) having a coordination number of 6. X is generally oxygen, although there are some important members of this group for which X is, for example, a halogen atom. The structure can be viewed as a network of corner-sharing oxygen octahedra, with B-ions lying at or near the centres of the octahedra, and A-ions occupying empty sites between the octahedra.

The onset of ferroelectric ordering in such crystals is accompanied by a variety of structural changes leading to new polar phases. One mechanism is the movement of B-ions inside the oxygen octahedra (e.g. Ti ions in $BaTiO_3$) along a specific crystallographic direction, which then becomes the polar axis. Such a displacement is also accompanied by changes in oxygen positions, and may even involve a disorder-order transition regarding site occupancy.

Alternatively, or even in addition to the above, the oxygen octahedra may rotate or distort to reduce the A-O distance (Glazer 1975).

(b) **Pyrochlore group**. Cadmium niobate ($Cd_2Nb_2O_7$) is a typical ferroelectric with a pyrochlore structure. A general structure $A_2B_2O_7$ of this kind has two types of oxygen atoms: those occurring as BO_6 octahedra, and others. The structure consists of corner-sharing BO_6 octahedra, forming -O-B-O- zigzag chains, with A-ions and the remaining oxygen atoms occupying sites between the octahedra.

(c) **Pseudoilmenite group**. Ilmenite is the mineral $FeTiO_3$. Some ferroelectrics with formula ABO_3 ($B = \mathrm{Nb, Ta}$) have a structure similar to ilmenite. $LiNbO_3$ is one such example, with space-group symmetry $R3c$. It can be viewed as a highly distorted perovskite structure. Along the polar c-axis, face-sharing oxygen octahedra repeat, with their centres occupied successively by an Nb ion, a vacancy, and an Li ion.

(d) **Tungsten-bronze group**. The general formula of structures belonging to this group is $A_5B_{10}O_{30}$ or $A_6B_{10}O_{30}$, with A = Pb, Sb, Li, Na, K, Ba, Sr, ..., and B = Nb, Ta, or W. They have a tunnel structure, and the A-atoms generally occupy sites in these tunnels or channels. The oxygen octahedra are quite highly distorted, and the channels may be 3-, 4-, or 5-sided. Examples are $Ba_{0.25}Sr_{0.75}Nb_2O_6$, $Ba_2NaNb_5O_{15}$, and $PbNb_2O_6$.

Ferroelectrics without Oxygen Octahedra

This class includes the nitrite group, the chalcohalide group, and the boracite group.

Some members of the nitrite group of ferroelectrics are $NaNO_2$ and KNO_2 (we may also include KNO_3 here).

The chalcohalide group has the general formula ABX, with A = Sb, Bi; B = S, Se; and X = Cl, Br, I. SbSI is a well-known member of this group.

Bunget & Popescu (1984) also include the chalcogenides and the halides under the chalcohalide category. Some of the relevant chalcogenides are: $Sn_2P_2S_6$, SbS_3, Ag_3AsS_3 (proustite), and Ag_3SbS_3 (pyrargyrite). The halides form quite large families. One of them has the general formula A_2BX_4, with typical members like Rb_2ZnCl_4 and $(NH_4)_2ZnCl_4$. Another family has the general formula ABF_4, with A = Ba and Sr, and B = Mn, Fe, Co, Ni, Mg, and Zn. $BaMnF_4$ is the best known member of this family.

Lastly we have the boracite group, which has the general formula $M_3^{2+}B_7O_{13}X^-$, with M = Mg, Ni, Co, Fe, Zn, Cd, Cu, and Cr, and X = Cl, Br, I. The mineral boracite has the general formula $Mg_3B_7O_{13}Cl$.

Members of the boracite group have the interesting feature that, for some of them (e.g. Ni-Cl boracite, Cr-Cl boracite and Mg-Cl boracite), a large number of ferroic properties (two or more out of ferroelectricity, ferroelasticity, ferromagnetism and ferrogyrotropy) coexist (Torre, Abrahams & Barns 1972; Toledano, Schmid, Clin & Rivera 1985; Ye, Burkhardt, Rivera & Schmid 1995; Castellanos-Guzman, Campa-Molina & Reyes-Gomez 1997).

Similarly, $BaMnF_4$ is a much investigated crystal because of its peculiar combination of ferroic properties (see Asahi et al. 1992).

SUGGESTED READING

H. D. Megaw (1957). *Ferroelectricity in Crystals.* Methuen, London.

F. Jona & G. Shirane (1962). *Ferroelectric Crystals.* Pergamon Press, Oxford.

I. Bunget & M. Popescu (1984). *Physics of Dielectrics.* Elsevier, Amsterdam.

Y. Xu (1991). *Ferroelectric Materials and Their Applications.* North-Holland, Amsterdam.

10.3 FERROELECTRIC PHASE TRANSITIONS

Ordinary dielectrics, described in §10.1, may be regarded as *linear dielectrics*, in the sense that, unless the electric field applied is exceedingly high, the response of the material is related linearly to the field (cf. Eq. 10.1.10). We also assumed in that section (except for the discussion leading to the Clausius-Mosotti equation in §10.1.6) that the inter-dipole interactions are negligible. One class of dielectrics for which this is certainly not the case are ferroelectrics. In them, there is a long-ranged cooperative interaction of the spontaneous polarization, generally extending over macroscopic distances. This interaction, or ordering, exists in opposition to the depolarizing fields and the thermal disordering processes, up to a temperature T_c, above which the disordering forces overcome the ordering forces. When this happens, there is a phase transition (a 'ferroelectric to paraelectric phase transition').

The disordered phase naturally has a higher point-group symmetry. If we start from such a phase of a dielectric, and cool the crystal through the ferroelectric phase transition, the crystal splits into domains to minimize the overall depolarizing field. And application of a sufficiently strong electric field to the ferroelectric phase can make some domains grow at the cost of others (cf. §10.7 below), giving rise to the well-known hysteretic behaviour (Fig. 1.2.1).

The following salient features of ferroelectrics emerge because of the facts outlined above:

(i) Ferroelectrics have a spontaneous-polarization component which can be reversed by the application of a strong enough electric field.

(ii) The spontaneous-polarization vector of a ferroelectric specimen has at least one component which is a *nonlinear* and a *double-valued* function of electric field (Fig. 1.2.1).

(iii) The long-ranged cooperative interaction of the reversible part of the spontaneous polarization in unit cells separated by macroscopic distances is overcome by thermal fluctuations at the ferroelectric to paraelectric phase transition.

The prototype symmetry for a ferroelectric phase transition may be either centrosymmetric or noncentrosymmetric. And in the latter case it may be either nonpolar or polar. We illustrate the three possibilities with examples:

(a) The cubic-to-tetragonal ferroelectric-ferroelastic transition in $BaTiO_3$ has a centrosymmetric prototype (of symmetry $m\bar{3}m$).

(b) KDP belongs to the ferroelectric-ferroelastic species $\bar{4}2mFmm2$. The prototype symmetry $\bar{4}2m$ is *nonpolar*, rather than centrosymmetric.

(c) A phase transition from a polar prototypic phase to another polar (ferroelectric) phase is also conceivable; for example for the Aizu species $mm2Fm$. In this case, the spontaneous polarization of the ferroelectric phase has a nonreversible part and a reversible part, the former corresponding to that which continues to be nonzero when the crystal passes from the ferroelectric phase to the prototypic phase on change of temperature. When an electric field is applied on the crystal in its ferroelectric phase for reversing the reversible component of the spontaneous polarization, the net result is a *reorientation*, rather than reversal, of the *total* spontaneous polarization (Fig. 10.3.1). Such a ferroelectric is called a *reorientable ferroelectric*.

In certain situations a reorientable ferroelectric phase can arise even from a nonplar prototype, an example being the ferroic species 32F2 (Aizu 1967). The three 2-fold axes in the prototype are perpendicular to the 3-fold axis, and only one of these survives in the ferroelectric phase. Consequently, the ferroelectric phase has three orientation states, related by the F-operations consisting of rotations by 0°, 120°, and 240° in the plane normal to the 3-fold axis of the prototype. Clearly, since a rotation by 180° is not an F-operation, only reorientation, and not reversal, of the spontaneous

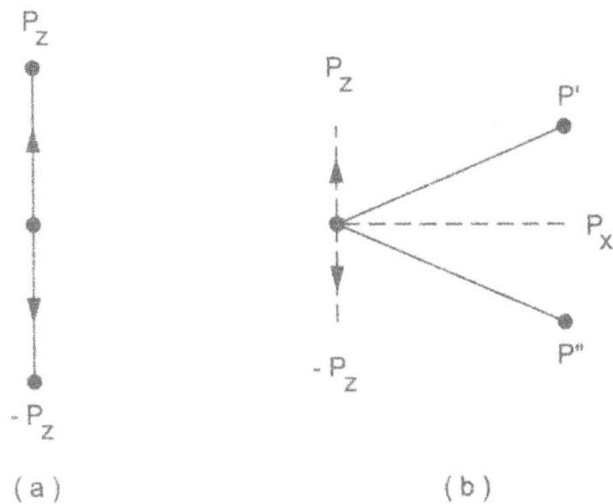

Figure 10.3.1: Illustrating the difference between a reversible ferroelectric (a), and a reorientable or divertible ferroelectric with a polar prototype (b).

polarization is possible.

There are a total of 33 species of reorientable ferroelectrics; these have been derived and tabulated by Aizu (1967).

The next question to consider is regarding the nature of the order parameter of a ferroelectric phase transition. It is a ferroic phase transition, and therefore at least one macroscopic tensor property coefficient must emerge in the lower-point-symmetry (i.e. ferroelectric) phase. There are three possibilities:

(i) Spontaneous polarization, a macroscopic tensor property, may itself be the order parameter. We then speak of a *proper* ferroelectric transition.

(ii) The order parameter is something other than spontaneous polarization, and the two have different symmetries. In this case the spontaneous polarization arises only because of its nonlinear coupling with the order parameter. This is a case of an *improper or faint* ferroelectric transition.

(iii) It can also happen that, though spontaneous polarization is not the order parameter, it has the same symmetry as the order parameter. Although there is a (bilinear) coupling between the two, the critical temperature dependence (responsible for the phase transition in the Landau picture) is carried by the order parameter (that is why it is called the order parameter), and not by the polarization. Such transitions are termed *pseudoproper* ferroelectric transitions (Dvorak 1974).

10.3.1 Proper Ferroelectric Phase Transitions

Ferroelectric phase transitions are a subclass of structural phase transitions, and therefore the general theories of phase transitions outlined in Chapter 5 are applicable to them. They are also a subclass of ferroic phase transitions; for them the change of point symmetry with respect to the prototype is such that the ferroic (ferroelectric) phase has a spontaneous-polarization component not present in the prototype. For a proper ferroelectric transition this component is the order parameter. Therefore the ferroelectric (polar) phase has the symmetry resulting from the intersection of the symmetry group of the prototype and the symmetry group of an appropriately oriented polar vector. Thence the number of ferroelectrically distinct orientation states in a proper ferroelectric is equal to the maximum number of orientation states possible (cf. Eq. 6.1.1). Such a phase is a *full ferroelectric* phase (cf. §6.2.2).

$BaTiO_3$ is a thoroughly investigated example of a proper ferroelectric. All the ferroic phase transitions that occur in it, with the cubic phase as the prototypic phase, are proper ferroelectric transitions (Devonshire 1949, 1951). They are also (weakly) first-order transitions. Since a polarization component is the order parameter, the Landau expansion (Eq. 5.3.9) can be written as

$$g = g_0 + \frac{a}{2}P^2 + \frac{b}{4}P^4 + \frac{c}{6}P^6 + \cdots, \tag{10.3.1}$$

where we have retained only even-power terms because $g(\mathbf{P}) = g(-\mathbf{P})$.

In the ferroelectric phase the order parameter is predicted to have the following temperature dependence by the Landau theory (Eq. 5.7.2):

$$P = \pm\left[-\frac{b}{2c}\left(1+\sqrt{1-\frac{4a'c}{b^2}(T-T_0)}\right)\right]^{1/2}, \tag{10.3.2}$$

and the inverse dielectric susceptibility in the ferroelectric phase is given by Eq. 5.7.5:

$$\chi_T^{-1} = \frac{b^2}{c}\sqrt{1-\frac{4a'c}{b^2}(T-T_0)}\left[\sqrt{1-\frac{4a'c}{b^2}(T-T_0)}+1\right] \tag{10.3.3}$$

Here T_0 is the stability limit of the prototypic phase. And the stability limit of the ferroelectric phase is given by Eq. 5.7.6:

$$T_0^- = T_0 + \frac{b^2}{4a'c} \tag{10.3.4}$$

The stability limits of the two phases do not coincide. The phase-transition temperature (T_c) is the temperature at which the two phases have the same free energy. This temperature is determined by Eq. 5.7.7.

At T_c the order parameter (spontaneous-polarization component P) changes discontinuously from a finite value, $(-(3b)/(4c))^{1/2}$, to zero. And the dielectric susceptibility has a finite value at this temperature.

By contrast, if the proper ferroelectric transition is a continuous transition, the Landau theory predicts (cf. Eqs. 5.3.27 and 5.3.28):

$$P = (a'/b)^{1/2}(T_c - T)^{1/2}, \tag{10.3.5}$$

$$\chi_T = C'/(T_c - T), \tag{10.3.6}$$

so that $\chi_T \to \infty$ as $T \to T_c$.

Moreover, for a continuous transition, the stability limits of the prototypic phase and the ferroelectric phase coincide at T_c; in this Landau-theoretic picture there is no temperature interval in which the two phases coexist.

The Elastic Ferroelectric

In this chapter we have not considered the nature of elastic anomalies in the vicinity of a proper ferroelectric transition. This problem has been analyzed in great detail by Devonshire (1949, 1951), and the treatment will not be repeated here. However, we must mention here an important distinction one makes in this context. This is regarding the nonpiezoelectric or piezoelectric nature of the prototype. The presence or absence of certain invariant terms in the Landau expansion depends on this distinction. If the prototype is nonpiezoelectric, a typical 'mixed' term in the Landau potential, which (along with other elastic-energy terms) can account for the broad features of the elastic behaviour, has the form $P_x P_y e_{xy}$. Such a phase transition is an improper or faint ferroelastic transition, (apart from being a proper ferroelectric transition). We return to this topic in Chapter 11.

10.3.2 Improper or Faint Ferroelectric Phase Transitions

In the example of $BaTiO_3$ mentioned above, the primary instability, namely the emergence of spontaneous polarization, is accompanied by a crystal-family-changing distortion of the crystal This is therefore not only a ferroelectric transition, but also a ferroelastic transition, although its ferroelastic nature is entirely because of a coupling of the spontaneous strain with the primary instability driving the transition, namely the spontaneous emergence of polarization. Such a transition is called a *proper ferroelectric improper (or faint) ferroelastic transition.*

Similarly, situations are possible (e.g. in the case of gadolinium molybdate (GMO)), wherein a spontaneous polarization arises as a result of its coupling with some other primary instability (order parameter) with geometrical or other transformation properties different from those of electric polarization. These are improper ferroelectric phase transitions.

One immediate consequence of such a situation is that domain pairs are possible which do not differ in any spontaneous-polarization component, although they differ with respect to the order parameter. Thus, unlike proper-ferroelectric phases, which are always full-ferroelectric phases, improper-ferroelectric phases can be *partial* ferroelectric phases (cf. §6.2.2). This is because, whereas a pair of variants *must* differ in the order parameter, there is no necessary condition that it must also differ in another, arbitrarily chosen, characteristic.

Another property of improper ferroelectrics is that if we consider the space groups S_0 and S of the prototypic and ferroelectric phases, then

$$S_0 \cap \Gamma_p \neq S, \qquad (10.3.7)$$

where Γ_p is the symmetry group of the reversible component of the spontaneous polarization. This again is only a manifestation of the fact that polarization is not the order parameter of this transition.

GMO is a better known example of a crystal which undergoes an improper ferroelectric transition. The prototype symmetry is $P\bar{4}2_1m$ (D^3_{2d}), which changes at $432K$ to $Pba2$ (C^8_{2v}) (Dvorak 1971; Levanyuk & Sannikov 1974). The order parameter in this case is a soft mode corresponding to the ($\frac{1}{2}\,\frac{1}{2}\,0$) point of the Brillouin zone of the prototype. It is thus an antiferrodistortive phase transition ($k \neq 0$) (cf. §5.4.2), entailing a two-fold increase in the volume of the primitive unit cell in going to the ferroic phase. Moreover, the point group $\bar{4}2m$ of the prototype has twice the order of the point group $mm2$ of the ferroic phase. Therefore, 2×2 or 4 distinct types of domains are possible in the ferroic phase. The order parameter of this transition is 2-dimensional. Let η_1 and η_2 be the two components. The four domain types correspond to the four combinations (η_1, η_2), $(\eta_1, -\eta_2)$, $(-\eta_1, \eta_2)$, and $(-\eta_1, -\eta_2)$. The spontaneous polarization P_3 is proportional to the product of the two components, so that only two distinct configurations are possible for it, namely $\eta_1\eta_2$ (or P_3) and $-\eta_1\eta_2$ (or $-P_3$). Thus there are domain types (e.g. (η_1, η_2) and $(-\eta_1, -\eta_2)$) which do not differ in spontaneous polarization.

Moreover, electric polarization, being a macroscopic tensor property, is invariant under a translation, whereas the order parameter, which corresponds to a zone-boundary soft mode (rather than a zone-centre soft mode), is not invariant under a translation operation. This fact also implies that we cannot explain the observed symmetry $Pba2$ (with a primitive unit cell twice as large as that of the prototype) as an intersection group of the prototype group $P\bar{4}2_1m$ and the symmetry group ∞m_z of the spontaneous-polarization component.

There are some other distinctive features of improper ferroelectrics which we can establish by carrying out a simple Landau-theoretic analysis. However, we must first do some additional spadework before writing the Landau expansion.

Two characteristics of the phase transition in GMO are representative of practically all improper ferroelectric transitions. One is that they are generally of the antiferrodistortive type (Toledano & Toledano 1976). The other is that their order parameter must have at least two components (Levanyuk & Sannikov 1974). Before going into the details of this we first describe the notion of the 'faintness index' of an improper ferroelectric transition.

Faintness Index

The onset of the order parameter at a phase transition can induce the onset of additional microscopic or macroscopic variables which do not necessarily belong to the IR of the prototype symmetry group to which the order parameter belongs. These are called faint variables (Aizu 1972b). An example is the onset of electrostrictive strain that accompanies the emergence of spontaneous polarization in $BaTiO_3$.

As mentioned above, improper ferroelectric transitions are generally driven by zone-boundary modes. Even when a phase transition is driven by a zone-centre mode, the properties corresponding to the faint variables will, by definition, be improper (or pseudoproper). What is more, even for a given physical property, not all its tensorial components may belong to the same IR (Aizu 1972c; Dvorak 1974). It is possible to split the macroscopic property tensor into a part that has the symmetry of the order parameter (the proper or pseudoproper part), and an improper part. Proper and improper parts of spontaneous polarization and spontaneous strain for ferroic transitions have been tabulated by Janovec et al. (1975). Toledano & Toledano (1976) have given a general treatment of order-parameter symmetries of nonferroelastic improper ferroelectric transitions.

A commonly used term for faint variables is "secondary order parameters", with the actual order parameter referred to as the "primary order parameter". We avoid such usage in this book, and take the view that *the* order parameter of a phase transition is one which carries the critical dependence on temperature (or on whatever is the controlling variable), and is exactly compatible with the observed symmetry change across the phase transition (cf. §5.3.2); its symmetry is neither less nor more than what is needed for this compatibility. The order parameter is the order parameter; there is nothing primary or secondary about it. The so-called secondary order parameters are faint variables which arise at a phase transition because of their coupling with the order parameter emerging at the phase transition as the instability driving the transition. Only in the case of a pseudoproper phase transition (see below) do we have a situation wherein more than one parameters have the symmetry expected of the order parameter. One has to then identify the primary instability among these, and call the other same-symmetry quantities arising at the phase transition as secondary order parameters. In fact, the order parameter in this case is a linear combination of all the same-symmetry parameters.

If a crystal is to undergo an improper ferroelectric transition, its Landau expansion must contain terms of coupling between polarization (which is the relevant faint variable in this case) and the order parameter. One can formulate a general coupling term as follows:

Let Γ be the IR of the prototype symmetry group according to which the order parameter transforms, and Γ' the IR according to which a polarization component (or any other faint variable of interest) transforms. With reference to Eq. 5.3.3, let η'_i and ϕ'_i be the quantities in Γ', corresponding to η_i and ϕ_i in Γ. A general coupling term then has the form

$$\eta^N \, \eta'^{n'} \sum_\alpha R_\alpha \, f_\alpha^{(N,n')}(\phi_i, \, \phi'_i) \tag{10.3.8}$$

Here $f^{(N,n')}$ is an invariant of order N and n' with respect to ϕ_i and ϕ'_i respectively.

In a coupling term the smallest value (say n) of N when $n' = 1$ is called the *faintness index* of that faint property (for which $n' = 1$) (Aizu 1972b, c, 1973b, d, 1974; Dvorak 1974).

For an improper ferroelectric, n is the lowest degree of the polynomial in the order-parameter components to which the spontaneous-polarization component couples.

According to Toledano & Toledano (1976), most of the ferroelectric transitions are of the proper type, and most of the nonferroelastic improper ferroelectric transitions are those for which faintness index is equal to 2 ($n = 2$), although the value $n = 3$ is also possible occasionally.

The versatile and comprehensive computer code ISOTROPY by Stokes & Hatch (1998) has been mentioned at several places in this book. Using this code, the characterization of a property (polarization, strain, etc.) as proper or improper, as well as the determination of its faintness index, can be made easily.

Dimensionality of the Order Parameter of an Improper Ferroelectric Transition

The dimensionality of the order parameter of an improper ferroelectric transition must be greater than unity (Levanyuk & Sannikov 1974; Gufan & Sakhnenko 1973).

This can be explained as follows.

In the Landau expansion the leading term responsible for improper-ferroelectric behaviour is linear in the spontaneous-polarization component (i.e. $n' = 1$ in expression 10.3.8). That is why the faintness index is defined with respect to it.

Let P be the spontaneous-polarization component emerging at the ferroelectric phase transition. If the order parameter has only one component, say η, the leading term would be of the form $P\eta^n$, n being the faintness index for ferroelectricity. If n is odd, $P\eta^n$ would have the same symmetry as $P\eta$, so that the latter term would also occur in the Landau expansion. But this is not possible because η and P do not have the same symmetry for an improper ferroelectric transition. An odd value of n is therefore ruled out.

If n is even, η^n would be invariant under all operations of the prototype group, and therefore, in $P\eta^n$, P also would be invariant, and more than one orientation states differing in P would not be possible. But this cannot be so for a ferroelectric phase, so an even value of n is also ruled out if the order parameter has only one component.

Thus the order parameter cannot be 1-dimensional for an improper ferroelectric transition. For nonferroelastic phase transitions of this type the order parameter is usually 2-dimensional, although a 3-dimensional order parameter is also encountered sometimes (Toledano & Toledano 1976).

Some Landau-Theory Results

We can now write a typical Landau expansion for an improper ferroelectric transition. The Landau free-energy density has three types of contributions: motive, interactive, and electric:

$$g = g_{motive} + g_{interactive} + g_{electric} \tag{10.3.9}$$

To concretize the meaning of this equation, we consider the specific example of GMO, and describe some salient features of the analysis carried out by Dvorak (1974) and Levanyuk & Sannikov (1974). This crystal belongs to the Aizu species $\bar{4}2mFmm2$. The polar axis is along 2_z, and normal to this direction the primitive unit cell has twice the area of the corresponding cell in the parent phase.

As discussed above, being an antiferrodistortive transition, it is necessarily an improper transition with respect to polarization or any other macroscopic (i.e. translationally invariant) property.

It may be mentioned here that the converse need not be true always: A ferrodistortive ferroelectric transition is not always a proper ferroelectric transition (see Indenbom 1960a, b).

In Eq. 10.3.9, g_{motive} stands for the order-parameter part of the Landau potential. The active IR is 2-dimensional. We denote by η_1 and η_2 the two components of the order parameter. Detailed analysis leads to the following expression:

$$g_{motive} = \frac{\alpha}{2}(\eta_1^2 + \eta_2^2) + \frac{\beta_1'}{4}(2\eta_1\eta_2)^2 + \frac{\beta_2'}{4}(\eta_1^2 - \eta_2^2)^2 + \frac{\beta_3'}{4}2\eta_1\eta_2(\eta_1^2 - \eta_2^2) + \frac{\gamma}{6}(\eta_1^2 + \eta_2^2)^3 \quad (10.3.10)$$

The $g_{interactive}$ part of the free energy pertains to 'mixed components', i.e. terms involving products of the order parameter and the spontaneous polarization P, each term being invariant under operations of the prototype symmetry group. Restricting ourselves to terms linear in P, a general expression for $g_{interactive}$ has the form

$$g_{interactive} = K\,P\,f(\eta_1, \eta_2) \quad (10.3.11)$$

For example, for g_{motive} defined by Eq. 10.3.10, the following expression has been arrived at by Dvorak (1971):

$$g_{interactive} = P\,[2a_1\eta_1\eta_2 + a_2(\eta_1^2 - \eta_2^2)] \quad (10.3.12)$$

The remaining term in Eq. 10.3.9 can be written as

$$g_{electric} = \frac{C}{2}P^2 - P\,E \quad (10.3.13)$$

Substituting Eqs. 10.3.10, 10.3.11 and 10.3.13 into Eq. 10.3.9, the spontaneous polarization for the improper ferroelectric phase is given by

$$E = \frac{\partial g}{\partial P} = 0 \simeq P_s C + K\,f(\eta_1, \eta_2), \quad (10.3.14)$$

so that

$$P_s = \frac{-K\,f(\eta_1, \eta_2)}{C} \quad (10.3.15)$$

The degree, n, of the functional $f(\eta_1, \eta_2)$ is the faintness index with respect to ferroelectricity. If γ_1, γ_2 are the basis vectors spanning the order-parameter space, we can write

$$f(\eta_1, \eta_2) = \eta^n f^{(n)}(\gamma_1, \gamma_2) \quad (10.3.16)$$

Therefore,

$$P_s \sim \eta^n \tag{10.3.17}$$

The higher the value of n, the smaller (or more faint) is the spontaneous polarization. This explains why n is called the faintness index.

An alternative (and better) term *faint ferroics* was coined by Aizu (1972b) for what are now generally known as improper ferroics.

Substituting from Eq. 5.3.27 into Eq. 10.3.17, we obtain the following temperature dependence for the faint variable P_s:

$$P_s \sim (T_c - T)^{n/2}, \quad T < T_c \tag{10.3.18}$$

Most frequently, $n = 2$, so that, for a continuous improper ferroelectric phase transition,

$$P_s \sim T_c - T \tag{10.3.19}$$

10.3.3 Pseudoproper Ferroelectric Phase Transitions

Improper ferroelectric phase transitions having faintness index equal to unity are called pseudoproper ferroelectric transitions (Dvorak 1974).

If Q denotes the order parameter of such a transition, the leading coupling term between the order parameter and electric polarization P has the bilinear (i.e. linear in both Q and P) form K_1QP, K_1 being the coupling constant.

If the coupling constant is large, it becomes difficult to distinguish between a pseudoproper and true-proper ferroelectric transition. Such is indeed the case for the transition in KDP (Dvorak 1972).

Although both Q and P have the same symmetry, it is Q which carries the critical temperature dependence, and is therefore the order parameter. P is taken as arising as a result of its coupling with the order parameter. A typical Landau expansion therefore has the following form (Dvorak 1974):

$$g = g_0 + \frac{a}{2}(T - T_0)Q^2 + \frac{b}{4}Q^4 + \frac{c}{6}Q^6 + K_1QP + \frac{\chi_0^{-1}}{2}P^2 \tag{10.3.20}$$

The inverse dielectric susceptibility for such a system can be shown to have the following temperature dependence:

$$\chi^{-1} = \chi_0^{-1} + \frac{K_1\chi_0^2}{a(T - T_c)}, \tag{10.3.21}$$

where

$$T_c = T_0 + K_1^2\chi_0/a \tag{10.3.22}$$

There is thus a shift of the transition temperature, the shift depending quite strongly on the magnitude of coupling between the order parameter and the polarization component.

If the coupling is large (as in KDP), there is not only a shift in the temperature at which the susceptibility diverges, the divergence behaviour itself is practically the same as in a true-proper ferroelectric transition.

If the coupling coefficient K_1 is small, as in $(NH_4)_2SO_4$ (Ikeda et al. 1973; Sawada et al. 1973), the divergence of susceptibility is insignificant; in fact the susceptibility is almost independent of temperature, as is typical of many *improper* ferroelectrics.

Thus, pseudoproper ferroelectric phase transitions may have features overlapping with both proper and improper ferroelectric phase transitions.

10.3.4 Ferroelectric Diffuse Transitions

A *sharp* ferroelectric *phase* transition occurs at a specific Curie temperature T_c, at which both $\epsilon'(\omega_0, T)$ and $\epsilon''(\omega_0, T)$ have their peak values, as expected from the Kramers-Kronig relations (cf. Eqs. 10.1.55 and 10.1.56).

By contrast, we may have a *diffuse* transition (which may not be a *phase* transition in the strict thermodynamic sense), as a function of temperature, such that there is no sharp rise in $\epsilon'(\omega_0, T)$ and $\epsilon''(\omega_0, T)$ at a specific temperature T_c. What we may have instead is a smeared or 'humped' curve for $\epsilon'(T)$, with a maximum occurring at, say, a temperature T'_m. Similarly, the smeared curve for $\epsilon''(T)$ may peak at a temperature T''_m. For a sharp phase transition, $T''_m = T'_m = T_c$. But for a diffuse transition, $T''_m \leq T'_m$. And if the crystal develops localized regions which have spontaneous polarization below a certain temperature T_p, it is a case of a *ferroelectric* diffuse transition (FDT).

A broad hump in the temperature variation of ϵ' may also be indicative of dipolar-glass behaviour, and we shall discuss this class of crystals in §10.4. Both FDTs and dipolar-glass transitions involve some kind of freezing of dynamical modes, but there is a qualitative difference between the two phenomena. We describe in this subsection some underlying features of the former.

We begin by considering the case of $BaTiO_3$, which exhibits a sharp-looking ferroelectric phase transition involving 'critical slowing down' (cf. §5.4.3) and 'critical freezing'.

In the cubic phase of this crystal, the Ti ion is nominally taken as sitting at the centre of the octahedral cage defined by oxygen atoms. In reality, however, even far above the cubic-tetragonal phase transition temperature T_c, this ion is found to be shifted along any of the eight $< 111 >$ directions (Comes, Lambert & Guinier 1968; Maglione & Jannot 1991). And these shifts are correlated in neighbouring unit cells, the correlations extending over about 100 Å in the $< 100 >$ directions. There are thus *polar clusters* even in the 'centrosymmetric' cubic phase, which is therefore centrosymmetric only on a crude enough scale. As the temperature T_c is approached from above, the correlation length becomes larger and larger, and the soft mode gets more and more overdamped, implying critical slowing down. Finally, critical freezing occurs at T_c.

This phase transition is predominantly of the displacive type, although there is also an order-disorder component near T_c (cf. §5.4.3). For displacive transitions (Lines & Glass 1977):

$$\epsilon(\omega) = \epsilon(\infty) + \frac{[\epsilon(0) - \epsilon(\infty)]\,\omega_0^2}{\omega_0^2 - \omega^2 + i\omega\gamma}, \tag{10.3.23}$$

where ω_0 is the resonance frequency for the damped harmonic oscillator representing the soft mode (cf. §5.4.3), and γ is the damping factor.

On the other hand, for order-disorder dynamics,

$$\epsilon(\omega) = \epsilon(\infty) + \frac{\epsilon(0) - \epsilon(\infty)}{1 + i\omega\tau}, \tag{10.3.24}$$

τ being the relaxation time.

In the case of $BaTiO_3$, the increasing overdamping of the soft mode on approaching T_c from above results in a *crossover* from resonant or displacive behaviour to relaxational or order-disorder behaviour (cf. Eq. 5.4.23). This occurs at a temperature as high as 100 K above T_c. It is presumably caused by an anharmonicity of the potential experienced by Ti ions at the centres of the TiO_6 groups; the Ti ions shift from the centres of the octahedra to any of the 8 equivalent positions along a $< 111 >$ direction. This can be looked upon as the formation of TiO_6 clusters (with a nonzero dipole moment), signalling some loss of atomic mobility, and a corresponding overdamping of the transverse-optic soft mode. With decrease of temperature there is a gradual increase in the life time of any particular orientation of these polar clusters. Schmidt (1990) refers to this as a phenomenon tending to *stabilize the cubic parent phase.* The cluster sizes increase on cooling, until the system freezes suddenly to the macroscopically polar phase of tetragonal symmetry (Blinc & Zeks 1974).

We must mention here that there is no general agreement yet about when is a transition sharp, and not diffuse (see Schmidt (1990) for a review). The rather sharp-looking transition in $BaTiO_3$ from cubic to tetragonal macroscopic symmetry occurs nominally at $T_c = 393K$. It is a discontinuous transition. Therefore the stability limits of the two phases do not coincide. In other words, there is a range of temperatures around T_c over which the two phases coexist (cf. §5.7.1). It is therefore a phase transition which starts at a temperature above T_c, say M_s, when the crystal is cooled from the cubic phase. As the temperature is lowered further, more and more of the cubic phase transforms to the tetragonal phase. The transformation is completed (or 'finished') at a temperature M_f ($M_s > T_c > M_f$).

As the crystal is cooled from a high temperature, *precursor effects* begin to appear even at temperatures far above T_c. At about $623K$, the temperature dependence of the refractive index (Burns & Dacol 1981), as well as of the reciprocal of the real part of the dielectric function (Kersten et al. 1988), begins to deviate from linearity, and the elastic compliance s_{11} begins to increase (Beige 1980).

At high enough temperatures there is good atomic mobility (for rotations or oscillations etc.) and the TiO_6 groups have octahedral symmetry. Had this continued down to T_c, the crystal would have undergone a normal displacive phase transition. What actually happens, however, at temperatures as high as $100K$ above T_c, is that there is a crossover from displacive to order-disorder behaviour. With decrease in temperature there is a gradual increase in the life time of any particular dipolar orientation, i.e. there is an increase in the time spent by the Ti ion at any of the 8 equivalent sites in the oxygen cage, and this life time gradually becomes larger than the characteristic lattice vibration time periods. There is thus an inelastic (and cooperative) coupling among the dipoles, setting in at temperatures far above T_c, which even affects the Ba positions. The larger and varied time scales introduced by this formation of clusters is a factor responsible for the coexistence of the paraelectric and ferroelectric phases over a range of temperatures around T_c.

The formation of dipolar clusters also results in an enhanced response to external electric fields. The response function (permittivity) gets an additional contribution from the tendency of the clusters to align as a whole along the applied field, with a concomitant movement of interphase boundaries.

Unlike the case of the phase transition in 'pure' $BaTiO_3$, FDTs with a large degree of difuseness generally occur in 'mixed crystals', in which crystallographically equivalent sites are occupied randomly by two or more types of cations. An example is that of the crystal $(Ba_{1-x}Sr_x)TiO_3$ (BST), with $x = 0.12$ (Tiwari, Singh & Pandey 1995; Singh & Pandey 1996; Singh, Singh, Prasad & Pandey 1996).

BST is a member of the class of mixed-crystal ferroelectrics with isovalent substitutions at 'A' and/or 'B' sites of the general perovskite structure ABO_3. This class can be represented by the general formula $(A_{1-x}A'_x)$ $(B_{1-y}B'_y)$ O_3, with A, A$'$ = Ba, Sr, Ca etc., and B, B$'$ = Ti, Zr, Sn, Hf etc. Here A and A$'$, and similarly B and B$'$, are isovalent atoms, which differ in size. This size difference causes local distortions of the lattice, with the concomitant possibility of local dipolar configurations.

Compositions of BST with $0 < x \leq 0.12$ behave almost like pure $BaTiO_3$ with respect to some properties. For example, $T''_m = T'_m$, and both T'_m and T''_m are independent of the frequency of the probing electric field (in the radio frequency range). Further, even for x as high as 0.08, $\epsilon'(T)$ shows a fairly sharp peak at T'_m. But for $x = 0.12$ this temperature dependence has a diffuse character, typical of an FDT. It has been argued by Singh et al. (1996) that the smeared dielectric behaviour for $x = 0.12$ is due to some peculiar order-parameter fluctuations occurring over a wide range of temperatures around T'_m.

Compositions of BST with $x > 0.12$ display dipolar-glass behaviour (§10.4).

Another member of the class of mixed ferroelectrics to which BST belongs is BCT [$(Ba_{1-x}Ca_x)TiO_3$]. It also exhibits FDT behaviour (Tiwari, Singh & Pandey 1994; Tiwari & Pandey 1994). When the ceramic of BCT is prepared by the conventional dry route, the solubility limit for introducing Ca in crystalline $BaTiO_3$ is $x = 0.12$ for a firing temperature of 1200^0C. However, a solubility limit of $x = 0.16$ is achieved if the so-called 'semiwet' method is adopted for preparing the material (Tiwari, Singh & Pandey 1994). And it is the ceramic prepared by the latter method which exhibits an FDT, the diffuseness increasing with increasing values of x. Further, it is found that, whereas $T''_m = T'_m$ for the former ceramic, $T''_m < T'_m$ for the latter (Tiwari & Pandey 1994).

We consider PZT [$Pb(Zr_{1-x}Ti_x)O_3$] next. The $x - T$ phase diagram of PZT has a nearly vertical 'morphotropic' phase boundary (MPB) near $x = 0.48$ (see, for example, Lines & Glass 1977). The room temperature structure of PZT has tetragonal symmetry for $x \geq 0.48$, and it is rhombohedral for $x \leq 0.470$ (Mishra, Singh & Pandey 1997). For $x = 0.475$ the rhombohedral phase (R phase) and tetragonal phase (T phase) coexist at room temperature. Thus the MPB represents a two-phase regime, and is therefore not sharp. It tilts towards the Zr-rich side as one goes up along the temperature axis in the phase diagram. The structure has the prototypic cubic symmetry at high temperatures.

Recently, a monoclinic ferroelectric phase has been discovered in $PbZr_{0.52}Ti_{0.48}O_3$ at about 250 K (Noheda, Cox, Shirane, Gonzalo, Cross & Park 1999).

The dielectric function $\epsilon(T)$ of PZT exhibits a rather sharp ferroelectric phase transition for the composition with $x = 0.515$, for a ceramic of very high density ($7.94 gcm^{-3}$) and reasonably large grain size (Mishra & Pandey 1997). A lower density ($7.82 gcm^{-3}$) results in a more diffuse transition.

The polar axis of the R phase of PZT has been conventionally believed to be along any of the $[111]_p$ axes of the cubic prototype. A recent neutron diffraction study has revealed the occurrence of random shifts of the order of 0.2 Å for the Pb ions along $< 100 >_p$ directions, superimposed on the shifts along $[111]_p$ (Corker, Glazer, Whatmore, Stallard & Fauth 1998). There is thus a domain-like 'local' structure, in which there are additional 'ordered' displacements of cations. This can be expected to contribute to the diffuseness of the FDT.

SUGGESTED READING

A. F. Devonshire (1949). Theory of barium titanate. Part I: *Phil. Mag.*, **40**, 1040. Part II: *Phil. Mag.*, **42**, 1065.

K. Aizu (1966). Possible species of ferroelectrics. *Phys. Rev.*, **146**, 423.

K. Aizu (1967). Ferroelectrics having an irreversible (but divertible) spontaneous polarization vector. *J. Phys. Soc. Japan*, **23**, 794.

A. P. Levanyuk & D. G. Sannikov (1974). Improper ferroelectrics. *Sov. Phys. - Usp.*, **17**, 199.

P. Toledano & J.-C. Toledano (1976). Order-parameter symmetries for improper ferroelectric nonferroelastic transitions. *Phys. Rev. B*, **14**, 3097.

M. E. Lines & A. M. Glass (1977). *Principles and Applications of Ferroelectrics and Related Materials.* Clarendon Press, Oxford.

H. T. Stokes & D. M. Hatch (1991). Coupled order parameters in the Landau theory of phase transitions in solids. *Phase Transitions*, **34**, 53.

G. Godefroy & B. Jannot (1992). Ferroelectric phase transitions and related phenomena. *Key Engg. Materials*, **68**, 81.

D. Pandey (1995). Diffuse transitions in mixed crystals. *Key Engg. Materials*, **101-102**, 177.

10.4 DIPOLAR GLASSES. RELAXOR FERROELECTRICS

Orientational glasses are crystals, some Wyckoff sites of which are associated randomly with dipole, quadrupole, or higher-order multipole moments which have orientational degrees of freedom, and which interact with one another sufficiently to undergo, below some freezing temperature T_f, a relaxational freezing into a state devoid of spatial long-range correlations. An article by Hochli, Knorr & Loidl (1990) provides an excellent review of the subject.

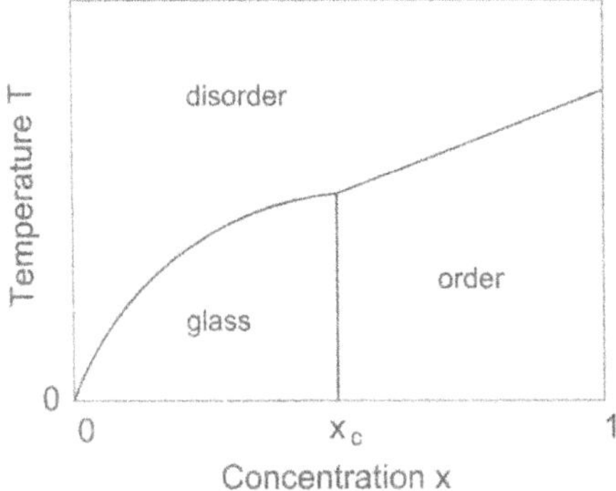

Figure 10.4.1: Typical phase diagram of a system like $KCl_{1-x}(OH)_x$. The Cl^- ions do not possess an orientational degree of freedom; only the dopant $(OH)^-$ ions do. Contrast this with Fig. 10.4.2, which is for a system with orientational degrees of freedom for *two* types of ions. [After Hochli et al. 1990.]

It is obvious from the above definition that orientational glasses must have several features in common with spin glasses and cluster glasses discussed in §9.2. As we shall see below, there is also a very important difference between spin glasses and the materials we discuss in this section. The difference has to do with the fact that spin-glass transitions, by and large, do not entail serious *structural* upheavals, whereas transitions in other glassy crystals do.

Orientational glasses may be dipolar glasses, or quadrupolar glasses, or both. We consider dipolar glasses here, and quadrupolar glasses in Chapter 11.

A typical example of a dipolar-glass crystal is that of $KCl_{1-x}(OH)_x$, i.e. KCl with some of the Cl^- ions substituted randomly by OH^- ions, resulting in dipole moments at and around the substituted sites. At $10K$ these dipoles can reorient rapidly, whereas at $0.1K$ they are frozen into a practically static configuration, although there is no net macroscopic polarization. Such systems have typically a phase diagram depicted in Fig. 10.4.1. For $x > x_c$ the interaction between the dopant ions becomes effective enough to produce long-range ordering (for $T < T_c$), resulting in a ferroelectric and/or ferroelastic phase.

We may also sometimes have a system $A_{1-x}B_x$, in which not only B but also A may be capable of orientational ordering (Fig. 10.4.2). A well-investigated example is that of the mixed crystal RADP (rubidium ammonium dihydrogen phosphate). The end member RDP of this mixed

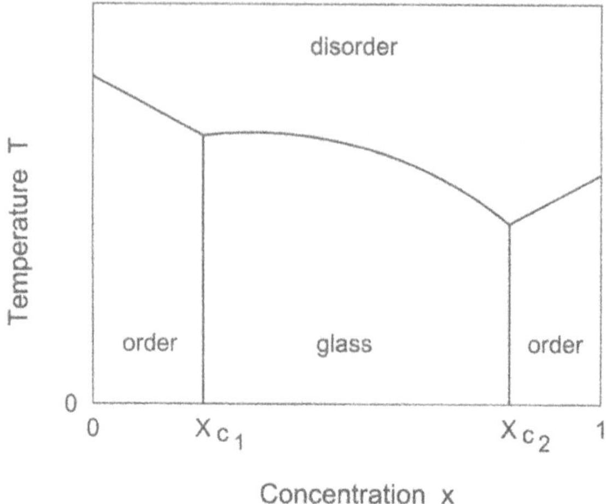

Figure 10.4.2: Schematic phase diagram for a system $A_{1-x}B_x$ such that both A and B are orientationally ordered crystals. For example, A may be a ferroelectric, and B an antiferroelectric (like in RADP). [After Hochli et al. 1990.]

crystal has a ferroelectric phase, whereas the other end member ADP has an antiferroelectric phase. Near $x = 0$ and $x = 1$ the ordering tendency prevails (for T less than the respective T_c). There is a certain minimum threshold of dopant concentration for glassy behaviour to occur, i.e. glassy behaviour exists only for $x_{c1} < x < x_{c2}$ (Fig. 10.4.2). This is a consequence of competing ferroelectric and antiferroelectric tendencies resulting in a disordered state, with degeneracy born out of frustration (cf. §9.2 for the corresponding situation in spin glasses). This quenched disorder is of a different nature from that existing for $T > T_c$ because of the dominant thermal fluctuations at high temperatures.

One of the characteristics of a ferroelectric with a glassy phase is the 'smearing' of the ferroelectric transition: There is no sharp phase transition at a particular temperature T_c. Instead, one observes a 'hump' in the temperature dependence of the dielectric susceptibility, and therefore speaks of a diffuse transition (cf. §10.3.4).

Orientational-glass behaviour in a variety of crystal families has been reviewed by Hochli et al. (1990).

We begin by describing the main classes of perovskite crystals which exhibit a glassy phase.

10.4.1 Classes of Glassy, Compositionally Modified, Ferroelectrics with Perovskite Type Structure

Perovskite structures have the composition ABO_3. Random substitution of a sufficiently large (but not too large) fraction of atoms by other suitable atoms can result in competing interactions, frustration, and consequent glassy behaviour.

Structures with Isovalent Substitution of 'A' and/or 'B' Sites

This is a class of mixed-crystal ferroelectrics with isovalent substitutions at 'A' and/or 'B' sites of the general perovskite structure ABO_3. The class can be represented by the general formula $(A_{1-x}A'_x)(B_{1-y}B'_y)O_3$, with A, A′ = Ba, Sr, Ca etc., and B, B′ = Ti, Zr, Sn, Hf etc. Here A and A′, and similarly B and B′, are isovalent atoms, which differ in sizes. This size difference causes local distortions of the lattice, with the concomitant possibility of local dipolar configurations.

As described in §10.3.4, BST ($Ba_{1-x}Sr_xTiO_3$) with $x \leq 0.12$ exhibits an FDT, but no glassy behaviour. However, for $x \geq 0.16$ the interaction among the dipolar clusters becomes strong enough for a dipolar glass state to ensue (Singh & Pandey 1996).

Structures with Offvalent Substitutions at 'A' and/or 'B' Sites

In this class, not only the sizes, but also the valencies of the substituent atoms are different from those in the host crystal. Consequently, the range of solubility can be quite limited, in general. Moreover, when A′ and/or B′ has a lower valency than A and B, respectively, an appropriate number of vacancies must be created at oxygen sites for maintaining charge neutrality. In the opposite case, when the substituent ion, say A′, has a higher valence than A, the structure has to alternate between A′-rich and A′-deficient regions.

The best known example of this class, which exhibits glassy behaviour for certain compositions, is $(Pb_{1-3x/2}La_x)(Zr_yTi_{1-y})O_3$ (or PLZT), the valence of Pb and La being 2 and 3 respectively.

Complex Compounds with Disorder at 'B' Sites

Unlike the solid solutions mentioned above, this category comprises *compounds*, with a disordered arrangement of B-site ions. The best known examples of this class are PMN ($Pb(Mg_{1/3}Nb_{2/3})O_3$), and PST ($Pb(Sc_{1/2}Ta_{1/2})O_3$). Another much-investigated material is 0.9PMN-0.1PT, i.e. PMN with 10% doping with lead titanate.

The term *relaxor ferroelectrics* is commonly used for this and other classes of ferroelectrics which exhibit a glassy-crystal phase. However, it must be emphasized that not all structural configurations in this class exhibit glassy or relaxor behaviour. For example, PST displays regular ferroelectric behaviour when in the phase in which the Sc and Ta ions are fully ordered. It is in the disordered phase that it undergoes a dipolar glass transition (Stenger, Scholten & Burggraaf 1979; Setter & Cross 1980).

10.4.2 Salient Features of Ferroelectric Crystals with a Dipolar-Glass Transition

Some features of ferroelectrics with a dipolar glass transition are similar to those which undergo ferroelectric diffuse transitions (§10.3.4). However, as we shall see here, there are some very important differences too (see the review by Pandey (1995) for a more detailed exposition).

Dielectric and Optical Properties

(**a**) Because translational periodicity is violated for a fraction of the Wyckoff sites, *all* properties of the crystal are affected to a small or large extent. In particular, the dielectric permittivity, instead of displaying a sharp peak around a phase-transition temperature T_c, exhibits a smeared-out variation with temperature (i.e. we have a diffuse transition). The temperature dependence of the real part of the permittivity has a humped appearance, with, say, T'_m as the temperature at which it has the maximum value ϵ'_m.

For $T > T'_m$, $\epsilon'(T)$ can be described by the following equation (Martirena & Burfoot 1974; Tiwari & Pandey 1994):

$$\frac{1}{\epsilon'(\omega, T)} = \frac{1}{\epsilon'_m(\omega, T)} + \frac{(T - T'_m)^\gamma}{C'} \tag{10.4.1}$$

The critical exponent γ ($1 \leq \gamma \leq 2$) is a measure of the degree of diffuseness or smearing. $\gamma = 1$ corresponds to Curie-Weiss behaviour (cf. Eq. 5.5.44). The value $\gamma = 2$ is expected for a 'quantum ferroelectric' (Rytz, Hochli & Bilz 1980). We consider quantum ferroelectrics in §10.5.

γ and C' are both found to increase with increasing diffuseness.

(**b**) In conventional ferroelectrics, the real part $\epsilon'(T)$ and the imaginary part $\epsilon''(T)$ of the dielectric permittivity generally exhibit a similar temperature dependence. For sharp phase transitions, both should peak at the same temperature, in accordance with the Kramers-Kronig relations (see Lines & Glass 1977). For a ferroelectric crystal displaying a diffuse transition the respective temperatures T'_m and T''_m at which they peak may be different (irrespective of the presence or absence of glassy behaviour), the difference increasing with increasing diffuseness; generally $T''_m(\omega) < T'_m(\omega)$ for a given frequency ω.

(**c**) Conventional ferroelectrics display a P versus E hysteresis loop (Fig. 1.2.1), which may be described as 'fat', in comparison to the slim hysteresis loops characteristic of ferroelectrics in a glassy phase. Moreover, these slim loops generally do not show saturation with electric field, unless the specimen crystal is at a temperature well below T'_m.

(**d**) For normal ferroelectrics with a sharp transition, $\epsilon'^{-1}(T_c) = \epsilon'^{-1}(T'_m) = P_s(T'_m) = 0$; i.e. the spontaneous polarization P_s drops to zero as the temperature of the crystal is raised to T_c $(= T'_m)$. By contrast, the (local) spontaneous polarization of a glassy ferroelectric becomes zero at a temperature T_p which is much higher than T'_m.

(**e**) Unlike FDTs, glass transitions in a ferroelectric exhibit a strong frequency dispersion, particularly in the RF region. Both ϵ' and the loss factor $\tan\delta$ exhibit this. And both $T'_m(\omega)$ and $T''_m(\omega)$ increase with increasing ω.

In the Debye model for dielectrics (see Jonscher 1983), the dipolar units are free to rotate, all units have the same (temperature independent) dipole moment, and there is no inter-dipole interaction. The dipolar relaxation in this model is purely a thermally activated process, with relaxation time τ given by an Arrhenius equation (cf. Eq. 9.2.30):

$$\tau = \frac{1}{\omega_0}\exp(k_B T_0/k_B T) = \frac{1}{\omega_0}\exp(T_0/T) \qquad (10.4.2)$$

The frequency ω_0 is the *attempt frequency* or the *Debye frequency*, and T_0 is the equivalent temperature of the activation energy.

In this model the permittivity is given by Eq. 10.1.75, which we rewrite as follows:

$$\epsilon'(\omega) = \epsilon'(\infty) + \frac{\Delta\epsilon'}{1+\omega^2\tau^2} \qquad (10.4.3)$$

Here $\Delta\epsilon'$ is the contribution of the freely reorientable dipoles to the static $(\omega = 0)$ dielectric function, and $\epsilon'(\infty)$ is the high-frequency dielectric function.

$\Delta\epsilon'$ has the following temperature dependence in the Debye model:

$$\Delta\epsilon' = C'/T \qquad (10.4.4)$$

Since the dipoles are non-interacting in this model, their motion can be frozen (apart from zero-point motion) only at $T = 0$, a fact correctly reflected by Eq. 10.4.2, according to which $\tau \to \infty$ as $T \to 0$. Thus the glass transition temperature, or the freezing temperature, T_f, is zero in this model:

$$T_f(\text{Arrhenius}) = 0 \tag{10.4.5}$$

For solids the Debye model is an unphysical model, though a useful idealization for reference purposes. The inter-dipole interaction in glassy solids is certainly nonzero, making them 'freeze' (i.e. undergo the glass transition) at a temperature T_f greater than the absolute zero. The relaxation time for solids has therefore been often modelled by a Vogel-Fulcher equation, i.e. Eq. 9.2.1 (Hochli et al. 1990; Viehland et al. 1990, 1991):

$$\tau = (\omega_0)^{-1} \exp[T_0/(T_m - T_f)], \tag{10.4.6}$$

implying that $\tau \to \infty$ (freezing) as $T_m \to T_f$; $T_f \neq 0$.

The temperature T_f is an empirical ('best-fit') Vogel-Fulcher temperature. For $T > T_f$ the dielectric response is predominantly fluctuation-driven, and for $T < T_f$ it is predominantly interaction-driven. However, below T_f the dipoles are not completely static and frozen. In any case, in dipolar cluster glasses T_f is found to depend on the range of ω values investigated, even down to the lowest practical probing frequencies ($\sim 1\,\text{mHz}$).

The ω-dependence of T_f reflects the fact that there is a whole range of activation energies and corresponding relaxation times, and that the fluctuation-driven and the interaction-driven regimes do not have a sharp boundary; they overlap over a considerable temperature range. We can interpret $(\omega_0)^{-1}$ in Eq. 10.4.6 as a typical *leading relaxation time.*

Eq. 10.4.6 does not always provide a satisfactory fit to the observed dielectric relaxation in glassy ferroelectrics, and other empirical relationships have been proposed. For example Cheng et al. (1997, 1998) used the following equation:

$$\tau = (\omega_0)^{-1} \exp(T_0/T'_m)^p, \tag{10.4.7}$$

where p is an adjustable parameter. It has been suggested by these authors that: $p^{-1} = 0$ for a normal ferroelectric; $p^{-1} = 1$ for a Debye medium; and $p^{-1} = 2$ for a glass. A relaxor ferroelectric like PMN-PT falls somewhere in-between a normal ferroelectric and a Debye medium.

(f) PMN has a cubic symmetry at high temperatures. For it, $T'_m = -5^0$C. If a field of 20 kV/m is applied to a crystal of it at, say, 90^0C in the [110] direction, and the system is cooled through T'_m down to, say, -80^oC, one observes a second peak in $\epsilon'(T)$ at $T'_t = -60^0$C, in addition to the larger peak at -5^0C (Arndt et al. 1988).

The main peak at T'_m and the additional peak at T'_t for the field-cooled (FC) crystal decrease in magnitude with increase of probing frequency.

On decreasing the temperature through T'_t there is a sudden appearance of a remanent (spontaneous) polarization at T'_t. We may also regard T'_t

as a *depolarization temperature*, above which the disordering forces (e.g. thermal fluctuations) win over the ordering forces; below T_t' the ordering forces establish a long-range order characteristic of a normal ferroelectric.

The peak in $\epsilon'(T)$ at T_t' is not present if the crystal is not under the influence of a poling field; i.e. if it is a zero-field-cooled (ZFC) specimen.

A similar behaviour is also observed for PLZT ceramic (Haertling & Land 1971a, b; Randall et al. 1987; Viehland et al. 1992). The X-ray diffraction pattern of the ZFC ceramic displays cubic symmetry down to the liquid-nitrogen temperature, whereas the FC ceramic has rhombohedral symmetry. This difference is attributed to a breakdown of ergodicity in the frozen state (cf. §9.2.5) (Hochli, Kofel & Maglione 1985).

(**g**) Refractive-index measurements made on PLZT as a function of temperature and composition provide evidence for local dipolar ordering, even at temperatures several hundred degrees above T_m' (Burns 1985; Burns & Dacol 1983). The macroscopic symmetry is cubic above T_c, and above a temperature T_d ($T_d \gg T_m'$), the refractive index varies linearly with temperature. However, at and below T_d a deviation from linearity is observed, which can be explained by postulating the formation of localized clusters of dipoles.

A similar behaviour is also exhibited by PMN.

Structural Properties

The most striking feature of ferroelectrics with a glassy phase is that the local-symmetry breaking temperature does not coincide with T_m'. In solid solutions like PLZT this temperature (T_d) is higher than T_m'. The same is also true about the complex perovskite compound PMN (Burns & Dacol 1983). As stated above, application of an electric field induces a structural change discernible by X-ray diffraction.

10.4.3 Spin Glasses vs. Dipolar Glasses

Dipolar glasses can be regarded as the electrical analogues of magnetic spin glasses described in §9.2. The main point of similarity is the random nature of the competing interactions in both cases (Brout 1965; Toulouse 1977).

Replacing the symbol S_i for spin by p_i for dipole moment, we can write by analogy with Eq. 9.2.5 the basic Edwards-Anderson type Hamiltonian for a dipolar glass as

$$\mathcal{H} = -\frac{1}{2}\sum_{i,j} J_{ij} p_i p_j + E \sum_i p_i, \tag{10.4.8}$$

where E denotes the external field, if any. We have chosen the Ising model, rather than the isotropic vector model.

Similarly, in analogy with Eq. 9.2.4 the following local order parameter can be defined:

$$q(T) = \lim_{t \to \infty} < \mathbf{p}_i^{(1)} \cdot \mathbf{p}_i^{(2)} > \qquad (10.4.9)$$

On cooling the crystal below a freezing temperature T_f, this parameter, which is zero above T_f, acquires a nonzero value.

Similarities apart, there is at least one important point of dissimilarity between spin glasses and orientational glasses. In spin glasses there is, relatively speaking, very little spin-lattice coupling. By contrast, in orientational glasses reorientation of moments (dipolar, quadrupolar, or higher-order) involves a movement of atoms, with the attendant effect on neighbouring atoms. Moment-lattice coupling is an important effect in the case of orientational glasses.

This moment-lattice coupling, unless it is very large, induces an ordering tendency in the crystal. The average interaction, $\bar{J}$, can be taken as a measure (or indicator) of this ordering tendency.

10.4.4 Dipolar-Glass Transitions vs. Ferroelectric Phase Transitions

Being a subset of structural transitions, dipolar-glass transitions are characterized by a strong coupling of the crystal lattice with local dipole moments. This is in contrast to what usually happens in spin-glass transitions and ferromagnetic or antiferromagnetic transitions.

We can make a comparative assessment of the various structural tendencies in terms of four parameters (Hochli et al. 1990): the pairwise interaction J; its average $\bar{J}$; its variance Var (J); and a moment-lattice coupling parameter (say K).

We first consider situations in which K is neither too small nor too large. Dipolar-glass transitions occur when J is small and Var (J) is large. A regular (conventional) ferroelectric phase transition is favoured when $\bar{J}$ is large and Var (J) is small. A small $\bar{J}$ and a small Var (J) correspond to a *superparaelectric* configuration; the term 'superparaelectric' can be understood by analogy with the term superparamagnetic, described in §9.2.10.

When K is small, the following four combinations are possible: superparamagnet ($\bar{J}$ small, Var (J) small); ferromagnet ($\bar{J}$ large, Var (J) small); spin glass (J small, Var (J) large); and canonical glass (large but ill-defined interaction).

Lastly, we consider a system for which J is small and Var (J) is large. When such a system is cooled from a high enough temperature, a spin-glass transition corresponds to a small K, a dipolar-glass transition to a medium

K, and a 'glassy crystal' (or strongly disordered) phase to a large K (larger than J and Var (J)).

To compare some typical features of dipole-glass transitions (DGTs) with those of ferroelectric phase transitions (FPTs) in the vicinity of the ordering temperature T_f, we first recall some theoretical results for the latter (cf. §10.3 and §5.3).

(a) Static linear susceptibility:

$$\chi_s = \frac{\partial P}{\partial E} \sim (T - T_f)^{-\gamma}, \qquad (10.4.10)$$

(b) Correlation length ($< P(0)P(x) >$):

$$\xi \sim |T - T_f|^{-\nu} \qquad (10.4.11)$$

(c) Time decay of the order parameter:

$$P(t) = P(0)\, e^{-\alpha t}, \quad T > T_f \qquad (10.4.12)$$

Typical values for critical exponents in the above equations are: $\gamma = 1$; $\nu = 0.5$; $\alpha \sim (T - T_f)^z$ near T_f, z being a dynamic exponent (Kadanoff et al. (1967); Hohenberg & Halperin (1977)).

Fig. 10.4.3 shows schematic plots of Eqs. 10.4.10-12. Also shown are the corresponding plots for dipolar-glass transitions (DGTs). The linear susceptibility for DGTs (the *Parisi susceptibility*) (Parisi 1979, 1983; Sompolinsky 1981; Fischer 1983) has the following temperature dependence near T_f (Fig. 10.4.3(a)):

$$\chi \sim (T - T_f)^{-1}, \quad T > T_f; \qquad (10.4.13)$$

$$\chi \sim \text{constant}, \quad T < T_f \qquad (10.4.14)$$

The spatial correlation length, ξ (Fig. 10.4.3(b)), for spin glasses is small, and has only a weak dependence on temperature (Carmesin & Binder 1988).

Lastly, the decay rate of the order parameter (Fig. 10.4.3(c)) has a complicated nonexponential character (Binder & Young 1986).

A universal theory of the dipolar-glass transition is not in sight yet (however, see Chamberlin 1998).

10.4.5 Relaxor Ferroelectrics

Although ferroelectric crystals with a glassy phase have several properties in common with spin glasses, important differences exist. The differences

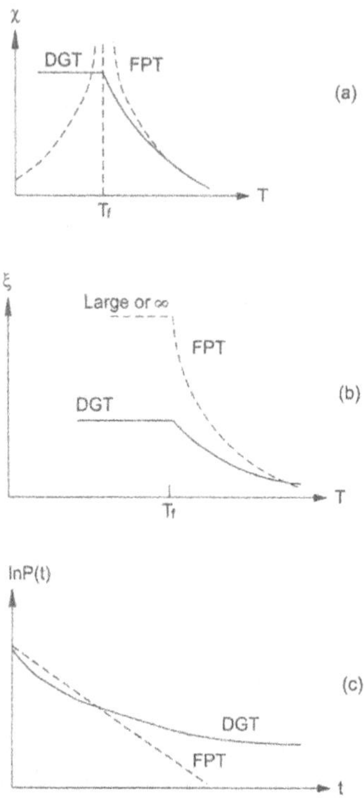

Figure 10.4.3: Comparative plots for ferroelectric phase transitions (FPTs, dashed lines) and dipolar-glass transitions (DGTs, solid lines) in the vicinity of the ordering temperature T_f: (a) linear (static) susceptibility; (b) correlation length; and (c) decay rate of order parameter. [Adapted from Hochli et al. 1990.]

arise because in ferroelectrics there are strong dipole-lattice coupling effects. These effects are not only of an electrical nature, but also have a very significant elastic or inelastic component. Because of the ordering effect of the dipole-lattice coupling, and the consequent formation of dipolar clusters, dipolar glasses are more akin to magnetic *cluster* glasses, than to magnetic *spin* glasses. And because of the rather strong local piezoelectric and electrostrictive effects, and the long-ranged nature of the mechanical interaction, dipolar (cluster) glass transitions also have much in common with martensitic phase transitions. This last aspect has emerged quite strongly from the work of Schmidt and coworkers (see Schmidt (1990) for an overview of this work).

Relaxor perovskite ferroelectrics like PMN thus present a fascinating (and technologically important) field of study in which concepts from diverse subjects have been invoked, although the overall situation is still far from clear. We discuss here some of the key aspects of their behaviour.

Relaxor ferroelectrics differ from conventional ferroelectrics in at least four important dielectric characteristics (Cross 1987, 1994): (i) the phase transition involved is diffuse, *and* the dielectric susceptibility is markedly dispersive below the temperature T'_m at which it peaks; (ii) T'_m itself is frequency-dependent, increasing with increasing frequency; (iii) the dielectric response above T'_m is not of the Curie-Weiss type; (iv) whereas the mean spontaneous polarization decays to zero at a temperature T_f (the freezing temperature) which is well below T'_m, the mean *square* polarization is nonzero even at temperatures 200 to 300 K above T'_m.

We still do not have a thorough understanding of the properties of relaxor ferroelectrics. Smolenskii & Agranovskaya (1958) explained the existence of a whole range of Curie points around T'_m in terms of coexisting phases having differing levels of composition and homogeneity. In terms of this *composition-variation model*, various static and dynamic properties of relaxors have been calculated using Rolov's (1965) method (see Pandey 1995). It turns out that the smearing or broadening of the transition temperature over such a wide range cannot be explained quantitatively in terms of Smolenskii's model. In any case the situation is far too complex to be explainable in terms of a single factor, namely local compositional inhomogeneity.

Several features of relaxor ferroelectrics are reminiscent of magnetic spin glasses and cluster glasses (§9.2). Cross (1987, 1994) advanced arguments in favour of a *superparaelectric-behaviour model*, by analogy with superparamagnetism (§9.2.10). The model moots the presence of small polar clusters, which undergo thermally-driven polarization reorientations among crystallographically equivalent directions.

It is pertinent to recall here that ferroelectric ordering is a cooperative effect. Therefore the involved energies scale with the volume of the ordered region, or the number of dipoles ordered. At sufficiently small volumes the barrier to polarization flipping becomes comparable to k_BT, and then the giant dipole moment of the entire cluster can reorient as a whole (i.e. coherently) from one direction to another equivalent direction.

The experimental evidence for the smallness of the dipolar clusters in PMN has been reported from transmission-electron microscopy (TEM) studies by Chen, Chan & Harmer (1989), Randall et al. (1990), and Bursill (1997). They observed $\frac{1}{2}[111]$ superlattice reflections, indicating the presence of short-range-ordered regions extending over 2-5 nm. What is more, heat treatment did not result in coarsening of these ordered regions, pointing to a certain *self-limiting* mechanism.

The self-limiting effect can be rationalized in terms of local charge imbalances (Chen et al. 1989). Although the global value of the Mg:Nb ratio in PMN is 1:2, the local value in a unit cell of an ordered cluster is 1:1. This results in a net charge for the unit cell, a situation that cannot persist over too large a number of contiguous unit cells. This is the factor preventing coarsening of these nonstoichiometrically ordered domains by heat treatment. Further evidence in support of this interpretation comes from the fact that if some of the Pb^{2+} ions on the A-sites are replaced by La^{3+}, there is an increase in the size of these clusters. And replacement by monovalent ions Li^{1+} has the opposite effect. Additional evidence for the validity of this line of reasoning has been reported recently by Gupta & Viehland (1997).

The case of another relaxor ferroelectric, namely $PbSc_{1/2}Ta_{1/2}O_3$ (PST), provides an interesting contrast. In this crystal there is no charge-imbalance for the unit cell, and heat-treatment does result in an increase in the degree of cooperative ordering (Setter & Cross 1980).

It is worth pointing out that, whereas the nanoclusters of various sizes in a PMN crystal have quite well-defined boundaries, as well as dynamic interactions among themselves, the edifice of the oxygen ions runs continuously across the entire crystal. The term *mictoelectric* may be appropriate for such a system, by analogy with mictomagnets (§9.2.10).

As the temperature of a relaxor ferroelectric like PMN is lowered from a value far above T'_m, there occurs a strong dispersion of the weak-field permittivity. As in the magnetic analogue (§9.2), a Vogel-Fulcher type relationship is observed in the temperature dependence of the relaxation frequency in the vicinity of the freezing temperature T_f (Viehland et al. 1990):

$$\omega = \frac{1}{\tau} = \frac{1}{\tau_0} e^{-E_a/k_B(T-T_f)} \tag{10.4.15}$$

The similarity of relaxor ferroelectrics with magnetic cluster glasses has been brought out by a large number of other studies as well, including the observation of a qualitative change in properties when a biasing field is applied. As the temperature is lowered below T'_m, more and more dipolar clusters undergo a glass-like freezing. They also have random orientations when cooled under zero field. Such randomly oriented dipoles generate a random internal field. Application of a field takes the system from a nonergodic (cf. §9.2.5) to a partially or a fully ergodic state, with the attendant formation of macroscopic domains (Imry & Ma 1975; Kleemann 1998). Several studies confirm this (Viehland et al. 1992; Ye & Schmid 1993; Park et al. 1995).

It is important to take note of two different types of polar clusters or domains in PMN: *frozen polar domains* (FPDs), and *dynamical polar domains*

(DPDs) (Qian & Bursill 1996a,b). Their existence and preponderance depends on temperature and on the degree of global charge compensation. At temperatures far above T'_m there is absence of order or coupling among the dipole moments. As the temperature is lowered, nanoclusters, with sizes limited by charge imbalance, form. Defects (e.g. chemical domain walls) act as pinning and aligning sites. The local fields at such sites, if they are strong enough, lead to the occurrence of FPDs in their vicinity. And in regions far from them, DPDs occur, the configurations of which fluctuate quite freely among the equivalent orientations available to them. In a certain finite temperature range the FPDs and the DPDs coexist. For temperatures below this range, as the cluster sizes increase with decreasing temperature, a percolation threshold is reached, such that the FPDs touch one another, and conventional ferroelectric behaviour sets in.

The models discussed above in this subsection are rather simplistic. A more realistic description of relaxor behaviour in PMN may be that in terms of the general Chamberlin theory outlined in §9.2.13. This is indicated by the experimental work of Kleemann (1998), who measured the time-decay of the polarization of zero-field-cooled PMN under isothermal conditions, in the temperature range 180-230 K. The polarization was deduced from birefringence data, via the relationship $\Delta n \sim < P >^2$. A rather small field of 1.2 kV/cm was applied parallel to the cubic [110] direction, and the time-decay of the induced birefringence was minitored after switching off the applied field. This was done at various fixed temperatures. The observations were found to be explained well by Chamberlin's theory. It is instructive to recapitulate some basic facts in this context.

When PMN is zero-field-cooled through the temperature T = 210K, there is no global breaking of the cubic symmetry; in fact, it is not a regular thermodynamic phase transition. Therefore, some authors refer to the high-temperature phase as the *ergodic phase*, and the low-temperature one as the *nonergodic phase.*

Owing to the chemical structure of this crystal, it has a quenched disorder, or a quenched random field configuration at all temperatures. Therefore, even for temperatures above T_m correlations exist between the fluctuations of the order parameter and those of the random electric fields in the crystal. This explains the occurrence of domain-like precursor phenomena in the ergodic phase (Glazounov, Tagantsev & Bell 1996). This phase also corresponds to a positive sign for the correlation coefficient C in Eq. 9.2.31.

In the nonergodic phase, C acquires a negative sign. According to the Chamberlin theory described in §9.2, there are only two 'universality classes' of slow-dynamics response: the KWW-like response class ($C > 0$), and the CvS-like response class ($C < 0$). There is thus a crossover from the KWW class to the CvS class when PMN enters the nonergodic phase.

Kleemann (1998) explains this as follows:

In the ergodic phase, as stated above, the nanometer sized polar domains or clusters have a dynamic existence because of their correlation with the quenched random fields. The probing electric field used by him causes a thermal-activation assisted reorientation of the cluster polarization, the relaxation of which is described by (cf. Eqs. 9.2.30 and 9.2.31)

$$\omega(s) = \omega(\infty)e^{-C/s}, \tag{10.4.16}$$

with $C > 0$ (CvS dynamics). This means that, contrary to commonsense expectation, the shortest relaxation times are expected for the *largest* clusters ($s \to \infty$) because, unlike small clusters, there are multiple ways for them for overcoming energy barriers.

By contrast, in the nonergodic phase there is a nearly frozen configuration of (fractal-like) domains. The small probing electric field therefore causes only a quasi-reversible movement of domain walls, the quenched random field providing a framework for the pinning forces which tend to pull the domain walls back to the zero-external-field configuration. This relaxation is predominantly of a non-activated nature, with relaxation rates still determined by the above equation, but with $C < 0$ (Glazounov, Tagantsev & Bell 1996).

Among the recent entrants to the growing family of relaxor ferroelectrics are: $Pb_{0.5}Ca_{0.5}TiO_3$ (Ranjan et al. 1997); and $Pb_{0.70}Ba_{0.30}ZrO_3$ (Pokharel et al. 1999).

10.4.6 Field-Induced Phase Transitions in Relaxor Ferroelectrics

By analogy with the theories of magnetic spin and cluster glasses, dipolar cluster glasses can also be expected to undergo a decrease in their nonergodic character in the presence of a biasing field. Moreover, changes in the magnitude and orientation of electric dipoles involve changes in the positional coordinates of ions, with the attendant strong (electric, as well as elastic) coupling with the crystal lattice. Therefore, tendency towards macroscopic ordering, which is anyway more prominent in dipolar and other orientational glasses compared to spin glasses, gets further accentuated by a biasing field, resulting in a structural phase transition at a low-enough temperature. When this happens, the global symmetry which is not lowered on the occurrence of the glass transition, gets lowered and the crystal behaves like a conventional ferroelectric. We have already described above (in §10.4.2) the occurrence of such a field-induced phase transition at a temperature T_t'.

As emphasized by Schmidt (1990), this situation is closely analogous to the well-known stress-induced martensitic phase transitions in some metallic alloys (cf. §11.5).

Cu – Zn – Al is a typical alloy which undergoes a martensitic phase transition, i.e. it has a high-temperature 'austenitic' phase and a low-temperature 'martensitic' phase: On cooling the austenitic phase, the martensitic phase starts forming at a temperature M_s. The proportion of the martensitic phase increases with cooling (in a 'burst-like' manner); this increase stops if the cooling is stopped. It is described as an *athermal transition*, not involving a diffusion of atoms. It has also been described as a *diffuse* transition in the sense that it extends over a whole range of temperatures, namely from M_s to M_f. M_f is the temperature at which the transformation process is finished (the transformation may not be 100%), and further cooling does not lead to an increase of the fraction of the martensitic phase.

Similarly the reverse transition from the martensitic phase to the austenitic phase starts at a temperature A_s on heating, and finishes at a temperature A_f.

The temperatures M_s, M_f, A_s and A_f can be altered by the application of an external stress field. Additives, as well as grain-size control, are also used for altering the range of temperatures over which the two phases coexist. In particular, one can *stabilize* the parent (austenitic) phase, i.e. one can prevent the occurrence of the transformation to the martensitic phase, by using suitable additives and/or manipulating the grain size. A fully stabilized system is also referred to as a *nontransforming system.*

The analogy with PMN, as also with certain compositions of PLZT, is obvious.

In PMN, there is no phase transition, i.e. no lowering of *global* symmetry, when only temperature is varied. Schmidt (1990) has therefore suggested that it should be described as a 'fully stabilized paraelectric'.

The effect of electric field on such a system is very different from that on a conventional ferroelectric. At temperatures below T_t' the field induces a transition to a phase of lower (rhombohedral) global symmetry, from the cubic-symmetry phase. This is reflected not only in the X-ray diffraction pattern, but also in the shape of the hysteresis loop, which acquires a square-shaped or 'fat' appearance, similar to that of a conventional ferroelectric.

Since the rhombohedral phase is a field-induced phase, it reverts to the cubic phase when the field passes through the zero value in a hysteresis-loop experiment. This transient occurrence of the nonpolar parent phase has been confirmed by birefringence measurements on another similar system, namely PLZT8/65/35 (Schmidt 1981). In it the polarization reversal occurs via a nonpolar, optically isotropic, state or phase.

Whereas in a conventional ferroelectric, polarization reversal is accompanied by movement of *domain* boundaries, in a relaxor ferroelectric the occurrence of the field-induced transition implies that the polarization reversal is accompanied by the creation and/or movement of *phase* boundaries.

As the temperature is raised from T_t' towards T_m', the hysteresis loop becomes slimmer and slimmer. In addition, no saturation of the polarization is observed even for fairly large-amplitude fields. Instead, the polarization continues to increase with the amplitude of the applied field.

SUGGESTED READING

H. Terauchy (1986). Dielectric glassy phases. *Phase Transitions*, **7**, 315.

A. P. Levanyuk & A. S. Sigov (1988). *Defects and Structural Phase Transitions.* Gordon & Breach, New York.

G. Schmidt (1990). Diffuse ferroelectric phase transitions in cubically stabilized perovskites. *Phase Transitions*, **20**, 127.

U. T. Hochli, K. Knorr & A. Loidl (1990). Orientational glasses. *Adv. Phys.*, **39**, 405.

E. Husson & A. Morell (1992). Ferroelectric materials with "ferroelectric diffuse phase transition". *Key Engg. Materials*, **68**, 217.

D. Viehland, J. F. Li, S. J. Jang, L. E. Cross & M. Wuttig (1992). Glassy polarization behaviour of relaxor ferroelectrics. *Phys. Rev. B*, **46**, 8013.

L. E. Cross (1994). Relaxor ferroelectrics: an overview. *Ferroelectrics*, **151**, 305.

D. Pandey (1995). Diffuse transitions in mixed ferroelectrics. *Key Engg. Materials*, **101-102**, 177.

Z.-Y. Cheng, R. S. Katiyar, X. Yao & A. Guo (1997). Dielectric behaviour of lead magnesium niobate relaxors. *Phys. Rev. B*, **55**, 8165.

H. Qian & L. A. Bursill (1996). Random-field Potts model for the polar domains of lead magnesium niobate and lead scandium tantalate. *Int. J. Mod. Phys. B*, **10**, 2027.

Z.-Y. Cheng, R. S. Katiyar, X. Yao & A. S. Bhalla (1998). Temperature dependence of dielectric constant of relaxor ferroelectrics. *Phys. Rev. B*, **57**, 8166.

W. Kleemann & E. K. Salje (1998) (Eds.). *Non-Exponential Relaxation and Rate Behaviour.* Special issue of *Phase Transitions*, **65**, 1-290.

10.5 QUANTUM FERROELECTRICS

Quantum ferroelectrics are ferroelectrics in which fluctuations of the electric polarization are governed by zero-point motion and other quantum-mechanical effects. Consequently, their critical exponents are different from those in the classical regime. The relevant temperature in this context is the saturation temperature, T_s, discussed in §5.4.5. Below this temperature, quantum effects become increasingly important as the temperature is lowered towards the absolute zero (Hayward & Salje 1998).

The most relevant quantity to investigate for a ferroelectric phase transition is the dielectric response. The effect of quantum phenomena on dielectric response was first studied by Barrett (1952), who derived an expression for the dielectric function of a system of harmonic oscillators obeying Bose statistics. His predictions were verified for $SrTiO_3$ by Sawaguchi, Kikuchi & Kodera (1952). Barrett's quantum-mechanical model was improved upon by Pytte (1972) by taking explicit account of the dipole-dipole interactions.

Interest in quantum ferroelectrics received a fresh impetus through the work of Oppermann & Thomas (1975) and Schneider, Beck & Stoll (1976), who employed renormalization-group techniques.

10.5.1 Displacive Limit of a Structural Phase Transition

The notion of a 'displacive limit' of a structural phase transition was introduced by Oppermann & Thomas (1975). The critical temperature T_c of a structural phase transition in general, and a ferroelectric phase transition in particular, is determined by a balance between two types of opposing influences. For $T > T_c > 0$ the disordering forces, which are predominantly of a thermal nature, prevail over long-ranged ordering forces like dipole-dipole coupling interactions. As the temperature is lowered towards T_c, the disordering tendencies are gradually overcome by the ordering interactions. The value of T_c for a crystal can be decreased by modifying the ordering interactions by parameters such as stress and composition. For a few crystals, conditions exist, or can be created by the influence of such parameters, such that $T_c = 0$. Such a critical temperature is called the displacive limit of a structural phase transition. In other words, the displacive limit is defined by that set of coupling parameters for which $T_c = 0$.

We have seen in Chapter 5 that critical fluctuations of thermal origin can become very dominant in the vicinity of T_c. But if $T_c = 0$, thermal fluctuations in the vicinity of T_c die out. One consequence of this is that, if we make only the classical calculations, i.e. if we do not take note of quantum fluctuations, a loss of universality is predicted for the critical phenomena (Morf, Schneider & Stoll 1977). This is because universality is mainly a result of a large order-parameter order-parameter correlation length, making a system practically independent of the microscopic details of the interactions involved.

However, quantum-mechanical fluctuations caused by zero-point motion reintroduce universality (but of a different class) (Morf et al. 1977). Consequences of quantum-mechanical fluctuations are very different from those of thermal fluctuations. For example, there is little or no dependence on temperature. That is, a quantum ferroelectric is characterized by the temperature independence of response functions (for T substantially below T_s). It follows that critical exponents for the displacive limit are not the same as those in a regime dominated by thermal fluctuations. There is thus a *crossover* of critical exponents when the displacive limit is approached.

For the lattice-dynamical model investigated by Morf et al. (1977), if a d-dimensional system has $T_c = 0$, then it corresponds to the Wilson fixed point of a $(d+1)$-dimensional system. In other words, when the effect of quantum fluctuations is included, the critical exponents for a d-dimensional system correspond to the Wilson exponents of a $(d+1)$-dimensional classical system. This amounts to an increase by unity of the effective dimensions of the lattice.

In particular, the predictions of the quantum-mechanical vector model near the displacive limit are about the same as those of the mean-field theory for $d = 4$.

Some important predictions of the model of Morf et al. (1977) at $T = 0$ are as follows:

$$P_s \sim (x - x_c)^{\beta_x}, \quad \beta_x = 1/2, \tag{10.5.1}$$

$$\epsilon^{-1} \sim (x - x_c)^{\gamma_x}, \quad \gamma_x = 1, \tag{10.5.2}$$

$$T_c \sim (x - x_c)^{1/\phi}, \quad \phi = 2 \tag{10.5.3}$$

Here x denotes the general coupling parameter, and x_c its limiting or critical value. In to Eq. 10.5.3, $\phi = 1$ for the classical case.

The term *quantum limit* is used in the literature in the context of quantum ferroelectrics (Rytz et al. 1980). Near $T = 0$, although the disordering influence of thermal fluctuations dies out, the quantum fluctuations may take over and may be so dominant for a system that it is prevented from undergoing a phase transition. It is then called an *incipient ferroelectric*. If, on the other hand, a critical composition x_c, or a critical pressure p_c, exists such that a ferroelectric phase transition does occur (at a low temperature) in spite of the quantum fluctuations, then this critical parameter defines the quantum limit for the system. That is, at the quantum limit the polar order has at least a marginal stability against quantum fluctuations.

It was believed at one stage that the number of different ferroelectric crystal systems for which quantum effects can be observed is not likely to be large (Hochli & Boatner 1979). However, the present thinking is different. We give an outline of the present viewpoint in the next subsection.

10.5.2 Modern Approach to Quantum Ferroelectrics

This subject is still in a stage of active development. Several aspects of the present position have been described in Kleemann & Salje (1998). Some conclusions from the work of Hayward & Salje (1998) are mentioned here.

A low-temperature extension of the Landau theory was outlined in §5.4.5 (Salje, Wruck & Thomas 1991). In particular, the saturation temperature T_s was introduced. This is the temperature characterising crossover, on cooling the crystal, from classical behaviour to quantum mechanical behaviour. Typically, the classical regime prevails for $T > 3T_s/2$, and the order parameter is totally saturated (i.e. is totally independent of temperature) for $T < T_s/2$ (Salje, Wruck & Thomas 1991). In the latter regime, quantum fluctuations are entirely responsible for determining the universality class of the system.

Eq. 5.4.26 is exact in the displacive limit defined in §10.5.1. For other situations, this equation still provides a good approximation for the solution of self-consistency equations (Salje, Wruck & Marais 1991). The formalism of Salje and coworkers has the merit that it is applicable to the entire relevant range of temperatures above and below T_s.

The value of T_s has been derived and tabulated by Hayward & Salje (1998) for a large number of crystals, not all of which are ferroelectrics. $T_s = 0$K for SbSI, but can be as high as 334 K for quartz.

The theory of Chamberlin (1998), described in §9.2.13, marks a definite advance in understanding the (slow) dynamics of quantum ferroelectrics, as also of several other condensed-matter systems.

We discuss some specific quantum ferroelectrics below.

10.5.3 Strontium Calcium Titanate

Bednorz & Muller (1984) established that $Sr_{1-x}Ca_xTiO_3$ (SCT) is a quantum ferroelectric, the critical exponents of which can be explained by an

XY model (cf. §5.5.2).

Pure $SrTiO_3$ is an incipient ferroelectric. It has cubic symmetry at room temperature, and undergoes, on cooling, an antiferrodistortive phase transition to a tetragonal phase of symmetry $I4/mcm$ at 105 K (Fleury & Worlock 1968). For this transition, T_s = 60 K (Hayward & Salje 1998). The unit cell of the tetragonal phase is rotated by 45^0 with respect to the cubic parent phase.

On further cooling, the relative dielectric function perpendicular to the tetragonal axis increases to a very high value, reaching 30,000 at about 3 K. At this temperature thermal fluctuations are very weak, and quantum fluctuations take over, so that the dielectric function becomes independent of temperature (see Bednorz & Muller (1984) for experimental data).

By analogy with $PbTiO_3$, this crystal should undergo a paraelectric-to-ferroelectric phase transition on cooling to 20 K (Cowley 1962). However, the dominant quantum fluctuations at such low temperatures stabilize the paraelectric phase. T_s = 20 K for this transition (Hayward & Salje 1998). It has turned out that application of the requisite amount of uniaxial stress can convert this incipient ferroelectric to a real, uniaxial ($n = 1$) ferroelectric.

Another method of inducing quantum-ferroelectric behaviour in this crystal is by replacing some of the Sr^{2+} ions randomly by Ca^{2+} ions. The mixed crystal $Sr_{1-x}Ca_xTiO_3$ is a quantum ferroelectric for $x_c < x < x_r$, where x_c = 0.002, and x_r = 0.016, with the spontaneous polarization, P_s, oriented along the pseudocubic directions [110] or $[1\bar{1}0]$ (Bednorz & Muller 1984; Bianchi, Dec, Kleemann & Bednorz 1995). Thus, in this range of x-values, SCT is an XY-ferroelectric, with a two-component order parameter ($n = 2$). It attains a very high peak value for the dielectric function ($\varepsilon(T) = 110,000$ at about 20 K) for $x = 0.0107$.

For $x > x_r$ the peak in $\epsilon(T)$, instead of being sharp as in a conventional ferroelectric, acquires a rounded appearance characteristic of a diffuse transition. Bednorz & Muller (1984) attributed this to the onset of a *random-field domain state* stipulated by Imry & Ma (1975). Although Sr^{2+} and Ca^{2+} ions have the same charge, their sizes are different (1.12 Å and 0.99 Å respectively). This results in the introduction of random strains locally, which couple with the electric polarization. Alternatively, some of the Ca^{2+} ions may also sit on Ti^{4+} sites, leading to local charge imbalance, which may be restored by the occurrence of vacancies on neighbouring oxygen sites. The random formation of such Ca^{2+}-vacancy dipoles results in a random electric field.

For such a random system the lower marginal dimensionality, d_l, is expected to be 4 (Imry & Ma 1975; Aharony 1978). Therefore, no ferroelectric ordering is expected for $d = 3$, but only a random-field domain state, as observed.

Another phase transition apparently occurs at T = 3.69 K in SCT with $x = 0.002$. In the recent work of Kleemann, Albertini, Chamberlin & Bednorz (1997) and Kleemann (1998), the real and imaginary parts of the dielectric function of this crystal were measured for the temperature range $1.5 \leq T \leq 15$ K and frequency range $10^{-3} \leq f \leq 10^7$ Hz. For the high-frequency regime, the system is found to relax in a CvS-like manner, with a positive correlation factor C. This factor reaches its maximum value of 68 at $T = 3.8$ K. Thus the high-frequency high-temperature ($T > 4$ K) response is characterized by the existence of an ergodic regime, involving activated flipping of clusters (just like in PMN).

At lower frequencies a crossover to nonactivated KWW-like behaviour appears to occur. This regime is characterized by a 3-dimensional percolation distribution function, n_s (cf. Eq. 9.2.29), with $C < 0$. Here also, the maximum magnitude of C, viz. $|C| = 172$, is observed near $T = 3.69$ K. For very low temperatures, the relaxation rates of large static domains acquire a surprisingly high value: $\omega_\infty \sim 2 \times 10^{-5}$ Hz. Dissipative quantum tunneling is conjectured to determine this (Kleemann, Albertini, Chamberlin & Bednorz 1997).

10.5.4 Potassium Tantalate Niobate

$KTa_{1-x}Nb_xO_3$ (KTN) is a much investigated quantum ferroelectric. Unlike SCT described above, it does not suffer from the complications of a structural phase transition before entering the quantum-ferroelectric regime. It is an $n = 3$ system, with $x_c = 0.008$. Classical ferroelectricity is exhibited by it for the entire range of compositions $0.008 \leq x \leq 1$. The composition with $x = 0.1$ exhibits the highest known value for the dielectric function ($\varepsilon(T_c) = 160,000$), which is obtained in this case at the cubic-to-tetragonal phase transition (Kind & Muller 1976).

As pointed out by Hochli, Weibel & Boatner (1977), KTN is the only known example of a cubic solid solution which exhibits a second-order ferroelectric phase transition the T_c of which can be varied continuously by varying x over a wide range.

Pure $KTaO_3$ ($x = 0$) does not display a ferroelectric phase transition, although there is a dielectric anomaly near $T = 0$ (Burkhard & Muller 1976).

$KTa_{0.95}Nb_{0.05}O_3$ is a ferroelectric, with $T_c \sim 60K$. And for $x \geq 0.05$, T_c increases almost linearly with x, reaching the value of 700 K for $KNbO_3$. The Nb concentration, x, serves as the control parameter (like high pressure) governing the quantum-ferroelectric behaviour.

For $x = x_c = 0.008$ the system becomes a quantum ferroelectric, with $T_c = 0$. The saturation temperature, T_s, estimated for KTN by Hayward

& Salje (1998), is 20 K. At $T = 0$ the spontaneous polarization varies as (Hochli et al. 1977):

$$P_s(x, 0) \sim (x - x_c)^{\beta_x}, \quad \beta_x = \frac{1}{2} \tag{10.5.4}$$

This is in agreement with the prediction that in the quantum regime the effective dimensionality of the lattice is increased by unity. (Morf et al. 1977). The upper marginal dimensionality, d_u, which sets the limit for mean-field behaviour, gets increased by unity when long-range interactions, as well as quantum-mechanical fluctuations, are incorporated in the calculation. Classical critical indices such as β_x in Eq. 10.5.4 are therefore applicable to quantum ferroelectrics down to $d = 2$ (instead of $d = 3$), except that logarithmic corrections are needed for $d = 2$. This is also true for the following:

$$\epsilon^{-1} \sim (x - x_c)^{\gamma_x}, \quad \gamma_x = 1, \tag{10.5.5}$$

$$T_c(x) \sim (x - x_c)^{1/\phi}, \quad \phi = 2 \tag{10.5.6}$$

Theoretical and experimental results are similar for both $KTa_{1-x}Nb_xO_3$ and $K_{1-y}Na_yTaO_3$, except that $x_c = 0.008$, whereas $y_c = 0.12$ (Hochli & Boatner 1979; Rytz, Hochli & Bilz 1980). At $y = y_c = 0.12$ the theoretical and experimental critical indices correspond to a Gaussian fixed point for a $d = 4$ isotropic system (Hochli & Boatner 1979).

Unlike SCT (§10.5.2), there is apparently no experimental evidence for the effect of random electric fields in KTN. Hochli & Boatner (1979) attribute this to the long range of the dipolar forces, which average out the local variations of the field.

10.5.5 Potassium Dihydrogen Phosphate

This well-known and popular nonlinear-optical crystal also exhibits quantum ferroelectric behaviour when subjected to a certain hydrostatic pressure. It is a uniaxial ferroelectric, for which T_c becomes zero at a pressure $p = p_c = 16.9$ kbar (Samara 1971, 1974, 1978; Nelmes et al. 1991). At the quantum limit, i.e. at $p = p_c$, the relation $\epsilon^{-1} \sim T^2$ is found to apply, in contrast to the relation $\epsilon^{-1} \sim T$ for a classical ferroelectric. The saturation temperature, T_s, for this crystal is 49 K (Hayward & Salje 1998).

SUGGESTED READING

U. T. Hochli & L. A. Boatner (1979). Quantum ferroelectricity in $K_{1-x}Na_x$ TaO_3 and $KTa_{1-y}Nb_yO_3$. *Phys. Rev. B*, **20**, 266.

J. G. Bednorz & K. A. Muller (1984). $Sr_{1-x}Ca_xTiO_3$: an XY quantum ferroelectric with transition to randomness. *Phys. Rev. Lett.*, **52**, 2289.

D. Rytz, U. T. Hochli & H. Bilz (1980). Dielectric susceptibility in quantum ferroelectrics. *Phys. Rev. B*, **22**, 359.

U. T. Hochli, K. Knorr & A. Loidl (1990). Orientational glasses. *Adv. Phys.*, **39**, 405.

E. K. H. Salje, B. Wruck & H. Thomas (1991a). Order-parameter saturation and low-temperature extension of Landau theory. *Z. Phys. B*, **82**, 399.

E. K. H. Salje, B. Wruck & S. Marais (1991b). Order parameter saturation at low temperatures: numerical results for displacive and O/D systems. *Ferroelectrics*, **124**, 185.

S. A. Hayward & E. K. H. Salje (1998). Low-temperature phase diagrams: nonlinearities due to quantum mechanical saturation of order parameters. *J. Phys.: Condens. Matter*, **10**, 1421.

W. Kleemann (1998). Correlated domain dynamics in relaxor ferroelectrics and random-field systems. *Phase Transitions*, **65**, 141.

R. V. Chamberlin (1998). Experiments and theory of the nonexponential relaxation in liquids, glasses, polymers and crystals. *Phase Transitions*, **65**, 169.

R. Ranjan, D. Pandey, V. Siruguri, P. S. R. Krishna & S. K. Paranjpe (1999). Novel features and phase transition behaviour of $(Sr_{1-x}Ca_x)TiO_3$: I. Neutron diffraction study. *J. Phys.: Condens. Matter*, **11**, 2233.

10.6 DOMAIN STRUCTURE OF FERROELECTRIC CRYSTALS

10.6.1 Domains in a Ferroelectric Crystal

A ferroelectric phase transition is characterized by the appearance of spontaneous polarization $\mathbf{P}_s$ (with respect to the prototype). Being a vector this polarization has a magnitude and a direction. This direction, when referred to the prototype, may or may not be equivalent to other directions (depending on the ferroic species involved). Also, this direction may be either along a polar or a nonpolar direction of the prototype. Thus there are four distinct possibilities to consider.

(i) $\mathbf{P}_s$ emerges along a unique but nonpolar axis of the prototype. An example is that of KDP. The Aizu species is $\bar{4}2mFmm2(p)$. The prototype or paraelectric point group is nonpolar. And the polar 2-fold axis of the ferroelectric phase coincides with the $\bar{4}$-axis of the prototype. The crystal has no preference for the two ends of the $\bar{4}$-axis when it makes the transition to the polar phase. Thus two orientation states are available, and the possibility exists for reversing the spontaneous polarization of a particular orientation state by applying a strong enough electric field in the opposite direction. And there is only one such axis along which this can happen. KDP is thus a *uniaxial ferroelectric.*

(ii) $\mathbf{P}_s$ develops along a nonunique and nonpolar axis of the prototype. This happens, for example, at the cubic-tetragonal transition in $BaTiO_3$. The three equivalent $< 100 >$ axes in the cubic phase give rise to a total of six possible orientations along which $\mathbf{P}_s$ can develop in the ferroelectric tetragonal phase, making it a *multiaxial ferroelectric* phase.

(iii) In the above two cases, $\mathbf{P}_s$ is zero in the prototypic phase. It can also happen that the prototype already has a spontaneous polarization (but only one orientation state), and an additional spontaneous polarization develops at the ferroelectric phase transition along a direction *different* from that of the polar axis of the prototype.

In terms of the distinction made by us between absolute and relative spontaneous polarization in §10.1.7, the absolute spontaneous polarization is nonzero in both the phases. The relative spontaneous polarization is zero in the prototypic phase and nonzero in the ferroelectric phase.

In this case, the absolute spontaneous polarization in the ferroelectric phase has a non-reversible part (which carried through across the transition from the prototypic phase), and a reversible part (which arose at the phase transition). When these two are superimposed and one looks at their vector sum, an applied electric field can only *reorient* the net spontaneous polarization (in discrete steps), rather than *reverse* it. One therefore speaks of a *reorientable ferroelectric* (cf. Fig. 10.3.1) (Shuvalov 1970).

Relative spontaneous polarization is always reversible, even for a reorientable ferroelectric, except for some trigonal or hexagonal prototype symmetries (cf. Aizu 1967).

Such a situation occurs in some boracites, wherein the prototype symmetry is $\bar{4}3m$ (T_d), and the ferroelectric-phase symmetry is $3m$ (C_{3v}), brought about by the emergence of a spontaneous polarization along any of the directions [111], [$\bar{1}$11], [1$\bar{1}$1], or [11$\bar{1}$]. (In fact there are 8 such directions, the other four being obtained by reversing these four.)

Several other such group-subgroup combinations have been tabulated by Blinc & Zeks (1974).

(iv) A fourth situation is conceivable in which additional spontaneous polarization develops at a phase transition *along* a *unique* polar axis. That is, in the lower-symmetry phase the spontaneous polarization has a non-reversible component carried over from the prototypic phase, and another non-reversible component along the *same* axis. However, this is not a ferroic (and therefore not a ferroelectric) phase transition because there is no change of point-group symmetry across it. The term *coelectric phase transitions* can be introduced for such phase transitions, by analogy with what Salje (1993a) calls coelastic phase transitions.

The various kinds of symmetry changes at ferroelectric phase transitions have been tabulated by Blinc & Zeks (1974).

10.6.2 Orientation of Walls Between Ferroelectric Domain Pairs

The orientation of the transition region (domain wall) between a ferroelectric domain pair is determined by two main factors: minimization of electrostatic energy and minimization of elastic energy. The latter comes from the fact that, associated with the relative spontaneous polarization is the relative spontaneous strain, arising from piezoelectric and/or electrostrictive effects. Generally, the elastic-energy term is more important than the electrostatic term because, even for a charged domain wall, the free carriers of charge (electrons) manage to achieve a high degree of effective charge neutrality (Arlt 1990).

A domain wall along which the mechanical or elastic-strain compatibility requirement is satisfied is called a *permissible wall* (Fousek & Janovec 1969; Wiesendanger 1973). This wall has such an orientation that components of spontaneous strain along it have the same value from the two domains separated by it.

Let a plane in the prototypic phase be specified by indices h, k, l (not necessarily the Miller indices) with reference to a standard Cartesian system of coordinates. Let $d\mathbf{s}(s_1, s_2, s_3)$ be an infinitesimal vector in the prototype, which changes (because of the real or notional phase transition from the prototypic phase) to $d^1\mathbf{s}$ and $d^2\mathbf{s}$ in the two ferroelectric domains separated by the plane (h, k, l). The condition of mechanical compatibility can be then expressed as (Fousek & Janovec 1969; Fousek 1971):

$$d^1s^2 - ds^2 = d^2s^2 - ds^2, \tag{10.6.1}$$

with

$$h\,ds_1 + k\,ds_2 + l\,ds_3 = 0 \tag{10.6.2}$$

Detailed analysis shows that there are two types of permissible domain walls: *W-walls* and *S-walls*. In addition, sometimes, no wall may exist which obeys the mechanical compatibility condition for permissibility; the term R-walls has been used by Fousek & Janovec (1969) for the walls that exist in such a case.

W-walls may be either W_f-walls or W_∞-walls:

If an F-operation for the domain pair is a 2-fold rotation or a mirror operation, then the domains are separated by a W_f-wall (perpendicular to the 2-fold axis, or parallel to the mirror plane). W_f-walls are prominent crystallographic planes; they have fixed and rational indices (h, k, l). Their orientation is independent of the spontaneous strain involved, and therefore does not vary with temperature.

If the F-operation is an inversion operation, symmetry considerations do not put any restriction on the orientation of a W-wall, and the symbol W_∞ is used in such a case. Other (physical) factors determine the orientations adopted by such walls.

S-walls have a temperature dependent orientation, and may have rational or irrational indices. Six categories (S_1 to S_6) for them have been identified by Fousek & Janovec (1969).

Excluding the case of W_∞-walls, permissible domain walls for a domain pair can occur in mutually perpendicular pairs. The following possible situations arise for the pairs: W_fW_f, W_fS_i, and S_iS_j.

The Wall Charge

The electric charge on a domain wall is determined by $\nabla \cdot \mathbf{P}_s$. If the mechanical compatibility requirement does not conflict with that of charge neutrality, a charge-neutral wall would have an orientation defined by a unit vector $\mathbf{n}$ normal to it such that

$$(\mathbf{P}_s - \mathbf{P}'_s) \cdot \mathbf{n} = 0, \tag{10.6.3}$$

where $\mathbf{P}_s$ and $\mathbf{P}'_s$ are the absolute spontaneous polarizations in the domains separated by the wall. A wall would be a charged wall if the component of polarization normal to it is discontinuous across the wall.

If (x, y, z) and (x', y', z') denote the direction cosines of the spontaneous-polarization vector in the two domains, the charge neutrality condition can be expressed as follows (Fousek 1971):

$$h(x - x') + k(y - y') + l(z - z') = 0 \qquad (10.6.4)$$

If the prototype has hexagonal or trigonal symmetry (for which 4-figure indices (hkil), (xyuz) and $(x'y'u'z')$ are used), the above condition changes to:

$$2(2h + k)(x - x') + 2(h + 2k)(y - y') + 3l(z - z') = 0 \qquad (10.6.5)$$

If this condition is not satisfied, the wall has a charge.

A W_∞-wall may have any orientation (so far as symmetry considerations are concerned), and may therefore be either charged or uncharged.

For a W_f-wall, if the F-operation is a mirror operation, the wall is charged. If the wall is perpendicular to a 2-fold axis of the prototype, it is charge-neutral.

For a mutually perpendicular W_f S_i pair of walls, if one is charged, the other is neutral.

Expected orientations of domain walls for all the 88 full-ferroelectric species have been derived and tabulated by Fousek (1971).

A head-to-tail configuration of the polarization vectors at a domain wall is energetically favourable, as it results in an uncharged wall. However, head-to-head and tail-to-tail arrangements have also been observed in $BaTiO_3$ and $KNbO_3$, with charge compensation by free electrons as a possible stabilization mechanism (Peng & Bursill 1983; Janovec & Dvorak 1986).

Coherent, Semicoherent, and Incoherent Boundaries

If all crystallographic planes run continuously from one side to the other across a domain boundary or a phase boundary, it is called a coherent boundary.

If some but not all crystallographic planes do so, it is a semicoherent boundary.

If no crystallographic plane runs continuously across the boundary, it is an incoherent boundary.

10.6.3 Thickness of Walls Between Ferroelectric Domain Pairs

The domain structure of ferroelectrics, which incorporates the thickness of domain walls, differs qualitatively from that of ferromagnetics. There are some basic reasons for this:

Firstly, in ferroelectrics there exist free carriers of charge, namely electrons, which can screen polarization; there are no magnetic monopoles, although electrons can still play the role of carriers of spin.

Secondly, electrostriction strains are one to two orders of magnitude larger than magnetostriction strains. Therefore, in ferroelectrics interactions with defects etc. generally have a more serious effect on the domain structure, compared to the situation in ferromagnets.

Thirdly, in ferroelectrics we do not have the equivalent of the quantum-mechanical exchange interaction responsible for long-range ordering in ferromagnets. Therefore, ferroelectric domain walls do not have to be thick, unlike Bloch walls in ferromagnets (§9.4.3). However, depending on the probe used for 'measuring' the thickness of a wall between ferroelectric domains, a distinction may have to be made between the 'effective' wall thickness and the 'real' wall thickness. All in all, it is a rather complex problem (Fousek 1992a), although important insights have been obtained in recent years (see Salje 1994).

X-ray diffraction has been used by several workers for determining the effective wall thickness in ferroelectrics and ferroelastics (see, e.g., Andrews & Cowley (1986); Wruck et al. (1994); Locherer et al. (1996)).

As the critical temperature T_c is approached, the effective wall thickness, W, tends to diverge (Salje 1994):

$$W = W_0\sqrt{T_c/(|T - T_c|)} \tag{10.6.6}$$

The parameter W_0 is a measure of the intrinsic wall thickness. It may be expected to be comparable to the lattice periodicity a_0. In real systems, however, wall bending, intersection of walls, defects, and several other factors can lead to a very different effective value for W_0. Andrews & Cowley (1986) arrived at a value $W_0 = 4.2a_0$ for KDP.

Similarly, Cao & Barsch (1990) determined $W_0 = 3.1a_0$ for $SrTiO_3$, using data from elastic properties, as well as curvatures of phonon-dispersion curves.

High-resolution electron microscopy (HREM) has emerged as an important experimental technique for determining the 'real' thickness of domain walls in ferroelectrics and other ferroics[1]. However, even here complications can arise from the charge carried by the probing beam of electrons. Moreover, wall thicknesses in the interior of a bulk specimen may be different from their values near the surface, or values for ultra-thin specimens (see, e.g., Lin & Bursill 1991; Bursill & Lin 1992).

[1]It must be mentioned here that, compared to electron diffraction, X-ray diffraction is better suited for obtaining certain types of information. X-ray diffraction gives information which is less localized, thus averaging the local structural parameters over fairly large volumes of the crystal. This enables one to perform a quantitative analysis of the strain fields associated with the microstructure (Locherer et al. 1996).

HREM reveals that for 180° domain walls, e.g. in $Ba_2NaNb_5O_{15}$ (BNN) crystals (Lin & Bursill 1991; Bursill & Lin 1992), the domain-wall thickness is virtually zero. The absolute spontaneous strain associated with the two domains separated by the wall is the same (so the relative spontaneous strain is zero), and the registry of the two crystal lattices is easy to achieve. The domain wall is elongated parallel to the polar axis, and sublattice steps and ledges occur.

When the F-operation between domains is other than a 180° rotation, misfit dislocations provide a possible mechanism for achieving registry of the two lattices at the domain wall (see Salje 1993a). The associated strain fields, as well as other factors like disorientations and warping of the wall lead to an increase in the *effective* or *apparent* wall thickness.

Mention must be made here of two basic models proposed for understanding domain-wall thicknesses, one due to Zhirnov (1959), and the other due to Kinase & Takahashi (1957). Zhirnov's model is of the Landau-Ginzburg type (cf. §5.5.2). For a scalar non-degenerate order parameter P one writes the following Landau-Ginzburg potential:

$$g = \frac{A}{2}(T - T_c)P^2 + \frac{B}{4}P^4 + \frac{G}{2}(\nabla P)^2 \tag{10.6.7}$$

The domain wall is assumed to exist perpendicular to the x-axis, with its mid-plane at $x = 0$. On solving the Euler-Lagrange equation

$$\frac{\partial}{\partial x}\left[\frac{\partial g}{\partial P}\right] - \frac{\partial g}{\partial P} = 0, \tag{10.6.8}$$

the following order-parameter profile is obtained (see Salje 1993a; Hatch et al. 1997):

$$P = P_0 \tanh\frac{x}{W}, \tag{10.6.9}$$

where W is the wall thickness, and P_0 is the value of the order parameter at large x. For a continuous phase transition, W is given by Eq. 10.6.6, with

$$W_0 = \frac{1}{2}\sqrt{\frac{G}{AT_c}} \tag{10.6.10}$$

This model predicts a domain-wall thickness of 5-20 Å for the 180° domains in $BaTiO_3$, and 50-100 Å for the 90° domains.

In Zhirnov's (1959) model the polarization vector does not rotate suddenly across the domain wall: rather it changes gradually. By contrast, there is an abrupt change in the direction of the polarization at the domain wall in the molecular model of Kinase & Takahashi (1957). Computation of the competing dipole-dipole interaction energy and the elastic energy in this model predicts a negligible thickness for the 180° walls in $BaTiO_3$.

More recently, Huang et al. (1997) have made a Landau-Ginzburg type calculation which predicts the occurrence of a structural transition within a 180° wall in a perovskite ferroelectric from an 'Ising-type' configuration to a 'Bloch-type' configuration on cooling, the latter being more stable in a wide temperature range.

Moving Domain Walls

So far we have considered widths of domain walls under static equilibrium conditions. Application of a driving field, e.g. during hysteresis-loop measurements, can make the domain walls move. In general the width of moving domain walls is greater than that at rest. Sidorkin (1997) has recently given a detailed theoretical treatment of this subject.

For thick domain walls one can work in the continuum approximation of the basic Zhirnov (1959) formalism by adding a kinetic-energy density term to the Landau-Ginzburg thermodynamic potential. For thin walls, due note has to be taken of the wall-surface energy.

SUGGESTED READING

J. Fousek & V. Janovec (1969). The orientation of domain walls in twinned ferroelectric crystals. *J. Appl. Phys.*, **40**, 135.

L. A. Shuvalov (1970). Symmetry aspects of ferroelectricity. *J. Phys. Soc. Japan*, **28** Suppl., 38.

J. Fousek (1971). Permissible domain walls in ferroelectric species. *Czech. J. Phys. B*, **21**, 955.

R. Blinc & B. Zeks (1974). *Soft Modes in Ferroelectrics and Antiferroelectrics.* North-Holland, Amsterdam.

G. Arlt (1990). Twinning in ferroelectric and ferroelastic ceramics: stress relief. *J. Mater. Sci.*, **25**, 2655.

Peng Ju Lin & L. A. Bursill (1991). HREM investigation of ferroelectric and ferroelastic domains in bronze type tunnel structures. *Phase Transitions*, **34**, 171.

J. Fousek (1992a). Domain investigations: A select review. In M. Liu et al. (Eds.), *ISAF92: Proc. 8th IEEE International Symposium on Applications of Ferroelectrics.* Greenville, S.C., U.S.A., Aug. 30 - Sept. 2, 1992. IEEE Catalog No. 92CH3080-9.

J. Fousek (1992b). Ferroelectric domains: Some recent advances. In N. Setter (Ed.), *Ferroelectric Ceramics*. Monte Verita, Birkhauser Verlag, Basel.

L. A. Bursill & Peng Ju Lin (1992). High resolution electron microscopy of electronic ceramic materials. *Key Engg. Materials*, **66-67**, 421. Also available in book form as J. Nowotny (Ed.), *Electronic Ceramic Materials*. Trans Tech Publications, Switzerland.

E. K. H. Salje (Ed.) (1994). *Mobile Domain Boundaries*. Special issue of *Phase Transitions*, **48**, 1-200.

X. R. Huang, X. B. Hu, S. S. Jiang & D. Feng (1997). Theoretical model of 180° domain-wall structures and their transformation in ferroelectric perovskites. *Phys. Rev. B*, **55**, 5534.

10.7 FERROELECTRIC DOMAIN SWITCHING

10.7.1 Kinetics of Domain Switching in Ferroelectrics

The time and field dependence of polarization reversal, or domain switching, in a ferroelectric is determined by a large number of complex processes, and cannot be described by a single model. The situation is complex, not only from the point of view of theory, but also of experiment; there are too many factors influencing any experimental measurement (Fatuzzo & Merz 1967; Lines & Glass 1977; Burfoot & Taylor 1979).

Merz (1954) was probably the first to conduct, on $BaTiO_3$, quantitative experiments to determine the switching time t_s, and its dependence on the applied electric field. The basic experiment was quite simple in design: Apply a step-function field and measure the displacement-current density as a function of time. For a strong-enough electric-field (at a given temperature) the direction of spontaneous polarization is reversed, and the displacement current is a measure of the rate of domain switching, assuming that the conduction current is negligible. For conducting specimens other techniques, which make a more direct measurement of the remanent polarization, have to be used (Chynoweth 1956; Husimi & Kataoka 1960; Ballman & Brown 1972).

For $BaTiO_3$ the observed kinetics of domain switching falls into two regimes, depending on whether the applied field E is less or more than about 10 kV/cm. For less than this value (Merz 1954):

$$i_{max} = i_0 \, e^{-\alpha/E}, \tag{10.7.1}$$

$$t_s = t_0 \, e^{\alpha/E} \tag{10.7.2}$$

Here i_{max} is the maximum value attained by the displacement current, before it starts decaying with time.

For $E > 10$ kV/cm (Stadler 1958):

$$t_s = k \, E^{-1.5} \tag{10.7.3}$$

A similar field dependence is observed for TGS crystals for $E > 20$ kV/cm (Fatuzzo & Merz 1959):

$$t_s = k \, E^{-1} \tag{10.7.4}$$

When ferroelectric domain switching occurs under the action of a driving field, the process involved may be either the nucleation and growth of regions with the new domain orientation, or, alternatively, a sideways movement of existing domain walls such that the domains oriented favourably with respect to the driving field grow at the cost of those less favourably oriented. In most situations, the latter mechanism is less likely, for the following reason (Landauer 1957): Unlike Bloch walls in ferromagnetic crystals, domain walls in ferroelectric crystals are usually only a few lattice-spacings thick. Therefore the energy required for a ferroelectric domain wall to move by one lattice spacing is not very different from the wall energy itself; i.e. it is a large quantity. By comparison, the energy gained by the process of domain-wall movement by one lattice spacing is not very large. It is therefore not a very probable mechanism of polarization reversal.

The more likely mechanism is similar to crystal growth (cf. §2.1), i.e. by nucleation and growth of regions with the new orientation of polarization (Miller & Weinreich 1960; Stadler & Zachmanidis 1963; Hayashi 1972). If the nucleation and growth occurs on existing domain walls, the whole process would mimic sideways motion of domain walls.

Crystal growth from a fluid phase is necessarily a first-order process. Therefore, as discussed in §2.1.1, nuclei of size greater than a critical value must form first, which can then grow into a larger crystal. The driving force for this process is provided by supercooling or supersaturation. The corresponding driving force for domain switching comes from the applied electric field. Unless the field applied is very large, nucleation of the new orientation state is expected to be only one lattice-spacing thick (see Lines & Glass 1977). According to Hayashi's (1972) model, 2-dimensional nuclei are formed on a 180° wall at a constant rate. They spread along the wall with different velocities in different directions. This growth of the domain wall parallel to its own plane can be quite fast because of the ledges and kinks available for the easy growth of regions with the new polarization orientation. When the entire surface of the wall is covered up by the new growth, the next stage is the formation and growth of another 2-dimensional nucleus (by thermal activation) on a flat *terrace* site (cf. §2.1.3). This is a slow and therefore the rate-determining process. Hayashi's (1972) analysis gives the following expression for the nucleation rate $1/\tau$:

$$1/\tau \sim e^{-(\delta/E)}, \qquad (10.7.5)$$

where δ is fairly independent of E. This equation describes correctly the observed switching rate $1/t_s$ (cf. Eq. 10.7.2).

The situation becomes more complex in the high-field regime, just as it does for the growth of a crystal from a highly supersaturated or highly supercooled fluid phase. Nuclei can now form even on top of other nuclei (see Chernov 1989). The corresponding situation in domain switching is that the apparent wall velocity now depends not only on the rate of nucleation, but also on the growth velocity of the already formed nuclei. Hayashi's (1972) analysis shows that, in the high-field regime, the exponential field dependence described by Eq. 10.7.5 is replaced by a power-law dependence. In fact, three regimes can be identified, each with its own field dependence for the wall velocity v:

(i) Thin wall; $E \sim 200$ V/cm:

$$v = k_1 E^{3/4} e^{-\delta/(2E)} \qquad (10.7.6)$$

(ii) Thick wall; step-like structure; $E \sim 1$ kV/cm:

$$v = k_2 E^{-1/2} e^{-\delta/E} \qquad (10.7.7)$$

(iii) Thin wall; $E > 30$ kV/cm:

$$v = k_3 E^n \qquad (10.7.8)$$

10.7.2 The Ferroelectric Hysteresis Loop

The characteristics of a typical hysteresis loop of a ferroelectric crystal (Fig. 10.7.1) are quite similar to those of a ferromagnetic crystal (Figs. 9.1.1 and 9.6.1). The coercive field E_c, the spontaneous polarization P_s, and the remanent polarization P_r have meanings similar to those of the corresponding ferromagnetic parameters (see, e.g., Lines & Glass 1977).

Fig. 10.7.1 also shows the D-E curve for a virgin (depolarized) specimen of a ferroelectric (curve OABC). Its interpretation is quite similar to that of the ferromagnetic analogue shown in Fig. 9.1.1. The segments OA, AB, and

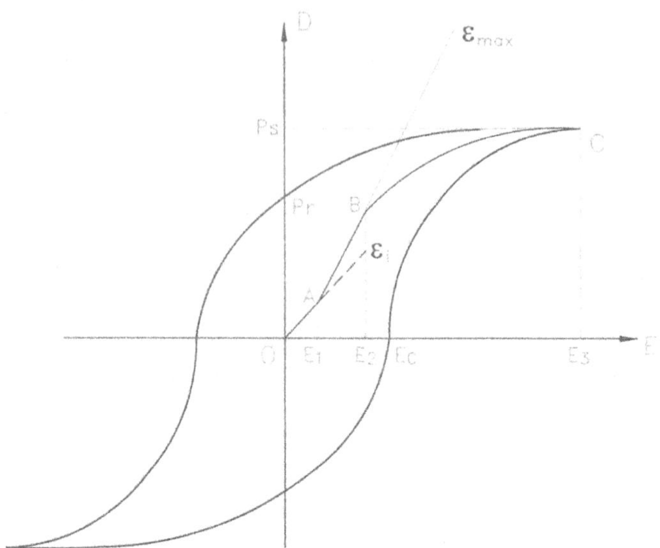

Figure 10.7.1: A typical ferroelectric hysteresis loop. As in the ferromagnetic analogue shown in Fig. 9.1.1 (and discussed in §9.1.1), the curve OABC is for a virgin specimen, having randomly oriented domains to start with. And like in Fig. 9.6.1 for a ferromagnetic crystal, the outermost envelope (on which the points marked E_c and P_r lie) is the so-called *limiting hysteresis loop.* The fields marked E_1, E_2 and E_3 are discussed in §14.2.1.

BC correspond, respectively, to reversible domain-wall bowing, irreversible domain-wall movement, and domain rotation. Also, by analogy with Fig. 9.1.1 again, the slope of the straight line defines the *initial permittivity* ϵ_i. And the inflexion point B corresponds to the maximum slope of the curve, and defines the *maximum permittivity*, ϵ_{max}.

SUGGESTED READING

E. Fatuzzo & W. J. Merz (1967). *Ferroelectricity.* North-Holland, Amsterdam.

M. E. Lines & A. M. Glass (1977). *Principles and Applications of Ferroelectrics and Related Materials.* Clarendon Press, Oxford.

J. C. Burfoot & G. W. Taylor (1979). *Polar Dielectrics and Their Applications.* Macmillan Press, London.

Chapter 11

FERROELASTIC CRYSTALS

We define ferroelastic crystals as those which undergo, or can be deemed to have undergone, at least one ferroelastic phase transition.

Ferroelastic phase transitions are a subset of ferroic phase transitions (cf. Fig. 5.2.2). Apart from the fact that they are nondisruptive phase transitions (NDPTs), they have the distinctive feature that they always involve a spontaneous shear distortion of the crystal lattice, such that there is a change of the *shape* (rather than the size) of the crystallographic unit cell. The case of the cubic-to-tetragonal phase transition in $BaTiO_3$ illustrates the point. It is a first-order phase transition, so that there is a change in the volume of the crystallographic unit cell. This fact is unimportant and irrelevant from the point of view of defining it as a ferroelastic phase transition.[1] What is relevant is that there is a change of the shape of the unit cell: from a cube to a square prism, brought about by an elongation of the cell edge along one of the basis vectors, say $\mathbf{a}_3$, and a reduction of the repeat distance along $\mathbf{a}_1$ and $\mathbf{a}_2$. Since the elongation can occur along any of the basis vectors $\mathbf{a}_1$, $\mathbf{a}_2$, $\mathbf{a}_3$, domain pairs can exist in the tetragonal phase, members of which differ in the direction chosen by the crystal for elongation. Such domain pairs are said to differ in relative spontaneous strain (cf. §11.1.3). If a phase transition results in the occurrence of at least one such domain pair, it is called a ferroelastic phase transition.

We begin this chapter by recapitulating some elastic properties of ordered crystals.

[1]We must emphsize that volume effects *are* important and relevant when it comes to determining the kinetics and energetics of a ferroelastic phase transition (Salje 1990).

11.1 SOME ELASTIC PROPERTIES OF ORDERED CRYSTALS

11.1.1 Strain, Stress, Compliance

A crystal is said to be deformed or strained when there is a change of the relative mean positions of its atoms. In crystal physics we replace the actual atomic structure of a crystal by a homogeneous continuum, which is anisotropic in general. In such a continuum we can define strain at a point in terms of the position vector $\mathbf{x}$ of that point.

As a result of the deformation or strain this position vector changes to, say, $\mathbf{x}'$. The vector

$$\mathbf{u} = \mathbf{x}' - \mathbf{x}, \tag{11.1.1}$$

called the *displacement vector*, is a measure of the displacement of the material point positioned initially at $\mathbf{x}$. The displacement vector can be expressed either as a function of the initial position vector $\mathbf{x}$ (the *Lagrangian form of strain*, or of the final position vector $\mathbf{x}'$ (the *Eulerian form of strain* (see, e.g., Mase 1970)). We shall adopt the Lagrangian form.

For defining the components of the strain tensor one introduces a Cartesian system of coordinates with respect to a certain point inside the continuum, taken as the origin O. Since the deformation or strain is defined in terms of *relative* distances between points, we can ignore any translation of the continuum as a whole. That is, we can assume without loss of generality that the origin O is not affected (moved) by the deformation. Therefore the components of the vectors $\mathbf{x}$, $\mathbf{x}'$ and $\mathbf{u}$ can be defined with respect to the *same* origin O.

Thence, in the Lagrangian formulation, components of $\mathbf{u}$ can be written as

$$u_i = u_i(x_1, x_2, x_3), \quad i = 1, 2, 3 \tag{11.1.2}$$

As shown in several standard texts (Nye 1957; Bhagavantam 1966), the state of deformation of a continuum can be described in terms of a symmetrical second-rank tensor, (e_{ij}), called the strain tensor. Depending upon the situation, several such tensors can be introduced.

When the relative displacements are vanishingly small, the relevant strain tensor is the *linear Lagrangian strain tensor*:

$$e_{ij} = \frac{1}{2}\left(\frac{\partial u_i}{\partial x_j} + \frac{\partial u_j}{\partial x_i}\right) \tag{11.1.3}$$

For larger displacements, one describes strain in terms of the *finite Lagrangian (or Green's) strain tensor*:

$$e_{ij} = \frac{1}{2}\left(\frac{\partial u_i}{\partial x_j} + \frac{\partial u_j}{\partial x_i} + \frac{\partial u_k}{\partial x_i}\frac{\partial u_k}{\partial x_j}\right) \tag{11.1.4}$$

We shall be concerned mainly with linear Lagrangian strain, defined by Eq. 11.1.3. It is a polar tensor of rank 2. Unlike the electric polarization tensor, the strain tensor has nonzero components even for centrosymmetric crystals.

When a crystal is deformed, its lattice parameters undergo changes. The lattice parameters are easily determined by an XRD or neutron diffraction experiment. It is therefore useful to derive the strain-tensor coefficients in terms of lattice parameters before and after a deformation. Such a derivation for several types of strain tensors has been carried out by Schlenker, Gibbs & Boisen (1978). We quote from their work for the linear Lagrangian strain tensor:

$$e_{11} = \frac{a'_1 \sin\beta' \sin\gamma'^*}{a_1 \sin\beta \sin\gamma^*} - 1, \tag{11.1.5}$$

$$e_{22} = \frac{a'_2 \sin\alpha'}{a_2 \sin\alpha} - 1, \tag{11.1.6}$$

$$e_{33} = \frac{a'_3}{a_3} - 1, \tag{11.1.7}$$

$$e_{12} = e_{21} = \frac{1}{2}\left[\frac{a'_2 \sin\alpha' \cos\gamma'^*}{a_2 \sin\alpha \sin\gamma^*} - \frac{a_1 \sin\beta \sin\gamma^*}{a_1 \sin\beta \sin\gamma^*}\right], \tag{11.1.8}$$

$$e_{13} = e_{31} = \frac{1}{2}\left[\frac{a'_1 \cos\beta'}{a_1 \sin\beta \sin\gamma^*} + \frac{\cos\gamma^*}{\sin\gamma^*} \cdot \left(\frac{a'_2 \cos\alpha'}{a_2 \sin\alpha} - \frac{a'_3 \cos\alpha}{a_3 \sin\alpha}\right) - \frac{a'_3 \cos\beta}{a_3 \sin\beta \sin\gamma^*}\right], \tag{11.1.9}$$

$$e_{23} = e_{32} = \frac{1}{2}\left[\frac{a'_2 \cos\alpha'}{a_2 \sin\alpha} - \frac{a'_3 \cos\alpha}{a_3 \sin\alpha}\right] \tag{11.1.10}$$

Here $(a_1, a_2, a_3, \alpha, \beta, \gamma)$ are the lattice parameters before the deformation of the crystal, and $(a'_1, a'_2, a'_3, \alpha', \beta', \gamma')$ are their values after the deformation. A * superscript stands for corresponding parameters of the reciprocal lattice. The Cartesian system of coordinates (for defining components of the strain tensor) is chosen as follows: The z-axis is taken along $\mathbf{a}_3$; the x-axis is taken along $\mathbf{a}_1^*$; and the y-axis is perpendicular to the z-axis and the x-axis, i.e. $\mathbf{a}_2$ is in the yz-plane.

Stress

Stress is force per unit area. A body in which one part exerts a force on neighbouring parts is said to be in a state of stress. Following Nye (1957), we can define stress at a point P in a material as follows: We consider an element of surface area δS passing through P, and imagine a unit vector l drawn perpendicular to this area. Let $\mathbf{f}\,\delta S$ be the force across this area. Then, it can be shown that (cf. Nye 1957), as $\delta S \to 0$, the components of force are related to the direction cosines of the area normal by

$$f_i = \sigma_{ij}\, l_j \tag{11.1.11}$$

The nine coefficients (σ_{ij}), which can be shown to form a tensor, define the *stress tensor.*

Imagine a cube embedded in the material, with its edges parallel to the Cartesian axis. It can be shown then that (σ_{ii}) are the *normal* components of stress, and $(\sigma_{ij},\ i \neq j)$ are the *shear* components.

If body torques are absent, the stress tensor has the following permutation symmetry:

$$\sigma_{ij} = \sigma_{ji} \tag{11.1.12}$$

Such a symmetric second-rank tensor can be represented by a surface called the *stress quadric*, defined by the relation

$$\sigma_{ij}\, x_i\, x_j = 1 \tag{11.1.13}$$

The stress tensor is a field tensor (cf. §3.1.3), and not a matter tensor. Therefore the symmetry of the stress quadric does not have to conform to the point-group symmetry of the material to which the stress is applied. Of course, the net symmetry of the stressed system (material plus applied stress) does depend on the initial symmetry of the material.

Through a coordinate transformation, the stress quadric can be referred to its principal axes, and, in the most general case, has the following form (*triaxial stress*):

$$(\sigma_{ij}) = \begin{bmatrix} \sigma_{11} & 0 & 0 \\ 0 & \sigma_{22} & 0 \\ 0 & 0 & \sigma_{33} \end{bmatrix} \tag{11.1.14}$$

Uniaxial Stress

A system is said to be under uniaxial stress when Eq. 11.1.14 has the following special form:

$$(\sigma_{ij}) = \begin{bmatrix} \sigma & 0 & 0 \\ 0 & 0 & 0 \\ 0 & 0 & 0 \end{bmatrix} \tag{11.1.15}$$

Biaxial Stress

This corresponds to the following representation:

$$(\sigma_{ij}) = \begin{bmatrix} \sigma_{11} & 0 & 0 \\ 0 & \sigma_{22} & 0 \\ 0 & 0 & 0 \end{bmatrix} \quad (11.1.16)$$

Compliance

Application of stress to a body produces strain in it, with a resultant change of shape and/or size. Below a certain *elastic limit*, and for sufficiently short durations of stress application, the body is able to return to its original shape and size when the stress is removed. Also, in the so-called *linear regime*, the strain is found to be linearly proportional to the applied stress (Hooke's law). In this regime, components of the strain tensor are related to those of the stress tensor through the following set of equations:

$$e_{ij} = s_{ijkl}\,\sigma_{kl} \quad (11.1.17)$$

This set of 9 equations can be inverted to yield the following:

$$\sigma_{ij} = c_{ijkl}\,e_{kl} \quad (11.1.18)$$

The coefficients (s_{ijkl}) in Eq. 11.1.17 define a fourth-rank tensor called the *elastic-compliance tensor*. Similarly, stress depends on strain through the 81 components of a fourth-rank *elastic stiffness tensor* (c_{ijkl}).

In the absence of body torques, the stress tensor in Eq. 11.1.17 is symmetric: $\sigma_{kl} = \sigma_{lk}$. Under these conditions,

$$s_{ijkl} = s_{ijlk} \quad (11.1.19)$$

Similarly, if conditions exist such that the strain tensor is also symmetric, we have

$$s_{ijkl} = s_{jikl} \quad (11.1.20)$$

Because of Eqs. 11.1.19 and 11.1.20, the number of nonzero independent components of the elastic-compliance tensor comes down from 81 to 36. This is intrinsic symmetry (§3.1.4).

The compliance tensor also possesses extrinsic symmetry. For example, if both the stress and the strain tensors can be assumed to be invariant under an inversion operation of the spatial coordinates, the compliance tensor would also have this symmetry. Inversion of coordinates therefore makes no difference to any component of the elastic-compliance tensor.

Other crystallographic operations, however, do impose restrictions on the components of the compliance tensor (in accordance with the Neumann theorem), bringing down the number of independent nonzero components. The results are tabulated in, for example, Nye (1957) and Sirotin & Shaskolskaya (1982).

Further restrictions are imposed by physical requirements. For example, the energy stored in a strained crystal must be positive definite if the configuration is to be stable one (see Nye 1957).

11.1.2 Absolute Spontaneous Strain

Strain in the absence of external stress is called spontaneous strain. This is also the meaning we assign in this book to *absolute* spontaneous strain. We append the adjective 'absolute' to make a distinction from what we call *relative* spontaneous strain in §11.1.3.

Strain must be defined with respect to a certain reference state in which all its components are zero. We can illustrate the notion of absolute spontaneous strain by considering the ferroelectric phase transition in TGS. The ferroic species involved is $2/mF2$. The emergence of spontaneous polarization P_2 in the polar phase of point-symmetry 2_y gives rise to a strain with reference to the prototypic phase (through the relation $e_{22} = Q_{2222}P_2^2$). This strain occurs in the absence of any applied stress, and is thus a spontaneous strain.

The ferroelectric phase of TGS has two orientation states, with space inversion as an F-operation that can map one state to the other. The two orientation states have parallel but oppositely oriented polarization vectors, and the F-operation reverses the orientation of this polarization when it maps one orientation state to the other.

The two states also possess (absolute) spontaneous strain, but the strain-tensor components are identical for them. This is because this strain varies as P^2, and not as $\mathbf{P}$. Since the two states do not differ in spontaneous strain, the polar phase of TGS is a nonferroelastic phase.

The F-operation is an inversion operation, and a strain tensor is invariant under this operation. Thus everything is self-consistent: The F-operation changes the sign of the polarization vector, as expected; and it does not alter any component of the spontaneous-strain tensor, again as expected.

The $R\bar{3}m \rightarrow R\bar{3}c$ phase transition in calcite provides an example of a nonferroelastic phase transition, which is nonferroelectric at the same time. In fact, it is a nonferroic phase transition because it does not involve any change of point-group symmetry. On account of ordering of the orientations of the carbonate groups, the symmetry changes from $R\bar{3}m$ to $R\bar{3}c$, accompanied by the emergence of absolute spontaneous strain. In this case there are no orientation states; only antiphase domains.

The need for making a distinction between absolute and relative spontaneous strain will become clear in the next section.

11.1.3 Relative Spontaneous Strain

The relative spontaneous strain tensor was introduced by Aizu (1970b) to meet the special requirements of ferroelastic state shifts. We consider the example of TGS again. As explained above, it is a nonferroelastic ferroelectric. Its two orientation states do not differ in spontaneous strain, so external stress cannot cause a state shift.

Other ferroic crystals in which the same situation occurs are SbSI (ferroic species $mmmFmm2$), and $LiNbO_3$ (ferroic species $\bar{3}mF3m$).

Clearly, the definition of spontaneous strain must be modified to exclude any orientational state shifts in such crystals from the ambit of ferroelastic state shifts.

Before describing Aizu's prescription, we draw attention to an additional problem which must also be attended to for defining spontaneous strain relevant to ferroelastic state shifts.

Strain, and therefore spontaneous strain, can be defined only with reference to some initial reference state. Suppose we define the reference state (with all elements of the spontaneous strain tensor equal to zero) at some particular temperature. The Curie temperature T_c can be a natural choice. However, because of thermal expansion, this reference state does not continue to have all its strain-tensor coefficients zero for $T > T_c$. This problem is more acute for ferroelasticity, and does not arise in the case of ferromagnetism. It also does not arise in the case of ferroelectrics with a nonpolar prototypic phase. This is a basic difference between spontaneous strain on one hand, and spontaneous polarization and spontaneous magnetization on the other. Given the point-group symmetry of a phase of a crystal, it is possible to tell whether or not this phase can have spontaneous polarization and/or spontaneous magnetization, and what are their magnitudes. The answer to this question does not require a reference to any prototype symmetry. For spontaneous strain, however, the questions cannot even be formulated adequately without reference to a prototype symmetry (Cracknell 1974).

Aizu (1970b) introduced a modified (i.e. relative) spontaneous-strain tensor which is appropriate for specifying ferroelastic state shifts uniquely. Aizu's formulation also includes a definition of *magnitude of spontaneous strain* for a full ferroelastic. All this is achieved by requiring that the magnitude of spontaneous strain be so defined as to be: (i) independent of the choice of the coordinate system; (ii) the same for all the orientation states of the full ferroelastic; and (iii) zero over the whole temperature range in the prototypic phase.

Consider a full ferroelastic with q orientation states $S_1, S_2, \cdots, S_q$. Let $\mathbf{e}(S_i)$ be the symmetry-adapted strain tensor for the orientation state S_i. The relative spontaneous strain tensor $\mathbf{e}_{(s)}(S_i)$ for this state is defined as (Aizu 1970b):

$$\mathbf{e}_{(s)}(S_i) = \mathbf{e}(S_i) - \frac{1}{q}\sum_{k=1}^{q} \mathbf{e}(S_k), \quad i = 1, 2, ..q \tag{11.1.21}$$

If the (i, j) element of $\mathbf{e}_{(s)}(S_k)$ is denoted by $e_{(s)ij}$, the magnitude e_s of relative spontaneous strain is defined by

$$e_s^2 = \sum_{i=1}^{3}\sum_{j=1}^{3} e_{(s)ij}^2 \tag{11.1.22}$$

These definitions satisfy all the three conditions stated above.

Application of Eq. 11.1.21 to the two orientation states of TGS gives us the result that the Aizu spontaneous strain for both the states is zero (even though they have nonzero absolute spontaneous strain).

For defining the Aizu (or relative) spontaneous strain for an orientation state, one takes the absolute spontaneous strain tensor for that state, and subtracts from it a tensor that is an average over the absolute spontaneous strain tensors of all the orientation states of the full ferroelastic. In other words, that part of the absolute spontaneous strain is subtracted that is common to all the orientation states.

In the case of TGS, the common part is equal to the full strain tensor itself, so the relative spontaneous strain part is zero, making it a nonferroelastic.

We consider the case of $SrTiO_3$ for illustrating the computation of the relative spontaneous strain tensor for a full ferroelastic. This crystal belongs to the ferroic species $m\bar{3}mF4/mmm$, with 48/16 or 3 orientation states. For any of the states, say S_1, the absolute spontaneous strain tensor can be written in the following form (which conforms to the symmetry $4/mmm$ of the ferroelastic phase):

$$\mathbf{e}(S_1) = \begin{pmatrix} e_{11} & 0 & 0 \\ 0 & e_{11} & 0 \\ 0 & 0 & e_{33} \end{pmatrix} \tag{11.1.23}$$

The absolute spontaneous strain tensors for the other two orientation states are as follows (Aizu 1970b; Wadhawan 1982):

$$\mathbf{e}(S_2) = \begin{pmatrix} e_{33} & 0 & 0 \\ 0 & e_{11} & 0 \\ 0 & 0 & e_{11} \end{pmatrix}, \tag{11.1.24}$$

$$\mathbf{e}(S_3) = \begin{pmatrix} e_{11} & 0 & 0 \\ 0 & e_{33} & 0 \\ 0 & 0 & e_{11} \end{pmatrix}, \tag{11.1.25}$$

The average or mean (absolute) strain tensor over the three states is

$$\mathbf{e}_m = \frac{1}{3}\begin{pmatrix} 2e_{11}+e_{33} & 0 & 0 \\ 0 & 2e_{11}+e_{33} & 0 \\ 0 & 0 & 2e_{11}+e_{33} \end{pmatrix} \tag{11.1.26}$$

By subtracting this tensor from the absolute spontaneous strain tensors given by Eqs. 11.1.23-25, we obtain the relative spontaneous strain tensors for the three states:

$$\mathbf{e}_{(s)}(S_1) = \begin{pmatrix} B & 0 & 0 \\ 0 & B & 0 \\ 0 & 0 & -2B \end{pmatrix}, \tag{11.1.27}$$

$$\mathbf{e}_{(s)}(S_2) = \begin{pmatrix} -2B & 0 & 0 \\ 0 & B & 0 \\ 0 & 0 & B \end{pmatrix}, \tag{11.1.28}$$

$$\mathbf{e}_{(s)}(S_3) = \begin{pmatrix} B & 0 & 0 \\ 0 & -2B & 0 \\ 0 & 0 & B \end{pmatrix}, \tag{11.1.29}$$

where

$$B = \frac{1}{3}(e_{11} - e_{33}) \tag{11.1.30}$$

The magnitude of relative spontaneous strain, as defined by Eq. 11.1.22, can be calculated from any of the Eqs. 11.1.27-29. In each case we get

$$e_s = \sqrt{6}\,|B| = \sqrt{2/3}\,|(e_{11} - e_{33})| \tag{11.1.31}$$

In the cubic phase, $e_{11} = e_{33}$, giving $B = 0$ over the entire temperature range of this prototypic phase.

Further, none of the nonzero elements of the relative spontaneous strain tensor has the same value in *all* the orientation states. These elements can thus be regarded as a kind of thermodynamic *state parameters* for full ferroelastics. They have been derived and tabulated for each of the 94 species of full ferroelastics by Aizu (1970b).

Salje (1993a) has given details for determining the elements of the spontaneous strain tensor in terms of the lattice parameters of a crystal.

11.1.4 Anelasticity

The movement or migration of atoms, defects (including domain boundaries and phase boundaries) through a material on application of stress depends not only on the stress applied, but also on time. The strain produced is not instantaneous, but lags behind the applied stress, and may approach the equilibrium value only asymptotically. The elastic (i.e. recoverable) dependence of strain on both stress and time is called anelasticity. *Elastic after-effect* is another term used, with practically the same meaning.

And the term *creep* is used for the time-dependent strain that is plastic or nonrecoverable.

Anelastic effects in a crystal involve two types of diffusion: diffusion of thermal energy, and diffusion of atoms, with corresponding thermal hysteresis and mechanical hysteresis effects.

Thermal hysteresis arises when the loading and/or unloading rates are too fast compared to processes such as thermal conduction.

Mechanical or elastic hysteresis is associated mainly with the diffusion rates of foreign atoms or other defects in the crystal, and depends strongly on temperature. In addition, there may also be a defect-independent ferroelastic response in certain crystals.

When a small compressive stress is applied to a defect-free nonferroelastic crystal, its length along the direction of the stress decreases, with a corresponding increase in lateral directions through the Poisson effect. When the stress is removed, such a crystal recovers its original shape quite quickly.

The situation changes very substantially if, for example, some interstitial atoms are present. Examples of such systems are: carbon atoms in a crystal of Fe; hydrogen atoms in a crystal of Nb; and basal-plane oxygen atoms in a $Y-Ba-Cu-O$ crystal (Wadhawan & Bhagwat 1989). When compressive stress is applied to such a crystal in a direction parallel to a unit-cell edge which has interstitial impurities, the rate of compressive deformation is affected by the rate at which the interstitial atoms can hop to interstitial sites in the plane normal to the direction of the compressive stress. Similarly, application of compressive stress in the lateral plane forces the interstitial impurities to hop back to their original sites (in a statistical sense). This is the essence of ferroelastic hysteresis described by Alefeld (1971) for some metal-hydrogen systems.

Snoek Relaxation and Gorsky Relaxation

Consider a crystal which is at equilibrium, and on which no stress or any other field has been applied. Let us assume that a small number of point defects (e.g. interstitial impurities or dopant atoms) have been introduced in the crystal. These point defects are assumed to be distributed randomly over crystallographically equivalent sites. Let us now apply a uniaxial stress. The crystal will respond by undergoing *stress-induced ordering* of these point defects by a mechanism whereby an impurity atom makes a specific choice from among the crystallographically equivalent sites available to it for occupation (Nowick & Heller 1963). Two main types of relaxation phenomena, with very different time scales, are involved, which we describe briefly.

A point defect in a crystal creates a strain field around it. The local distortion produced in the lattice can be visualized as an *elastic dipole* described by a polar second-rank tensor (Zener 1948; Nowick & Heller 1963; Kroner 1964; Alefeld 1971).

The concept of the elastic dipole has been introduced by analogy with electric dipoles. In fact, they are more similar to magnetic dipoles, than to electric dipoles. But there are important differences among all three. While electric monopoles (charges) exist, magnetic and elastic monopoles do not. The elastic dipole is described by a polar second-rank tensor. By contrast, the electric dipole is described by a polar first-rank tensor, and the magnetic dipole by an axial first-rank tensor. Whereas electric and elastic dipoles possess time-inversion symmetry, magnetic dipoles do not. There are important consequences of all these differences. For example, since both stress and strain are second-rank tensors, they are connected by a fourth-rank tensor, namely the compliance tensor. The higher ranks of the tensors involved result in a much richer and more complex mechanical behaviour compared to the electric and magnetic analogues.

The existence of a point defect and the elastic field associated with it usually lowers the local symmetry of the crystal lattice. Being a symmetric second-rank polar tensor, the general representation surface often associated with the strain tensor is an ellipsoid. Under the action of external stress, the elastic dipoles get reoriented, typically over a picosecond time scale. This *orientational relaxation* is called Snoek relaxation (Snoek 1941, 1942).

By and large, interstitial defects cause an expansion of the crystal. By contrast, the effect of substitutional defects on the density of the crystal may be quite small. The actual situation, of course, is a complex interplay of these two types of effects. If there is a net expansion or contraction, then an external stress field, producing a gradient in this dilatation or contraction, results in *diffusional relaxation* or Gorsky relaxation (Gorsky 1935; Alefeld, Volkl & Schaumann 1970).

Whereas a lowering of the local site symmetry around the defect site is a prerequisite for the occurrence of the Snoek effect, this is not the case for the Gorsky effect.

The Snoek and Gorsky relaxations can be distinguished experimentally on the basis of the time scales involved. To take the example of hydrogen in Nb, the typical time for the Snoek hopping of hydrogen atoms under stress is of the order of $10^{-12} - 10^{-11}$ seconds. On the other hand, Gorsky diffusion over a crystal size of 1 mm needs $10^{12} - 10^{13}$ hoppings of a point defect, which works out to a relaxation time of the order of 10 to 100 seconds.

Lattice Gas

Another relevant concept in the present context is that of a lattice gas (Lee & Yang 1952). A lattice gas is a gas of particles constrained to exist only on lattice sites. Hydrogen atoms in a crystal of Nb can be viewed as constituting a lattice gas (Alefeld et al. 1969). Their presence and mutual interactions modify the compliance tensor of the host crystal. They also display critical phenomena and phase transitions, just as ordinary gases do at high-enough densities and low-enough temperatures.

A *phase separation* occurs below a critical temperature. That is, there occur regions of increased hydrogen concentration (positive spontaneous strain), coexisting with regions of reduced hydrogen concentration (negative spontaneous strain).

SUGGESTED READING

J. F. Nye (1957). *Physical Properties of Crystals: Their Representation by Tensors and Matrices. The Clarendon Press, Oxford.*

A. S. Nowick & W. R. Heller (1963). Anelasticity and stress-induced ordering of point defects in crystals. *Adv. Phys.*, **12**, 251.

H. W. Hayden, W. G. Moffatt & J. Wulff (1965). *The Structure and Properties of Materials. Vol. III: Mechanical Behaviour.* Wiley, New York.

G. Alefeld, G. Schaumann, J. Tretkowski & J. Volkl (1969). Ferroelasticity of niobium due to hydrogen as a lattice gas. *Phys. Rev. Lett.*, **22**, 697.

K. Aizu (1970b). Determination of state parameters and formulation of spontaneous strain for ferroelastics. *J. Phys. Soc. Japan*, **28**, 706.

S. Dattagupta & R. Ranganathan (1984). Linear response analysis of Gorsky relaxation of light interstitials in the presence of ordering. *J. Phys. F: Metal Phys.*, **14**, 1417.

E. K. H. Salje (1993a). *Phase Transitions in Ferroelastic and Co-elastic Crystals.* Student edition. Cambridge University Press, Cambridge.

11.2 STRUCTURAL CLASSIFICATION OF FERROELASTICS

Ferroic phase transitions can be either ferroelastic or nonferroelastic (cf. §5.2). Ferroelastic phase transitions involve a distortion of the shape of the crystallographic unit cell; e.g. a cube changing into a rhombus. What kind of crystal structures are more prone to spontaneous ferroelastic distortions? Which ferroelastic phases have large magnitudes of spontaneous strain? There are no systematic answers to such questions yet, although some progress has been made in this direction.

Our understanding of the corresponding questions in the case of ferroelectric crystals is somewhat better. Abrahams & Keve (1971) conducted a fairly comprehensive survey of ferroelectric crystal structures and arrived at some general conclusions. For example, they could classify a large number of ferroelectrics into three categories: one-dimensional, two-dimensional, and three-dimensional, depending on whether the atomic displacement vectors (with reference to the prototypic structure; cf. §8.4) responsible for the net spontaneous polarization are confined predominantly to one direction, one plane, or not confined to any of these. They observed that the spontaneous polarization is the largest for the first category, and smallest for the third.

A different approach for a structural classification of ferroelastic crystals could be in terms of the chemical groupings of the atoms involved, to see if the relevant *elastoactive atomic groupings* can be located. A good attempt in this direction was made by Dudnik & Kiosse (1983), although much more needs to be done. Progress in this area can enable us to tailor-make ferroelastics for specific applications.

We summarize here the work of Dudnik & Kiosse (1983), whom we shall call DK for short. Their work deals only with inorganic nonferroelectric ferroelastics.

DK identify 15 *structural families of ferroelastics* according to their structural type. In addition, they mention 8 individual ferroelastics which cannot be assigned to these 15 families, and, when more examples are available, they may become representatives of additional structural families. These 8 crystals are: H_3BO_3, Sb_5O_7I, KIO_2F_2, $Mg_2B_2O_5$, PtGeSe, $K_2Ba(NO_2)_4$, $ZnGeF_6.6H_2O$, and $KClO_3$.

Two of these, namely H_3BO_3 (Wadhawan 1978b) and $KClO_3$ (Wadhawan 1980) have the special feature that the magnitude of spontaneous strain is extremely large for them. Another crystal for which this is the case is 9-hydroxy-1-phenalenone (Svensson & Abrahams 1984). Stress-induced creation and/or movement of domain walls is naturally more difficult in such crystals, particularly in thick specimens. It is possible that, in addition to genuine ferroelastic switching, the phenomenon of slip may also be involved, making the experimental study of their domain-wall movement a very complicated proposition.

DK conclude from their crystal-chemical survey that an elastoactive grouping of atoms can normally be identified in each of the 15 ferroelastic families. These groupings of atoms can easily change their orientation and shape under an applied stress. They usually include the anion or cation complexes which are dominant in determining the packing of the building blocks of the crystal structure; hence their key role in determining the ferroelastic response to the applied stress.

Most of the 15 ferroelastic families and some of the 8 other special types of pure ferroelastic crystals can be assigned to one or the other of five crystal-chemical classes. Very briefly, these classes are as follows:

1. Ferroelastics containing isolated or 'zero-dimensional' (mostly tetrahedral) anionic complexes.

2. Ferroelastics with anion or cation complexes connected to each other and forming infinite chains, bands, or layers.

3. Ferroelastics with anion complexes connected by hydrogen bonds into chains, bands or layers.

4. Ferroelastics with a 3-dimensional framework formed by corner-sharing anionic octahedral complexes.

5. Ferroelastic molecular-ionic crystals.

Class 1 incorporates seven of the 15 ferroelastic families: palmierites; fergusonites; teilorites; langbeinites; complex cyanides; and double trigonal molybdates and tungstates. We mention one example from each of these seven families in the same sequence: $Pb_3(PO_4)_2$, $BiVO_4$, K_2CrO_4, $LiTlSO_4$, $K_2Mn_2(SO_4)_3$, $K_2Hg(CN)_4$, and $NaFe(MoO_4)_2$.

Class 2 covers four of the 15 families: pentaphosphates (e.g. LaP_5O_{14}), fresnoites (e.g. $Ba_2TiGe_2O_8$), $M_4A(XO_4)_3$ (e.g. $K_4Zn(MoO_4)_3$), and ditellurites (e.g. $SrTe_2O_5$). This class also includes Sb_5O_7I.

Class 3 incorporates the alkali trihydro-selenites (e.g. $KH_3(SeO_3)_2$), and also H_3BO_3.

Class 4 comprises the perovskite-family ferroelastics, an example being $LaAlO_3$. Elpasolites like $Cs_2NaBiCl_6$ are taken as a subfamily of this family.

Lastly, Class 5 consists of the calomel family, represented by Hg_2Cl_2.

There is some difficulty in putting compounds of the type $Pb_8X_2O_{13}$, with $X = P, V, As$, into any of the above classes. They can perhaps be assigned to Class 1.

SUGGESTED READING

S. C. Abrahams & E. T. Keve (1971). Structural basis of ferroelectricity and ferroelasticity. *Ferroelectrics*, **2**, 129.

E. F. Dudnik & G. A. Kiosse (1983). The structural peculiarities of some pure ferroelastics. *Ferroelectrics*, **48**, 33.

11.3 FERROELASTIC PHASE TRANSITIONS

Ferroelastic phase transitions are ferroic phase transitions involving a change of the crystal family. Our crystallographic classification of phase transitions, described in §5.2, divides all ferroic transitions into two classes: ferroelastic and nonferroelastic-ferroic, thus assigning a pride of place to ferroelastic transitions (compared to other ferroic transitions like ferroelectric or ferromagnetic). There is a fundamental reason for doing this, which is connected with the nature of critical fluctuations in the vicinity of such transitions (Bruce 1976; Cowley 1976; Folk, Iro & Schwabl 1976a, b, 1979). For a continuous or quasi-continuous phase transition the fluctuations of the order parameter become increasingly large and correlated as T_c is approached. The fluctuations are 'premonitory' in nature, in the sense that small regions of the crystal are locally transformed into the new phase even before the critical point is actually reached. It is as if the structure has a premonition of the phase it is heading for. If homogeneous strain is the order parameter (as for proper ferroelastic phase transitions), the appearance of these fluctuating regions of the new (ferroelastic) phase results in a lattice mismatch between the new phase and the old phase, which has a curbing effect on the critical fluctuations. [Thus, these fluctuations are not only premonitory, they are also suicidal !] Moreover, the strain fields are of a long-ranged nature. For all these reasons, ferroelastic phase transitions, particularly proper ferroelastic transitions, are well-described by mean-field theories like the Landau theory (Cowley 1976; Salje 1991). This is irrespective of whether or not the transition is also concurrently a ferroelectric transition. Thus, ferroelastic phase transitions are a class apart, as reflected in our classification.

11.3.1 True-Proper and Pseudoproper Ferroelastic Phase Transitions

Ferroelastic phase transitions with homogeneous strain as the order parameter, or with homogeneous strain having the same symmetry as the actual order parameter, are called proper ferroelastic phase transitions.

If the order parameter can be identified with a component of the strain tensor, we speak of a *true-proper ferroelastic transition.*

If the primary instability driving the proper ferroelastic transition is some parameter other than homogenous strain, but the strain emerging at the phase transition has the same symmetry as the order parameter, we speak of a *pseudoproper ferroelastic transition.*

If a true-proper ferroelastic transition is driven by a lattice-dynamical soft mode, the soft mode is a zone-centre ($k = 0$) acoustic phonon.

When an optical phonon mode softens, its frequency tends to zero. But the frequency of an acoustic phonon is anyway zero at $k = 0$. What then is the meaning of softening of such a mode? The softening of such a mode means that its *velocity* ($\partial\omega/\partial k$) tends to zero at the ferroelastic transition (Rehwald 1973; Cowley 1976). Stated differently, although the frequency of an acoustic phonon mode is zero *at* $k = 0$, when such a mode softens, its frequency becomes zero even in the neighbourhood of $k = 0$.

True-Proper Ferroelastic Transitions and Critical Fluctuations

A ferroelastic phase transition involves a lattice mismatch between the parent phase and the daughter phase.

Pretransitional (or premonitory) fluctuations set in even before the critical temperature T_c is actually reached. Because of the lattice mismatch the critical fluctuations tend to be self-curbing (suicidal), making critical phenomena less prominent for true-proper ferroelastic transitions. If we add to this the fact that strain fields are long-ranged fields, mean-field theories like the Landau theory can be expected to be more applicable to them than to other types of phase transitions (Patashinskii & Pokrovskii 1979). This is indeed found to be the case by detailed analyses (Cowley 1976; Folk, Iro & Schwabl 1976a, b, 1979; Als-Nielsen & Birgeneau 1977; Salje 1991).

As stated earlier, softening of an acoustic mode (at $k = 0$) means that its velocity becomes vanishingly small. If we couple this to the fact that acoustic velocities in crystals are generally direction-dependent, we arrive at the possibility that mode softening may occur in certain selected directions only. In fact, it is possible to split the d-dimensional acoustic wavevector

k into an m-dimensional *soft* component and a $(d - m)$-dimensional *hard* component. Renormalization-group considerations then lead to the following expression for the upper marginal dimensionality (Folk et al. 1976a, b):

$$d_u = 2 + \frac{m}{2} \tag{11.3.1}$$

We recall from §5.5.5 that if the dimensionality d of a system is such that $d > d_u$, results of the Landau theory are valid, and there is no need for a recourse to a renormalization-group theory treatment. If $d = d_u$, the Landau theory works in a 'marginal' way, its results requiring only minor 'logarithmic' corrections (Wegner & Riedel 1973; Stephen, Abrahams & Straley 1975).

According to Eq. 11.3.1, for systems with $d = 3$ a mean-field theory should work well for a proper ferroelastic phase transition if $m = 1$; and it should require only small (logarithmic) corrections if $m = 2$.

The case $m = 0$ corresponds to no critical fluctuations with wavelengths less than the dimensions of the specimen crystal, and we can expect completely classical (mean-field) behaviour. And $m = 3$ (isotropic elastic crystal) will certainly entail nonclassical critical behaviour.

The case $m = 1$ corresponds to *one-dimensional soft sectors*; i.e. the wavevector of an acoustic soft mode is confined to specific directions in reciprocal space. Similarly, $m = 2$ means *two-dimensional soft sectors*, with the wavevector of the soft mode confined to specific planes of reciprocal space. Only those wavevector components are 'relevant' for the critical phenomena which approach the static spontaneous strain in the limit $k \to 0$. This restricts greatly the size of the correlated region in reciprocal space, thus bringing down the value of d_u for proper ferroelastic phase transitions.

True-Proper Ferroelastic Transitions

For these transitions, onset of homogeneous strain itself is the primary instability driving the transition. For quite some time, hardly any temperature-induced transition of this type could be identified, and the only known example was the pressure-induced transition in paratellurite, TeO_2 (Peercy & Fritz 1974; Worlton & Beyerlein 1975; Uwe & Tokumoto 1979). However, some genuine, temperature-induced, transitions of this type do occur in a few crystals. These crystals are: lithium ammonium tartrate monohydrate (LAT-monohydrate) (Sawada, Udagawa & Nakamura 1977; Sawada & Nakamura 1985); and 2,2,6,6-tetramethyl piperidino oxy(tanane) (Sawada & Nakamura 1985).

Since critical fluctuations are expected to be unimportant for true-proper ferroelastic transitions, the critical exponents predicted by the Landau theory apply. In particular, the order-parameter critical exponent β

has the value $\frac{1}{2}$ for a continuous transition of this type.

Since a component of the strain tensor is the order parameter, if extensive thermodynamic variables other than that conjugate to the spontaneous strain are held constant, the generalized susceptibility corresponding to the relevant strain component, i.e. a principal component of the elastic compliance tensor, is predicted by the Landau theory to diverge at T_c, with $\gamma = 1$ in Eq. 5.5.44. In fact, observation of such a divergence constitutes one part of the proof for the occurrence of such a transition. The other part of the proof is to demonstrate that there is no other (primary) instability (e.g. softening of an optical mode coupled to the acoustic mode) which forces the compliance coefficient(s) to diverge (see Sawada & Nakamura 1985, and Bulou et al. 1992).

Pseudoproper Ferroelastic Transitions

Most of the known proper ferroelastic transitions are of the pseudoproper variety, rather than of the true-proper variety. For the former, the order parameter has the same symmetry as the homogeneous strain arising at the transition, but the order parameter is not homogeneous strain itself. In many cases the order parameter is an optical soft mode, and the spontaneous strain arises because of its bilinear coupling with the order parameter; the faintness index is equal to unity.

Aizu (1971) divided ferroelastic transitions into the *elastic type* and the *optical type*. The former can now be identified with true-proper ferroelastic transitions. They represent "elastically soft and optically hard" behaviour near T_c. If they are continuous transitions, at least one sound velocity always passes through zero at T_c (Aubry & Pick 1971).

Ferroelastic transitions of the optical type occur in crystals which are "optically soft and elastically hard". We can regard them as a subclass of pseudoproper ferroelastic transitions.

The following crystals have been investigated for the occurrence of a pseudoproper ferroelastic transition in them:

$BiVO_4$ (David 1983; David & Wood 1983; Bulou et al. 1992).

$LaNbO_4$ (Brixner et al. 1977; Mariathasan, Finger & Hazen 1985).

LaP_5O_{14} and other rare-earth pentaphosphates (Weber, Tofield & Liao 1975; Toledano, Errandonea & Jaguin 1976; Errandonea 1980; Schwabl 1980).

$Na_5Al_3F_{14}$ (see Bulou et al. 1992).

Nb_3Sn (Schwabl 1980).

NaOH (Schwabl 1980).

In – Tl alloy (Liakos & Saunders 1982).

KDP (Brody & Cummins 1968).

$DyVO_4$, $TbVO_4$, $TmVO_4$ (Sandercock 1972).

$KH_3(SeO_3)_2$ (Shuvalov, Ivanov & Sitnik 1967; Toledano & Toledano 1980).

$CaCl_2$, $CaBr_2$ (Unruh 1993, 1995).

Betaine fumarate (Unruh 1995).

Betaine borate (Unruh 1995).

For $BiVO_4$, $LaNbO_4$, LaP_5O_{14} and $Na_5Al_3F_{14}$, softening of an optical phonon mode provides the order parameter, and the spontaneous strain arises as a result of its bilinear coupling with the optical mode.

However, other primary instabilities are also possible. In KDP the phase transition is triggered by the ordering of protons. In $DyVO_4$ and other rare-earth vanadates the primary cause of the transition is the cooperative Jahn-Teller effect (see Gehring & Gehring 1975).

As in the case of pseudoproper ferroelectric phase transitions (§10.3.3), the presence of a bilinear coupling term between the strain and the order parameter in the Landau potential for a ferroelastic transition has the effect of shifting the transition temperature, the shift depending on the square of the coupling coefficient (see Bulou et al. (1992) for a detailed discussion). If the coupling coefficient is small, it becomes difficult to distinguish between a true-proper and a pseudoproper ferroelastic transition.

The case of $LaNbO_4$ presents a peculiar situation. Here, although there is an optical soft mode of the same symmetry as the spontaneous strain emerging at the phase transition, the optical mode undergoes only a small degree of softening.

11.3.2 Improper Ferroelastic Phase Transitions

Ferroelastic phase transitions in which the strain-tensor component emerging at the transition does not have the same symmetry as the order parameter are called improper ferroelastic transitions. The spontaneous strain arises as a result of its coupling with the order parameter, and the faintness index (§10.3.2) is equal to or greater than 2.

Since homogeneous strain is a macroscopic parameter, its symmetry in a ferroelastic phase cannot be the same as that of the order parameter of any antiferrodistortive (i.e. $k \neq 0$) ferroelastic transition. Therefore all antiferrodistortive ferroelastic transitions are necessarily improper. Even when $k = 0$ for a ferroelastic transition, it would still be improper-ferroelastic if the order parameter is a component of a tensor of rank other than 2, i.e. if the tensor rank is different from that of the strain tensor (Toledano & Toledano 1980; Izyumov & Syromyatnikov 1990). The cubic-tetragonal phase transition in $BaTiO_3$ is an example of this, wherein the order parameter (P_3) is a component of a tensor (polarization) of rank 1.

Most of the ferroelastic transitions are of the improper type (Toledano & Toledano 1980). Those which are not simultaneously ferroelectric, i.e. the so-called purely ferroelastic transitions, have been analyzed by Toledano & Toledano (1980), who have also tabulated a large number of crystals in which they occur. Additional examples of improper ferroelastics are: $LaAlO_3$, $RbCaF_3$, $KMnF_3$, $RbAlF_4$, $TlAlF_4$, ReO_3, MF_3 (with M = Al, Ga, Cr, ..), $CoZrF_6$, $ZnZrF_6$ (see Bulou et al. 1992), $Rb_3H(SeO_4)_2$ (Schranz et al. 1991) and GMO (see Bulou et al. (1992) for a large list of references on this well-investigated improper ferroelectric-ferroelastic).

We dwell on the example of GMO to illustrate a number of typical features of improper ferroelastics.

GMO undergoes an antiferrodistortive (cell-doubling) phase transition at 432 K, the symmetry group of the crystal changing from $P\bar{4}2_1m$ (D^3_{2d}) to $Pba2$ (C^8_{2v}) (Borchardt & Bierstedt 1966, 1967). The order parameter for this transition is a doubly degenerate soft mode corresponding to the $(\frac{1}{2}, \frac{1}{2}, 0)$ point, or the M-point, of the Brillouin zone of the parent phase (Axe, Dorner & Shirane 1971; Dvorak, Axe & Shirane 1972; Dvorak 1974).

Since there is a change of the crystal family from tetragonal to orthorhombic, the transition is ferroelastic.

And since it entails cell doubling, the soft mode involved is a zone-boundary mode, making it an improper ferroelastic transition.

Since the point group of the ferroic phase is a polar group, it is also a ferroelectric (*improper* ferroelectric) transition. What lends further support to such a conclusion is the absence of large dielectric anomalies in the vicinity of the transition (Cross, Fouskova & Cummins 1968; Levanyuk & Sannikov 1970; Pytte 1970; Dvorak 1971a, b).

Because of the improper character of the transition not only with respect to ferroelectricity, but also with respect to ferroelasticity, the elastic response also shows a non-divergent behaviour in the vicinity of the transition.

For an antiferrodistortive transition, all macroscopic physical properties emerging in the ferroic phase as a result of the transition, i.e. arising as a result of their coupling with the order parameter, are faint variables. For most of the improper ferroelastic transitions the faintness index for strain is expected to have the value 2 (from group-theoretic analysis), although the values 3 and 4 are also possible in a few cases (Toledano & Toledano 1980; Aizu 1973b; Janovec, Dvorak & Petzelt 1975).

A faintness index of 2 means that the spontaneous strain is coupled to the square of the order parameter. Since the critical exponent β for the order parameter has the value $\frac{1}{2}$ in the Landau theory, this coupling implies that the strain varies linearly with temperature for $T < T_c$.

Antiferrodistortive phase transitions are not necessarily driven by long-ranged interactions (because the wavevector **k** is not zero for them). One would therefore expect that critical phenomena would be more important for them, and thence for improper ferroelastic transitions. However, the work of Salje (1990, 1991) indicates that practically all ferroelastic transitions are well-described by mean-field theories. This is related to the long-ranged nature of strain fields.

SUGGESTED READING

W. Rehwald (1973). The study of structural phase transitions. *Adv. Phys.*, **22**, 721.

R. A. Cowley (1976). Acoustic phonon instabilities and structural phase transitions. *Phys. Rev. B*, **13**, 4877.

R. Folk, H. Iro & F. Schwabl (1976). Critical statics of elastic phase transitions. *Z. Physik B*, **25**, 69.

F. Schwabl (1980). Elastic phase transitions. *Ferroelectrics*, **24**, 171.

J.-C. Toledano & P. Toledano (1980). Order parameter symmetries and free energy expansions for purely ferroelastic transitions. *Phys. Rev. B*, **21**, 1139.

J. K. Liakos & G. A. Saunders (1982). *Phil. Mag. A*, **46**, 217.

A. Sawada & T. Nakamura (1985). Intrinsic ferroelastic transition from a piezoelectric paraelastic phase. In G. W. Taylor et al. (Eds.), *Piezoelectricity*. Gordon & Breach, London. This paper discusses the physical origin of a proper ferroelastic transition.

E. K. H. Salje (1990, 1993a). *Phase Transitions in Ferroelastic and Co-elastic Crystals.* Cambridge University Press, Cambridge. This is a relatively advanced text, laying less emphasis on the symmetry aspects of ferroelastic transitions, and more on the real structure of ferroelastic phases of crystals.

E. K. H. Salje (1992). Application of Landau theory for the analysis of phase transitions in minerals. *Phys. Reports*, **215**, 49.

A. Bulou, M. Rousseau & J. Nouet (1992). Ferroelastic phase transitions and related phenomena. *Key Engg. Materials*, **68**, 133.

W. Schranz (1994). Static and dynamic properties of the order-disorder phase transition in KSCN and related crystals. *Phase Transitions*, **51**, 1.

H.-G. Unruh (1995). Soft modes at ferroelastic phase transitions. *Phase Transitions*, **55**, 155.

W. Schranz, M. Fally & D. Havlik (1998). Non-exponential relaxation in macroscopic susceptibilities. *Phase Transitions*, **65**, 27.

11.4 QUADRUPOLAR GLASSES

The general term 'orientational glasses' (Sullivan et al. 1978) is used for dipolar, quadrupolar, and higher multipole-moment glasses which are the electrical and/or mechanical analogues of magnetic spin glasses. We have already considered spin glasses in §9.2, and dipolar glasses in §10.4. We take a brief look at quadrupolar glasses in this section.

Quadrupolar glasses are mixed crystals with a random distribution of local strain fields, which undergo a glass transition at a certain low temperature. At high temperatures the quadrupolar moments associated with the molecules or molecular ions can rotate quite freely. At low-enough temperatures they undergo a transition to a phase which has orientational order.

Examples of such host crystals are KCN, N_2 and ortho-H_2. Their long-range ordering can be disturbed by introducing atoms which either differ in size (e.g. Na^+ ions replacing some of the K^+ ions in KCN), or result in the replacement of some of the quadrupolar units by non-quadrupolar units. Examples of the latter kind are: random replacement of some of the CN^- ions in KCN by Br^- ions; random replacement of N_2 by Ar; and random replacement of ortho-H_2 by para-H_2. We shall concentrate on KCN because it is also a ferroelastic crystal.

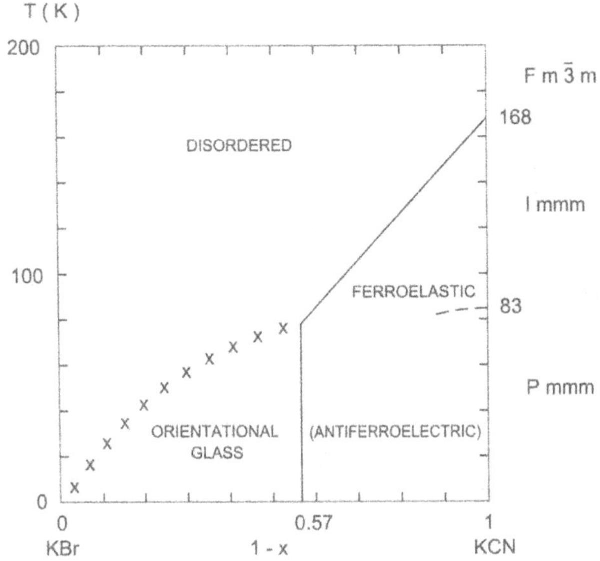

Figure 11.4.1: Phase diagram for $K(CN)_{1-x}Br_x$. [After Anderson (1985).]

The CN^- molecular ions in KCN have a small (electric) dipole moment, and the dipole-dipole interaction is quite weak compared to the elastic quadrupole interaction. The mixed crystal $K(CN)_{1-x}Br_x$ has been investigated very extensively because the rod-shaped CN^- ion has a size similar to that of the Br^- ion, and therefore good-quality crystals can be grown easily over the entire range of compositions $0 \leq x \leq 1$.

Fig. 11.4.1 shows the phase diagram for this system. Pure KCN has cubic symmetry $Fm\bar{3}m$ above 168 K (Durant et al. 1980). The phases below 168 K are ferroelastic. The symmetry of the phase between 168 K and 83 K is $Immm$. At 83 K the crystal undergoes an antiferroelectric phase transition, with symmetry changing to $Pmmm$.

The transition from $Fm\bar{3}m$ to $Immm$ symmetry is the result of aligning of the axes of the CN^- groups by the elastic quadrupolar interaction, without introduction of any electric dipolar order. The transition from $Immm$ to $Pmmm$ symmetry is the result of antiferroelectric ordering of the CN^- groups.

We next see the effect of randomly substituting Br^- ions at CN^- sites. In $K(CN)_{1-x}Br_x$, so long as $(1-x) > 0.57$, the crystal still possesses long-range order. Below this limit, and for sufficiently low temperatures (cf. Fig. 11.4.1), a glassy phase is more stable.

This glassy phase (or quadrupolar-glass phase) exhibits the usual universal properties of a glasses, like the linear temperature dependence of specific heat, and T^2 dependence of thermal conductivity at low temperatures (cf. §9.2.4).

Sethna et al. (1984) gave a mean-field theory for the ferroelastic phase transition in pure KCN. This theory was extended to the glassy state of $K(CN)_{1-x}Br_x$ for explaining some of its properties in terms of the quadrupolar interaction (Sethna & Chow 1985). The quadrupole is defined in terms of the vector **n** pointing from the C to the N atom in $(CN)^-$. Experiments indicate that 180° flipping of the $(CN)^-$ dipoles dominates the dielectric-loss mechanism even in the glassy phase (Bhattacharya et al. 1982; Birge et al. 1984), implying that the difference between the **n** and -**n** configurations can be ignored, and the Hamiltonian must be invariant under an inversion of the vector **n**.

In the mean-field approximation, a cyanide ion experiences an effective quadrupolar field **Q** due to long-ranged elastic forces transmitted by lattice strains. This field is defined by (Chowdhury 1986):

$$Q^i_{\alpha\beta} = [n^i_\alpha n^i_\beta - \frac{1}{3}\delta_{\alpha\beta}] \tag{11.4.1}$$

The Hamiltonian is expressed as

$$\mathcal{H} = -\frac{1}{2}\sum Q^i_{\alpha\beta}\, Q^j_{\alpha\beta}\, c_i c_j, \tag{11.4.2}$$

where c_i and c_j are the occupation probabilities for the sites i and j.

In spite of the fact that this mean-field theory ignored frustration (§9.2.6), it could explain correctly the distribution of barrier heights hindering the rotation of the cyanide ions.

A model incorporating frustration was formulated by Kanter & Sompolinsky (1986). Several other models have been reviewed by Binder & Reger (1992). The problem is inherently very complex, and a general theory does not exist yet, although several useful results have been obtained by adopting a mean-field approach; this approach is not too bad because of the generally long range of the elastic interaction.

Although analogies have been drawn from the Sherrington-Kirkpatrick-Parisi formalism developed for spin glasses (§9.2.7-8), one basic difference is that, whereas spin-glass systems possess inversion symmetry for the spins, the orientational glasses lack this symmetry. The effect of competing interactions leading to frustration is also very different here. The frustration effect in quadrupolar glasses generally decays rapidly with the length of the frustration loop (cf. Fig. 9.2.6), and is thus not present at all length scales.

On the whole the mean-field approximation has a better validity for quadrupolar glasses, than for spin glasses.

SUGGESTED READING

J. P. Sethna & K. S. Chow (1985). Microscopic theory of glassy disordered crystals: $(KBr)_{1-x}(KCN)_x$. *Phase Transitions*, **5**, 317.

U. T. Hochli, K. Knorr & A. Loidl (1990). Orientational glasses. *Adv. Phys.*, **39**, 405.

K. Binder & J. D. Reger (1992). Theory of orientational glasses: Models, concepts, simulations. *Adv. Phys.*, **41**, 547.

W. Schranz, M. Fally & D. Havlik (1998). Non-exponential relaxation in macroscopic susceptibilities. *Phase Transitions*, **65**, 27.

11.5 MARTENSITIC PHASE TRANSITIONS

11.5.1 General Features

Typically, a martensitic phase transition is a diffusionless, first-order, structural phase transition in which the kinetics and morphology are dominated by the strain energy arising from 'shear-like' displacements of atoms. Spontaneous shear strain is an essential feature of martensitic transitions. Since spontaneous shear strain is also central to ferroelastic phase transitions, it is natural to wonder about the relationship, or the degree of overlap, between these two kinds of phase transitions (Wadhawan 1985). It turns out that some types of martensitic transitions are of the nondisruptive type (cf. §5.2), and therefore are no different from ferroelastic transitions. The rest are of the disruptive type, and cannot be equated with ferroelastic transitions. This situation is depicted in our Venn-Euler diagram for phase transitions (Fig. 5.2.2), wherein a certain (small) subset of martensitic transitions is shown to overlap with a subset of ferroelastic transitions.

In martensitic transitions there is a definite orientational relationship between the parent phase and the daughter phase. There is a definite planar interface (called the *habit plane*) between the two phases. Theories of martensitic transitions lay great stress on the notion of *invariant-plane strain*, as the habit plane is found to remain invariant with respect to spontaneous shear and rotation relative to the crystallographic axes of the two phases (Roytburd 1978, 1993; Wayman 1981).

The martensitic phase transition has been exploited for several decades for the quench-hardening of steels. The basic transition, on cooling, is from the parent phase *austenite* to the daughter phase *martensite*. The transition involves both a change of density, and a shearing of the lattice.

Martensitic transitions have been investigated in a large number of other alloys also, and, more recently, even in some nonmetallic systems. The terms 'austenitic' and 'martensitic' are commonly used in a generic sense, to apply to all transitions similar to the austenite-martensite transition in steel, referring respectively to the high-temperature phase and the low-temperature phase.

In the metals and alloys investigated the austenitic phase commonly has bcc symmetry, which is a more open structure compared to the close-packed fcc or hcp structures.

Spontaneous strain plays a central role in determining all the main features of martensitic phase transitions. Such transitions can be therefore divided into three types: M_1, M_2 and M_3, depending on the magnitude of the spontaneous strain (Krumhansl 1989; Lindgard & Mouritsen 1990; Izyumov, Laptev & Syromyatnikov 1994). Type M_1 transitions involve small spontaneous strains and are nondisruptive. The other two types are disruptive, with M_3 involving very large strains, and M_2 involving moderate strains. The M_1 type are the same as quasi-continuous ferroelastic transitions.

The classical Landau theory can be applied to M_1 transitions (Horovitz, Murray & Krumhansl 1978; Cao & Barsch 1990), and not at all to the M_3 type. With due care the M_2 type is sometimes amenable to such a treatment (Gunton & Saunders 1973; Madhava & Saunders 1976).

Fig. 11.5.1 shows schematically what happens to properties like electrical resistance, or change of length, or change of volume, when an austenitic phase of an alloy is cooled through the martensitic phase transition.

At a temperature M_s, called the *martensite-start temperature*, the martensitic phase starts forming. On further cooling, more and more of martensite forms, till a temperature called *martensite-finish temperature*, M_f, is reached. Typically, no diffusion of atoms is involved, and the transition is *athermal*. If we dwell at some temperature between M_s and M_f, the fraction of martensite formed stops changing. The transformed fraction increases further if the cooling is resumed. For the M_3 type transition, because of the large strains involved, the transformed fraction increases (on cooling) predominantly in a burst-like manner, by the sudden appearance of new plate-like regions of martensite, rather than by growth of the existing regions. The difference $M_s - M_f$ is typically about 20 K, and in this temperature range the austenitic phase and the martensitic phase coexist.

When the martensitic phase is heated, the austenitic phase starts forming at a temperature A_s (*austenite-start temperature*), which is higher than M_s. On further raising of temperature, the fraction of the austenitic phase continues to increase till a temperature A_f, called *austenite-finish temper-*

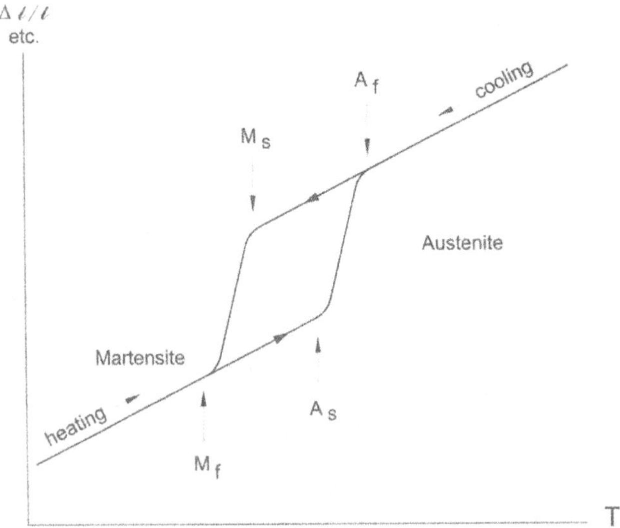

Figure 11.5.1: Typical plot of the temperature variation of some property like fractional change of length ($\Delta l/l$), or volume ($\Delta V/V$), or electrical resistance R, for a thermally driven martensitic phase transition. [After Wayman (1992).]

ature, is reached.

Except for the M_1-type martensitic transitions, the occurrence of a lattice-dynamical soft mode may not be invoked, in general, for explaining their mechanism. Since they occur at high temperatures and involve large displacements of atoms, anharmonic effects are expected to play a dominant role.

Lindgard & Mouritsen (1990) explained the occurrence of martensitic transitions in M_2 and M_3 systems in terms of an interplay between two fluctuating strain components, the fluctuations being caused by the anharmonicity of the interactions involved. This two-strain theory also explained the occurrence of pre-transition phenomena at temperatures far above M_s. Such precursor phenomena (e.g. appearance of distorted clusters of the martensitic phase) are easier to appreciate in terms of critical fluctuations heralding *continuous* phase transitions. Their occurrence for a strongly discontinuous transition, involving atomic displacements comparable to unit-cell dimensions, is indeed a peculiar feature of M_2 and M_3 type martensitic transitions. They are explained as metastable formations (rather than critical fluctuations) resulting from the strong anharmonicity of the operative interactions.

11.5.2 Pseudoelasticity and Pseudoplasticity

Pseudoelasticity

Since strain energy plays a dominant role in the occurrence and kinetics of a martensitic phase transition, it is only natural that application of mechanical stress has a dramatic effect on the entire process.

Assuming that the martensitic phase has a higher density than the austenitic phase, stress has the effect of raising M_s to a higher value, say M_s^s. We thus have a stress-induced martensitic transition, starting at a temperature M_s^s. Because of this transition, the strain produced in the specimen is much more (as high as 16%), than what can be expected from a purely elastic response. And when the stress is released, the reverse phase transition occurs, and the highly deformed specimen rebounds to its original shape and size. The terms *pseudoelasticity* or *superelasticity* are used for such behaviour (see, for example, Boyko, Garber & Kossevich 1994).

Since this process occurs as a consequence of the stress-induced phase transformation, one speaks of *pseudoelasticity by transformation* (Fig. 11.5.2 (a)). The term superelasticity is used for it because of the very large elastic response to stress.

If the external stress is applied at a temperature at which the system is already in the martensitic phase, we have the phenomenon of *pseudoelasticity by reorientation*, which is no different from ferroelastic switching (Fig. 11.5.2(b)).

Pseudoplasticity

The word 'elasticity' has two different connotations, depending on the context. We may use it to specify the nature of an interaction. For example, an interaction may be elastic (or mechanical), rather than, say, electric or magnetic.

Alternatively, we may wish to make a distinction between elasticity and plasticity. If a body changes its shape and size when a stress is applied, and bounces back to its original shape and size when the stress is removed, we speak of elastic behaviour (which may be linear or nonlinear). If the property of bouncing back or regaining the original configuration on release of external stress is missing, the response is said to be plastic, rather than elastic.

The term pseudoelasticity has been used for the two systems shown in Fig. 11.5.2 because in each case, on application of stress, the new configuration is surrounded by the old configuration, which provides a restoring

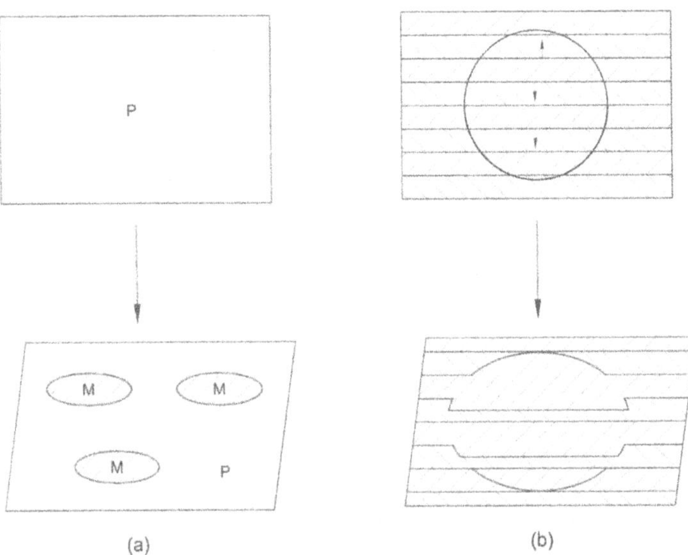

Figure 11.5.2: (a) Pseudoelasticity by transformation (or superelasticity), and (b) pseudoelasticity by reorientation (or ferroelasticity).

force, roughly proportional to the volume fraction of the new configuration. When the external stress is removed, the restoring force causes the material to exhibit the 'bouncing back', or 'elastic', or 'rubber-like' feature (Lieberman et al. 1975).

By contrast, Fig. 11.5.3 depicts situations in which there is a single, planar interface between the initial and the transformed or reoriented configuration, and there is no rubber-like or bouncing-back action expected when the external stress is released. We therefore have the phenomena of *pseudoplasticity by transformation* (part (a)), and *pseudoplasticity by reorientation* (or 'ferroplasticity') (part (b)) (Warlimont 1976).

In physics (though not so much in physical metallurgy) the term ferroelasticity is frequently used to cover both ferroelasticity and ferroplasticity.

11.5.3 Crystallographic Reversibility of a Phase Transition

When a symmetry operation is lost at a phase transition, the phase in which this symmetry element is absent develops domain types or variants related by this symmetry operation. In other words, starting from a single variant in the initial phase, one ends up with two or more variants in the

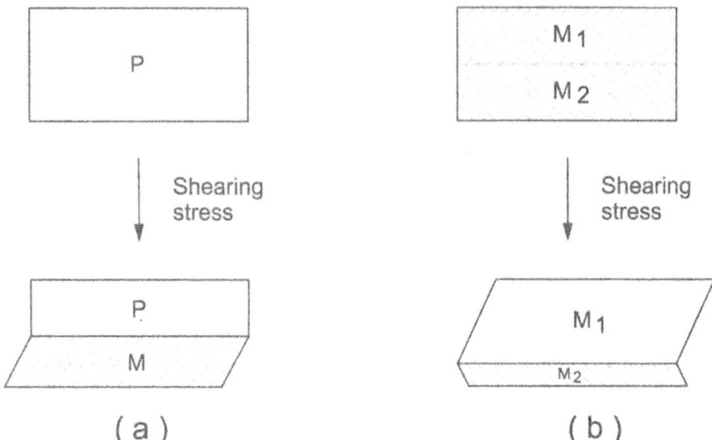

Figure 11.5.3: (a) Pseudoplasticity by transformation, and (b) pseudoplasticity by reorientation (also called 'ferroplasticity', or 'plasticity related to domain-wall movement').

final phase. If this final phase is made to undergo the reverse transition, the initial single-variant configuration may or may not be recovered. A phase transition is said to be crystallographically reversible if, on going through the forward and reverse transitions, one ends up with the same configuration of domain states one started with (Portier & Gratias 1982). The concept of crystallographic reversibility is particularly relevant in the context of the shape-memory effect associated with martensitic phase transitions (§11.5.4). We consider here an example of a crystallographically reversible phase transition. However, it is not a martensitic transition because there is no spontaneous strain involved.

The group-tree formalism described in §7.5 is ideally suited for analyzing the crystallographic reversibility of a phase transition. We apply it here to the transition $Fm\bar{3}m \leftrightarrow Pm\bar{3}m$, which occurs in Au_3Cu (cf §7.4.4).

The forward transition is brought about by merely cooling the crystal, with no anisotropic external influence present, so that the solicitation symmetry (g) is that of a sphere, i.e. SO_3. Therefore the group tree is as shown in Fig. 11.5.4(a). We see that $n_{01} = 4$, implying that, starting from a single variant in the initial phase, we end up with 4 variants in the final phase, described by Eq. 7.4.8.

We now consider the reverse transformation, depicted in Fig. 11.5.4(b). Since $n_{10} = 1$, we conclude that all the four variants of the ordered phase end up in the same unique variant of the disordered (prototypic) phase.

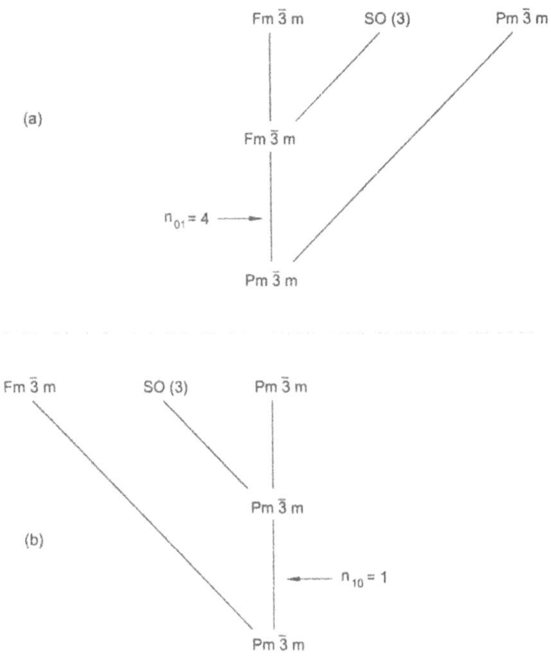

Figure 11.5.4: Group trees for the forward (a), and reverse (b), phase transition in the alloy Au_3Cu.

The phase transition from the disordered phase to the ordered phase and back is therefore crystallographically reversible.

11.5.4 Shape-Memory Effect

One-Way Shape-Memory Effect

The shape-memory effect (SME) is usually associated with a martensitic phase transition, although it can also occur in other situations. A material is said to exhibit the one-way SME if, after it has been pseudoplastically deformed, it recovers its original shape on heating slowly through the martensitic phase transition to the higher-symmetry phase.

In a typical SME cycle (Fig. 11.5.5(a)), a specimen is first cooled to the martensitic phase, resulting in the appearance of domain structure, possibly having a *self-accommodating* 'diamond' morphology for minimizing the overall strain energy. The specimen is then subjected to a deforming stress. The deformation is abnormally high for the stress applied, because of the property of pseudoelasticity by reorientation (cf. §11.5.2). This

Figure 11.5.5: The typical SME cycle. (a) One-way SME; (b) two-way SME.

deformation reduces the number of domains. Ideally, for a perfect SME, one would like this number to be just unity.

When the stress is removed, and the specimen heated back through the phase transition to the parent phase, it reverts to its initial shape, as if it remembers its initial shape in spite of the large deformation it went through.

If the deforming stress is applied at a temperature *above* M_s, then pseudoplasticity by transformation, as well as pseudoelasticity by transformation, also contribute to the set of processes responsible for the SME (Warlimont 1976).

Strains as high as 10% can be recovered in this manner in some systems exhibiting the one-way SME (Warlimont 1992).

There is a range of temperatures over which the austenitic and the martensitic phases coexist, with phase boundaries separating them. Further, there are domain boundaries within the martensitic phase. For good shape recovery it is necessary that the phase boundaries, as well as the domain boundaries, are created, moved, and/or annihilated in a *reversible* manner. A number of contributing factors must exist if this is to happen to a substantial extent (Nakanishi 1975; Wasilewski 1975).

For example, plastic deformation, which is by and large an irreversible process involving creation and/or movement of dislocations, should not occur. [By contrast, pseudoplastic deformation *can* be a reversible process in certain situations, involving movement of phase boundaries and domain boundaries.] Thus a high yield strength of the parent phase is desirable from this point of view. But this requirement clashes with the need for a low driving force for the nucleation and/or growth of the martensitic phase inside the austenitic phase. The main contribution to the enthalpy barrier opposing the nucleation of the martensitic plates in the parent matrix comes from a term of the form Ce_m^2, where C is the appropriate shear elastic stiffness constant of the parent phase, and e_m is the effective martensitic strain. If C is small, even a low value of applied stress can induce the nucleation of the martensitic phase, but a low C also means a low yield strength, with the attendant increased possibility of plastic deformation involving dislocations. These two conflicting requirements can be met by applying the deforming stress at a temperature only slightly above M_s (Wasilewsky 1975).

A vanishingly small value of C indicates the presence of a proper ferroelastic phase transition. All materials for which this is the case are thus candidates for the occurrence of the SME (Nakanishi 1975). A case in point is that of the ferroelastic superconductor Y – Ba – Cu – O (Somayazulu, Rao & Wadhawan 1989), for which the SME was anticipated from such considerations, and then demonstrated (Tiwari & Wadhawan 1991).

Another important factor influencing efficient shape recovery in SME systems is that of the degree of crystallographic reversibility of the phase-transition process (cf. §11.5.3). An ideal situation for the reversible movement of all interfaces involved in the SME cycle would be one in which, starting from a single domain or crystal of the parent phase, one ends up with the same single domain on completing the shape-memory cycle; i.e. $n_{10} = 1$ in Fig. 11.5.4(b). If it is not possible to make $n_{10} = 1$, one should at least try to make it as small as possible. We illustrate this with an example.

In the alloy NiAl (Chakravorty & Wayman 1976) there is a martensitic phase transition, with point-group symmetry changing from $m\bar{3}m$ to $4/mmm$ on cooling through the transition, but the 4-fold axis of the martensitic phase does not coincide with any 4-fold axis of the parent phase. Therefore, in terms of the notation used in §7.5 for constructing a group tree, $H_0 = G_0 = m\bar{3}m$, $G_1 = 4/mmm$, and $N_{01} = \bar{1}$. The group tree therefore looks as in Fig. 11.5.6(a) for the martensitic transition brought about by cooling, i.e. under the influence of only a scalar, namely temperature (with nominal symmetry $(g) = SO_3$). We find that $n_{01} = 24$, meaning that as many as 24 domain types can exist in the martensitic phase.

What happens when we heat this system back to the parent phase? Fig. 11.5.6(b) describes it pictorially. We end up with 8 variants. The fact that $n_{10} = 8$, rather than 1, means that if one were to start with a

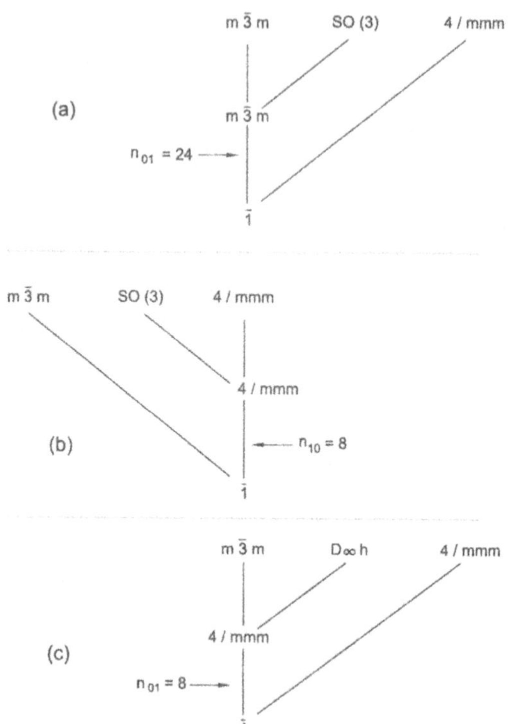

Figure 11.5.6: Group trees for the martensitic phase transition in NiAl. (a) Phase transition on cooling. (b) Reverse phase transition on heating. (c) Same as (a), except that the martensitic transition on cooling occurs in the presence of a uniaxial stress parallel to a 4-fold axis of the cubic phase.

single-domain crystal of the NiAl alloy in the austenitic phase and cool it to the martensitic phase, and then heat it back to the austenitic phase, one would not recover a single-domain crystal, and the transition is therefore not crystallographically reversible.

But we have not yet considered the effect of uniaxial stress on this system (the stress may be either internal, or externally applied). Let us assume that the stress is directed along a 4-fold axis of the cubic phase, i.e. we take the solicitation to have the symmetry

$$(g) = D_{\infty h[100]} \tag{11.5.1}$$

We assume that the stress does not alter the nature of the phase transition; i.e. $G_1 = 4/mmm$, as before. The group tree can now be constructed as in Fig. 11.5.6(c), with $n_{01} = 8$. Thus the application of the specified stress brings down the number of variants in the martensitic phase from 24 (as in Fig. 11.5.6(a)) to 8.

Choice of a different direction for the application of uniaxial stress (or a suitable combination of stresses) could perhaps bring down this number still further.

For efficient shape recovery in the one-way SME in a material, the ideal goal to be achieved (through a judicious choice of internal or external stress patterns) is the formation of only a single variant in the martensitic phase, having just one interface with the parent phase.

In the case of the $Y-Ba-Cu-O$ ceramic, crystallographic reversibility was demonstrated by a group-tree analysis by Tiwari & Wadhawan (1991).

It is pertinent to note here that the one-way SME occurs not only in alloys, but also in insulators like PLZT (Kimura et al. 1981; Wadhawan et al. 1981) and $Y-Ba-Cu-O$ (Tiwari & Wadhawan 1991). The maximum recoverable strain in these and other ceramics is in the range of quarter of a percent to half a percent (also see Virkar et al. 1991).

Compared to alloys, the occurrence of the SME in ferroelectric insulators offers the possibility of using electric field as an additional control parameter. The first such study was on PLZT (Wadhawan et al. 1981). Its composition x/65/35 is close to the morphotropic phase boundary (Haertling & Land 1971; Haertling 1971). It is both a ferroelectric and a ferroelastic. In fact, it is a relaxor ferroelectric in a certain composition range. For $x > 4.5$ it undergoes a phase transition (Keve & Annis 1973) which is similar to a martensitic phase transition. Above T_c the material behaves almost like a normal paraelectric. However, the dielectric behaviour immediately below T_c is not that of a normal ferroelectric, but rather that of a 'quasiferroelectric', in which there are polar distorted microregions shorter than the wavelength of light. Application of an electric field between T_c and a certain lower temperature T_p changes the microdomains to macrodomains. If the electric field is switched off the macrodomains revert back to randomly oriented and distributed microdomains, giving a macroscopically nonferroelectric state. The observation of slim ferroelectric hysteresis loops is evidence for this (Carl & Geisen 1973). When the electric field is applied at a temperature below T_p, stable macrodomains, indicative of long-ranged order, are induced and the material behaves like a normal ferroelectric and ferroelastic, exhibiting fat hysteresis loops. For temperatures between T_c and T_p, microdomains of polar short-range order coexist with the paraelectric matrix, just as martensite coexists with austenite over a certain temperature range.

The effect of electric field on the SME in PLZT of composition 6.5/65/35 was studied by Wadhawan et al. (1981). The modification introduced by the electric field, applied in conjunction with bending stress on thin bars of the ceramic, could be interpreted in terms of the above-mentioned model. In the temperature region in which the microdomains exist, the system is 'electrically soft' but 'mechanically hard', and the load does not produce a large bending. However, when the electric field is switched on, it changes the microdomains to macrodomains, and the material becomes mechanically soft also; consequently the same load produces an additional amount of bending.

Two-Way Shape-Memory Effect

In the two-way SME (Fig. 11.5.5(b)) the specimen, after a repeated process of "training", can be made to remember its shape in both the phases (Schroeder & Wayman 1977). What the training achieves is crystallographic reversibility, with either phase as the starting phase (Guenin 1989).

The usual procedure is to train or programme a specimen by thermomechanical treatment. When in the martensitic phase, it is deliberately constrained during heating to the austenitic phase, so as to suppress the usual one-way shape memory. This results in a movement of defects to certain sites so as to generate built-in microstresses in the austenitic phase, which programme the specimen to behave as in a stress-induced martensitic transition. In other words, the internal microstresses favour the formation of only (or mainly) a single variant of the martensite when cooled to the martensitic phase, resulting in macroscopic spontaneous deformation of the specimen on cooling. After a large number of such training cycles the built-in stresses correspond to a solicitation such that crystallographic reversibility is attained with either phase as the starting phase. In particular, during the heating part of the two-way shape-memory cycle, the original shape is regained through the usual shape-memory process involving a phase transition to the parent phase.

Guenin (1989) has adduced evidence which points to the formation of oriented structural defects like dislocations in the austenitic phase. Such defects act as internal biasing fields, resulting in the nucleation and/or growth of specific variants when the system is cooled to the martensitic phase.

11.5.5 Falk's Universal Model for Shape-Memory Alloys

The Landau theory of phase transitions was originally formulated for continuous transitions, whereas martensitic transitions, even of the M_1 type, are discontinuous transitions. One has to therefore employ Devonshire's (1954) generalization of the theory for dealing with them. Further, in writing a typical Landau expansion we assume that the order parameter is a slowly varying function of distance, whereas for a martensitic transition there is a range of temperatures in which the system has phase boundaries and domain boundaries at which there is a sharp variation of the order parameter. One must therefore include terms in the Landau expansion which involve spatial derivatives of the order parameter, in the spirit of Ginzburg's (1955, 1961) extension of the Landau theory (cf. §5.5.2). Such a treatment for martensitic transitions was given in the pioneering work of Falk (1980, 1982a, b, 1983, 1984).

Falk (1980, 1982b) proposed a simple, but fairly universal, model Landau expansion, taking a suitably defined strain as the order parameter. This model explains a large number of features for a whole group of alloys exhibiting the SME. This calculation is the mechanical counterpart of Devonshire's (1954) treatment of discontinuous ferroelectric phase transitions driven by temperature and electric field. The martensitic alloys NiTi, NiAl, CuZn, CuAlNi, AgCd, and $CuAuZn_2$ are treated together in this theory, ignoring the differences in the stacking sequence of their close-packed layers.

After a suitable rescaling, the following *ansatz* is made for the Landau free energy:

$$f(e,t) = f_0(t) + (t + \frac{1}{4})e^2 - e^4 + e^6 + e'^2 \qquad (11.5.2)$$

Here f, e, e' and t stand, respectively, for scaled values of free-energy density, shear strain, strain gradient, and temperature. It is a universal equation for a large number of alloys. The properties of specific materials enter through the scaling constants. It is remarkable that such a simple one-component one-dimensional model leads to qualitative agreement with numerous experiments on shape-memory alloys: stress-strain curves exhibiting elasticity, ferroelasticity and superelasticity in the appropriate temperature regions; the mode softening; the SME; the occurrence of stress-induced and temperature-induced phase transitions; and the latent heat of the transition.

SUGGESTED READING

R. J. Wasilewski (1975). In J. Perkins (Ed.), *Shape Memory Effects in Alloys*. Plenum Press, New York.

H. Warlimont (1976). Shape memory effects. *Mater. Sci. & Engg.*, **25**, 139.

C. M. Wayman (1981). Martensitic transformations: An overview. In *Proc. Int. Conf. on Solid-Solid Transformations*. Carnegie-Melon University, Pittsburgh, Aug. 10-14, 1981. Published by the Metals Society of the

AIME, USA.

G. Guenin (1989). Martensitic transformation: Phenomenology and the origin of the two-way memory effect. *Phase Transitions*, **14**, 165.

R. Tiwari & V. K. Wadhawan (1991). Shape-memory effect in the Y-Ba-Cu-O ceramic. *Phase Transitions*, **35**, 47.

J. Pouget (1991). Lattice model for nonlinear patterns in ferroelastic-martensitic materials. *Phase Transitions*, **34**, 105.

Yu. A. Izyumov, V. M. Laptev & V. N. Syromyatnikov (1994). Phenomenological theory of martensitic and reconstructive phase transitions. *Phase Transitions*, **49**, 1.

V. S. Boyko, R. I. Garber & A. M. Kossevich (1994). *Reversible Crystal Plasticity*. American Institute of Physics, New York.

P. Entel, K. Kadau, R. Meyer, H. C. Herper, M. Schroter & E. Hoffmann (1998). Large-scale molecular-dynamics simulations of martensitic nucleation and shape-memory effects in transition metal alloys. *Phase Transitions*, **65**, 79.

11.6 DOMAIN STRUCTURE OF FERROELASTIC CRYSTALS

11.6.1 Domains in Ferroelastic Crystals

The number of variants in the ferroelastic phase of a crystal is given by the ratio of the orders of the point group of the prototype and the point group of the ferroelastic phase in question.

For a full ferroelastic, any domain pair is a ferroelastic domain pair. For a partial ferroelastic, a domain pair may be either ferroelastic or non-ferroelastic.

Whereas the possible number of variants in a ferroelastic phase can be determined from group-theoretical considerations, the actual number of domains in a given specimen depends on the history of the specimen and on external conditions. For example, the number of domains can be reduced by cooling the crystal through the ferroelastic phase transition under a uniaxial stress.

11.6.2 Suborientation States

Consider a ferroelastic domain pair having a common domain wall. Because the two domains differ in spontaneous strain, they must undergo a small mutual rotation to make physical contact at the domain wall. The term disorientations is used for such rotations (cf. §7.1.4). They depend on sample history and external conditions.

Working under the parent-clamping approximation (PCA) (§5.1.2) amounts to ignoring the disorientations. If we do not ignore them, an F-operation for a ferroelastic domain pair is no longer exactly equal to a symmetry operation of the prototype group. Moreover, the difference between the two operations can occur in more than one symmetry-equivalent ways, as exemplified by the case of the ferroelastic superconductor Y–Ba–Cu–O (Wadhawan 1988). In other words, disorientations can have (and do have) symmetry-equivalent counterparts, as a result of which the actual number of orientation states in a ferroelastic phase can be more than that calculated under the PCA. The term *suborientation states* (as against orientation states) has been introduced to take cognisance of this possibility (Boulesteix 1984; Shuvalov, Dudnik & Wagin 1985; Dudnik & Shuvalov 1989).

Because of the complex nature of strain fields in a ferroelastic specimen, a rigorous group-theoretical enumeration of suborientation states is a very difficult, if not impossible, task.

11.6.3 Double Ferroelasticity

It has been argued by Guymont (1981, 1991) that it is not necessary to refer to a supergroup prototype symmetry for defining spontaneous strain and, in a more general sense, for rationalizing the domain structure of a ferroic phase of a crystal: it is enough that the nondisruption condition (§5.1.1) be satisfied, so that one can make a meaningful statement about lost or gained symmetry operators at the phase transition. For any phase transition, the lost symmetry operators determine the variants.

Consider two phases of symmetry G and H. And let I be their intersection group:

$$I = G \cap H \tag{11.6.1}$$

The symmetry operators lost at the phase transition $G \rightarrow H$ are $(G - I)$, and the number of variants is G/I $(= q_{12}$, say).

Let these variants be called S_1, S_2, ... $S_{q_{12}}$. One can choose any of them, say S_1, as the 'initial' variant, and define a spontaneous-strain tensor for it, symmetry-adapted with respect to the group I. The relative spontaneous strain can then be defined for each of the q_{12} variants, as discussed in §11.1.3. The phase of symmetry H is then a full or partial ferroelastic phase, depending on whether all the variants have, or do not have, distinct relative spontaneous strain tensors. No reference to a prototype supergroup is required for this procedure.

Now consider the reverse transition $H \to G$. The symmetry operators lost in this are $(H - I)$, and the number of variants is H/I ($= q_{21}$, say). Let these variants be called V_1, V_2, .. $V_{q_{21}}$. We can once again choose any of them, say V_1, as the initial variant, and assign a spontaneous-strain tensor to it, symmetry-adapted for the symmetry group I. After this the relative spontaneous strain tensor can be computed for each of the variants. If there is at least one pair of variants or domains, say (V_1, V_2), which is a ferroelastic domain pair, the phase G is a ferroelastic phase, just as phase H is a ferroelastic phase if at least one domain pair, say (S_1, S_2), is a ferroelastic domain pair. Thus, in principle, ferroelastic state shifts can be expected in both the phases. The term 'double ferroelasticity' was coined by Guymont (1981, 1991) for this property.

Alternatively, if possible, both G and H can be regarded as subgroups of a certain suitably chosen prototype symmetry group, and such an approach may end up providing a more comprehensive explanation of the totality of the observed domain structure than that possible from Guymont's prescription.

Guymont was not able to identify a real-life example of double ferroelasticity, although he constructed theoretical examples. A possible actual example was suggested by Wadhawan (1991), namely the NiAl alloy. However, in view of the rather strict definition of ferroic phase transitions adopted in this book (in terms of the nondisruption condition) it appears that the phase transition in NiAl cannot qualify to be called a ferroic phase transition, and is therefore also not a ferroelastic phase transition. The reasons are as follows.

As depicted in Fig. 11.5.6, for this system $G = m\bar{3}m$ and $H = 4/mmm$. But the 4-fold axis of the phase with symmetry H does not coincide with any symmetry axis of the phase with symmetry G, and therefore $I = \bar{1}$ (Fig. 11.5.6(a)). The point groups G, H, and I are of orders 48, 16 and 2. Therefore 24 orientational variants are possible in the phase H, and 8 orientational variants are possible in phase G, purely on the basis of symmetry operators lost in going from one phase to the other. Although one can work out relative-strain tensors for the two phases, the phase transition is probably reconstructive, and therefore disruptive. In our scheme of things, only nondisruptive phase transitions can be ferroic, and thence ferroelastic if they entail a change of crystal family.

11.6.4 Orientation of Walls Between Ferroelastic Domain Pairs

The strain compatibility criterion used by Fousek & Janovec (1969) for determining the orientations of walls between ferroelectric domain pairs (§10.6.2) was employed by Sapriel (1975) for all the 94 full-ferroelastic species of crystals. According to this criterion, a permissible domain wall between a contiguous domain pair can exist only along that plane which, as a result of the ferroelastic phase transition, undergoes equal deformation in the two domains separated by it.

Let e_{ij} and e'_{ij} denote the components of the relative spontaneous strain tensor for the two contiguous orientation states. Then, Sapriel's (1975) formulation leads to the following equation determining the orientations of permissible domain walls:

$$\sum_{i,j}(e_{ij} - e'_{ij})x_i x_j \;=\; 0 \tag{11.6.2}$$

Here x_i (i=1, 2, 3) are the coordinates of a point on the domain wall, measured with respect to an origin on the mid-plane of the wall, in a Cartesian frame fixed in the prototype.

Eq. 11.6.2 is a second-degree equation, representing a conical surface. It splits into a product of two linear equations (representing two planar surfaces or walls) if the following condition for permissibility of the walls is satisfied (Sapriel 1975):

$$\det|(e_{ij}) - (e'_{ij})| \;=\; 0 \tag{11.6.3}$$

The relative spontaneous strain tensor is defined (§11.1.3) by subtracting out the average value of absolute spontaneous strain from the strain tensors of all the orientation states. Therefore, (e_{ij}) and (e'_{ij}) in Eq. 11.6.3 are traceless matrices.

When we deal with absolute strain tensors, rather than relative strain tensors, even when the two individual strain matrices are not traceless, the difference may still be so. The tracelessness of the strain-difference matrix is, in fact, the second part of Sapriel's strain-compatibility condition:

$$\mathrm{Tr}\,[(e_{ij}) - (e'_{ij})] \;=\; 0 \tag{11.6.4}$$

The fact that the matrix for the strain-difference tensor in Eq. 11.6.3 is traceless leads to the result that the two permissible planar walls must be mutually perpendicular.

If Eqs. 11.6.3 and 11.6.4 are not satisfied, and the two domains under consideration happen to be contiguous, the wall between them is said to be an *impermissible wall*. It is likely to be nonplanar, stressed, and/or diffuse. It may also be less mobile, or even immobile.

Two types of permissible domain walls were defined by Fousek & Janovec (1969): W_f-walls and S-walls (§10.6.2). The corresponding nomenclature introduced by Sapriel (1975) is: *W-walls* and *W′-walls*.

In addition, W_∞-walls have the same meaning in the work of Fousek & Janovec (1969) and Sapriel (1975).

We illustrate the use of this formalism by considering the case of $Pb_3(PO_4)_2$ discussed by Sapriel (1975). It belongs to the pure ferroelastic Aizu species $\bar{3}mF2/m$, for which the number of orientation states is 3 (say S_1, S_2, S_3). The monoclinic ferroelastic phase is assumed to have its 2-fold axis along the y-axis of the Cartesian frame of reference chosen in the prototype. Therefore, for the 'initial' state S_1 the symmetry-adapted relative spontaneous strain tensor has the representation (Aizu 1970b):

$$\mathbf{e}_{(s)}(S_1) = \begin{pmatrix} -A & 0 & C \\ 0 & A & 0 \\ C & 0 & 0 \end{pmatrix} \qquad (11.6.5)$$

The following three can be taken as a representative set of F-operations for this species: rotations by 0^0, 120^0, and -120^0, resulting in orientation states S_1, S_2, S_3, respectively. The relative spontaneous strain tensors for S_2 and S_3 therefore have the following forms (Aizu 1970b):

$$\mathbf{e}_{(s)}(S_2) = \begin{pmatrix} \frac{1}{2}A & \frac{\sqrt{3}}{2}A & -\frac{1}{2}C \\ \frac{\sqrt{3}}{2}A & -\frac{1}{2}A & \frac{\sqrt{3}}{2}C \\ -\frac{1}{2}C & \frac{\sqrt{3}}{2}C & 0 \end{pmatrix}, \qquad (11.6.6)$$

$$\mathbf{e}_{(s)}(S_3) = \begin{pmatrix} \frac{1}{2}A & -\frac{\sqrt{3}}{2}A & -\frac{1}{2}C \\ -\frac{\sqrt{3}}{2}A & -\frac{1}{2}A & -\frac{\sqrt{3}}{2}C \\ -\frac{1}{2}C & -\frac{\sqrt{3}}{2}C & 0 \end{pmatrix} \qquad (11.6.7)$$

Let a, b, c, and β denote the lattice parameters of the monoclinic phase at room temperature. Then A and C are given by (Toledano et al. 1975):

$$A = (y - z/\sqrt{3})/2b, \qquad (11.6.8)$$

$$C = (z + 3a\cos\beta)/6a\sin\beta \qquad (11.6.9)$$

We now have all the information needed for writing down the equations determining the orientations of permissible domain walls. For example, for walls between S_2 and S_3, Eq. 11.6.2 takes the form

$$xyA + yzC = 0, \qquad (11.6.10)$$

which splits into two linear equations:

$$y = 0, \tag{11.6.11}$$

$$xA + zC = 0 \tag{11.6.12}$$

We notice that Eq. 11.6.11 represents a W-wall, whereas Eq. 11.6.12 represents a W'-wall. The orientation of the latter depends on lattice parameters, and therefore varies with temperature. The two walls are mutually perpendicular.

The spontaneous deformation of the lattice does not affect the z-axis. The angle made by the W'-wall with this axis is given by $\tan\theta = -c/a$. Substituting $A = 21.8 \times 10^{-3}$ and $C = 6.6 \times 10^{-3}$ into this equation (Toledano et al. 1975), one gets $\theta = 17^0$, in agreement with experiment (Brixner et al. 1973).

Sapriel's formulation is based on several simplifying assumptions, which must be dealt with, wherever necessary:

(a) It ignores disorientations (§7.1.4 and 11.6.2), i.e. it works under the parent-clamping approximation (PCA).

(b) The magnitude of spontaneous strain is assumed to be small. Corrections have to introduced if this is not the case.

(c) If the ferroelastic domain pair is also a ferroelectric domain pair, the electrostatic-energy contribution to the free energy is zero only for electrically neutral walls. For a charged wall the optimum orientation can be different from that obtained from Sapriel's formalism.

(d) The presence of twinning dislocations and the interactions among them can result in wedge-shaped or lenticular domains. This requires a more elaborate calculation of the shape and orientation of the domain walls involved (Bornarel 1972; Bornarel & Lajzerowicz 1972; Fousek, Glogarova & Kursten 1976; Salje 1993a).

11.6.5 Phase Boundaries and Polydomain Phases in Ferroelastics

As discussed in §5.3.13, for a discontinuous phase transition there is a range of temperatures in which both the parent phase and the daughter phase are stable, i.e. the two phases can coexist. Naturally, there are walls or boundaries between them. The nature, orientation, and location of these phase boundaries has a role to play in determining the domain structure of the daughter phase at temperatures below the stability limit of the parent phase (Roytburd 1974). It is important to understand the phase boundaries if one wishes to understand and control the final domain structure of the daughter phase or the ferroic phase.

The phase boundaries have a symmetry aspect and a structural aspect. We have already considered the symmetry aspect in a general way in §8.1.1. We now discuss the structural aspect, in which strain considerations play a dominant role, not only for ferroelastics but also for ferroelectrics.

There have been at least three approaches to the problem of determining the orientation and location of phase boundaries in ferroelastics (see Dec (1993) for a review). In an increasing order of complexity and sophistication these are: the modified Sapriel formulation (Boulesteix et al. 1986); the Metrat formulation (Metrat 1980); and the Roytburd formulation (see Roytburd (1993) for an update on this approach).

We have described in §11.6.4 Sapriel's (1975) theory of domain-wall orientations in full ferroelastics. Sapriel adopted the criterion that the orientation of the wall separating a ferroelastic domain pair must be such as to make it strain-free. Boulesteix et al. (1986) applied the same criterion for determining the orientations of walls which separate an orientation state of the ferroelastic phase, not from another orientation state of the same phase, but from the prototypic phase.

The relative spontaneous strain for the prototypic phase is zero, by definition. Therefore, if we equate to zero one of the two strain-tensor matrices in Eq. 11.6.2, say (e'_{ij}), we get the following condition for a strain-free heterophase interface:

$$\sum_{ij} e_{ij}\, x_i\, x_j \;=\; 0 \qquad (11.6.13)$$

As in Sapriel's approach for ferroelastic domain pairs, solutions of Eq. 11.6.13, corresponding to mutually perpendicular phase boundaries, are determined when the following condition is satisfied (cf. Eq. 11.6.3):

$$\det |(e_{ij})| \;=\; 0 \qquad (11.6.14)$$

This extension of Sapriel's theory suffers from the flaw that it assumes that there is no change of specific volume (or density) at the phase transition.

We turn to the Metrat (1980) approach next. Here one takes note of the fact that phase boundaries may be between phases, one or both of which may be twinned. For example, in $KNbO_3$ (which undergoes the same sequence of phase transitions as $BaTiO_3$), the tetragonal phase is twinned because it is a derivative structure of the cubic phase. And the next phase on cooling, namely the orthorhombic phase, is twinned because some rotational symmetry elements are lost in going from the tetragonal symmetry to the orthorhombic symmetry. The twinning at each phase transition occurs to minimize the overall energy, particularly the strain energy.

Because of the twinning at each ferroelastic phase transition (which is sample dependent), one defines a mean spontaneous strain for the tetragonal phase (M) and for the orthorhombic phase (N). Both M and N are defined with respect to the prototypic cubic phase of lattice parameter $a_0 = V^{1/3}$:

$$M = \sum m_i \Phi_i M_i, \tag{11.6.15}$$

$$N = \sum n_i \theta_i N_i, \tag{11.6.16}$$

where m_i, n_i are the relative fractions of the individual domains; M_i, N_i are the respective spontaneous strains; and Φ_i, θ_i their rotations with respect to the cubic phase.

Choosing, as is usual, a common Cartesian system of coordinates, (x_1, x_2, x_3), in the cubic phase (in which all spontaneous strains are zero, by definition), the strain compatibility condition leads to the following equations for the phase boundary between the two twinned phases:

$$D_{ij} x_i x_j = 0, \tag{11.6.17}$$

$$D_{ij} = N_{ik} N_{jk} - M_{ik} M_{jk} \tag{11.6.18}$$

Eq. 11.6.17 reduces to the product of two plane-surface solutions,

$$(h_1 x + k_1 y + l_1 z)(h_2 x + k_2 y + l_2 z) = 0, \tag{11.6.19}$$

if the following conditions are obeyed (for $i \neq j$):

$$\det |D_{ij}| = 0, \tag{11.6.20}$$

$$D'^2_{ij} \equiv (D^2_{ij} - D_{ii} D_{jj}) > 0 \tag{11.6.21}$$

The direction cosines of the plane-normals are given by (Metrat 1980):

$$h_1 = \frac{D_{11}}{D_1}, \; k_1 = \frac{D_{12} + D'_{12}}{D_1}, \; l_1 = \frac{D_{13} \pm D'_{13}}{D_1}, \tag{11.6.22}$$

where

$$D_1^2 = D_{11}^2 + (D_{12} + D'_{12})^2 + (D_{13} \pm D'_{13})^2; \tag{11.6.23}$$

and

$$h_2 = \frac{D_{11}}{D_2}, \; k_2 = \frac{D_{12} - D'_{12}}{D_2}, \; l_2 = \frac{D_{13} \mp D'_{13}}{D_2}, \tag{11.6.24}$$

where

$$D_2^2 = D_{11}^2 + (D_{12} - D'_{12})^2 + (D_{13} \mp D'_{13})^2 \tag{11.6.25}$$

An illustration of the Metrat-type analysis is provided by the work of Topolov et al. (1990) on twinning in the in the orthorhombic phases of $PbHfO_3$.

Polydomain Phases

The importance of twinning in determining the heterophase and homophase boundaries in ferroelastics was recognized in its most general form in the work of Roytburd (1974, 1983, 1993; also see Roytburd & Yu 1994). Roytburd developed the concepts of *elastic domains* and *polydomain phases.* The original motivation for this work was to deal with the polydomain structure related to martensitic phase transitions. In most cases such transitions give rise to B-twins (§7.4.3), rather than S-twins associated with ferroelastic phase transitions. However, the theory is general enough to cover both types (see Roytburd 1978).

A ferroelastic phase transition is characterized by a spontaneous deformation of the crystal lattice, resulting in a change of the crystal family. Consequently the emergence and propagation of the ferroelastic phase in the restricted volume of the parent matrix creates a strain field. Beyond a certain amount of growth of the ferroelastic phase the strain energy in the surrounding matrix becomes so large that the system has to reduce it by opting for a self-induced ferroelastic switching of the daughter phase to a domain of opposite spontaneous strain. The growth of the new ferroelastic domain again causes an increasing amount of strain in the surrounding matrix. When this strain becomes unbearable the system opts for another spontaneous ferroelastic switching, this time back to the original state of spontaneous starin. This can occur repeatedly, so that what we have is a *polysynthetic twin*, or a *polytwin*, which is a plane-parallel plate comprising alternating plane-parallel domains.

The same reasoning can be applied, not only to the homophase interfaces within the ferroelastic phase, but also to the phase boundaries. In the vicinity of the ferroelastic phase transition the system is in a state of upheaval (e.g. because of lattice-dynamical mode softening), and mechanical stress (internal or external) may easily suppress the phase transition, or shift or extend the temperature range over which it occurs. One is thus led to the concept of *polydomain phases*, which has turned out to be a fundamental concept in the theory of ferroic phase transitions in crystals (Roytburd 1993).

The basic idea in this theory is that, although there may be a discontinuous change of lattice parameters at the martensitic and/or ferroelastic phase transition, there must be a continuity of the two lattices at the phase boundary. It is assumed further that, during the phase transition, there is neither a creation nor a movement of dislocations.

Similarly, one imposes the condition of lattice continuity at domain boundaries as well.

In continuation of this basic assumption, and in keeping with experimental observations on a large number of systems, one postulates that the polytwin and/or the polydomain phase is the result of self-induced deformation twinning, subject to the constraint that the overall strain is an *invariant-plane strain*. What the latter means is that, not only does the product phase or domain meet the parent phase or domain in a coherent manner (i.e. across a continuous lattice), the intervening plane is also an undistorted and unrotated plane (on an average). Such a plane is known as an *invariant plane* (Weschler, Lieberman & Read 1953; Bowles & Mackenzie 1954).

Over the years there has been a change in the perception regarding the physics of the processes leading to the formation of polytwins. Rather than treating polytwinning as a process of deformation twinning, it is now considered as an integral part of the process of phase transition. In other words, a theory of a ferroelastic phase transition must take explicit note of the formation of structural domains at the free-energy-minimization stage itself. The structural domains (and, for that matter, even ferroelectric and ferromagnetic domains) must be viewed as so configured as to minimize the overall free energy (Barsch & Krumhansl 1984). This configuration has very different properties from those of a single-domain state, and the special term *elastic domains* has been coined by Roytburd (1974) to emphasize their different identity.

Since the basic physics of coherent phase boundaries and domain boundaries is not very different (Boulesteix et al. 1988), the terms 'elastic domains' and 'polydomain phases' are used interchangeably sometimes.

Virtual phase is another relevant term. A virtual phase can exist only as a part of a polydomain structure (Roytburd 1971).

An elastic domain, or a polydomain phase, or a virtual phase, has properties which depend on sample history. The existence of interfaces provides a large number of extra degrees of freedom, not available for single-domain states (Khachaturyan, Shapiro & Semenovskaya 1991). These extra degrees of freedom also interact in a highly nonlinear way with external fields, making such systems strong candidates for smart-structure applications (§14.3). The shape-memory effect is the best known example of this.

Let us call the parent (higher point-group symmetry) phase as Phase 1, and the daughter phase as Phase 2, with $\mathbf{e}_s$ as the spontaneous strain of a single-domain state of Phase 2. There can, in general, be a change of lattice-translation vectors at the ferroelastic phase transition. Therefore, to achieve continuity of the lattices at the interfacial or phase-boundary region, both phases must undergo an additional spontaneous deformation. Let the strain in Phase 1 at the interface be $\mathbf{e}_1$, and let $\mathbf{e}_2$ be the additional strain occurring in Phase 2 for achieving matching of lattice parameters at the interface.

In spite of these additional spontaneous strains, the orientations of the two lattices may not match at the interface. One can treat Phase 1 as the reference phase, and introduce the necessary rotational strain $\mathbf{w}$ in Phase 2 only.

Thus the total spontaneous strain of Phase 2 is:

$$\mathbf{D}_2 = \mathbf{e}_s + \mathbf{e}_2 + \mathbf{w} \tag{11.6.26}$$

And for Phase 1:

$$\mathbf{D}_1 = \mathbf{1} + \mathbf{e}_1 \tag{11.6.27}$$

A coherent (or lattice-matched) phase boundary implies that if we consider an arbitrary vector $\mathbf{R}$ on the invariant plane, the change in this vector because of the lattice strains is the same for Phase 1 and Phase 2:

$$(\mathbf{D}_1 - \mathbf{D}_2) \cdot \mathbf{R} = 0 \tag{11.6.28}$$

Let $\mathbf{n}$ denote a unit vector along the normal to the interface. Since $\mathbf{R}$ is a vector in the interface, we have $\mathbf{R} \cdot \mathbf{n} = 0$.

It helps to visualize the invariant plane as 'vertical', so that $\mathbf{n}$ is 'horizontal' for this plane. There is a *transition layer* of finite thickness (with the invariant plane located near the half-thickness point), such that the lattice planes on the two sides of the invariant plane are displaced in the vertical direction by amounts proportional to their distance from the invariant plane. Let the vector $\mathbf{s}$ denote this displacement; $\mathbf{s} = 0$ for the invariant plane, and this vector points in opposite directions on the two sides of this plane.

The condition of continuity of the two lattices (Eq. 11.6.28) implies that the net strain $(\mathbf{D}_2 - \mathbf{D}_1)$ at any lattice plane parallel to the invariant plane should be proportional to the distance of this plane from the invariant plane; i.e.

$$(\mathbf{D}_2 - \mathbf{D}_1)_{ij} = s_i\, n_j, \quad i, j = 1, 2, 3 \tag{11.6.29}$$

The strain $s_i n_k$ is the invariant-plane strain. It has a pure-strain part, and a rotation part:

$$\Delta e_{ij} = \frac{1}{2}\,(s_i n_j + n_i s_j), \tag{11.6.30}$$

$$w_{ij} = \frac{1}{2}\,(s_i n_j - n_i s_j) \tag{11.6.31}$$

In some ferroelastic phase transitions the relative spontaneous strain $\mathbf{e}_s$ itself is an invariant-plane strain; i.e. $e_2 = 0$ in Eq. 11.6.26 and $e_1 = 0$ in Eq. 11.6.27. Then

$$e_{sij} = \frac{1}{2}\,(s_i n_j + n_i s_j), \tag{11.6.32}$$

and there is a lattice plane which is identical for the parent phase and the daughter phase. This happens in the case of $Pb_3(VO_4)_2$, $Pb_3(PO_4)_2$ (Shuvalov et al. 1985;), and $NaNbO_3$ (Dec & Kwapulinski 1989).

In a more general case, i.e. when the relative spontaneous strain is not an invariant-plane strain, a combination of two mechanisms can operate: (a) both the parent and the daughter lattices undergo additional strains to achieve their gradual matching on the two sides of the invariant plane; (b) the daughter phase breaks into a twinned structure to effectively become an elastic domain. This elastic domain has a mean spontaneous strain $\bar{\mathbf{e}}_s$, which plays the role of the relative spontaneous strain in Eq. 11.6.32:

$$\bar{\mathbf{e}}_s(\alpha) \simeq (1-\alpha)\mathbf{e}_{s1} + \alpha\mathbf{e}_{s2}, \tag{11.6.33}$$

where $(1-\alpha)$ and α are the volume fractions of the two domain states comprising the elastic domain. Several examples of such situations, including a comparison between theory and experiment, have been discussed by Dec (1993).

The concepts of optimum domain fraction α and effective (or average) macroscopic spontaneous strain $\bar{\mathbf{e}}_s$ are capable of further generalization, in the sense that they can be treated as *thermodynamic variables* determining the actual evolution of a phase transition in a solid (Roytburd 1978). The contributions of electric and/or magnetic fields (internal or external) to the overall free energy can also be included. A general observation is that the elastic interaction dominates over the electric and magnetic interactions.

The system can minimize its free energy not only by creating elastic (parallel-plate) domains, but also by creating *higher-order domain structures* (see Roytburd 1993, 1994). The polytwin described above is the lowest-order domain structure. In spite of the 'tuning' possible by a self-variation of the domain fraction α, the overall stress caused by the occurrence of the ferroic phase transition may still be high. The system can lower it further by forming *hierarchical domain structures* (2nd order, 3rd order, etc.), with polytwins of a lower order serving effectively as 'single domains' for the domain structure of the next higher order. Roytburd (1993, 1994) cites the presence of disorientations w (Eq. 11.6.31) as one of the causative factors for the occurrence of higher-order domain structures.

11.6.6 Some Further Aspects of the Effect of Long Ranged Elastic Interaction on Domain Structure

The microstructure of ferroelastic crystals is determined by a variety of interacting factors, and therefore a number of context-dependent approaches have been adopted for understanding this complex problem. In

§11.6.5 we gave an outline of Roytburd's approach in terms of hierarchies of domain structure. Another approach, which has resulted in good progress and insight, particularly during the last decade and a half, is that based on treating explicitly the anisotropic and long-ranged strain interaction as the main causative factor determining the kinetics and other characteristics of a structural phase transition. In recent years this approach has been spearheaded by Khachaturyan and coworkers (see Khachaturyan, Semenovskaya & Long-Qing Chen (1994) for a recent review), and by Salje and coworkers (see Salje (1995) and Bratkovsky, Heine & Salje (1996) and the references therein).

The basic idea is as follows: Structural changes occurring in unit-cell i result in a local stress field, which affects even a distant cell j through a 'knock-on effect', i.e. a transmission of stress through successive unit cells. The effective elastic interaction $J(r_{ij})$ over long distances in the crystal has a strongly anisotropic or angle-dependent part (resulting in a few *soft directions* parallel to domain walls), and a so-called *Zener-Eshelby interaction* J_z of infinite range (Zener 1948b; Eshelby 1956).

One of the most striking manifestations of the elastic interaction in some crystals is the observation of (cross-hatched) *tweed patterns* when the crystal is quenched through a ferroelastic phase transition. Even for a discontinuous ferroelastic phase transition a dense mass of *embryos* of the ferroelastic phase is present as thermodynamic fluctuations (usually concentration inhomogeneities), and this happens even at temperatures far above the transition temperature T_c.

The occurrence of significant precursor effects in the parent phase, even for discontinuous phase transitions, is peculiar to solid-solid transitions. Generally the kinetics of the processes is quite slow (although there can also be a fast component), and precursor tweed patterns have been observed even above T_c in crystals such as Y – Ba – Cu – O (Schmahl et al. 1989). It has been postulated that the presence of some kind of static disorder or inhomogeneity (e.g. due to dopants or defects) is a necessary prerequisite for the existence of tweed embryos in the parent phase. In fact such systems have several features reminiscent of quadrupolar glasses (Kartha et al. 1991).

The present understanding is that tweed embryos are present as thermal fluctuations at temperatures even well above T_c. As T_c is approached from above, the embryos become more and more prominent, and freeze into metastable structures on quenching the specimen crystal through T_c. The metastable tweed pattern then sharpens and coarsens with the passage of time (Salje 1993b; Khachaturyan et al. 1994).

The Elastic Interaction and Structural Phase Transitions

The preponderance of the strain coupling over the Coulomb interaction and the van der Waals interaction in determining most of the important features of a structural phase transition has been brought out in detail by Khachaturyan et al. (1994), and by Bratkovsky et al. (1995). We describe here some key features of their work.

The atomic structure of a crystal is determined by the manner in which its atoms, or groups of atoms, pack themselves in a lowest-free-energy configuration. Irrespective of whether or not the crystal is also a ferroelectric and/or a ferromagnetic, when the atomic configuration changes in any unit cell (as a function of temperature, pressure, etc.), the dimensions of the unit cell change. That is, a spontaneous shear strain or volume strain develops. This local displacement field is bound to influence the neighbouring unit cells through a 'push-pull' mechanism. The deformation in these latter cells, in turn, influences *their* neighbours, and so on. It follows that there is an inherent ordering interaction, $J(\mathbf{r}_{ij})$, in the crystal, mediated entirely by strain, irrespective of the presence or absence of other ordering interactions like the magnetic exchange interaction or the electric dipolar interaction.

Depending on the symmetry of $J(\mathbf{r}_{ij})$, this elastic interaction may lead to either ferroelastic or antiferroelastic ordering. Ferroelastic ordering is induced by a field which falls off as r^{-3}. And the strain mediating antiferroelastic ordering decays with distance as r^{-5}.

At short distances the function $J(\mathbf{r}_{ij})$ has a very complicated angular and spatial dependence. At large distances it has the following form for the ferroelastic case:

$$J_{ferro}(\mathbf{r}) \sim J_s + J_z, \tag{11.6.34}$$

where

$$J_s = \frac{\sum_l A_l Y_{lm}}{r^3}, \tag{11.6.35}$$

$$J_z = Z/N \tag{11.6.36}$$

J_z is the Zener-Eshelby interaction term; it is a constant, N being the number of unit cells in the specimen crystal. Y_{lm} are spherical harmonics of order l, the $l = 4$ term being the most dominant. Ferroelastic ordering ensues if the local stress field has the symmetry of a macroscopic shear strain.

For antiferroelastic ordering the following functional dependence applies:

$$J_{antiferro} \sim \frac{\sum_l A_l Y_{lm}(\theta, \phi)}{r^5} \tag{11.6.37}$$

The J_z term in Eq. 11.6.34, being a constant, has an infinite range. Each unit cell makes a contribution of the order of $1/N$ to the total ferroelastic shear, so that contributions from all the N cells add up to a constant value, which determines the temperature T_c of the ferroelastic phase transition, as well as the enthalpy of the transition.

It is the J_s term in Eq. 11.6.34 which determines the existence of metastable tweed textures.

Ferroic-Phase Morphology as an Internal Thermodynamic Parameter

The **r**-dependence of the strain interaction expressed by Eq. 11.6.35 is similar to dipole-dipole interaction between electric or magnetic finite elements. The dominant interaction in all these situations is strongly anisotropic, and decays as $1/r^3$. This similarity of the main underlying interaction for all the three types of primary ferroics (ferromagnetics, ferroelectrics and ferroelastics) is quite deep, and warrants further elaboration:

Following Khachaturyan et al. (1994), we begin by noting that the total strain energy per unit volume of a ferroelastic crystal is proportional to Ce_s^2, where C is a typical shear modulus, and e_s represents the effective spontaneous strain at the temperature of interest. This energy density has the dimensionality [energy / L^3]. For estimating the *total* energy of the specimen crystal we must multiply it by a characteristic quantity having a dimensionality [L^3]. Now, a *finite*-ranged interaction is characterized by a certain length parameter, namely the range of the interaction. By contrast, an infinite-ranged interaction like the elastic interaction does not have such a characteristic length. The only parameter available to us, which has the dimension [L^3], is the volume V of the daughter (ferroelastic) phase. Therefore the total strain energy of the fully or partially transformed specimen crystal has the form

$$E_{strain} = \alpha C e_s^2 V, \tag{11.6.38}$$

where α is a dimensionless proportionality coefficient, the value of which can be obtained by detailed calculations (Khachaturyan 1967).

The elastic interaction has two main characteristics, namely its infinite range and its strong anisotropy. These make the volume-dependent part of the strain energy a nonlinear function of the geometrical configuration of the phase boundaries and the domain boundaries.

The nonlinear volume dependence of the strain-energy part of the total Hamiltonian has a serious consequence for the simple Landau-theoretic treatment of phase transitions given in Chapter 5. The Landau expansion of the free-energy *density* (Eq. 5.3.9) is written with the implicit assumption that the total free energy can be obtained by multiplying the free-energy density by the volume of the specimen. We now find that the volume dependence of the free energy is not linear.

Another consequence of the dependence of α in Eq. 11.6.38 on the volume fraction and the shape of the transformed phase is that it is essential to include in the Landau expansion terms depending on the gradient of the order parameter. In other words, we must work with the Landau-Ginzburg thermodynamic potential (Eq. 5.5.4), rather than the Landau potential (Eq. 5.3.9), in spite of the fact that critical fluctuations of the order parameter are, as a rule, less important when the dominant interaction mediating the phase transition is a long-ranged interaction.

As pointed out by Khachaturyan et al. (1994), the $1/r^3$ dependence of the elastic interaction expressed by Eq. 11.6.35, and the dependence of the factor α in Eq. 11.6.38 on the volume fraction and geometrical configuration of the ferroelastic phase, points to a general similarity between (proper) ferroelastic and other (proper) primary-ferroic transitions, namely ferroelectric and ferromagnetic transitions. All three types are influenced by long-ranged dipole-dipole interactions, and all three result in a ferroic phase which must split into domains to minimize the overall interaction energy. In fact, α in Eq. 11.6.38 can be regarded as the mechanical analogue of the demagnetization factor in ferromagnets and the depolarization factor in ferroelectrics.

A realistic theory of a primary-ferroic phase transition must recognize the morphology of the emerging ferroic phase as an internal thermodynamic parameter (which evolves with time and temperature as the new phase emerges and grows) to minimize the overall free energy (Khachaturyan 1983; Salje 1993a).

The Four Categories of Strain Coupling

Depending on the symmetry of the local displacement field, and whether or not strain-compatibility conditions are satisfied at the domain walls, four types of strain coupling have been identified (Bratkovsky et al. 1995).

If the symmetry of the local displacement field is not that of a macroscopic strain, $J(\mathbf{r})$ results in an antiferroelastic ordering. This category of strain coupling has been therefore called *Type AF*. There is no macroscopic deformation of the specimen in this case.

When the symmetry of the local displacement field is that of a macroscopic strain, $J(\mathbf{r})$ is ferroelastic. Two possibilities arise. For a particular domain pair the wall separating them may or may not obey the two Sapriel (1975) conditions for strain compatibility (Eqs. 11.6.3 and 11.6.4). If they do not, the ferroelastic walls between them are said to be of the 'non-Sapriel (ns) type', or *Type F(nS)*. In such a case the two domains may differ in volume strain.

When the Sapriel conditions are fulfilled, the ferroelastic strain tensors may be either shear strain tensors (*Type F(xy)*), or they may be of the type

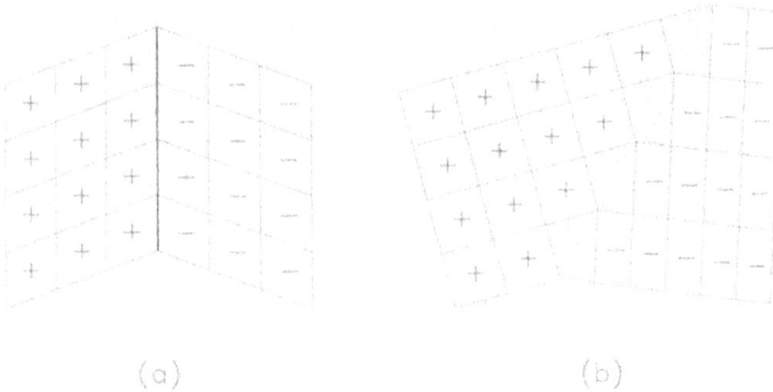

Figure 11.6.1: Domain wall for Type F(xy) strain (a), and Type F(xx-yy) strain (b). In (a) there is no additional distortion at the domain wall, whereas in (b) there is bound to be a distortion of the unit cells in the vicinity of the wall. [After Bratkovsky et al. 1995]

encountered in, say, a cubic-to-tetragonal transition (*Type F(xx-yy)*).

Transitions of the ferroelastic type $F(xy)$ are characterized by an xy-strain, i.e. $e_{xy} \neq 0$ for them (Fig. 11.6.1(a)). For their case, creation of domain walls hardly costs any energy. Therefore arrays of such walls, with e_{xy} changing sign at each successive wall, can form easily to minimize the overall macroscopic strain. As discussed in connection with the Sapriel formulation, such permissible walls can occur in mutually perpendicular pairs. Sets of mutually perpendicular walls of this type have the appearance of a tweed pattern. The tetragonal-to-orthorhombic phase transition in Co-doped Y-Ba-Cu-O, brought about by varying the concentration of Co, is an example of this situation (see Salje 1993a).

The ferroelastic type $F(xx - yy)$ arises when the spontaneous strain e_{xx} is equal to $-e_{yy}$ (Fig. 11.6.1(b)). In this case, although a coherent domain boundary is possible, the two domains do not match exactly at the boundary. There is a transition layer spread over a few unit cells around the boundary to accommodate the lattice mismatch. Naturally, creation of such walls costs a substantial amount of energy.

Lastly, in the non-Sapriel ferroelastic case F(nS), the Sapriel conditions for permissible domain walls are not satisfied, and the wall need not be a coherent, planar, wall. Examples of types of spontaneous strains for which the Sapriel conditions are not obeyed are $(xy+yz+zx)$ and $(2zz-xx-yy)$.

Conserved and Nonconserved Order Parameters. Kinetics

Structural phase transitions need not always occur under equilibrium conditions. When a nonequilibrium process is involved, a 'kinetic order parameter', η_{kin}, must be introduced for dealing with the question of the rate of the phase transition. Such a parameter is a measure of the 'kinetic deformation pattern', in the same way as the order parameter η defined in §5.3.2 is a measure of the static or equilibrium deviation of the structure in the ferroic phase from that in the prototypic phase. It is found for several systems (see Salje 1992) that η_{kin} and η describe the same structural configuration, so that one can often drop the subscript 'kin'. However, this is not the case in general (Dattagupta et al. 1991b).

A nonequilibrium state tends towards equilibrium with the passage of time. To describe the time evolution of $\eta(t)$, one introduces a *heat bath*, to which the state variables are coupled. The heat bath enables thermal fluctuations to occur, as well as an exchange of energy with the system, resulting in an evolution towards equilibrium.

For certain situations (see Salje 1993b) a constraint on the kinetic rate law is demanded by the invariance of the chemical composition. In such a condition a change (or 'spin flip' in the language of the Ising model) of the order parameter in some part of the system must be compensated by an opposite change ('spin flop') nearby. One then speaks of a *conserved order parameter*, and *Kawasaki dynamics* (Kawasaki 1966).

In certain other situations, a spin flip can occur almost independently of other spin variables, and we then have a *nonconserved order parameter*, obeying *Glauber dynamics* (Glauber 1963).

Within the approximations of the linear response theory (LRT; cf. §E.3.1), an ordering process associated with a structural phase transition, and occurring throughout the specimen at the same time, can be described by the so-called *Landau-Khalatnikov equation* (Machlup & Onsager 1953a,b; Landau & Khalatnikov 1954; Tani 1969):

$$\frac{d\eta}{dt} = -\Gamma \frac{\partial g}{\partial \eta} \tag{11.6.39}$$

Here η is the order parameter, g the free-energy density, and Γ a set of 'kinetic coefficients' which are assumed to be only weakly dependent on temperature. This equation from irreversible thermodynamics expresses the fact that the rate at which the order parameter tends to approach the equilibrium value is large if the thermodynamic force $\partial g/\partial \eta$ is large. It describes the relaxation of states close to equilibrium into the equilibrium state.

Ginzburg & Landau (1958) went beyond the simple LRT, and formulated a nonlinear equation for describing the kinetics of systems having nonconserved order parameters. Similarly, Cahn & Hilliard (1958, 1959) formulated the corresponding equation for conserved kinetics (see below). The *Ginzburg-Landau equation* and the *Cahn-Hilliard equation* can be expressed together as follows (see Patashinskii & Pokrovskii 1979; Landau & Lifshitz 1980; Dattagupta et al. 1991b; Bratkovsky, Heine & Salje 1996):

$$\frac{\partial \eta}{\partial t} = -\nu \frac{\delta g_L(\eta)}{\delta \eta}, \tag{11.6.40}$$

with

$$\nu = -\nu_c \nabla^2 + \nu_n \tag{11.6.41}$$

Here ν_c and ν_n are the kinetic coefficients for the conserved and nonconserved order parameters respectively.

It has been found that, in general, one can identify the driving force for the kinetic process with the excess free energy of the phase transition involved (Malcherek, Salje & Kroll 1997).

A general rate equation, covering a large variety of kinetics of order-disorder situations, and allowing for both conserved and nonconserved kinetics, was put forward by Salje (1988). The *Salje equation* can be written as

$$\frac{\partial \eta_{\mathbf{k}}}{\partial t} = -\frac{1}{\tau k_B T} [1 - (\xi_c^2/\xi^2) \exp(\xi^2 k^2/2)] \left(\frac{\partial g}{\partial \eta(\mathbf{r})}\right)_{\mathbf{k}} \tag{11.6.42}$$

In this equation the order parameter η can be inhomogeneous in general, and $\eta_{\mathbf{k}}$ denotes its $\mathbf{k}$th Fourier component. τ denotes some fundamental time scale of the rate process, and the free-energy density g is a function of $\eta(\mathbf{r})$ and temperature T. ξ_c reflects the extent of conservation of the order parameter. The ratio ξ_c^2/ξ^2 is a measure of the mixing of the two extreme situations corresponding to pure Kawasaki dynamics (conserved order parameter) and pure Glauber dynamics (nonconserved order parameter).

For $\xi_c = 0$ the Salje equation can be shown to reduce to the Ginzburg-Landau equation for Glauber dynamics.

Similarly, putting $\xi_c^2/\xi^2 = 1$ and making suitable approximations reduces the Salje equation to the following:

$$\frac{\partial \eta_k}{\partial t} = -\frac{1}{\tau k_B T} (k^2 \xi^2/2) \left(\frac{\partial g^K(\eta, T)}{\partial \eta(\mathbf{r})}\right), \tag{11.6.43}$$

where $g^K(\eta, T)$ is the (kinetic) Gibbs-energy density for Kawasaki dynamics. This is the Cahn-Hilliard equation (mentioned above) for a system described by a conserved order parameter.

Partially Conserved Order Parameter

A situation wherein $0 < \xi_c/\xi < 1$ corresponds to a *partially conserved order parameter* (Salje 1988, 1992, 1993b; Dattagupta et al. 1991b). Broadly speaking, what may happen is that the order parameter is nonconserved in small 'mesoscopic' parts of the crystal, but is conserved globally through suitable exchanges among these small regions. The 'nonconserved part' of the local rate equation may be written as follows (Salje 1992):

$$\left[\frac{d\eta(\mathbf{r})}{dt}\right]_{\text{local}} = -\frac{1}{\tau_1 k_B T}\frac{\partial g(\mathbf{r})}{\partial \eta}, \tag{11.6.44}$$

with g having the Ginzburg-Landau form.

The effect of exchange between two regions with different order parameters $\eta(\mathbf{r})$ and $\eta(\mathbf{r}')$ is incorporated by assuming that, to lowest order, the rate of flow of the order parameter from one region to the other is proportional to the difference between the driving forces at $\mathbf{r}$ and $\mathbf{r}'$ (Marais, Salje & Heine 1991; Marais & Salje 1991):

$$\left[\frac{\partial\eta(\mathbf{r})}{\partial t}\right]_{\mathbf{r}\to\mathbf{r}'} = \left(\frac{\tau_2^{-1}}{k_B T}\right)\left[\frac{\partial g(\mathbf{r}')}{\partial \eta} - \frac{\partial g(\mathbf{r})}{\partial \eta}\right] \tag{11.6.45}$$

Here τ_2 is a 'flow constant'. An integration over $\mathbf{r}'$ provides the total change of η at $\mathbf{r}$.

The final equation for the rate law for the 'mixed' case, derived by Marais & Salje (1991), is

$$\frac{\partial\eta(\mathbf{r})}{\partial t} = -\frac{\tau_1^{-1}+\tau_2^{-1}}{k_B T}\left(1 - \frac{\tau_2^{-1}}{\tau_1^{-1}+\tau_2^{-1}}\frac{\sinh(\xi\nabla)}{\xi\nabla}\right)\left(\frac{\partial g(\mathbf{r})}{\partial \eta}\right) \tag{11.6.46}$$

Here ξ denotes a characteristic length scale of the exchange kinetics.

The τ-dependent terms in Eq. 11.6.46 are seen to form an operator determining the extent of conservation of the order parameter. $\tau_2^{-1}/(\tau_1^{-1}+\tau_2^{-1})$ relates to the ratio ξ_c^2/ξ^2 of the length scales in Eq. 11.6.42. Up to second order, Eq. 11.6.46 is the same as Eq. 11.6.42. Both are Salje equations, in that both scale the degree of conservation of the order parameter.

Going back to the simplistic Landau-Khalatnikov equation (Eq. 11.6.39), it is worthwhile to emphasize that, in the light of the brief description given above, the kinetic coefficients Γ are functions of the order parameter in the case of partial conservation.

New developments continue to occur in this topical area of research, a flavour of which can be found in Salje (1999).

Pattern Formation

Crystals with partially conserved order parameters often involve local correlations among ordering processes. Salje (1993b) and coworkers have investigated 'framework structures' like the mineral cordierite in this context, and certain general trends have emerged from their work. It is found that a mixing of nonconserved and conserved kinetics results in *bifurcation behaviour*: uniform states are formed mainly for the nonconserved order parameter, and periodic patterns (like tweed texture) are formed when there is a sufficient contribution from the conserved order parameter. As transient states these patterns may decay on approach to equilibrium. In the case of conserved order parameters this decay process relates to coarsening of the texture.

Detailed calculations show that for F(xy) and F(xx-yy) type of ferroelastic transitions described earlier in this subsection (which allow the formation of permissible domain walls) the kinetics is such that the fluctuations, which occur even at temperatures far above T_c, freeze into a metastable tweed structure on quenching to a temperature below T_c, and this metastable structure coarsens slowly with the passage of time.

The coarsening process often changes the tweed pattern to a *needle-domain pattern* or a *stripe pattern*. The physical origin of the tendency towards the formation of needle domains is the need for the domains of the ferroelastic crystal to undergo disorientations (§7.1.4) to achieve registry of the lattices across a domain wall.

11.6.7 Ferrielastics and Their Domain Structure

The notion of ferrielasticity has been introduced by analogy with ferrimagnetism. A crystal is said to be in a ferrielastic phase if it satisfies three conditions (Sawada 1990): (i) It has spontaneous strain; (ii) there are two or more equivalent orientation states; and (iii) it is possible to identify two contributions to spontaneous strain which are of *opposite* signs and *different* magnitudes.

A ferrielastic exhibits an anomalous temperature dependence of spontaneous strain not found in ordinary ferroelastics (Sawada 1990).

Some members of the family of crystals $[A(CH_3)_4]_2XBr_4$ have been identified as ferrielastics (Sawada, Matsumoto & Tanaka 1993; Tanaka et al. 1995; Sawada, Watanabe & Tanaka 1997). Here A stands for N^{5+} or P^{5+}, and X for Co^{2+} or Mn^{2+}.

Crystals of $[N(CH_3)_4]_2XBr_4$, with X = Co^{2+} or Mn^{2+}, are ferrielastics. By contrast, $[P(CH_3)_4]_2XBr_4$, with X = Co^{2+} or Mn^{2+}, are ferroelastics (rather than ferrielastics) because their two sublattice strains have the *same* sign. The ferrielastic members of the family exhibit the so-called *square domains*. These domains are delimited by walls which show a characteristic change in brightness while passing through a so-called *compensation temperature* T_z. At $T = T_z$ the two sublattice strains cancel or compensate each other completely.

Like the analogy between ferrielasticity and ferrimagnetism, the concept of *antiferroelasticity* has also been introduced by analogy with the magnetic counterpart (Aizu 1969b). Bratkovsky et al. (1995) have discussed some structural mechanisms responsible for antiferroelastic ordering.

SUGGESTED READING

J. Sapriel (1975). Domain-wall orientations in ferroelastics. *Phys. Rev. B*, **12**, 5128.

G. Metrat (1980). Theoretical determination of domain structure at transitions from twinned phase: Applications to the tetragonal-orthorhombic transition of $KNbO_3$. *Ferroelectrics*, **26**, 801.

A. G. Khachaturyan (1983). *The Theory of Structural Transformations in Solids.* Wiley, New York.

C. Boulesteix (1984). A survey of domains and domain walls generated by crystallographic phase transitions causing a change of the lattice. *Phys. Stat. Solidi* (a), **86**, 11.

L. A. Shuvalov, E. F. Dudnik & S. V. Wagin (1985). Domain structure geometry of real ferroelastics. *Ferroelectrics*, **65**, 143.

C. Boulesteix, B. Yangui, M. Ben Salem, C. Manolikas & S. Amelinckx (1986). The orientation of interfaces between a prototype phase and its ferroelastic derivatives: theoretical and experimental study. *J. Physique*, **47**, 461.

V. K. Wadhawan (1988). Epitaxy and disorientations in the ferroelastic superconductor $YBa_2Cu_3O_{7-x}$. *Phys. Rev. B*, **38**, 8936.

G. Arlt (1990). Twinning in ferroelectric and ferroelastic ceramics: stress relief. *J. Mater. Sci.*¡ **25**, 2655.

M. Guymont (1991). Ferroelasticity in non-group-subgroup transitions. Possibility of double ferroelasticity. *Phase Transitions*, **34**, 135.

E. K. H. Salje (1993a). *Phase Transitions in Ferroelastic and Coelastic Crystals.* Student edition. Cambridge Univ. Press, Cambridge.

A. L. Roytburd (1993). Elastic domains and polydomain phases in solids.

Phase Transitions, **45**, 1.

J. Dec (1993). Paraelastic-ferroelastic interfaces. *Phase Transitions*, **45**, 35.

M. H. Yoo & M. Wuttig (Eds.) (1994). *Twinning in Advanced Materials*. The Minerals, Metals & Materials Society, Pennsylvania.

A. M. Bratkovsky, E. K. H. Salje, S. C. Marais & V. Heine (1995). Strain coupling as the dominant interaction in structural phase transitions. *Phase Transitions*, **55**, 79.

11.7 FERROELASTIC DOMAIN SWITCHING

The switching of one ferroelastic domain to another, under the action of an appropriate driving field like uniaxial stress, occurs through the movement of existing domain walls or by the creation and movement of new ones. We consider some aspects of the statics and dynamics of such domain switching in this section.

11.7.1 The Optimum Switching Configuration

Consider a ferroelastic domain pair $\{S_1, S_2\}$. Let us imagine that the entire specimen is initially in one of the domain states, say S_1. The *critical stress* for this domain pair can be roughly defined as the minimum uniaxial stress needed for converting a measurable portion of S_1 to S_2 across a specified domain wall. Such a definition cannot be very precise because it ignores the time factor.

It should be remembered that, in general, the critical stress for $S_1 \rightarrow S_2$ may not be the same as that for $S_2 \rightarrow S_1$ (see Wadhawan 1982). Also, as indicated in the above definition, since more than one types of domain walls may be possible between S_1 and S_2, the critical stress is not only domain-pair specific, it is also domain-wall specific.

The relative spontaneous strain for a domain pair may be either a shear strain or a tensile strain (with respect to a coordinate system chosen in the prototype in accordance with the usual conventions). For a pure shear strain the needed optimum geometrical configuration for applying the critical stress is quite straightforward to determine (see Salje 1993a). The same is also true for purely tensile strain for a simple system like the tetragonal phase of $BaTiO_3$ (see Wadhawan 1982).

The situation can become somewhat complicated for a tristable ferroelastic like $Pb_3(PO_4)_2$, for which the relative spontaneous strain for any of the three ferroelastic domain pairs has both a diagonal (tensile) part and a nondiagonal (shear) part. For a ferroelastic domain pair in any crystal, and for any general applied-stress configuration, the free-enthalpy difference is given by Eq. 6.2.18:

$$-\Delta g = \Delta e_{(s)ij}\, \sigma_{ij} \tag{11.7.1}$$

$Pb_3(PO_4)_2$ belongs to the Aizu species $\bar{3}mF2/m$, with three orientation states. Consider two of these states, say S_1 and S_2. What is the optimum direction for applying compressive stress so that the least amount of stress can effect ferroelastic switching from S_1 to S_2 ?

Let (l_1, l_2, l_3) be the direction cosines of the direction along which the stress is applied:

$$\sigma_{ij} = l_i l_j \sigma \tag{11.7.2}$$

Then Eq. 11.7.1 takes the following form (Wadhawan 1982):

$$\frac{\Delta g}{\sigma} = \frac{3}{2}a(l_2^2 - l_1^2) - \sqrt{3}al_1l_2 - \sqrt{3}cl_2l_3 + 3cl_3l_1, \tag{11.7.3}$$

where a and c are values of $e_{(s)22}$ and $e_{(s)13}$ respectively for orientation state S_1. The method of Lagrange multipliers can be used, subject to the constraint

$$l_1^2 + l_2^2 + l_3^2 = 1, \tag{11.7.4}$$

for determining the maximum positive value of $\Delta g/\sigma$. The final result is that the direction requiring the least amount of compression for achieving ferroelastic switching from S_1 to S_2 is that along (74.13°, 160°, 81.48°) in the standard Cartesian frame fixed in the trigonal prototypic phase (Wadhawan 1982). This calculation has been made under the parent-clamping approximation; i.e. it ignores disorientations.

11.7.2 Plasticity Related to Ferroelastic Domain Switching

When stress is applied to a crystal, it deforms. The deformation or strain may be elastic or plastic. The strain induced by the stress is said to be elastic if it becomes zero on removal of stress.

Plastic strain, on the other hand, does not become zero when the applied stress is removed. This happens if the atoms or molecules of the object are displaced by the applied stress to new surroundings from which they cannot bounce back to their old places when the external stress is no longer present. There is usually a change in the coordination numbers of at least a fraction of the atoms if the material exhibits plastic behaviour.

Plasticity related to ferroelastic domain switching is also called pseudoplasticity (§11.5.2) (Warlimont 1976). To put the subject of pseudoplastic behaviour in a proper perspective, we first summarize the salient features of 'true' plastic behaviour.

Consider a crystal that is not in a ferroelastic phase. This can happen when, for example, the crystal is in a prototypic phase. When the deforming stress applied to it exceeds the elastic limit, it deforms plastically. The plastic response depends on factors such as the magnitude and direction of the applied load, the temperature, and the rate of loading. And at the atomic-structure level, some of the mechanisms of plastic deformation are: slip, mechanical twinning, and diffusional creep (Klassen-Neklyudova 1963, 1964).

Now consider a crystal in a ferroelastic phase, on which a deforming stress is applied. For a ferroelastic domain pair in it, if the applied stress is higher than the critical stress (defined in §11.7.4 below), plastic deformation (or rather pseudoplastic deformation) by ferroelastic domain switching would take place. And the critical stress is generally an order of magnitude smaller than the stress required for the creation and movement of dislocations. Further, the critical stress approaches the value zero as the temperature rises towards T_c. This is because the contribution to the free enthalpy (Eq. 11.7.1) by the term involving spontaneous strain drops to zero at the phase transition to the prototypic phase.

If there is a large symmetry descent in going from the prototypic symmetry to the symmetry of the ferroelastic phase in question, a large number of domain types become available, which further favours pseudoplastic deformation over plastic deformation.

Thus, in full ferroelastics, particularly when they are in polycrystalline (ceramic) form, pseudoplastic deformation may be the only mechanism by which they get deformed permanently. On the other hand, in a single-crystal partial ferroelastic, domain pairs are possible for which relative spontaneous strain is zero, and their deformation at large applied loads is governed by ordinary plastic processes.

11.7.3 Mobility and Thickness of Domain Boundaries in Ferroelastics

It is clear from the discussion in this section, and that in §11.6, that the question of the thickness and the kinetics of motion of domain boundaries and phase boundaries in the vicinity of a ferroelastic phase transition is linked intimately with the presence of a highly complex, long-ranged, and anisotropic elastic interaction. Some of the rate equations used for building up theories of the kinetics and the microstructure related to the ferroelastic phase transition were described in §11.6.6. The free-energy expansion in such equations includes gradient terms for the order parameter (Gordon 1986, 1991; Salje 1993a; Semenovskaya & Khachaturyan 1991, 1992; Chen, Wang & Khachaturyan 1992). This is because a moving domain wall implies a local perturbation of energy: The system relaxes towards equilibrium via a kinetic pathway determined by the thermodynamic driving force $\partial g \,/\, \partial \eta$.

A central result of these theories is that a suitably defined order parameter has the following profile across the moving domain wall (Collins et al. 1979; Salje 1993a):

$$\eta(r,t) \;=\; \eta_0 \tanh \frac{r - vt}{\omega(1 - \frac{v^2}{c_0^2})^{1/2}} \tag{11.7.5}$$

Here 2ω is the thickness of the wall when its speed v is zero, and c_0 is some characteristic speed. η_0 is the value of the order parameter at the mid-plane of the wall.

Thus the wall thickness varies with its velocity. Also, it increases very rapidly when the critical temperature T_c is approached (Salje 1993a).

Tweed structure, i.e. a transient or metastable collection of aligned microdomains of different orientation states, is expected to be a common feature of structural phase transitions. Its formation provides a partial reduction of the overall strain energy. On further cooling below T_c, the tweed structure generally evolves (coarsens) into a polytwin structure consisting of domains of alternating signs of the spontaneous shear strain. However, if there exists a substantial concentration of impurities or other defects, they may act as *pinning sites*, preventing the relaxation of the tweed texture to a self-accommodating polytwin pattern.

The tweed pattern appears as a precursor even for $T > T_c$, and its coarsening with decreasing temperature has many features in common with spin glasses (Kartha et al. 1991).

The question of the 'real' thickness of domain walls separating ferroelastic domain pairs is debatable. Moreover, different techniques tend to measure different quantities, partly because of the possible presence of disorientations, as well as strain fields. Electron microscopy probes smaller regions than X-ray diffraction, and indicates smaller wall thicknesses than X-ray diffraction (Locherer et al. 1996). Typial values reported for ferroelastic wall thicknesses at low temperatures are 1-10 nm, and follow a temperature dependence described by the Ginzburg-Landau theory (Locherer et al. 1996; Salje et al. 1999).

11.7.4 The Ferroelastic Hysteresis Loop

In the beginning of this chapter, ferroelasticity was defined by us in terms of a ferroelastic phase transition. This ensures that the crystal can exist in at least two equivalent orientation states which differ in relative spontaneous strain. And since in our scheme of things only nondisruptive phase transitions (NDPTs) can possibly be ferroelastic, the switchability of one ferroelastic orientation state to another under the action of a suitable uniaxial stress can be taken for granted. This is because NDPTs do not involve serious structural upheavals, and therefore enthalpy barriers between ferroelastic domain pairs can be expected to be low. Thus hysteretic behaviour in the ferroelastic phase is a consequence of the ferroelastic phase transition.[2]

Stress-strain hysteresis loops (Fig. 1.2.1) characteristic of ferroelastic materials are considerably more difficult to record and interpret than their ferromagnetic and ferroelectric analogues. This is because whereas magnetization, magnetic field, polarization, and electric field are all tensors of rank 1, strain and stress are tensors of rank 2, a fact which increases the complexity of the problem for the mechanical case. Moreover, mechanical fields are more difficult to generate, apply, reverse in sign, and measure, compared to magnetic and electric fields. For such reasons, whereas ferromagnetic and ferroelectric loop tracers are commonplace, there has not been much progress in the development of experiments for generating ferroelastic hysteresis loops routinely.

Fig. 11.7.1 shows a ferroelastic hysteresis curve for $Pb_3(P_{0.8}V_{0.2}O_4)_2$ (Salje & Hoppman 1976; Salje 1993a).

The origin of this plot has been *chosen* so as to achieve symmetry with respect to positive and negative applied stresses. The loss of the 3-fold axis of symmetry in going from the prototype to the ferroelastic phase results in three possible domain types in the ferroelastic phase. After the transition to the ferroelastic phase, a virgin sample on which no external stress has been applied may have all three of the orientation states in equal proportions, so that the net macroscopic strain is zero. On application of a positive uniaxial stress, some domains grow at the cost of others, and the specimen

[2]Earlier authors (e.g. Salje 1993a) have taken stress-strain hysteresis as the *defining* feature of a ferroelastic material. For phases resulting from NDPTs, this approach and our approach are equivalent. However, differences can arise when one is dealing with a *disruptive* phase transition. For reasons discussed in §5.1, we do not deal with disruptive phase transitions in this book. For phase transitions which violate the nondisruption condition it is meaningless to talk in terms of lost or gained symmetry operators. Consequently the very meaning of orientation states becomes qualitatively different in such cases. The situation is similar for ferroelectrics. The occurrence of hysteresis is often taken as a defining feature of a ferroelectric material, although this approach can run into problems when one is dealing with a relaxor ferroelectric.

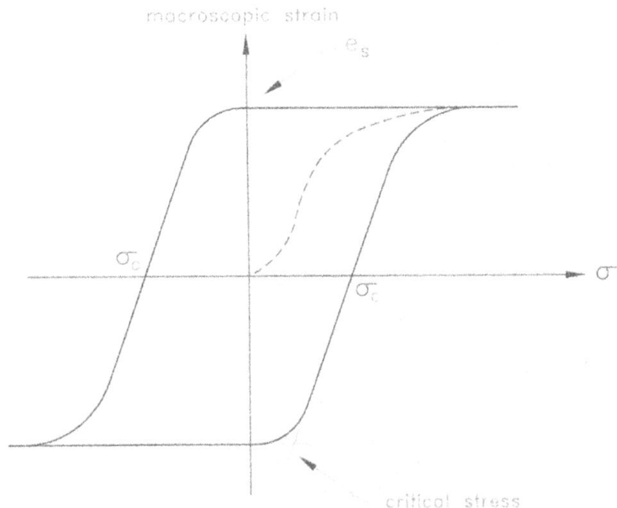

Figure 11.7.1: Ferroelastic hysteresis loop for $Pb_3(P_{0.8}V_{0.2}O_4)_2$. [Adapted from Salje 1993a.]

acquires a nonzero positive macroscopic strain (dashed line in Fig. 11.7.1). On the application of a large enough stress the macroscopic strain reaches a saturation value, beyond which the increase is of a nonferroelastic nature.

When the stress is gradually reduced, the macroscopic strain also decreases, but has a nonzero value at $\sigma = 0$. This is the *spontaneous macroscopic strain.*

The macroscopic strain can be brought to zero by applying a negative stress σ_c, called the *coercive stress.*

A further increase in the magnitude of the negative stress results in a negative value of the macroscopic strain, till it reaches a saturation value with respect to ferroelastic switching.

On reversing the change in the negative stress, the magnitude of the negative macroscopic strain also decreases, but has a negative nonzero value (equal in magnitude to the spontaneous macroscopic strain) at $\sigma = 0$.

Application of too small a positive stress to the specimen does not produce any pseudoplastic strain, but only a reversible elastic strain. At a certain critical value of the applied stress, called the *critical stress,* a measurable pseudoplastic change in the domain structure occurs.

The whole hysteretic cycle can be retraced repeatedly.

SUGGESTED READING

V. K. Wadhawan (1982). Ferroelasticity and related properties of crystals. *Phase Transitions*, **3**, 3.

A. G. Khachaturyan (1983). *The Theory of Structural Transformations in Solids.* Wiley, New York.

E. K. H. Salje (1993a). *Phase Transitions in Ferroelastic and Co-elastic Crystals* (Student edition). Cambridge University Press, Cambridge.

V. S. Boyko, R. I. Garber & A. M. Kossevich (1994). *Reversible Crystal Plasticity.* American Institute of Physics, New York.

K. R. Locherer, S. A. Hayward, P. J. Hirst, J. Chrosch, M. Yeadon, J. S. Abell & E. K. H. Salje (1996). X-ray analysis of mesoscopic structures. *Phil. Trans. R. Soc. Lond. A*, **354**, 2815.

Chapter 12

SECONDARY AND HIGHER-ORDER FERROICS

Ferromagnetics, ferroelectrics, and ferroelastics are the three types of primary ferroics. These were discussed in the previous three chapters. In this chapter we consider secondary and higher-order ferroics. There are six types of secondary ferroics: ferrobielectrics, ferrobimagnetics, ferrobielastics, ferroelastoelectrics, ferromagnetoelectrics, and ferromagnetoelastics. Still higher order ferroics can be defined by referring to Eq. 6.2.18.

12.1 SECONDARY AND HIGHER ORDER FERROIC PHASE TRANSITIONS

Ferroic phase transitions entail the emergence of at least one macroscopic tensor property coefficient. If this property is spontaneous polarization, spontaneous magnetization, or spontaneous strain, at least one domain pair would differ in any of these, and the ferroic phase is a primary ferroic phase. In a non-primary ferroic phase, no domain pair differs in any of these *spontaneous* quantities, but there is at least one domain pair for which an appropriate combination of applied fields can *induce* a difference with respect to one of these properties. Another applied field can then cause ferroic domain switching through one of these induced differences.

Nonprimary nonmagnetic ferroic phase transitions have been analyzed exhaustively, within the framework of the Landau theory, by Toledano & Toledano (1977). We shall refer to this work in subsequent sections of this chapter. Some general results and conclusions are mentioned here.

It has been demonstrated by these authors that both proper and improper transitions of this type are possible. For the former, only 2-dimensional and 3-dimensional order parameters are found, whereas for the latter the order parameters may be 2-, 3-, 4- or 6-dimensional.

The following interesting observation was also made in this work: Most of the actual examples of improper nonferroelastic transitions seem to violate the Lifshitz condition (§5.3.9), whereas most of the improper ferroelastic transitions respect this condition.

More recently, Janovec, Richterova & Litvin (1993) have given an analysis of possible F-operations in nonferroelastic phases. They have shown that there are 48 such operations in all. They have also expressed these operations in terms of dichromatic point groups.

SUGGESTED READING

K. Aizu (1973a). Second-order ferroic state shifts. *J. Phys. Soc. Japan*, **34**, 121.

R. E. Newnham & L. E. Cross (1974a & b). Symmetry of secondary ferroics. *Mat. Res. Bull.*, **9**, 927 and 1021.

P. Toledano & J.-C. Toledano (1977). Order-parameter symmetries for the phase transitions of nonmagnetic secondary and higher-order ferroics. *Phys. Rev. B*, **16**, 386.

R. E. Newnham, S. Trolier-McKinstry & J. R. Giniewicz (1993). Piezoelectric, pyroelectric and ferroic crystals. *J. Mater. Edu.*, **15**, 189.

V. Janovec, L. Richterova & D. B. Litvin (1993). Nonferroelastic twin laws and distinction of domains in nonferroelastic phases. *Ferroelectrics*, **140**, 95.

12.2 FERROBIELECTRICS AND FERROBIMAGNETICS

For a ferrobielectric crystal there exists at least one domain pair such that, when referred to a common system of coordinate axes, the dielectric permittivity tensor differs in at least component for the individuals comprising the domain pair (cf. §6.2.2). This tensor is a polar i-tensor of rank 2. So is the strain tensor. Therefore all ferroelastics are potential ferrobielectrics.

A ferrobimagnetic crystal is defined similarly in terms of the magnetic permeability tensor, which is also a polar i-tensor of rank 2. Therefore all ferroelastics are potential ferrobimagnetics also.

Thus, *all ferroelastics are potential ferrobielectrics and ferrobimagnetics, and vice versa.*

If a ferroelectric state shift is possible for a given domain pair, the free-energy difference for this domain pair is (cf. §6.2.18) $\Delta P_{(s)i} E_i$; i.e. it depends on the first power of the electric field. The corresponding free-energy difference for a ferrobielectric domain pair is $\frac{1}{2}\Delta\epsilon_{ij} E_i E_j$; i.e. it depends on the second power of the electric field. Thus, if a state shift is possible through both these terms, the ferroelectric state shift would occur for a lower electric field, masking the observation of ferrobielectric switching. Ferrobielectric state shifts can therefore be observed with less difficulty for nonferroelectric domain pairs only.

Similar considerations apply to the mutual relationship between ferromagnetic and ferrobimagnetic state shifts.

Although ferrobielectric domain pairs do not differ in *spontaneous* polarization, field-induced switching for them can be understood in terms of an *induced* difference in polarization as follows:

$$-\Delta g = \frac{1}{2}\Delta\epsilon_{ij} E_i E_j = \frac{1}{2}\Delta P_j^{induced} E_j \qquad (12.2.1)$$

The difference in the permittivity tensors of the two domain states makes them respond differently to the same applied electric field, thus creating an induced polarization difference $\Delta P_j^{induced}$. The second component of the electric field, namely E_j, if large enough, can effect a ferrobielectric state shift.

Similarly, for a ferrobimagnetic domain pair:

$$-\Delta g = \frac{1}{2}\Delta\mu_{ij} H_i H_j = \frac{1}{2}\Delta M_j^{induced} H_j \qquad (12.2.2)$$

$NaNbO_3$ and $SrTiO_3$ are likely crystals which may exhibit ferrobimagnetic domain switching (Newnham & Cross 1974a). $SrTiO_3$ is ferroelastic below 110 K. Its 90° domains have a free-energy difference proportional to $(\epsilon_{33} - \epsilon_{11})E^2$, offering the possibility of ferrobielectric switching.

NiO exhibits both ferroelastic and ferrobimagnetic domain switching (Slack 1960; Roth 1960; Newnham & Cross 1974a). It belongs to the ferroic species $m\bar{3}mF\bar{3}m$. A magnetic field of 5000 Oersteds is sufficient to cause ferrobimagnetic movement of domain walls in a well-annealed specimen of this antiferromagnetic crystal.

SUGGESTED READING

R. E. Newnham & L. E. Cross (1974a). Symmetry of secondary ferroics. I. *Mat. Res. Bull.*, **9**, 927.

12.3 FERROBIELASTICS

In the beginning of Chapter 11 we gave the example of the cubic-to- tetragonal phase transition in $BaTiO_3$ to illustrate a ferroelastic phase transition. We now consider the case of the $\beta \to \alpha$ transition in quartz (Salje 1992) to illustrate a nonferroelastic ferrobielastic transition. A comparison of the two cases can be quite instructive.

β-quartz has point-group symmetry 622, whereas α-quartz has 32. The ratio of the orders of the two point groups is 2, so that α- quartz can exist in two orientation states (the so-called Dauphiné twins). The phase transition entails change of lattice parameters (Carpenter et al. 1998), so that the two orientation states have nonzero absolute spontaneous strain. However, they have zero *relative* spontaneous strain because the tensor coefficients describing absolute spontaneous strains of the two states are identical (cf. Eq. 11.1.21).

By contrast, in the case of $BaTiO_3$ the c-axis of the tetragonal phase can point along any of the three mutually perpendicular directions of the basis vectors spanning the cubic phase, so that it is possible to identify three types of domain pairs which differ in relative spontaneous strain.

The orientation states in a ferrobielastic domain pair differ in elastic compliance. Therefore an applied stress can induce a strain difference between the two orientation states, on which stress can act to cause ferrobielastic switching and domain wall motion through the following driving force (Eq. 6.2.18):

$$-\Delta g = \frac{1}{2}\Delta s_{ijkl}\,\sigma_{ij}\,\sigma_{kl} = \frac{1}{2}\Delta e_{kl}^{induced}\,\sigma_{kl} \qquad (12.3.1)$$

The elastic compliance tensor is a fourth-rank tensor. So is the photoelastic tensor, although the two do not have the same intrinsic symmetry (see, for example, Sirotin & Shaskolskaya 1982). If a domain pair is ferrobielastic, but not simultaneously ferroelastic, the optical indicatrices on the two sides of the domain wall are identical in all respects, and the two domain states are not easily distinguishable under the conventional optical microscope. Application of a uniaxial stress induces a strain difference, and, through the photoelastic effect, also a difference in the orientations of the two optical indicatrices, making the two domain states optically distinguishable. However, unlike for a ferroelastic domain pair, when the stress is removed the optical distinction disappears.

If for a domain pair both ferroelastic and ferrobielastic state shifts are allowed by symmetry, the ferroelastic shift is likely to occur at a considerably lower applied stress, and will therefore mask the occurrence of the ferrobielastic state shift. A purely ferrobielastic state shift is therefore best observed for a nonferroelastic domain pair.

There are only five nonferroelastic ferrobielastic species (Newnham & Cross 1974b; Toledano & Toledano 1977): $4/mmmF4/m$; $\bar{3}mF\bar{3}$; $6/mF\bar{3}$; $6/mmmF\bar{3}m$; and $6/mmmF\bar{3}$.

In addition, there are 10 ferroic species which are simultaneously ferrobielastic and ferroelastoelectric: $\bar{4}2mF\bar{4}$; $4mmF4$; $4/mmmF\bar{4}$; $3/mF3$; $6F3$; $\bar{6}m2F32$; $622F32$; $6mmF3m$; $6mmF3$; and $6/mmmF32$. The most important and best-investigated crystal in this category is quartz, which belongs to the species 622F32.

Although ferrobielastic phase transitions have been predicted for several crystals (Newnham & Cross 1974b), only two purely ferrobielastic transitions and two ferrobielastic-ferroelastoelectric transitions are actually known to occur (Toledano & Toledano 1977). The two purely ferrobielastic crystals are NbO_2 and $LaCoO_3$. For the former the symmetry change is from $P4_2/mnm$ to $I4_1/a$. It is an improper ferrobielastic transition, mediated by a 4-dimensional order parameter (Shapiro et al. 1974). However, a more complex sequence of phase transitions in this crystal has been considered by Toledano & Toledano (1977).

For $LaCoO_3$ the symmetry change is from $R\bar{3}c$ to $R\bar{3}$. The transition is proper ferrobielastic, involving a 1-dimensional order parameter (Raccah & Goodenough 1967).

SiO_2 (quartz) and $AlPO_4$ are crystals which have phases which are simultaneously ferrobielastic and ferroelastoelectric. The two crystals are also isomorphous in the prototypic as well as the ferroic phases, undergoing the phase transition $P6_{2,4}22 \rightarrow P3_{1,2}21$ (Bachheimer & Dolino 1975; Lang, Datars & Calvo 1969) [1]. The transition is proper ferrobielastic-ferroelastoelec-tric, involving a 1-dimensional order parameter.

Domain Walls in Ferrobielastics

Quartz is the main ferrobielastic for which some information regarding domain walls is available (Newnham & Cross 1974b; Laughner 1982; Newnham, Trolier-McKinstry & Giniewicz 1993). Whereas for a purely ferro-

[1] It may be mentioned here that SiO_2 also has a high-temperature phase called cristobalite. It is stable between 1743 K and the melting point 2001 K. It has cubic symmetry, $Fd\bar{3}m$, and undergoes a (metastable) first-order phase transition at about 493 K to a tetragonal phase of symmetry $P4_32_12$ or $P4_12_12$ (Hatch & Ghose 1991). Similarly, $AlPO_4$ has a high-temperature (cristobalite) cubic phase of symmetry $F\bar{4}3m$, which undergoes a phase transition to an orthorhomic phase of symmetry $C222_1$ (Hatch, Ghose & Bjorkstam 1994)

elastic domain pair the orientation of domain walls is determined by the strain-compatibility condition, with the strain difference coming from the relative spontaneous strains of the two states, there is no spontaneous-strain difference present in the case of a purely ferrobielastic domain pair. Only *induced* strain differences can occur for the latter case.

For the ferroic species $622F32$ to which quartz belongs, there are only two orientation states or variants, the F-operation relating them being a 180° rotation about the [001] (or x_3) trigonal axis. The two variants are nothing but the Dauphiné twins which occur in α-quartz.

Under this F-operation all elements of the elastic-compliance tensor of α-quartz remain invariant except s_{1123} (and those related to s_{1123} through the symmetry operations of the crystal class 32 to which α-quartz belongs). Referred to a common Cartesian frame of reference, s_{1123} has opposite signs in the two variants of α-quartz.

When the only external field applied is uniaxial stress σ, the enthalpy barrier between the two variants (with contributions from all elastic-compliance coefficients related to s_{1123} by symmetry) has the following form (Indenbom 1960a; Anderson, Newnham & Cross 1977):

$$\Delta g = {}^2g - {}^1g \sim -4s_{1123}(\sigma_{11}\sigma_{23} - \sigma_{22}\sigma_{23} + 2\sigma_{12}\sigma_{13}), \qquad (12.3.2)$$

where ${}^1g$ and ${}^2g$ are the free-energy densities for Variant 1 and Variant 2 (cf. Eq. 6.2.18).

Thus $(\sigma_{11}\sigma_{23} - \sigma_{22}\sigma_{23} + 2\sigma_{12}\sigma_{13})$ is the stress-component combination for achieving ferrobielastic switching in α-quartz. Details of this ferroic switching have been discussed in several publications (Aizu 1973a; Newnham & Cross 1974b; Anderson et al. 1977; Bertagnolli, Kitinger & Tichy 1979; Laughner, Wadhawan & Newnham 1981; Wadhawan 1982; Laughner 1982; Newnham, Trolier-McKinstry & Giniewicz 1993).

The composition plane separating unstressed Dauphiné twins of quartz is usually of an irregular shape, as there is no strain compatibility condition to satisfy when there is no strain difference between these N-twins. The situation, however, is different for the laboratory conditions created for studying the creation and/or movement of this interface. Application of stress induces a strain difference in the two domain stats, and this difference has the following dependence on applied stress (Newnham & Cross 1974b): $\Delta e_{11} = 4s_{1123}\sigma_{23}$; $-\Delta e_{22} = 4s_{1123}\sigma_{23}$; $\Delta e_{33} = 0$; $\Delta e_{23} = 4s_{1123}(\sigma_{11} - \sigma_{22})$; $\Delta e_{31} = 8s_{1123}\sigma_{12}$; and $\Delta e_{31} = 8s_{1123}\sigma_{31}$. We notice that $\Delta e_{33} = 0$. There is thus a tendency expected for the domain walls to be parallel to the x_3-axis or the optic axis. This is indeed observed to be the case.

These domain walls, though generally parallel to the optic axis of quartz, are seldom planar (unlike the situation for permissible domain walls in ferroelastics).

The shapes of domains created in α-quartz by stress depend on applied stress, defects, and sample history. They may be either in the form of thin stripes or thick wedges running parallel to the optic axis, or large-volume diffuse twins (Laughner 1982; Newnham, Trolier-McKinstry & Giniewicz 1993). The magnitude of the coercive stress can be brought down by applying the stress slowly, and also by repeated thermal cycling. This indicates the thermally activated nature of the process of ferrobielastic switching in this crystal. A correlation is also observed between the coercive stress and the shapes of the created domains.

SUGGESTED READING

M. V. Klassen-Neklyudova (1964). *Mechanical Twinning of Crystals.* Consultants Bureau, New York.

R. E. Newnham & L. E. Cross (1974b). Symmetry of secondary ferroics. II. *Mat. Res. Bull.*, **9**, 1021.

P. Toledano & J.-C. Toledano (1977). Order-parameter symmetries for the phase transitions of nonmagnetic secondary and higher-order ferroics. *Phys. Rev. B*, **16**, 386.

G. Dolino & J. P. Bachheimer (1985). Effect of the $\alpha - \beta$ transition on the mechanical properties of quartz. In G. W. Taylor et al. (Eds.), *Piezoelectricity.* Gordon & Breach, U.K.

12.4 FERROELASTOELECTRICS

A ferroic phase of a crystal is said to be ferroelastoelectric if there exists a domain pair for which the piezoelectric-tensor difference Δd_{ijk} is nonzero in the following equation (cf. Eq. 6.2.18):

$$-\Delta g = \Delta d_{ijk} \, E_i \, \sigma_{jk} \tag{12.4.1}$$

If such a domain pair is contiguous, i.e. if we have an orientational twin (cf. §7.1.6), the possibility exists that we can make one twin grow at the cost of the other ('ferroelastoelectric switching') by a *simultaneous* application of electric and mechanical fields of appropriate directions and magnitudes.

Piezoelectric and Electrostriction Tensors Revisited

We stated the usual definitions of piezoelectric and electrostriction tensors in §10.1.4. All the noncentrosymmetric crystal classes except 432 exhibit the piezoelectric effect. If we consider a nonferroelastic piezoelectric crystal under the influence of an electric field alone, Eq. 6.2.12 reduces to

$$e_{ij} = d_{kij}\, E_k \tag{12.4.2}$$

This equation embodies the converse piezoelectric effect, the direct piezoelectric effect being the generation of electric polarization (or electric displacement, as in Eq. 6.2.10) on application of mechanical stress:

$$D_i = d_{ijk}\, \sigma_{jk} \tag{12.4.3}$$

The piezoelectric effect is a *linear* effect. A quadratic effect, called electrostriction, is present in all crystals, including centrosymmetric crystals. If we are dealing with a linear dielectric, i.e. if the polarization **P** (and thence the electric displacement **D**) is proportional to the first power of the electric field **E**, the electrostriction tensor can be defined through Eq. 10.1.31:

$$e_{ij} = M_{ijkl}\, E_k\, E_l \tag{12.4.4}$$

However, ferroelectrics, particularly relaxor ferroelectrics (§10.4.5), are far from being linear dielectrics, and for them the **M**-tensor in Eq. 12.4.4 is a constant only for very small values of the applied electric field. For higher values the **P**(**E**) curve (or the **D**(**E**) curve) shows a saturation effect, implying that the coefficients M_{ijkl} in Eq. 12.4.4 have field-dependent values.

It is found experimentally that, even for relaxor ferroelectrics (e.g. for 0.9PMN-0.1PT (Cross et al. 1980; Nomura & Uchino 1985; Newnham 1990)), P^2 varies linearly with strain. Therefore the following is considered a better equation for defining electrostriction (than Eq. 12.4.4):

$$e_{ij} = Q_{ijkl}\, D_k\, D_l \tag{12.4.5}$$

If we make a similar change in the definition of the piezoelectric tensor, and also ignore the formal distinction between direct and converse piezoelectric effects, we can replace Eq. 12.4.2 by

$$e_{ij} = d_{ijk}\, D_k \tag{12.4.6}$$

Combining Eqs. 12.4.6 and 12.4.5,

$$e_{ij} = d_{ijk}D_k + (Q_{ijkl}D_l)D_k = (d_{ijk} + Q_{ijkl}D_l)D_k \tag{12.4.7}$$

We can interpret this equation in two ways. We can regard the first term as coming from the true piezoelectric effect, and the second from the false or indirect piezoelectric effect. The second term may also be viewed as a correction to be applied to the linear piezoelectric term when the electric field is not vanishingly small (Nye 1957).

Alternatively, we may invoke the Curie principle and examine the question of whether or not the electric field is lowering the net symmetry of the system. For example, even a centrosymmetric crystal will effectively behave as a noncentrosymmetric system while under the influence of an electric field, resulting not only in the modification of those tensor coefficients which are already nonzero, but also in the emergence of new tensor coefficients which are zero when there is no anisotropic external influence present. These new coefficients (e.g. in the second term in Eq. 12.4.7 in the case of a centrosymmetric crystal) are the result of what is called a *morphic effect* (Nye 1957).

The Electromechanical Order of a Ferroic State Shift

Aizu's (1972a) concept of the electromechanical order of a state shift for a nonmagnetic crystal was described in §6.2.2. Out of a total of 773 possible ferroic species, 212 are nonmagnetic. Aizu derived and tabulated the electrical, mechanical, and electromechanical orders of state shifts for each of the 212 nonmagnetic species. Several theorems were also established in this context, and some general conclusions were drawn. We mention a few of them here.

The concept of the order of a ferroic state shift is fairly analogous to that of the order of a structural phase transition. The higher the order of a state shift, the more subtle is the nature of the change in the atomic configuration. Also, since higher-order state shifts can be effected only by higher powers of the driving field(s), such state shifts are likely to be masked by the occurrence of lower order state shifts of the same nature.

It follows from Aizu's work that the electromechanical order of a state shift can never be higher than the electrical order and/or the mechanical order of the same state shift. When it is lower, it is reasonable to expect that a suitable combination of electrical and mechanical fields may be more effective in inducing such a state shift than either field by itself.

Aizu (1972a) also brought out the point that ferroic crystals belonging to the same species not only have the same number of orientation states, but also the same orders of state shifts.

Some Ferroelastoelectric Crystals

The number of nonmagnetic ferroelastoelectric species which are neither ferroelastic, nor ferroelectric, nor ferrobielastic, is 15 (Aizu 1972a; Newnham & Cross 1974b; Toledano & Toledano 1977). We list them here, along with the number (n_{os}) of orientation states in each, as well as the orders of electrical (n_e), mechanical (n_m), and electromechanical (n_{em}) orders of state shifts in each.

Species	n_{os}	n_e	n_m	n_{em}
$mmmF222$	2	3	∞	2
$4/mF\bar{4}$	2	3	∞	2
$4/mmmF422$	2	>4	∞	2
$4/mmmF\bar{4}2m$	2	3	∞	2
$\bar{3}mF32$	2	3	∞	2
$6/mF\bar{6}$	2	3	∞	2
$\bar{6}m2F\bar{6}$	2	3	>2	2
$6mmF6$	2	>4	>2	2
$6/mmmF\bar{6}$	4	3	>2, ∞	2
$6/mmmF\bar{6}m2$	2	3	∞	2
$6/mmmF622$	2	>4	∞	2
$m3F23$	2	3	∞	2
$432F23$	2	3	>2	2
$m\bar{3}mF\bar{4}3m$	2	3	∞	2
$m\bar{3}mF23$	4	3, >4	>2, ∞	2, >2

For most of these, inversion is the only F-operation, in which case, if ferroelastoelectric switching can be effected in a crystal belonging to any of these species, there would be a concomitant reversal of handedness or chirality (the 'ferrogyrotropic effect'; cf. §6.3). A state shift with inversion as an F-operation is electrically odd-order and mechanically ∞th order (Aizu 1972a).

Three confirmed cases of purely ferroelastoelectric phase transitions have been discussed by Toledano & Toledano (1977). These occur in FeS (Townsend et al. 1976); $CsCuCl_3$ (Hirotsu 1975); and NH_4Cl (Wang & Wright 1974).

FeS belongs to the Aizu species $6/mmmF622$, and the phase transition is improper ferroelastoelectric. This strongly discontinuous transition violates the Lifshitz condition (Toledano & Toledano 1977).

NH_4Cl, belonging to the species $m\bar{3}mF\bar{4}3m$, undergoes a proper ferroelastoelectric phase transition involving a 1-dimensional order parameter. The possibility of ferroelastoelectric poling in this crystal, requiring a combination of electric and mechanical poling fields, was first discussed by Newnham (1974) and Newnham & Cross (1974b), and demonstrated by Mohler & Pitka (1974).

More recently, $(C_2H_5NH_3)_2ZnCl_4$ has been shown to undergo a purely ferroelastoelectric phase transition (Tello et al. 1994). The Aizu species in this case is $mmmF222$. Since the prototypic and the ferroic phases belong to the same crystal family (orthorhombic), and there is also a loss of inversion symmetry in going to the ferroic phase, this is a nonferroelastic ferrogyrotropic phase transition.

We have discussed in §12.3 the 10 nonmagnetic species which are simultaneously ferrobielastic and ferroelastoelectric. Two known examples of crystals belonging to such species are quartz and $AlPO_4$. They belong to the same species $622F32$. There are only two orientation states, and the F-operation reverses the sign of the x_1-axis. The two orientation states therefore differ, not only in the sign of s_{1123}, but also of the piezoelectric coefficient d_{111}. When such a crystal is under the influence of not only uniaxial stress, but also electric field $\mathbf{E}$, Eq. 12.3.2 gets modified to (Laughner, Newnham & Cross 1979):

$$\Delta g = -4s_{1123}(\sigma_{11}\sigma_{23} - \sigma_{22}\sigma_{23} + 2\sigma_{12}\sigma_{13}) - d_{111}(E_1\sigma_{11} - E_1\sigma_{22} - 2E_2\sigma_{12}) \tag{12.4.8}$$

Ferroelastoelectric switching in α-quartz was demonstrated by Laughner et al. (1979) by choosing $\sigma = \sigma_{11}$ and $E = E_1$.

SUGGESTED READING

K. Aizu (1972a). Electrical, mechanical and electromechanical orders of state shifts in nonmagnetic ferroic crystals. *J. Phys. Soc. Japan*, **32**, 1287.

P. Toledano & J.-C. Toledano (1977). Order-parameter symmetries for the phase transitions of nonmagnetic secondary and higher-order ferroics. *Phys. Rev. B*, **16**, 386.

G. W. Taylor et al. (Eds.) (1985). *Piezoelectricity.* Gordon & Breach, U. K.

12.5 FERROMAGNETOELASTICS

In §6.2.2 the possible presence of a ferromagnetoelastic state shift for an orientational twin was defined in terms of the relation

$$-\Delta g = \Delta Q_{ijk} H_i \sigma_{jk} \tag{12.5.1}$$

Such state shifts require the simultaneous presence of a magnetic field and a uniaxial stress of appropriate directions and magnitudes. The overcoming of the enthalpy barrier for domain switching in this case can be understood either in terms of an induced strain difference on which the applied uniaxial stress acts, or in terms of an induced magnetization difference on which the applied magnetic field acts. That is, Eq. 12.5.1 can be rewritten either as

$$-\Delta g = \Delta e_{jk}^{induced} \sigma_{jk}, \tag{12.5.2}$$

or as

$$-\Delta g = \Delta M_i^{induced} H_i \tag{12.5.3}$$

The piezomagnetic tensor (Q_{ijk}) is a third-rank axial tensor. There are 66 magnetic point groups for which this tensor is nonzero (Birss 1964). If in a piezomagnetic phase an orientational twin differs only in the sign of the magnetic spin, it is reasonable to assume that ferroic switching can certainly be effected, as the enthalpy barriers for spin flipping are generally small compared to those for nonmagnetic (i.e. structural) ferroic switching. Since ferromagnetoelastic domain switching requires the simultaneous application of magnetic field and uniaxial stress, the presence of ferromagnetism and/or ferroelasticity may often mask the observation of ferromagnetoelastic switching. 31 of the 66 piezomagnetic crystal classes are pyromagnetic (and therefore potentially ferromagnetic). The remaining 35 are antiferromagnetic. One should look for the ferromagnetoelastic effect in those piezomagnetic classes out of these 35 which are the result of (real or hypothetical) nonferroelastic phase transitions. Since ferroelastic phases cannot have cubic or hexagonal symmetry (because for them a crystallographic prototype cannot be defined), one can look for a more easily observable ferromagnetoelastic effect in those 35 antiferromagnetic crystal classes which are cubic or hexagonal. There are 17 of them:

$6'$, $\bar{6}'$, $6'/m'$, 622, $6'22'$, $6mm$, $6'mm'$, $\bar{6}m2$, $\bar{6}'m'2$, $\bar{6}'m2'$, $6/mmm$, $6'/m'mm'$, 23, $m3$, $4'32$, $\bar{4}'3m'$, and $m\bar{3}m'$.

Ferromagnetoelastic switching can also be observed in the remaining 18 antiferromagnetic piezomagnetic classes, provided that either the concerned phase is not simultaneously ferroelastic, or, if it is ferroelastic, it is not a *full* ferroelastic (so that one can look for the ferromagnetoelastic effect in nonferroelastic domain pairs). These 18 classes are:

222, $mm2$, mmm, 32, $3m$, $\bar{3}m$, $4'$, $\bar{4}'$, $4'/m$, 422, $4'22$, $4mm$, $\bar{4}mm'$, $\bar{4}2m$, $\bar{4}'2m'$, $\bar{4}'2'm$, $4/mmm$, and $4'/mmm'$.

CoF_2 and MnF_2 (belonging to the magnetic crystal class $4'/mmm'$) (Borovik-Romanov 1959) and $FeCO_3$ (magnetic symmetry $\bar{3}$m) (Borovik-Romanov 1960; Pickart 1960) are piezomagnetic crystals for which coefficients of the piezomagnetic tensor were determined experimentally. Ferromagnetoelastic switching in $FeCO_3$ was indeed effected by Borovik-Romanov et al. (1962) by applying fields H_1 and σ_{23} below the antiferromagnetic phase transition temperature, i.e. below 30 K (also see Newnham & Cross 1974a).

SUGGESTED READING

R. E. Newnham & L. E. Cross (1974a). Symmetry of secondary ferroics. I. *Mat. Res. Bull.*, **9**, 927.

12.6 FERROMAGNETOELECTRICS

The magnetoelectric effect is the induction of magnetization by an electric field. The linear part of the effect is described by (cf. Eq. 6.2.11):

$$B_i = \alpha_{ij} E_j \tag{12.6.1}$$

Alternatively, it can be described as the induction of electric displacement by a magnetic field (cf. Eq. 6.2.10):

$$D_i = \alpha_{ij} H_j \tag{12.6.2}$$

Ferromagnetoelectric switching in a ferroic crystal is determined by the following free-energy density for a relevant orientational twin (cf. Eq. 6.2.18):

$$-\Delta g = \Delta\alpha_{ij}\, E_i\, H_j, \tag{12.6.3}$$

or

$$-\Delta g = \Delta D_j^{induced}\, H_j, \tag{12.6.4}$$

or

$$-\Delta g = \Delta B_i^{induced}\, E_i \tag{12.6.5}$$

The choice of crystals in which the magnetoelectric effect may be observed is limited by the following two factors: (a) The magnetoelectric effect is absent in all crystals the symmetry groups of which include time-inversion and/or space-inversion operations. In particular, crystals containing antitranslations in their Shubnikov groups are excluded. These are the so-called Type II antiferroelectrics (Sirotin & Shaskolskaya 1982; Landau, Lifshitz & Pitaevskii 1984). (b) The manifestation of the magnetoelectric effect requires the simultaneous application of appropriate magnetic and electric fields. Therefore it is more difficult to observe for those domain pairs which are ferromagnetic and/or ferroelectric.

58 of the 90 magnetic crystallographic point groups permit magnetoelectricity (Birss 1964). If we exclude from these the 18 pyromagnetic classes, we are left with 40 magnetoelectric classes in which to look for the ferromagnetoelectric switching effect. These have been listed by Newnham & Cross (1974a). The list can be further shortened by excluding ferroelectric crystals, or rather ferroelectric domain pairs.

Bertaut & Mercier (1971) tabulated magnetoelectric coefficients for about 20 materials, the most important among these being Cr_2O_3. Magnetoelectricity in this crystal was predicted by Dzyaloshinskii (1959), and measurements were first carried out by Astrov (1960).

When a crystal undergoes a ferromagnetoelectric phase transition, the various domain states in the ferroic phase can, in principle, occur with equal probability. If a judicious application of electric and magnetic fields can achieve preponderance of one domain type over others, it demonstrates the phenomenon of 'poling' through ferromagnetoelectric switching. This was achieved for Cr_2O_3 by Shtrikman & Treves (1963) (also see Newnham & Skinner 1976).

SUGGESTED READING

A. J. Freeman & H. Schmid (Eds.) (1975). *Magnetoelectric Interaction Phenomena in Crystals.* Gordon & Breach, New York.

Yu. I. Sirotin & M. P. Shaskolskaya (1982). *Fundamentals of Crystal Physics.* Section 73. Mir Publishers, Moscow.

L. D. Landau, E. M. Lifshitz & L. P. Pitaevskii (1984). *Electrodynamics of Continuous Media.* Pergamon Press, Oxford.

D. B. Litvin, V. Janovec & S. Y. Litvin (1994). Nonferroelastic magnetoelectric twin laws. *Ferroelectrics*, **162**, 275.

12.7 TERTIARY FERROICS

Tertiary ferroicity is difficult to demonstrate because of its possible masking by the presence of corresponding primary and secondary effects.

According to the symmetry analysis carried out by Aizu (1972a) for nonmagnetic crystals, there are only four ferroic species, out of a total of 212, for which no electrical, mechanical, and electromechanical state shifts of order less than 3 can occur. The number of orientation states is only 2 for all of them. For three, namely $6/mmmF6/m$, $\bar{4}3mF23$, and $m\bar{3}mFm3$, the electrical order of state shifts is 5 or higher, the mechanical order is 3 or higher, and the electromechanical order is 3 or higher. The fourth ferroic species, namely $m\bar{3}mF432$, also has the same characteristics, except that, with space inversion as the F-operation, the mechanical order of state shifts is infinity.

Tertiary or higher-order ferroic behaviour should be looked for in these four species. No actual tertiary-ferroic switching appears to have been demonstrated yet because of the high coercive fields involved, and also because of the low interest in such systems. However, Newnham & Cross (1974b) and Amin & Newnham (1980) have discussed a number of candidate crystals.

It follows from Aizu's (1972a) analysis that ferrotrielectric, and even ferroquadrielectric, behaviour is not possible in any crystal. ferrotrielastic Ferrotrielastic behaviour is possible for crystals belonging to the three species $6/mmmF6/m$, $\bar{4}3mF23$, and $m\bar{3}mFm3$.

The mineral elpasolite (K_2NaAlF_6) belongs to the Aizu species $m\bar{3}mF$ $m3$. Its two orientation states differ in 2 of the 8 independent third-order elastic constants. It is a potential ferrotrielastic.

Another such potential ferrotrielastic identified by Amin & Newnham (1980) is cadmium chlorapatite, $Cd_5(PO_4)_3$. It belongs to the species $6/mmmF6/m$. Its two orientation states differ in the signs of C_{111112} and C_{112331}.

SUGGESTED READING

A. Amin & R. E. Newnham (1980). Tertiary ferroics. *Phys. Stat. Solidi* (a), **61**, 215.

Chapter 13

POLYCRYSTAL FERROICS AND COMPOSITE FERROICS

A crystal, by definition, is an infinite object; otherwise the points of its underlying lattice would not be strictly equivalent. Real crystals, however, are never infinite. We consider size effects in ferroic crystals in this chapter. Also discussed is the ferroic behaviour of polycrystals and composites, which are characterized, not only by a surface, but also by *interfaces* which can be of a more complex nature (e.g. due to the greater likelihood of aggregation of inclusions and pores in their vicinity) than the homophase and heterophase interfaces encountered in single-crystal ferroics.

We first consider free particles (crystallites) of ferroic materials, and then their incorporation in polycrystals (ceramics and polycrystalline alloys) and composites.

We do not consider other types of defects in ferroics in this book (except what was discussed in §5.9, and during the discussion on spin glasses etc.), and instead refer the reader to a review article by Hilczer (1995).

13.1 SIZE EFFECTS IN FERROIC MATERIALS

13.1.1 General Considerations

The presence of a surface introduces a surface-energy term in the total free energy of a ferroic or nonferroic crystal. The relative importance of this term increases as the size of the crystal is decreased, entailing an increase in the surface-to-volume ratio.

Apart from the dominance of the surface-energy term at small sizes, new effects can arise due to the truncation of the infinite lattice. For example, for predominantly covalent crystals, the lattice parameters increase with decreasing size in the nanometer regime, thus simulating negative pressure.

From a practical or experimental point of view, another factor which assumes greater importance as crystal sizes decrease is the role played by surface impurities like adsorbed phases or oxide layers, etc.

Several types of size effects must be distinguished. One is the occurrence of very small self-limiting regions in large single crystals due to causes such as charge imbalance (as in PMN crystals), or very dilute levels of impurities or dopants (as in spin glasses, etc.). We have already considered these in Chapters 9, 10, and 11, and will be mentioned here only for comparison purposes, or for drawing analogies.

The second type of small-size effects occur when sizes of isolated single crystals (i.e. crystallites or particles) are made smaller and smaller. These will be discussed in §13.1.2, 13.1.3 and 13.1.4.

The third type of small-size effects occur in ceramics and alloys with very small grains. These will be taken up in §13.2.

The fourth class of systems in which characteristic size effects can be important and interesting are nanocomposites. We discuss these in §13.3.6. Even thin films can be treated as belonging to this class because they are normally in contact with a second phase, namely the substrate.

The physics of clusters and nanophase materials is already a vast subject (Multani & Wadhawan 1990). Here we can devote space only for some general remarks.

13.1.2 Size Effects in Ferromagnetic Powders

This topic has been reviewed in substantial detail by Multani et al. (1990) and Haneda & Morrish (1990).

In §9.2.10 we described the occurrence of superparamagnetism, as also a specific blocking temperature, for small magnetic particles embedded in rock materials. The basic phenomenon is the same even for free crystallites or particulate matter.

As the size of ferromagnetic crystals is reduced, the saturation magnetization M_s is found to decrease. Coey (1971) investigated γ-Fe_2O_3 particles in the 6.5 nm size range by Mossbauer spectroscopy, and found the the presence of a noncollinear spin structure. More extensive studies have since established the occurrence of similar noncollinear spin arrangements in small crystals of CrO_2, $NiFe_2O_4$, and $CoFe_2O_4$ (see Haneda & Morrish

1990). It has been also established that, at least in the case of $CoFe_2O_4$ particles, there is a core with a collinear spin arrangement, coexisting with outer surface layers in which the spins have a noncollinear (canted) configuration. The presence of the latter naturally results in a lower overall value for M_s.

Such a shell structure for the spins is not observed in fine particles of Fe, FeNi, and FeCo alloys (Morrish & Pollard 1986), although M_s decreases for them also with decreasing particle size. For example, $M_s = 130$ emu/g for Fe particles in the 6 nm size range (Birringer et al. 1986), whereas this value is 220 emu/g for bulk α-Fe.

The ferromagnetic or ferrimagnetic transition temperature T_c is also found to decrease with the size of the crystallite. T_c for Ni falls by about 40 K for 70 nm particles (Valiev et al. 1989).

The lowering of T_c with particle size can be put to practical use. This is best exemplified by the case of barium ferrite ($BaFe_{12}O_{19}$), a hexagonal hard ferrite. It finds applications not only in permanent magnets, but is also being developed as a recording medium and a magneto-optic medium (see Haneda & Morrish 1990 for a review). Among other things, smaller particle sizes mean greater information-storage densities.

The occurrence of a ferromagnetic phase transition entails the formation of a magnetic domain structure in the ferromagnetic phase, at least for a bulk crystal. For sufficiently small particle sizes the formation of domain walls is not favoured, and the whole particle becomes a single-domain particle. An example is that of Fe particles suspended in mercury. For them, the crystallite size below which a single-domain configuration is favoured is about 23 nm (Kneller & Luborsky 1963). And it is in the region of 28 nm for particles of $Fe_{0.4}Co_{0.6}$ (Kneller & Luborsky 1963).

Speaking in general terms, as the size of a ferromagnetic crystallite is reduced, four successive stages of behaviour can be distinguished (Newnham, Trolier-McKinstry & Ikawa 1990):

(1) Polydomain; (2) Single-domain; (3) Superparamagnetic; and (4) Paramagnetic.

Stage 2 is a consequence of the fact that creation of a domain wall costs energy, and below a certain volume-to-surface ratio the trade-off between this cost and the benefit of decreasing the demagnetizing field gets reversed (Kittel 1946).

Stage 3 was discussed in §9.2.10. A typical dimension below which it sets in is 20 nm. This transition to the superparamagnetic state or phase also entails an enhancement of overall symmetry. The basic reason for this, as mentioned in §9.2.10, is this: Because of the cooperative nature of ferromagnetic ordering, the ordering energies involved scale with volume. With particle sizes as small as 20 nm or less, the energy barrier for a collective spin flip (for the entire crystallite) from one easy direction of magnetization to another becomes comparable to thermal-fluctuation energies (Bean & Jacobs 1956). The result is a zero-magnetization, single-domain particle with an extremely high magnetic permeability.

Stage 4 arises because the number of atoms or molecules involved is so small that no substantial long-ranged cooperative ordering, characteristic of a ferromagnetic phase, is possible.

13.1.3 Size Effects in Ferroelectric Powders

The review article by Multani et al. (1990) provides a wealth of information on this topic.

Like the four successive states of a ferromagnetic crystallite, induced by a decrease in the particle size (described above in §13.1.2), there are, in general, four successive size-induced states or phases in a ferroelectric powder: (1) Polydomain → (2) Single-domain → (3) Superparaelectric → (4) Paraelectric.

As a rule, electro-elastic coupling effects are stronger than magneto-elastic coupling effects. Therefore, sample history is more important in determining small-size effects in powders of ferroelectrics than in powders of ferromagnetics.

Ishikawa, Yoshikawa & Okada (1988) and Lee, Halliyal & Newnham (1988) investigated size effects in $PbTiO_3$ powders, and found that its tetragonal form is stable down to 20 nm sizes. The corresponding critical size for $BaTiO_3$ is 1200 Å; for sizes smaller than this, its tetragonal phase at room temperature reverts to the cubic phase (see Newnham & Trolier-McKinstry 1990). However, the behaviour depends strongly on residual strains.

Small-size effects in $BaTiO_3$ and $PbTiO_3$ have been investigated more recently by Bursill et al. (1997) and Jiang, Peng & Bursill (1998a, b). For $BaTiO_3$ the properties of the ultrafine powder were found to depend on the method of preparation. The $BaTiO_3$ nanoparticles prepared by the sol gel (SG) method had a critical size (about 130 nm) below which the occurrence of the tetragonal-cubic ferroelectric phase transition is suppressed. By contrast, the nanoparticles prepared by the steric acid gel (SAG) method remained cubic for all the sizes studied, a behaviour attributed to (non-surface) 'chemical effects' like intergrowth defects (alternating polytypic layers of cubic and hexagonal $BaTiO_3$), vacancies, and charged and uncharged dopants.

Nanocrystals of $PbTiO_3$ prepared by the SG method were found to remain tetragonal down to 25 nm, although the c/a ratio went on decreasing. As revealed also by the Raman scattering work of Ishikawa (1988), the soft mode responsible for the ferroelectric phase transition disappears altogether for particle sizes less than 25 nm, thus indicating the disappearance of the phase transition.

The ferroelectric transition temperature T_c for $BaTiO_3$ was found to decrease with decreasing particle size below 4200 nm, down to 110 nm. The corresponding sizes for $PbTiO_3$ were observed to be 90 nm and 25 nm.

Effect of particle size on the ferroelectric behaviour of PZT powders was investigated by Mishra & Pandey (1995, 1997). It was concluded that, for both tetragonal and rhombohedral compositions, the ferroelectric transition is suppressed below a certain size due to the presence of intense depolarization fields resulting from the single-domain configuration at small sizes (also see Arlt & Pertsev (1991)). An earlier study by Srinivasan et al. (1984) on $Pb(Zr_{0.510}Ti_{0.490})O_3$ particles, prepared by a sol gel method resulting in particle sizes of the order of 16 nm, indicated the absence of any broad (6 atomic percent) morphotropic phase boundary between tetragonal and rhombohedral phases. The material exhibited features of a near-amorphous phase.

Effect of particle size on the ferroelectric phase transition in nanocrystals of $PbSc_{1/2}Ta_{1/2}O_3$ (PST) has been studied recently by Park, Knowles & Cho (1998) for particles in the 160-10 nm size range. The tetragonality ratio c/a decreases monotonically with particle size, becoming unity at 273 K for a mean particle size of 53 nm. For sizes less than 20 nm, and down to 10 nm, the tetragonality of the crystal structure persists, even though there is no peak in the temperature dependence of either the dielectric function or the DSC curve.

13.1.4 Size Effects in Ferroelastic Powders

By analogy with ferromagnetic and ferroelectric powders discussed above, four regimes of the size dependence of the ferroelastic properties crystals could be envisaged:

(i) In large ferroelastic crystals (typically, larger than 1 micron), the usual domain structure occurs, as also the characteristic stress-strain hysteretic behaviour.

(ii) Reduction in size may lead to a single-domain ferroelastic phase (typically for sizes between 1000 nm and 100 nm). Although the overall process involves several contributions to the free energy and its minimization, the

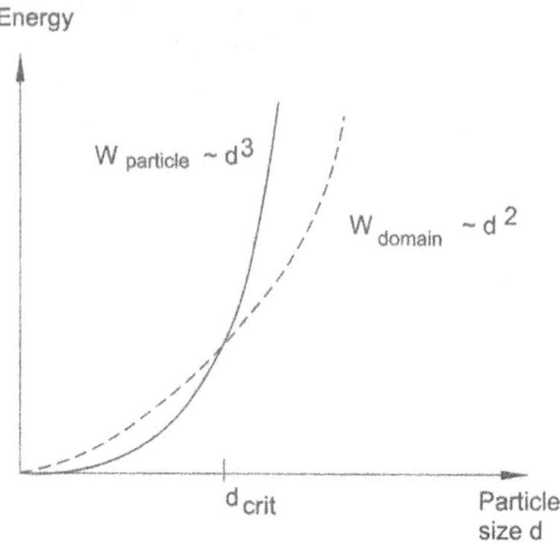

Figure 13.1.1: Competition between the homogeneous elastic-energy term (which varies as d^3) and the domain-wall energy term (which varies as d^2). At $d = d_{crit}$ the two curves cross, and for particle sizes less than d_{crit} spontaneous creation of domain walls is not favoured energetically. [Adapted from Arlt (1990).]

two main terms are the volume term and the surface term (the latter including the domain-wall term also). For a particles size d, the volume term scales as d^3, and the domain wall term as d^2 (Fig. 13.1.1). The size, d_{crit}, at which the two terms match (they have opposite signs) is the size below which a configuration devoid of domain walls is favoured.

(iii) At still smaller sizes a *superparaelastic* phase may ensue. Such a phase may exhibit zero macroscopic spontaneous starin, but very large elastic compliance (Newnham & Trolier-McKinstry 1990).

(iv) For particle sizes approaching molecular dimensions, all vestiges of cooperative ferroelastic ordering may disappear, and the system may revert to a higher-symmetry paraelastic phase.

For ferromagnetic crystals the demagnetization field and its mimimization plays a major role in determining the domain structure, and for small enough crystal sizes other forces become dominant and a single-domain state is favoured. A similar role is played by the depolarization field in the case of a ferroelectric. There is no analogous mechanical field for a *free* ferroelastic particle (Eshelby 1961).

However, if the same ferroelastic crystallite is in an anisotropic environment (e.g. in a ceramic), its spontaneous deformation at a ferroelastic phase transition would create a strain field in the surrounding material, and we would have *mechanical depolarization* (see Arlt 1990).

SUGGESTED READING

M. Multani, P. Ayyub, V. Palkar & P. Guptasarma (1990). Limiting long-range-ordered solids to finite sizes in condensed-matter physics. *Phase Transitions*, **24-26**, 91.

K. Haneda & A. H. Morrish (1990). Mossbauer spectroscopy of magnetic small particles with emphasis on barium ferrite. *Phase Transitions*, **24-26**, 661.

D. Pandey, Neelam Singh & S. K. Mishra (1994). Effect of particle size on ferroelectric transitions. *Indian J. Pure & Appl. Phys.*, **32**, 616.

A. Aharony (1996). *Introduction to the Theory of Ferromagnetism.* Clarendon Press, Oxford.

13.2 POLYCRYSTAL FERROICS

The use of ferroic and other single crystals in devices is generally limited by considerations of feasibility, cost, size, and shape. Very often, polycrystals, particularly ceramics, provide an attractive alternative.

In a ceramic the various crystallites or grains are separated by grain boundaries, and the intergrain bonding is comparable in strength and nature to intragrain bonding. In the preparation of a ceramic (which is generally an oxide), one normally starts with a powder which is often a collection of individual and unbonded crystallites. Their fusion and the resultant formation of grains and grain boundaries results in a material (the *polycrystal*) which has properties that are influenced very substantially by the sizes and shapes of the grains, and by the nature of the grain boundaries; i.e. by the microstructure and the nanostructure. A grain boundary is a region of the polycrystal that acts as a transition layer between two grains, and thus has a structure, composition, and purity level different form that of the grains separated by it.

In a ferroic polycrystal the domain structure inside the crystallites or grains is determined by a number of factors, particularly the grain size and shape. The ferroic domains in a polycrystal can be induced to have a preferred orientation or texture by the process of poling, i.e. by applying a field strong enough to cause domain-wall motion.

13.2.1 Polycrystal Ferromagnetics

Magnetic Ceramics

Magnetic ceramics have been reviewed by Goldman (1988, 1990) and Valenzuela (1994). Most of them contain Fe_2O_3 as the main oxide, although some contain the oxides of Mn, Cr or Al. The term 'ferrites' applies to most of the magnetic ceramics, and they are generally ferrimagnetic, rather than ferromagnetic.

There are three main categories of magnetic ceramics: magnetic spinels, magnetoplumbites, and magnetic garnets.

Spinel is the name of the nonmagnetic mineral $MgO.Al_2O_3$. Magnetic spinels have the composition $MO.Fe_2O_3$, where M = Mn^{2+}, Ni^{2+}, Cu^{2+}, Co^{2+}, Fe^{2+}, Zn^{2+}, or their combinations; even monovalent ions like Li^+ occur in some magnetic spinels. Similarly, the Fe^{3+} ion may sometimes occur in combination with other trivalent ions. Spinels have a pseudocubic chemical unit cell with 8 formula units in it. The true (magnetic) symmetry is certainly not cubic; otherwise ferroic properties cannot arise.

Magnetoplumbites have the general formula $MO.6Fe_2O_3$, with M = Ba, Sr or Pb. The Fe ions may sometimes be replaced partially by Al, Ga, Cr or Mn ions. The crystal structure has pseudohexagonal symmetry.

The best known magnetic garnet is YIG (yttrium iron garnet), having the formula $3Y_2O_3 . 5Fe_2O_3$. The general formula is $3M_2O_3 . 5Fe_2O_3$, or $M_3Fe_5O_{12}$. All the metal ions are trivalent. The crystal structure has a dodecahedral configuration.

The ferrites allow a very high degree of compositional tunability to suit specific applications.

Permeability of Magnetic Ceramics

For a metallic or insulator magnetic polycrystal to exhibit high permeability, it is necessary that the domain walls in it should move easily under an applied magnetic field. In the case of ferromagnetic metals, very high permeability can be achieved because the grain boundaries are relatively thin and domain walls can generally move right across them from one grain to another. Pinning centres such as pores are also quite negligible for them. The situation is different in ferrites. For them the domain walls are thicker, pinning centres like impurities, pores and inclusions abound, and chemical inhomogeneity can be high. If the average grain size is small, it further leads to a lowering of the permeability.

Resistivity of Magnetic Ceramics

Ferrites score over metallic magnets in applications where high resistivity is a necessity. This happens when prevention of eddy-current losses is a major consideration, as in high-frequency applications.

In ferrites, the resistivity depends on factors such as the iron content, and the nature and number-density of grain boundaries. Addition of CaO has been found to increase the resistivity.

Polycrystalline Magnetic Alloys

Properties of practically all the magnetic materials, both alloys and insulators, originate from three elemental constituents, namely Fe, Co, and Ni, which are ferromagnetic at room temperature. We have already considered the magnetic insulators, i.e.the ferrites. We now focus on the magnetic alloys.

Compared to ferrites, magnetic alloys have large values of saturation magnetization, M_s, as also higher T_c values. This is because in metallic systems there are no intervening oxygen atoms to 'dilute' the magnetic moment per unit volume. For the same reason, i.e. because of the closer distances between the interacting spins, strong direct-exchange ordering interactions are established, which can withstand the disordering thermal forces up to higher temperatures compared to the oxides.

The relative permeability of a ferromagnet is high if its M_s/K ratio is high, where K denotes the overall anisotropy (magnetocrystalline anisotropy, shape anisotropy, etc.). For metallic ferromagnets, not only is M_s high in general, even K can be made small through several techniques (see Valenzuela 1994). Therefore, compared to magnetic ceramics, extremely high values of permeability (approaching 2×10^6) have been achieved in magnetic alloys.

As mentioned earlier, an interesting feature of grain boundaries in magnetic polycrystals is that the domain walls in the magnetic grains can be generally made to move past the grain boundaries from one grain to another. This is consistent with the high permeabilities exhibited by such materials.

The element Fe is the basic component used for making all soft ferromagnetic alloys. These alloys can be tailor-made to exhibit a very wide range of properties, keeping in view the intended applications.

The most widely used soft magnetic alloy is Fe – Si. Some other important alloys are Fe – Al, Fe – Ni, Fe – Co, and Fe – Pt. For a good review of their properties the book by Valenzuela (1994) should be consulted.

13.2.2 Polycrystal Ferroelectrics

Ferroelectricity can occur in only 10 of the 32 crystal classes. These 10 polar classes (1, 2, 3, 4, 6, m, $mm2$, $3m$, $4mm$, $6mm$), being noncentrosymmetric, form a subset of the 21 noncentrosymmetric classes. All these 21 classes, except one, are piezoelectric, the exception being the class 432. Thus all ferroelectric classes are piezoelectric as well, although the converse is not true (cf. Fig. 10.1.1). Piezoelectrics, particularly from the applications point of view, have a close relationship with ferroelectrics.

Further, since all the ferroelectrics with which we deal here are insulators (we do not discuss semiconductor ferroelectrics due to limitations on space), the term 'polycrystalline ferroelectrics' becomes synonymous with 'ceramic ferroelectrics'.

Most of the ceramic ferroelectrics are used either as piezoelectrics, or, if made transparent, as optical electroceramics.

Piezoceramics

When a ceramic of a piezoelectric crystal is prepared, initially its grains are oriented randomly, making it an effectively centrosymmetric material. Therefore, it cannot exhibit the piezoelectric effect unless it is poled. The process of poling amounts to switching some domains into directions that are more favourably inclined with respect to the poling field. Naturally, this switching cannot occur unless the crystalline species has alternate orientation states available to it, to which an unfavourably oriented domain can switch, i.e. unless the material is a ferroelectric. Thus ceramics of only those piezoelectrics can be poled which are ferroelectric as well.

Nonferroic piezoelectrics cannot be poled.

If a material is simultaneously ferroelectric and ferroelastic, then even uniaxial stress can be used for poling its ceramic so that it exhibits the piezoelectric effect.

The most important piezoceramic is PZT ($PbZr_xTi_{1-x}O_3$). It has the same perovskite structure as $BaTiO_3$ (Clarke & Glazer 1974; Corker, Glazer, Whatmore, Stallard & Fauth 1998). In its $T - x$ phase diagram there occurs a 'morphotropic' (temperature independent) phase boundary (MPB) near $x = 0.53$. It is a nearly vertical line parallel to the T-axis, and extends to a temperature above which the structure acquires the prototypic cubic symmetry. At room temperature, and near the MPB, the Zr-rich side has tetragonal symmetry. The material displays a very high dielectric permittivity for compositions near the MPB. Also, the electromechanical coupling coefficient, k, is very large for compositions near the MPB, implying a large conversion of the input electrical (mechanical) energy to mechanical (electrical) energy in transducer applications of the material (Mishra, Pandey & Singh 1996; Mishra, Singh & Pandey 1997; Mishra & Pandey 1997). The large electromechanical response in the MPB region is attributed to the presence of the tetragonal to rhombohedral phase transition nearby, and the consequent peak in the dielectric response (Karl & Hardtl 1971; Wersing 1981; Cross 1993; Mishra & Pandey 1997).

Compared to piezoelectric single crystals, piezoceramics can be produced more easily, at a much lower cost, and in a variety of shapes and sizes (see, for example, Pohanka & Smith 1988). PZT has the highest d_{33} coefficient of all the known single-phase materials, a typical value for a properly fabricated and well-poled specimen being about $400 \times 10^{-12} C/N$.

Optical Electroceramics

The use of transparent ferroelectric single crystals like KDP, $BaTiO_3$, and GMO for electro-optical applications is limited by considerations of cost, size and shape. During the late 1960s, processes were developed for producing optically transparent electroceramics. In 1971, Haertling & Land reported the fabrication of 100% transparent lanthanum-doped PZT (PLZT) ceramic (not counting the 18% reflection losses). With the use of broadband antireflection coatings, 98% transmission can be obtained these days (see Haertling (1980) for a review).

In PLZT some of the Pb^{2+} sites are occupied by La^{3+} ions. It can be therefore represented by the formula $Pb_{1-3y/2}La_y(Zr_zTi_{1-z})O_3$. The short-hand notation used for specifying this composition is $y/z/(1-z)$, where $z/(1-z)$ is the Zr/Ti ratio, and y is the atomic percentage of La^{3+} ions.

Depending on the composition, as well as their intended application, PLZT ceramics have been divided into three main categories in terms of their relevant electro-optic characteristics: 'memory', 'linear', and 'quadratic'.

The Memory Characteristic. A typical composition of PLZT displaying memory characteristics is 8/65/35, with an average grain size of 2 microns. By successively applying electric pulses in two mutually perpendicular directions, two optically distinct states of the ceramic, corresponding to different orientations of the ferroelectric domains, can be established. The ceramic exhibits memory in the sense that the domain orientations are stable, i.e. they do not revert back to their original states when there is no external field present.

The Linear Characteristic. For applications based on the Pockels effect a suitable composition of PLZT is 8/60/40. It has tetragonal structure, and high coercivity. The ceramic shows excellent electro-optical linearity for dc fields ranging from -1.4 MV/m to 2.0 MV/m, with an effective Pockels coefficient, r_c, of about 100 pm/V. This is 6 times larger than that for single-crystal $LiNbO_3$. An alternative composition, 8/65/35, has an even higher value of 520 pm/V, but this is at the expense of linearity. Grain size has a strong effect on the linearity of the response.

The Quadratic Characteristic. Compositions of PLZT close to the paraelectric-ferroelectric phase boundary, particularly 8.8/65/35, 9.5/65/35, and 8/70/30, are popular for Kerr-effect applications. These compositions fall in the relaxor-ferroelectric regime, and undergo field-induced phase transitions. At room temperature the gross point-group symmetry is cubic (the same as the actual symmetry of the prototypic phase), and only a slim hysteresis loop is observed (§10.4). Application of electric field induces a transition to the optically anisotropic rhombohedral or tetragonal 'normal' ferroelectric phase (with a fat hysteresis loop). The optical anisotropy increases quadratically with electric field.

These compositions are also strongly ferroelastic, and even exhibit the shape-memory effect (SME). The SME in alloys cannot be manipulated substantially by electric fields because of the high electrical conductivity. This limitation is not there for an insulator like PLZT. The effect of electric field on the SME in 6.5/65/35 PLZT was demonstrated by Wadhawan et al. (1981). It would be interesting to produce the SME in such a material by purely electrical means, without any externally applied mechanical stress.

The ferroelastic nature of the quadratic compositions of PLZT enables their biasing and poling by mechanical means. This fact is exploited in applications like image storage (Haertling 1980).

The composition 8.2/70/30 is antiferroelectric, but because of its high degree of 'electric softness' it becomes ferroelectric on application of a field greater than 1 MV/m. The hysteresis loop is therefore very narrow near the origin, and wide for large fields. Concomitant to this field-induced transition to the ferroelectric phase is the development of birefringence of the order of 0.05, and a marked increase in the scattering of light. This is particularly pronounced for grain sizes in the 10-15 microns range. This scattering effect is employed, in the longitudinal mode, in device applications for reducing the light flux over angles of about 3°.

13.2.3 Polycrystal Ferroelastics

The subject of polycrystalline ferroelastics has developed mostly in the context of the SME exhibited by several alloy systems. Among the insulators there are systems which are simultaneously ferroelastic and ferroelectric, and perhaps the best examples of these are the PLZT compositions with the quadratic characteristic (§13.2.2).

Among the polycrystalline materials investigated which are purely ferroelastic, i.e. which are not simultaneously ferroelectric and/or ferromagnetic, an interesting case is that of the Y – Ba – Cu – O ceramic. SME in it was demonstrated by Tiwari & Wadhawan (1991). Among the reasons cited by these workers for expecting the SME in it was the fact that, at least for the geometry adopted by them for conducting the experiment, the important requirement of crystallographic reversibility (cf. §11.5.3) was indeed met. Whereas the actual crystalline symmetry of the tetragonal and orthorhombic phases of this material is $4/mmm$ and mmm, the symmetry of the ceramic is ∞/mm in the parent (nominally tetragonal) phase. This enhanced symmetry of the polycrystal, compared to the single crystal, is partly instrumental in ensuring crystallographic reversibility of the phase transition for the geometry of the system investigated.

SUGGESTED READING

L. M. Levinson (Ed.) (1988). *Electronic Ceramics.* Marcel Dekker, New York.

R. E. Newnham (1989). Electroceramics. *Rep. Prog. Phys.*, **52**, 123.

G. Arlt (1990). Twinning in ferroelectric and ferroelastic ceramics: stress relief. *J. Mater. Sci.*, **25**, 2655.

Y. Xu (1991). *Ferroelectric Materials and Their Applications.* North-Holland, Amsterdam.

M. J. Mayo (1993). Synthesis and applications of nanocrystalline ceramics. *Materials & Design*, **14**, 323.

R. Valenzuela (1994). *Magnetic Ceramics.* Cambridge University Press, Cambridge.

S. Li, J. A. Eastman, R. E. Newnham & L. E. Cross (1996). Susceptibility of nanostructured ferroelectrics. *Jpn. J. Appl. Phys.*, **35**, L502.

13.3 COMPOSITES WITH AT LEAST ONE FERROIC CONSTITUENT

How much finer things are in composition than alone.

Ralph Waldo Emerson, *Journals*

A composite is a material made up of two or more submaterials (or constituents, or phases) which are strongly and intimately bonded together.

From the applications point of view, composites can be branded as either *structural* or *nonstructural*.

Re-enforced cement concrete (RCC), particle board, and plywood are examples of structural composites. Compared to nonstructural composites, the field of structural composites is old and highly developed (for a recent review, see Hanson (1995)).

In this section we focus, by and large, on nonstructural composites.

13.3.1 General Considerations

Sometimes an application of a material may have conflicting requirements. For example, we may wish to make a transducer material which has a large d_{33} or d_{31} piezoelectric coefficient, low density, and high mechanical flexibility. Although a poled PZT ceramic is strongly piezoelectric, it is also very brittle and dense. Compared to it, a polymer like PVF_2 (polyvinylidene difluoride) is flexible and light-weight, but not strongly piezoelectric. If we combine the two, and make a carefully patterned composite (the pattern depending on the intended application), the composite material can not only meet the conflicting requirements, it may even have enhanced properties. In fact, a composite may sometimes even have *new* properties, not present in any of its constituent phases separately.

The freedom provided by the large number of ways in which two or more constituent materials can be configured together to design a composite with various connectivities and symmetries offers fascinating possibilities. And very often the ideas for the design come from how nature does it in forming the tremendous variety of crystals from the 100-odd elements. *Crystals are nature's own composites*, although on the very small atomic scale (Newnham & Trolier-McKinstry 1990).

13.3.2 Sum, Combination, and Product Properties of Composites

Sum Properties

A trivial example of a sum property of a composite is its density, which is the average (arithmetic mean) of the densities of its constituent materials.

In §3.1.7 we defined a tensor property **T** as relating a force **X** to a response **Y** through a linear constitutive relation **Y** = **TX**. The property **T** can be described as an X-Y effect.

Consider a composite which has two constituents, each with its own X-Y effect. The proportionality tensors **T** are different for the two constituents. Together the two materials produce a net X-Y property for the composite (they may produce other properties also, some of them new). Such a property is a sum property (van Suchtelen 1972). It is obtained as a weighted sum of the T-tensors for the constituent phases. The weighted summation is not always an easy task, as the sum property may depend critically on the mutual geometrical configurations of the constituent materials (see Hale 1976).

Examples of sum properties include electrical and thermal resistivity, dielectric permittivity, thermal expansion, and elastic compliance (Hale 1976).

Electrical resistivity is a good example for visualizing how a sum property depends on the geometrical arrangement in the composite. One extreme is when the constituent phases are aligned parallel to the probing field, and the other extreme is when they are perpendicular.

The resistivity $\bar{\rho}$ of the composite can be expressed as

$$\bar{\rho}^n = V_1 \rho_1^n + V_2 \rho_2^n + \cdots, \qquad (13.3.1)$$

where V_1, V_2 are the volume fractions of the two component phases; ρ_1, ρ_2 are their resistivities; $n = 1$ for a series configuration of the resistances; and $n = -1$ for parallel mixing.

In a general situation the two may occur together, and the sum property lies between the arithmetic mean and the geometric mean, although its realistic calculation can be a highly nontrivial task, particularly if the two constituents have widely different properties.

When a sum rule similar to Eq. 13.3.1 is written for capacitances, or dielectric permittivities, the values of n get interchanged for series and parallel configurations.

Combination Properties

Consider a property that is a combination of two properties. For example, the acoustic speed v of a wave travelling along a long thin rod depends on Young's modulus E and density ρ: $v = (E/\rho)^{1/2}$. When a composite is made from two or more materials, the mixing rule for their Young's moduli may not be the same as the mixing rule for their densities (Newnham 1985, 1986, 1988). The mixing rule for the combination property can then become a very complicated affair. It can even happen that, unlike for sum properties, the value of the combination property may lie *outside* the range defined by the values of the property for the pure phases.

Newnham (1986) cites experimental evidence for an actual example of this. For a composite made by embedding oriented steel filaments in an epoxy matrix, the wave velocity v_T for a wave travelling transverse to the direction of orientation of the steel filaments is measured to be less than the value of this velocity for both epoxy and steel. This is attributed to different effects of volume fraction on stiffness and density.

How strongly the mixing rules can depend on direction is well illustrated by the same example, wherein it is found that, unlike the transverse velocity v_T, the longitudinal velocity v_L does behave like a sum property, implying that for v_L the stiffness and density follow the same mixing rule.

Product Properties

Consider a diphasic composite, with an X-Y effect in Phase 1, and a Y-Z effect in Phase 2. The two phases are so configured that when the response Y occurs in Phase 1 on application of force X, this response gets transferred to, i.e. it acts as a force on, Phase 2 (albeit with some dissipation or damping factor). The X-Y effect and the Y-Z effect are then said to be coupled, and the composite exhibits an X-Z effect. The last-mentioned effect is an example of a product property (van Suchtelen 1972).

As an example, suppose Phase 1 is magnetostrictive, and Phase 2 is piezoelectric. Then X is magnetic field, and Y is the magnetostrictive strain produced in Phase 1 by the magnetic field. The strain Y can result in a pressing of Phase 2, leading to the creation of an electric field (Z) through the inverse piezoelectric effect. The net result is a magnetoelectric effect. And this can happen even when neither Phase 1 nor Phase 2 are individually magnetoelectric.

This is the remarkable thing about product properties: They can arise in a composite even when they cannot occur in the constituent phases for reasons such as symmetry restrictions.

In the above example the coupling is mechanical. Several examples of possible electrical, optical, magnetic, thermal and chemical couplings have been considered and tabulated by van Suchtelen (1972) and Hale (1976).

The X-Y effect in bulk Phase 1 is described by a tensor T_1, and the Y-Z effect in bulk Phase 2 is described by a tensor T_2:

$$T_1 = \frac{\partial Y}{\partial X}, \tag{13.3.2}$$

$$T_2 = \frac{\partial Z}{\partial Y} \tag{13.3.3}$$

Let T_3 be the tensor describing the product property X-Z. Then

$$T_3 = \frac{\partial Z}{\partial X} = k_c k_s T_1 T_2 \tag{13.3.4}$$

Here k_c and k_s are factors of magnitude less than unity.

k_c is a measure of the coupling efficiency between the two phases. If this coupling is poor, only a small fraction of the effect Y in Phase 1 would be transmitted to Phase 2.

The factor k_s is a structural coupling factor, which depends not only on the volume fractions of the two phases, but also on their geometrical layout.

Situations exist wherein these coupling factors are complex, rather than real. They may even be tensor quantities (van Suchtelen 1972).

It is also possible that the X-Z effect occurs as a net effect from more than one Y-effects, so that Eq. 13.3.4 has to be generalized to

$$T_3 = \frac{\partial Z}{\partial X} = k_c \sum_i k_{si} T_{1i} T_{2i} \tag{13.3.5}$$

13.3.3 Symmetry of Composites

The directional symmetry of any macroscopic property of a material is influenced by its point-group symmetry. Single crystals can be normally taken as homogeneous media for describing this macroscopic symmetry. The corresponding proposition for composites has to be examined carefully because of the various length scales involved. The dimensions of the phases in a composite may be anywhere between a few nanometers to a few centimeters. Therefore, for assigning a point-group symmetry to such a diphasic or multiphasic object one must choose a length scale which is large enough to produce a meaningful and realistic average over a sufficiently representative volume of the composite. When we speak of the point-group symmetry of a composite we implicitly assume that this aspect has been taken care of.

The phases or materials which constitute a composite have symmetries of their own. In addition, and equally importantly, they have a specific rotational and translational relationship with one another. The individual symmetries, as well as the mutual disposition, together define the net point-group symmetry of the composite, in accordance with the Curie-Shubnikov principle (cf. Appendix C).

Having chosen the submaterials that would constitute a composite, the designer still has at his disposal the choice of their mutual spatial relationship. The net point-group symmetry, and thence the symmetry of the macroscopic properties of the composite, can be altered by altering this spatial relationship. This offers a remarkable degree of flexibility in developing new materials for specific applications.

The Curie-Shubnikov principle, which determines the symmetry of a composite, has a dissymmetrization aspect, and sometimes also a symmetrization aspect. In other words, the symmetry of a composite may be either equal to or higher than the common minimum symmetry. And the common minimum symmetry, defined by the intersection group G_d in Eq. C.1.3, is never higher than (in fact, it is mostly lower than) the symmetries of the individual components of the composite (cf. Eq. C.1.4).

It is also a self-evident proposition that if the symmetry decreases, the number of nonzero independent coefficients of any tensor property increases. And if the symmetry increases, this number goes down. A familiar example is that of the dielectric permittivity tensor. It has only one independent component for crystals belonging to the cubic crystal classes, two independent components for trigonal, tetragonal and hexagonal classes, and three components for orthorhombic, monoclinic and triclinic classes (under appropriate similarity transformations, where needed).

Translated to the field of composites (the design of which can be chosen by human and other living beings, rather than by nature alone), such considerations lead to two remarkable generalizations:

(a) *It is possible, in principle, to generate new property coefficients by so designing a composite as to decrease its overall symmetry appropriately; i.e. by effecting dissymmetrization.*

(b) *It is possible, in principle, to get rid of some undesirable properties by appropriately increasing the symmetry of a composite; i.e. by effecting symmetrization.*

Considerable help can be obtained in achieving these objectives by invoking the Hermann theorem of crystal physics (cf. §3.3).

We illustrate these ideas by considering two examples.

The first example illustrates the consequences of dissymmetrization in a composite, namely creating a magnetoelectric material comprising two

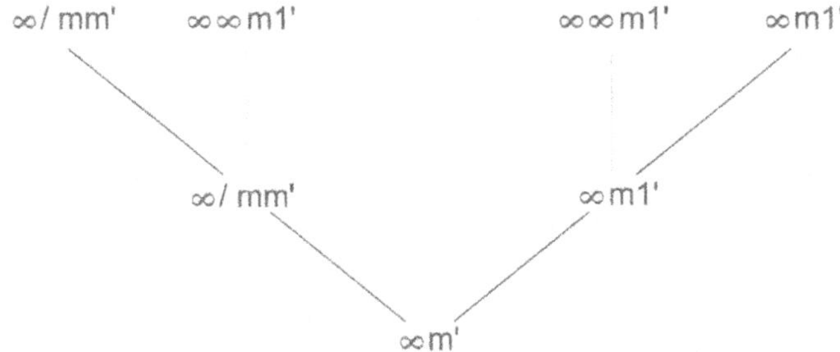

Figure 13.3.1: Group-tree depicting the various dissymmetrization processes occurring in a polycrystalline $BaTiO_3 - (Co, Ti)Fe_2O_4$ composite, poled and magnetized by parallel electric and magnetic fields.

submaterials neither of which can exhibit magnetoelectricity individually. Such a composite material was developed at the Philips Laboratories by combining $BaTiO_3$ and $(Co, Ti)Fe_2O_4$ (van den Boomgaard et al. 1974, 1978; van Run et al. 1974; also see Bracke & van Vliet 1981, and Hanumaiah et al. 1994). The first material is a ferroelectric, and the second a ferrimagnetic.

$BaTiO_3$ has point-group symmetry $4mm1'$ in the single-crystal state. In the composite it occurs as an electrically poled ceramic, the symmetry of which is the same as that of a cone, namely $\infty m1'$ (see the right-hand part of the group tree depicted in Fig. 13.3.1).

The single-crystal symmetry of $(Co, Ti)Fe_2O_4$ is $4/mm'm'$. In the composite, under the action of an applied magnetic field, it acquires a symmetry which is described by the magnetic Curie group ∞/mm' (cf. the left-hand part of Fig. 13.3.1) (Newnham, Skinner & Cross 1978; Newnham et al. 1980; Newnham 1985).

In accordance with the Curie principle of superposition of dissymmetries, the net symmetry of the composite comprising the two submaterials is given by the intersection group G_d:

$$G_d = \infty m1' \cap \infty/mm' = \infty m' \tag{13.3.6}$$

The symmetry group $\infty m'$ allows the magnetoelectric effect, whereas the component groups $\infty m1'$ and ∞/mm' do not (Newnham 1985).

Physically, the processes involved are as follows. Application of a magnetic field does nothing significant to the $BaTiO_3$ ceramic, but it causes a magnetostrictive strain in the magnetically poled ceramic of the spinel $(Co, Ti)Fe_2O_4$. The dimensional changes caused by this strain get transmitted to the poled $BaTiO_3$ piezoelectric ceramic, and an electric voltage arises in it through the piezoelectric effect. Thus an XY property (magnetostriction) couples with a YZ property (piezoelectricity) to give the product property XZ, namely magnetoelectricity.

What is equally striking is the magnitude of the magnetoelectric effect produced. It is a hundred times more than that in Cr_2O_3.

We now consider an example of a composite which has been designed to possess *higher* macroscopic symmetry than its building blocks, with the objective of eliminating some undesirable properties through the process of symmetrization. The example is that of plywood. It is a structural composite, but similar configurations have also been adopted for making nonstructural ceramics (e.g. fibre-reinforced polymer-matrix laminated composites designed for embedded-sensor applications (Hansen 1995)).

Cross-ply plywood is made of an odd number of plies of wood, bonded together so that the fibre axes of successive plies are at right angles (Stavsky & Hoff 1969; Countryman, Carney & Welsh 1969; Hansen 1995). The outermost plies are called *faces*, and the term *centres* is used for plies which have their fibre axes parallel to those in the faces. And the plies with fibre axes at right angles to these are called *cores*. Thus cross-ply plywood can be viewed as made of two interpenetrating objects, one consisting of all the centres (symmetry $G_1 = mmm$), and the other consisting of all the cores (symmetry $G_2 = mmm$).

The remarkable thing about such a design is that, whereas each ply, if alone, can warp under changes of temperature and humidity, the composite cross-ply structure does not.

This and some other properties of plywood can be understood in terms of the Curie-Shubnikov principle and the Hermann theorem of crystal physics (Wadhawan 1987a).

G_1 and G_2 are the same symmetry group mmm. The intersection group G_d is also mmm. The symmetry of the overall object is obviously higher than G_d, and we have therefore to identify a suitable symmetrizer (cf. §C.2) to describe correctly the enhanced or extended symmetry of cross-ply plywood. The symmetrizer can be readily seen to be $g_2 = 4_z$, so that the symmetry of this plywood configuration is (cf. Eq. C.2.2)

$$G_s = (mmm) \cup 4_z(mmm) = 4/mmm \tag{13.3.7}$$

With reference to the Hermann theorem, this corresponds to $N = 4$ (cf. §3.3). This means that, on a sufficiently macroscopic scale, the system has an axis of 4-fold rotational symmetry at the point-group level. It follows that if we consider any tensor property of the material of rank r, then for $r < N$, i.e. for $r < 4$, the effective value of N is ∞ for that tensor property. Thermal expansivity is one such property. For it, $r = 2$. Since $N > 2$ for cross-ply plywood, the Hermann theorem predicts that $N = \infty$ for the thermal expansivity of this composite material. That is, we get isotropic thermal expansion (or zero warping) in the plane normal to the pseudo-4-fold axis of cross-ply plywood. This is also known as *transverse isotropy*: We say that cross-ply plywood displays transverse isotropy with respect to thermal expansivity.

Since $N = \infty$ so far as thermal expansivity of cross-ply plywood is concerned, the number of independent components of this tensor is the same for the symmetries ∞/mm and $4/mmm$; this number is 2, the two independent components being α_{11} and α_{33} (see, e.g., Sirotin & Shaskolskaya 1982).

By contrast, if we consider a single ply , its point-group symmetry is mmm, implying that $N = 2$. Therefore, since r is only equal to N, and not less than N, Hermann theorem does not predict transverse isotropy for thermal expansivity. Thus warping can occur (i.e. it is not forbidden by symmetry) in a single ply of wood on variation of temperature.

Next we consider a higher-rank property for the same composite, namely the elastic-stiffness tensor, for which $r = 4$. Since $r = N$, the Hermann theorem does not predict transverse isotropy of elastic stiffness. This is indeed verified by detailed calculation. The number of independent elastic constants for a material of point-group symmetry $4/mmm$ is 6, compared to the number 5 for the symmetry class ∞/mm (Stavsky & Hoff 1969; Sirotin & Shaskolskaya 1982). Thus isotropy of elastic behaviour is not present for the basal plane of cross-ply plywood.

What should we do to achieve this transverse isotropy for this laminated composite ? Thanks to the Hermann theorem, the answer is immediate: Make N greater than 4 (Wadhawan 1987a).

A convenient choice for N is 6, resulting in the so-called hex-ply structure. It amounts to stacking and bonding laminae such that the angle between successive pairs increases in steps of 60°. The basic 'unit cell' of this *angle-ply* plywood would thus comprise any three successive laminae or plies with grains or fibres making angles of 0°, 60° and 120° with some direction in the basal plane. Such a structure has a gross point-group symmetry 622. It follows from the Hermann theorem that, since $N = 6$, any tensor property of rank 5 or less ($r \leq 5$) would behave as if $N = \infty$ for it. In particular, the elastic stiffness tensor (for which $r = 4$) would behave as if the point-symmetry of the composite is $\infty 2$. Thus pseudo-isotropy in the transverse plane (the basal plane perpendicular to the 6-fold axis) would be achieved.

For the limit group $\infty 2$, as well as for the group 622, the number of independent components of the elastic-stiffness tensor is 5, whereas it is 6 for the symmetry group $4/mmm$ of cross-ply plywood. Thus symmetry enhancement (in this case in a specific plane) in going from $4/mmm$ to 622 not only results in a reduction of the number of independent elastic coefficients (from 6 to 5), but also in the elimination of an undesirable property, namely non-isotropy of elastic behaviour in the principal plane of plywood.

The photoelastic tensor is also a 4th-rank tensor, although it is not identical to the elastic-stiffness tensor. Several *nanocomposites* can be designed for optical applications (see Newnham, Trolier-McKinstry & Ikawa 1990). In a nanocomposite, at least one of the phases has at least one dimension in the nanometer range (i.e. much below the wavelength of light). The easily accessible fabrication technology for making flat or laminated nanocomposites (stretching, rolling, or poling of thin layers followed by bonding) usually results in orthorhombic symmetry (meaning $N = 2$), and the consequent optically biaxial behaviour for the individual layers. Transverse isotropy, which amounts to uniaxial optical behaviour, should be achievable by adopting a $4/mmm$ configuration similar to cross-ply plywood. But if it is desired that the laminated nanocomposite should not only be optically uniaxial, but also possess transverse isotropy of photoelastic behaviour, then the hexply configuration is necessary (and $N = 8$ would be even better). Such a system will be largely free from vibration-induced fluctuations of birefringence in the basal plane. Here is a solution looking for a problem !

13.3.4 Connectivity of Composites

Compared to single-phase materials, the power and promise of composites lies, not only in the gross symmetry that can be designed for achieving dissymmetrization or symmetrization as desired, but also in the flexibility available in choosing the connectivities of the phases or submaterials present.

Connectivity of a submaterial in a composite has been specified as a *number* (Newnham, Skinner & Cross 1978). It is the number of dimensions in which the submaterial is self-connected.

In the 3-dimensional world there are ten possible connectivities for a diphasic composite: 0-0, 1-0, 2-0, 3-0, 1-1, 2-1, 3-1, 2-2, 3-2 and 3-3 (however, see below).

Imagine an epoxy matrix in which small particles of another material are dispersed so that they do not touch one another. This is an example of 3-0 connectivity, because one of the phases, namely the epoxy, is self-connected in all the three dimensions, whereas the other phase (comprising the dispersed particles) is not self-connected in any dimension or direction.

Several other examples have been discussed by Newnham, Skinner & Cross (1978), Pilgrim, Newnham & Rohling (1987), and Newnham & Trolier-McKinstry (1990).

The 10 types of connectivity listed above do not exhaust all possibilities. Consider the case of 3-1 connectivity, exemplified by a composite in which rods of the poled piezoelectric ceramic PZT are embedded in a polymer matrix. Contrast this with another composite, made by taking a rectangular block of poled PZT ceramic, drilling parallel holes in it across one of the three pairs of parallel faces, and filling these holes with a polymer (Klicker, Biggers & Newnham 1981; Safari, Newnham, Cross & Schulze 1982). Clearly, the properties of these two composites are markedly different. A convention has therefore been introduced by writing the connectivity of the 'active' phase first (Pilgrim, Newnham & Rohling 1987). Taking PZT as the active phase, and the polymer as the inert one, the connectivity of the first composite is 1-3, and that of the second is 3-1.

Such considerations lead to the identification of six additional connectivities in diphasic composites in three dimensions: 0-1, 0-2, 0-3, 1-2, 1-3, and 2-3, giving a total of 16.

We described sum properties in §13.3.2. The averaging formula to be used for calculating a sum property of a composite depends in a crucial manner on the nature of the connectivity pattern, the series and parallel configurations for resistances and capacitances being familiar examples of this.

The possible number of distinct connectivity patterns shoots up sharply (from 16 to 64) in going from diphasic to triphasic composites (Pilgrim et al. 1987).

The question of connectivity is linked to other aspects such as relative sizes and orientations of the phases involved, and the characteristic length of the probing field with respect to the length scales of the phases (Newnham & Trolier-McKinstry 1990).

A nomenclature scheme for composites, as also a classification for them, has been attempted by Pilgrim et al. (1987) (also see Newnham & Trolier-McKinstry 1990). Unfortunately the entire problem is extremely complex, and one has to resort to notions such as 'precedence rules' based on parameters such as: (i) unique desired property; (ii) desired property coefficient in a shared property; (iii) tensor order of coefficient or property; (iv) volume fraction; (v) weight fraction; (vi) formula weight or repeat-unit weight.

This part of materials science awaits a qualitative breakthrough. Presumably, a combination of group theory and graph theory will lead to it.

13.3.5 Transitions in Composites

The fact that composite materials offer a large number of degrees of freedom for manipulation is reflected in the variety of transitions that are possible in them (Newnham & Trolier-McKinstry 1990):

(i) **Phase transitions in specific phases of the composite**. These are regular phase transitions involving a change of the thermodynamic state, and brought about by changes in scalar influences like temperature and hydrostatic pressure.

(ii) **Connectivity transitions**. We enumerated 16 connectivity classes in §13.3.4. Connectivity transitions amount to going from one such class to another.

(iii) **Field-induced transitions**. Poling of a ferroic composite is a typical example of this. After poling the composite often acquires the macroscopic symmetry of the poling field. Newnham & Trolier-McKinstry (1990) call them 'symmetry transitions'. However, since symmetry changes can also occur due to regular thermodynamic phase transitions, or even due to connectivity transitions, we prefer to call them transitions in composites induced by external non-scalar fields.

(iv) **Combination transitions**. It is conceivable that the above three types of transitions do not always occur only one at a time, and two or more of them may occur *simultaneously*.

For ferroic phase transitions the compact and informative Aizu symbol was described in §6.1.1. Something similar has been attempted for the above transitions in composites. An important input into the Aizu symbol is the information about the prototype symmetry. One could similarly define the 'initial connectivity'. Pilgrim et al. (1987) define it as the connectivity at 0 K in the absence of any external tensor fields (at zero or ambient hydrostatic pressure).

In the Aizu symbol for ferroic transitions, the letter F (for 'ferroic') separates the symbol for the prototype symmetry on its left and the symbol for the ferroic symmetry on its right. For the transitions in composites one can similarly choose a capital letter (T for temperature, P for pressure, E for electric field, etc.) for the force field responsible for the transition. A superscript can be attached to it to specify the nature of the transition (p for a thermodynamic phase transition, c for a connectivity transition, and f for a field-induced symmetry change).

A large number of examples for this notation have been described by Pilgrim et al. (1987) and Newnham & Trolier-McKinstry (1990). We select a few involving ferroic submaterials.

Consider a composite formed by liquid-phase sintering of PZT particles. The isolated PZT particles constitute the active phase and the 3-dimensionally connected phase separating these particles is the passive phase. The connectivity class is thus 0-3, and the symmetry of the composite is $\infty\infty m$. Application of a poling electric field under suitable conditions forces the ferroelectric PZT particles to have a preferred direction for the spontaneous polarization, reducing the overall symmetry to ∞m. The poling is thus an example of a field-induced symmetry change in the composite. The symbol for this transition is

$$\infty\infty m\ 0-3\ E^f\ \infty m\ 0-3$$

Similarly, magnetic field can be used for magnetizing a porous film of γ-Fe_2O_3. The pores constitute the inert phase. The symbol for this transition caused in the composite by the magnetic field is

$$\infty\infty m\ 3-0\ H^f\ \infty/mm'\ 3-0$$

To illustrate a connectivity transition we consider the example of PZT particles in a polymer matrix. To achieve poling of such a composite, two additional steps are taken. One is to introduce very fine particles of carbon (of size much smaller than that of PZT particles) into the polymer matrix, and the other is to apply hydrostatic pressure to make the PZT particles touch one another in three dimensions. Let us denote by letters A, B and C the three phases, namely carbon, PZT and polymer.

Phase A is entirely within Phase C, and the particles of Phase A are much smaller than those of Phase B. Phases A and C can therefore be taken as forming a *quasi-composite* (Pilgrim et al. 1987), with connectivity (0-3). So far as its relationship with Phase B is concerned, it behaves like a single-phase material, in which Phase B is embedded with zero connectivity. The connectivity of the pseudo-diphasic composite is thus 0-3. To indicate that the second phase in this is not really a single phase, but rather a quasi-composite, the overall connectivity is written as 0-3(0-3).

For an appropriately chosen volume fraction for Phase B, application of adequate hydrostatic pressure P can make the PZT particles touch one another in three dimensions, changing he connectivity of the composite to 3-3(0-3) (Sa-gong et al. 1985, 1986).

Now imagine the following processes: We apply an electric poling field E and a hydrostatic pressure P, and after the poling has been achieved, E and P are removed. The following self-explanatory symbol describes the two transitions effected in the composite:

$$\infty\infty m\ 0-3\ (0-3)\ P^cE^f\ \infty m\ 0-3\ (3-3)\ P^c\ \infty m\ 0-3\ (0-3)$$

13.3.6 Ferroic Nanocomposites

Ferromagnetic Nanocomposites

Ferrofluids (ferromagnetic fluids) are a striking example of how nanocrystals of a ferromagnetic material, when dispersed in a fluid, can constitute a composite that has a remarkable range of applications. The magnetic particles used have typically a size of 10 nm. At this small size they exhibit superparamagnetism, with the attendant phenomenally large permeability (cf. §13.1.2). This size is also small enough to let Brownian motion prevent their settling down under gravity, thus ensuring colloidal stability.

For preserving 0-3 connectivity (i.e. for preventing clustering), the actual composite configuration in a ferrofluid either includes a polymer coating for the nanoparticles, or has a like charge on the surface of the particles so that they repel one another. Such measures prevent their agglomeration over long periods of time (Berkovsky 1978; Bacri et al. 1988).

Ferrofluids find applications as various types of seals in the computer industry, in gas lasers, motors, blowers, clean-room robotics, etc. (Rosenweig 1982).

Another major type of ferromagnetic nanocomposites are those used in the recording industry (including computer memories). The need for going to nanosizes comes from the fact that the smaller the size of the particles, the greater is the density of information storage. Nanoparticles of γ-Fe_2O_3 are generally used. Their powder is dispersed in a binder and coated on a tape or a disc (Camras 1988). This is again a 0-3 connectivity configuration.

Some of the more exotic connectivity patterns evolved for improved performance as magnetic recording media have been reviewed by Newnham, Trolier-McKinstry & Ikawa (1990).

Ferroelectric Nanocomposites

Electric analogues of ferrofluids, i.e. ferroelectric fluids, have been discussed by Bachmann & Barner (1988). Milled particles of $BaTiO_3$, having an average size of 10 nm and dispersed in a mixture of heptane and oleic acid, exhibit a permanent dipole moment. Another nanocomposite investigated in this connectivity class (i.e. 0-3) comprises 20 nm $PbTiO_3$ particles dispersed in a polymer matrix (Lee, Halliyal & Newnham 1988). Poling of the nanoparticles could be achieved by using strong enough electric fields, and the composite was demonstrated to exhibit piezoelectricity.

A strong motivation for decreasing (towards the nanometer regime) the size of the ferroelectric particles used in composites is provided by their expected nonlinear-optical and electro-optical applications. Use of small enough particles is anticipated to result in acceptable levels of optical transparency and homogeneity.

Ferroelastic Nanocomposites

The best investigated example of this class of composites is that of submicron sized particles of zirconia dispersed in a major phase like alumina, with the purpose of enhancing the fracture-toughness of the latter. The basic mechanism of how this toughening occurs will be described in §14.2.4.

SUGGESTED READING

J. van Suchtelen (1972). Product properties: A new application of composite materials. *Philips Research Reports*, **27**, 28.

D. K. Hale (1976). The physical properties of composite materials. *J. Mater. Sci.*, **11**, 2105.

B. Berkovsky (1978) (Ed.). *Thermomechanics of Magnetic Fluids.* Hemisphere Publishing Corporation, Washington.

J. van Suchtelen (1980). Non-structural applications of composite materials. *Ann. Chim. Fr.*, **5**, 139.

Tsu-Wei Chou, R. L. McCullough & R. B. Pipes (1986). Composites. *Sci. Amer.*, **255**, 166.

R. E. Newnham (1986). Composite electroceramics. *Ann. Rev. Mater. Sci.*, **16**, 47.

V. K. Wadhawan (1987a). The generalized Curie principle, the Hermann theorem, and the symmetry of macroscopic tensor properties of composites. *Mat. Res. Bull.*, **22**, 651.

R. E. Newnham & S. Trolier-McKinstry (1990a). Crystals and composites. *J. Appl. Cryst.*, **23**, 447.

R. E. Newnham & S. Trolier-McKinstry (1990b). Structure-property relationships in ferroic nanocomposites. *Ceramic Transactions*, **8**, 235.

R. E. Newnham, S. Trolier-McKinstry & H. Ikawa (1990). Multifunctional ferroic nanocomposites. *J. Appl. Cryst.*, **23**, 447.

J. R. Hanson (1995). Introduction to advanced composite materials. In E. Udd (Ed.), *Fiber Optic Smart Structures.* Wiley, New York.

Mel M. Schwartz (1997). *Composite Materials, Vol. 1: Properties, Nondestructive Testing, and Repair.* Prentice Hall, New Jersey.

Chapter 14

APPLICATIONS OF FERROIC MATERIALS

What I am advocating is that we realise how much we owe to society. It keeps us - and if I look around myself I find that it keeps us in luxury - for doing what we want to do anyway, for doing what gives us most pleasure. I believe that we should show in return, some helpfulness and be less than annoyed if one of our conclusions or discoveries finds a practical application.

E. P. Wigner (1989)

In this chapter we first recapitulate (in §14.1) the salient features of ferroic materials. Such a review provides a basis for grouping their applications into five types (called Types A to E). The applications of ferroic materials are then described very briefly and selectively in §14.2. Use of ferroic materials in smart structures is considered separately in §14.3, after describing the basics of the concept of smart structures and materials.

14.1 SALIENT FEATURES OF FERROIC MATERIALS

We have defined a ferroic material as one which undergoes, or can be realistically thought of as capable of undergoing, at least one ferroic phase transition.

We have defined a ferroic phase transition as any. *nondisruptive* phase transition involving a change of the point-group symmetry of the underlying crystal structure.

Phase transitions in single crystals or polycrystals have been conventionally thought of as being either temperature-driven, or pressure-driven, or composition-driven; i.e. the control parameter has been usually taken as either T, or p, or x. All three of these are *always present*: Any material is at some temperature T, under some hydrostatic pressure p, and has some composition represented notionally by the symbol x. All these control parameters are *scalars*, and have the maximal directional symmetry embodied in the point group $\infty\infty 11'$. Changes of point-group symmetry at ferroic phase transitions can be therefore discussed without taking explicit note of the (high) symmetry of any of these three scalar parameters.

When we consider the situation in a relaxor ferroelectric like PMN, we find that, if it happens to be a system which has a fully stabilized cubic phase (cf. §10.3.4 and 10.4.6), there is no temperature-induced ferroic phase transition at all, in the sense that there is no change of the global point-group symmetry, and yet PMN is widely perceived as a ferroelectric material.

We observe next that this situation arises because of the conventional definition of a phase transition in terms of T, p or x only. The basic reason for a phase transition to occur in a material is that there is a lowering of the free energy as a result of it. But the free energy can be a function, not only of the conventional scalar fields T, p and x, but also of other fields like electric field, magnetic field, or uniaxial stress. A phase transition driven by any of these tensor fields is as much a phase transition as that driven by a scalar field.

One could therefore define a ferroic phase transition as any nondisruptive phase transition brought about by a scalar or tensor control parameter, and entailing a change of point-group symmetry.

Application (or superposition) of a tensor field on a crystal lowers the net symmetry of the total system (crystal plus external field) in accordance with the Curie principle. This lowering of symmetry is not at all problematic in identifying the additional change of symmetry at a tensor-field-induced phase transition because the lowering of symmetry caused by the superimposed field *before* the occurrence of the phase transition is present as an extra influence for the daughter phase also. For example, it does not affect (gets cancelled out in) the definition of the relative spontaneous polarization or strain of a domain state. What is important for the definition of, for example, a ferroelectric domain pair is that the members of the pair *differ* in spontaneous polarization. The part in which they do not differ (and this includes the polarization *induced* by the external field) has no significant role to play in domain switching, or in the movement of a phase boundary when the tensor-field-induced phase transition occurs.

Thus, so far as ferroic domain structure and the movement of domain boundaries and phase boundaries are concerned, field-induced phase transitions are hardly any different from phase transitions induced by a scalar parameter[1].

In view of the above, any bulk ferroic crystal (*including* a relaxor ferroic) can be thought of as having three characteristic features:

(A) Existence of ferroic orientation states, resulting from long-ranged ordering with respect to at least one macroscopic tensor property.

(B) Existence of domain boundaries and phase boundaries in the ferroic phase, which disappear in the prototypic phase, and which can be moved by applying a field, the movement becoming easier in the close vicinity of the ferroic phase transition.

(C) Enhancement of certain properties in the vicinity of the ferroic phase transition; some of these properties *must* be macroscopic tensor properties because, by definition, there is a change of point-group symmetry at a ferroic phase transition.

We are particularly interested in the *macroscopic* tensor properties which get enhanced near the ferroic phase transition, because these are the properties which determine the practical applications of ferroic materials in the laboratory or in the industry.

14.1.1 Existence of the Ferroic Orientation State

The fact that there occurs a ferroic phase transition in a material, leading to the formation of domains in the ferroic phase, underscores the fact that there is something about the crystal structure and about the nature of interactions among atoms constituting the structure, which permits or promotes long-ranged cooperative ordering.

For this reason, the occurrence of proper ferroelectric phase transitions in, say, $BaTiO_3$ can be interpreted as a proclivity of its crystal structure to attain states (orientation states) with high polarizabilities (through co-operative processes).

Several ferroelectrics find applications as converters of laser frequencies because of their large-magnitude nonlinear optical (NLO) properties. Generally speaking, large NLO polarizabilities can be expected in those

[1]It is important to realize, however, that whereas the domain-structure statics may not depend significantly on whether a ferroic phase transition is caused by a scalar field or a tensor field, the actual 'reaction pathway' and the kinetics for the occurrence of the transition may become radically different in the presence of an external anisotropic influence. This is particularly true for spin-glass systems and their electrical and mechanical counterparts.

materials which have large *linear* polarizabilities (Lines & Glass 1977).

14.1.2 Mobility of Domain Boundaries and Phase Boundaries

Any ferroic phase of a crystal has at least two orientation states because its point-group symmetry is a proper subgroup of the prototype point group. Therefore the occurrence, or the possibility of occurrence, of boundaries which separate such domain states is an essential characteristic of ferroic phases. And these boundaries can move under the action of appropriate nonscalar fields.

The magnitude of such a field required for moving a domain boundary decreases, and the mobility of the domain boundary increases, as one approaches the critical temperature T_c; beyond it the ferroic phase is no longer the least-free-energy phase. This is because of a lowering of the enthalpy barrier between contiguous domain pairs as the critical point is approached.

In addition to the occurrence of domain boundaries at all temperatures below T_c, phase boundaries can exist in a certain temperature range. The lower bound of this range is the temperature which is the stability limit of the prototype, and the upper bound is the temperature which is the stability limit of the ferroic phase (§5.3.13). The existence of phase boundaries is particularly important for ferroics with a diffuse phase transition, or a field-induced phase transition.

14.1.3 Enhancement of Certain Macroscopic Properties Near a Ferroic Phase Transition

Not only do new tensor-property coefficients arise in the ferroic phase because of the reduction of the point-group symmetry of the prototype, certain macroscopic properties also become large in the vicinity of the ferroic phase transition.

A tensor property **T** is defined as relating a force (or control parameter) **X** to a response **Y** through a linear constitutive relation (cf. Eq. 3.1.35):

$$\mathbf{Y} = \mathbf{T}\mathbf{X} \tag{14.1.1}$$

In particular, in the context of a ferroic phase transition, **X** can be any of the six control parameters: T, p, x, **E**, **H**, $\boldsymbol{\sigma}$. Thus a number of macroscopic tensor properties or response functions ($= \partial \boldsymbol{Y}/\partial \boldsymbol{X}$) can be defined by taking $\boldsymbol{X}$ as any of these six parameters. Which of them becomes large near the transition depends on the proper or improper nature of the ferroic transition with respect to a given property; it also depends on the extent of coupling (quantified by the faintness index) between the order parameter and the macroscopic property in question.

Two broad subcategories may be introduced here. The controlling field may either be a scalar ($\boldsymbol{T}$, $\boldsymbol{p}$, or $\boldsymbol{x}$), or a nonscalar ($\mathbf{E}$, $\mathbf{H}$, $\boldsymbol{\sigma}$).

14.1.4 A Comparative Analysis of the Properties of Ferroic Materials

Ferroic materials have certain similarities and certain differences. The differences arise mainly because of the fundamental differences in the nature of the interactions driving the ferroic transition.

Similarities

Ferroic materials are characterized by the occurrence of long-ranged ordering below a critical point; moreover, this (nondisruptive) ordering is necessarily accompanied by a change of the point-group symmetry of the material. Often the change of point-group symmetry may tend to be masked, either because the material is a ceramic, or because it splits into a domain structure to minimize the overall demagnetization, depolarization, and/or elastic energy.

The dipole-dipole part of the interaction in all the three primary ferroics, namely ferromagnetics, ferroelectrics, and ferroelastics, is fairly long-ranged, decaying with distance as $1/r^3$. Because of this long-ranged nature of the interaction, a splitting into domains must occur (unless prevented by size effects), not only because symmetry considerations allow this, but also because otherwise very high demagnetization, depolarization, or elastic fields would develop.

The availability of two or more orientation states in a ferroic makes it possible to pole it, i.e. change its domain structure by a suitable external field such that either only one domain, or very few domains, are present.

The change of the point-group symmetry, built into the definition of a ferroic phase transition, ensures that at least one new macroscopic tensor property coefficient arises in the ferroic phase.

The adopted definition of a ferroic phase transition also ensures that at least one macroscopic response function is large in the vicinity of the transition. Even when a ferroic phase transition is improper with respect to ferromagnetism, ferroelectricity or ferroelasticity, the corresponding magnetic, electric or elastic susceptibility generally increases to some extent at the transition (because there is a coupling of the spontaneous magnetization, polarization or strain with the order parameter, and the susceptibility associated with the order parameter necessarily becomes large in the vicinity of the ferroic transition).

One of the most important areas of applications of ferroic materials is in smart structures, some of the commonly used ferroics for this being PZT/PLZT, TERFENOL – D ($Tb_{1-x}Dy_xFe_2$), PMN, and NITINOL (Ni – Ti). Newnham (1997, 1998) has pointed out an important similarity among these materials: Most of them undergo at least two phase transitions, and, together, the two transitions encompass both atomic ordering and atomic displacements. PZT has a partially ordered cubic phase, and rhombohedral and tetragonal quasi-ordered phases near the morphotropic phase boundary. TERFENOL – D undergoes a paramagnetic-ferrimagnetic phase transition, followed by another transition whereby the magnetic spins reorient along different directions. In PMN there is a diffuse transition from a partially ordered cubic phase to a relaxor phase, followed by a phase transition to a regular ferroelectric phase. In NITINOL there is a partially ordered cubic structure, which transforms to an ordered martensitic phase over an extended temperature range, involving complex structural changes and coexisting phases.

Differences

There are several differences between ferromagnets and other types of ferroic materials. One is that a purely ferromagnetic phase transition is generally not a *structural* phase transition, in the sense that there is hardly any change in the crystal structure when the spins get ordered.

The largely non-structural nature of ferromagnetic transitions means that they can be expected to be always nondisruptive. By contrast, we have to make a careful distinction between disruptive and nondisruptive structural phase transitions, and insist that only the latter can be ferroelectric, ferroelastic etc. (provided they also entail a change of point-group symmetry).

The ferromagnetic exchange interaction is of quantum-mechanical origin, and is short-ranged. Moreover, magnetic monopoles do not exist. The situation is markedly different for the electric analogue.

Further, magnetostriction is generally a much weaker effect compared to electrostriction. The former is generally at the ppm level, whereas the latter can be as large as 1%, or even more.

Because of the exchange interaction, ferromagnetic domain walls are very thick. By contrast, ferroelectric and ferroelastic domain walls can be thin or thick, depending on the degree of coherence between the two lattices meeting at the wall.

Because of the long-ranged nature of electric and elastic interactions, proper ferroelectric and proper ferroelastic phase transitions are generally well described by a mean-field theory like the Landau-Ginzburg theory, and the critical region is very small (Salje 1993a). This is generally not the case

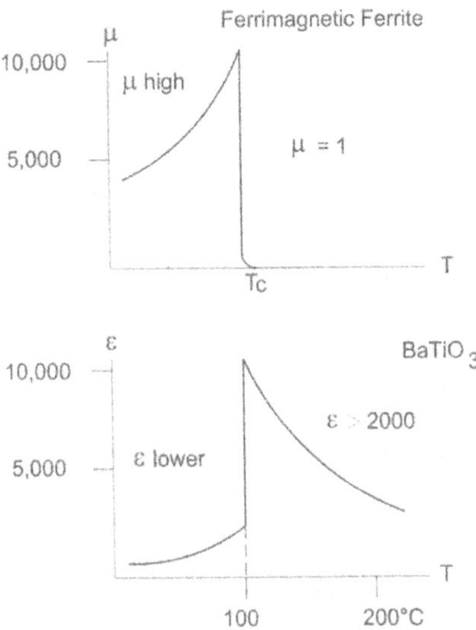

Figure 14.1.1: Comparison of Curie-Weiss behaviour in a typical ferrimagnetic ferrite with that in a $BaTiO_3$ type ferroelectric. [After Cross (1996).]

for ferromagnetic phase transitions.

Although a Curie-Weiss type dependence on temperature is expected for the generalized susceptibilities of both proper ferromagnets and proper ferroelectrics (and also proper ferroelastics), an important difference must be recognized. Fig. 14.1.1 shows this.

Part (a) of the figure shows the typical relative permeability versus temperature curve for a ferrimagnetic ferrite, and Part (b) a relative permittivity versus temperature curve for a $BaTiO_3$ type ferroelectric.

In the case of the ferroelectric the cooperative long-ranged ordering is predominantly of the dipole-dipole type, and not very strong. Therefore, even moderate electric fields can influence the ordering substantially. This fact is reflected in the large value of the electric Curie constant C in the Curie-Weiss equation ($\varepsilon = C/(T - T_c)$). The consequent large value of the permittivity at T_c continues to be large for wide temperature intervals on both sides of T_c.

By contrast, since ferromagnetic or ferrimagnetic ordering is mainly through the very strong exchange interaction (and not so much through the much weaker magnetic dipole-dipole interaction), the magnetic Curie constant is much smaller, resulting in a very small permeability in the paramagnetic phase.

An idea of how strong the spontaneous magnetic ordering is can be had by making a rough estimate of the Weiss molecular field H_m from the equation $\mu_B H_m \sim k_B T_c$. Taking $T_c = 1000$ K gives $H_m \sim 10^7$ gauss. This corresponds to a molecular field constant $\gamma = H_m/M \sim 10^7/10^3 \sim 10^4$. This is to be compared to the much smaller value of $4\pi/3$ for the Lorentz factor for a simple dipole-dipole interaction in a ferroelectric.

SUGGESTED READING

L. E. Cross (1996). Ferroelectric ceramics: Materials and application issues. *Ceramic Transactions*, **68**, 15.

R. E. Newnham (1997). Molecular mechanisms in smart materials. *MRS Bulletin*, May 1997.

14.2 APPLICATIONS

Applications of specific categories of ferroic materials have been reviewed in several books and articles, including the following:

Ferromagnetics: Wohlfarth (1980b); Goldman (1990); Valenzuela (1994).

Ferroelectrics: Lines & Glass (1977); Cross & Hardtl (1980); Herbert (1982); Jain (1988); Levinson (1988); Moulson & Herbert (1990); Xu (1991); Swartz & Wood (1992); Cross (1995).

Ferroelastics: Wadhawan (1982).

Secondary Ferroics: Quartz, which is a ferrobielastic, as well as a ferroelastoelectric, is the most important secondary ferroic from the point of view of applications (Momosaki & Kogure 1982; Brice 1985; Besson, Groslambert & Walls 1985; Ward 1989).

In view of the above-mentioned published material, our discussion of the applications of ferroic materials is of a highly selective nature. In §14.1 we described some *common* features of ferroics. These can form a possible basis for focusing on their applications based on each such main characteristic. Such a grouping of applications is not an easy task, as several applications are based on more than one ferroic characteristic. Nevertheless, such a grouping helps bring out the nature of the main property used in a particular application, and we adopt it here.

The role of ferroics in smart structures is discussed separately in §14.3.

14.2.1 Applications Related to the Existence of the Ferroic Orientation State

A primary ferroic phase transition is characterized by long-ranged cooperative ordering. Applications of several ferroic materials, particularly ferroelectrics, are based mainly on this fact. We label such applications as *Type A applications.*

A.1. Electrooptic Modulators

Several ferroelectric crystals find applications as electrooptic modulators. Examples of such crystals exploiting the Pockels effect are: KD_2PO_4 (DKDP); $LiNbO_3$; $LiTaO_3$; and $Sr_{0.75}Ba_{0.25}Nb_2O_6$ (SBN). And KTN (cf. §10.5.3) is a well-known ferroelectric used for electrooptic modulation through the Kerr effect.

These crystals have significantly large nonlinear polarizabilities. However, there is another factor which contributes to the large electrooptic effect, and which is the direct offshoot of the existence of ferroelectric orientation states or domains. Rubidium hydrogen selenate ($RbHSeO_4$) provides a dramatic example of this (Salvestrini et al. 1994). This crystal is a ferroelectric-ferroelastic (Pietraszko et al. 1979; Suzuki, Osaka & Makita 1979). It exhibits what is called a *giant Pockels effect.*

To understand what this means, and why it arises, we begin by considering a widely used NLO crystal like KDP. It is a paraelectric at room temperature, and becomes a ferroelectric only below a certain low temperature (123 K). Thus its room-temperature applications do not make use of any ferroic property, and if the applied electric field is not too large, its dielectric response is linear: $D_i = \epsilon_{ij} E_j$. For large applied fields, as when a laser beam is incident on the crystal, nonlinear response becomes significantly large:

$$D_i = \epsilon_{ij} E_j + r_{ijk} E_j E_k + R_{ijkl} E_j E_k E_l + \ldots \qquad (14.2.1)$$

The tensor (r_{ijk}) determines the Pockels effect, and (R_{ijkl}) the Kerr effect.

No domain effects are involved in this, and the substantial NLO effects are genuinely because of high electric fields, and because of large nonlinear polarizabilities.

Now consider a crystal, not necessarily KDP, in a ferroelectric phase, with the associated domain structure. On the basis of the typical hysteresis loop drawn in Fig. 10.7.1 we can expect four types of dielectric response, in increasing order of the applied electric field.

(a) **Domain-wall bowing.** When the field applied is less than E_1, the dielectric response is linear and reversible. The domain walls tend to be pinned at their existing locations by imperfections in the crystal, and under the action of the small applied field only a slight bowing of the walls occurs. Assuming that several ferroelectric domains, with different allowed directions for the spontaneous polarization are present, the bowing of the domain walls amounts to a change in the relative volumes of the domains. Although the net dielectric response is a complicated function of the actual domain structure, it is intuitively clear that it is likely to be a considerably larger effect than the response of a nonferroelectric or single-domain crystal to a small electric field.

(b) **Domain-wall movement**. For $E_1 < E < E_2$, the applied field is strong enough to make the domain walls move (in a largely irreversible manner) after releasing them from their pinning sites. The optical indicatrix has different orientations in different domains. The movement of domain walls is such as to make one particular orientation of the indicatrix preponderate over the others. The overall optical effect of this is a large effective rotation of the 'average' orientation of the indicatrix, with a concomitant large Pockels effect.

$RbHSeO_4$ crystals are ferroelectric-ferroelastic at room temperature, with triclinic pseudo-orthorhombic symmetry. To minimize the overall strain energy the ferroelastic domain structure is mainly that of slabs having alternating signs of spontaneous strain. The 'anomalous' effect of electric field on light deflected from the domain walls has been investigated by Tsukamoto & Futama (1993), and attributed to domain-wall motion. The coercive field (E_c in Fig. 10.7.1) is 75 V/mm. Salvestrini et al. (1994) measured, as a function of electric field, the total phase shift introduced by a polydomain crystal of $RbHSeO_4$ for a laser beam. The fields applied were in the range of E_c. It was observed that the total phase shift varies linearly with the applied field for $E \leq E_2$ (cf. Fig. 10.7.1). This linear variation is presumably caused by domain-wall movement.

(c) **Domain rotation** . For $E = E_2$ practically all the domain walls have been obliterated, and for larger values of E up to E_3 the main dielectric response is by the rotation of the single ferroelectric domain towards the direction of the applied field. There is a corresponding rotation of the optical indicatrix, clearly with an abnormally large Pockels effect.

(d) **Nonferroelectric nonlinear dielectric response**. For $E > E_3$ the dielectric response is genuinely nonferroic. It is also markedly nonlinear because of the high value of the field applied and is described by Eq. 14.2.1.

It is clear from the above description that the giant Pockels effect in

$RbHSeO_4$ crystals, and presumably in other ferroelectric crystals, is a consequence of the existence of ferroelectric orientations states (domains). It is also influenced by the actual domain structure, and the underlying processes are quite complicated, requiring further investigation. What is quite striking is that the effective Pockels effect observed in this material is about 600 times larger than in KDP (Salvestrini et al. 1994).

A.2. Laser Frequency Converters

It is not an accidental matter that many of the best known crystals used for laser-frequency conversion are ferroelectrics (e.g. $LiNbO_3$; KDP and analogues; $KTiOPO_4$ (KTP) and analogues). An obvious reason for this is that, being ferroelectrics, they satisfy the condition of being noncentrosymmetric crystals. Another reason is that crystals with large linear polarizabilities are also usually the crystals with large nonlinear polarizabilities (Lines & Glass 1977).

A.3. Permanent Magnets

The long-range ordering of magnetic moments in a ferromagnet can result in a large (and spontaneous) magnetic moment per unit volume. For making permanent magnets an additional requirement is that the net magnetization should not decrease much with the passage of time. One way of ensuring this is to have a grain size smaller than about 1 micron, so that no domain wall movement can occur (because no domain walls are sustained by the small grains (cf. §13.1.2)).

In order that devices based on permanent magnets be as small and as efficient as possible, it is necessary that the material used has a large remanent induction B_r, a large coercive field H_c, and a large energy product $(BH)_{max}$ (§9.4.4).

The coercive field can be increased by creating strong impediments to the movement of domain walls. This is done by introducing inhomogeneities into the polycrystal. An early example of this is the introduction of 1 wt. % of C in Fe. Although grain boundaries also provide some opposition to domain wall movement, their role can be further enhanced by a finely dispersed nonmagnetic second phase which tends to dwell in the grain-boundary regions.

Alloys of Fe, Ni, Co and Al ('Alnico alloys') constitute around 6% of the total world production of hard magnetic materials (Valenzuela 1994). They have a high value of T_c ($\sim 1070K$), and therefore are used in high-temperature applications.

$SmCo_5$ has exceptionally hard magnetic properties (Strnat 1988). It has very high magnetocrystalline anisotropy. Another similar material is

Sm_2Co_{17} (see Valenzuela 1994). Because of their very high BH-product they are used in miniaturized motors and actuators etc.

The high cost of Sm and Co has led to the development of a cheaper magnet material, namely $Nd_2Fe_{14}B$.

Ferrites offer a better alternative to alloy magnets in many applications, in spite of the fact that the highest relative permeability attained for them ($\sim$ 100,000) is much lower than what has been possible with the metallic systems. The high resistivity of ferrites is a definite advantage in many situations. The main applications of ferrites are in loudspeakers, dc motors, and stepping motors.

A.4. Small-Signal Applications of Ferrites

For small magnetic fields the magnetization curve of a ferromagnet has a constant slope (Fig. 9.1.1). Conversely, small ac signals are transformed linearly to a magnetic flux by such materials. Because of their higher resistivities, ferrites are preferred over magnetic alloys for high-frequency applications based on this linear response function. Ferrite cores are used extensively in antennas etc.

As the ferromagnetic phase transition is approached, not only does the permeability become high, it also varies nonlinearly with temperature. This can be undesirable in certain applications in which the temperature may vary. Therefore, additives are introduced in systems like Ni – Zn ferrites to smear out the ferromagnetic phase transition.

A.5. Ferrites in SMPS

Another application of ferrite cores is in switched-mode power supplies used for computers and peripherals etc. (see Valenzuela 1994). Compact and efficient power supplies can be made by using ferrite-core transformers operating at $\sim$ 25 kHz. Mn – Zn ferrites, having several additives which bring down the coercive field H_c, have been used.

14.2.2 Applications Exploiting the Mobility of Domain Boundaries and Phase Boundaries

A ferroic phase transition results in transformation twinning or domain structure in the ferroic phase. The domains can be made to shrink or expand by a suitably configured driving field. This fact, when exploited for single-crystal ferroics, can lead to their *detwinning*. The corresponding term generally used for ceramic ferroics is *poling*; it means achieving a preferred orientation for the domains in the various grains of the ferroic ceramic.

It is important to remember that only ferroic ceramics can be poled. Grains of nonferroic materials do not have available to them the choice of switching to alternative orientation states, because none exist.

We call applications exploiting the moveability of domain boundaries and phase boundaries in ferroics as *Type B applications.*

B.1. Detwinning of Ferroic Crystals

Crystals of the ferroelectric $LiNbO_3$, which find extensive applications in nonlinear optics, are generally required to be used in a single-domain state. But when they are grown by the Czochralski technique from the melt, they undergo a phase transition to the ferroelectric phase on their way to cooling to room temperature. Therefore, at room temperature they possess the domain structure expected for the ferroic species $\bar{3}mF3m$. To obtain single-domain crystals, one cools the crystal from its paraelectric phase under the action of an electric field so that only one domain direction is favoured during and after the ferroelectric transition.

Similarly, ferroelastic crystals are detwinned by applying uniaxial stress. Several examples have been reviewed by Wadhawan (1982). These include $SmAlO_3$, $CsFeF_4$, $LaFeO_3$, Mg – Cl boracite, and tris-sarcosine calcium chloride.

Uniaxial stress can be used for detwinning a ferrobielastic crystal, the best-investigated example being that of α-quartz (see Klassen-Neklyudova 1964).

B.2. Poling of Ferroic Polycrystals, and Periodic Domain Inversion of Ferroic Crystals

A ferroic ceramic is essentially isotropic to begin with (because of the random orientations of its grains). Consequently it displays little or no directional properties like piezoelectricity or pyroelectricity. An electric field is used for poling ferroelectric ceramics. Poling of electroceramics like PZT is routinely done for using them, say, as piezoelectric elements in gas lighters.

Similarly one can pole a ferroelastic ceramic by applying uniaxial stress. The availability of ferroelasticity as a ferroic property extends considerably the range of materials which can be poled. This is because, whereas ferroelectricity can occur only in polar noncentrosymmetric crystal classes, ferroelasticity can occur even in centrosymmetric crystal classes.

Similar considerations apply to secondary ferroics also (Newnham & Skinner 1976). NH_4Cl is a ferroelastoelectric. That means that appropriate mechanical and electric fields must be applied *together* for effecting domain switching, and thence poling, in it. This has been demonstrated (see Newnham & Skinner 1976).

We now turn to poling or domain switching of ferroic single crystals.

Periodic domain inversion (PDI) of ferroelectric crystals like lithium niobate, lithium tantalate, and KTP for achieving 'quasi phase matching' (QPM) conditions for the frequency doubling of laser radiation offers interesting possibilities. One produces a periodic array of ferroelectric domains, with successive domains having opposite directions of spontaneous polarization. This results in a periodic correction for the crystal dispersion, leading to phase-velocity matching between the fundamental and the second harmonic (if a right repeat distance is chosen) (Armstrong, Bloembergen, Ducuing & Pershan 1962; Fejer 1994).

A number of techniques have been used for introducing PDI in ferroelectrics (see Byer (1992) for a review). For bulk crystals, PDI is achieved during the process of crystal growth itself. This is done, for example, by 'laser-heated pedestal growth', or by off-axis rotation of the seed crystal in the so-called Czochralski method of crystal growth from the melt. For wafer crystals, on the other hand, one of the techniques used combines photolithography with electric poling, using a suitable mask (see Hu, Thomas & Webjorn 1996). Another technique, used by Lim, Fejer & Byer (1989) for a lithium niobate wafer, involved photolithographically patterned diffusion of Ti. Electron beam writing is one more approach, adopted for lithium niobate and lithium tantalate by Ito, Takyu & Inaba (1991), for obtaining periodic inversion of ferroelectric domains.

A typical repeat distance for the modulation is 5 microns. By choosing a suitable repeat distance, QPM can be obtained over a wide range of wavelengths of the laser radiation. Much of the work so far has been on KTP (Thomas & Glazer 1991; Hu, Thomas & Webjorn 1995; Hu, Thomas, Gupta & Risk 1995), and lithium niobate (Kitaoka, Mizuuchi, Yokoyama, Yamamoto, Narumi & Kato 1999; Fujimura, Suhara & Nishihara 1999).

An important configuration, which is the ferroelastic counterpart of the ferroelectric poling described above for achieving a periodically reversed domain structure, is that of neodymium pentaphosphate (NPP), NdP_5O_{14}. This crystal undergoes periodic ferroelastic switching very readily (Weber, Tofield & Liao 1975; Huang, Jiang, Hu, Xu, Zeng, Feng & Wang 1995). Meeks & Auld (1985) used this property for creating regularly spaced domain walls in NPP, and the spacing of the walls was tunable from 70 microns to 0.5 micron. The result was a tunable optical grating. What is more, since NPP is also a low-threshold laser-host crystal, one can develop a laser with a built-in, instantly tunable, optical grating (Meeks, Auld & Newnham 1985).

B.3. Magnetic Recording

As implied by the existence of the ferromagnetic (or ferrimagnetic) hysteresis loop, a ferromagnetic material has at least two stable remanent states, which can form the basis of information storage schemes employing binary logic. Ferrite core memories were used extensively in the early 1970s for this purpose. Although semiconductor memories are currently used in computers, the ferrite core memories have the advantage that they are nonvolatile; i.e. each remanent state is stable by itself, not requiring the use of an external field for keeping the system in that state.

Ferrites are used for reading and writing of information on tapes and discs, the basic processes used being the same for audio, video and computer recording. Apart from the conventional methods using magnetic fields produced by currents, magneto-optical recording, as well as optical readout, are also used. Faraday rotation and Kerr effect are employed. This results in higher data storage, faster access times, and virtually absent wear and tear (see Valenzuela 1994).

B.4. Thin-Film Ferroelectric Memories

A very active area of current research is that of thin-film integrated ferroelectrics, particularly with the objective of developing high-density information storage systems for use in 'smart cards' (Cross & Trolier-McKinstry 1997; Auciello, Scott & Ramesh 1998; Kingon 1999). The basic job is to integrate ferroelectric memory elements with silicon-based IC chips.

Typical ferroelectrics used are PZT and SBT ($SrBi_2Ta_2O_3$). A recent entrant to the fray is BLT ($Bi_{3.25}La_{0.75}Ti_3O_{12}$), i.e. La-doped bismuth titanate (BTO) (Park et al. 1999). The 'up' and 'down' orientation states of a ferroelectric domain provide the basis for the binary-code memories. The configuration is intrinsically nonvolatile because all orientation states are equally stable, and thus do not require the application of an external biasing field for the domain to remain in that state.

The use of ferroelectrics also offers the possibility of high dielectric constants, thus reducing the sizes of the capacitor elements made from them and used in what are commonly known as DRAMs (dynamic random-access memories).

Assuming that the ferroelectric domain walls move at about the speed of sound, they can move across a film of thickness 1 micron in about 1 nanosecond. Sustained research has, in fact, pushed the switching time of what are called NVFRAMs (nonvolatile ferroelectric RAMs) into the picosecond regime. Theoretical considerations show that, in very small capacitors, the ferroelectric switching time is determined, not by the speed of movement of the domain wall, but by the time needed for the new ferroelectric domain to nucleate. Using experimental data for nucleation rates for PZT, Scott (1998) calculated the ultimate switching speed for PZT capacitors to be 600 ± 200 picoseconds. This compares well with the actual value of about 900 picoseconds.

Low-density NVFRAMs are now being produced commercially, whereas use of high-density configurations in devices still needs additional research for achieving short-term and long-term reliability, freedom from excessive fatigue, lower-temperature deposition, and overcoming of the problems associated with compatibility between the ferroelectric film and the silicon substrate.

The presently used DRAMs in computers are based on the silicon technology, and suffer from the problems of large size, a not-very-large number of write cycles, and rather long write times. Thin-film ferroelectric memories can solve these problems, but then some new problems crop up which need to be attended to. SBT requires a rather high processing temperature, which can degrade the silicon IC chip with which it has to be integrated. PZT, when deposited on the commonly used Pt electrodes, suffers from a decrease of the effective spontaneous polarization when switched (i.e. read and written) in a DRAM a large number of times. The use of BLT appears to overcome these problems to a substantial extent. Films of BLT, deposited on metal electrodes, are reported by Park et al. (1999) to be: free from polarization fatigue; integrable with the rest of the device at temperatures of the order of 650^0C; and having a larger remanent polarization than SBT films.

Kingon (1999) has discussed the prospects of ferroelectric thin-film memories providing a better alternative to the existing materials and technologies. Thin films of the relaxor material PMN-PT may offer some attractive possibilities in this regard (Maria, Hackenberger & Trolier-McKinstry 1998).

The following additional applications, exploiting the switchability of ferroelastic domains and the consequent mobility of the domain walls, have been discussed by Wadhawan (1982): micropositioner with a memory; variable acoustic delay line; tailored domain patterns for resonator applications; focusing acoustic transducers; moving line source of light; optical shutter and colour modulator; page composer.

14.2.3 Applications Using Enhanced Macroscopic Properties near the Ferroic Phase Transition

We call them *Type C applications*.

For a proper ferroic phase transition, the generalized susceptibility corresponding to the order parameter becomes large, this susceptibility being a macroscopic physical property. Other properties coupled to this susceptibility also become large.

Moreover, if we consider, for example, a proper ferroelectric transition, since the spontaneous polarization rises rapidly with temperature just below T_c, the material exhibits a large pyroelectric effect. Similarly for proper ferroelastic phase transitions.

Even for an improper ferroic transition, since the transition is ferroic there must be at least one macroscopic tensor property coefficient which becomes nonzero below T_c. Although the faintness index for the concerned property is 2 or higher, the coupling of this property with the order parameter generally results in an enhancement of some macroscopic properties in the vicinity of T_c.

C.1. Pyroelectric Detectors

Several ferroelectrics find applications as pyroelectric detectors. This is because all ferroelectrics are pyroelectrics as well, and also because the spontaneus polarization of a ferroelectric generally shows a strong temperature sensitivity just below T_c. This subject has been reviewed by Lines & Glass (1977).

C.2. Applications in Capacitors

$BaTiO_3$ is a well-known example of a ferroelectric which finds applications in capacitors (Cross & Hardtl 1980).

Relaxor ferroelectrics are another class of ferroics which offer attractive possibilities for applications in capacitors. They not only have high dielectric constants, but also a diffuse transition. They can thus serve as fairly temperature-stable, high-volume-efficiency, capacitors.

C.3. Acousto-optic Modulators

An ultrasonic beam can set up a strain modulation, and therefore a refractive-index modulation, in a crystal. Such a crystal can then serve as an optical grating. A variety of light modulators can be made from this acousto-optic set-up, making possible both amplitude modulation and frequency modulation. Crystals of As_2S_3 are often used for this purpose.

Several figures of merit have been introduced for comparing the modulation efficiency of different crystals, including the following (Dixon 1967):

$$M_1 = \frac{n^7 p^2}{\rho v} \tag{14.2.2}$$

Here n is the average refractive index, p an appropriate elasto-optic coefficient, ρ the density, and v the relevant acoustic velocity. Apart from the general requirement of large n and small ρ, ferroelastic crystals present some interesting possibilities, particularly in the vicinity of the ferroelastic phase transition. In the vicinity of this transition, even a small stress can produce a large strain, and therefore a large change of birefringence, implying a large p in Eq. 14.2.2. The situation becomes particularly favourable if the transition is a proper or a pseudoproper ferroelastic transition. In this case, some acoustic velocity necessarily tends to zero as the transition point is approached. And a vanishingly small v in Eq. 14.2.2 means a very large increase in M_1. $BiVO_4$, a pseudoproper ferroelastic, is a promising material in this context.

The phase diagram of $Pb_3(P_{1-x}V_xO_4)_2$ has a large ferroelastic regime (Hodenberg & Salje 1977), and its ferroelastic phase transition can be tuned to the temperature of application by varying x. For example, for $x = 0.21$ the phase transition occurs at room temperature (Wadhawan & Glazer 1981). The figures of merit, evaluated by Salje (1976), are as large as those for the widely used As_2S_3.

14.2.4 Applications Involving Field-Induced Phase Transitions

We classify them as *Type D applications.*

D.1. Transformation Toughening of Materials

A material is said to be tough if it resists fracture by resisting the propagation of cracks. Zirconia (ZrO_2) is the best known example of a material in which the occurrence of a ferroic phase transition is exploited for increasing the fracture toughness, either of the zirconia ceramic itself, or of the matrix material to which it is added as a second phase. We focus on this material here for illustrating the processes involved in the transformation toughening of materials.

Zirconia is a material of strategic importance, having a very high melting point, high chemical stability, and very low thermal conductivity. It undergoes the following sequence of phase transitions (Subbarao, Maity & Srivastava 1974; Subbarao 1990; Nagarajan & Rao 1993):

$$\text{Monoclinic} \overset{1443\text{K}}{\rightarrow} \text{Tetragonal} \overset{2643\text{K}}{\rightarrow} \text{Cubic} \overset{2953\text{K}}{\rightarrow} \text{Melt}$$

The monoclinic-tetragonal (or m-t) transition involves a change of crystal family, and is also fairly nondisruptive. It is therefore a ferroelastic phase transition. So is the tetragonal-cubic (t-c) transition.

The m-t transition shows considerable thermal hysteresis: the m $\rightarrow$ t transition occurs at 1443 K, whereas the t $\rightarrow$ m transition occurs at 1123 K. There is thus a large temperature range in which the two phases are simultaneously stable. The stability limits of the two phases also depend on grain size.

The strongly first-order t-m transition not only involves a substantial volume change, the volume *increases*, rather than decreases, on cooling across the transition.

The t-c ferroelastic transition is quite mild, with hardly any volume change.

The phase transitions involved can be suppressed fully or partially, either by decreasing the grain size, or by putting additives like Y_2O_3, CeO_2, MgO, or CaO.

The cubic and tetragonal phases can be stabilized to room temperature by decreasing the grain sizes to sub-micron values, or by the addition of lower-valent oxides.

We consider the t-m transition to explain the basic transformation-toughening mechanism. If t-ZrO_2 has been made stable at room temperature, it means that the t-m phase transition has been arrested. The central idea employed in transformation toughening is that this arrested phase transition can be made to occur by mechanical stress. The phase transition is ferroelastic (involving a shear strain of about 8%), as well as strongly first-order, with an increase of volume on entering the m-phase. When a crack tends to propagate through such a material, the large stress field at the tip of the crack induces the occurrence of the t-m transition in the region around the tip. The resulting 5.6% volume increase leads to the formation of several small cracks in front of the larger crack. This causes a blunting of the main crack, and an increased absorption of internal stress per unit crack propagation. As a supplementary mechanism, ferroelastic switching (and also slip) in the m-phase absorbs the *shear* stress at the crack front. This is a good example of *ferroelastic switching as a stress-accommodating mechanism*. The end result is that a seemingly brittle ceramic like zirconia acquires a fracture toughness comparable to that of a metal.

It also turns out that addition of the metastable t-phase of zirconia to any ceramic matrix can toughen the latter. Some of the ceramics toughened by this stress-induced transformation mechanism in t-ZrO_2 are Al_2O_3, cordierite, mullite, and the apatites.

In all, there are three main types of transformation-toughened configurations involving zirconia.

(a) **Partially stabilized zirconia** (PSZ). This consists of fine (< 0.1 micron) t-zirconia inclusions dispersed in a c-zirconia matrix (Trefilov 1995), and possesses high fracture toughness and good resistance to thermal shock. The partial stabilization of the t-phase is achieved by mixing small amounts of additives like MgO, CaO, or Y_2O_3.

(b) **Tetragonal zirconia polycrystals** (TZP). These consist mostly of sub-micron-sized particles of t-zirconia.

(c) **Zirconia toughened ceramics** (ZTC). Zirconia toughened alumina (ZTA) is a typical example of this. The t-zirconia particles occur both within the alumina grains, as well as in the inter-grain regions. It is at the latter sites that the toughening mechanism operates. ZTCs like corderites, mullite, silica and apatite find several applications, including those as extremely low-porosity bioceramics.

The high fracture toughness of zirconia, coupled with its very low thermal conductivity, makes it a very important refractory material. Even a 1-mm thick coating of zirconia on a metal results in a 300 K drop in temperature across the coating. Its use in hybrid engines with improved fuel efficiencies is thus very promising.

D.2. Kerr-Effect Applications of PLZT

The 'quadratic' characteristic of certain relaxor-ferroelectric compositions of PLZT derives from field-induced phase transitions (cf. §13.2.2). The following field dependence of the effective electro-optic coefficient (the R-coefficient) responsible for the Kerr effect has been derived by Haertling & Land (1971a):

$$R = -\frac{2\bar{\Delta n}}{n^3 E^2} \tag{14.2.3}$$

Here $\bar{\Delta n}$ is the birefringence, and n the refractive index. In the absence of the electric field the macroscopic symmetry is cubic and the material is non-birefringent. Thus the birefringence is caused entirely by the field. The R-coefficient is very large because its origin lies in the field-induced phase transition in the relaxor ferroelectric (Meitzler & O'Bryan 1973; Carl & Geisen 1973; Keve & Bye 1975), rather than in the usual modification of the optical indicatrix by the strong electric field present in a laser beam in the normal manifestation of the Kerr effect. The R-coefficient for 8.5/65/35 PLZT is $38.60 \times 10^{-16}\, m^2/V^2$, compared to only $0.17 \times 10^{-16}\, m^2/V^2$ for single crystals of $KTa_{0.65}Nb_{0.35}O_3$.

The ferroelastic nature of these ceramics can be used to advantage in certain situations. Application of uniaxial stress can be used for poling the domains along a direction that would result in maximum possible change of birefringence with electric field (see Lines & Glass 1977).

D.3. Shape-Memory-Effect Applications

The shape-memory effect exhibited by several ferroelastic alloys (cf. §11.5.4) finds a number of applications (cf. Perkins 1975; Schetky 1979; Wayman 1980). One of the most striking of these is as actuators in smart structures, and we shall discuss this separately in §14.3.

Other applications include a variety of electrical and mechanical connectors, an integrated-circuit package, and a heat engine exploiting the shape change on thermal cycling (Owen 1975).

14.2.5 Applications Involving Transport Properties

These are *Type E applications.*

E.1. Photoferroelectric Applications

Ferroelectric crystals like $LiNbO_3$ can be doped with ions such as Cr^{3+} so that they can absorb photons in the wavelength range in which they are transparent (Glass & Anston 1972). Absorption of photons results in release of charge carriers to the conduction band, where they wander along the polar axis (under the action of the internal electric field of the ferroelectric) till they are localized by trapping centres such as impurities, defects, etc. The net result is a space-charge field, and thence an additional dipole moment. There is also a concomitant local change of birefringence through an internal electro-optic effect. This is known as the *photorefractive effect*, and the host crystal is called a photoferroelectric (Fridkin 1979; Sturman & Fridkin 1992).

Several applications of such materials, including those for holography, have been discussed by Lines & Glass (1977).

E.2. Photoferroelastics

A phenomenon corresponding to the photorefractive effect, but in a non-ferroelectric ferroelastic, namely Sb_5O_7I, was reported by Fridkin et al. (1981). This crystal belongs to the ferroic species $6/mF2/m$. Photons of an appropriate energy induce a change in the birefringence of the crystal when they are absorbed by it. Such a crystal is called a photoferroelastic.

The explanation of the effect is similar to that in a photoferroelectric, except that now there is no internal electric field available for transporting the charge carriers freed by the incident photons to trapping centres. Fridkin et al. (1981) postulated an interaction between the free charge carriers and a zone-boundary acoustic mode. There is thus a *photodeformation* through the acousto-optic effect.

E.3. Photostrictors

In a photoferroelectric, the photovoltaic effect and the resultant photorefractive effect, are not the only consequences of release and transportation of free charge carriers along the polar axis. The space-charge field resulting from the trapping of these charge carriers also results in a change of dimensions via internal piezoelectric and electrostrictive effects (Brody 1983; Uchino, Miyazawa & Nomura 1983; Uchino, Aizawa & Nomura 1985; Sada, Inoue & Uchino 1987; Tanimura & Uchino 1988). The term photostriction has been used for this phenomenon in which illumination of a material induces strain in it.

The photovoltaic part of the photostriction effect is explained by treating the material as a semiconductor, with a certain band gap. However, this effect is very different from that in a conventional p-n junction of a solar cell, in that, because it occurs in a ferroelectric with a very high built-in biasing field along the polar axis, the voltage generated is much greater than the band-gap value.

So far, most of the work on the development of photostrictors has been done by Uchino and coworkers (see Uchino 1996). Much of their work has been on PLZT ceramic, doped with various materials, notably WO_3. The PLZT composition chosen is close to the morphotropic phase boundary of PZT, with 0.5/52/48 a typical choice. The band gap is of the order of 3.3 eV. Therefore much of the work has been carried out using the 380 nm line from a UV lamp, with a typical intensity of 10 mW/cm^2.

The photovoltage reaches a value of several kV/cm, and the photocurrent under illumination is a few nano-amperes. By adopting a so-called 'bimorph configuration', displacements of the order of 150 microns can be achieved at the tip of a 20 mm long, 0.35 mm thick, bimorph (Uchino 1996).

Several applications of the photostrictor effect in PLZT have been conceived, and some of them demonstrated. The most attractive thing about such a system is that it can serve as a *remote-control actuator*, requiring neither electric lead wires, nor even an electrical circuit. The absence of lead wires makes it particularly attractive for applications in micro-robotics. Other applications are in photo-driven relays and micro-walking devices.

Another area of applications envisaged by Uchino is in optical telecommunications as optical telephones or *photophones*. It is anticipated that the information technology of the 21st century would have the following major components: solid-state lasers as light sources, optical fibers as the media for transferring information, and a photo-acoustic device based on photostrictors as the optical telephone.

A limitation at present is the slow response times (build-up times and decay times) of the processes involved. In a photo-driven relay devised by Uchino (1996), the typical delay time was 1-2 seconds, in spite of the fact that a dual-beam method was used for avoiding the time delay that would normally occur in the 'off' process due to the low dark conductivity of the material.

The solution to this problem may lie in fabricating a composite, rather than working with a single-phase ceramic.

SUGGESTED READING

S. L. Swartz & V. E. Wood (1992). Ferroelectric thin films. *Condensed Matter News*, **5**, 4.

K. Uchino (1996). New applications of photostriction. *Innovations in Mater. Sci.*, **1** (1), 11.

R. Ramesh (Ed.) (1997). *Thin Film Ferroelectric Materials and Devices.* Kluwer, Dordrecht.

L. E. Cross & S. Trolier-McKinstry (1997). Thin-film integrated ferroelectrics. In *Encyclopedia of Applied Physics*, Vol. 21. Wiley-VCH Verlag, GmBH.

K. Uchino (1997). *Piezoelectric Actuators and Ultrasonic Motors.* Kluwer, Dordrecht.

O. Auciello, J. F. Scott & R. Ramesh (1998). The physics of ferroelectric memories. *Physics Today*, July 1998 issue, p. 22.

14.3 FERROIC MATERIALS IN SMART STRUCTURES

> *. . . it is reasonable to assume that highly integrated smart structures very similar to the biological model will be technologically possible in the near future. These smart structures will need to interface with human beings. Given the level of their anticipated sophistication and adaptive abilities, they will appear as living conscious entities to the majority of those people interacting with them. This will be the case in spite of the fact that they will not meet the formal requirements of either life or consciousness.*
>
> W. B. Spillman (1992)

This is the last section of the main text of the book, and in it we shall find a reference to, and use of, many of the key concepts and results described in the book.

In prehistoric times materials were used in the form in which they occurred in nature. The next stage was the deliberate design of materials (e.g. alloys, ceramics, composites) with certain desirable but *fixed* properties. Progress in physics, chemistry and materials science has now given us the capability to design and fabricate materials or structures which can *adapt* their properties in a pre-conceived and useful manner to changes encountered by them in environmental conditions. An additional recent development has been the evolution of the field of artificial intelligence, which enables us to introduce a modicum of *learning* into the design of adaptive materials or structures, so that they can respond to a situation in a way similar to that of a moderately smart or intelligent living being. Some of the basic characteristics of an intelligent living being are: sensing, actuation, control, and learning (Ahmad et al. 1990; Davidson 1992; Coghlan 1992; Knowles 1992).

Smart materials or structures may be formally defined as *materials or structures with an ability to respond in a pre-designed useful manner to changing environmental conditions* (Thompson, Gandhi & Kasivisvanathan 1992). There are two broad categories of them: *passively smart*, and *actively smart*, materials or structures (Newnham 1991; Newnham & Ruschau 1991a, b; Newnham 1997). Both have a sensing characteristic and a (predesigned) response characteristic. The difference lies in the fact that actively smart materials or structures are connected to an external power source, field, or feedback system (and a control module) (Fig. 14.3.1), which is designed to enhance and/or control their response, and usually results in an enhanced actuator characteristic of the assembly.

14.3.1 Smart Systems, Structures, and Materials

It is necessary to make a distinction between smart systems, smart materials, and smart structures.

A three-term (PID) temperature controller is an example of a smart *system*. It consists of complex circuitry, using a large number of diverse components, and it becomes nonfunctional if sliced arbitrarily into two or more parts.

By contrast, a thermistor (a resistor the resistivity of which is a predesigned useful function of temperature) is a smart *material*. If it is cut into two or more parts, each part is still a thermistor which can act as a (passively) smart material, adjusting its electrical resistance autonomously, in a predesigned manner, against variations of temperature.

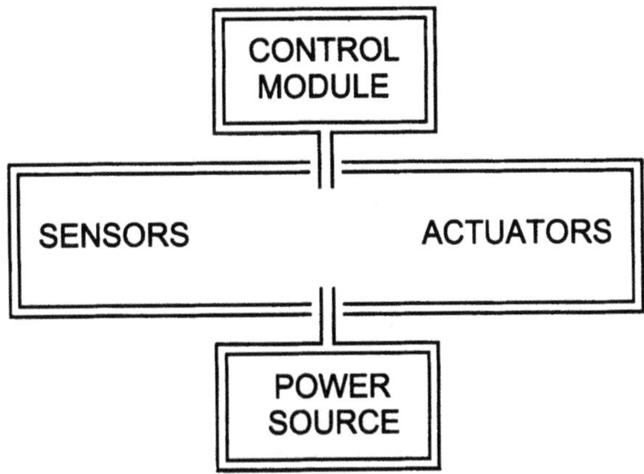

Figure 14.3.1: The essential parts of an actively smart structure.

Smart *structures* can be regarded as intermediate between smart materials and smart systems. In a smart-structure composite the sensor elements (e.g. optical fibers) and actuator elements (e.g.thin shape-memory-alloy wires) are embedded all over the composite at the fabrication stage itself. With progressive micro-miniaturization of the components of smart composites, the distinction between smart structures and smart materials may become more and more blurred. In any case, the former is a more general term than the latter.

In what follows, we shall not be discussing smart systems at all. We shall use the term 'smart structures' to cover even smart materials, where applicable.

Biomimetics

In their quest for designing novel smart structures, scientists and engineers tend to fall back again and again on how Nature has been doing it for biological systems, which have evolved through natural selection for millions of years. In biomimetics one aims at mimicking Nature for developing smart or even intelligent systems, structures, or materials, and it is now a science in its own right.

Spillman (1992) has put forward the hypothesis that *the most efficient smart structure (biological or not) for a given purpose should have a* self-similar *level of functioning at every hierarchical level of its organization; in other words, it should have a* fractal *character.*

In Fig. 14.3.1 the control module and the power source have been shown as residing outside the body of the smart structure. As the subject of smart structures advances further, a stage may come when the control module will acquire a distributed character, and will become largely a part of the smart structure or material (Helferty, Boussalis & Wang 1992).

Depending on their type and level of sophistication, smart structures may possess one or more of the following characteristics (Takagi 1989; Newnham 1994): sensing; actuation; selectivity; shapeability; self recovery; self repair; stability, including multistability; standby phenomena; and switchability.

14.3.2 Passively Smart Structures

The smartness of a passively smart structure does not require an external power source or active feedback mechanism for coming into play. For this reason, they are, in general, closer to being smart *materials*, rather than smart systems. Also, for the same reason, their performance is generally of a less dramatic nature and, although they find a large number of applications, we hardly notice them as doing something particularly smart !

The thermistor was mentioned above as an example of a passively smart material. The ceramic varistor is another example. Its electrical resistivity decreases rapidly, and highly nonlinearly, on application of high voltages. A zinc oxide varistor can act as a lightning protector. When struck by lightning, its resistance falls to a very low value, and the current is bypassed to the ground. The highly nonlinear I-V characteristic is thus the standby protection property of this passively smart material. It offers an alternative to the conventional copper-based lightning protector in that its resistivity at low voltages is much higher.

Other examples of passively smart materials are optical limiters and photochromic glasses. A rather elaborately investigated case is that of the 'novelty filter' PHOTOGREY, an optical glass developed by Corning Glass Works (cf. Spillman 1992). Its transmissivity varies nonlinearly with the intensity of the light falling on it. Thus it automatically reduces the variations of the intensity of light passing through it.

A solution of fullerene in, say, toluene serves a similar purpose. It acts as a passively smart material because, for high input fluences of optical radiation, the output fluence saturates to a constant value (Tutt & Kost 1991; Mishra, Rawat & Mehendale 1997).

Many other examples have been described by Newnham (1990).

14.3.3 Actively Smart Structures

Actively smart structures and materials possess both sensor and actuator functions, and involve external biasing or feedback (Fig. 14.3.1). They thus have an externally aided, usually nonlinear, self-tunability feature with respect to one or more of the macroscopic tensor properties of the material(s) involved. Ferroic materials, which have several highly nonlinear macroscopic properties in the vicinity of the ferroic phase transition, are obviously relevant in this context, although they are not the only possible choice. Electrorheological (ER) fluids are some of the other materials used in actively smart structures (Gandhi, Thompson & Choi 1989).

Adaptive learning can also be incorporated in the device applications of actively smart materials by the use of fast, real-time, information processing arrangements involving neural networks (Grossman et al. 1989).

Among the options for embedded sensors in actively smart structures are piezoelectric ceramics, resistive strain gauges, and optical fibres (Bowden, Fanucci & Nolet 1989).

Optical fibres are particularly popular as embedded sensors because they can be used in large lengths in extremely small diameters, and they do not undergo structural degradation during embedding because of their ability to withstand the high temperatures and pressures necessary for the fabrication of the smart composite. They also possess very high response rates and sensitivities. Since they are sensitive along their entire length, they are very well suited for real-time detection of structural changes at any or all points in the composite structure.

At present the most popular actuator option is that of shape-memory alloys, particularly NITINOL[2]. We shall discuss them in some detail in §14.3.5.

A large shape-memory effect (SME) can arise if mediated by a martensitic or ferroelastic phase transition. The SME has been observed not only in metallic systems, but also in PLZT (Wadhawan et al. 1981; Schmidt 1990) and $Y - Ba - Cu - O$ (Tiwari & Wadhawan 1991). GMO and its terbium and dysprosium analogues also probably display the SME (Virkar et al. 1991). Although the recoverable shape strain of nonmetallic SME materials is only of the order of 0.5% (or less), their high electrical resistivity can be an advantage over shape-memory alloys (SMAs). In any case, the figure 0.5% compares favourably with the recommended maximum strain of 1% for SMAs used for *cyclic* applications for 100,000 or more cycles (Stoeckel & Simpson 1992). Apparently, no investigations have yet been carried out on nonmetallic SME materials for repeated applications involving a large number of cycles.

Among the other options for actuators in smart structures are piezoelectric ceramics like PZT, and large-electrostriction relaxor ferroelectrics like PMN (Varadan et al. 1992). The basic mechanism here is the genera-

[2]Mention must be made here of another very promising actuator material for applications in which the controlling field is a magnetic field, namely the alloy TERFENOL – D (with a typical composition $Tb_{0.3}Dy_{0.7}Fe_2$) (Chaudhry & Rogers 1995). It has excellent properties for fast, heavy-load, actuator applications (see du Tremolet de Lacheisserie 1993, page 353, for details).

tion of strain by the application of voltage. This strain is inherently smaller than that involved in shape-memory alloys. Thus, piezoelectric and electrostrictive actuators are considered suitable for high-frequency low- and medium-stroke applications, although the total stroke can be enhanced by fabricating multilayer actuators (Uchino 1992; Sugawara et al. 1992).

SMAs are ideal for low-frequency high-stroke applications (cf §14.3.5). The low-frequency restriction is imposed by the slowness with which the SMA wire embedded in the composite can cool back to the martensitic phase, after actuator function has been realized through shape recovery by heating to the austenitic phase. Another problem is the fatigue of SMA wires, especially in high-strain configurations.

14.3.4 Tuning of Properties of Ferroics by External Fields

If a material is to respond in a variable manner to suit the changing environmental conditions, it should be able to tune its relevant property automatically. The presence of an external biasing field, as in the case of actively smart materials, often results in a larger range of tunability of the concerned property than is possible for passively smart materials (for which no biasing field is applied). A nonlinear dependence of the property on the biasing field is usually advantageous for achieving self-tunability of the property.

For a concrete discussion of this question we consider the strain tensor **e** and the electric displacement vector **D** of a crystal placed in a nonmagnetic environment:

$$e = \boldsymbol{s}^{E,T}\boldsymbol{\sigma} + \boldsymbol{d}^{T}\boldsymbol{E} + \boldsymbol{\alpha}^{E}\mathrm{T} \tag{14.3.1}$$

$$\boldsymbol{D} = \boldsymbol{d}^{T}\boldsymbol{\sigma} + \boldsymbol{\epsilon}^{\sigma,T}\boldsymbol{E} + \boldsymbol{p}^{\sigma}\mathrm{T} \tag{14.3.2}$$

The total strain **e** has contributions from the stress field $\boldsymbol{\sigma}$, the electric field **E** (through the piezoelectric effect), and thermal expansion (Nye 1957). Similarly, **D** has, respectively, piezoelectric, dielectric, and pyroelectric contributions.

Now suppose the crystal chosen is such that it undergoes a proper ferroelastic phase transition at a temperature not far from the device-application temperature. Two aspects of such crystals are very important from the point of view of their applications in actively smart structures. One is that some component (or a combination of components) of the elastic compliance tensor ($\mathbf{s}^{\mathbf{E},T}$ in Eq. 14.3.1) becomes a highly nonlinear and strong function of temperature in the vicinity of the transition temperature. This feature makes it easy to tune the elastic (or rather pseudoplastic) response of the material by applying an external stress field.

The other aspect of the ferroelastic transition is that application of a stress field leads to a shifting of the transition temperature. This has other concomitant effects like the occurrence of a 'mechanical' shape memory effect (superelasticity), governed by the formation of stress-induced martensite (see, e.g. Guenin 1989; Wayman 1992; Stoeckel & Simpson 1992).

The above remarks about ferroelastic materials also apply, *mutatis mutandis*, to other primary ferroics. The dielectric response function (denoted by ϵ in Eq. 14.3.2) of a proper ferroelectric phase of a material becomes large and highly nonlinear in the vicinity of the transition temperature T_c. And T_c can be shifted by applying an electric field (Jiang 1992).

Eqs. 14.3.1 and 14.3.2 also have other features relevant to smart-material applications of ferroics in the vicinity of T_c. For example, although the piezoelectric tensor $\mathbf{d}$ is zero for a centrosymmetric crystal, it can acquire nonzero (even large) *field-tunable* components in some materials (notably relaxor ferroelectrics) because of its coupling with the primary instability driving the phase transition. PMN and PLZT are the best-investigated examples of this. The ferroelectric hysteresis loop for them does not disappear suddenly at a single temperature, but rather decreases gradually in area with increasing temperature. What is even more relevant for the present discussion is the fact that there is no change in the gross point-group symmetry of the material in the Curie range of temperatures. For example, for PMN the gross point-group symmetry at room temperature is $m\bar{3}m$, the same as that of the paraelectric high-temperature phase (Prokhorov & Kuz'minov 1990). Since this is a centrosymmetric point group, all components of the $\mathbf{d}$ tensor must be zero. And yet the material can exhibit a nonzero and large piezoelectric response when subjected to an external electric biasing field (e.g. $d_{33} = 130$ pC/N for $E_3 = 3.7$ kV/cm).

The mechanical analogues of this in metallic systems are the alloys that exhibit diffuse martensitic transitions. Their analogous relevant features are cluster formation and coexisting phases.

In nonmetallic ferroelastics a well-known example of this type is that of the solid solution $(KBr)_{1-x}(KCN)_x$ (Hochli, Knorr & Loidl 1990; Loidl 1991). For $x < 0.6$ the orientational disorder of the dumbbell-shaped $(CN)^-$ ion (which carries both an elastic dipole and an electric dipole) is frozen-in and transitions to orientational-glass states occur.

PLZT is another extensively investigated material, which becomes very compliant in the vicinity of the diffuse ferroic transition (Meitzler & O'Bryan 1973). It also exhibits the shape-memory effect (Wadhawan et al. 1981).

14.3.5 Applications of Ferroic Materials in Smart Structures

Ferroic materials find applications as both sensors and actuator in smart structures.

Piezoelectric Ceramics as Sensors

Although optical fibres are the most popular choice at present as sensors in smart structures, piezoelectric ceramics are also strong contenders for this application, especially if they can be developed in the form of thin fibres for easy and extended embeddability in the composite structure; otherwise their usual brittleness can create problems in certain applications. The vicinity of a ferroic phase transition at the temperature of application gives enhanced sensitivity. Their sensor action is through the production of voltage under deformation. This is largely a reversible process, which makes their use very attractive.

Tunable Transducer

The concept of developing a fully tunable composite transducer has been described by Newnham (1991). It can tune its sensor and actuator functions, and thence act as an actively smart structure. The tuning is achieved by making available external electrical and mechanical biasing fields.

It is constructed from rubber and a relaxor ferroelectric, both capable of highly nonlinear behaviour even under moderate external fields: rubber has a highly nonlinear response to mechanical stress, and a relaxor ferroelectric like PMN has a highly nonlinear electrical and electromechanical response to electric field. Therefore the biasing enables the composite structure to tune its property coefficients in response to a changing environment. The properties which can be tuned include resonant frequency, acoustic and electrical impedance, damping factors, and electromechanical coupling coefficients.

An important property of a transducer is its resonant frequency f:

$$f = \frac{1}{2t}\sqrt{c/\rho} \tag{14.3.3}$$

Here t is the relevant thickness, c the stiffness coefficient, and ρ the density. Rubber has the special property that its stiffness increases dramatically under stress. Thus, mechanical biasing stress can be used for tuning the stiffness, and therefore the resonant frequency f, of a composite in which rubber is one of the constituents.

In the experiments conducted in the laboratory of Prof. Newnham (1991), a multilayer laminate comprising alternating steel shims and rubber layers, each of thickness 0.1 mm, was subjected to various amounts of compressive stress, and measurements of Young's modulus E were made. It was found that the stiffness c quadrupled from 600 to 2400 MN/cm^2 (core-sponding to a doubling of the resonant frequency f) for a biasing stress of 200 MN/cm^2.

Rubber is not a piezoelectric material. Therefore, to make a transducer, one has to design a composite using rubber and, say, poled PZT ceramic. Such a composite was made by Newnham and coworkers. It consisted of thin rubber layers, PZT, and metallic head and tail masses, this triple sandwich being held together by a stress bolt. At low stress rubber is very soft, and effectively isolates the resonating PZT member from the metallic head and tail masses. At high stress the rubber stiffens, leading to a large coupling between PZT and the metal pieces. The result is that the radial resonant frequency doubles (from 19 to 37 kHz) when the bias stress is changed from 20 to 100 MPa. Moreover, as the rubber stiffens under stress, the mechanical quality factor Q increases from 11 to 34.

In this arrangement only a mechanical tuning capability is present (from rubber), and no tunability comes from PZT. The reason is that the piezoelectric strain produced in PZT is a *linear* function of the applied electric field, and the piezoelectric coefficient is a constant, not tunable by an electric biasing field. To achieve electrical tunability, we must replace PZT by a relaxor ferroelectric like PMN.

PMN is a strongly electrostrictive material. Not only is the mechanical response to electric field large, it is also *nonlinear* (varying as square of the applied field). Thus it has property coefficients which are tunable by an external electric field. Compared to PZT, PMN has the following features to offer: (i) The macroscopic strain produced in PMN by electric fields is comparable to that in PZT. At zero biasing field (at room temperature) the material exhibits no piezoelectric effect. Measurements carried out on $Pb(Mg_{0.3}Nb_{0.6}Ti_{0.1})O_3$ (i.e. on PMN – PT) under a biasing field of 3.7 kV/cm gave a value of 1300 pC/N for d_{33}. (ii) There is practically no hysteresis problem to cope with. (iii) Even the dielectric susceptibility of PMN can be tuned by applying a dc biasing field.

All this implies that the electromechanical coupling coefficient, k, can be tuned by the electric field. So also the electrical impedance, because the field-induced polarization saturates at high field values.

Thus a tunable transducer can be mae out of rubber and PMN, exploiting the elastic nonlinearity and the piezoelectric nonlinearity. By controlling the mechanical and electrical biasing fields though a negative feedback configuration of this actively smart composite, all the main parameters of the transducer can be tuned, namely resonant freqency f, the acoustic impedance Z_A, the electric impedance Z_E, the inverse mechanical damping factor Q, and the electromechanical coupling factor k.

Shape-Memory Alloys in Smart Structures

Shape-memory alloys (SMAs) have several attractive features which make them very suitable in a variety of smart-structure applications (Davidson 1992; Rogers 1992; Stoeckel & Simpson 1992). Much of the work in this area has been done on NITINOL. Pseudoplastic strains as large as 7 to 8% can be completely recovered in this material on heating across the martensite-austenite phase transition. The transition temperature can be tailored to lie anywhere between 0^oC and 100^oC. The thermal hysteresis of the transition during heating and cooling can be made less than 5^oC (Grossman et al. 1989). Wang (1992) has reported a thermal hysteresis as small as 0 to 1^oC. This very narrow thermal hysteresis occurs in some binary and ternary Ni – Ti alloys possessing the premartensitic 'R-phase' (Stoeckel & Simpson 1992). Whereas the martensitic and the austenitic phases have yield strengths of 80 and 620 MPa respectively, a rather large stress of 700 MPa is generated if the deformed martensitic phase is physically prevented from regaining its shape in the high-temperature austenitic phase. Not only is the stress generated large, and the available stroke length large, the stress is also *constant* during the stroke (Wayman 1992). The Young's modulus of the high-temperature phase is about four times that of the low-temperature martensitic phase. Thus a spring made of this material can change its spring constant by a factor of four on being heated to the upper phase. SMAs like Ni-Ti can thus find the following two types of uses in smart structures: (i) application of forces and torques; and (ii) variation of material properties like stiffness.

Correspondingly, there are two broad categories of applications envisaged for SMAs (Davidson 1992). Both involve the use of SMA wires of diameter about 200-500 microns, embedded in a pseudoplastically elongated state in the smart composite, and prevented from recovering their normal (memorized) length during fabrication. When an electric current (usually in the form of pulses) is passed to heat such a wire to its high-temperature phase, two things can happen. If an SMA wire is configured not to coincide with the neutral axis of the structure, the recovery force generated by the phase transition leads to a bending of the structure in a pre-designed way. If, on the other hand, the SMA wire is along the neutral axis, there is generated a uniformly distributed stress along the length of the structure. Creation of a state of residual strain leads to a concomitant tuning of the stiffness, and of the natural frequency of vibration. If the wire is in the form of a coil, a change of length of the coil also occurs (Furuya & Shimada 1991).

For many of the applications of SMAs in smart structures the one-way SME is not useful, as it is a one-time operation only. For automatic repeated or cyclic operations the SMAs have to be trained for the two-way SME (Guenin 1989; Maclean et al. 1992; cf. §11.5.4).

A variety of applications of SMAs in smart-structure and smart-system configurations have been either investigated or actually demonstrated. We survey some of them here.

SMAs as sensor elements. The use of SMA wires as sensors in smart structures has been described by Baz et al. (1992a). These sensors undergo stretching when the beam they are embedded in is deflected by a load. They generate a signal proportional to the stretching, the signal being then used for sending electrical current through another set of SMA wires for appropriate actuator response.

Robotics. SMAs can convert thermal energy to mechanical energy through the shape-recovery process on heating. Their use in robotics is therefore only to be expected. What has been achieved till now is the development of smart *systems*, rather than smart structures, employing SMAs as actuators. Both 'biased' and 'differential' actuators have been developed (Furuya & Shimada 1991). The heating of the SMA wire or coil is usually of the resistive type, obtained by passing a train of electrical pulses. The response speed is limited by the rate at which the system can cool back to the martensitic phase.

Applications in space technology. SMAs can contribute to a large number of applications in space technology (Schetky 1991). An example of an actively smart structure is provided by the folding-box type protective shroud based on SMAs (Schetky 1991). On being heated by solar energy in outer space, the SMA actuator converts itself from a stowed to a fully deployed (unfolded) shape, thus providing protective shielding to the satellite.

In the zero-gravity environment of outer space, there is no damping possible from gravitational forces. Smart structures offer perhaps the only way out for the control of vibrations, and for achieving an accurate pointing of, for example, high-gain antennas. SMAs have been envisaged as actuators in truss members of large, high-precision, space structures (Spillman 1992). In the design of the truss, optical fibres are integrated into it to act as sensors. Vibration levels sensed by these are analyzed in the control module of the truss, which then provides electrical power to the SMA actuators to ensure that the vibrations are reduced to acceptable levels (Spillman 1992). According to Wada et al. (1990), future NASA missions will require large space systems (including optical interferometers) to be positioned and controlled with sub-micron accuracies. Incorporation of suitable adaptive structures into a truss is regarded as the main possible solution to the problem of accurate positioning and orientation.

Active control of buckling of composite beams. Baz et al. (1992b) have introduced and investigated the idea of impregnating flexible fibreglass composite beams with NITINOL wires, so that the buckling characteristics of such beams can be controlled dynamically by exploiting the large recovery force generated in the SMA wires when they are fed thermal energy for transforming to the austenitic phase. The NITINOL wires are embedded in vulcanized rubber sleeves situated along the neutral axis of the beam. A non-contact sensor monitors any buckling of the beam, and activates the heating of the NITINOL wires, which in turn prevent the buckling from taking place. In the configuration investigated by Baz et al. (1992b), the critical buckling load could be increased by a factor of three compared to the uncontrolled beam. Such an arrangement enables one to achieve a higher performance-to-weight ratio, without compromising on the stability of the beam.

Although Baz et al. (1992b) studied the use of only the *thermally* activated SME, it would be interesting to examine the use of the mechanically induced SME for the same purpose. This, if feasible, would have some additional advantages: no sensor for buckling would be needed; no heating arrangement would be required; and since there would be no heating, there would be no softening of the surrounding matrix.

The smart traversing beam. The design of long-span support bridges requires traversing beams that are light in weight, have high strength, and do not deflect excessively under moving loads. Baz et al. (1992a) have demonstrated the feasibility of fabricating such beams from smart composites, using NITINOL wires for countering autonomously the deflection produced in the beam by moving loads. The NITINOL wires are embedded parallel to, but not coinciding with, the neutral axis of the beam. When a deflection is sensed, the control system sends electrical current through an appropriate number of NITINOL wires to heat them to the austenitic phase. The large recovery forces generated by this action counteract the tendency of the beam to deflect. However, such an arrangement is not very effective if the deflecting load moves very fast on the bridge.

Compliant wing sections for controlling the flight of aerodynamic and hydrodynamic vehicles. Biological structures (birds, fish) have articular bone or flexible cartilage configurations. Movement is achieved by contraction of muscles. Analogies with these have spurred activity for the design of compliant wing sections for the adaptive control of surfaces, with SMA wires as the actuating elements (Maclean et al. 1992; Beauchamp et al. 1992). In the studies carried out by Beauchamp et al. (1992), the SMA actuators were situated *outside* the wing (or foil, or fin). Two sets of SMA wires were fixed to the opposite sides of the trailing edge of the foil, the leading edge being fixed to a post. The two sets of SMA wires provided actuation towards their respective sides when any one set was heated to the austenitic phase. Thus this arrangement employed the one-way SME, the restoration of shape on cooling being provided by the spring metal backbone of the foil or wing. The dynamically varying wing section was found to produce a higher lift force, and a lower flow separation, compared to the rigid wing.

The smart artificial muscle. Modelling studies have been conducted by Thursby et al. (1989) on an idealized neuro-muscular functional unit, with fibre-optic sensors, SMA actuators, and an artificial neural network for a 'brain'. The typical muscle can be modelled as a bundle of matter having a combination of series and parallel connections through individual fibres (tendons etc.). The interaction with the nervous system is through these fibres, which also act as sensors and actuators. In the artificial smart muscle, SMA wires correspond to these fibres. In the biological system each muscle fibre is actuated separately by an alpha motor neuron. A problem presents itself here, in that it is very difficult to activate each SMA wire separately by electrical means in the artificial muscle. Moreover, these wires, being made of metallic alloys, have low resistivity. This makes it very impractical to achieve I^2R heating individually for a complicated and large network of SMA wires connected in series and parallel configurations. Thursby et al. (1989) and Grossman et al. (1989) have therefore experimented with *laser-heat activated SMA structures*. This provides some other advantages also (see below). It was demonstrated that artificial neural networks can be trained to control the functions of such a muscle system.

Laser-heat activated SMA structures. As described by Grossman et al. (1989), the use of electrical heating of SMA wires in conventional smart composites produces a nonlinear all-or-none type of response. The nonlinearity stems partially from the fact the resistance of the wires is temperature dependent, and therefore the temperature change produced per unit voltage is not constant. These authors have explored the possibility of using pulsed laser heating as an alternative strategy. Optical fibres embedded in the same composite in which the SMA wires are embedded are used for transporting the laser energy for a very localized heating of the SMA actuators. The optical fibres also act as sensors. Feasibility studies were carried out with high-power CO_2 and frequency-doubled Nd:YAG laser radiation, but eventually the inherently small-sized diode lasers will be embedded in the smart composite itself. With the use of appropriate techniques involving ultra-fast real-time data processing with neural networks, they were able to achieve a *linear* control of individual SMA segments, without overburdening the integrity of the composite.

The parallel-processing architecture of the neural network results in high speed. The network also 'learns' by example during 'training', and then attempts to apply the memorized algorithms to unfamiliar inputs.[3] This also contributes to high overall speed of response. The output from the network is fed as control signals for the laser supplying heat power to the SMA actuators.

Use of laser heating by Grossman et al. (1989) for activating a large array of SMA wires solves the three main problems encountered when electrical heating is employed:

(i) By using very thin optical fibres (with a diameter of the order of 100 microns) for carrying laser heat to any specific SMA segment, the degradation of the strength of the composite is minimized.

(ii) By using separate optical fibres for the activation of each SMA segment, the problem of crosstalk or any other interference is overcome.

(iii) The amount of optical power required per unit rise of temperature is constant. This greatly simplifies the control-circuit requirements.

Grossman et al. (1989) have used a three-layer *perceptron* neural net-

[3]A two-minute introduction to neural networks is offered here for the totally uninitiated. Neural networks are a set of interconnected 'neurons' working concurrently (see, for example, Neelakanta & de Groff 1994). They can learn system-dynamics without requiring *a priori* information regarding the system structure (Thursby, Grossman & Yoo 1990; Grossman & Thursby 1995). They are based on models of the brain and its behaviour. A neural network develops *associations* between objects pertaining to a given problem domain. It consists of three basic elements: processing elements (PEs) and their connections; a method to train the network to solve certain problems (*learning*); and a method to recall the information from the network. A PE is a building block of a neural network. A neural network consists of one or more PEs connected together. Each connection is associated with a numeric value called the *connection weight*, that forms a memory unit of the network. PEs are usually divided into disjoint subsets called *layers*. Functionally, each PE forms a weighted sum of the inputs impinging on it, using a *sum function*. This sum is then transformed by a *transfer function* to a value that is fed to an *output function* to produce the output of that PE. PEs in the same layer have the same sum, transfer and output functions. 'Learning' is a process of adjusting the connection weights in order to make the network develop correct associations between objects concerning some application. The learning is done by a *learning rule* or function, there being a single rule for an entire layer.

work for establishing the proof-of-concept of the laser activated SMA structure. Data from the optical-fibre sensors are fed to the neural network, which calculates the strains and the signals to be sent to individual SMA actuators for corrective action. The training of the neural network is done off-line for determining the optimum weights.

General conclusions. Although the subject of smart materials and structures is in its infancy, extensive research going on in a large number of laboratories all over the world can be expected to result in great breakthroughs and rapid progress. Commercially successful application of SMAs in smart structures will require further work in certain areas. We list three of these here.

(a) For the successful application of laser-activated SMA structures, the transfer functions will have to be determined extensively between the input optical intensity, pulse shape, and pulse width on the one hand, and mechanical and thermal properties on the other (Grossman et al. 1989).

(b) Friend (1992) has drawn attention to the problem of prediction of strain trajectories during *partial* actuation of SMAs in the continuous-mode operation, and the problem of long-term stability of the actuation strain. The long-term stability is good if the SMA actuator does not have to work against any biasing load. But in most real-life applications the SMA is embedded in the smart-structure composite, and has to do work. Therefore the recovery response of the SMA is not constant, and has to be predicted by sophisticated modelling calculations. Further, as pointed out by Friend (1992) again, biased as well as partial-cycle actuation will result in complex trajectories. This would necessitate the development of very complex control algorithms for modelling the complicated stress-strain behaviour realistically. Work in this direction is already in progress in some laboratories (Maclean, Patterson & Misra 1990).

(c) For cyclic applications of SMAs, the usable limits on the maximum recoverable strain and on the maximum generated stress have considerably lower values than what are available for no-load and noncyclic operations. For NITINOL the recommended maximum strain drops to 2% from 8%, and the maximum stress drops to 140 MPa from 650 MPa, when the number of cycles is around 10,000 (Stoeckel & Simpson 1992). Further research can hopefully push forward these limits on performance.

What we are going to see in the near future is the coming together of three *mega-technologies*, namely advanced materials, information technology, and biotechnology, for the evolution of biomimetic materials, structures, and systems. Ferroic materials in general, and shape-memory alloys in particular, clearly have a crucial role to play in this scenario.

SUGGESTED READING

I. Ahmad, A. Crowson, C. A. Rogers & M. Aizawa (Eds.) (1990). *U.S.-Japan Workshop on Smart / Intelligent Materials and Systems*, March 19-23, 1990, Honolulu, Hawaii. Technomic Pub. Co., Lancaster.

R. O. Claus (Ed.) (1991). *Proc. Conf. on Optical Fiber Sensor-Based Smart materials and Structures*, April 3-4, 1991, Blacksburg, Virginia. Technomic Pub. Co., Lancaster.

R. E. Newnham (1991). Tunable transducers: Nonlinear phenomena in electroceramics. In *Chemistry of Electronic Ceramic Materials*, Special Publication 804, National Institute of Standards & Technology. (Proceedings of the International Conference held in Jackson, WY, August 17-22, 1990. Issued January 1991.)

G. J. Knowles (Ed.) (1992). *Active Materials and Adaptive Structures.* IOP Publishing, Bristol.

B. Culshaw, P. T. Gardiner & A. M. Donach (Eds.) (1992). *First European Conference on Smart Structures and Materials.* IOP Publishing, Bristol. (SPIE Vol. 1777.)

E. Udd (Ed.) (1995). *Fiber Optic Smart Structures.* Wiley, New York.

R. E. Newnham (1997). Molecular mechanisms in smart materials. *MRS Bulletin*, May 1997.

R. E. Newnham (1998). Phase transformations in smart materials. *Acta Cryst.*, **A54**, 729.

Chapter 15

EPILOGUE

The trouble with facts is that there are so many of them.

Samuel McChord Crothers

I had two objectives to meet in the writing of this text. One was to attempt a unified and reasonably self-contained account of the physics and applications of ferroic materials (single crystals, ceramics, composites). The other was to try to make it easier for the beginner to comprehend the concepts and jargon used in the research papers on the subject.

To achieve the first task, I have taken the help of symmetry considerations. Although symmetry arguments seldom provide numbers for experimental verification, they have great unifying and systematizing power. Prototype symmetry, and the spontaneous breaking of the point-group part of it at a ferroic phase transition, is the central notion in the physics of ferroic materials. The definition of prototype symmetry needed to be made more precise, which I have done in §5.1.3.

The second objective made it necessary for me to opt for greater breadth than depth in the treatment of the topics discussed. In view of the wide diversity of the concepts to be covered, I was compelled to be brief. I have no regrets about this because there is an important pedagogic principle involved in the approach adopted by me: I believe that, often, the best way to explain a complex or 'advanced' topic to a student is by describing it *briefly*. Once the student has understood the basic idea, and also learnt the jargon used by experts, there should be no difficulty in going deeper into the topic by consulting the SUGGESTED READING material.

Quite deliberately, I have often discussed topics which fall outside the purview of ferroic phase transitions and ferroic materials. This has been done to place the subject of ferroic materials in a proper perspective, and to induce the reader either to draw analogies, or to appreciate the difference between ferroic and nonferroic behaviour. The similarity between crystal growth and domain-boundary (and phase-boundary) movement is an example of the former type. And the distinction between structural extended defects and compositional extended defects (Chapter 8) is an example of the latter type.

As is natural when one attempts the first comprehensive and connected treatment of a subject, gaps in the overall growth of the subject become visible. The most serious gap I noticed was that regarding the very definition of a ferroic phase transition (and thence a ferroic material). I have provided my answer to it (in §5.1) by invoking Guymont's nondisruption condition.

Another matter of definition, which still needs some detailed analysis, is that of the choice of a control parameter for defining a phase transition in a crystalline material, particularly a phase transition involving a change of symmetry. As discussed in §14.1, there is no strong reason for choosing only temperature, hydrostatic pressure and/or composition (all scalars) as the control parameters for defining a symmetry-changing phase transition. For domain-structure systematics etc., anisotropic influences like electric field, magnetic field and/or uniaxial stress can be equally valid thermodynamic control parameters for inducing a symmetry-changing phase transition. In other words, for several purposes, field-induced phase transitions are no different from transitions effected by scalar control parameters.

The *thermodynamic* definition of a phase transition in a crystal runs into difficulties when the number N of unit cells in it is not infinite (see, for example, Privman & Fisher 1983). The problem becomes increasingly serious as we go down to nanometer scales. Under the circumstances, a *symmetry*-based definition can become more meaningful. In §5.6.1 I quoted Landau's celebrated statement ('*symmetry cannot change continuously*'), which Anderson (1981) has called *the First Theorem of condensed matter physics*. One can dwell on this theorem to make the point that even a crystal that is not infinite can be ascribed a definite symmetry, *at least over a typical length scale*. So long as this typical length scale encompasses a few repeat distances, there is a specific symmetry assignable to that part of the crystallite. And the symmetry-based definition of a ferroic phase transition is simply that there should be *a change of point-group symmetry over the specified length scale* [1] (in a nondisruptive manner).

Admittedly, such a definition can run into problems for a system with nanodomains with, say, thermally induced movement of domain boundaries

---

[1]The specification of a length scale for defining the symmetry of a crystal is nothing unusual. Even an infinite crystal has Wyckoff site symmetries not all of which are the same as the symmetry of the entire crystal. The symmetry of a crystal cannot be defined for sizes smaller than than that of its Wigner-Seitz cell.

and phase boundaries. One has to then specify not only a length scale, but also a time scale, for assigning the point-group symmetry.

Nanocrystals do not have the same crystal structure throughout: As one approaches the surface, the lattice parameters may tend to change (usually increase) gradually. But so long as there is a core for which a unit cell can be identified, and this unit cell repeats itself a few times, there is a phase with a well-defined symmetry, which can therefore undergo a transition to another phase of different symmetry, under appropriate conditions.

Application of even a vanishingly small anisotropic field changes the symmetry of a crystallite; this is *not* a phase transition, but only a process of dissymmetrization. As this external field is increased in magnitude, the net symmetry (of crystal plus field) may suddenly change at some value of the field. *This* is a phase transition.

Thus symmetry considerations help reduce the ambiguity present in the thermodynamic definition of a phase transition in small crystals.

Certainly, important new features can arise sometimes when a phase transition is induced by a non-scalar field. An example is the effect of the presence of an external magnetic field on the degree of degeneracy of the ground state of a frustrated system like a spin glass (cf. §9.2.6).

Although suggestions for further reading have been given at the end of almost all the sections, I advise the reader to pay particular attention to the following texts:

An authoritative recent book on the theory of ferromagnetism is that by Aharony (1996). Also recommended strongly are the books by Jiles (1991) and Valenzuela (1994) on the subject of magnetic materials.

For ferroelectrics, the book by Lines & Glass (1977) continues to be my favourite. A useful addition to the basic literature is the book by Xu (1991), which covers several materials-science aspects of ferroelectrics. Pandey's (1995) review article on diffuse transitions in mixed ferroelectrics provides a good starting point for the newcomer to this topical field of research.

Salje's (1993a) book on ferroelastics is a welcome addition to the student-oriented literature on this topic. Salje has also guest-edited some very useful special issues of the journal PHASE TRANSITIONS, on this subject (cf. Salje (1994, 1995, 1999); Kleemann & Salje (1998)).

I have not given space to the experimental techniques used for studying ferroic materials. The reason is that practically all the usual techniques for investgating condensed matter are also relevant for ferroic materials. In this regard there is nothing special about them.

Some pioneering work on nonstructural composites was done by van Suchtelen (1972, 1976) (also see Hale 1976). As explained in §13.3, there is tremendous scope for exploiting the flexibility offered by composites for designing new systems, incorporating ferroics, with specific end-uses in mind. Applications of ferroic materials in smart structures are examples of this.

Because of the constraints on time and space, some topics had to be left out. Ferroelectric liquid crystals is one such topic. Another is that of polymer ferroelectrics. Ferroic behaviour of quasicrystals will also be interesting to explore. For example, if a quasicrystal undergoes a transition to a phase of ordinary crystallographic symmetry, the noncrystallographic symmetry operators lost at the phase transition (e.g. 5-fold symmetry rotations) can be expected to leave behind their signatures as operators which map one ferroic domain to another (under the parent-clamping approximation). Of course, this would be in addition to the effect of loss (if any) of the normal crystallographic symmetry operators.

> *Yes, there is a Nirvana; it is in leading your sheep to a green pasture, and in putting your child to sleep, and in writing the last line of your poem.*
>
> Kahlil Gibran

Appendix A

SET THEORY

The first man who noticed the analogy between a group of seven fishes and a group of seven days made a notable advance in the history of thought.

A. N. Whitehead

This appendix serves mainly as a prelude to Appendix B on group theory. We introduce a few concepts, definitions, and notation.

Sets. A set is defined as a unification of well defined objects of thought into a whole.

If an object x belongs to a set S, it is called an *element* of S, or a *member* of S. This is denoted by $x \in S$. The contrary statement (x not an element of S) is abbreviated as $x \notin S$.

If P is the statement which defines the elements of a set, the set is denoted by $\{x \,|\, x \text{ satisfies } P\}$, where the vertical line is read as "such that".

A set without any elements is called an *empty set*, or a *null set*, and is denoted by $\emptyset$.

A is called a *subset* of B (denoted by $A \subset B$) if every element of A is also an element of B. If $A \subset B$, then $B \supset A$, the latter statement meaning "B contains A" or "B is a superset of A".

Two sets A and B are said to be *equal* ($A = B$) if, for every x, $x \in A$ and $x \in B$. Otherwise, $A \neq B$.

If $A \subset B$ and $A \neq B$, A is a *proper subset* of B. If a subset A of B may or may not be a proper subset, one writes $B \subseteq A$.

Algebra of Sets. The *union* (or *sum*) $A \cup B$ and the *intersection* (or *product*) $A \cap B$ of two sets A and B are defined as follows:

$$A \cup B = \{x \,|\, x \in A \text{ or } x \in B\}; \tag{A.0.4}$$

$$A \cap B = \{x \mid x \in A \text{ and } x \in B\}; \tag{A.0.5}$$

If A and B have no common elements, they are called *disjoint sets*. For such sets $A \cap B = \emptyset$.

Union and intersection obey the following two distribution laws:

$$A \cup (B \cap C) \;=\; (A \cup B) \cap (A \cup C), \tag{A.0.6}$$

$$A \cap (B \cup C) \;=\; (A \cap B) \cup (A \cap C) \tag{A.0.7}$$

If $\{A_\alpha\}$ is a nonempty collection of sets, then

$$\cup A_\alpha \;=\; \{x \mid x \text{ belongs to at least one } A\}; \tag{A.0.8}$$

$$\cap A_\alpha \;=\; \{x \mid x \text{ belongs to every } A\}; \tag{A.0.9}$$

For two sets A and B, the *relative complement* of B in A is the set $A - B$, defined as

$$A - B \;=\; \{x | x \;\in\; A \text{ but } x \;\notin\; B\} \tag{A.0.10}$$

Some authors use the notation $A \setminus B$ in place of $A - B$ for the difference between the sets A and B. Example: $\{1, 2, 3\} \setminus \{2, 3, 4\} = \{1\}$.

If $B \subset A$, then the set $B^c \;=\; A \setminus B$ is called the *complement* of B to A. We note that $B \cup B^c \;=\; A$, and $B \cap B^c \;=\; \emptyset$.

Often it is useful to define the largest set relevant for a given physical problem, and then consider its subsets. Such a set, $\mathcal{U}$, is called the *universe* or the *universal set*. An example of $\mathcal{U}$ in plane geometry is the set of all points in the plane.

The complement of A in $\mathcal{U}$ is denoted by A'. For two sets A and B the following results are self-evident:

$$(A')' \;=\; A \tag{A.0.11}$$

$$\mathcal{U}' \;=\; \emptyset, \quad \emptyset' \;=\; \mathcal{U} \tag{A.0.12}$$

$$A \cap A' \;=\; \emptyset, \quad A \cup A' \;=\; \mathcal{U} \tag{A.0.13}$$

$$A \subset B \;\text{ if and only if }\; B' \subset A' \tag{A.0.14}$$

$$(A \cup B)' \;=\; A' \cap B' \tag{A.0.15}$$

$$(A \cap B)' \;=\; A' \cup B' \tag{A.0.16}$$

Venn-Euler Diagrams. Venn-Euler diagrams are illustrative diagrams for expressing relationships between sets. One represents a set by a simple plane area, usually bounded by a circle. Fig. A.0.1 shows the Venn-Euler diagram for sets $A = \{a, b, c, d\}$, $B = \{c, d, e, f\}$, and illustrates graphically the fact that $A \cap B = c, d$.

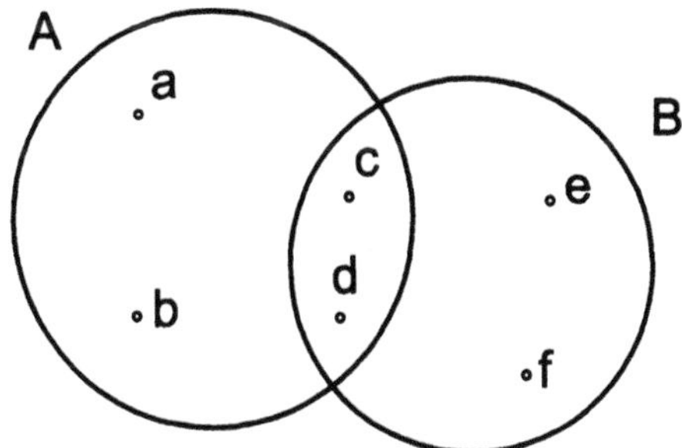

Figure A.0.1: The Venn-Euler diagram for two sets A and B with a nonempty intersection set.

Cartesian Product. For two sets A and B, the cartesian-product set $A \times B$ is formed by taking all the *ordered pairs* (p, q), where $p \in A$ and $q \in B$. A plane is an example of a cartesian product of two sets of points on a line.

Point Fields, Functions, Mappings. A point field is a set of elements called *points.*

A mapping or a *correspondence* from a point field A to a point field B is said to be defined if, for every point p in A a point p' is associated in B: $p' \in B \; \forall \; p \in A$ (where the symbol $\forall$ denotes "for every"). $p' = f(p)$ is called the *image* of p. Such an assignment or mapping also defines the *function* f, and is sometimes written as $f: \; A \xrightarrow{f} B$.

A is the *domain* of f, and B is its *co-domain.*

Operators. Transformations. If the domain and the co-domain of a function f is one and the same set A, f is said to be an operator or a transformation on A.

One-to-One Mappings. f is called a one-to-one function or mapping of a set A to a set B if no two different elements of A are mapped to the same element of B.

Onto Mappings. If in a mapping by f from set A to set B, every element of B is the image of at least one element of A, f is an onto mapping or function. In an onto mapping the image of A is the whole of B ($f(A) = B$). A is said to be mapped *into* B if $f(A) \subset B$.

Equivalent Sets. Set A is equivalent to B ($A \sim B$) if a function or mapping exists from A to B which is both one-to-one and onto.

Infinite Sets. A set is infinite if it is equivalent to one of its proper subsets. Otherwise it is a finite set.

Denumerable Sets. Consider the set of natural numbers, $N = \{1, 2, 3, ...\}$. Any set D is a denumerable set if it is equivalent to the set N.

Countable Sets. A set is countable if it is either finite or denumerable. A set is non-denumerable or non-countable if it is infinite and if it is not equivalent to the set N of natural numbers.

Permutations. A transformation defined on a finite point field is a permutation.

SUGGESTED READING

S. Lipschutz (1981). *Theory and Problems of Set Theory and Related Topics.* Schaum's Outline Series, McGraw-Hill, Singapore.

Appendix B

GROUP THEORY

The Book of Nature is written in mathematical characters.

Galileo Galilei, in *Il Saggiatore*

Group theory provides the mathematical language for describing the symmetries of physical systems. The most important link between symmetry and physics is through the theory of representations of groups, specially the *Wigner theorem.* According to this theorem, *the physical parameters describing the properties of a system are transformed according to the irreducible representations of the symmetry group of the system.*

We summarize here the basic concepts and definitions of group theory, and state (mostly without proof) its important theorems and results relevant for the description of ferroic phase transitions and ferroic materials. Many of the concepts described here are illustrated during their actual use in various chapters of the book.

B.1 ABSTRACT GROUP THEORY

Sets with Algebraic Structure

If a law of composition (or combination, or multiplication) is defined for a set so that any two elements of the set can be combined or multiplied to give another element of the set, the set is said to have *algebraic structure*, as well as the property of *closure.* An example is the set of all positive and negative integers (including zero), with subtraction as the law of composition. We notice that in this example the law of composition is not *associative* [e.g. $(16 - (-9)) - 12 \neq 16 - (-9 - 12)$].

Groups

A group $G(e, a, b, c, ...)$ is a set of *distinct* elements with algebraic structure, the law of composition for which is associative, and which includes an *identity element*, as well as the *inverse* of every element.

The identity element, e, is that which has the property that, for every element a of the group,

$$ea = ae = a \tag{B.1.1}$$

The inverse b of any element a has the property that

$$ab = ba = e \tag{B.1.2}$$

Naturally, if b is the inverse of a, then a is the inverse of b.

The inverse of an element a is denoted by a^{-1}. Therefore, if $a^{-1} = b$, then $b^{-1} = a$.

The number of elements in a group is called its *order*. A group of finite order is called a *finite group*. The order of a group G is conventionally denoted as either g or $|G|$.

Some examples of groups follow:

(i) The set of all integers, $I(..-3, -2, -1, 0, 1, 2, ..)$, with ordinary summation as the law of composition. The identity element for this group is 0, and the inverse of any element n is the integer $-n$.

(ii) The set of all nonsingular square matrices of order n, with ordinary matrix multiplication as the law of composition.

(iii) The set $(0, 1, 2, 3, ..n-1)$ of n integers under the law of addition *modulo*(n). If, for example, $n = 5$, then: $3 + 4 = 2$, $4 + 4 = 3$, $4 + 2 = 1$, etc. This group is referred to as the Z_n *group* (see below).

Abelian Groups

A group is said to be Abelian or commutative if all its elements commute with one another; i.e. $ab = ba$ for all a and b belonging to the group.

Groups of Transformations

A transformation (such as rotation, reflection, translation, permutation) which leaves a physical system invariant is called a *symmetry transformation* of the system. That the set of all such transformations forms a group (called the *symmetry group* of the system) is reasonably straightforward to visualize. Any two symmetry transformations, performed in succession,

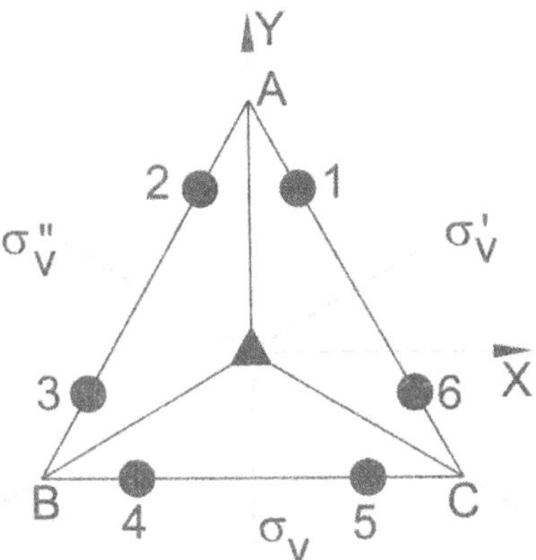

Figure B.1.1: Symmetry elements of an equilateral triangle in 2-dimensional space. [After Ludwig & Falter 1988.]

keep the system invariant, thus satisfying the closure requirement. The identity transformation corresponds to no transformation at all, and is a member of the set. For every symmetry transformation there obviously exists the inverse transformation also. Finally, successive transformations obey the associative law. This will become more clear when we discuss an example below (that of a triangle).

Symmetry Group of an Equilateral Triangle

Let us consider the symmetry group of an equilateral triangle (Fig. B.1.1). If the triangle is rotated by an angle $2\pi/3$ about an axis perpendicular to its own plane, and passing through its centroid, it is transformed back into itself. This axis is said to be an axis of 3-fold symmetry (or a triad), and the symmetry operation just described is denoted by C_3. We follow the convention of taking counter-clockwise rotations as positive. Two successive operations of C_3 (denoted by C_3^2) also leave the triangle invariant. Three such successive operations amount to a total rotation of 2π, bringing the triangle to its original configuration, and thus amounting to an identity operation: $C_3^3 = e$.

In addition to these three rotational operations of symmetry, there are three reflection operations also, σ_v, σ_v', σ_v'', in Fig. B.1.1, which bring the triangle back into coincidence with itself. These mirror planes of symmetry pass through the 'vertical' axis C_3; hence the subscript v in σ_v etc.

The operations of the six symmetry elements of the equilateral triangle can be shown conveniently in terms of the mappings of the points 1,2,..6

marked on the triangle in Fig. B.1.1.

$$e \quad : \quad 1 \to 1 \quad 2 \to 2 \quad 3 \to 3 \quad 4 \to 4 \quad 5 \to 5 \quad 6 \to 6$$

$$\begin{array}{lll} C_3 & : & 1 \to 3 \to 5 \to 1 \quad 2 \to 4 \to 6 \to 2 \\ C_3^2 & : & 1 \to 5 \to 3 \to 1 \quad 2 \to 6 \to 4 \to 2 \end{array}$$

$$\begin{array}{lllll} \sigma_v & : & 1 \leftrightarrow 2 & 3 \leftrightarrow 6 & 4 \leftrightarrow 5 \\ \sigma_v' & : & 1 \leftrightarrow 6 & 2 \leftrightarrow 5 & 3 \leftrightarrow 4 \\ \sigma_v'' & : & 1 \leftrightarrow 4 & 2 \leftrightarrow 3 & 5 \leftrightarrow 6 \end{array}$$

This set of mappings also enables us to see easily the effect of successive operations. For example, C_3 carries point 1 to 3, and σ_v' carries 3 to 4. Thus the combined effect of these two symmetry operations, denoted by the product $C_3\sigma_v'$, is to take point 1 to point 4. And the operation $1 \to 4$ is also achieved by σ_v''. Thus $C_3\sigma_v' = \sigma_v''$.

We can satisfy ourselves that every conceivable product of the elements in the set $\{e, C_3, C_3^2, \sigma_v, \sigma_v', \sigma_v''\}$ is again a member of this set. In fact, this set constitutes a group, denoted conventionally by the symbol C_{3v} or $3m$. Table B.1.1 gives what is called the *group multiplication table* for the group, namely the results of successive applications (or products ab) of all possible pairs of symmetry elements (a, b) of the group.

We notice that the group C_{3v} is not an Abelian group. For example, $C_3\sigma_v = \sigma_v''$, whereas $\sigma_v C_3 = \sigma_v'$, so that the products are not always commutative. We also notice that each element of the group appears once, and only once, in each row and column of the group table.

Generators of a Finite Group. Cyclic Groups

One can generate all the elements of a group starting from a certain minimum number of properly chosen elements of the group. The smallest set of such elements, whose powers and products can generate the entire group, comprises the generators of the group.

A nontrivial example of a single element generating the entire group is provided by the case where the element a is such that $a^n = e$, where n is the smallest positive integer satisfying this condition. Since a is an element of the group, so are $a^2, a^3, ..$ etc.; we must stop at $a^n = e$. Any power of a higher than n does not give another distinct element because $a^{n+k} = a^k$. The generator a thus generates a group $\{a, a^2, a^3, ..a^{n-1}, e\}$ of order n. Any group generated by a single element is called a cyclic group. All cyclic groups are Abelian, but the converse is not necessarily true.

Table B.1.1: Multiplication table for elements of the group C_{3v}.

a \ b	e	C_3	C_3^2	σ_v	σ_v'	σ_v''
e	e	C_3	C_3^2	σ_v	σ_v'	σ_v''
C_3	C_3	C_3^2	e	σ_v''	σ_v	σ_v'
C_3^2	C_3^2	e	C_3	σ_v'	σ_v''	σ_v
σ_v	σ_v	σ_v'	σ_v''	e	C_3	C_3^2
σ_v'	σ_v'	σ_v''	σ_v	C_3^2	e	C_3
σ_v''	σ_v''	σ_v	σ_v'	C_3	C_3^2	e

Conjugate Elements and Classes

If for two elements b and c of a group, an element a exists such that

$$a^{-1}ba = c, \tag{B.1.3}$$

the elements b and c are said to be conjugate elements. The above equation represents the *similarity transformation* of b by a.

From Table B.1.1 we can pick up examples of conjugate elements. For instance, $C_3^2\sigma_v = \sigma_v'$. Multiplying both sides of this equation by C_3 from the right, and noting that $C_3^2 = C_3^{-1}$, we get $C_3^{-1}\sigma_v C_3 = \sigma_v' C_3$. Table B.1.1 can be consulted again to see that $\sigma_v' C_3$ is nothing but σ_v'', so that

$$C_3^{-1}\sigma_v C_3 = \sigma_v'' \tag{B.1.4}$$

Thus σ_v and σ_v'' are conjugate elements.

It can be shown readily that if an element b is conjugate to c as well as to d, then c and d are conjugate elements; in fact it means that b, c and d are all conjugate to one another. This suggests the possibility of splitting a group into sets such that all elements of a set are conjugate to one another, but no two elements belonging to different sets are conjugate to each other.

Such sets of elements of a group are called *conjugacy classes*, or simply *classes*.

The identity element e of a group always constitutes a class by itself, because for any element a belonging to the group, $a^{-1}ea = e$.

It is easily verified by consulting Table B.1.1 for the group C_{3v} considered above that it is composed of the classes (e), (C_3, C_3^{-1}), and $(\sigma_v, \sigma_v', \sigma_v'')$.

Subgroups

A subset F of the elements of a group G which satisfies all the criteria for defining a group, and which has the same law for the products of elements as the larger group, is called a subgroup of G.

The group C_{3v} considered above has the following nontrivial or *proper* subgroups: $\{e, C_3, C_3^{-1}\}$, $\{e, \sigma_v\}$, $\{e, \sigma_v'\}$, $\{e, \sigma_v''\}$. The whole group, as well as the group $\{e\}$, are also its subgroups. These are called the *improper* or trivial subgroups.

We notice that all the proper subgroups of C_{3v} are cyclic, although the whole group is not cyclic.

Cosets of a Subgroup

Let G be a group of order g, and $H\{h_1, h_2, ..h_h\}$ one of its proper subgroups of order h. Since H is a proper subgroup of G, there exists at least one element $a \in G$ which does not belong to H. The set $(ah_1, ah_2, ...ah_h)$ is called the *left coset* of H by a (and is denoted by aH). Similarly, the set $(h_1a, h_2a, ...h_ha)$ is the *right coset* of H by a.

A coset is not always a group. For example it cannot always have the identity element; the identity element occurs only in the subgroup H, and not in other cosets.

In general the left coset aH and the right coset Ha are not identical. However, they are not disjoint.

The element a is a *coset representative.* Its choice is not unique; any member of the coset can be the coset representative.

Two left cosets aH and bH are either identical, or have no element in common (i.e. are disjoint sets). Otherwise, some h_i and h_j can be found such that $ah_i = bh_j$, in which case $a = bh_jh_i^{-1}$, implying that a is a member of the coset bH, contrary to the definition of a coset. [a, being a coset representative for aH, cannot also be a member of another coset bH.] Similarly, two right cosets are either identical, or disjoint.

The number of elements in a coset is equal to the order of the proper subgroup H.

Lagrange Theorem for Subgroups

According to this theorem the order of any subgroup H of G is a divisor of the order g of G. We have seen above that any two distinct cosets of a group are always disjoint, and that in each coset the number of elements is equal to the order of the subgroup. It follows that the elements of G can be split into an integral number, n, of disjoint sets; therefore $g = nh$, where h is the order of H. The number n is called *the index of H in G*.

Normal Subgroups

Consider a subgroup $H \subset G$, and some element $s \in G$. If the left and right cosets of H with respect to all $s \in G$ are the same, H is said to be a normal or *invariant* subgroup of G. It follows from this definition that every element of the set sH is equal to some element of Hs, i.e. $sh_i = h_j s$, or

$$h_i = s^{-1} h_j s \tag{B.1.5}$$

But this is just the conjugation relation between h_i and h_j, and implies that, if an element h_i belongs to a normal subgroup H, then all elements conjugate to h_i also belong to H. In other words, *a normal subgroup consists of entire classes of the larger group G*.

The converse can also be shown to be true, namely, if a subgroup H consists of entire classes of G, it must be a normal subgroup of G. For example, $\{e, C_3, C_3^{-1}\}$ is a normal subgroup of the group C_{3v}, but $\{e, \sigma_v\}$ is not.

Conjugate Subgroups

Consider a subgroup H of a group G. Another subgroup H' of G is said to be conjugate or *equivalent* to H if there exists an element $g \in G$ such that $H' = g^{-1} H g$. One can thus divide all subgroups of a group into *classes of conjugate subgroups*, or *conjugacy classes of subgroups*.

Conjugacy classes may contain different numbers of subgroups. The conjugacy class of a normal subgroup contains only one subgroup.

Product Groups

Consider two groups H ($h_1 = e, h_2, ...h_h$) and K ($k_1 = e, k_2, ...k_k$) of orders h and k respectively, such that they are disjoint except for the identity element e, and such that every element of H commutes with every element of K. One can define an *outer direct product group*, G, of order hk ($G = H \otimes K$), with (h_i, k_j) as a typical element, and with the following multiplication law among its elements:

$$(h_i, k_j) \cdot (h_k, k_l) = (h_i h_k, k_j k_l) \tag{B.1.6}$$

The group H is isomorphic to (h_i, e), and the group K is isomorphic to (e, k_j). It follows from this definition that H and K are normal subgroups of G.

If $H = K$ ($= G$, say), we get a special product group with its elements satisfying the relation

$$(h_i, h_i) \in G \otimes G \tag{B.1.7}$$

This group is isomorphic to G, and is called *the inner direct product group of* G.

If G is a group with subgroups H and K such that $H \cap K = \{e\}$ and:

$$k_j H = H k_j \text{ for all } k_j \in K \text{ and all } h_i \in H, \tag{B.1.8}$$

and

$$\text{for all } p \in G \text{ we have } p = h_i k_j, \tag{B.1.9}$$

then G is called the *semi-direct product group* of H and K:

$$G = H \circledS K \tag{B.1.10}$$

We notice that, in this case, whereas H is an invariant subgroup of G, K may not be so, in general.

Factor Groups

Let us take the normal subgroup $\{e, C_3, C_3^{-1}\}$ of C_{3v}, and form all possible *distinct* cosets from it by using the elements of C_{3v}. In fact we can form only one such distinct coset, namely $K_2 = (\sigma_v, \sigma_v', \sigma_v'')$. The use of other possible elements, namely σ_v' and σ_v'', for forming more cosets simply gives K_2 with an altered sequence for its elements. We reserve the symbol K_1 for the initial set itself: $K_1 = (e, C_3, C_3^{-1})$. (We ignore the fact that K_1 is actually a group, and not just a set.)

It turns out that K_1 and K_2 can be treated as elements of a group $\mathcal{K}$ called the factor group of G with respect to the normal subgroup K_1. For this group, the product of two elements (cosets) is obtained by multiplying each element of the first coset with every element of the other coset, *counting repeated elements only once*:

$$K_1 K_2 = (e, C_3, C_3^{-1})(\sigma_v, \sigma_v', \sigma_v'') \tag{B.1.11}$$

$$= (\sigma_v, \sigma_v', \sigma_v'', \sigma_v'', \sigma_v, \sigma_v', \sigma_v', \sigma_v'', \sigma_v) \tag{B.1.12}$$

$$\rightarrow (\sigma_v, \sigma_v', \sigma_v'') = K_2 \tag{B.1.13}$$

In fact, K_1 plays the role of the identity element for the factor group.

A general definition can be given as follows: If H is a normal subgroup of G, the set of all the distinct cosets of H in G, together with the coset multiplication rule given above, is called the factor group or the *quotient group* of G with respect to H. It is denoted by $\mathcal{K} = G/H$ (G over H). The order k of $\mathcal{K}$ can be readily shown to be g/h.

Mappings Between Groups. The Kernel

A multiplication table, like Table B.1.1, contains all the information about the algebraic or analytical structure of a group. Two groups having the same multiplication table, and therefore the same structure, are said to be *isomorphic* to each other. The correspondence between the elements of two groups can be discussed in a more general manner through the language of mappings.

Any mapping between sets with an algebraic structure which satisfies the requirement $f(b)f(a) = f(ba)$ is called a *homomorphic mapping* or a *homomorphism*. Similarly, a mapping from a group G_1 to a group G_2 is a homomorphism if the group structure is preserved in the mapping, that is, if to each element a in G_1 there corresponds a unique element $\phi(a)$ in G_2 such that $\phi(ba) = \phi(b)\phi(a)$, and if the mapping is defined for all the elements of G_1.

The set of all elements of G_1, which are mapped onto the unit element of G_2, is called the *kernel* of the mapping.

In a homomorphism it is possible that several elements of G_1 are mapped to the same element of G_2 (e.g. $\phi(a) = \phi(b)$ when $a \neq b$). When a mapping between two groups is homomorphic, one-to-one, and onto, it is an *isomorphism*. Two isomorphic groups have the same multiplication table, and are completely equivalent. In fact they comprise one and the same abstract group.

Intersection Groups

Consider two groups G_1 and G_2. A group I comprising only of elements common to both G_1 and G_2 is called an intersection group. Symbolically,

$$I = G_1 \cap G_2 \tag{B.1.14}$$

I is a subgroup of both G_1 and G_2.

The Z_n Group

The group of integers modulo n, under addition, is called the Z_n group.

The Ising model is an example of Z_2 symmetry.

A clock model, with vectors limited to n equally spaced points on a 2-dimensional unit circle, has Z_n symmetry.

B.2 LINEAR SPACES AND OPERATORS

The application of abstract group theory to physical problems involves the use of representations of groups in physical spaces. The most used spaces are the linear spaces or vector spaces.

Fields

Consider a set of elements $F(a, b, c, ...)$, for which two binary operations are defined, namely an *addition* (denoted by $+$) and a *multiplication* (denoted by $.$). F is a field if: (a) it is an Abelian group under addition, with zero (0) as the identity element; and (b) all its nonzero elements also form an Abelian group under multiplication, with unity as the identity element.

Common examples of fields are: the set of all real numbers ($\mathcal{R}$); and the set of all complex numbers ($\mathcal{C}$). The elements of a field are called *scalars*.

Vector Spaces

A set of elements $\{u, v, w, ..\}$ (e.g. vectors, points, matrices, functions) is called a *linear space* or *vector space* $\mathcal{L}$ over the field $F(a, b, c, ..)$ of scalars if the following two conditions are satisfied for all $u, v, .. \in \mathcal{L}$ and all $a, b, c, .. \in F$:

(a) $\mathcal{L}$ is an Abelian group ($\mathcal{L}$,+) under addition, with 0 as the identity element.

(b) An operation called *scalar multiplication* exists for combining any scalar of the field F and any element of $\mathcal{L}$ such that, for every $u, v \in \mathcal{L}$ and $a, b \in F$ we have:

$$a(u + v) = au + av \in \mathcal{L}, \tag{B.2.1}$$

$$(a + b)u = au + bu \in \mathcal{L}, \tag{B.2.2}$$

$$a(bu) = (ab)u, \tag{B.2.3}$$

$$1u = u, \tag{B.2.4}$$

$$0u = 0 \tag{B.2.5}$$

It is usually not necessary to distinguish between the scalar zero and the element 0 of the vector space.

The elements of a vector space are called *vectors*.

If the vector space is defined over the field of real numbers, it is called a *real vector space*; if defined over a field of complex numbers, it is a *complex vector space.*

Inner-Product Space. Norm

If a vector space $\mathcal{L}$ satisfies the following conditions, it is called an inner product space, or a *metric vector space*: For every pair of elements $u, v \in \mathcal{L}$ there exists a unique scalar (u, v) in the field F such that

$$(u, v) = (v, u)^*, \tag{B.2.6}$$

$$(au, bv) = a^* b(u, v), \tag{B.2.7}$$

$$(w,\ au + bv) = a(w, u)\ +\ b(w, v), \tag{B.2.8}$$

where the * denotes 'complex conjugate'.

Ordinary three-dimensional space, with the familiar rule for the scalar product of two vectors, is an example of an inner-product space.

Consider a set of n-tuplets of numbers $u = (u_1, u_2, ...u_n)$, $v = (v_1, v_2, ...$ $v_n)$, ... over a field to which the scalars u_i belong. If, for example, the field concerned is the complex field $\mathcal{C}$, this set of n-tuplets constitutes a vector space over $\mathcal{C}$. This vector space would qualify to be called an inner-product space if we define the scalar product of any two of the n-tuplets, say u and v, as a complex number given by

$$(u, v) = u_i^*\, v_i \tag{B.2.9}$$

[Here we follow the convention that a summation from $i = 1$ to n over the repeated index i is implied.]

If $v = u$, we have

$$(u, u) \;=\; |u_i|^2 \tag{B.2.10}$$

The nonnegative square root of this number is called the norm (or *length*) of the vector u.

Cauchy Sequence

A sequence is an infinite set of numbers c_1, c_2, ... c_n, ... associated with each positive integer n according to some specified rule. This sequence is said to be *convergent*, with a limit c, if, for every real positive number ϵ, howsoever small, there exists a positive integer N such that, for every $n > N$, we have $|c_n - c| < \epsilon$.

A sequence is a Cauchy sequence if, for every ϵ, an N can be found such that, for any two integers $n > N$ and $m > N$, we have $|c_n - c_m| < \epsilon$. Every convergent sequence is a Cauchy sequence, and vice versa.

A sequence can be defined, not only in terms of real or complex numbers, but also, for example, in terms of vectors in 2, 3, or n-dimensional space.

Hilbert Space

If every Cauchy sequence of elements belonging to an inner-product space $\mathcal{L}$ has a limit which also belongs to $\mathcal{L}$, we have a *complete* space.

Any complete inner product space is a Hilbert space.

As a counter example, the space of all rational numbers is not a complete space because one can construct a Cauchy sequence in this space with an *irrational* limit.

In quantum mechanics the state of a system is described by a vector in an appropriate Hilbert space.

Basis Vectors

In ordinary 3-dimensional space ($\mathcal{R}_3$) any vector can be defined in terms of its three 'components', namely vectors along three non-coplanar (and pairwise noncollinear) axes. One can similarly define an n-dimensional vector space $\mathcal{L}_n$ such that the whole of this space is *spanned* by a set of n similarly chosen basis vectors $x_1, x_2, ..x_n$.

Any two vectors x_i and x_j of $\mathcal{L}_n$ are said to be *linearly independent* if it is impossible to find a scalar c such that $x_i - cx_j = 0$.

This equation can be generalized to arrive at the following condition: A set of m vectors (x_i) of $\mathcal{L}_n$ is a linearly independent set if and only if the equation

$$a_i x_i = 0 \tag{B.2.11}$$

is satisfied only when all the coefficients a_i in the summation are zero.

Any set of n linearly independent vectors $(x_1, x_2, ..x_n)$ in $\mathcal{L}_n$ is called a *complete set.* Such a set, the choice of which is not unique, constitutes the basis vectors of the space. Any vector $\mathbf{u}$ in $\mathcal{L}_n$ can be written as a sum of component vectors parallel to the complete set of basis vectors:

$$\mathbf{u} = u_i x_i \tag{B.2.12}$$

It is usually advantageous to choose basis vectors $\{\mathbf{e}_i\}$ with unit norm, rather than choosing basis vectors $\{x_i\}$ with an arbitrary norm. Further, it is usually very convenient to choose vectors which are orthogonal to one another. Basis vectors that are orthogonal, as well as of unit magnitude (*orthonormal basis vectors*), satisfy the condition

$$(\mathbf{e}_i, \mathbf{e}_j) = \delta_{ij}, \tag{B.2.13}$$

where δ_{ij} is the Kronecker delta function.

Consider two vectors

$$\mathbf{u} = u_i e_i \tag{B.2.14}$$

and

$$\mathbf{v} = v_i e_i \tag{B.2.15}$$

in $\mathcal{L}_n$. The scalar product of these vectors is given by

$$(\mathbf{u}, \mathbf{v}) = (\mathbf{v}, \mathbf{u})^* = u_i^* v_i \tag{B.2.16}$$

Linear Operators

Consider two vector spaces $\mathcal{L}$ and $\mathcal{L}'$ over the same field F. A mapping $T : \mathcal{L} \rightarrow \mathcal{L}'$ is called a *linear transformation* or a *linear operator* if for all $\mathbf{u}, \mathbf{v} \in \mathcal{L}$ and all $c \in F$, we have

$$T(\mathbf{u} + \mathbf{v}) = T\mathbf{u} + T\mathbf{v} \tag{B.2.17}$$

and

$$T(c\mathbf{u}) = cT\mathbf{u} \tag{B.2.18}$$

Thus a linear operator A is defined over $\mathcal{L}$ by associating a vector $|\phi>$ to each vector $|\psi>$ of $\mathcal{L}$: $|\phi>= T|\psi>$. In quantum mechanics the dynamical variables of a system are described by linear operators in Hilbert space.

Vectors in a vector space can be specified in terms of their components, with basis vectors serving as the 'coordinate axes'. Eqs. B.2.15 serve as an example of this. When operators act on a vector space, we have two choices regarding their action. We may treat the operator as an *active operator*, which leaves the basis vectors (and therefore the coordinate axes) fixed, and maps all other vectors to their new images. Alternatively, we can treat it as a *passive operator*, which simply maps the basis vectors to their respective images, leaving other vectors unchanged; its effect on the other vectors of $\mathcal{L}_n$ is then felt through the changes it causes in their components referred to the transformed basis vectors. In this book, unless stated specifically to the contrary, we treat the operators as active operators.

Isometric Mappings

Coordinate transformations or mappings which preserve distances are called isometric mappings, or *isometries*.

Affine Mappings

Mappings which preserve parallelism of lines, but may or may not preserve distances, are called affine mappings.

Unitary and Orthogonal Operators, and Their Representations

Let us consider a passive operator T, which transforms the set of basis vectors $(\mathbf{e}_i)$ in $\mathcal{L}_n$ over the field $\mathcal{C}$ to a new set $(\mathbf{e}'_i)$:

$$\mathbf{e}'_i = T\mathbf{e}_i = \mathbf{e}_j T_{ji}, \quad i = 1 \text{ to } n \tag{B.2.19}$$

Here T_{ji} are scalars in $\mathcal{C}$, denoting the components of $\mathbf{e}'_i$ along $\mathbf{e}_j$. If the operator T is such that it transforms one set of orthonormal basis vectors to another, it is called a *unitary operator*. Such operators keep invariant the norms and scalar products of the vectors of $\mathcal{L}_n$.

The scalars T_{ji} in Eq. B.2.19 can be determined by forming the scalar product of $\mathbf{e}_j$ and $\mathbf{e}_i$:

$$(\mathbf{e}_j, \mathbf{e}'_i) = (\mathbf{e}_j, T\mathbf{e}_i) = (\mathbf{e}_j, \mathbf{e}_j T_{ji}) \tag{B.2.20}$$

Since the $(\mathbf{e}_i)$ are an orthonormal basis (cf. Eq. B.2.13), we get

$$(\mathbf{e}_j, T\mathbf{e}_i) = T_{ji} \tag{B.2.21}$$

T_{ji} is referred to as the *matrix element* of the operator T between $\mathbf{e}_j$ and $\mathbf{e}_i$. And the square matrix $[T_{ij}]$ of order n gives a representation of the operator T in the basis $(\mathbf{e}_i)$ in $\mathcal{L}_n$.

We have assumed here that $\mathcal{L}_n$ is defined over the field $(\mathcal{C})$ of complex numbers. If it is defined, instead, over a field $\mathcal{R}$ of real numbers, T is an *orthogonal* operator, rather than a unitary operator.

Symmetry Transformation Operators, and Their Effect on Functions

As discussed in Appendix A, a function or a mapping is defined by associating a point $f(\mathbf{r})$ in a point field B for every point $\mathbf{r}$ in a point field A. A *transformation* $\mathbf{R}$ takes every point $\mathbf{r}$ in A to another point in A:

$$\mathbf{r}' = \mathbf{R}\,\mathbf{r} \tag{B.2.22}$$

Correspondingly, the function $f(\mathbf{r})$ is transformed to, say, $\mathbf{R}_f\, f(\mathbf{r}')$. The operators $\mathbf{R}$ and $\mathbf{R}_f$ act on different spaces; $\mathbf{R}$ on coordinate space, and $\mathbf{R}_f$ on function space.

$\mathbf{R}_f$ is a symmetry transformation if

$$\mathbf{R}_f\, f(\mathbf{r}') = f(\mathbf{r}) \tag{B.2.23}$$

Eq. B.2.22 can be rewritten as

$$\mathbf{r} = \mathbf{R}^{-1}\,\mathbf{r}' \tag{B.2.24}$$

Substitution of $\mathbf{r}$ from Eq. B.2.24 into Eq. B.2.23 gives

$$\mathbf{R}_f\, f(\mathbf{r}') = f(\mathbf{R}^{-1}\,\mathbf{r}') \tag{B.2.25}$$

Since both $\mathbf{r}$ and $\mathbf{r}'$ are defined in the same point field A, we can drop the primes in this equation:

$$\mathbf{R}_f\, f(\mathbf{r}) = f(\mathbf{R}^{-1}\,\mathbf{r}) \tag{B.2.26}$$

Symmetry transformations in crystals are described by the Seitz operator $\{\mathbf{R}|\mathbf{t}\}$ (Eq. 2.2.11 of Chapter 2), where $\mathbf{R}$ is a rotation (or rotation-inversion) operator and $\mathbf{t}$ is a translation operator. Eq. B.2.26 has therefore to be generalized to

$$\{\mathbf{R}_f|\mathbf{t}\}\, f(\mathbf{r}) = f(\{\mathbf{R}|\mathbf{t}\}^{-1}\,\mathbf{r}) \tag{B.2.27}$$

On using Eq. 2.2.14 this becomes

$$\{\mathbf{R}_f|\mathbf{t}\}\, f(\mathbf{r}) = f(\mathbf{R}^{-1}\mathbf{r} - \mathbf{R}^{-1}\mathbf{t}) \tag{B.2.28}$$

B.3 REPRESENTATIONS OF FINITE GROUPS

Matrix Representations of Groups

We have seen in §B.2 that the square matrix $[T_{ij}]$ of order n provides a *representation* of the operator T in the basis $(\mathbf{e}_i)$ in linear vector space $\mathcal{L}_n$ of dimensionality n. We also know that the set of all nonsingular, square, distinct matrices of any order forms a group, with ordinary matrix multiplication as the law of composition for the group. The use of matrices for representations of groups is therefore only natural.

Formally, a representation $\Gamma(G)$ of a finite group $G = \{e, a, b, c..\}$ of order g is a homomorphic mapping of the elements of G onto a group of nonsingular linear operators T that map a linear space $\mathcal{L}_n$ onto itself:

$$\Gamma(G):\ \ s \in G \rightarrow T(s) \tag{B.3.1}$$

The space $\mathcal{L}_n$ is the *space of the representation* (rep) of G, and n is the *dimension of the representation*. By introducing a basis, the operators $T(s)$ can be defined by their matrix representation given by Eq. B.2.21.

Consider a set of nonsingular matrices $\Gamma = \{\Gamma(e), \Gamma(b), ..\}$, all of the same order, such that, if $ab = c$ is in G, then

$$\Gamma(a)\Gamma(b) = \Gamma(c) \tag{B.3.2}$$

is in the set Γ. Such a set of matrices forms a representation of G. The dimension of the representation is equal to the order of the matrices.

A matrix representation is, in general, a homomorphic mapping $G \to \Gamma$; that is, a many-to-one correspondence of the elements of G onto the elements of the set of matrices Γ. If the correspondence is isomorphic (one to one), the representation is said to be *true* or *faithful*. In this case the set of matrices also constitutes a group Γ which is isomorphic to the group G.

The Identity Representation

The simplest representation of a group is that homomorphic mapping in which all elements of the group are mapped to unity (a unit matrix of order 1). This is called the unit representation, or the identity representation.

The basis functions $\mathbf{e}_i$, in terms of which the matrices of a representation are defined (Eq. B.2.21) always include one function which is invariant under all the operations of the group. This single basis function gives the identity representation.

The identity representation is a 1-dimensional representation. Therefore all its characters are equal to unity (cf. Eq. B.3.10 below).

Inequivalent Representations

Consider two matrix representations Γ_1 and Γ_2 of a group G:

$$\Gamma_1 = \{\Gamma_1(e), \Gamma_1(a), \Gamma_1(b), ..\},$$

$$\Gamma_2 = \{\Gamma_2(e), \Gamma_2(a), \Gamma_2(b), ..\}$$

If a nonsingular matrix S can be found such that

$$\Gamma_1(a) = S^{-1}\,\Gamma_2(a)\,S \tag{B.3.3}$$

for all $a \in G$, then Γ_1 and Γ_2 are called equivalent representations of G. Two representations that are not equivalent are called inequivalent representations. Representations with different dimensions are necessarily inequivalent.

The matrix S in Eq. B.3.3 is said to perform an *equivalence transformation*, or a *similarity transformation*.

Unitary Representations

A matrix T is said to be unitary if $TT^\dagger = T^\dagger T = E$ (a unit matrix).

If all the matrices of a representation are unitary, it is called a unitary representation.

A representation of a finite group can be always brought into a unitary form by means of an equivalence transformation.

Reducible and Irreducible Representations. Invariant Subspaces

Consider two matrix representations Γ_1 and Γ_2 of a group G. From these, one can construct a new representation of a larger dimension:

$$\Gamma(G_i) = \begin{bmatrix} \Gamma_1(G_i) & 0 \\ 0 & \Gamma_2(G_i) \end{bmatrix} \tag{B.3.4}$$

This representation clearly satisfies the condition expressed by Eq. B.3.2. It is the direct sum of the representations Γ_1 and Γ_2:

$$\Gamma = \Gamma_1 \oplus \Gamma_2, \tag{B.3.5}$$

and its dimension is the sum of the dimensions of Γ_1 and Γ_2.

A representation having the block-diagonal structure of Eq. B.3.4, or which can be cast into such a structure through a similarity transformation, is called a reducible representation.

If no similarity transformation exists which can cast a representation into the block-diagonal form, the representation is said to be an irreducible representation.

A reducible matrix representation can be *decomposed* or *reduced* into a direct sum of irreducible representations. When this is done through a similarity transformation like Eq. B.3.3, the underlying basis also undergoes a linear transformation (cf. Eq. B.2.19). The structure of Eq. B.3.4 implies that the new basis must divide itself into subsets or subspaces, such that each of the subspaces possesses the closure property under the operations of the group G. Thus the basis spanning a reducible matrix representation can always be split into a direct sum of *invariant subspaces.*

In contrast to this, the basis spanning an irreducible representation is already an invariant subspace (of itself), and cannot be decomposed further; it is an *irreducible invariant subspace.*

The Great Orthogonality Theorem

Let us consider all the inequivalent irreducible representations (IIRs) of a group G of order g. Let $\Gamma_{ij}^{(\mu)}(s)$ denote a typical matrix element of the μth IIR, with n_μ as the dimension of this IIR. For a given set of values for μ, i, and j, and with s running over all the elements of the group, the matrix elements can be regarded as the components of a vector in a g-dimensional space. The orthogonality theorem expresses the orthogonality relations satisfied by such vectors.

According to this theorem, for any two IIRs $\Gamma(\mu)$ and $\Gamma(\nu)$ of dimensions n_μ and n_ν, the following relation must hold:

$$\sum_{s\in G} \Gamma_{ij}^{(\mu)}(s)\, \Gamma_{kl}^{(\nu)}(s) = \frac{g}{n_\mu}\, \delta_{\mu\nu}\, \delta_{ik}\, \delta_{jl} \tag{B.3.6}$$

The following are three special cases of Eq. B.3.6:

$$\sum_s \Gamma_{ij}^{(\mu)}(s)\,\Gamma_{ij}^{(\nu)}(s) = 0 \quad \text{if } \mu \neq \nu; \tag{B.3.7}$$

$$\sum_s \Gamma_{ij}^{(\mu)}(s)\,\Gamma_{kl}^{(\mu)}(s) = 0 \quad \text{if } i \neq k \text{ and/or } j \neq l; \tag{B.3.8}$$

$$\sum_s \Gamma_{ij}^{(\mu)}(s)\,\Gamma_{ij}^{(\mu)}(s) = \frac{g}{n_\mu} \tag{B.3.9}$$

Eq. B.3.7 states that any two vectors in the g-dimensional space are orthogonal to each other if they are taken from matrices of two different representations.

Eq. B.3.8 states that even when the vectors are from the same representation, they are orthogonal to each other if they are taken from different sets of elements in the matrices of this representation.

Eq. B.3.9 expresses the fact that the length of any such vector is $\sqrt{g/n_\mu}$.

Characters of a Representation

The matrices of a representation of a group, for a given vector space, are not unique. They depend on the choice of basis vectors, and even on the order in which these basis vectors are chosen. However, such representation matrices are related to one other through similarity transformations, since they are all defined for the same vector space which has the property of closure. Since the trace of a matrix is invariant under a similarity transformation, it follows that the traces of all the matrices of a representation can uniquely characterize a representation, no matter what set of basis vectors is chosen.

The trace of a matrix $\Gamma^{(\mu)}(s)$ is called its character, $\chi_\mu(s)$:

$$\chi_\mu(s) = \sum_{i=1}^{n_\mu} \Gamma_{ii}^{(\mu)}(s) \tag{B.3.10}$$

The character of any element of a group is the same in two equivalent representations. The character is invariant under equivalence.

Since the elements in a class of a group are related by similarity transformations, it follows that the character $\chi_\mu(s)$ is the same for all elements s in a given class.

The First Orthogonality Relation for Characters

Let C_s denote the class to which the elements s belong, and let g_s be the number of such elements $\sum_s g_s = g$. Then the following relation can be shown to hold:

$$\sum_{C_s=1}^{n_c} g_s \chi_\nu^*(C_s)\, \chi_\mu(C_s) \;=\; g\, \delta_{\nu\mu}, \tag{B.3.11}$$

where n_c is the number of classes the group has.

The Second Orthogonality Relation for Characters

Let n_r denote the total number of IIRs of a group G, and C_a and C_b two such IIRs. Then the following orthogonality relation holds:

$$\sum_{\mu=1}^{c} \chi_\mu^*(C_a)\, \chi_\mu(C_b) \;=\; \frac{g}{g_a}\delta_{ab} \tag{B.3.12}$$

Relationship between n_r and n_c

With reference to Eq. B.3.11 we can treat

$$[\sqrt{g_1}\chi_\mu(C_1),\ \sqrt{g_2}\chi_\mu(C_2)\ \ldots\sqrt{g_{n_c}}\chi_\mu(C_{n_c})]$$

as a vector in n_c-dimensional vector space. Eq. B.3.11 can be then thought of as an inner product of such vectors. There is one such vector for each μ, so that there are n_r vectors in all. Since an n_c-dimensional space cannot have more than n_c mutually orthogonal vectors, we must have

$$n_r \;\leq\; n_c \tag{B.3.13}$$

Similarly, with reference to Eq. B.3.12 we can regard

$$[\chi_1(C_i),\ \chi_2(C_i),\ \ldots, \chi_{n_r}(C_i)]$$

as a vector in n_r-dimensional space. The number of such vectors is n_c, and we must have

$$n_c \;\leq\; n_r \tag{B.3.14}$$

Eqs. B.3.13 and B.3.14 lead to the result

$$n_r \;=\; n_c \tag{B.3.15}$$

Thus, *the number of IIRs a group can have is equal to the number of classes in it.*

The Character Table

The character table is an arrangement of n_r^2 $(= n_c^2)$ characters $\chi^\mu(C_i)$ in a table with n_r rows and n_c columns. There are as many rows in this table as there are IIRs, and there are an equal number of columns, one for each class of the group.

Relationship between the Dimensions of the IIRs of a Group and Its Order

Characters of reducible representations are called *compound* characters, and those of irreducible representations are called *simple* or *primitive characters.* Characters being a class property, for a given IR the maximum number of distinct primitive characters is equal to the number of classes of the group, namely n_c. And the number of IIRs is also equal to n_c (Eq. B.3.15). There are thus n_c^2 primitive characters $\chi^{(\mu)}(C_i)$ in all. The character table is a $n_c \times n_c$ matrix of such primitive characters. If the primitive characters are weighted by the factors $\sqrt{g_i}/\sqrt{g}$, the matrix with matrix-elements $\sqrt{g_i/g}\chi^{(\mu)}(C_i)$ is unitary.

We denote the classes of the group G by $C_1, C_2, ...C_{n_c}$, and the IIRs by $\Gamma_1, \Gamma_2, ...\Gamma_{n_c}$. The symbol C_1 is normally reserved for the class of the identity element g_1 or e. And Γ_1 is used for denoting the identity or trivial representation of G. Thus, $\chi_\mu(C_1)$ is equal to the trace of $\Gamma_\mu(g_1)$. And the latter is a $\mu \times \mu$ unit matrix. Therefore,

$$\chi_\mu(C_1) = n_\mu \tag{B.3.16}$$

We write Eq. B.3.12 for the case $C_a = C_b = C_1$:

$$\sum_{\mu=1}^{n_r} \chi_\mu^*(C_1)\,\chi_\mu(C_1) = \frac{g}{g_1}\delta_{11} = g \tag{B.3.17}$$

Substitution of Eq. B.3.16 into Eq. B.3.17 results in

$$\sum_{\mu=1}^{n_r} n_\mu^2 = g \tag{B.3.18}$$

Thus, *the sum of the squares of the dimensions of all the IIRs of a group is equal to the order of the group.*

Subduced Representations

When a field is applied to a system, its symmetry is, in general, reduced (in accordance with the Curie principle). Symmetry may also be reduced through spontaneous symmetry breaking, as, for example, when a crystal undergoes a phase transition on change of its temperature. The resultant symmetry group, H, is frequently a subgroup of the initial group G.

Naturally, the representations $D(G)$ of G are also representations of H, though not, in general, *irreducible* representations (even when they are irreducible for G). These representations are called subduced representations of G on H.

On subduction, the representation space $L^{(\mu)}$ of G decomposes into irreducible subspaces for H.

Induced Representations

Consider a group G having H as one of its proper subgroups. The subgroup H, being a group, has its own irreducible representations. Let $D^{(\lambda)}$ be one such irreducible representation, of dimension d, and with basis functions $\phi_1, \phi_2, ...\phi_d$. Then, for any element $s \in H$, we have

$$s\phi_\nu = \sum_\mu \phi_\mu D^{(\lambda)}_{\mu\nu}(s) \tag{B.3.19}$$

Since H is a proper subgroup of G, say of index k, the following coset-decomposition can be written (§B.1):

$$G = r_1 H + r_2 H + ... + r_k H \tag{B.3.20}$$

We can choose one of the coset-representatives, say r_j, and define a new set of basis functions, $\{\phi_{j\nu}\}$, as follows:

$$\phi_{j\nu} = r_j \phi_\nu, \quad j = 1, 2, ...k; \quad \nu = 1, 2, ...d \tag{B.3.21}$$

This set can be used as the basis for a new, kd-dimensional, representation of G. It is readily verified that for any element $p \in G$, we have

$$p\phi_{j\nu} = \sum_{i\mu} \phi_{i\mu} D_{i\mu, j\nu}(p) \tag{B.3.22}$$

where

$$D_{i\mu,j\nu}(p) = \delta_{ij}(p)\, D^{(\lambda)}_{\mu\nu}(r_i^{-1} p r_j), \tag{B.3.23}$$

$$\delta_{ij}(p) = 1 \text{ when } pr_j \in r_i H, \tag{B.3.24}$$

$$\delta_{ij}(p) = 0 \text{ when } pr_j \notin r_i H \tag{B.3.25}$$

Eqs. B.3.23-25 define what is called the induced representation of $D^{(\lambda)}$ onto G.

Symmetric and Antisymmetric Product Representations

Consider two representations $\Gamma^{(\mu)}$ and $\Gamma^{(\nu)}$ of a group G, of dimensions μ and ν respectively. The basis functions $x_1, x_2, .., x_\mu$ for $\Gamma^{(\mu)}$ are transformed as

$$G x_j = \sum_i x_i D^{(\mu)}_{ij} \tag{B.3.26}$$

Similarly, the basis functions $y_1, y_2, .., y_\nu$ for the representation $\Gamma^{(\nu)}$ transform as

$$G y_l = \sum_i y_k D^{(\nu)}_{kl} \tag{B.3.27}$$

Therefore the products $x_j y_l$ will be transformed as

$$G(x_j y_l) = G x_j \, G y_l = \sum_{ik} x_i y_k [D_{ij}^{(\mu)} \, D_{kl}^{(\nu)}] \tag{B.3.28}$$

We can define a new set of 'product' matrices as follows:

$$[D^{(\mu x \nu)}(G)]_{ik,jl} = D_{ij}^{(\mu)}(G) \, D_{kl}^{(\nu)}(G) \tag{B.3.29}$$

The rows and columns of these new matrices are described by double indices. It follows from the structure of Eq. B.3.28 that the set of functions $\{x_j y_l\}$ can form a basis for a new representation, called the *direct-product representation*; it can be verified that this set of product matrices forms a group, and that it satisfies all the requirements for a representation also. It is a $\mu\nu$-dimensional representation. Product representations are, in general, reducible.

Of special interest to us (in the context of the Landau theory of phase transitions) is the case when $\Gamma^{(\mu)} = \Gamma^{(\nu)} = \Gamma^{(\alpha)}$ (say). In other words, one can construct squares (and cubes, etc.) of a given representation. From Eq. B.3.28 we can obtain the following relation:

$$G\,(x_j y_l + x_l y_j) = \sum_{ik} (x_i y_k + x_k y_i) \, D_{ij}^{(\alpha)} \, D_{kl}^{(\alpha)} \tag{B.3.30}$$

This implies that the functions $(x_j y_l + x_l y_j)$ possess the property of closure, and can be used as basis functions for representations of G. The representation $[\Gamma^{(\alpha)}]^2$ based on these functions is called the *symmetric product representation.*

Similarly, an *antisymmetric product representation*, $\{\Gamma^{(\alpha)}\}^2$, can be defined, with the set $(x_j y_l - x_l y_j)$ serving as basis functions.

B.4 SOME CONTINUOUS GROUPS

Discrete and Continuous Groups

If the number of elements of a group is denumerably infinite, it is called a discrete infinite group. If the number of its elements is nondenumerably infinite, it is a continuous group.

Consider the set of all real numbers. This set is called a continuous group of order 1 because only one parameter, say x, is sufficient to specify any real number in the interval $[-\infty, \infty]$.

The Euclidean Group E(3)

The set of all distance-preserving geometrical operations (translations, rotations, inversions) in three-dimensional space constitutes a group called the Euclidean group $E(3)$.

The Translation Group T(3)

The set of all translational operations in three-dimensional space constitutes a group called the translation group $T(3)$.

Consider two successive translations defined by the vectors $\mathbf{a}_1$ and $\mathbf{a}_2$, with $\mathbf{t}_{\mathbf{a}_1}$ and $\mathbf{t}_{\mathbf{a}_2}$ as the corresponding transformation operators. The composition law in $T(3)$ is

$$\mathbf{t}_{\mathbf{a}_1}\mathbf{t}_{\mathbf{a}_2} = \mathbf{t}_{\mathbf{a}_1+\mathbf{a}_2} \tag{B.4.1}$$

Thus $T(3)$ is a commutative group.

The Three-Dimensional Rotation-Inversion Group O(3)

The group of all 3×3 orthogonal matrices is a continuous group called the orthogonal group $O(3)$. Such matrices represent orthogonal (length-preserving) transformations in three-dimensional, real, vector space. Therefore the group of such transformations (also denoted by $O(3)$) is isomorphic to the group of the orthogonal matrices. These transformations involve rotations, space-inversions, or their combinations. $O(3)$ is therefore the rotation-inversion group in three dimensions.

The Three-Dimensional Rotation Group SO(3)

Any orthogonal matrix R has the property that

$$R\,R^T = R^T\,R = E, \tag{B.4.2}$$

where R^T is the transpose of matrix R, and E is the unit matrix. The matrices R and R^T have the same determinant, (det R), and therefore the above equation yields $(\det R)^2 = 1$, or

$$\det R = \pm 1 \tag{B.4.3}$$

The matrices of the group $O(3)$ therefore divide themselves into two sets: those with $\det R = +1$, which correspond to *proper* rotations, and those with $\det R = -1$, corresponding to *improper* rotations. The first set constitutes a group, $SO(3)$, *the special orthogonal group*, in three dimensions.

The improper rotations can be viewed as products of proper rotations with the space-inversion operation, and the matrix for the space-inversion operation is $J = -E$. The space-inversion matrix J and the identity matrix E comprise a group with two elements. The space-inversion operation commutes with all the rotation operations, and therefore one can split $O(3)$ into a direct product:

$$O(3) = SO(3) \otimes (E, J) \tag{B.4.4}$$

The Space-Time Rotation-Inversion Group O′(3)

In general, the three-dimensional space under consideration may not be invariant to the time-inversion operation e'. There can then be three types of inversion operations: the space inversion operation i; the time inversion operation e'; and the total inversion operation i'. The group

$$E_0 = \{e, i, e', i'\} \tag{B.4.5}$$

is called *the full inversion group.* It has the following proper subgroups: the space inversion group J, the time inversion group E', and the total inversion group J':

$$J = C_i = \bar{1} = \{e, i\} \tag{B.4.6}$$

$$E' = C_1' = 1' = \{e, e'\} \tag{B.4.7}$$

$$J' = C_i'(C_1) = \bar{1}' = \{e, i'\} \tag{B.4.8}$$

If we make a further separation into proper and improper rotations, we obtain the following relations, the validity of which is self-evident:

$$SO'(3) = SO(3) \otimes E', \tag{B.4.9}$$

$$S\tilde{O}(3) = SO(3) \otimes J' \tag{B.4.10}$$

$$O'(3) = SO(3) \otimes E_0 = O(3) \otimes E' \tag{B.4.11}$$

The group $O'(3)$ is the general space-time rotation-inversion group. *Any magnetic point group is a subgroup of this group* (Opechowski 1974; Ascher 1966).

The O_n Group

The symmetry group of unit vectors on the surface of a sphere in n-dimensional space is called the O_n group.

The XY-model is an example of O_2 symmetry, and the Heisenberg model is an example of O_3 symmetry.

SUGGESTED READING

N. Hamermesh (1964). *Group Theory.* Addison-Wesley, Reading.

W. Opechowski & R. Guccione (1965). In G. T. Rado & H. Suhl (Eds.), *Magnetism*, Vol. IIA, Chap. 3. Academic Press, New York.

W. Ludwig & C. Falter (1988). *Symmetries in Physics.* Springer-Verlag, Berlin.

T. Inui, Y. Tanabe & Y. Onodera (1990). *Group Theory and Its Applications in Physics.* Springer-Verlag, Berlin.

Appendix C

THE CURIE SHUBNIKOV PRINCIPLE

> *. . we must admit that without the help of symmetry and invariance principles we would not have been able to obtain even the probably approximate laws of nature which we now use and which had such a large effect on our ways of life.*
>
> E. P. Wigner (1984)

Symmetry considerations have played, and continue to play, a vital role in our quest for discovering and understanding the laws of nature. There is a hierarchical progression in our knowledge of the universe: from events to laws of nature, and from laws of nature to symmetry or invariance principles (Wigner 1967).

When two or more systems or fields are superimposed, the net symmetry of the composite system is, more often than not, lower than that of any of the superimposed components. This lowering of symmetry is referred to as the process of *dissymmetrization*, the net symmetry being the highest symmetry common to all the components, taking due account of the mutual orientation and placement of the components.

The opposite process, that is, the process by which the net symmetry is higher than the highest common symmetry, is called symmetrization (Shubnikov & Koptsik 1974).

In modern science, Pierre Curie is believed to be amongst the earliest of physicists to have seriously considered the causes of symmetry in nature. His principle of superposition of dissymmetries, to be discussed here, was enunciated in a lecture "Sur la Symétrie" given in Paris at the French

Mineralogical Society meeting on 13th of November 1884 (Curie 1884). His ideas were consolidated further in a paper published ten years later (Curie 1894). Amongst the contemporaries of Curie was Neumann, who proposed in 1885 the first symmetry prescription for finding the number of independent elastic constants of crystals. Neumann's student Voigt introduced into physics the concept of tensors.

Another pioneer in this field was Minnigerode, who stated the following principle (Minnigerode 1884): "The group of the structure of a crystal is contained in the group of each of its physical properties" (quoted by Brandmuller (1986)).

Nearer to our times, Shubnikov (1951) systematized the application of symmetry considerations to crystal physics and physical crystallography, and made a number of original contributions (for reviews, see Koptsik (1968, 1983), and Shubnikov & Koptsik (1974)). The generalized form of the Curie principle, which we propose to call the Curie-Shubnikov principle, owes its present statement to Shubnikov.

C.1 THE CURIE PRINCIPLE. DISSYMMETRIZATION

The effect truly exists beforehand in its cause.

The ancient Sankhya system
of Hindu philosophy.

Following Shubnikov & Koptsik (1974), we state the Curie principle of superposition of dissymmetries as follows:

When several phenomena of different origin are superimposed in one and the same system, their dissymmetries are summed. There only remain the symmetry elements common to each phenomenon taken separately.

The term *dissymmetry* may be simply interpreted as "the absence of symmetry". But it has a more precise meaning, which we now explain.

The notion of a universal set was stated in Appendix A. The extension of this notion to symmetry groups is obvious. Let us use the symbol $\tilde{G}$ for the *universal group* (also called the *embracing group* or the *fundamental group*). It is the highest-symmetry group relevant for a given problem. The introduction of this group is necessary for defining precisely the dissymmetry of an object or phenomenon. If one is dealing with, for example, the classical tensor properties of nonmagnetic perfect crystals, one can take

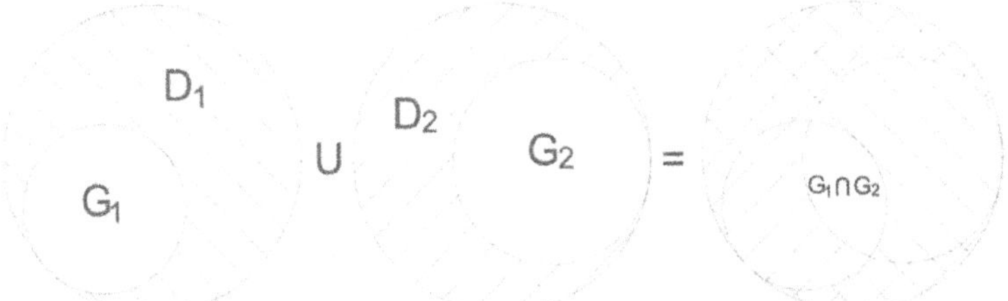

Figure C.1.1: Venn-Euler diagrams illustrating the Curie principle of addition of dissymmetries.

the universal group $\tilde{G}$ as $O(3)$, the orthogonal group for all rotations and inversions in three-dimensional space (§B.4).

If an object or phenomenon or field has a symmetry group G, its dissymmetry D is simply the complement of G in $\tilde{G}$ (cf. Appendix A):

$$D = \tilde{G} \setminus G \tag{C.1.1}$$

The summing of dissymmetries, referred to in the above statement of the Curie principle, for objects or phenomena having symmetry groups G_1, G_2, ..., can thus be expressed as the following union of sets:

$$D_1 \cup D_2 \cup ... = \tilde{G} \setminus G_1 \cup \tilde{G} \setminus G_2 \cup ... \tag{C.1.2}$$

Fig. C.1.1 shows the Venn-Euler diagrams for this equation for a case involving only two groups G_1 and G_2.

The shaded areas in the first two diagrams define D_1 and D_2, and when the two shaded areas are added up, as in the third diagram, what is left unshaded is the symmetry group, say G_d, which describes the symmetry of the composite system obtained by superimposing the systems of symmetries G_1 and G_2. G_d, called the intersection group, is the largest common subgroup of G_1 and G_2, taking due account of the mutual disposition of the symmetry elements of G_1 and G_2. The result can be generalized to any number of superimposed symmetries:

$$G_d = G_1 \cap G_2 \cap ... = \cap_i G_i \tag{C.1.3}$$

Naturally, the intersection group G_d cannot be of a higher order than that of any of the component groups G_1, G_2, .. :

$$G_d \subseteq G_i \tag{C.1.4}$$

This relation embodies the statement of the *Neumann theorem* of crystal physics (Nye 1957):

The symmetry G_i possessed by any macroscopic physical property of a crystal cannot be lower than the point-group symmetry of the crystal.

The point group (G_d) of the crystal is simply that resulting from the intersection of the symmetry groups of all the macroscopic physical properties possessed by the crystal.

We see that although Neumann's work was independent of that of Curie, the Curie principle happens to be more general, and implicitly includes the Neumann theorem.

Shubnikov & Koptsik (1974) also state what they call *the Neumann-Minnigerode-Curie (or NMC) principle* as follows:

$$G_{object} \subseteq G_{property} \tag{C.1.5}$$

If an object is to possess a certain physical property (that is, if this property is not to be prohibited by the symmetry of the object), it is necessary that the symmetry group of the object be at least a subgroup of the group of symmetry operations of the physical property.

Recently, Rosen (1995) has given "derivations" of a number of symmetry principles, the main one being what he calls *the symmetry principle*:

The symmetry group of the cause is a subgroup of the symmetry group of the effect.

That is:

$$G_{cause} \subseteq G_{effect} \tag{C.1.6}$$

A somewhat less precise, but more handy, version of this principle is as follows (Rosen 1995):

The effect is at least as symmetric as the cause.

Clearly, Eq. C.1.5 represents only a special case of Eq. C.1.6.

Since the effect cannot be less symmetric than the cause, the symmetry principle can be used for defining a lower bound on the symmetry of the effect. Rosen (1995) refers to this as the *minimalistic use of the symmetry principle.*

Again, since the cause cannot have a higher symmetry than the effect, the symmetry principle provides an upper bound on the symmetry of the cause. This is the *maximalistic use of the symmetry principle.*

The statement of the Curie principle expressed by Eq. C.1.3 is based on certain implicit assumptions. The relaxation of these assumptions leads to successive generalizations and modifications of the statement of the principle (Koptsik 1983; Brandmuller 1986). We discuss one such generalization in §C.2.

We conclude this section by presenting a variation on the way the Curie principle is stated sometimes. Imagine a situation in which all the G_is in Eq. C.1.3 are either absent, or isomorphic to the universal group $\tilde{G}$, except one, say G_1. In that case, Eq. C.1.3 reduces to $G_d = G_1$. This would seem to suggest that the effect (represented by the group G_d) has the same symmetry as the cause (represented by G_1), a statement that should not be taken too literally. The actual statement attributed to Curie is the following (Curie 1894b; Ascher 1977): 'The characteristic symmetry of a phenomenon is the maximal symmetry compatible with the phenomenon. A phenomenon can exist in a medium which has its symmetry or that of one of the subgroups of the characteristic symmetry'.

Many authors take this as the main statement of the Curie principle (Boccara 1981; Sirotin & Shaskolskaya 1982; Senechal 1990), and state the principle as follows:

> *When certain causes produce certain effects, the elements of symmetry in the causes ought to reappear in the effects produced. When certain effects reveal a certain assymetry, this asymmetry must be found in the causes which gave birth to them.*

Naturally, when several causes are present simultaneously, only those elements of symmetry can "reappear" which are common to all the causes. But in every case, "the effect truly exists beforehand in its cause".

C.2 THE CURIE SHUBNIKOV PRINCIPLE. SYMMETRIZATION

One type of situation in which Eq. C.1.3 may not be valid is when the groups G_1, G_2, ... are equivalent subgroups of $\tilde{G}$. This is easier to comprehend when the superposition is of tangible geometrical objects, rather than fields or phenomena. To be specific, if we superimpose geometrical objects that are *equal*, it is possible to construct composite systems that have *higher* symmetries than the symmetry G_d predicted by Eq. C.1.3. Fig. C.2.1 illustrates this for the case of two equal isosceles triangles.

Consider an isosceles triangle, as shown in Fig. C.2.1(a). The origin of the coordinate axes, as well as their orientation with respect to the geometrical figures shown in parts a, b, and c of Fig. C.2.1, is fixed, and is as shown. Fig. C.2.1(b) shows another isosceles triangle, of the same size as in Fig. C.2.1(a), but with orientation as shown. Let G_1 and G_2 denote the symmetry groups for the two triangles.

The group G_1 has only two elements, an identity element and a mirror plane of symmetry normal to the y-axis $[G_1 = (1, m_y)]$. In the International

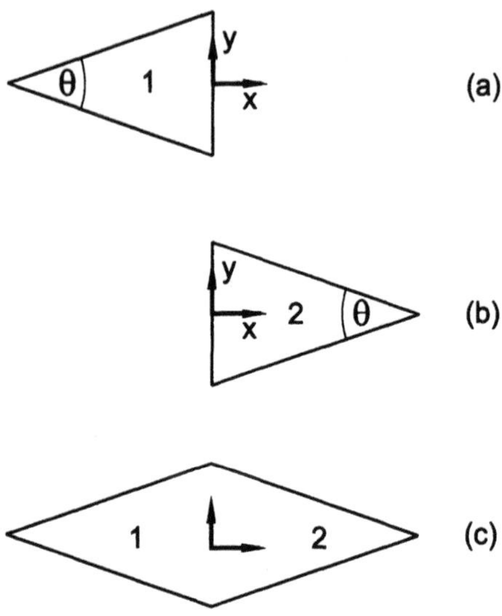

Figure C.2.1: An illustration of symmetrization.

crystallographic notation, this group is denoted by the symbol m_y, so that $G_1 = m_y$. G_2 also has the same two symmetry elements: $G_2 = (1, m_y)$, so that $G_2 = m_y$.

We now create a composite figure from these two identical (equal) triangles by joining them so that they are coplanar and share an arm as shown in Fig. C.2.1(c). The resulting figure is a rhombus. The rhombus has an extra mirror plane of symmetry (m_x) normal to the x-axis, which its two component triangles do not have. This violates Eq. C.1.3, because this equation predicts

$$G_d = m_y \cap m_y = m_y, \tag{C.2.1}$$

whereas the actual symmetry group, say G_s, of the composite figure (the rhombus) is $(1, m_y, m_x, i)$, where i denotes the inversion operation (the rhombus is a centrosymmetric figure).

Thus when two equal isosceles triangles are combined in the manner shown in Fig. C.2.1, the resultant figure has a higher symmetry than that given by the intersection group G_d. This is an example of symmetrization ($G_s \supset G_d$).

The reason for the occurrence of symmetrization lies in the fact that the net symmetry of a composite object consisting of equal parts includes even those symmetry operations which map one part to *another*, whereas for a composite object made up from unequal parts, only those symmetry operations comprise the net symmetry group G_d which simultaneously map each part onto *itself*. Symmetrization occurs in the example of Fig. C.2.1 because the mirror operation m_x can map Triangle 1 onto Triangle 2, and Triangle 2 onto Triangle 1, as a consequence of the fact that the triangles are equal in size (apart from having a specifically chosen relative disposition). By contrast, operations of the intersection group G_d can only map Triangle 1 into itself, and, simultaneously, Triangle 2 into itself.

Shubnikov has generalized the group-theoretical formulation of the Curie principle to include the possibility of symmetrization also, so that the generalized principle is seldom violated (at least for systems which do not interact so strongly as to alter their individual symmetries). According to the generalized principle (which we call the "Curie-Shubnikov principle"):

The symmetry group G_s of a composite object, or phenomenon, or field, is given by the union of G_d (defined by the original statement of the Curie principle) with a set of *symmetrizers* M (Shubnikov & Koptsik 1974):

$$G_s = G_d \cup M = \cap_i G_i \cup M \tag{C.2.2}$$

Thus G_s is an *extended group*.

In G_d, the subscript d denotes dissymmetrization, and in G_s, s denotes symmetrization.

To illustrate the meaning of M, we go back to the example of Fig. C.2.1. To obtain G_s correctly from G_d, we choose M as

$$M = (m_x, i) \tag{C.2.3}$$

Since $m_y m_x = i$ (that is, a reflection operation m_x, followed by a reflection operation m_y, is equal to an inversion operation through the origin), we can rewrite Eq. C.2.3 as

$$M = (1, m_y)\, m_x = G_d\, m_x \tag{C.2.4}$$

The resultant symmetry G_s, according to Eq. C.2.2 is

$$G_s = G_d \cup G_d m_x = (1, m_y) \cup (m_x, i) = mm \tag{C.2.5}$$

The extended group G_s thus describes correctly the symmetry of the rhombus.

The general expression given by Shubnikov & Koptsik (1974) for the symmetrizer M is

$$M = G_d g_2 \cup G_d g_3 \cup \ldots G_d g_j, \tag{C.2.6}$$

where $\{g_2, g_3, \ldots g_j\}$ are the representative elements of an appropriate system of cosets. For more details the book by Shubnikov & Koptsik (1974) should be consulted, as also a paper by Vlachavas (1986).

Further generalizations and modifications of the Curie principle have been made by a number of workers. One such modification attempts to remove the restriction that the superimposed phenomena or fields are of the non-interacting type (in the sense that formation of the composite system does not alter the inherent symmetries of the individual components) (Shubnikov & Koptsik 1974; Koptsik 1983).

Another generalization extends the domain of validity of the principle even to dynamical effects like infrared absorption and the Raman effect (Brandmuller 1986).

An extension to magnetic groups, as well as to the Curie limiting groups, is also discussed by Brandmuller (1986).

Rosen (1995) discusses the following: the equivalence principle; the symmetry principle; the equivalence principle for processes; the symmetry principle for processes; the general symmetry evolution principle; and the special symmetry evolution principle.

C.3 LATENT SYMMETRY

The universe is full of magical things patiently waiting for our wits to grow sharper.

Eden Phillpotts

In Fig. C.2.1(a), for any value of θ between 0 and π ($0 < \theta < \pi$), the symmetry group of the isosceles triangle is m_y. This is true even when $\theta = \pi/2$. However, for $\theta = \pi/2$ the composite figure obtained by combining two such triangles is not a rhombus, but a square. The symmetry group (G_s) of a square is $4mm$, a group of order 8, and not 4 (as for the group mm for the rhombus). Thus, when $\theta = \pi/2$, it is not sufficient to choose the symmetrizer defined by Eq. C.2.3. Two additional symmetrizers must be chosen to construct the group $4mm$ from the intersection group G_d (Wadhawan 1987a).

Fig. C.3.1 shows a plot of the order n of the group G_s as a function of θ. We see singular behaviour at $\theta = \pi/2$. This happens in spite of the fact that the point-group symmetry of the triangle is the same ($G_1 = m_y$) for all values of θ between 0 and π. *It is as if the isosceles triangle with $\theta = \pi/2$ has, in addition to its explicit symmetry m_y, a certain latent symmetry which manifests itself only when two such triangles are brought together to form a composite figure, namely the square.* Conventional group theory is not adequate for taking account of this.[1]

[1]The act of choosing two additional symmetrizers for obtaining the symmetry group $4mm$ as an extended group from m_y is an act of hindsight. We already know the final answer, and therefore go looking for a rationalization of the additional symmetry

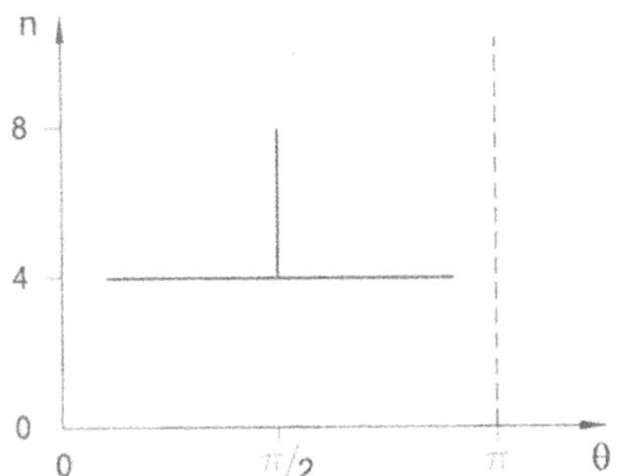

Figure C.3.1: Variation of the order of the extended group G_s with the angle θ of the isosceles triangle drawn in Fig. C.2.1(a).

It is conceivable that latent symmetry, which we have illustrated here with only a geometrical example, may also be important when certain *fields* are superimposed in the physical world.

SUGGESTED READING

A. V. Shubnikov & V. A. Koptsik (1974). *Symmetry in Science and Art.* Plenum Press, New York.

V. A. Koptsik (1983). Symmetry principle in physics. *J. Phys. C: Solid State Phys.*, **16**, 23.

J. Brandmuller (1986). An extension of the Neumann-Minnigerode-Curie principle. *Comp. & Maths. with Appls.*, **12B**, 97.

J. Rosen (1995). *Symmetry in Science. An Introduction to the General Theory.* Springer-Verlag, Berlin. A good modern book on the foundations of the role of symmetry in science. Logic and philosophy are combined to present "derivations" of a number of symmetry principles. The bibliography is particularly valuable for the serious student of symmetry.

---

enhancement occurring for a specific value of θ, namely $\pi/2$. The isosceles triangle has only the symmetry m_y even when $\theta = \pi/2$. It is when we divide it into two parts (by the mirror line m_y) that we find that each part now has a symmetry m_{xy} (but each part now does not have the original symmetry m_y of the entire triangle !). The operator m_{xy}, of course, is one of the many symmetrizers one can choose for explaining the symmetry of the square. Group theory does not forewarn us about the latent symmetry which comes out into the open only for the composite object.

Appendix D

THE FOURIER TRANSFORM

Two functions $F(\mathbf{r})$ and $F(\mathbf{k})$ are said to be Fourier transforms of each other if

$$F(\mathbf{r}) = \int_{-\infty}^{\infty} F(\mathbf{k}) \exp(i\mathbf{r}.\mathbf{k})\, d\mathbf{k}, \tag{D.0.1}$$

and

$$F(\mathbf{k}) = \frac{1}{(2\pi)^3} \int_{-\infty}^{\infty} F(\mathbf{r}) \exp(-i\mathbf{r}.\mathbf{k})\, d\mathbf{r} \tag{D.0.2}$$

If $\mathbf{r}$ denotes vectors in real space, then $\mathbf{k}$ corresponds to vectors in reciprocal space.

Of special interest is the situation where $F(\mathbf{k}) = 1$ in Eq. D.0.1. For simplicity, we consider the problem in one dimension by replacing $\mathbf{r}$ by x and $\mathbf{k}$ by k:

$$F(x) = \int_{-\infty}^{\infty} \exp(ikx)\, dk \tag{D.0.3}$$

This is a divergent integral. We therefore evaluate it first over a finite interval $(-q, q)$, and then take the limit $q \to \infty$:

$$F(x) = \lim_{q\to\infty} F_q(x), \tag{D.0.4}$$

where

$$F_q(x) = \int_{-q}^{q} \exp(ikx)\, dk = \frac{\exp(iqx) - \exp(-iqx)}{ix} = 2\frac{\sin qx}{x} \tag{D.0.5}$$

We define at this stage the *Dirac delta function*, $\delta(x)$, as follows: If we have any continuous, differentiable, function $f(x)$, the Dirac delta function is that for which

$$\int_{-\infty}^{\infty} f(x')\,\delta(x')\,dx' \;=\; f(0) \tag{D.0.6}$$

Let us now consider the integral

$$I \;=\; \lim_{q\to\infty} \int_{-\epsilon}^{\epsilon} f(x)\,\frac{\sin qx}{x}\,dx, \tag{D.0.7}$$

where $(-\epsilon, \epsilon)$ is a vanishingly small interval around the origin. Over this small interval, $f(x)$ can be replaced by the constant value $f(0)$.

We introduce a change of variable, $qx = y$, so that $x = y/q$ and $dx/x = dy/y$. Then

$$I \;=\; f(0)\,\lim_{q\to\infty} \int_{-\epsilon q}^{\epsilon q} \frac{\sin y}{y}\,dy \;=\; -2f(0) \int_{0}^{\infty} \frac{\sin y}{y}\,dy \tag{D.0.8}$$

By contour integration, the integral on the right-hand side can be evaluated as $\pi/2$, so that

$$I \;=\; \pi\, f(0) \tag{D.0.9}$$

Thus, up to the constant π, the function $(\sin qx/x)$ in Eq. D.0.7 satisfies the definition of the Dirac delta function (Eq. D.0.6). Therefore, Eqs. D.0.3-5 can be combined to yield the following result:

$$\frac{1}{2\pi} \int_{-\infty}^{\infty} \exp(ikx)\,dk \;=\; \delta(x) \tag{D.0.10}$$

Thus the Fourier transform of a constant is a delta function. Conversely, the Fourier transform of a delta function is a constant.

Next we consider the Fourier transform of the exponential function e^{-cx}, where c is a positive constant.

We begin by noting that the Fourier transform of this function cannot be defined because $e^{-cx} = \infty$ for $x = -\infty$ (cf. Eq. D.0.2). But we can define the Fourier transform of $e^{-c|x|}$:

$$F(k) \;=\; \frac{1}{2\pi} \int_{-\infty}^{\infty} e^{-c|x|+ikx}\,dx \tag{D.0.11}$$

Since $|x| = x$ for $x > 0$, and $|x| = -x$ for $x < 0$, we can write

$$2\pi\,F(k) \;=\; \int_{0}^{\infty} e^{-cx+ikx}\,dx \;+\; \int_{-\infty}^{0} e^{cx+ikx}\,dx \tag{D.0.12}$$

We introduce a change of variable, $x = -y$, for the second integral:

$$2\pi\,F(k) \;=\; \int_{0}^{\infty} e^{-cx+ikx}\,dx \;+\; \int_{0}^{\infty} e^{-cy-iky}\,dy \tag{D.0.13}$$

On carrying out these integrations we get

$$2\pi F(k) = \left.\frac{e^{-cx+ikx}}{-c+ik}\right|_0^\infty - \left.\frac{e^{-cy-iky}}{c+ik}\right|_0^\infty \tag{D.0.14}$$

Since $e^{-cx+ikx} = e^{-cx}(\cos kx + i\sin kx)$, and a sine or cosine function cannot have a magnitude exceeding unity, the total term is equal to zero for $x = \infty$. Therefore,

$$2\pi F(k) = \frac{1}{c-ik} + \frac{1}{c+ik} = \frac{2c}{c^2+k^2}, \tag{D.0.15}$$

or

$$F(k) = \frac{c}{\pi(c^2+k^2)} \tag{D.0.16}$$

Such a function is called a *Lorentzian.* Thus the Fourier transform of the function $e^{-c|x|}$ is a Lorentzian.

The Fourier transform of the function e^{-cr}/r is of interest in statistical mechanics:

$$f(k) = \frac{1}{(2\pi)^3}\int_{r=0}^{\infty}\int_{\theta=0}^{\pi}\int_{\phi=0}^{2\pi}\frac{e^{-cr}}{r}\, e^{i\mathbf{k}.\mathbf{r}}\, r^2\, dr\, \sin\theta\, d\theta\, d\phi \tag{D.0.17}$$

We can choose the polar axis of coordinates along $\mathbf{k}$. Then

$$f(k) = \frac{2\pi}{(2\pi)^3}\int_0^\infty e^{-cr} r\, dr \left[\frac{-e^{ikr\cos\theta}}{ikr}\right]_0^\pi, \tag{D.0.18}$$

which simplifies to

$$f(k) = \frac{1}{(2\pi)^2 ik}\int_0^\infty dr\, [e^{-r(c-ik)} - e^{-ir(c+ik)}], \tag{D.0.19}$$

or

$$f(k) = \frac{1}{2\pi^2(c^2+k^2)} \tag{D.0.20}$$

We redefine the constant c as follows:

$$\xi \equiv 1/c \equiv \sqrt{c_1/c_2} \tag{D.0.21}$$

[In §5.5.2, ξ is identified with the order parameter correlation length.] Then

$$\frac{e^{-cr}}{r} = \frac{e^{-r/\xi}}{r}, \tag{D.0.22}$$

and its Fourier transform is

$$f(k) = \frac{c_1}{2\pi^2(c_2+c_1k^2)} \tag{D.0.23}$$

The Convolution Theorem

If we have two functions $B(\mathbf{r})$ and $L(\mathbf{r})$ in real space, their convolution $C(\mathbf{r})$ is defined as

$$C(\mathbf{r}) \equiv B(\mathbf{r}) * L(\mathbf{r}) = L(\mathbf{r}) * B(\mathbf{r}) = \int B(\mathbf{r}-\mathbf{r}')\, L(\mathbf{r}')\, d\mathbf{r}' \quad \text{(D.0.24)}$$

Let us compute the Fourier transform of the convolution function $C(\mathbf{r})$:

$$\int_{-\infty}^{\infty} C(\mathbf{r})\, e^{-i\mathbf{r}.\mathbf{k}}\, d\mathbf{r} = \int_{-\infty}^{\infty}\int_{-\infty}^{\infty} B(\mathbf{r}-\mathbf{r}')\, L(\mathbf{r}')\, e^{-i\mathbf{r}.\mathbf{k}}\, d\mathbf{r}'\, d\mathbf{r}$$

$$= \int_{-\infty}^{\infty} d\mathbf{r}'\, L(\mathbf{r}')\, e^{-i\mathbf{r}'.\mathbf{k}} \int_{-\infty}^{\infty} B(\mathbf{r}-\mathbf{r}')\, e^{-i(\mathbf{r}-\mathbf{r}').\mathbf{k}}\, d\mathbf{r} = L(\mathbf{k})\, B(\mathbf{k}) \quad \text{(D.0.25)}$$

where $L(\mathbf{k})$ is the Fourier transform of $L(\mathbf{r})$, and $B(\mathbf{k})$ is the Fourier transform of $B(\mathbf{r})$.

Similarly, if we compute the Fourier transform of the product $B(\mathbf{r})L(\mathbf{r})$, we get

$$\int_{-\infty}^{\infty} B(\mathbf{r})L(\mathbf{r})e^{-i\mathbf{r}.\mathbf{k}}\, d\mathbf{r} = \int_{-\infty}^{\infty} B(\mathbf{r}')L(\mathbf{r}')e^{-i\mathbf{r}'.\mathbf{k}}\, d\mathbf{r}'$$

$$= \int_{-\infty}^{\infty} B(\mathbf{r}') \int_{-\infty}^{\infty} L(\mathbf{k}')e^{i\mathbf{r}'.\mathbf{k}'}\, d\mathbf{k}'\, e^{-i\mathbf{r}'.\mathbf{k}}\, d\mathbf{r}'$$

$$= \int_{-\infty}^{\infty} L(k') \int_{-\infty}^{\infty} B(\mathbf{r}')e^{i\mathbf{r}'.(\mathbf{k}'-\mathbf{k})}\, d\mathbf{r}'\, d\mathbf{k}'$$

$$= \int_{-\infty}^{\infty} L(\mathbf{k}')B(\mathbf{k}-\mathbf{k}')\, d\mathbf{k}' = B(\mathbf{k}) * L(\mathbf{k}) \quad \text{(D.0.26)}$$

Eqs. D.0.25 and D.0.26 represent a general result known as the convolution theorem:

If two functions are convoluted in one space, the result in the other space is the product of their respective Fourier transforms.

The convolution of a function $B(\mathbf{r})$ with a delta function $\delta(\mathbf{r}-\mathbf{r}_0)$ is

$$B(\mathbf{r}) * \delta(\mathbf{r}-\mathbf{r}_0) = \int_{-\infty}^{\infty} B(\mathbf{r}')\, \delta(\mathbf{r}-\mathbf{r}'-\mathbf{r}_0)\, d\mathbf{r}' = B(\mathbf{r}-\mathbf{r}_0) \quad \text{(D.0.27)}$$

The result is thus the same function, but with the origin shifted to $\mathbf{r}_0$. Convolution can thus be used as a mathematical device for reproducing a function at one or more positions. A diffraction grating provides a simple example of this. It can be regarded as the convolution of one of its elements (corresponding to $B(\mathbf{r})$ in Eq. D.0.24) with a function repeating the ideal point grating (a set of equally spaced points). The convolution function $C(\mathbf{r})$ in Eq. D.0.24 corresponds to repeating the grating element at each of the equally spaced points of the function $L(\mathbf{r})$. The result is a set of equally spaced elements. Similarly, the diffraction pattern of the grating can be regarded as the product of two functions, namely the diffraction pattern of a single element and the diffraction pattern of the set of equally spaced ideal points.

SUGGESTED READING

J. W. Goodman (1968). *Introduction to Fourier Optics.* McGraw-Hill, New York. Reissued in 1988.

Ya. B. Zeldowich & A. D. Myskis (1976). *Elements of Applied Mathematics.* Mir Publishers, Moscow.

J. F. James (1995). *A Student's Guide to Fourier Transforms, with Applications in Physics and Engineering.* Cambridge University Press.

Appendix E

THERMODYNAMICS AND STATISTICAL MECHANICS

In this Appendix we summarize some relevant concepts, definitions, and results of thermodynamics and statistical mechanics.

E.1 THERMODYNAMICS

The Moving Finger writes; and, having writ,
Moves on: nor all thy Piety nor Wit
Shall lure it back to cancel half a line,
Nor all thy Tears Wash out a Word of it.

Omar Khayyam

(Translated by Edward Fitzgerald)

Thermodynamics aims at describing the bulk behaviour of macroscopic systems in terms of only a few measurable *thermodynamic parameters.* In the case of a gas or a liquid, for example, such parameters are pressure p, volume V, and temperature T.

If these parameters are independent of time, the system is said to be in a *steady state.*

If, in addition, there is no macroscopic flow of heat or particles through the system, it is said to be in *equilibrium.*

A *state function* is any property which, in equilibrium, depends only on the thermodynamic parameters, rather than on the history of the system.

State parameters may be either *extensive* or *intensive.* The former (like internal energy or entropy) are proportional to the size of the system, and the latter (like pressure and temperature) are independent of the size of the system.

The state functions of a system are not entirely independent of one another. They are connected by *equations of state.* The Boyle's law for an ideal gas is an example of an equation of state connecting p, V and T. Similarly the Curie law is an equation of state for a paramagnetic crystal connecting magnetic field H, magnetization M per unit volume, and temperature:

$$M - \frac{CH}{T} = 0, \tag{E.1.1}$$

where C is the Curie constant.

The *first law of thermodynamics* expresses the conservation of energy, and also takes note of two types of energy, namely heat energy E and work energy W:

$$dU = dQ - dW \tag{E.1.2}$$

Here dQ is the amount of heat energy added to the system, and dW the amount of work done by the system. The law states that their difference must appear completely as the change of the internal energy of the system.

A change in the state functions of a system comprises a *thermodynamic process.*

Such a process is a *quasistatic process* if it occurs infinitely slowly, and is thus close to equilibrium at all times.

It is a *reversible process* if it is quasistatic and its path in thermodynamic space can be reversed exactly. Otherwise it is an *irreversible* process.

If no heat is exchanged with the surroundings, i.e. if $dQ = 0$, the process is said to be an *adiabatic process.*

Entropy S is introduced as an extensive state parameter through the *second law of thermodynamics.* The law states that if heat dQ is added to a system at temperature T, then

$$dQ \leq T\,dS, \tag{E.1.3}$$

where the equality sign can apply only for a reversible process.

Real-life processes are irreversible.

E.1.1 Thermodynamic Potentials

A variety of thermodynamic potentials can be defined. The term 'potential' is used by analogy with potential energy in mechanics, which is a measure of the work obtainable from a system. For example, for an adiabatic process ($dQ = 0$), internal energy $U(S, V)$ plays the role of a potential: the decrease of internal energy is equal to the maximum amount of work obtainable through an adiabatic process, the maximum corresponding to reversible processes. This can be shown as follows.

We first write Eq. E.1.2 for a reversible process:

$$\Delta U = (\Delta Q)_{rev} - (\Delta W)_{rev} \tag{E.1.4}$$

If the process is adiabatic ($\Delta Q = 0$) but not reversible, we have (since ΔU is a perfect differential)

$$\Delta U = -(\Delta W)_{irrev} \tag{E.1.5}$$

Eliminating ΔU from Eqs. E.1.4 and E.1.5, and using Eq. E.1.3, we get

$$(\Delta W)_{rev} - (\Delta W)_{irrev} = \int T\,dS \geq 0 \tag{E.1.6}$$

Thus the decrease in internal energy corresponds to the maximum value of ΔW obtainable through an adiabatic process; this maximum occurs when the process is reversible. In this sense U (per unit volume) plays the role of a 'potential'.

Other thermodynamic potentials are defined for other modes of extracting work from a system. Before describing these, we first generalize our description to include the possibility of a change in the number of particles of the system.

For a reversible process Eqs. E.1.3 and E.1.2 yield

$$dU = T\,dS - p\,dV \tag{E.1.7}$$

Each of the terms in this equation has dimensions of energy or work, and the last term corresponds to mechanical work, namely the work that the system does on expansion at constant pressure. If electric field, magnetic field, and/or uniaxial stress are also present, additional work terms must be included. Thus, in general:

$$dU = T\,dS + X_i\,dx_i, \tag{E.1.8}$$

where X is a *generalized force*, and x the *generalized displacement* conjugate to it.

Similarly, if the system has N_j molecules of type j, and this number can vary,

$$dU\,(S,\,\{x_j\},\,\{N_j\}) = T\,dS + X_i\,dX_i + \mu_j\,dN_j \tag{E.1.9}$$

Here μ_j is the *chemical potential*, defined by

$$\mu_j = \partial U \,/\, \partial N_j \tag{E.1.10}$$

An often-used thermodynamic potential is the *Helmholtz potential* (or the Helmholtz free energy per unit volume), A. It is defined as

$$A = U - TS \tag{E.1.11}$$

It is a state function, and dA is a perfect differential:

$$dA = dU - T\,dS - S\,dT, \tag{E.1.12}$$

which, on using Eq. E.1.9, becomes

$$dA = -S\,dT + X_i\,dx_i + \mu_j\,dN_j \tag{E.1.13}$$

We have seen above that U plays the role of a potential for extracting work out of the system by an adiabatic process. Similarly A plays the role of a potential for extracting work through an *isothermal* ($dT = 0$) process. This can be proved as follows:

Substituting Eq. E.1.2 in E.1.12,

$$dA = dQ - dW - T\,dS - S\,dT \tag{E.1.14}$$

If the process is reversible ($dQ = T\,dS$) and isothermal ($dT = 0$), this becomes

$$-dA = (dW)_{rev} \tag{E.1.15}$$

If the process is irreversible (and isothermal), the second law of thermodynamics tells us that $dQ - TdS \leq 0$, and therefore

$$-dA \geq (dW)_{irrev} \tag{E.1.16}$$

In either case $-dA$ is the maximum amount of work that can be extracted from the system at constant temperature.

Gibbs potential (or Gibbs free-energy density), G, is another important thermodynamic potential. When the only 'force' present is hydrostatic pressure, it is defined as

$$G = U - TS + pV = A + pV \tag{E.1.17}$$

It is a state function, and in differential form

$$dG = dU - T\,dS - S\,dT + p\,dV + V\,dp \tag{E.1.18}$$

On using the first and the second law,

$$dG = (dQ - T\,dS) - (dW - p\,dV) + V\,dp - S\,dT \tag{E.1.19}$$

If the process is reversible ($dQ = T\,dS$), isobaric ($dp = 0$), and isothermal ($dT = 0$),

$$-dG = dW - p\,dV \tag{E.1.20}$$

Thus $-dG$ is a measure of the maximum work that can be extracted from the system at constant pressure and temperature.

A *spontaneous process* is one that occurs by itself, without any change in the external fields applied on the system. We note from Eq. E.1.19 that if for such a process pressure and temperature are constant and $dW - p\,dV = 0$, and $dQ - T\,dS \leq 0$, then G can only decrease. The system tends towards a configuration whereby G is minimized.

In the context of ferroic materials it is relevant to define a *generalized Gibbs potential*, for which the pV term is often omitted, and terms corresponding to electric, magnetic, and non-hydrostatic stress contributions are included (cf. Eq. 6.2.1):

$$G = U - TS - E_iD_i - H_iB_i - \sigma_{ij}e_{ij} \tag{E.1.21}$$

E.1.2 Homogeneous Functions

A function $f(x)$ is a homogeneous function of degree k if, for an arbitrary scale factor b

$$f(bx) = \frac{1}{b^k} f(x) \tag{E.1.22}$$

Consider one of the thermodynamic potentials, e.g. internal energy U. It is a function of extensive parameters S, V, and N. Let the functional relationship be of the form

$$U = f(S, V, N) \tag{E.1.23}$$

Suppose the size of the system increases by a factor b. Then

$$bU = f(bS, bV, bN) \tag{E.1.24}$$

Thus, although the internal energy increases by a factor b, the functional relationship, f, is not affected, and U is a homogeneous function of S, V, and N of degree -1.

Eq. E.1.24 is true for any value of b, and therefore also for $b = V^{-1}$. Therefore

$$U = V f(S/V, N/V) \tag{E.1.25}$$

It can be shown similarly that other thermodynamic potentials also have a trivial dependence on one of the extensive parameters. This leads to the introduction of *thermodynamic-potential densities*; e.g.

$$u = \left.\frac{\partial U}{\partial V}\right|_{S,N} \tag{E.1.26}$$

Other examples are discussed by Chaikin & Lubensky (1995).

E.2 EQUILIBRIUM STATISTICAL MECHANICS

There is something fascinating about science. One gets such wholesale returns of conjecture out of such a trifling investment of fact.

Mark Twain

Critical phenomena (cf. §5.5) occurring in the vicinity of ferroic and other phase transitions involve interaction among a large number of particles, and therefore statistical mechanics is the appropriate theoretical apparatus for dealing with them. Concepts from statistical mechanics are also invoked when one deals with response functions and susceptibilities of materials.

E.2.1 Microcanonical Ensemble

An adequate description of the motion of particles of a system with, say, N particles and s degrees of freedom is provided by Hamilton's equations of motion in terms of generalized coordinates $q_1, q_2, ...q_s$ and the conjugate generalized momenta $p_1, p_2, ...p_s$. Together these $2s$ parameters define a phase space, such that any specific microscopic state of the assembly of particles is represented by a *phase point* or *representative point* in this space. As the system evolves with time, this point traces a *phase line* or *phase trajectory*. In accordance with the requirements of causality, each representative phase point develops out of the phase point preceding it in time.

Because of the large number of particles involved, it is neither practical nor particularly useful to solve the large number of equations of motion. It would be much better to relate the microscopic parameters to a few macroscopic or average parameters which can be determined experimentally. This was achieved by Gibbs by introducing the notion of the *ensemble*.

Gibbs replaced the time-dependent phase line by a fictitious and complete phase line existing at one particular time. Each point on the latter corresponds to a separate system with the same macroscopic parameters like energy E, volume V, and number of particles N, but different possible microscopic states. And all such points are accorded equal weight, implying that all microscopic states with the same macroscopic parameters (E, V, N) are equally probable. This imagined collection of similar, noninteracting, systems, all existing at the same time, is referred to as the *Gibbs ensemble*. Members of the ensemble, all of which correspond to the same N, V and E, are called *elements* of the ensemble.

In the Gibbs formalism, calculation of macroscopic properties is carried out under the assumption that the time average of a property at equilibrium is the same as the *ensemble average*, which is the instantaneous average over the entire statistical ensemble. This assumption is known as the *ergodicity hypothesis*. In other words, we assume that all microscopic states of the system (for a constant energy) are equally probable and accessible.

Since all accessible microscopic states are included, the number of elements in phase space is extremely large, and therefore can be taken as changing *continuously* from one region of phase space to a neighbouring region. An ensemble can therefore be characterized by a *distribution function*, D, which defines the number density with which the phase points are distributed in phase space.

The ensemble average of a quantity $R(q,p)$ is defined as

$$\bar{R} = \frac{1}{M}\int R(q,p)\, D(q,p,t)\, d\Gamma, \tag{E.2.1}$$

where M is the total (and very large) number of phase points:

$$M = \int D\, d\Gamma \tag{E.2.2}$$

One can define a (normalized) *density function* or *probability density*, ρ, as

$$\rho = D/M, \tag{E.2.3}$$

in terms of which the ensemble average of R is

$$\bar{R} = \int R\,\rho\, d\Gamma \tag{E.2.4}$$

For a conservative system the energy E is a constant of motion:

$$E(q_1, q_2, \ldots q_N, p_1, p_2, \ldots p_N) = \text{constant} \tag{E.2.5}$$

Therefore in the $2N$-dimensional phase space it defines a $(2N-1)$-dimensional hyperspace called the *energy surface* or the *ergodic surface*.

Consider an ensemble described by the ergodic surface defined by Eq. E.2.5. Such an ensemble satisfies the following condition for statistical equilibrium for any density function which is a function of E alone:

$$\left(\frac{\partial \rho}{\partial t}\right)_{q,p} = 0 \tag{E.2.6}$$

The simplest possible choice is:

$$\rho(E) = \text{constant} \times \delta(E - E_0) \tag{E.2.7}$$

The ensemble described by such a density function is called the *microcanonical ensemble.*

There is always an uncertainty ΔE in the specification of the energy. Therefore, rather than an ergodic surface, there is really an *ergodic shell* bounded by surfaces corresponding to energies E and $E+\Delta E$. The density function is constant inside the ergodic shell, and zero outside it, and the distribution of phase points in the shell is uniform.

The connection between the microscopic ensemble and thermodynamics is provided by *entropy.* Let the volume occupied by the microscopic ensemble in phase space be denoted by $\Gamma(E)$:

$$\Gamma(E) = \int_{E<\mathcal{H}(q,p)<E+\Delta} \rho(q,p)\, dq^s\, dp^s, \tag{E.2.8}$$

where $\rho = 1$ inside the ergodic shell, and zero outside it.

If we denote the volume occupied by the ergodic surface of energy E by $\Omega(E)$, then

$$\Gamma(E) = \Omega(E+\Delta) - \Omega(E) = g(E)\,\Delta, \tag{E.2.9}$$

where the *density of states*, $g(E)$, is defined by

$$g(E) = \partial\Omega(E)/\partial E \tag{E.2.10}$$

The statistical-mechanical entropy is defined as

$$S(E,V) = k_B \ln \Gamma(E) \tag{E.2.11}$$

It can be demonstrated that this entropy is the same as thermodynamic entropy, and k_B can be identified with the Boltzmann constant.

The fact that the thermodynamic entropy is related to $\Gamma(E)$, and thence to the number of microscopic states accessible to the system in equilibrium, is known as *the Boltzmann principle.*

A changeover from classical considerations to quantum statistical mechanics amounts to taking note of the fact that the product $\Delta q \Delta p$ cannot be less than a value of the order of the Planck's constant h. The phase space is then imagined as divided into elementary cells of volume h^N.

The assertion that each accessible microscopic state is equally probable ("principle of equal probability") enables us to deal with *composite* systems. Consider two systems (N_1, V_1, E_1) and (N_2, V_2, E_2). When they are isolated from each other and from other systems, the volumes occupied by their microscopic ensembles in their respective phase spaces are $\Gamma_1(E_1)$ and $\Gamma_2(E_2)$. Now suppose they are brought in thermal contact to constitute a single, isolated, composite system in equilibrium. The total energy is consequently $E_1 + E_2$. For this composite system the microscopic ensemble is described by the density function

$$\begin{aligned} \rho_{12}(E) &= \text{constant for } E_1 + E_2 \leq E \leq E_1 + E_2 + 2\Delta; \\ &= 0 \text{ otherwise} \end{aligned} \tag{E.2.12}$$

The phase space volume occupied by the composite system is the product

$$\Gamma_{12}(E_1, E_2) = \Gamma_1(E_1)\Gamma_2(E_2) \tag{E.2.13}$$

E.2.2 Canonical Ensemble

The microcanonical ensemble provides a suitable description for an *isolated* system, which does not exchange energy or number of particles with the surroundings. In the context of phase transitions it is more appropriate to use the *canonical ensemble*, wherein the system can vary its temperature by exchange of energy with a (much larger) heat bath, or heat reservoir.

The system of interest and the reservoir with which it is in thermal equilibrium can be together viewed as an isolated, constant-energy, system describable as a microcanonical ensemble.

Let N_r and N_t be the number of particles in the reservoir and the system of interest respectively; $N_r \gg N_t$. The total energy ($E = E_r + E_t$) is assumed to lie within a shell defined by the ergodic surfaces corresponding to energies E and $E + 2\Delta$.

Let $\Gamma_r(E_r)$ be the volume occupied by the reservoir in its own phase space. The probability of finding the system of interest in a state within $dp_t\, dq_t$ of the point (p_t, q_t) in phase space, irrespective of the state the reservoir is in, is proportional to $dp_t\, dq_t\, \Gamma_r(E_r)$. Since $E_r = E - E_t$, the density function in the Γ space for the system of interest is

$$\rho(p_t, q_t) \propto \Gamma_r(E - E_t) \tag{E.2.14}$$

We now make use of Eq. E.2.11 for the present problem to obtain

$$k_B \ln \Gamma_r(E - E_t) = S_r(E - E_t) \tag{E.2.15}$$

If the fluctuations in energy are not too strong, one can write

$$S_r(E-E_t) = S_r(E) - E_t \left.\frac{\partial S_r(E_r)}{\partial E_r}\right|_{E_r=E} + \dots \simeq S_r(E) - \frac{E_t}{T}, \quad \text{(E.2.16)}$$

where T can be identified with the temperature of the reservoir. Eq. E.2.15 can thus be written as

$$\Gamma_r(E-E_t) = e^{S_r(E-E_t)/k_B} = e^{S_r(E)/k_B}\, e^{-E_t/(k_B T)} \quad \text{(E.2.17)}$$

The first factor on the right-hand side can be treated as constant because the heat reservoir is much larger than the system of interest.

Substituting Eq. E.2.17 in E.2.14,

$$\rho(p_t, q_t) = C\, e^{-E_t/k_B T}, \quad \text{(E.2.18)}$$

where C is a constant to be determined through normalization.

The energy of the system can be expressed in terms of the Hamiltonian $\mathcal{H}_t(p_t, q_t)$. We can also drop the subscript 't' to obtain, finally,

$$\rho(p,q) = C\, e^{-\mathcal{H}(p,q)/k_B T} = C\, e^{-\beta\mathcal{H}(p,q)}, \quad \text{(E.2.19)}$$

where $\beta = 1/(k_B T)$.

The factor $e^{-\beta\mathcal{H}}$ is called the *Boltzmann factor.*

In the above definition of the density function of a system in equilibrium with a large heat reservoir, the presence of the latter is felt only through the temperature T. The ensemble defined by the density function given by Eq. E.2.19 is called the *canonical ensemble.*

E.2.3 Partition Function

The volume occupied by an ensemble in phase space is called the partition function of the ensemble, and is usually denoted by $\mathcal{Z}(N, V, T)$:

$$\mathcal{Z}(N,V,T) = \frac{1}{N!\, h^s} \int e^{-\beta\mathcal{H}(p,q)}\, d^s p\, d^s q \quad \text{(E.2.20)}$$

The factor h^s is introduced to make the partition function a dimensionless quantity. And the factor $N!$ is necessary for "correct Boltzman counting" (see, e.g., Plischke & Bergersen (1994), p. 33, for a discussion of these two factors).

Since the constant C in Eq. E.2.19 is determined by equating the integral of the density function to unity, the density function can be written finally as

$$\rho(N,V,T) = e^{-\beta\mathcal{H}(p,q)}/\mathcal{Z}(N,V,T) \quad \text{(E.2.21)}$$

All thermodynamics functions of the system can be determined once the following equation is established:

$$\mathcal{Z}(N, V, T) = e^{-A(V,T)}, \tag{E.2.22}$$

where $A = U - TS$ is the Helmholtz free energy.

The ensemble average of a macroscopic quantity B (having a discrete set of states labelled by s) in the canonical ensemble is thus given by

$$< B >= \frac{\sum_s B\rho}{\sum_s \rho} = \frac{\sum_s Be^{-\beta\mathcal{H}}}{\sum_s e^{-\beta\mathcal{H}}} \tag{E.2.23}$$

E.2.4 Quantum Statistical Mechanics

In quantum mechanics observables are associated with Hermitian operators operating on a suitable Hilbert space (cf. §B.2), and a state of a system is a vector $|\psi>$ in the same space. The wave function can be written as a linear superposition of a complete set of orthonormal stationary wave functions $\{\phi_n\}$:

$$\psi = \sum_n a_n\phi_n, \tag{E.2.24}$$

with $< \phi_j|\phi_k >= \delta_{jk}$. Here n stands collectively for a set of quantum numbers, and $|a_n|^2$ is the probability that the system exists in a state with quantum numbers n.

The average of a large number of measurements of a macroscopic observable R is given by

$$< R >= \frac{\overline{(\psi, R\psi)}}{\overline{(\psi, \psi)}} = \frac{\sum_m \sum_n \overline{(a_n, a_m)}(\phi_n, R\phi_m)}{\sum_n \overline{(a_n, a_n)}}, \tag{E.2.25}$$

where $\overline{(a_n, a_m)}$ is the time average of the scalar product (a_n, a_m).

We consider a system which, although not completely isolated, interacts with the surroundings so weakly that its energy is nearly constant. We can choose the set $\{\phi_n\}$ such that its members are eigenstates of the Hamiltonian $\mathcal{H}$ of the system:

$$\mathcal{H}\phi_n = E_n\phi_n \tag{E.2.26}$$

In quantum statistical mechanics we make the "equal probability" postulate stated above and the "random phase" postulate, as a result of which we have $\overline{(a_n, a_m)} = \delta_{nm}$ for $E < E_n < E + \Delta$, and zero for energy values outside this interval. The total wavefunction (Eq. E.2.24) can therefore be written as

$$\psi = \sum_n b_n\phi_n, \tag{E.2.27}$$

where $|b_n|^2 = 1$ for $E < E_n < E + \Delta$, and $|b_n|^2 = 0$ for energy outside this interval.

The result of measuring R is therefore expected to be

$$< R > = \frac{\sum_n |b_n|^2 (\phi_n, R\phi_n)}{\sum_n |b_n|^2} \tag{E.2.28}$$

We next consider the question of the representation of an operator R. This is done by defining its matrix elements with respect to a complete set of eigenstates. For this, one introduces the so-called *density matrix*, which is defined keeping in mind the fact that only the square-moduli $|b_n|^2$ appear in the definition of $< R >$ in Eq. E.2.28:

$$\rho_{mn} \ = \ (\phi_n, \rho\phi_m) \ = \ \delta_{mn}|b_n|^2 \tag{E.2.29}$$

The density operator ρ operates on state vectors in Hilbert space, and can be represented as

$$\rho \ = \ \sum_n |\phi_n > \ |b_n|^2 \ < \phi_n| \tag{E.2.30}$$

Eq. E.2.8 can be rewritten in terms of the density matrix as follows:

$$< R > = \frac{\sum_n (\phi_n, R\rho\phi_n)}{\sum_n (\phi_n, \rho\phi_n)} \ = \ \frac{\mathrm{Tr}(R\,\rho)}{\mathrm{Tr}\rho} \tag{E.2.31}$$

Here Tr stands for "trace". Since the trace of a matrix is not altered by a similarity transformation, the merit of Eq. E.2.31 is that it is independent of the choice of representation for the operator R.

For the microscopic ensemble the density matrix is

$$\rho_{mn} \ = \ \delta_{mn}|b_n|^2, \tag{E.2.32}$$

where the $|b_n|^2$ have constant nonzero values if the energy eigenvalues lie between E and $E + \Delta$; and zero otherwise.

For the canonical ensemble the density matrix is

$$\rho_{mn} \ = \ \delta_{mn} e^{-\beta E_n}, \tag{E.2.33}$$

and the density operator is

$$\rho \ = \ \sum_n |\phi_n > e^{-\beta E_n} \ < \phi_n| \ = \ e^{-\beta\mathcal{H}} \sum_n |\phi_n >< \phi_n| \ = \ e^{-\beta\mathcal{H}}, \tag{E.2.34}$$

where we have made use of the completeness property of eigenstates.

The partition function (cf. Eq. E.2.20) is given by

$$\mathcal{Z}(N, V, T) \ = \ \mathrm{Tr}\,\rho \ = \ \sum_n e^{-\beta E_n}, \tag{E.2.35}$$

where we have replaced the phase-space integration in Eq. E.2.20 by a summation over the states of the system:

$$\frac{1}{N!\,h^s}\int dp\,dq \rightarrow \sum_n \tag{E.2.36}$$

We can finally write the ensemble average of R in the canonical ensemble as

$$< R > = \frac{\mathrm{Tr}(R\,e^{-\beta\mathcal{H}})}{\mathcal{Z}} \tag{E.2.37}$$

E.2.5 Fluctuations

The properties of a system fluctuate about their mean values. The same is therefore true about the elements of an ensemble.

Let P_i be the probability or density function of finding a system in the state i, and M_i the value of some property of interest when the system is in that state. Obviously, $\sum_i P_i = 1$. The *mean value* of the property is defined as

$$\bar{M} = \sum_i P_i M_i \tag{E.2.38}$$

A measure of the fluctuations is provided by the *mean-square deviation*:

$$\overline{(\delta M)^2} = \overline{(M - \bar{M})^2} = \sum_i P_i (M_i - \bar{M})^2$$

$$= \sum P_i M_i^2 - 2\bar{M}\sum P_i M_i + \bar{M}^2 = \overline{M^2} - \bar{M}^2 \tag{E.2.39}$$

The square root of this is the *standard deviation*, ΔM.

In statistical mechanics the averages $\overline{M^2}$ and $\bar{M}^2$ in Eq. E.2.39 are obtained as ensemble averages. We illustrate this for the case of energy fluctuations in the canonical ensemble, for which the probability P is given by the density function defined by Eq. E.2.21, and the partition function can be written as

$$\mathcal{Z} = \sum_i e^{-\beta E_i} \tag{E.2.40}$$

The mean energy is

$$< E > = \sum_i P_i E_i = \frac{\sum_i E_i e^{-\beta E_i}}{\mathcal{Z}} = -\frac{\partial \mathcal{Z}/\partial\beta}{\mathcal{Z}} \tag{E.2.41}$$

Similarly,

$$< E^2 > = \sum_i P_i E_i^2 = \frac{\partial^2 \mathcal{Z}/\partial\beta^2}{\mathcal{Z}} \tag{E.2.42}$$

From Eq. E.2.41,

$$-\frac{\partial < E >}{\partial \beta} = \frac{1}{\mathcal{Z}}\frac{\partial^2 \mathcal{Z}}{\partial \beta^2} - \frac{1}{\mathcal{Z}^2}\left(\frac{\partial \mathcal{Z}}{\partial \beta}\right)^2, \tag{E.2.43}$$

which on using Eqs. E.2.41, 42 and 39 becomes

$$-\frac{\partial < E >}{\partial \beta} = < E^2 > - < E >^2 = < (\delta E)^2 > \tag{E.2.44}$$

The left-hand side of this equation is related to specific heat at constant volume:

$$C_V \equiv \left(\frac{\partial < E >}{\partial T}\right)_V = \left(\frac{\partial < E >}{\partial \beta}\right)_V \frac{\partial \beta}{\partial T} = \left(\frac{\partial < E >}{\partial \beta}\right)_V (-k_B\beta^2)$$

$$= k_B\beta^2 < (\delta E)^2 > = \frac{< (\delta\, E)^2 >}{k_B T^2} \tag{E.2.45}$$

Energy fluctuations can be defined as $\Delta E/ < E >$:

$$\frac{\Delta E}{< E >} \equiv \frac{(< (\delta E)^2 >)^{1/2}}{< E >} = \frac{(k_B T^2 C_V)^{1/2}}{< E >} \tag{E.2.46}$$

This relation connects C_V, a response function (cf. Eq. E.3.39 below), with the mean-square fluctuation of the energy function, and is an example of a general theorem called the *fluctuation dissipation theorem* (cf. §E.3).

Both C_V and $< E >$, being extensive parameters, scale as N. Therefore the energy fluctuations scale as $(1/N)^{1/2}$. If N is of the order of the Avogadro number, the fluctuations in the energy, occurring because the system described by the canonical ensemble is in contact with a heat reservoir, are normally very small, and then there is hardly any difference between the canonical and the microcanonical ensembles.

E.2.6 Correlation Functions

Intrinsic or spontaneous fluctuations exist even when a system is at equilibrium. Because of these fluctuations one employs statistical or ensemble averages for defining any correlation between the value of a property at a point $\mathbf{r}$ at time t and the value of the same or a different property at another point $\mathbf{r}'$ at a later time t'. A variety of correlation functions can be defined.

In the context of phase transitions in ferroic materials we are interested in the correlation functions connected with the order parameter. Let $m(\mathbf{r})$

denote its density. The macroscopic order parameter associated with a ferroic phase transition is then the ensemble average M:

$$M = \left\langle \int m(\mathbf{r})\, d\mathbf{r} \right\rangle \tag{E.2.47}$$

A measure of the correlation between $m(\mathbf{r})$ and $m(0)$ is provided by the spatial order parameter-order parameter correlation function $\Gamma(\mathbf{r})$:

$$\Gamma(\mathbf{r}) = < m(\mathbf{r}) m(0) > - < m(\mathbf{r}) >< m(0) > \tag{E.2.48}$$

If the system is homogeneous (i.e. invariant under translation), we have $< m(\mathbf{r}) >=< m(0) >$, and Eq. E.2.47 becomes

$$\Gamma(\mathbf{r}) = < m(\mathbf{r}) m(0) > - < m(0) >^2 \tag{E.2.49}$$

We denote the Fourier transform of $m(\mathbf{r})$ by $m(\mathbf{k})$, and of $\Gamma(\mathbf{r})$ by $\Gamma(\mathbf{k})$. As shown in Appendix D, the Fourier transform of a constant is a delta function. Therefore, taking the Fourier transform of both sides in Eq. E.2.49,

$$\Gamma(\mathbf{k}) = < m(\mathbf{k})\, m(0) > - < m(0) >^2 (2\pi)^3 \delta(\mathbf{k}) \tag{E.2.50}$$

By definition, the order parameter is zero for temperatures above the critical temperature:

$$< m(0) > = 0, \quad T > T_c \tag{E.2.51}$$

We also have, by definition,

$$m(\mathbf{r}) = \frac{1}{(2\pi)^3} \int m(\mathbf{k})\, e^{i\mathbf{k}.\mathbf{r}}\, d\mathbf{k}, \tag{E.2.52}$$

so that

$$m(0) = \frac{1}{(2\pi)^3} \int m(\mathbf{k})\, d\mathbf{k} \tag{E.2.53}$$

We note further that for two wavevectors $\mathbf{k}$ and $\mathbf{k}'$,

$$< m(\mathbf{k}) m(\mathbf{k}') > = (2\pi)^3\, \delta(\mathbf{k} + \mathbf{k}')\, |m(\mathbf{k})|^2 \tag{E.2.54}$$

Substituting Eqs. E.2.51, 53 and 54 into E.2.50 we obtain

$$\Gamma(\mathbf{k}) = < |m(\mathbf{k})|^2 > \tag{E.2.55}$$

E.3 NONEQUILIBRIUM STATISTICAL MECHANICS

E.3.1 Linear Response Theory

We now consider time-dependent phenomena in systems that are not at equilibrium, but close to equilibrium. For such systems linear response theory (LRT) is applicable.

For probing a system with a typical spectroscopic technique, we may either subject it to some force, and study its *response*, or we may keep the force 'on' for a long time and then remove it to study its *relaxation* to a state of equilibrium. In the LRT we assume that the disturbance caused to the system by the probing force is negligibly small, so that first-order perturbation theory is applicable for interpreting the spectroscopic data.

LRT provides the necessary link between experiment and correlation functions. This link can be worked out by applying the techniques of statistical mechanics to nonequilibrium systems. Of special importance in this connection is the *fluctuation dissipation theorem* (see below).

E.3.2 Time Correlation Functions

The discussion here, and in the following sections, follows the work of Kubo et al. (1985).

A time correlation function between dynamical variables $A(t_0)$ and $B(t_0 + t)$ is defined as the ensemble average

$$C_{AB}(t) \; = < A(t_0) \, B(t_0 + t) > \qquad \text{(E.3.1)}$$

It is independent of t_0 for *stationary processes.* (A stationary process is one which occurs under stationary conditions like constant temperature, pressure, electric field, etc.)

A special case of the above function is the following *autocorrelation function:*

$$C_{AA} \; = < A(t_0) \, A(t_0 + t) > \qquad \text{(E.3.2)}$$

This function is a measure of the time (the *correlation time*) over which the variable A retains its own memory till it is wiped out by fluctuations.

To gain a somewhat deeper understanding of time correlation functions, we begin with the equation of motion of a particle, as specified by Newton's second law:

$$m \frac{d\mathbf{u}}{dt} \; = \mathbf{F} \qquad \text{(E.3.3)}$$

Here $\mathbf{F}$ stands for the total force experienced by the particle.

A common example of the motion of a particle manifesting fluctuations is that of Brownian motion. For this case the force has two types of contributions:

$$\mathbf{F} \; = \; -m\gamma\mathbf{u} + \mathbf{R}(t) \tag{E.3.4}$$

The first term, proportional to the velocity $\mathbf{u}$ of the particle, represents the *frictional force*, with $m\gamma$ as the frictional coefficient. The second term stands for the *random force*, arising because of the random collisions of the molecules of the liquid with the particle under observation. Therefore the equation of motion is

$$m\,\frac{d\mathbf{u}}{dt} \; = \; -m\gamma\mathbf{u} + \mathbf{R}(t) \tag{E.3.5}$$

An equation of motion like this, with a random-force component, is referred to as a *Langevin equation* (Coffey, Kalmykov & Waldron 1996).

For dealing with real-life situations it is necessary to generalize this equation in at least two ways. One is the introduction of a possible external force $\mathbf{X}(\mathrm{t})$. The other is to abandon the assumption that the friction is determined only by the instantaneous velocity of the Brownian particle, and to recognize its dependence on velocities at *all* times previous to t. The generalized Lagevin equation then reads

$$\frac{d\mathbf{u}(t)}{dt} \; = \; -\int_{-\infty}^{t} \gamma(t-t')\,\mathbf{u}(t')\,dt' + \frac{\mathbf{R}(t)}{m} + \frac{\mathbf{X}(t)}{m} \tag{E.3.6}$$

The external force can be an arbitrary function of time, as well as a combination of various forces. The essence of the LRT is the assumption that the external force disturbs the equilibrium of the system only slightly. This assumption has two consequences: The effect is *linearly* related to the cause, and the various causes (as also the corresponding effects) can be summed by *linear superposition.*

In particular, $\mathbf{X}(t)$ can be expressed as a Fourier series, and the effects of the various harmonics can be summed by linear superposition. It is therefore sufficient to consider the effect of any one of the Fourier components, in particular the fundamental component, and assume that $\mathbf{X}(t)$ has the form

$$\mathbf{X}(t) \; = \; \mathbf{X}_0 \cos\omega t \; = \; \Re\{\mathbf{X}_0 e^{i\omega t}\} \tag{E.3.7}$$

The average velocity produced by this force is

$$< \mathbf{u}(t) > = \; \Re\{\mu(\omega)\,\mathbf{X}_0\, e^{i\omega t}\} \tag{E.3.8}$$

Here $\mu(\omega)$ is the *complex mobility*, related to the frictional coefficient through

$$\mu(\omega) \; = \; \frac{1}{m}\,\frac{1}{i\omega + \gamma(\omega)}, \tag{E.3.9}$$

where

$$\gamma(\omega) = \int_0^\infty \gamma(t)\, e^{-i\omega t}\, dt \tag{E.3.10}$$

Since analysis in terms of harmonic components is very common in LRT, it is instructive to state a theorem in this context.

Let a stationary process $z(t)$ be sampled over a time interval $0 \le t \le T$. We can write

$$z(t) = \sum_{n=-\infty}^{\infty} a_n\, e^{i\omega_n t}, \tag{E.3.11}$$

where the frequencies of the harmonics are given by

$$\omega_n = 2\pi n/T, \quad n = 0, \pm 1, \pm 2, ... \tag{E.3.12}$$

The average strength of the amplitudes of the harmonics is defined in terms of the mean-square averages of their real and imaginary parts:

$$< |a_n|^2 > = < |a'_n|^2 > + < |a''_n|^2 > \tag{E.3.13}$$

The average intensity $I(\omega)$ observable for a frequency window $\Delta\omega$ is

$$I(\omega)\, \Delta\omega = \sum_{\omega_n \in \Delta\omega} < |a_n|^2 > \tag{E.3.14}$$

The interval between neighbouring frequencies is $2\pi/T$, so that there are $\Delta\omega/(2\pi/T)$ of them in the band $\Delta\omega$. Therefore the intensity spectrum at any frequency ω is given by

$$I(\omega) = \lim_{T\to\infty} \frac{T}{2\pi} < |a_m|^2 > \tag{E.3.15}$$

This spectrum can be obtained in terms of the time correlation function $\phi(t)$ for the process $z(t)$ by using the *Wiener-Khintchine theorem* (Wiener 1930; Khintchine 1934). The correlation function is

$$\phi(t) = < z(t_0)\, z(t_0 + t) > \tag{E.3.16}$$

And the theorem states that

$$I(\omega) = \frac{1}{2\pi} \int_{-\infty}^{\infty} \phi(t)\, e^{-i\omega t}\, dt \tag{E.3.17}$$

E.3.3 Fluctuation Dissipation Theorem

Going back to the generalized Lagevin equation, Eq. E.3.5, we note that it is a linear equation and thus can be subjected to harmonic analysis. One of the results of such an analysis is the following (Kubo et al. 1985):

$$\mu(\omega) = \frac{1}{m} \frac{1}{i\omega + \gamma(\omega)} = \frac{1}{k_B T} \int_0^\infty < u(t_0) u(t_0 + t) > e^{-i\omega t}\, dt \tag{E.3.18}$$

It gives a general expression for the complex mobility in terms of the Fourier-Laplace transform of the correlation function of velocity. It is an instance of the *fluctuation dissipation theorem of the first kind* (cf. Kubo et al. 1985). It states that the response function of a system to an external influence is dictated by correlations between thermal fluctuations occurring at different times in the system *in the absence of the external influence.* In other words, the progress towards equilibrium of a macroscopic nonequilibrium system is decided by the same laws which determine the regression of spontaneous microscopic fluctuations in an equilibrium state of the system.

The theorem relates equilibrium fluctuations to dissipation in the linear regime. Several other examples or variations of this theorem are encountered in this book.

E.3.4 Response Function

Application of an electric or magnetic field to a material induces a response in the form of induced electric or magnetic polarization. Similarly application of mechanical stress induces a strain, and application of a temperature gradient or a concentration gradient induces a flow of heat or mass. If the fields applied are sufficiently small, the response is linearly related to the field. If not, one must admit dependence on higher powers of the forces.

The term *generalized force* is appropriate when we do not wish to specify the nature of the force field.

Similarly the term *generalized displacements* can be applied to describe the changes in the atomic positions, magnetic dipoles, or charge clouds that occur on the application of a generalized force.

Generalized currents are the electric, thermal or other currents caused by the generalized forces, and reflect the tendency of the system to change from the initial equilibrium state to a new equilibrium state.

Let us denote by $X_\nu(t)$ the generalized forces, and by $B_\mu(t)$ the generalized displacements or currents. Then, within the domain of the LRT, the following linear relations exist by a superposition of the various effects:

$$B_\mu(t) - B_\mu^{eq} = \sum_\nu L_{\mu\nu} X_\nu(t) \tag{E.3.19}$$

Here $B_\mu^{eq} = B_\mu(0)$, and $L_{\mu\nu}$ are the so-called *kinetic coefficients.*

This equation can be generalized or improved by recognizing that an effect at any time t may be the result of all the relevant causes at all times $t' < t$. This is especially important for forces (causes) that vary so rapidly that the displacements or currents (effects) lag behind them. Therefore we must integrate over all the time-variations of the forces for times prior to the instant of observation of the effect:

$$B_\mu(t) - B_\mu^{eq} = \sum_\nu \chi_{\mu\nu}^\infty X_\nu(t) + \int_{-\infty}^t dt' \sum_\nu \Phi_{\mu\nu}(t-t')\, X_\nu(t') \quad \text{(E.3.20)}$$

Here the first term on the right-hand side denotes effects the delay in the appearance of which can be neglected, so that they are practically "instantaneous".

As a concrete example, let us consider a situation in which a pulsed force of type α is applied at time $t = t_1$ to a system which was in equilibrium before the application of the force:

$$X_\nu(t) = \delta_{\nu\alpha}\, \delta(t - t_1) \quad \text{(E.3.21)}$$

Substituting this in Eq. E.3.19 yields

$$B_\mu(t) - B_\mu^{eq} = \chi_{\mu\alpha}^\infty\, \delta(t-t_1) + \Phi_{\mu\alpha}(t-t_1)\, \theta(t-t_1), \quad \text{(E.3.22)}$$

where θ is the *Heavyside unit step function*:

$$\begin{aligned} \theta(x) &= \int_{-\infty}^\infty dx'\, \delta(x') = 1 \quad \text{for } x > 0, \\ &= 0 \quad \text{for } x < 0 \end{aligned} \quad \text{(E.3.23)}$$

The force pulse is centered around $t = t_1$. The first term in Eq. E.3.22 corresponds to the 'instantaneous' (pulsed) response at $t = t_1$. It is zero for $t > t_1$. Therefore

$$B_\mu(t) - B_\mu^{eq} = \Phi_{\mu\alpha}(t - t_1) \quad \text{for } t > t_1 \quad \text{(E.3.24)}$$

This equation provides a physical interpretation for the function Φ. For $t \gg t_1$ the system will return once again to equilibrium. Φ is therefore called the *aftereffect function*, or *the response function* (Kubo et al. 1985).

E.3.5 Relaxation

Let us consider a system which is under the action of an external force X_α for such a long time that an equilibrium situation prevails. At time t_1 we remove the force suddenly. That is,

$$\begin{aligned} X_\nu(t) &= \delta_{\nu\alpha} \quad \text{for } t < t_1, \\ &= 0 \quad \text{for } t > t_1 \end{aligned} \quad \text{(E.3.25)}$$

To determine the effect of this action, we substitute Eq. E.3.25 into E.3.20. We get

$$B_\mu(t) = B_\mu^{eq} + \chi_{\mu\alpha}^\infty + \psi_{\mu\alpha}(0) \quad \text{for } t < t_1, \quad \text{(E.3.26)}$$

$$B_\mu(t) \;=\; B_\mu^{eq} + \psi_{\mu\alpha}(t - t_1) \quad \text{for } t > t_1, \tag{E.3.27}$$

where

$$\psi_{\mu\alpha}(t) \;\equiv\; \int_t^\infty dt'\, \Phi_{\mu\alpha}(t') \tag{E.3.28}$$

The last equation defines what is called *the relaxation function*: As $t \to \infty$, $\psi_{\mu\alpha}(t) \to 0$, and from Eq. E.3.27, $B_\mu(t)$ relaxes to the equilibrium value B_μ^{eq}.

Eq. E.3.28 also defines the relationship between the relaxation function and the response function, which can be re-expressed as

$$\Phi_{\mu\nu}(t) \;=\; -\frac{d\psi_{\mu\nu}(t)}{dt} \tag{E.3.29}$$

E.3.6 Generalized Susceptibility

We consider next the response of the system to a force that is neither a delta-function of time nor a 'down-step' function defined by Eq. E.3.25, but a general function of time. Such a function can be written as a Fourier integral:

$$X_\nu(t) \;=\; \frac{1}{2\pi}\int_{-\infty}^{\infty} X_{\nu,\omega}\, e^{-i\omega t}\, d\omega \tag{E.3.30}$$

We assume that the LRT is applicable, so that the effects of the various harmonics can be combined by linear superposition. Therefore it is sufficient to work with only a single harmonically varying force.

Eq. E.3.30 can be Fourier-inverted to yield

$$X_{\nu,\omega} \;=\; \int_{-\infty}^{\infty} X_\nu(t)\, e^{i\omega t}\, dt \tag{E.3.31}$$

Equations similar to E.3.30 and E.3.31 can also be written for the generalized displacements or currents:

$$B_\mu - B_\mu^{eq} \;=\; \frac{1}{2\pi}\int_{-\infty}^{\infty} B_{\mu,\omega}\, e^{-i\omega t}\, d\omega, \tag{E.3.32}$$

$$B_{\mu,\omega} \;=\; \int_{-\infty}^{\infty} [B_\mu(t) - B_\mu^{eq}]\, e^{i\omega t}\, dt \tag{E.3.33}$$

In addition we rewrite Eq. D.0.10 of Appendix D in terms of the changed notation:

$$\delta(x) \;=\; \frac{1}{2\pi}\int e^{i\omega x}\, d\omega \tag{E.3.34}$$

If we now substitute Eq. E.3.20 into E.3.33, and use E.3.30, we get the Fourier transform of E.3.20:

$$B_{\mu,\omega} \;=\; \sum_\nu \chi_{\mu\nu}(\omega)\, X_{\nu,\omega}, \tag{E.3.35}$$

where we have introduced the following complex function:

$$\chi_{\mu\nu}(\omega) = \chi^{\infty}_{\mu,\nu} + \int_0^{\infty} \Phi_{\mu\nu}(t)\, e^{i\omega t}\, dt \tag{E.3.36}$$

The linear relation expressed by Eq. E.3.35 makes us identify $\chi_{\mu\nu}(\omega)$ as the (complex) *generalized susceptibility.*

It is related to the response function Φ through Eq. E.3.36.

And its relationship to the relaxation function ψ can be established by substituting Eq. E.3.29 into E.3.36, integrating by parts, and using the fact that $\psi_{\mu\nu}(t \to \infty) = 0$:

$$\chi_{\mu\nu} = \chi^{\infty}_{\mu\nu} + i\omega \int_0^{\infty} \psi_{\mu\nu}(t)\, e^{i\omega t}\, dt \tag{E.3.37}$$

As a special case, it follows from this equation that

$$\chi_{\mu\nu}(0) = \chi^{\infty}_{\mu\nu} + \psi_{\mu\nu}(0) \tag{E.3.38}$$

The static generalized susceptibility is thus a real function.

We also note from Eq. E.3.35 that in the zero-frequency limit,

$$B_{\mu,0} = \sum_{\nu} \chi_{\mu\nu}(0)\, X_{\nu,0} \tag{E.3.39}$$

When only one type of generalized force, say μ, is present (or dominant), then, in the static limit, we can define the *differential susceptibility*:

$$\chi_{\mu} = \frac{\partial B_{\mu}}{\partial X_{\mu}} \tag{E.3.40}$$

We recall some familiar examples of generalized susceptibility:

Specific heat at constant volume:

$$C_V = \left(\frac{\partial Q}{\partial T}\right)_V = T\left(\frac{\partial S}{\partial T}\right)_V \tag{E.3.41}$$

Specific heat at constant pressure:

$$C_p = \left(\frac{\partial Q}{\partial T}\right)_p = T\left(\frac{\partial S}{\partial T}\right)_p \tag{E.3.42}$$

Isothermal compressibility:

$$K_T = -\frac{1}{V}\left(\frac{\partial V}{\partial p}\right)_T \tag{E.3.43}$$

Adiabatic compressibility:

$$K_S = -\frac{1}{V}\left(\frac{\partial V}{\partial p}\right)_S \tag{E.3.44}$$

Finally, we indicate the important relationship between the static generalized susceptibility χ and the order parameter autocorrelation function $\Gamma(r)$ (cf. Eq. E.2.49).

The canonical-ensemble average of a quantity B is given by Eq. E.2.23. In the presence of a generalized force X, with the corresponding generalized displacement x, the Hamiltonian changes appropriately and the ensemble average is given by

$$< B > = \frac{\sum_s B\, e^{-\beta(\mathcal{H}-Xx)}}{\sum_s e^{-\beta(\mathcal{H}-Xx)}} \tag{E.3.45}$$

We differentiate this equation with respect to X to obtain the generalized differential susceptibility:

$$\chi = \frac{\partial < B >}{\partial X} = \beta\,[< Bx > - < B >< x >] \tag{E.3.46}$$

To extract a familiar example from this we identify X with magnetic field h, and both B and x with magnetic moment M. Then χ is magnetic susceptibility, and we have shown here that

$$\chi = \frac{\partial M}{\partial h} = \beta\,[< M^2 > - < M >^2] \tag{E.3.47}$$

This can be expressed in terms of the magnetic-moment density $m(r)$ (cf. Eq. E.2.47) as

$$\chi = \beta V \int dr\,[< m(r)m(0) > - < m(0) >^2], \tag{E.3.48}$$

which, on using the definition of the spatial autocorrelation function $\Gamma(r)$ (Eq. E.2.49) becomes

$$\chi = \beta V \int dr\,\Gamma(r) \tag{E.3.49}$$

This equation is one more example of the fluctuation dissipation theorem. It describes the response χ of the system to a perturbing field h in terms of the correlations among the spontaneous fluctuations of the system existing when there is no perturbation present.

SUGGESTED READING

R. Kubo, M. Toda & N. Hashitsume (1985). *Statistical Physics II: Nonequilibrium Statistical Mechanics*, second edition. Springer-Verlag, Berlin.

D. Chandler (1987). *Introduction to Modern Statistical Mechanics.* Oxford University Press, New York.

S. Dattagupta (1987). *Relaxation Phenomena in Condensed Matter Physics.* Academic Press, New York.

K. Huang (1987). *Statistical Mechanics.* Wiley, New York.

REFERENCES CITED

Abrahams, S. C. (1971). *Mat. Res. Bull.*, **6**, 881.

Abrahams, S. C. & E. T. Keve (1971). *Ferroelectrics*, **2**, 129.

Aharony, A. (1978). *Solid State Comm.*, **28**, 667.

Aharony, A. (1996). *Introduction to the Theory of Ferromagnetism.* Clarendon Press, Oxford.

Ahmad, I., A. Crowson, C. A. Rogers & M. Aizawa (Eds.) (1990), *U.S.-Japan Workshop on Smart/Intelligent Materials and Systems*, March 19-23, 1990, Honolulu, Hawaii. Technomic Pub. Co., Lancaster.

Aizu, K. (1962), *Rev. Mod. Phys.*, **34**, 550.

Aizu, K. (1964a). *Phys. Rev. A*, **133**, 1350.

Aizu, K. (1964b). *Phys. Rev. A*, **133**, 1584.

Aizu, K. (1964c). *Phys. Rev. A*, **134**, 701.

Aizu, K. (1966). *Phys. Rev.*, **146**, 423.

Aizu, K. (1967). *J. Phys. Soc. Japan*, **23**, 794.

Aizu, K. (1969a). *J. Phys. Soc. Japan*, **27**, 387.

Aizu, K. (1969b). *J. Phys. Soc. Japan*, **27**, 1171.

Aizu, K. (1970a). *Phys. Rev. B*, **2**, 754.

Aizu, K. (1970b). *J. Phys. Soc. Japan*, **28**, 706.

Aizu, K. (1970c). *J. Phys. Soc. Japan*, **28**, 717.

Aizu, K. (1971). *J. Phys. Chem. Solids*, **32**, 1959.

Aizu, K. (1972a). *J. Phys. Soc. Japan*, **32**, 1287.

Aizu, K. (1972b). *J. Phys. Soc. Japan*, **33**, 629.

Aizu, K. (1972c). *J. Phys. Soc. Japan*, **33**, 1390.

Aizu, K. (1973a). *J. Phys. Soc. Japan*, **34**, 121.

Aizu, K. (1973b). *J. Phys. Soc. Japan*, **35**, 180. Also see *J. Phys. Soc. Japan* (1973), **35**, 951 and 1567 for errata.

Aizu, K. (1973d). *J. Phys. Soc. Japan*, **35**, 691.

Aizu, K. (1974). *J. Phys. Soc. Japan*, **36**, 1273.

Aizu, K. (1975). *J. Phys. Soc. Japan*, **38**, 1592.

Aizu, K. (1977). *J. Phys. Soc. Japan*, **42**, 424.

Aizu, K. (1978). *J. Phys. Soc. Japan*, **44**, 683.

Aizu, K. (1979). *J. Phys. Soc. Japan*, **46**, 384.

Ajayan, P. M. & L. D. Marks (1990). *Phase Transitions*, **24-26**, 229.

Alario-Franco, M. A. (1987). *Cryst. Latt. Def. & Amorph. Mat.*, **14**, 357.

Alefeld, G. (1971). In R. E. Mills (Ed.), *Critical Phenomena in Alloys, Magnets and Superconductors*. McGraw-Hill, New York.

Alefeld, G., G. Schaumann, J. Tretkowski & J. Volkl (1969). *Phys. Rev. Lett.*, **22**, 697.

Alefeld, G., J. Volkl & G. Schaumann (1970). *Phys. Stat. Solidi*, **37**, 337.

de Almeida, J. R. & D. J. Thouless (1978). *J. Phys. A*, **11**, 983.

Als-Nielsen, J. & R. J. Birgeneau (1977). *Amer. J. Phys.*, **45**, 554.

Amelinckx, S., G. van Tendeloo, D. van Dyck & J. van Landuyt (1989). *Phase Transitions*, **16/17**, 3.

Amin, A. & R. E. Newnham (1980). *Phys. Stat. Sol.* (a), **61**, 215.

Anderson, A. C. (1985). *Phase Transitions*, **5**, 301.

Anderson, J. S. (1973). *J. Chem. Soc. Dalton*, 1107.

Anderson, J. S. & B. G. Hyde (1965). *Bull. Soc. Chim. Fr.*, 1215.

Anderson, J. S. & B. G. Hyde (1967). *J. Phys. Chem. Solids*, **28**, 1393.

Anderson, S. F. & P. W. Anderson 91975). *J. Phys. F: Metal Phys.*, **5**, 965.

Anderson, P. W. (1958). In *Proc. Conf. Phys. of Dielectrics*. Academy of Sciences, Moscow.

Anderson, P. W. (1960). In G. I. Skanavi (Ed.), *Fizika Dielektrikov*. Akad. Nauk SSSR, Moscow.

Anderson, P. W. (1981). In N. Boccara (Ed.), *Symmetries and Broken*

Symmetries in Condensed Matter Physics. IDSET, Paris.

Anderson, P. W. (1984). *Basic Notions of Condensed Matter Physics*. Addison-Wesley, California.

Anderson, P. W. (1988a). *Phys. Today*, **41**(1), 9.

Anderson, P. W. (1988b). *Phys. Today*, **41**(3), 9.

Anderson, P. W. (1988c). *Phys. Today*, **41**(6), 9.

Anderson, P. W. (1988d). *Phys. Today*, **41**(9), 9.

Anderson, P. W. (1989a). *Phys. Today*, **42**(7), 9.

Anderson, P. W. (1989b). *Phys. Today*, **42**(9), 9.

Anderson, P. W. (1990). *Phys. Today*, **43**(3), 9.

Anderson, P. W., B. I. Halperin & C. M. Varma (1972). *Phil. Mag.*, **25**, 1.

Anderson, T. L., R. E. Newnham & L. E. Cross (1977). *Proc. 21st Annual Symp. on Frequency Control*, 171.

Andersson, S. & B. G. Hyde (1974). *J. Solid State Chem.*, **9**, 92.

Andrews, T. (1869). *Phil. Trans. R. Soc.*, **159**, 575.

Andrews, S. R. & R. A. Cowley (1986). *J. Phys. C: Solid State Phys.*, **19**, 615.

Arlt, G. (1990). *J. Mater. Sci.*, **25**, 2655.

Arlt, G. & N. A. Pertsev (1991). *J. Appl. Phys.*, **70**, 2283.

Armstrong, J. A., N. Bloembergen, J. Ducuing & P. S. Pershan (1962). *Phys. Rev.*, **127**, 1918.

Asahi, T., M. Tomizawa & J. Kobayashi (1992). *Phys. Rev. B*, **45**, 1971.

Ascher, E. (1966a). *Helv. Phys. Acta*, **39**, 40.

Ascher, E. (1966b). *Helv. Phys. Acta*, **39**, 466.

Ascher, E. (1966c). *Physics Letters*, **20**, 352.

Ascher, E. (1977). *J. Phys. C: Solid State Phys.*, **10**, 1365.

Ascher, E. & J. Kobayashi (1977). *J. Phys. C: Solid State Phys.*, **10**, 1349.

Astrov, D. N. (1960). *Sov. Phys. - JETP*, **38**, 984.

Aubry, S. & R. Pick (1971). *J. de Physique*, **32**, 657.

Auciello, O., J. F. Scott & R. Ramesh (1998). The physics of ferroelectric memories. *Physics Today*, July 1998 issue, p. 22.

Axe, J. D., B. Dorner & G. Shirane (1971). *Phys. Rev. Lett.*, **26**, 519.

Bachheimer, J. P. & G. Dolino (1975). *Phys. Rev. B*, **11**, 3195.

Bachmann, R. & K. Barner (1988). *Solid State Comm.*, **68**, 865.

Bacri, J.-C., R. Perzynski & D. Salin (1988). *Endeavor*, New Series, **12**(2), 76.

Baker, M. & B. G. Hyde (1978). *Phil. Mag.*, **38**, 615.

Ballman, A. A. & H. Brown (1972). *Ferroelectrics*, **4**, 189.

Barrett, J. H. (1952). *Phys. Rev.*, **86**, 118.

Barrett, W. F., W. Brown & R. A. Hadfield (1900). *Sci. Trans. Roy. Dublin Soc.*, **7**, 67.

Barrow, J. D. (1988). *The World Within the World.* Oxford University Press, Oxford.

Barsch, G. R. & J. A. Krumhansl (1984). *Phys. rev. Lett.*, **53**, 1069.

Bausch, R. (1972). *Z. Physik*, **254**, 81.

Baz, A., J. Ro, S. Poh & Gilheany (1992a). In G. J. Knowles (Ed.) (1992).

Baz, A., J. Ro, M. Mutua & Gilheany (1992b). In G. J. Knowles (Ed.) (1992).

Bean, C. P. & J. D. Livingston (1956). *J. Appl. Phys.*, **27**, 1448.

Bean, C. P. & J. D. Livingston (1959). *J. Appl. Phys.*, **30**, 120S.

Beauchamp, C. H., R. H. Nadolink & L. M. Dean (1992). In G. J. Knowles (Ed.) (1992).

Bednorz, J. G. & K. A. Muller (1984). *Phy. Rev. Lett.*, **52**, 2289.

Beige, H. (1980). *Acta Phys. Slov.*, *30*, 71.

Belyi, V. N. (1982). *Sov. Phys. Crystallogr.*, **27**, 516.

Bell, A. E. & A. D. Kaplin (1975). *Contemp. Phys.*, **16**, 375.

Ben Salem, M. & B. Yangui (1995). *Key Engg. Materials*, **101-102**, 61.

Bendersky, L. A., J. W. Cahn & D. Gratias (1989). *Phil. Mag.*, **60**, 837.

Benedeck, G., T. P. Martin & G. Pacchioni (Eds.) (1988). *Elemental and Molecular Crystals.* Springer-Verlag, Berlin.

Bennema, P. (1993). In D. T. J. Hurle (Ed.), *Handbook of Crystal Growth.* North-Holland, Amsterdam. Vol. 1, Chap. 7.

Bennema, P. & J. P. van der Eerden (1987). In I. Sunagawa (Ed.), *Mor-*

phology of Crystals, Part A. Terra Scientific, Tokyo.

Berkovsky, B. (1978) (Ed.). *Thermomechanics of Magnetic Fluids.* Hemisphere Publishing Corporation, Washington.

Bertagnolli, E., E. Kittinger & J. Tichy (1979). *J. Appl. Phys.*, **50**, 6267.

Bertaut, F. & F. Forrat (1956). *C. R. Acad. Sci.*, **242**, 382.

Bertaut, E. F. & M. Mercier (1971). *Mat. Res. Bull.*, **6**, 907.

Bessada, C., A. H. Fuchs, J. J. Pinvidic & H. Szwarc (1981). In J. Lascombe (Ed.), *Dynamics of Molecular Crystals.* Elevier, Amsterdam.

Besson, R. J., J. M. Groslambert & F. L. Walls (1985). In G. W. Taylor et al. (Eds.), *Piezoelectricity.* Gordon & Breach, U. K.

Bhagavantam, S. (1966). *Crystal Symmetry and Physical Properties.* Academic Press, London.

Bhagwat, K. V., R. Subramanian & V. K. Wadhawan (1983). *Phase Transitions*, **4**, 19.

Bhagwat, K. V., V. K. Wadhawan & R. Subramanian (1986). *J. Phys. C: Solid State Phys.*, **19**, 345.

Bhat, H. L. (1985). *Prog. Cryst. Growth & Charac.*, **11**, 57.

Bhattacharya, S., S. R. Nagel, L. Fleishman & S. Susman (1982). *Phys. Rev. Lett.*, **48**, 1267.

Bialas, H. & G. Schauer (1982). *Phys. Stat. Solidi (a)*, **72**, 679.

Bianchi, U, J. Dec, W. Kleemann & J. G. Bednorz (1995). *Phys. Rev. B*, **51**, 8737.

Binder, K. (1979). In T. Riste (Ed.), *Ordering in Strongly Fluctuating Condensed Matter Systems.* Plenum Press, New York.

Binder, K. (1980). In E. G. D. Cohen (Ed.), *Fundamental Problems in Statistical Mechanics.* North-Holland, Amsterdam.

Binder, K. & J. D. Reger (1992). *Adv. Phys.*, **41**, 547.

Binder, K. & A. P. Young (1986). *Rev. Mod. Phys.*, **58**, 801.

Birge, N. O., Y. H. Jeong, S. R. Nagel, S. Bhattacharya & S. Susman (1984). *Phys. Rev. B*, **30**, 2306.

Birman, J. L. (1966). *Phys. Rev. Lett.*, **17**, 1216.

Birman, J. L. (1978). In P. Kramer & A. Rieckers (Eds.), *Group Theoretical Methods in Physics.* Springer-Verlag, Berlin.

Birman, J. L. (1982). *Physica*, **114A**, 564.

Birringer, R., U. Herr & H. Gleiter (1986). *Suppl. Trans. Jpn. Inst. Metals*, **27**, 43.

Birss, R. R. (1964). *Symmetry and Magnetism*. North-Holland, Amsterdam.

Blinc, R. & B. Zeks (1974). *Soft Modes in Ferroelectrics and Antiferroelectrics*. North-Holland, Amsterdam.

Blinc, R. & A. P. Levanyuk (Eds.) (1986). *Incommensurate Phases in Dielectrics*. North-Holland, Amsterdam.

Bobeck, A. H. (1967). *Bell Syst. Tech J.*, **46**, 1901.

Boccara, N. (1968). *Annals of Phys.*, **47**, 40.

Boccara, N. (Ed.) (1981). *Symmetries and Broken Symmetries in Condensed Matter Physics*. IDSET, Paris.

Bollmann, W. (1970). *Crystal Defects and Crystalline Interfaces*. Springer-Verlag, Berlin.

Bollmann, W. (1982). *Crystal Lattices, Interfaces, Matrices: An Extension of Crystallography*. Published by the author himself (Geneva).

Bonin, K. D. & V. V. Kresin (1997). *Electric-Dipole Polarizabilities of Atoms, Molecules and Clusters*. World Scientific, Singopore.

Borchardt, H. J. & P. E. Bierstedt (1966). *Appl. Phys. Lett.*, **8**, 50.

Borchardt, H. J. & P. E. Bierstedt (1967). *J. Appl. Phys.*, **38**, 2057.

Bornarel, J. (1972). *J. Appl. Phys.*, **43**, 845.

Bornarel, J. & J. Lajzerowicz (1972). *Ferroelectrics*, **4**, 177.

Borovik-Romanov, A. S. (1959). *JETP*, **36**, 1954.

Borovik-Romanov, A. S. (1960). *JETP*, **38**, 1088.

Borovik-Romanov, A. S., G. G. Aleksanjan & E. G. Rudashevsky (1962). *Proc. Int. Conf. on Magnetism and Crystallography*, Kyoto.

Boulesteix, C. (1983). In K. Gschneider & L. Eyring (Eds.), *Handbook on the Physics and Chemistry of Rare Earths*, Vol. 5, Chap. 44. North-Holland, Amsterdam.

Boulesteix, C. (1984). *Phys. Stat. Solidi* (a), **86**, 11.

Boulesteix, C. (Ed.) (1992). *Diffusionless Phase Transitions and Related Structures in Oxides*. Trans Tech Publications, Zurich.

Boulesteix, C. (Ed.) (1995). *Diffusionless Phase Transitions in Oxides.* Trans Tech Publications, Zurich.

Boulesteix, C., M. Ben Salem, B. Yangui, Z. Kang & L. Eyring (1988). *Phys. Stat. Solidi* (a), **107**, 469.

Boulesteix, C., B. Yangui, M. Ben Salem, C. Manolikas & S. Amelinckx (1986). *J. Physique*, **47**, 461.

Bowden, M. L., J. P. Fanucci & S. C. Nolet (1989). *Proc. SPIE (Fiber Optic Smart Structures and Skins II)*, **1170**, 180.

Bowles, J. S. & J. K. Mackenzie (1954). *Acta Metall.*, **2**, 129.

Boyko, V. S., R. I. Garber & A. M. Kossevich (1994). *Reversible Crystal Plasticity.* American Institute of Physics, New York.

Bracke, L. P. M. & R. G. van Vliet (1981). *Int. J. Electronics*, **51**, 255.

Bradley, C. J. & A. P. Cracknell (1972). *The Mathematical Theory of Symmetry in Solids.* Clarendon Press, Oxford.

Bragg, W. L. & E. J. Williams (1934). *Proc. Roy. Soc. London A*, **145**, 699.

Bragg, W. L. & E. J. Williams (1935a). *Proc. Roy. Soc. London A*, **151**, 540.

Bragg, W. L. & E. J. Williams (1935b). *Proc. Roy. Soc. London A*, **152**, 231.

Brandmuller, J. (1986). *Comp. & Maths. with Appls.*, **12B**, 97.

Bratkovsky, A. M., E. K. H. Salje, S. C. Marais & V. Heine (1995). *Phase Transitions*, **55**, 79.

Bratkovsky, A. M., V. Heine & E. K. H. Salje (1996). *Phil. Trans. R. Soc. Lond.*, **354**, 2875.

Brewster, D. (1824). *Edinbg. J. Sci.*, **1**, 208.

Brice, J. C. (1985). *Rev. Mod. Phys.*, **57**, 105.

Brixner, L. H., P. E. Biersted, W. F. Jaep & J. R. Barkley (1973). *Mater. Res. Bull.*, **8**, 497.

Brixner, L. H., J. F. Whitney, F. C. Zumsteg & G. A. Jones (1977). *Mater. Res. Bull.*, **12**, 17.

Brockman, F. G., van der Heide & M. W. Louwerse (1969). *Philips Tech. Reports*, 30, 323.

Brody, E. M. & H. Z. Cummins (1968). *Phys. Rev. Lett.*, **21**, 1263.

Brody, P. S. (1983). *Ferroelectrics*, **50**, 27.

Brophy, J. H., R. M. Rose & J. Wulff (1965). *The Structure and Properties of Materials. Vol. II: Thermodynamics of Structure.* Wiley, New York.

Brout, R. (1965). *Phys. Rev. Lett.*, *14*, 176.

Brout, R. & H. Thomas (1967). *Physics*, **3**, 317.

Brown, M. E. & M. D. Holingsworth (1995). *Nature*, **376**, 323.

Brown, W. F. (1962). *Magnetostatic Principles in Ferromagnetism.* North-Holland, Amsterdam.

Bruce, A. D. (1976). *Ferroelectrics*, **12**, 21.

Bruce, A. D., R. A. Cowley & A. F. Murray (1978). *J. Phys. C: Solid State Phys.*, **11**, 3591.

Bulou, A., M. Rousseau & J. Nouet (1992). *Key Engg. Materials*, **68**, 133.

Bunget, I. & M. Popescu (1984). *Physics of Dielectrics.* Elsevier, Amsterdam.

Burfoot, J. C. & G. W. Taylor (1979). *Polar Dielectrics and Their Applications.* Macmillan Press, London.

Burkhard, H. & K. A. Muller (1976). *Helv. Phys. Acta*, **49**, 725.

Burns, G. (1985). *Phase Transitions*, **5**, 261.

Burns, G. & F. H. Dacol (1981). *Ferroelectrics*, **37**, 661.

Burns, G. & F. H. Dacol (1983). *Phys. Rev. B*, **28**, 2527.

Burns, G. & A. M. Glazer (1990). *Space Groups for Solid State Scientists*, second edition. Academic Press, London.

Bursill, L. A. (1997). *Ferroelectrics*, **191**, 129.

Bursill, L. A. & Peng Ju Lin (1992). *Key Engg. Materials*, **66-67**, 421.

Bursill, L. A., B. Jiang, J. L. Peng, T. L. Ren, W. L. Zhong & P. L. Zhang (1997). *Ferroelectrics*, **191**, 281.

Burton, W. K., N. Cabrera & F. C. Frank (1951). *Philos. Trans. R. Soc. London*, **A243**, 299.

Busch, G. (1938). *Helv. Phys. Acta*, **11**, 269.

Busch, G. & P. Scherrer (1938). *Naturwissenschaft*, **23**, 737.

Busch, G. (1991). *Condensed Matter News*, **1**, 20.

Byer, R. L. (1992). In S. Miyata, *Nonlinear Optics.* Elsevier, Amsterdam.

Cady, W. G. (1946). *Piezoelectricity.* McGraw-Hill, New York.

Cahn, J. W. (1977). *Acta Met.*, **25**, 721.

Cahn, J. W. & J. E. Hilliard (1958). *J. Chem Phys.*, **28**, 258.

Cahn, J. W. & J. E. Hilliard (1959). *J. Chem Phys.*, **31**, 688.

Cahn, J. W. & G. Kalonji (1981). In *Proc. Int. Conf. on Solid-Solid Transformations*, Carnegie-Melon University, Pittsburgh, Aug. 10-14, 1981. Published by the Metals Society of the AIME, USA.

Cahn, R. W. (1954). *Adv. Phys.*, **3**, 363.

Campbell, I. A. & S. Senoussi (1992). *Phil. Mag.*, **65**, 1267.

Camras, M. (1988). *Magnetic Recording Handbook.* Van Nostrand, New York.

Canella, V. & J. A. Mydosh (1972). *Phys. Rev. B*, **6**, 4220.

Canella, V., J. A. Mydosh & J. I. Budnick (1971). *J. Appl. Phys.*, **42**, 1689.

Carmesin, H. O. & K. Binder (1988). *J. Phys. A*, **21**, 4053.

Cao, W. & G. R. Barsch (1990). *Phys. Rev. B*, **41**, 4334.

Carl, K. & K. Geisen (1973). *Proc. IEEE*, **61**, 967.

Carpenter, M. A., E. K. H. Salje, A. Graeme-Barber, B. Wruck, M. T. Dove & K. S. Knight (1998). *American Mineralogist*, **83**, 2.

Castellanos-Guzman, A. G., J. Campa-Molina & J. Reyes-Gomez (1997). *Ferroelectrics*, **190**, 1.

Chaikin, P. M. & T. C. Lubensky (1995). *Principles of Condensed Matter Physics.* Cambridge University Press, Cambridge.

Chakravorty, S. & C. M. Wayman (1976). *Met. trans.*, **A7**, 555 and 569.

Chamberlin, R. V. (1993). *Phys. Rev. B*, **48**, 15638.

Chamberlin, R. V. (1994). *J. Appl. Phys.*, **76**, 6401.

Chamberlin, R. V. (1996). *Europhys. Lett.*, **33**, 545.

Chamberlin, R. V. (1998). *Phase Transitions*, **65**, 169.

Chamberlin, R. V. & D. N. Haines (1990). *Phys. Rev. Lett.*, **65**, 2197.

Chandler, D. (1987). *Introduction to Modern Statistical Mechanics.* Oxford University Press, New York.

Chaudhry, Z. & C. Rogers (1995). In E. Udd (Ed.), *Fiber Optic Smart Structures.* Wiley, New York.

Chavan, S. A., J. V. Yakhmi & I. K. Gopalakrishnan (1995). *Mater. Sci. & Engg.*, **C3**, 175.

Chelkowski, A. (1980). *Dielectric Physics.* Elsevier, Amsterdam.

Chen, Chinh-Wen (1986). *Magnetism and Metallurgy of Soft Magnetic Materials.* Dover Publishers, New York.

Chen, J., H. M. Chan & M. P. Harmer (1989). *J. Amer. Ceram. Soc.*, **72**, 593.

Chen, T. P., F. R. Chen, Y. C. Chuang, Y. D. Guo, J. G. Peng, T. S. Huang & L. J. Chen (1992). *J. Cryst. Growth*, **118**, 109.

Chen, L.-Q., Y. Wang & A. G. Khachaturyan (1992). *Phil. Mag. Lett.*, **65**, 15.

Cheng, Z.-Y., R. S. Katiyar, X. Yao & A. Guo (1997). *Phys. Rev. B*, **55**, 8165.

Cheng, Z.-Y., R. S. Katiyar, X. Yao & A. S. Bhalla (1998). *Phys. Rev. B*, **57**, 8166.

Chern, Mao-Jin & R. A. Phillips (1972). *J. Appl. Phys.*, **43**, 496.

Chernov, A. A. (1984). *Crystal Growth.* Springer-Verlag, Berlin.

Chernov, A. A. (1989). *Contemp. Phys.*, **30**, 251.

Chernov, A. A. (1998). *Acta Cryst.*, A**54**, 859.

Chikazumi, S. (1991). In Y. Ishikawa & N. Miura (Eds.), *Physics and Engineering Applications of Magnetism.* Springer-Verlag, Berlin.

Chittipedi, S., K. R. Cromak, J. S. Miller & A. J. Epstein (1987). *Phys. Rev. Lett.*, **58**, 2695.

Chou, Tsu-Wei, R. L. McCullough & R. B. Pipes (1986). *Sci. Amer.*, **255**, 166.

Chowdhury, D. (1986). *Spin Glasses and Other Frustrated Systems.* Princeton University Press, Princeton.

Chynoweth, A. G. (1956). *J. Appl. Phys.*, **27**, 78.

Clark, A. E. (1980). In E. P. Wohlfarth (Ed.), *Ferromagnetic Materials*, Vol. 1. North-Holland, Amsterdam.

Clarke, R. & A. M. Glazer (1974). *J. Phys. C: Solid State Phys.*, **7**, 2147.

Clarke, R. & A. M. Glazer (1976). *Ferroelectrics*, **14**, 695.

Claus, R. O. (Ed.) (1991). *Proc. Conf. on Optical Fiber Sensor-Based Smart materials and Structures*, April 3-4, 1991, Blacksburg, Virginia.

Technomic Pub. Co., Lancaster.

Cochran, W. (1959). *Phys. Rev. Lett.*, **3**, 412.

Cochran, W. (1960). *Adv. Phys.*, **9**, 387.

Cochran, W. (1973). *The Dynamics of Atoms in Crystals.* Arnold, London.

Coey, J. M. D. (1971). *Phys. Rev. Lett.*, **27**, 1140.

Coey, J. M. D. (1978). *J. Appl. Phys.*, **49**(3), 1646.

Coffey, W. T., Yu. P. Kalmykov & J. T. Waldron (1996). *The Langevin Equation.* World Scientific, Singapore.

Coghlan, A. (1992). *New Scientist*, 4 July 1992, page 27.

Collins, M. A., A. Blumen, J. F. Currie & J. Ross (1979). *Phys. Rev. B*, **19**, 3630.

Comes, R., M. Lambert & A. Guinier (1968). *Solid State Comm.*, **6**, 715.

Comes, R., M. Lambert & A. Guinier (1970). *J. Phys. Soc. Japan*, **28**, Suppl., 195.

Corker, D. L., A. M. Glazer, R. W. Whatmore, A. Stallard & F. Fauth (1998). *J. Phys.: Condens. Matter*, **10**, 6251.

Cotton, F. A. (1971). *Chemical Applications of Group Theory*, second edition. Wiley, New York.

Countryman, D. R., J. M. Carney & J. L. Welsh (1969). In A. G. H. Deitz (Ed.), *Composite Engineering Materials.* M.I.T. Press, Cambridge, Massachusetts.

Cowley, R. A. (1962). *Phys. Rev. Lett.*, **9**, 159.

Cowley, R. A. (1976). Phys. Rev. B, **13**, 4877.

Cracknell, A. P. (1972). *Acta Cryst.*, **A28**, 597.

Cracknell, A. P. (1974). *Adv. Phys.*, **23**, 673.

Cross, L. E. (1987). *Ferroelectrics*, **76**, 241.

Cross, L. E. (1993). *Ferroelectric Ceramics, Tutorial Reviews*: Theory, Processing and Applications. Birkhauser, Basel.

Cross, L. E. (1994). *Ferroelectrics*, **151**, 305.

Cross, L. E. (1995). *Japanese J. Appl. Phys.*, **34**, 2525.

Cross, L. E. (1996). *Ceramic Transactions*, **68**, 15.

Cross, L. E., A. Fouskova & S. E. Cummins (1968). *Phys. Rev. Lett.*, **21**, 812.

Cross, L. E. & K. H. Hardtl (1980). *Encyclopedia of Chemical Technology*, **10**, 1.

Cross, L. E., S. J. Jang, R. E. Newnham, S. Nomura & K. Uchino (1980). *Ferroelectrics*, **23**, 187.

Cross, L. E. & R. E. Newnham (1974). *Ferroic Crystals for Electrooptic and Acoustooptic Applications.* Penn State University, Pennsylvania. Appendices VII and VIII.

Cross, L. E. & S. Trolier-McKinstry (1997). Thin-film integrated ferroelectrics. In *Encyclopedia of Applied Physics*, Vol. 21. Wiley-VCH Verlag, GmBH.

Cuevas, A. Gomez, J. M. Perez Mato, M. J. Tello, G. Madariaga, J. Fernandez, Lopez Echarri, F. J. Zuniga & G. Chapuis (1984). *Phys. Rev. B*, **29**, 2655.

Cullity, B. D. (1972). *Introduction to Magnetic Materials.* Addison-Wesley, Reading.

Culshaw, B., P. T. Gardiner & A. McDonach (Eds.) (1992). *First European Conference on Smart Structures and Materials.* IOP Publishing, Bristol. (SPIE Vol. 1777.)

Cummins, S. E. (1970). *Ferroelectrics*, **1**, 11.

Curie, J. (1889). *Ann. Chim. Phys.*, **6**, 244.

Curie, J. & P. Curie (1980a). *C. R. Acad. Sci. Paris*, **91**, 294.

Curie, J. & P. Curie (1980b). *C. R. Acad. Sci. Paris*, **91**, 383.

Curie, P. (1884). *Soc. Mineralog. France Bull., Paris*, **7**, 418.

Curie, P. (1894a). *J. de Physique*, **3**, 393.

Curie, P. (1894b). *J. de Physique*, **5**, 289.

Dattagupta, S. (1981). *Bull. Mater. Sci.*, **3**, 133.

Dattagupta, S. (1987). *Relaxation Phenomena in Condensed Matter Physics.* Academic Press, New York.

Dattagupta, S., V. Heine, S. Marais & E. K. H. Salje (1991a). *J. Phys.: Condens. Matter*, **3**, 2963.

Dattagupta, S., V. Heine, S. Marais & E. K. H. Salje (1991b). *J. Phys.: Condens. Matter*, **3**, 2975.

Dattagupta, S. & R. Ranganathan (1984). *J. Phys. F: Metal Phys.*, **14**, 1417.

David, W. I. F. (1983). *J. Phys. C: Solid State Phys.*, **16**, 5093.

David, W. I. F., A. M. Glazer & A. W. Hewat (1979). *Phase Transitions*, **1**, 155.

David, W. I. F. & I. G. Wood (1983). *J. Phys. C: Solid State Phys.*, **16**, 5149.

Davidson, R. (1992). *Materials & Design*, **13**, 87.

Debye, P. (1945). *Polar Molecules*. Dover, New York.

Dec, J. (1993). *Phase Transitions*, **45**, 35.

Dec, J. & J. Kwapulinski (1989). *Phase Transitions*, **18**, 1.

Dekker, A. J. (1957). *Solid State Physics*. Prentice Hall, New York.

Deonarine, S. & J. L. Birman (1983). *Phys. Rev. B*, **27**, 4261.

Devarajan, V. & A. M. Glazer (1986). *Acta Cryst.*, **A42**, 560.

Devonshire, A. F. (1949). *Phil. Mag.*, **40**, 1040.

Devonshire, A. F. (1951). *Phil. Mag.*, **42**, 1065.

Devonshire, A. F. (1954). *Adv. Phys.*, **3**, 85.

Diep, H. T. (Ed.) (1994). *Magnetic Systems with Competing Interactions (Frustrated Spin Systems)*. World Scientific, Singapore.

Dillon, J. F. (1957). *Phys. Rev.*, **105**, 759.

Dimitrakopulos, G. P. & Th. Karakostas (1996). *Acta Cryst.*, **A52**, 62.

Dixon, R. W. (1967). *J. Appl. Phys.*, **38**, 5149.

Dmitriev, V. & P. Toledano (1994). *Phase Transitions*, **49**, 57.

Dolino, G. (1988). In S. G. Ghose, J. M. D. Coey & E. K. H. Salje (Eds.), *Structural and Magnetic Phase Transitions in Minerals*. Springer- Verlag, Berlin.

Dolino, G. & J. P. Bachheimer (1985). In G. W. Taylor et al. (Eds.), *Piezoelectricity*. Gordon & Breach, U. K.

Dolino, G & P. Bastie (1995). *Key Engg. Mater.*, **101-102**, 285.

Doni, E. G., G. L. Bleris, Th. Karakostas, J. G. Antonopoulos & P. Delavignette (1985). *Acta Cryst.*, **A41**, 440.

Donnay, G. & J. D. H. Donnay (1974). *Canadian Mineralogist*, **12**, 422.

Durand. D., X. Scarvada Do Carmo, A. Anderson & F. Luty (1980). *Phys. Rev. B*, **22**, 4005.

Dorner, B., J. D. Axe & G. Shirane (1972). *Phys. Rev. B*, **6**, 1950.

Dove, M. T., A. P. Giddy & V. Heine (1992). *Ferroelectrics*, **136**, 33.

Dudnik, E. F. & G. A. Kiosse (1983). *Ferroelectrics*, **48**, 33.

Dudnik, E. F. & L. A. Shuvalov (1989). *Ferroelectrics*, **98**, 207.

Dvorak, V. (1970). *J. Phys. Soc. Japan*, **28**, 252.

Dvorak, V. (1971a). *Phys. Stat. Solidi*, **B45**, 147.

Dvorak, V. (1971b). *Phys. Stat. Solidi*, **B46**, 763.

Dvorak, V. (1974). *Ferroelectrics*, **7**, 1.

Dvorak, V. (1978). *Czech. J. Phys.*, **B28**, 989.

Dvorak, V., A. P. Levanyuk & D. G. Sannikov (1975). *Izv. Akad. Nauk*, **39**, 659.

Dzyaloshinskii, I. E. (1959). *Sov. Phys. JETP*, **37**, 881.

Dzyaloshinskii, I. E. (1964). *Sov. Phys. JETP*, **19**, 960.

Dzyaloshinskii, I. E. (1964a). *Zh. Eksp. Teor. Fiz.*, **46**, 1352.

Dzyaloshinskii, I. E. (1964b). *Zh. Eksp. Teor. Fiz.*, **47**, 336.

Dzyaloshinskii, I. E. (1964c). *Zh. Eksp. Teor. Fiz.*, **47**, 992.

Echt, O., K. Sattler & E. Recknagel (1981). *Phys. Rev. Lett.*, **47**, 1121.

Edwards, S. F. & P. W. Anderson (1975). *J. Phys. F: Metal Phys.*, **5**, 965.

Eshelby, J. D. (1956). *Solid State Physics*, **3**, 79.

Eshelby, J. D. (1961). In *Progress in Solid Mechanics*, Vol. 2, pp. 89-140. North-Holland, Amsterdam.

Einstein, A. (1910). *Annln Phys*, **33**, 1275.

Elliot, R. J. (1983). In M. Ausloos & R. J. Elliot (Eds.), *Magnetic Phase Transitions*. Springer-Verlag, Berlin.

Entel, P., K. Kadau, R. Meyer, H. C. Herper, M. Schroter & E. Hoffmann (1998). *Phase Transitions*, **65**, 79.

Enz, U. (1982). In E. P. Wohlfarth (Ed.), *Ferromagnetic Materials*, Vol. 3. North-Holland, Amsterdam.

Errandonea, G. (1980). *Phys. Rev. B*, **21**, 5221.

Evarestov, R. A. & V. P. Smirnov (1993). *Site Symmetry in Crystals. Theory and Applications*. Springer-Verlag, Berlin.

Fabry, J., V. Petricek, I. Cisarova & J. Kroupa (1997). *Acta Cryst. B*, **53**,

272.

Falk, F. (1980). *Acta Met.*, **28**, 1773.

Falk, F. (1982a). *J. de Physique*, **43**, C4-3.

Falk, F. (1982b). *J. de Physique*, **43**, C4-203.

Falk, F. (1983). *Z. Physik B*, **51**, 177.

Falk, F. (1984). *Z. Physik B*, **54**, 159.

Fatuzzo, E. & W. J. Merz (1959). *Phys. Rev.*, **116**, 61.

Fatuzzo, E. & W. J. Merz (1967). *Ferroelectricity*. North-Holland, Amsterdam.

Fayard, M., R. Portier & D. Gratias (1981). In N. Boccara (Ed.), *Symmetries and Broken Symmetries in Condensed Matter Physics*. IDSET, Paris.

Fejer, M. M. (1994). *Physics Today*, May issue, page 25.

Fichtner, K. (1986). *Comp. & Maths. with Appls.*, **12B**, 751.

Fieschi, R. & F. G. Fumi (1953). *Il Nuovo Cimento*, **10**, 865.

Fischer, K. H. (1983). *Phys. Stat. Solidi* (b), **116**, 357.

Fischer, K. H. & J. A. Hertz (1991). *Spin Glasses*. Cambridge University Press, Cambridge.

Fischmeister, H. F. (1985). *J. de Physique*, Colloque C4, **46**, 3.

Fisher, M. E. (1974). *Rev. MOd. Phys.*, **46**, 597.

Fisher, M. E. (1998). *Rev. Mod. Phys.*, **70**, 653.

Fleury, P. A. & J. M. Worlock (1968). *Phys. Rev. B*, **174**, 613.

Folk, R., H. Iro & F. Schwabl (1976a). *Phys. Lett.*, **57A**, 112.

Folk, R., H. Iro & F. Schwabl (1976b). *Z. Physik*, **B25**, 69.

Folk, R., H. Iro & F. Schwabl (1979). *Phys. Rev. B*, **20**, 1229.

Fousek, J. (1971). *Czech. J. Phys. B*, **21**, 955.

Fousek, J. (1992a). In M. Liu et al. (Eds.), ISAF92: *Proc. 8th IEEE International Symposium on Applications on Ferroelectrics*. Greenville, SC, USA, Aug. 30 - Sept. 2, 1992. IEEE Catalog No. 92CH3080-9.

Fousek, J. (1992b). In N. Setter (Ed.), *Ferroelectric Ceramics*, Monte Verita, Birkhauser Verlag, Basel.

Fousek, J., M. Glogarova & Kursten (1976). *Ferroelectrics*, **11**, 469.

Fousek, J. & V. Janovec (1969). *J. Appl. Phys.*, **40**, 135.

Freeman, A. J. & H. Schmid (1975) (Eds.). *Magnetoelectric Interaction Phenomena in Crystals.* Gordon & Breach, London.

Fridkin, V. M. (1979). *Photoferroelectrics.* Springer-Verlag, Berlin.

Fridkin, V. M., N. A. Korchagina, M. A. Kosonogov, R. M. Magomadov, A. I. Rodin, E. D. Rogach & K. A. Verchovskaya (1981). *Ferroelectrics*, **31**, 15.

Friedel, G. (1926). *Lecons de Cristallographie.* Berger-Levrault, Paris. Reprint published by Blanchard, Paris.

Friedel, J. (1981). In N. Boccara (Ed.), *Symmetries and Broken Symmetries in Condensed Matter Physics.* IDSET, Paris.

Friend, C. M. (1992). *Proc. SPIE (Smat Materials and Structures)*, **1777**, 181.

Frohlich, H. (1949). *Theory of Dielectrics.* Clarendon Press, Oxford.

Fujimura, M., T. Suhara & H. Nishihara (1999). *Bull. Mater. Sci.*, **22**, 413.

Fulcher, G. S. (1925). *J. Amer. Ceram. Soc.*, **8**, 339.

Fumi, F. G. (1952a). *Acta Cryst.*, **5**, 44.

Fumi, F. G. (1952b). *Phys. Rev.*, **86**, 561.

Fumi, F. G. (1952c). *Acta Cryst.*, **5**, 691.

Furuya, Y. & H. Shimada (1991). *Materials & Design*, **12**, 21.

Futama, H. & R. Pepinsky (1962a). *Bull. Amer. Phys. Soc.*, **7**, 177.

Futama, H. & R. Pepinsky (1962b). *J. Phys. Soc. Japan*, **17**, 725.

Haertling, G. H. (1971). *J. Amer. Ceram. Soc.*, **54**, 303.

Haertling, G. H. & C. E. Land (1971). *J. Amer. Ceram. Soc.*, **54**, 1.

Hayward, S. A. & E. K. H. Salje (1998). *J. Phys.: Condens. Matter*, **10**, 1421.

Gabay, M. & G. Toulouse (1981). *Phys. Rev. Lett.*, **47**, 201.

Gandhi, M. V., B. Thompson & S. B. Choi (1989). *J. Compoite Mate.*, **23**, 1232.

Gaugain, J. M. (1856a). *C. R. Acad. Sci. Paris*, **42**, 1264.

Gaugain, J. M. (1856b). *C. R. Acad. Sci. Paris*, **43**, 916.

Gehring, G. A. & K. A. Gehring (1975). *Rep. Prog. Phys.*, **38**, 1.

Gibbs, J.W. (1876). *The Scientific Papers (1961).* Dover, New York. Vol. 1.

Ginzburg, V. L. (1945). *Zh. Eksp. Teor. Fiz.*, **15**, 739.

Ginzburg, V. L. (1946). *J. Phys. U.S.S.R.*, **10**, 107.

Ginzburg, V. L. (1949). *Zh. Eksp. Teor. Fiz.*, **19**, 36.

Ginzburg, V. L. (1955). *Nuovo Cimento*, **2**, 1234.

Ginzburg, V. L. (1961). *Sov. Phys. Solid State*, **2**, 1824.

Ginzburg, V. L. & L. D. Landau (1958). *Zh. Eksp. Teor. Fiz.*, **20**, 1064.

Giordano, N. & W. P. Wolf (1977). *Phys. Rev. Lett.*, **39**, 342.

Givargizov, E. I. (1991). *Oriented Crystallization on Amorphous Substrates.* Plenum Press, New York.

Glass, A. M. & D. H. Anston (1972). *Opt. Commun.*, **5**, 45.

Glauber, R. J. (1963). *J. Math. Phys.*, **4**, 294.

Glazer, A. M. (1975). *Acta Cryst.*, **A31**, 756.

Glazer, A. M. (1988). In E. K. H. Salje (Ed.), *Physical Properties and Thermodynamic Behaviour of Minerals.* D. Reidel Publishing Co.

Glazer, A. M. & K. Stadnicka (1986). *J. Appl. Cryst.*, **19**, 108.

Glazer, A. M., K. Stadnicka & S. Singh (1981). *J. Phys. C: Solid State Phys.*, **14**, 5011.

Glazounov, A. E., A. K. Tagantsev & A. J. Bell (1996). *Phys. Rev. B*, **53**, 11281.

Globus, A. (1977). *J. de Physique* (Paris), **C1-38**, C1-1.

Godefroy, G. & B. Jannot (1992). *Key Engg. Materials*, **68**, 81.

Goldman, A. (1988). In L. M. Levinson (Ed.), *Electronic Ceramics.* Marcel Dekker, New York.

Goldman, A. (1990). *Modern Ferrite Technology.* Van Nostrand Reinhold, New York.

Goldrich, F. E. & J. L. Birman (1968). *Phys. Rev.*, **167**, 528.

Goldstone, J. (1961). *Nuovo Cim.*, **19**, 154.

Gordon, A. (1986). *Physica*, **138B**, 239.

Gordon, A. (1991). *Phys. Lett. A*, **154**, 79.

Gorsky, W. S. (1935). *Phys. Z. Sowjet.*, **8**, 443, 562.

Goss, N. P. (1935). *Trans. Amer. Soc. Metals*, **23**, 511.

Granicher, H. & K. A. Muller (1971). *Mat. Res. Bull.*, **6**, 977.

Gratias, D., R. Portier, M. Fayard & M. Guymont (1979). *Acta Cryst.*, **A35**, 885.

Gratias, D. & R. Portier (1982). *J. de Physique*, Colloque C6, **43**, 15.

Gratias, D. & A. Thalal (1988). *Phil. Mag. Lett.*, **57**, 63.

Griffiths, R. B. (1970). *Phys. Rev. Lett.*, **24**, 715.

Gross, D. J. (1995). *Physics Today*, **48**, 46, December issue.

Grossman, B. et al. (1989). *Proc. SPIE (Fiber Optic Smart Structures and Skins II)*, **1170**, 123.

B. G. Grossman & M. H. Thursby (1995). In E. Udd (Ed.), *Fiber Optic Smart Structures.* Wiley, New York.

Guenin, G. (1989). *Phase Transitions*, **14**, 165.

Guimaraes, D. M. C. (1979a). *Acta Cryst.*, **A35**, 108.

Gumlich, E. & P. Goerens (1912). *Trans. Faraday Soc.*, **8**, 98.

Gunton, D. J. & G. A. Saunders (1973). *Solid State Comm.*, **12**, 569.

Gupta, S. M. & D. Viehland (1997). *J. Amer. Ceram. Soc.*, **80**, 477.

Gupta, T. K., F. F. Lange & J. H. Bechtold (1978). *J. Mater. Sci.*, **13**, 1464.

Guymont, M. (1978). *Phys. Rev. B*, **18**, 5385.

Guymont, M. (1981). *Phys. Rev. B*, **24**, 2647.

Guymont, M. (1991). *Phase Transitions*, **34**, 135.

Guymont, M., D. Gratias, R. Portier & M. Fayard (1976). *Phys. Stat. Solidi* (a), **38**, 629.

Haas, C. (1965). *Phys. Rev.*, **140**, A863.

Haberland, H. (Ed.) (1994). *Clusters of Atoms and Molecules.* Springer-Verlag, Berlin.

Haertling, G. H. & C. E. Land (1971a). *J. Amer. Ceram. Soc.*, **54**, 1.

Haertling, G. H. & C. E. Land (1971b). *J. Amer. Ceram. Soc.*, **54**, 303.

Haertling, G. H. (1988). In L. M. Levinson (Ed.), *Electronic Ceramics.* Marcel Dekker, New York.

Hahn, T. (Ed.) (1992). *International Tables for Crystallography. Vol. A:*

Space-Group Symmetry. Kluwer Acad. Publishers, Dordrecht.

Hahn, T. & H. Wondratschek (1994). *Symmetry of Crystals: Introduction to International Tables for Crystallography Vol. A*. Heron Press, Sofia.

Hale, D. K. (1976). *J. Mater. Sci.*, **11**, 2105.

Hall, E. O. (1954). *Twinning and Diffusionless Transformations in Metals*. Butterworths, London.

Halperin, B. H. & C. M. Varma (1976). *Phys. Rev. B*, **14**, 4030.

Hamano, K., Y. Ikeda, T. Fujimoto, K. Ema & S. Hirotsu (1980). *J. Phys. Soc. Japan*, **49**, 2278.

Hammermesh, N. (1964). *Group Theory*. Addison-Wesley, Reading.

Haneda, K. & A. H. Morrish (1990). *Phase Transitions*, **24-26**, 661.

Hanson, J. R. (1995). In E. Udd (Ed.), *Fiber Optic Smart Structures*. Wiley, New York.

Hanumaiah, A., T. Bhimasankaram, S. V. Suryanarayana & G. S. Kumar (1994). *Bull. Mater. Sci.*, **17**, 405.

Hartman, P. (1987). In I. Sunagawa (Ed.), *Morphology of Crystals, Part A*. Terra Scientific, Tokyo.

Hatch, D. M. & W. Cao (1999). *Ferroelectrics*, **222**, 1.

Hatch, D. M. & S. Ghose (1991). *Phys. Chem. Minerals*, **17**, 554.

Hatch, D. M., S. Ghose & J. L. Bjorksam (1994). *Phys. Chem. Minerals*, **21**, 67.

Hatch, D. M., R. A. Hatt & H. T. Stokes (1997). *Ferroelectrics*, **191**, 29.

Hatch, D. M., P. Hu, A. Saxena & G. R. Barsch (1996). *Phys. Rev. Lett.*, **76**, 1288.

Hatch, D. M., J. S. Kim, H. T. Stokes & J. W. Felix (1986). *Phys. Rev. B*, **33**, 6196.

Hatch, D. M. & H. T. Stokes (1986). *Phase Transitions*, **7**, 87.

Hatch, D. M. & H. T. Stokes (1988). *Isotropy Subgroups of the 230 Crystallographic Space Groups*. World Scientific, Singapore.

Hatch, D. M., H. T. Stokes & R. M. Putnam (1987). *Phys. Rev. B*, **35**, 4935.

Hatch, D. M., H. T. Stokes, K. S. Aleksandrov & S. V. Misyul (1989). *Phys. Rev. B*, **39**, 9282.

Hatt, R. A. & D. M. Hatch (1999). *Ferroelectrics*, **226**, 61.

Hayashi, M. (1972). *J. Phys. Soc. Japan*, **33**, 739.

Hayden, H. W., W. G. Moffatt & J. Wulff (1965). *The Structure and Properties of Matter. Vol. III: Mechanical Behaviour*. Wiley, New York.

Hayward, S. A. & E. K. H. Salje (1996). *Am. Mineral.*, **81**, 1332.

Hayward, S. A. & E. K. H. Salje (1998). *J. Phys.: Condens. Matter*, **10**, 1421.

Heesch, H. (1929). *Z. Kristallogr.*, *71*, 95.

Heesch, H. (1930). *Z. Kristallogr.*, *73*, 725.

Heine, V. (1960). *Group Theory in Quantum Mechanics*. Pergamon Press, New York.

Heisenberg, W. (1928). *Z. Physik*, **49**, 619.

Helferty, J. J., D. Boussalis & S. J. Wang (1992). In G. J. Knowles (Ed.) (1992).

Herbert, J. M. (1982). *Ferroelectric Transducers and Sensors*. Gordon & Breach, New York.

Hermann, C. (1929). *Z. Kristallogr.*, **69**, 533.

Hermann, C. (1934). *Z. Kristallogr.*, **89**, 32.

Hermelbracht, K. & H. G. Unruh (1970). *Z. Angew. Phys.*, **28**, 285.

Herring, C. (1942). *J. Franklin Inst.*, **233**, 525. For a correction see R. J. Elliot (1954), *Phys. Rev.*, **96**, 280.

Hilczer, B. (1995). *Key Engg. Materials*, **101-102**, 95.

Hilpert, S. (1909). *Ber. Deutsch. Chem. Ges. Bd 2*, **42**, 2248.

Hirotsu, S. (1975). *J. Phys. C: Solid State Phys.*, **8**, L12.

Hoare, M. R. (1973). In I. Prigogine (Ed.), *Advances in Chemical Physics*. Wiley, New York. Vol. 40.

Hoare, M. R. & P. Pal (1972). *J. Cryst. Growth*, **17**, 77.

Hochli, U. T. & L. A. Boatner (1979). *Phys. Rev. B*, **20**, 266.

Hochli, U. T., K. Knorr & A. Loidl (1990). *Adv. Phys.*, **39**, 405.

Hochli, U. T., P. Kofel & M. Maglione (1985). *Phys. Rev. B*, **32**, 4546.

Hochli, U. T., H. E. Weibel & L. A. Boatner (1977). *Phys. Rev. Lett.*, **39**, 1158.

Hodenberg, R. V. & E. K. H. Salje (1977). *Mat. Res. Bull.*, **12**, 1029.

Hohenberg, P. C. & B. I. Halperin (1977). *Rev. Mod. Phys.*, **49**, 435.

Holt, D. B. & D. M. Wilcox (1971). *J. Cryst. Growth*, **9**, 193.

Horovitz, B., J. L. Murray & J. A. Krumhansl (1978). *Phys. Rev. B*, **13**, 3549.

Hosoya, M. (1977). *J. Phys. Soc. Japan*, **42**, 399.

Hu, Z. W., P. A. Thomas M. C. Gupta & W. P. Risk (1995). *Appl. Phys. Lett.*, **66**, 13.

Hu, Z. W., P. A. Thomas & J. Webjorn (1995). *J. Phys. D*, **28**, A189.

Hu, Z. W., P. A. Thomas & J. Webjorn (1996). *J. Appl. Cryst.*, **29**, 279.

Huang, K. (1987). *Statistical Mechanics.* Wiley, New York.

Huang, X. R., X. B. Hu, S. S. Jiang & D. Feng (1997). *Phys. Rev. B*, **55**, 5534.

Huang, X. R., S. S. Jiang, X. B. Hu, X. Y. Xu, W. Zeng, D. Feng & J. Y. Wang (1995). *Phys. Rev. B*, **52**, 9932.

Hurd, C. M. (1982). *Contemp. Phys.*, **23**, 469.

Husimi, K. & K. Kataoka (1960). *Rev. Sci. Instrum.*, **31**, 418.

Husson, E. & A. Morell (1992). *Key Engg. Materials*, **68**, 217.

Hyde, B. G., A. N. Bagsaw, S. Andersson & M. O'Keeffe (1974). *Ann. Rev. Mat. Sci.*, **4**, 43.

Iizumi, M. & K. Gesi (1977). *Solid State Comm.*, **22**, 37.

Imry, Y. & S.-K. Ma (1975). *Phys. Rev. Lett.*, **35**, 1399.

Indenbom, V. L. (1960a). *Sov. Phys. Cryst.*, **5**, 106.

Indenbom, V. L. (1960b). *Izv. Akad. Nauk SSSR, Ser. Fiz.*, **24**, 1180.

Inui, T., Y. Tanabe & Y. Onodera (1990). *Group Theory and Its Applications in Physics.* Springer-Verlag, Berlin.

Ishibashi, Y. & V. Dvorak (1978). *J. Phys. Soc. Japan*, **44**, 32.

Ishikawa, K. (1998). *Phys. Rev. B*, **37**, 5852.

Ishikawa, K., K. Yoshikawa & N. Okada (1988). *Phys. Rev. B*, **37**, 5852.

Ising, E. (1925). *Z. Physik*, **31**, 253.

Ito, H., C. Takyu & H. Inaba (1991). *Electron. Lett.*, **27**, 1221.

Iwasaki, H. & K. Sugi (1971). *Appl. Phys. Lett.*, **19**, 92.

Izyumov, Yu. A., V. M. Laptev & V. N. Syromyatnikov (1994). *Phase Transitions*, **49**, 1.

Izyumov, Yu. A. & V. N. Syromyatnikov (1990). *Phase Transitions and Crystal Symmetry*. Kluwer, Dordrecht.

Izyumov, Yu. A., V. M. Laptev & V. N. Syromyatnikov (1994). *Phase Transitions*, **49**, 1.

Jahn, H. A. (1949). *Acta Cryst.*, **2**, 33.

Jain, J. D. (1988). *IETE Technical Review (India)*, **5**, 351.

James, J. F. (1995). *A Student's Guide to Fourier Transforms, with Applications in Physics and Engineering*. Cambridge University Press.

Janner, A. & T. Janssen (1979). *Physica*, **99A**, 47.

Janovec, V. (1972). *Czech J. Phys.*, **B22**, 974.

Janovec, V. (1976). *Ferroelectrics*, **12**, 43.

Janovec, V. (1981). *Ferroelectrics*, **35**, 105.

Janovec, V. (1983). *Phys. Lett.*, **99A**, 384.

Janovec, V. & V. Dvorak (1986). *Ferroelectrics*, **66**, 169.

Janovec, V., V. Dvorak & J. Petzelt (1975). *Czech. J. Phys.*, **B25**, 1362.

Janovec, V., G. Godefroy & L. R. Godefroy (1984). *Ferroelectrics*, **53**, 333.

Janovec, V., L. Richterova & D. B. Litvin (1992). *Ferroelectrics*, **126**, 287.

Janovec, V., L. Richterova & D. B. Litvin (1993). *Ferroelectrics*, **140**, 95.

Janovec, V., W. Schranz, H. Warhanek & Z. Zikmund (1989). *Ferroelectrics*, **98**, 171.

Jaric, M. V. (1981). *Phys. Rev. B*, **23**, 3460.

Jaric, M. V. (1982a). *Physica*, **114A**, 550.

Jaric, M. V. (1982b). *Phys. Rev. B*, **25**, 2015.

Jaric, M. V. & J. L. Birman (1977). *Phys. Rev. B*, **16**, 2564.

Jerphagnon, J. & R. Chemla (1976). *J. Chem Phys.*, **65**, 1522.

Jerphagnon, J., D. Chemla & R. Bonneville (1978). *Adv. Phys.*, **27**, 609.

Jiang, Q. (1992). In G.J. Knowles (Ed.) (1992).

Jiang, B, J. L. Peng & L. A. Bursill (1998a). *Ferroelectrics*, **207**, 445.

Jiang, B, J. L. Peng & L. A. Bursill (1998b). *Ferroelectrics*, **207**, 587.

Jiles, D. (1991). *Introduction to Magnetism and Magnetic Materials.* Chapman & Hall, London.

Jona, F. & G. Shirane (1962). *Ferroelectric Crystals.* Pergamon Press, Oxford.

Jonscher, A. K. (1983). *Dielectric Relaxation in Solids.* Chelsea Dielectrics Press, London.

Joshi, A. W. (1982). *Elements of Group Theory for Physicists*, third edition. Wiley Eastern, New Delhi.

Juretschke, H. J. (1974). *Crystal Physics: Macroscopic Physics of Anisotropic Solids.* Benjamin, London.

Kadanoff, L. P. (1966). *Physics (NY)*, **2**, 263.

Kadanoff, L. P., W. Gotze, D. Hamblen, R. Hecht, E. A. S. Lewis, V. V. Palciauskas, M. Rayl, J. Swift, D. Aspens & J. Kane (1967). *Rev. Mod. Phys.*, **39**, 395.

Kaku, M. & J. Thompson (1997). *Beyond Einstein: The Cosmic Quest for the Theory of the Universe.* Oxford University Press, Oxford.

Kalonji, G. (1982). *Symmetry Principles in the Physics of Crystalline Interfaces.* Ph.D. thesis, M.I.T.

Kalonji, G. (1985). *J. de Physique*, Colloque C4, **46**, 249.

Kalonji, G. & J. W. Cahn (1982). *J. de Physique*, **43**, C6-25.

Kanamori, J. (1963). In G. T. Rado & H. Suhl (Eds.), *Magnetism*, Vol. 1. Academic Press, New York.

Kanter, I. & H. Sompolinsky (1986). *Phys. Rev. B*, **33**, 2073.

Karl, K. & K. H. Hardtl (1971). *Phys. Stat. Sol. (a)*, **8**, 87.

Kartha, S., T. Castan, J. A. Krumhansl & J. P. Sethna (1991). *Phys. Rev. Lett.*, **67**, 3630.

Kasuya, T. (1956). *Prog. Theor. Phys.*, **16**, 45.

Kawasaki, K. (1966). *Phys. Rev.*, **145**, 224.

Kay, H. F. & P. Vousden (1949). *Phil. Mag.*, **40**, 1019.

Kersten, O., A. Rost & G. Schmidt (1988). *Ferroelectrics*, **80**, 995.

Keve, E. T. & A. D. Annis (1973). *Ferroelectrics*, **5**, 77.

Keve, E. T. & K. L. Bye (1975). *J. Appl. Phys.*, **46**, 810.

Khachaturyan, A. G. (1967). *Sov. Phys. - Solid State*, **8**, 2163.

Khachaturyan, A. G. (1983). *Theory of Structural Transformations in Solids.* Wiley, New York.

Khachaturyan, A. G., S. M. Shapiro & S. Semenovskaya (1991). *Phys. Rev. B*, **43**, 10832.

Khachaturyan, A. G., S. Semenovskaya & Long-Qing Chen (1994). In M. H. Yoo & M. Wuttig (Eds.), *Twinning in Advanced Materials.* The Minerals, Metals & Materials Society, Pennsylvania.

Khintchine, A. I. (1934). *Math. Ann.*, **109**, 604.

Kim, J. S., H. T. Stokes & D. M. Hatch (1986). *Phys. Rev. B*, **33**, 6210.

Kimberling, C. H. (1972). *Amer. Math. Monthly*, **79**, 136.

Kimoto, K. & I. Nisida (1977). *J. Phys. Soc. Japan*, **42**, 2071.

Kimura, T., R. E. Newnham & L. E. Cross (1981). *Phase Transitions*, **2**, 113.

Kinase, W. & H. Takahashi (1957). *J. Phys. Soc. Japan*, **12**, 464.

Kind, R. & K. A. Muller (1976). *Commun. Phys.*, **1**, 223.

Kingon, A. (1999). *Nature*, **401**, 658.

Kitayoka, Y., K. Mizuuchi, T. Yokoyama, K. Yamamoto, K. Narumi & M. Kato (1999). *Bull. Mater. Sci.*, **22**, 405.

Kittel, C. (1946). *Phys. Rev.*, **70**, 965.

Kittel, C. (1949). *Rev. Mod. Phys.*, **21**, 541.

Kittel, C. (1951). *Phys. Rev.*, **82**, 729.

Kittel, C. (1966). *Introduction to Solid State Physics*, 3rd edition. Wiley, New York.

Klassen-Neklyudova, M. V. (1963). *Plasticity of Crystals.* Consultants Bureau, New York.

Klassen-Neklyudova, M. V. (1964). *Mechanical Twinning of Crystals.* Consultants Bureau, New York.

Kleemann, W. (1998). *Phase Transitions*, **65**, 141.

Kleemann, W., A. Albertini, R. V. Chamberlin & J. G. Bednorz (1997). *Europhys. Lett.*, **37**, 145.

Kleemann, W. & E. K. H. Salje (1998) (Eds.). *Non-Exponential Relaxation and Rate Behaviour.* Special issue of *Phase Transitions*, **65**, 1-290.

Klicker, K. A., J. V. Biggers & R. E. Newnham (1981). *J. Amer. Cer.*

Soc., **64**, 5.

Kneller, E. F. & F. E. Luborsky (1963). *J. Appl. Phys.*, **34**, 656.

Knowles, G. J. (Ed.) (1992). *Active Materials and Adaptive Structures.* IOP Publishing, Bristol.

Kobayashi, J. (1991). *Phase Transitions*, **36**, 95.

Kobayashi, J. (1992). *Condensed Matter News*, **1** (6), 17.

Kobayashi, J., T. Asahi, S. Takahashi & A. M. Glazer (1988). *J. Appl. Cryst.*, **21**, 479.

Kobayashi, J., Y. Enomoto & Y. Sato (1972). *Phys. Stat. Solidi* (b), **50**, 335.

Kobayashi, J. & Y. Uesu (1983). *J. Appl. Cryst.*, **16**, 204.

Kobayashi, J., Y. Uesu & H. Takehara (1983). *J. Appl. Cryst.*, **16**, 212.

Kobayashi, J. & N. Yamada (1962). *J. Phys. Soc. Japan*, **17**, 876.

Kohlrausch, R. (1854). *Pogg. Ann. Phys.*, **91**, 56.

Konak, C., V. Kopsky & F. Smutny (1978). *J. Phys. C: Solid State Phys.*, **11**, 2493.

Kopsky, V. (1976). *J. Magn. & Mag. Materials*, **3**, 201.

Kopsky, V. (1979a). *Acta Cryst.*, **A35**, 83.

Kopsky, V. (1979b). *Acta Cryst.*, **A35**, 95.

Kopsky, V. (1982). *Group Lattices, Subduction of Bases, and Fine Domain Structures for Magnetic Crystal Point Groups.* Academia, Prague.

Kopsky, V. & D. G. Sannikov (1977). *J. Phys. C: Solid State Phys.*, **10**, 4347.

Koptsik, V. A. (1968). *Sov. Phys. - Cryst.*, **12**, 667.

Koptsik, V. A. (1975). *Krist. and Tech.*, **10**, 231.

Koptsik, V. A. (1983). *J. Phys. C: Solid State Phys.*, **16**, 23.

Koster, G. F. (1957). In Seitz & Turnbull (Eds.), *Solid State Physics*, **5**, 174. Academic Press, New York.

Kotru, P. N. & K. K. Raina (1982). *Crystal Res. & Technol.*, **17**, 1077.

Kotru, P. N., S. K. Kachroo & K. K. Raina (1985). *Crystal Res. & Technol.*, **20**, 27.

Kovalev, O. V. (1993). *Representations of the Crystallographic Space Groups: Irreducible Representations, Induced Representations and Corepresen-*

tations, 2nd edition. Edited by H. T. Stokes & D. M. Hatch. Gordon & Breach, USA.

Kroner, E. (1964). *Phys. Kondens. Materie*, **2**, 262.

Krumhansl, J. A. (1989). DOE-LINE Workshop, Oct. 24-28, Berkely, CA.

Krumhansl, J. A. (1998). *Phase Transitions*, **65**, 109.

Kubo, R., M. Toda & N. Hashitsume (1985). *Statistical Physics II: Nonequilibrium Statistical Mechanics*, second edition. Springer-Verlag, Berlin.

Kumaraswamy, K. & N. Krishnamurthy (1980). *Acta Cryst.*, **A36**, 760.

Lahti, P. M. (Ed.) (1999). *Magnetic Properties of Organic Materials.* Marcel Dekker, New York.

Landau, D. P., B. E. Keen, B. Schneider & W. P. Wolf (1971). *Phys. Rev. B*, **3**, 2310.

Landau, L. D. (1937a). *Phys. Z. Sowjet*, **11**, 26 (in Russian). For an English translation, see D. ter Haar (Ed.), *Collected Papers of L. D. Landau.* Gordon & Breach, New York (1965).

Landau, L. D. (1937b). *JETP*, **7**, 19.

Landau, L. D. (1937c). *Phys. Z. Sowjet*, **11**, 545.

Landau, L. D. (1937d). *JETP*, **7**, 627.

Landau, L. D. & I. M. Khalatnikov (1954). *Dokl. Akad. Nauk SSSR*, **96**, 469.

Landau, L. D. & E. M. Lifshitz (1958). *Statistical Physics.* Pergamon Press, Oxford. Third edition published as Lifshitz & Pitaevsky (1980), Part 1.

Landau, L. D., E. M. Lifshitz & L. P. Pitaevsky (1984). *Electrodynamics of Continuous Media*, second edition. Pergamon Press, New York.

Landauer, R. (1957). *J. Appl. Phys.*, **28**, 227.

Lang, R., W. R. Datars & C. Calvo (1969). *Phys. Lett.*, **A30**, 340.

Lange, S. B. (1974). *Sourcebook of Pyroelectricity.* Gordon & Breach, London.

Laughner, J. W. (1982). Ph.D. thesis. The Pennsylvania State University.

Laughner, J. W., R. E. Newnham & L. E. Cross (1979). *Phys. Stat. Sol.* (a), **56**, K83.

Laughner, J. W., V. K. Wadhawan & R. E. Newnham (1981). *Ferroelectrics*, **36**, 439.

Laves, F. (1975). *Acta Cryst.*, **A31**, S9.

Lavrencic, B. B. & T. S. Shigenari (1973). *Solid State Comm.*, **13**, 1329.

Lebedev, N. I., A. P. Levanyuk & A. S. Sigov (1984). *Ferroelectrics*, **55**, 241.

LeCraw, R. C., E. G. Spencer & C. S. Porter (1958). *Phys. Rev.*, **110**, 1311.

Lederman, F. L., M. B. Salamon & L. W. Schaklette (1974). *Phys. Rev. B*, **9**, 2981.

Lee, M., A. Halliyal & R. E. Newnham (1988). *Ferroelectrics*, **87**, 71.

Lee, T. D. & C. N. Yang (1952). *Phys. Rev.*, **87**, 410.

Leeuw, F. H. de, R. van den Doel & U. Enz (1980). *Rep. Prog. Phys.*, **43**, 689.

Levanyuk, A. P. & D. G. Sannikov (1969). *Sov. Phys. JETP*, **28**, 134.

Levanyuk, A. P. & D. G. Sannikov (1970). *Fiz. Tverd. Tela*, **12**, 812.

Levanyuk, A. P. & D. G. Sannikov (1971a). *Sov. Phys. Solid State*, **12**, 2418.

Levanyuk, A. P. & D. G. Sannikov (1971b). *Sov. Phys. JETP*, **33**, 600.

Levanyuk, A. P. & D. G. Sannikov (1974). *Sov. Phys. Usp.*, **17**, 199.

Levanyuk, A. P. & D. G. Sannikov (1975). *Sov. Phys. Solid State*, **17**, 327.

Levanyuk, A. P. & D. G. Sannikov (1976). *Sov. Phys. Solid State*, **18**, 245.

Levanyuk, A. P. & A. S. Sigov (1988). *Defects and Structural Phase Transitions.* Gordon & Breach, New York.

Levin, K., C. M. Soukoulis & G. S. Grest (1979). *J. Appl. Phys.*, **50**, 1695.

Levinson, L. M. (Ed.) (1988). *Electronic Ceramics.* Marcel Dekker, New York.

Li, S., J. A. Eastman, R. E. Newnham & L. E. Cross (1996). *Jpn. J. Appl. Phys.*, **35**, L502.

Liakos, J. K. & G. A. Saunders (1982). *Phil. Mag. A*, **46**, 217.

Lieberman, D. S., M. A. Schmerling & R. S. Karz (1975). In J. Perkins (Ed.), *Shape-Memory Effects in Alloys.* Plenum Press, New York.

Lifshitz, E. M. (1941). *Sov. Phys.-JETP*, **11**, 255.

Lifshitz, R. (1997). *Rev. Mod. Phys.*, **69**, 1181.

Lifshitz, E. M. & L. P. Pitaevski (1980). *Statistical Physics*, third edition.

Pergamon Press, Oxford.

Lim, E. J., M. M. Fejer & R. L. Byer (1989). *Electron. Lett.*, **25**, 174.

Lindgard, P.-A. & O. G. Mouritsen (1990). *Phys. Rev. B*, **41**, 68.

Lines, M. E. & A. M. Glass (1977). *Principles and Applications of Ferroelectrics and Related Materials.* Clarendon Press, Oxford.

Lipschutz, S. (1981). *Theory and Problems of Set Theory and Related Topics.* Schaum's Outline Series, McGraw-Hill, Singapore.

Lipson, H. & I. Taylor (1958). *Fourier Transforms and X-Ray Diffraction.* G. Bell, London.

Litvin, D. B. (1982). *J. Math. Phys.*, **23**, 337.

Litvin, D. B. (1984). *Acta Cryst. A*, **40**, 255.

Litvin, D. B., V. Janovec & S. Y. Litvin (1994). *Ferroelectrics*, **162**, 275.

Litvin, D. B., J. N. Kotzev & J. L. Birman (1982). *Phys. Rev. B*, **26**, 6947.

Litvin, D. B., S. Y. Litvin & V. Janovec (1995). *Acta Cryst.*, **A51**, 524.

Litvin, D. B. & T. R. Wike (1991). *Character Tables and Compatibility Relations of the Eighty Layer Groups and Seventeen Plane Groups.* Plenum Press, New York.

Locherer, K. R., S. A. Hayward, P. J. Hirst, J. Chrosch, M. Yeadon, J. S. Abell & E. K. H. Salje (1996). *Phil. Trans. R. Soc. Lond.*, **354**, 2815.

Loidl, A. (1991). *Phase Transitions*, **34**, 225.

Lomont, J. S. (1964). *Applications of Finite Groups.* Academic Press, New York.

Ludwig, W. & C. Falter (1988). *Symmetries in Physics.* Springer-Verlag, Berlin.

Lydanne, R. H., R. G. Sachs & E. Teller (1941). *Phys. Rev.*, **59**, 673.

Lyubarskii, G. Ya. (1960). *The Application of Group Theory in Physics.* Pergamon Press, Oxford.

Lyuksyutov, I., A. G. Naumovets & V. Pokrovsky (1992). *Two-Dimensional Crystals.* Academic Press, New York.

Ma, S. K. (1976). *Modern Theory of Critical Phenomena.* Benjamin / Cummings, Reading, Massachusetts.

Machlup, S. & L. Onsager (1953a). *Phys. Rev.*, **91**, 1505.

Machlup, S. & L. Onsager (1953b). *Phys. Rev.*, **91**, 1512.

Mackay, A. L. (1962). *Acta Cryst.*, **15**, 916.

Maclean, B. J., B. F. Carpenter, J. L. Draper & M. S. Misra (1992). In G. J. Knowles (Ed.) (1992).

Maclean, B. J., G. J. Patterson & M. S. Misra (1990). In Ahmad et al. (Eds.).

Madhava, M. R. & G. A. Saunders (1976). *Solid State Comm.*, **19**, 791.

Maglione, M. & B. Jannot (1991). *Phase Transitions*, **33**, 23.

Magneli, A. (150). *Arkiv Kemi*, **1**, 513.

Makita, Y., A. Sawada & Y. Takagi (1976). *J. Phys. Soc. Japan*, **41**, 167.

Malcherek, T., E. K. H. Salje & H. Kroll (1997). *J. Phys.: Condens. Matter*, **9**, 8075.

Malozemoff, A. P. & J. C. Slonczewski (1979). *Physics of Magnetic Domain Walls in Bubble Materials.* Academic Press, New York.

Mannheim, P. D. (1986). *Comp. & Maths. with Appls.*, **12B**, 169.

Marais, S. & E. K. H. Salje (1991). *J. Phys.: Condens. Matter*, **3**, 3667.

Marais, S., E. K. H. Salje & V. Heine (1991). *Phys. Chem. Min.*, **18**, 180.

Maria, J.-P., W. Hackenberger & S. Trolier-McKinstry (1998). *J. Appl. Phys.*, **84**, 5147.

Mariathasan, J. W. E., L. W. Finger & R. m. Hazen (1985). *Acta Cryst. B*, **41**, 179.

Martin, T. P. (1988). In Benedeck et al. (Eds.), *Elemental and Molecular Clusters.* Springer-Verlag, Berlin.

Mase, G. E. (1970). *Theory and Problems of Continuum Mechanics.* McGraw-Hill, New York.

Mason, W. P. (1947). *Phys. Rev.*, **72**, 854.

Mason, W. P. (1949). *Piezoelectric Crystals.* Van Nostrand, New York.

Matthias, B. T. (1949). *Phys. Rev.*, **75**, 1771.

Matthias, B. T. & J. P. Remeika (1949). *Phys. Rev.*, **76**, 1886.

Mayo, M. J. (1993). *Materials & Design*, **14**, 323.

Meeks, S. W. & B. A. Auld (1985). *Appl. Phys. Lett.*, **47**, 102.

Meeks, S. W., B. A. Auld & R. E. Newnham (1985). *Jap. J. Appl. Phys.*, **24.2**, 658.

Megaw, H. D. (1945). *Nature*, **155**, 484.

Megaw, H. D. (1952). *Acta Cryst.*, **5**, 739.

Megaw, H. D. (1957). *Ferroelectricity in Crystals.* Methuen, London.

Megaw, H. D. (1973). *Crystal Structures, A Working Approach.* Saunders, New York.

Meitzler, A. H. & H. M. O'Bryan (1973). *Proc. IEEE*, **61**, 959.

Mermin, N. D. (1991). In D. P. DiVincenzo & P. Steinhardt (Eds.), *Quasicrystals: The State of the Art.* World Scientific, Singapore.

Merz, W. J. (1949). *Phys. Rev*, **76**, 1221.

Merz, W. J. (1954). *Phys. Rev B*, **95**, 690.

Metrat, G. (1980). *Ferroelectrics*, **26**, 801.

Michel, L. (1981). In N. Boccara (Ed.), *Symmetries and Broken Symmetries in Condensed matter Physics.* IDSET, Paris.

Miller, J. S., A. J. Epstein & W. M. Reiff (1988). *Science*, **240**, 40.

Miller, S. C. & W. F. Love (1967). *Tables of Irreducible Representations of Space Groups and Corepresentations of Magnetic Space Groups.* Pruet Press.

Miller, R. C. & G. Weinreich (1960). *Phys. Rev.*, **117**, 1460.

Milovsky, A. V. & O. V. Kononov (1985). *Mineralogy.* Mir Publishers, Moscow.

Minnigerode, B. (1884). *Nachr. Akad. Wiss. Gottingen, Math-phys. / Klasse Iia*, **184**, 195.

Mirman, R. (1999). *Point Groups, Space Groups, Crystals, Molecules.* World Scientific, Singapore.

Mishima, T. (1932). *Iron Age*, **130**, 346.

Mishra, S. K. & D. Pandey (1995). *J. Phys.: Condens. Matter*, **7**, 9287.

Mishra, S. K. & D. Pandey (1996). *Appl. Phys. Lett.*, **69**, 1707.

Mishra, S. K. & D. Pandey (1997). *Phil. Mag. B*, **76**, 227.

Mishra, S. K., D. Pandey & A. P. Singh (1996). *Appl. Phys. Lett.*, **69**, 1707.

Mishra, S. K., A. P. Singh & D. Pandey (1997). *Phil. Mag. B*, **76**, 213.

Mishra, S. R., H. S. Rawat & S. C. Mehendale (1997). *Appl. Phys. Lett.*, **71**(1), 46.

Miyake, S. & R. Ueda (1946). *J. Phys. Soc. Japan*, **1**, 32.

Miyake, S. & R. Ueda (1947). *J. Phys. Soc. Japan*, **2**, 93.

Mohamad, I. J., L. Zammit-Mangion, E. F. Lambson & G. A. Saunders (1982). *J. Phys. Chem. Solids*, **43**, 749.

Mohler, E. & R. Pitka (1974). *Solid State Comm.*, **14**, 791.

Momosaki, E. & S. Kogure (1982). *Ferroelectrics*, **40**, 203.

Mookerjee, A. (1979). *Pramana - J. Phys.*, **14**, 11.

Mookerjee, A. & S. B. Roy (1983). *J. Phys. F*, **13**, 1945.

Mookerjee, A. & S. B. Roy (1984). *J. Phys. F*, **14**, 2714.

Morf, R., T. Schneider & E. Stoll (1977). *Phys. Rev. B*, **16**, 462.

Moria, T. (1963). In G. T. Rado & H. Suhl (Eds.), *Magnetism*, Vol. 1. Academic Press, New York.

Morrish, A. H. & R. J. Pollard (1986). *Adv. Ceram.*, **16**, 393.

Mort la Bresque (1987). *Mosaic*, **18**, 3.

Mott, N. F. & E. A. Davis (1979). *Electronic Processes in Non-Crystalline Materials.* Oxford University Press, Oxford.

Moulson, A. J. & J. M. Herbert (1990). *Electroceramics: Materials, Properties, Applications.* Marcel & Dekker, New York.

Mueller, H. (1935). *Phys. Rev.*, **47**, 175.

Mueller, H. (1940a). *Phys. Rev.*, **57**, 829.

Mueller, H. (1940b). *Phys. Rev.*, **58**, 565.

Mueller, H. (1940c). *Phys. Rev.*, **58**, 805.

Mueller, H. (1940d). *Ann. N. Y. Acad. Sci.*, **40**, 321.

Muller, K. A. (1981). In K. A. Muller & H. Thomas (Eds.), *Phase Transitions - I.* Springer-Verlag, Berlin.

Mukamel, D. & M. V. Jaric (1984). *Phys. Rev. B*, **29**, 1465.

Multani, M., P. Ayyub, V. Palkar & P. Guptasarma (1990). *Phase Transitions*, **24-26**, 91.

Multani, M. S. & V. K. Wadhawan (Eds.) (1990). *Physics of Clusters and Nanophase Materials. Phase Transitions*, **24-26**, 1-834.

Multi, A. R. & D. B. Holt (1972). *J. Mater. Sci.*, **7**,694.

Mutaftschiev, B. (1993). In D. T. J. Hurle (Ed.), *Handbook of Crystal Growth.* Elsevier. Vol. 1, Chap. 4.

Mydosh, J. (1992). *Spin Glasses.* Taylor & Francis, London.

Mydosh, J. A. & G. J. Nieuwenhuys (1980). In E. P. Wohlfarth (Ed.), *Ferromagnetic Materials*, Vol. 1. North-Holland, Amsterdam.

Nagarajan, V. S. & K. J. Rao (1993). *J. Mater. Chem.*, **3**, 43.

Naish, V. E. (1963). *Izv. Akad. Nauk SSSR, Ser. Fiz.*, **27**, 1496.

Nakanishi, N. (1975). In J. Perkins (Ed.), *Shape Memory Effect in Alloys.* Plenum Press, New York.

Néel, L. (1948). *Ann. de Phys.*, **3**, 137.

Néel, L. (1949). *Ann. Geophys.*, **5**, 99.

Neelakanta, P. S. & D. F. de Groff (1994). *Neural Network Modeling: Statistical Mechanics and Cybernetic Perspectives.* CRC Press, London.

Nelmes, R. J., M. I. McMahon, R. O. Piltz & N. G. Wright (1991). *Ferroelectrics*, **124**, 355.

Nesbitt, E. A., H. J. Williams, J. H. Wernick & R. C. Sherwood (1961). *J. A. P.*, **32**, 342 S.

Newnham, R. E. (1974). *Amer. Mineralogist*, **59**, 906.

Newnham, R. E. (1985). *J. Mater. Edu.*, **7**, 605.

Newnham, R. E. (1986). *Ann. Rev. Mater. Sci.*, **16**, 47.

Newnham, R. E. (1988). *Crystallogr. Rev.*, **1**, 253.

Newnham, R. E. (1989). *Rep. Prog. Phys.*, **52**, 123.

Newnham, R. E. (1990). *Ferroelectrics*, **102**, 1.

Newnham, R. E. (1991). National Institute of Standards Special Publication 804: *Chemistry of Electronic Ceramic Materials.* Proceedings of International Conference, held in Jackson, WY, Aug. 17-22, 1990. Issued Jan. 1991.

Newnham, R. E. (1994). *Electroceramics II*, **3**, 1771. Deutsche Keramische Gesellschaft Koln, Germany.

Newnham, R. E. (1997). *MRS Bulletin.* May 1997.

Newnham, R. E. (1998). *Acta Cryst.*, A**54**, 729.

Newnham, R. E., L. J. Bowen, K. A. Klicker & L. E. Cross (1980). *Materials in Engg.*, **2**, 93.

Newnham, R. E. & L. E. Cross (1974a). *Mat. Res. Bull.*, **9**, 927.

Newnham, R. E. & L. E. Cross (1974b). *Mat. Res. Bull.*, **9**, 1021.

Newnham, R. E. & G. R. Ruschau (1991a). *Earth & Mineral Sci.*, **60**, 27.

Newnham, R. E. & G. R. Ruschau (1991b). *J. Amer. Ceram. Soc.*, **74**, 463.

Newnham, R. E. & D. P. Skinner (1976). *Mat. Res. Bull.*, **11**, 1273.

Newnham, R. E., D. P. Skinner & L. E. Cross (1978). *Mat. Res. Bull.*, **13**, 525.

Newnham, R. E. & S. Trolier-McKinstry (1990a). *J. Appl. Cryst.*, **23**, 447.

Newnham, R. E. & S. Trolier-McKinstry (1990b). *Ceramic Transactions*, **8**, 235.

Newnham, R. E., S. Trolier-McKinstry & J. R. Giniewicz (1993). *J. Mater. Edu.*, **15**, 189.

Newnham, R. E., S. Trolier-McKinstry & H. Ikawa (1990). *Mat. Res. Soc. Symp. Proc.*, **175**, 161.

Nielsen, J. W. (1976). *IEEE-MAG*, **12**, 327.

Noether, E. (1918). *Nachrichten Gesell. Wissenschaft. Gottingen*, **2**, 235.

Noheda, B. D. E. Cox, G. Shirane, J. A. Gonzalo, L. E. Cross & S.-E. Park (1999). *Appl. Phys. Lett.*, **74**, 2059.

Nomura, S. & K. Uchino (1985). In G. W. Taylor et al. (Eds.), *Piezoelectricity.* Gordon & Breach, U. K.

Nowick, A. S. (1995). *Crystal Properties via Group Theory.* Cambridge University Press, Cambridge.

Nowick, A. S. & W. R. Heller (1963). *Adv. Phys.*, **12**, 251.

Nowotny, J. (1992) (Ed.), *Electronic Ceramic Materials.* Trans Tech Publications, Switzerland.

Nye, J. F. (1976). *Physical Properties of Crystals.* Clarendon Press, Oxford. Corrected version of the first edition published in 1957.

Onsager, L. (1944). *Phys. Rev.*, **65**, 117.

Opechowski, W. (1974). *Int. J. Magn.*, **5**, 317.

Opechowski, W. (1977). In R. T. Sharp & B. Colman (Eds.), *Group Theoretical Methods in Physics.* Academic Press, New York.

Opechowski, W. & R. Guccione (1965). In G. T. Rado & H. Suhl (Eds.), *Magnetism.* Academic Press, New York. Vol. IIA, Chap. 3.

Opermann, R. & H. Thomas (1975). *Z. Physik B*, **22**, 387.

Overhauser, A. W. (1978). *Adv. Phys.*, **27**, 343.

Owen, W. S. (1975). In J. Perkins (Ed.), *Shape Memory Effects in Alloys.* Plenum Press, New York.

Palmer, R. G. (1982). *Adv. Phys.*, **31**, 669.

Pandey, D. (1995). *Key Engg. Materials*, **101-102**, 177.

Pandey, D., Neelam Singh & S. K. Mishra (1994). *Indian J. Pure & Appl. Phys.*, **32**, 616.

Paquet, D. & J. Jerphagnon (1980). *Phys. Rev. B*, **21**, 2962.

Parisi, G. (1979). *Phys. Rev. Lett.*, **43**, 1754.

Parisi, G. (1980a). *J. Phys. A*, **13**, 1101.

Parisi, G. (1980b). *J. Phys. A*, **13**, 1887.

Parisi, G. (1980c). *J. Phys. A*, **13**, L115.

Parisi, G. (1980d). *Phil. Mag. B*, **41**, 677.

Parisi, G. (1980e). *Phys. Rep.*, **67**, 97.

Parisi, G. (1983). *Phys. Rev. Lett.*, **50**, 1946.

Park, J.-H., B.-K. Kim, K.-H. Song & S. J. Park (1995). *Mater. Res. Bull.*, **30**, 435.

Park, Y., K. M. Knowles & K. Cho (1998). *J. Appl. Phys.*, **83**, 5702.

Parlinski, K., Z. Q. Li & Y. Kawazoe (1999). *Phase Transitions*, **67**, 681.

Park, B. H., B. S. Kang, S. D. Bu, T. W. Noh, J. Lee & W. Jo (1999). *Nature*, **401**, 682.

Patashinski, A. Z. & V. I. Pokrovski (1979). *Fluctuation Theory of Phase Transitions.* Pergamon Press, Oxford.

Pauthenet, R. (1956). *C. R. Acad. Sci.*, **242**, 1859.

Peercy, P. S. & I. J. Fritz (1974). *Phys. Rev. Lett.*, **32**, 466.

Peierls, R. E. (1955). *Quantum Theory of Solids.* Oxford University Press, London.

Peng, J. L. & L. A. Bursill (1983). *Phil. Mag. A*, **48**, 251.

Peng, J. L. & L. A. Bursill (1991). *Phase Transitions*, **34**, 171.

Pepinsky, R. (1962). *Bull. Amer. Phys. Soc.*, **7**, 177.

Pepinsky, R. (1963). *Proc. Symp. on Ferromagnetism & Ferroelectricity.* State Press, Leningrad.

Perelomova, N. V. & M. M. Tagieva (1983). *Problems in Crystal Physics, with Solutions.* Mir Publishers, Moscow.

Perkins, J. (1975) (Ed.), *Shape Memory Effects in Alloys.* Plenum Press, New York.

Pfeuty, P. & G. Toulouse (1977). *Introduction to the Renormalization Group and to Critical Phenomena.* Wiley, New York.

Pickart, S. J. (1960). *Bu.. Amer. Phys. Soc.*, **5**, 357.

Pietraszko, A., A. Waskovska, S. Olejnik & K. Lukaszewicz (1979). *Phase Transitions*, **1**, 99.

Pilgrim, S. M., R. E. Newnham & L. L. Rohling (1990). *Mat. Res. Bull.*, **22**, 677.

Pimpinelli, A. & J. Villain (1998). *Physics of Crystal Growth.* Cambridge University Press, Cambridge.

Pina, C. M., U. Becker, P. Risthaus, D. Bosbach & A. Putnis (1998). *Nature*, **395**, 483.

Pine, A. S. (1970). *Phys. Rev. B*, **2**, 2049.

Plischke, M. & B. Bergersen (1994). *Equilibrium Statistical Mechanics*, second edition. World Scientific, Singapore.

Pohanka, R. C. & P. L. Smith (1988). In L. M. Levinson (Ed.), *Electronic Ceramics.* Marcel Dekker, New York.

Pokharel, B. P., R. Ranjan, D. Pandey, V. Siruguri & S. K. Paranjpe (1999). *Appl. Phys. Lett.*, **74**, 756.

Pond, R. C. & W. Bollmann (1979). *Trans. Roy. Soc.*, **292**, 449.

Pond, R. C. & D. S. Vlachavas (1983). *Proc. R. Soc. Lond. A*, **386**, 95.

Pond, R. C. & P. E. Dibley (1990). *Colloque de Physique*, **51**, Colloque C1, supplement 1, C1.

Portier, R. & D. Gratias (1982). *J. de Physique*, **43**, C4-17.

Portigal, D. L. & E. Burstein (1968). *Phys. Rev.*, **170**, 673.

Potts, R. B. (1952). *Proc. Camb. Phil. Soc.*, **48**, 106.

Pouget, J. (1991). *Phase Transitions*, **34**, 105.

Privman, V. & M. E. Fisher (1983). *J. Stat. Phys.*, **33**, 385.

Prokhorov, A. M. & Yu. S. Kuz'minov (1990). *Ferroelectric Crystals for Laser Radiation Control.* Adam Hilger, Bristol.

Pumphrey, P. H. & K. M. Bowkett (1971). *Scripta Metall.*, **5**, 365.

Pumphrey, P. H. & K. M. Bowkett (1972). *Scripta Metall.*, **6**, 31.

Pytte, E. (1970). *Solid State Commun.*, **8**, 2101.

Pytte, E. (1972). *Phys. Rev. B*, **5**, 3758.

Qian, H. & L. A. Bursill (1996a). *Int. J. Mod. Phys. B*, **10**, 2007.

Qian, H. & L. A. Bursill (1996b). *Int. J. Mod. Phys. B*, **10**, 2027.

Raccah, P. M. & J. B. Goodenough (1967). *Phys. Rev.*, **155**, 932.

Ramachandran, G. N. & S. Ramaseshan (1961). In S. Flugge (Ed.), *Handbuch der Physik*, Springer-Verlag, Berlin, Vol. 25/1, pp. 1-217.

Raman, C. V. & T. M. K. Nedungadi (1940). *Nature*, **145**, 147.

Ramesh, R. (Ed.) (1997). *Thin Film Ferroelectric Materials and Devices.* Kluwer, Dordrecht.

Randal, C., D. Barber, R. Whatmore & P. Groves (1987). *Ferroelectrics*, **76**, 311.

Randall, C. A., A. D. Hilton, D. J. Barber & T. R. Shrout (1990). *J. Mater. Sci.*, **25**, 3461.

Ranganathan, S. (1966). *Acta Cryst.*, **21**, 197.

Ranganathan, S. & B. Roy (1992). In S. Ranganathan et al. (Eds.), *Interfaces: Structure and Properties.* Oxford & IBH Pub. Co., New Delhi.

Ranjan, R., N. Singh, D. Pandey, V. Siruguri, P. S. R. Krishna, S. K. Paranjpe & A. banerjee (1997). *Appl. Phys. Lett.*, **70**, 3221.

Ranjan, R., D. Pandey, V. Siruguri, P. S. R. Krishna & S. K. Paranjpe (1999). *J. Phys.: Condens. Matter*, **11**, 2233.

Rao, C. N. R. (1984). *Solid State Chemistry. Perspective Report Series* - 1, Indian Nat. Sci. Acad., New Delhi.

Rao, C. N. R. & K. J. Rao (1978). *Phase Transitions in Solids: An Approach to the Study of the Chemistry and Physics of Solids.* McGraw-Hill, New York.

Redfern, S. A. T. & P. F. Schofield (1996). *Phase Transitions*, **59**, 25.

Rehwald, W. (1973). *Adv. Phys.*, **22**, 721.

Renou, A. & M. Gillet (1981). *Surf. Sci.*, **106**, 27.

Riley, S. J. (1990). *Phase Transitions*, **24-26**, 271.

Rogers, C. A. (1992). *Proc. SPIE (Smart Materials and Structures)*, **177**,

163.

Rolov, B. N. (1965). *Sov. Phys. - Solid State*, **6**, 1676.

Rosen, J. (1995). *Symmetry in Science. An Introduction to the General Theory.* Springer-Verlag, Berlin.

Rosenweig, R. E. (1982). *Sci. Amer.*, October issue, p. 136.

Roth, W. L. (1960). *J. Appl. Phys.*, **31**, 2000.

Roytburd, R. L. (1971). *Sov. Phys. - Solid State*, **13**, 1523.

Roytburd, A. L. (1974). *Sov. Phys. - Usp.*, **17**, 326.

Roytburd, A. L. (1978). In H. Ehrenreich, F. Seitz & D. Turnbull (Eds.), *Solid State Phys.*, **33**, 317. Academic Press, New York.

Roytburd, A. L. (1983). *Izv. Akad. Nauk SSSR*Ser. Phys., **47**, 435.

Roytburd, A. L. (1993). *Phase Transitions*, **45**, 1.

Roytburd, A. L. & Y. Yu (1994). In M. H. Yoo & M. Wuttig (Eds.), *Twinning in Advanced Materials.* The Minrals, Metals & Materials Society, Pennsylvania.

Ruderman, M. A. & C. Kittel (1954). *Phys. Rev.*, **96**, 99.

Rytz, D., U. T. Hochli & H. Bilz (1980). *Phys. Rev. B*, **22**, 359.

Sada, T., M. Inoue & K. Uchino (1987). *J. Cer. Soc. Japan*, **5**, 545.

Safari, A., R. E. Newnham, L. E. Cross & W. A. Schulze (1982). *Ferroelectrics*, **41**, 197.

Sa-gong, G., A. Safari, S. J. Jang & R. E. Newnham (1985). *Ferroelectrics*, **45**, 131.

Sa-gong, G., A. Safari & R. E. Newnham (1985). *Proc. 6th IEEE Int. Symp. on Applns. of Ferroelectrics*, pp. 281-284.

Saint-Gregoire, P. (1995). *Key Engg. Materials*, **101-102**, 237.

Saito, Y. (1996). *Statistical Physics of Crystal Growth.* World Scientific, Singapore.

Saksena, B. D. (1940). *Proc. Indian Acad. Sci.*, **A12**, 93.

Salje, E. K. H. (1976). *Phys. Stat. Sol.* (a), **33**, K165.

Salje, E. K. H. (1988). *Phys. Chem. Mineral.*, **15**, 336.

Salje, E. K. H. (1990, 1993a). *Phase Transitions in Ferroelastic and Coelastic Crystals* (Student edition published in 1993). Cambridge University Press, Cambridge.

Salje, E. K. H. (1991). *Phase Transitions*, **34**, 25.

Salje, E. K. H. (1992). *Phys. Reports*, **215**, 49.

Salje, E. K. H. (1993b). *J. Phys.: Condens. Matter*, **5**, 4775.

Salje, E. K. H. (1994) (Ed.), *Mobile Domain Boundaries.* Special issue of *Phase Transitions*, **48**, 1-200.

Salje, E. K. H. (1995a) (Ed.), *Kinetics of Ferroelastic Phase Transitions.* Special issue of *Phase Transitions*, **55**, 1-244.

Salje, E. K. H. (1995b). *Eur. J. Mineral.*, **7**, 791.

Salje, E. K. H. (1999) (Ed.), *Strain Near Structural Phase Transitions.* Special issue of *Phase Transitions*, **67**, 539-808.

Salje, E. K. H., M. T. Dove, I. Tsatskis, K. Locherer & J. Chrosch (1999). *Phase Transitions*, **67**, 539.

Salje, E. K. H. & G. Hoppman (1976). *Mat. Res. Bull.*, **11**, 1545.

Salje, E. K. H., B. Wruck & H. Thomas (1991a). *Z. Phys. B*, **82**, 399.

Salje, E. K. H., B. Wruck & S. Marais (1991b). *Ferroelectrics*, **124**, 185.

Salje, E. K. H., U. Bismayer, B. Wruck & J. Hensler (1991c). *Phase Transitions*, **35**, 61.

Salvestrini, J. P., M. D. Fontana, M. Aillerie & Z. Czapla (1994). *Appl. Phys. Lett.*, **64**, 1920.

Samara, G. A. (1971). *Phys. Rev. Lett.*, **27**, 103.

Samara, G. A. (1974). *Ferroelectrics*, **7**, 221.

Samara, G. A. (1978). *Ferroelectrics*, **20**, 87.

Sandercock, J. R. (1972). *Solid State Comm.*, **11**, 729.

Sangwal, K. & R. Rodriguez-Clemente (1991). *Surface Morphology of Crystalline Solids.* Trans-Tech Pub., Zurich.

Sannikov, D. G. (1993). *Crystallogr. Rep.*, **38** (4), 577.

Sano, M., K. Ito & K. Nagata (1986). *Jap. J. Appl. Phys.*, **25**, 627.

Sapriel, J. (1975). *Phys. Rev. B*, **12**, 5128.

Sawada, A. (1990). *J. Phys. Soc. Japan*, **60**, 3593.

Sawada, A., Y. Ishibashi & Y. Takagi (1977). *J. Phys. Soc. Japan*, **43**, 195.

Sawada, A., Y. Makita & Y. Takagi (1976). *J. Phys. Soc. Japan*, **41**, 174.

Sawada, A., H. Matsumoto & K. Tanaka (1993). *Ferroelectrics*, **140**, 245.

Sawada, A. & T. Nakamura (1985). In G. W. Taylor et al. (Eds.), *Piezoelectricity*. Gordon & Breach, London.

Sawada, A., M. Udagawa & T. Nakamura (1977). *Phys. Rev. Lett.*, **39**, 829.

Sawada, A., Y. Watanabe & K. Tanaka (1997). *Ferroelectrics*, **191**, 141.

Sawaguchi, E., Y. Akishige & M. Kobayashi (1985). *J. Phys. Soc. Japan*, **54**, 480.

Sawaguchi, E., A. Kikuchi & Y. Kodera (1962). *J. Phys. Soc. Japan*, **17**, 1666.

Saxena, A., G. R. Barsch & D. M. Hatch (1994). *Phase Transitions*, **46**, 89.

Schetky, L. M. (1979). *Sci. Amer.*, **241**(5), 68.

Schetky, L. M. (1991). *Materials & Design*, **12**, 29.

Schlenker, J. L., G. V. Gibbs & M. B. Boisen (1978). *Acta Cryst. A*, **34**, 52.

Schmidt, G. (1990). *Phase Transitions*, **20**, 127.

Schmidt, G., H. Arndt, G. Borchhardt, J. V. Cieminski, T. Petzsche, K. Borman, A. Sternberg, A. Zirnite & V. A. Isupov (1981). *Phys. Stat. Solidi* (a), **63**, 501.

Schmidt, V. S., A. B. Wertern & A. G. Baker (1976). *Phys. Rev. Lett.*, **37**, 839.

Schmahl, W. W., A. Putnis, E. K. H. Salje, P. Freeman, A. Graeme-Barber, R. Jones, K. K. Singh, J. Blunt, P. P. Edwards, J. Lorem & K. Mirza (1989). *Phil. Mag. Lett*, **60**, 241.

Schneider, T., H. Beck & E. Stoll (1976). *Phys. Rev. B*, **13**, 1123.

Schottky & Wagner (1931). *Z. Phys. Chem. B*, **11**, 163.

Schranz, W. (1994). *Phase Transitions*, **51**, 1.

Schranz, W., P. Dolinar, A. Fuith & H. Warhanek (1991). *Phase Transitions*, **34**, 189.

Schranz, W., M. Fally & D. Havlik (1998). *Phase Transitions*, **65**, 27.

Schroeder, T. A. & C. M. Wayman (1977). *Acta Met.*, **25**, 1375.

Schwabl, F. (1980). *Ferroelectrics*, **24**, 171.

Schwartz, Mel M. (1997). *Composite Materials, Vol. 1: Properties, Non-destructive Testing, and Repair.* Prentice Hall, New Jersey.

Schweda, E. (1992). *Key. Engg. Materials*, **68**, 187.

Scott, J. F. (1998). *Ferroelectric Reviews*, **1**, 1.

Semenovskaya, S. & A. G. Khachaturyan (1991). *Phys. Rev. Lett.*, **67**, 2223.

Semenovskaya, S. & A. G. Khachaturyan (1992). *Phys. Rev. B*, **46**, 6511.

Senechal, M. (1986). *Comp. & Maths. with Appls.*, **12B**, 565.

Senechal, M. (1990). *Crystalline Symmetries: An Informal Mathematical Introduction.* Adam Hilger, Bristol.

Seitz, F. (1936). *Ann. Math. Stat.*, **37**, 17.

Sethna, J. P. & K. S. Chow (1985). *Phase Transitions*, **5**, 317.

Sethna, J. P., S. R. Nagel & T. V. Ramakrishnan (1984). *Phys. Rev. Lett.*, **53**, 2489.

Setter, N. & L. E. Cross (1980). *J. mater. Sci.*, **15**, 2478.

Shafranovsky, I. I. (1968). *Lectures on Crystal Morphology.* Vysshaya Shkola, Moscow. In Russian.

Shapiro, S. M., J. D. Axe, G. Shirane & P. M. Raccah (1974). *Solid State Comm.*, **15**, 377.

Sheftal, N. N. (1966a). In A. V. Shubnikov & N. N. Sheftal (Eds.), *Growth of Crystals*, Vol. 4. Consultants Bureau, New York.

Sheftal, N. N. (1966b). In A. V. Shubnikov & N. N. Sheftal (Eds.), *Growth of Crystals*, Vol. 4. Consultants Bureau, New York.

Sheftal, N. N. (1976). In N. N. Sheftal (Ed.), *Growth of Crystals*, Vol. 10. Consultants Bureau, New York.

Sherrington, D. & S. Kirkpatrick (1975). *Phys. Rev. Lett.*, **35**, 1972.

Shirane, G., S. Hoshino & K. Suzuki (1950). *Phys. Rev.*, **80**, 1105.

Shtrikman, S. & D. Treves (1963). *Phys. Rev.*, **130**, 986.

Shubnikov, A. V. (1951). *Symmetry and Antisymmetry of Finite Figures.* USSR Academy of Sciences, Moscow.

Shubnikov, A. V. (1960). *Principles of Optical Crystallography.* Consultants Bureau, New York.

Shubnikov, A. V., N. V. Belov, and others (1964). *Colored Symmetry.*

Pergamon Press, Oxford.

Shubnikov, A. V. & V. A. Koptsik (1974). *Symmetry in Science and Art.* Plenum Press, New York.

Shuvalov, L. A. (1970). *J. Phys. Soc. Japan*, **28** Suppl., 38.

Shuvalov, L. A. (Ed.) (1988). *Modern Crystallography IV. Physical Properties of Crystals.* Springer-Verlag, Berlin.

Shuvalov, L. A., K. A. Aleksandrov & I. S. Zheludev (1959). *Kristallografia*, **4**, 130.

Shuvalov, L. A., E. F. Dudnik, V. A. Nepochatenko & S. V. Wagin (1985). *Izv. Akad. Nauk SSSR Ser. Fiz.* , **49**, 297 (in Russian).

Shuvalov, L. A., E. F. Dudnik & S. V. Wagin (1985). *Ferroelectrics*, **65**, 143.

Shuvalov, L. A. & N. R. Ivanov (1964). *Kristallografia*, **9**, 363.

Shuvalov, L. A., N. R. Ivanov & T. K. Sitnik (1967). *Sov. Phys. Crystallogr.*, **12**, 315.

Sidorkin, A. (1997). *Ferroelectrics*, **191**, 109.

Singh, N. & D. Pandey (1996). *J. Phys.: Condens. Matter*, **8**, 4269.

Singh, N., A. P. Singh, Ch. D. Prasad & D. Pandey (1996). *J. Phys.: Condens. Matter*, **8**, 7813.

Sirotin, Yu. I. & M. P. Shaskolskaya (1982). *Fundamentals of Crystal Physics.* Mir Publishers, Moscow.

Slack, G. A. (1960). *J. Appl. Phys.*, **31**, 1571.

Slater, J. C. (1941). *J. Chem. Phys.*, **9**, 16.

Slater, J. C. (1950). *Phys. Rev.*, **78**, 748.

Smolenskii, G. A. & A. I. Agranovskaya (1958). *Sov. Phys. - Tech. Phys.*, **3**, 1380.

Snoek, J. L. (1941). *Physica*, **8**, 711.

Snoek, J. L. (1942). *Physica*, **9**, 862.

Somayazulu, M. S., S. M. D. Rao & V. K. Wadhawan (1989). *Mat. Res. Bull.*, **24**, 795.

Sompolinsky, H. (1981). *Phys. Rev. Lett.*, **47**, 935.

Speiser, A. (1927). *Theorie der Gruppen von Endlichen Ordnung.* Springer, Berlin.

Spencer, E. G., R. C. LeCraw & F. Reggia (1956). *Proc. IRE*, **44**, 790.

Spillman, W. B. (1992). *Proc. SPIE (Smart Materials and Structures)*, **1777**, 97.

Srinivasan, T. P. (1988). *J. Phys. C: Solid State Phys.*, **21**, 4207.

Srinivasan, M. R., P. Ayyub, M. S. Multani, V. R. Palkar & R. Vijayaraghavan (1984). *Phys. Lett.*, **A101**, 435.

Stadler, H. L. (1958). *J. Appl. Phys.*, **29**, 1485.

Stadler, H. L. & P. J. Zachmanidis (1963). *J. Appl. Phys.*, **34**, 3255.

Standards on Piezoelectric Crystals (1949). *Proc. Institute of Radio Engineers*, **37**, 1378.

Stanley, H. E. (1971). *Introduction to Phase Transitions and Critical Phenomena.* Oxford University Press, New York.

Stark, J. P. (1988). *Phys. Rev. B*, **38**, 1139.

Stavsky, Y. & N. J. Hoff (1969). In A. G. H. Deitz (Ed.), *Composite Engineering Materials.* M.I.T. Press, Cambridge, Massachusetts.

Stein, D. L. (Ed.) (1992). *Spin Glasses and Biology.* World Scientific, Singapore.

Stenger, C. G. F., F. L. Scholten & A. J. Burggraaf (1979). *Solid State Commun.*, **32**, 989.

Stephen, M. J., E. Abrahams & J. P. Straley (1975). *Phys. Rev. B*, **12**, 256.

Stoeckel, D. & J. Simpson (1992). In G. J. Knowles (Ed.) (1992).

Stokes, H. T. & D. M. Hatch (1984). *Phys. Rev. B*, **30**, 4962.

Stokes, H. T. & D. M. Hatch (1988). *Isotropy Subgroups of the 230 Crystallography Space Groups.* World Scientific, Singapore.

Stokes, H. T. & D. M. Hatch (1991). *Phase Transitions*, **34**, 53.

Stokes, H. T. & D. M. Hatch (1998). Computer code ISOTROPY. Website http://www.physics.byu.edu/ stokesh/isotropy.html

Stokes, H. T., D. M. Hatch & H. M. Nelson (1993). *Phys. Rev. B*, **47**, 9080.

Stokes, H. T., D. M. Hatch & J. D. Wells (1991). *Phys. Rev. B*, **43**, 11010.

Stokes, H. T., J. S. Kim & D. M. Hatch (1987). *Phys. Rev. B*, **35**, 388.

Strnat, K. J. (1988). In E. P. Wohlfarth & K. H. J. Buschow (Eds.),

Ferromagnetic Materials, Vol. 4. North-Holland, Amsterdam.

Strnat, K. J., G. J. Hoffer, W. Ostertag & I. C. Olson (1966). *J. A. P.*, **37**, 1252.

Stryjewski, E. & N. Giordano (1977). *Adv. Phys.*, **26**, 487.

Sturman, B. I. & V. M. Fridkin (1992). *The Photovoltaic and Photorefractive Effects in Noncentrosymmetric Materials.* Gordon & Breach, Philadelphia.

Subbarao, E. C. (1990). *Ferroelectrics*, **102**, 267.

Subbarao, E. C., H. S. Maity & K. K. Srivastava (1974). *Phys. Stat. Sol.* (a), **21**, 9.

Suga, H. & S. Seki (1974). *J. Non-cryst. Solids*, **16**, 171.

Suga, H. & S. Seki (1981). *Faraday Discussions*, 69.

Sugano, S., Y. Nishina & S. Ohnishi (Eds.) (1991). *Microclusters.* Springer-Verlag, Berlin.

Sugawara, Y., K. Onitsuka, S. Yoshikawa, Q. C. Xu, R. E. Newnham & K. Uchino (1992). In G. J. Knowles (Ed.) (1992).

Sullivan, N. S., M. Devoret, B. P. Cowan & C. Urbina (1978). *Phys. Rev. B*, **17**, 5016.

Sunagawa, I. (1987). In I. Sunagawa (Ed.), *Morphology of Crystals*, Part B. Terra Sci. Pub. Co., Tokyo.

Sutton, A. P. & R. W. Balluffi (1996). *Interfaces in Crystalline Materials.* Clarendon Press, Oxford.

Sutton, M. & R. L. Armstrong (1982). *Phys. Rev. B*, **25**, 1813.

Suzuki, S., T. Osaka & Y. Makita (1979). *J. Phys. Soc. Japan*, **47**, 1741.

Svensson, C. & S. C. Abrahams (1980). *J. Appl. Cryst.*, **17**, 459.

Swartz, S. L. & V. E. Wood (1992). *Condensed Matter News*, **1**, 4.

Synkers, M., P. Delavignette & S. Amelinckx (1971). *Phys. Stat. Solidi* (b), **48**, K1.

Tagantsev, A. K. & A. E. Glazounov (1998). *Phase Transitions*, **65**, 117.

Takagi, T. (1989). *The Concept of Intelligent Materials and the Guidelines for R & D Promotion.* Japan Science & Technology Agency Report. Tsukuba Science City, Japan.

Tanaka, K., T. Shimada, Y. Nishihata & A. Sawada (1995). *J. Phys. Soc. Japan*, **64**, 146.

Tani, K. (1969). *J. Phys. Soc. Japan*, **26**, 93.

Tanimura, M. & K. Uchino (1988). *Sensors & Mater.*, **1**, 47.

Tavger, B. A. (1958). *Sov. Phys. - Cryst.*, **3**, 341.

Tavger, B. A. & V. M. Zaitsev (1956). *Zh. Eksp. Teor. Fiz.*, **30**, 564.

Taylor, G. W., J. J. Gagnepain, T. R. Meeker, T. Nakamura & L. A. Shuvalov (Eds.) (1985). *Piezoelectricity*. Gordon & Breach, U.K.

Tello, M. J., A. Lopez-Echarri, J. Zubillaga, I. Ruiz-Larrea, F. J. Zuniga, G. Madariaga & A. Gomez-Cuevas (1994). *J. Phys.: Condens. Matter*, **6**, 6751.

Temkin, D. E. (1964). In N. N. Sirota (Ed.), *Mechanism and Kinetics of Crystallization*. Nauka i Tekhnika, Minsk.

Terauchi, H. (1986). *Phase Transitions*, **7**, 315.

Theumann, W. K. & R. Koberle (Eds.) (1990). *Neural Networks and Spin Glasses*. World Scientific, Singapore.

Thomas, H. (1971). In E. J. Samuelsen, E. Andersen & J. Feder (Eds.), *Structural Phase Transitions and Soft Modes*. Universitetsforlaget, Oslo.

Thompson, B. S., M. V. gandhi & S. Kasivisvanathan (1992). *Materials & Design*, **13**, 3.

Thouless, D. J., P. W. Anderson & R. G. Palmer (1977). *Phil. Mag.*, **35**, 593.

Thrasher, M. A., A. R. Shahin, P. H. Meckl & J. D. Jones (1992). *Proc. SPIE (Smart Materials and Structures)*, **1777**, 197.

Thursby, M. H., B. G. Grossman, T. Alavie & K. S. Yoo (1989). *Proc. SPIE (Fiber Optic Smart Structures and Skins II)*, **1170**, 316.

Thursby, M. H., B. G. Grossman & K. S. Yoo (1990). In Ahmad et al. (Eds.).

Tiller, W. A. (1991a). *The Science of Crystallization: Microscopic Interfacial Phenomena*. Unievrsity Press, Cambridge.

Tiller, W. A. (1991b). *The Science of Crystallization: Macroscopic Phenomena and Defect Generation*. University Press, Cambridge.

Tiwari, R. & V. K. Wadhawan (1991). *Phase Transitions*, **35**, 47.

Tiwari, V. S. & D. Pandey (1995). *J. Am. Ceram. Soc.*, **77**, 1819.

Tiwari, V. S., Neelam Singh & D. Pandey (1994). *J. Am. Ceram. Soc.*, **77**, 1813.

Tiwari, V. S., Neelam Singh & D. Pandey (1995). *J. Phys.: Condens. Matter*, **7**, 1441.

Toda, M., R. Kubo & N. Saito (1992). *Statistical Physics I: Equilibrium Statistical Mechanics*, second edition. Springer-Verlag, Berlin.

Toledano, J. C. (1979a). *J. Solid State Chem.*, **27**, 41.

Toledano, J.-C., G. Errandonea & J. P. Jaguin (1976). *Solid State Commun.*, **20**, 905.

Toledano, J.-C. & P. Toledano (1980). *Phys. Rev. B*, **21**, 1139.

Toledano, J.-C., L. Pateau, J. Primot, J. Aubree & D. Morin (1975). *Mater. Res. Bull.*, **10**, 103.

Toledano, P. & J.-C. Toledano (1977). *Phys. Rev. B*, **16**, 386.

Toledano, P. (1992). *Key Engg. Materials*, **68**, 1.

Toledano, P. & V. Dmitriev (1993). *Condensed Matter News*, **2**(5), 9.

Toledano, P. & V. Dmitriev (1995). *Key Engg. Materials*, **101-102**, 311.

Toledano, P. & V. Dmitriev (1997). *Ferroelectrics*, **191**, 85.

Toledano, P., H. Schmid, M. Clin & J.-P. Rivera (1985). *Phys. Rev. B*, **32**, 6006.

Toledano, P. & J. C. Toledano (1976). *Phys. Rev. B*, **14**, 3097.

Toledano, P. & J. C. Toledano (1982). *Phys. Rev. B*, **25**, 1946.

P. A. Thomas & A. M. Glazer (1991). *J. Appl. Cryst.*, **24**, 968.

Topolov, V. Yu., L. E. Balyunis, A. V. Turik & O. E. Fesenko (1990). *Ferroelectrics*, **110**, 41.

Torre, L. P., S. C. Abrahams & R. L. Barns (1972). *Ferroelectrics*, **4**, 291.

Toulouse, G. (1977). *Commun. Phys.*, **2**, 115.

Townsend, M. G., J. R. Gosselin, R. J. Tremblay & A. H. Webster (1976). *J. Phys. (Paris)*, Suppl. **10**, C4-11.

Trefilov, V. I. (1995) (Ed.), *Ceramic and Ceramic-Matrix Composites*. Chapman & Hall, London.

du Tremolet de Lacheisserie, E. (1993). *Magnetostriction: Theory and Applications of Magnetoelasticity*. CRC Press, London.

Tsukamoto, T. & H. Futama (1993). *Phase Transitions*, **45**, 59.

Tsukamoto, T., M. Komukae, S. Suzuki, H. Futama & Y. Makita (1983). *J. Phys. Soc. Japan*, **52**, 3966.

Tutt, L. W. & A. Kost (1991). *Nature*, **356**, 225.

Uchino, K. (1992). *Proc. 1st European Conf. on Smart Structures & Materials*, Section 5. Glasgow.

Uchino, K. (1996). *Innovations in Mater. Res.*, **1**, 11.

Uchino, K. (1997). *Piezoelectric Actuators and Ultrasonic Motors.* Kluwer, Dordrecht.

Uchino, K., M. Aizawa & S. Nomura (1985). *Ferroelectrics*, **64**, 199.

Uchino, K., Miyazawa, Y. & S. Nomura (1983). *Jpn. J. Appl. Phys.*, **22**, Suppl. 22-2, 102.

E. Udd (Ed.) (1995). *Fiber Optic Smart Structures.* Wiley, New York.

Unruh, H.-G. (1993). *Phase Transitions*, **45**, 77.

Unruh, H.-G. (1995). *Phase Transitions*, **55**, 155.

Uwe, H. & H. Tokumoto (1979). *Phys. Rev. B*, **19**, 3700.

Vainshtein, B. K. (1981). *Modern Crystallography I. Symmetry of Crystals: Methods of Structural Crystallography.* Springer-Verlag, Heidelberg.

Vainshtein, B. K. & A. A. Chernov (Eds.) (1988). *Modern Crystallography.* Nova Science Publishers, Commack, New York.

Valasek, J. (1920). *Phys. Rev.*, **15**, 537.

Valasek, J. (1921). *Phys. Rev.*, **17**, 475.

Valenzuela, R. (1994). *Magnetic Ceramics.* Cambridge University Press, Cambridge.

Valiev, R. Z., R. R. Mulyukov, K. Y. Mulyukov, V. I. Navykov & L. I. Trusov (1989). *Pisma v Zhurnal Tekhnicheskoi Fiz.*, **15**, 78.

van den Boomgaard, J., D. R. Terrell, R. A. J. Born & H. F. J. J. Giller (1974). *J. Mater. Sci.*, **9**, 1705.

van den Boomgaard, J., D. R. Terrel & R. A. J. Born (1978). *J. Mater. Sci.*, **13**, 1538.

van der Eerden, J. P. (1993). In D. T. J. Hurle (Ed.), *Handbook of Crystal Growth.* North-Holland, Amsterdam. Vol. 1, Chapter 6.

van der Waals, J. D. (1873). Ph.D. thesis, University of Leiden.

van Run, A. M. J. G., D. R. Terrell & J. H. Scholing (1974). *J. Mater. Sci.*, **9**, 1710.

van Suchtelan, J. (1972). *Philips Research Reports*, **27**, 28.

van Suchtelan, J. (1980). *Ann. Chim. Fr.*, **5**, 139.

van Tendeloo, G. & S. Amelinckx (1974). *Acta Cryst.*, **A30**, 431.

van Vlack, J. H. (1947). *Annales de l'Institut Henri Poincaré*, **10**, 57.

Varadan, V. V., L. C. Chin & V. K. Varadan (1992). *Proc. SPIE (Smart Materials and Structures)*, **1777**, 1.

Venkataraman, G., D. Sahoo & V. Balakrishnan (1989). *Beyond the Crystalline State: An Emerging Perspective.* Springer-Verlag, Berlin.

Vere, A. W. (1987). *Crystal Growth: Principles and Progress.* Plenum Press, New York.

Verma, A. R. (1951a). *Nature*, **167**, 939.

Verma, A. R. (1951b). em Phil. Mag., **42**, 1005.

Verma, A. R. (1953). *Crystal Growth and Dislocations.* Butterworths, London.

Viehland, D., S. J. Jang, L. E. Cross & M. Wuttig (1990). *J. Appl. Phys.*, **68**, 2916.

Viehland, D., J. F. Li, S. J. Jang, L. E. Cross & M. Wuttig (1992). *Phys. Rev. B*, **46**, 8013.

Virkar, A. V., J. F. Sue, P. Smith, K. Mehta & K. Prettyman (1991). *Phase Transitions*, **35**, 27.

Vlachavas, D. S. (1984). *Acta Cryst.*, **A40**, 213.

Vogel, H. (1921). *Z. Phys.*, **22**, 645.

Wada, B. K., J. L. Fanson, G. S. Chen & C. P. Kuo (1990). In Ahmad et al. (Eds.) (1990).

Wadhawan, V. K. (1978a). *Current Science*, **47**, 534.

Wadhawan, V. K. (1978b). *Mat. Res. Bull.*, **13**, 1.

Wadhawan, V. K. (1979). *Acta Cryst. A*, **35**, 629.

Wadhawan, V. K. (1980). *Acta Cryst. A*, **36**, 851.

Wadhawan, V. K. (1982). *Phase Transitions*, **3**, 3.

Wadhawan, V. K. (1985). *Mater. Sci. Forum*, **3**, 91.

Wadhawan, V. K. (1987a). *Mat. Res. Bull.*, **22**, 651.

Wadhawan, V. K. (1987b). *Phase Transitions*, **9**, 297.

Wadhawan, V. K. (1988). *Phys. Rev. B*, **38**, 8936.

Wadhawan, V. K. (1989). *Ferroelectrics*, **97**, 171.

Wadhawan, V. K. (1991). *Phase Transitions*, **34**, 3.

Wadhawan, V. K. (1997). *Acta Cryst. A*, **53**, 546.

Wadhawan, V. K. (1998). *Phase Transitions*, **64**, 165.

Wadhawan, V. K. & K. V. Bhagwat (1989). *Phase Transitions*, **19**, 27.

Wadhawan, V. K. & C. Boulesteix (1992). *Key Engg. Materials*, **68**, 43.

Wadhawan, V. K. & A. M. Glazer (1981). *Phase Transitions*, **2**, 75.

Wadhawan, V. K. & A. M. Glazer (1989). *Phys. Rev. B*, **39**, 9631.

Wadhawan, V. K., M. C. Kernion, T. Kimura & R. E. Newnham (1981). *Ferroelectrics*, **37**, 575.

Wadhawan, V. K. & M. S. Somayazulu (1986). *Phase Transitions*, **7**, 59.

Wadsley, D. A. (1964). In L. Mandelcorn (Ed.), *Nonstoichiometric Compounds.* Academic Press, New York.

Wainer, E. & S. Solomon (1942). *Titanium Alloys Manufacturing Co. Reports 8 and 9.*

Wang, F. E. (1992). In G. J. Knowles (Ed.) (1992).

Wang, F. Y. (1973). In H. Herman (Ed.), *Treatise on Materials Science and Technology.* Academic Press, New York.

Wang, C. H. & R. B. Wright (1974). *J. Chem. Phys.*, **60**, 849.

Ward, R. W. (1989). *Proc. 38th Ann. Freq. Control Symp.*, pp. 22-31.

Warlimont, H. (1976). *Mater. Sci. & Engg.*, **25**, 133.

Wasilewski, R. J. (1975). In J. Perkins (Ed.), *Shape-Memory Effects in Alloys.* Plenum Press, New York.

Waterman, P. C. (1959). *Phys. Rev.*, **113**, 1240.

Wayman, C. M. (1980). *J. Metals*, **32**, 129.

Wayman, C. M. (1981). *Proc. Int. Conf. on Solid State Transformations.* Carnegie-Melon Univ., Pittsburgh, Aug. 10-14, 1981. Published by the Metals Society of the AIME, USA.

Wayman, C. M. (1992). In G. J. Knowles (Ed.), *Active Materials and Adaptive Structures.* IOP Publishing, Bristol.

Weber, L. (1929). *Z. Kristallogr.*, **70**, 309.

Weber, H. P., B. C. Tofield & P. F. Liao (1975). *Phys. Rev. B*, **11**, 1152.

Wegner, F. J. & E. K. Riedel (1973). *Phys. Rev. B*, **7**, 248.

Weiss, P. (1907). *J. Phys.*, **6**, 661.

Weissman, M. B. (1998). In A. P. Young (Ed.), *Spin Glasses and Random Fields.* World Scientific, Singapore.

Wersing, W. (1981). *Ferroelectrics*, **37**, 611.

Weschler, M. S., D. S. Lieberman & T. A. Read (1953). *Trans. AIME*, **197**, 1503.

Whatmore, R. W., R. Clarke & A. M. Glazer (1978). *J. Phys. C: Solid State Phys.*, **11**, 3089.

White, R. M. & T. H. Geballe (1979). *Long Range Order in Solids.* Suppl. 15 to *Solid State Phys.*. Academic Press, New York.

Widom, B. (1965). *J. Chem. Phys.*, **43**, 3898.

Wiener, N. (1930). *Acta Math.*, **55**, 117.

Wiesendanger, E. (1973). *Czech. J. Phys. B*, **23**, 91.

Wigner, E. P. (1959). *Group Theory and Application to Atomic Spectra.* Academic Press, New York.

Wigner, E. P. (1967). *Symmetries and Reflections: Scientific Essays of E. P. Wigner.* Indian University Press, Bloomington and London.

Wigner, E. P. (1984). In W. W. Zachary (Ed.), *Proc. XIII Int. Colloq. on Group Theoretical Methods in Physics*, Maryland, U. S. A. (21-25 May 1984). World Scientific, Singapore.

Wigner, E. P. (1989). In H. A. Bethe et al., *From a Life of Physics.* World Scientific, Singapore.

Wilson, K. G. (1977). *Scientific American*, **241**, 140.

Wohlfarth, E. P. (1977). *Physica*, **86-88B**, 852.

Wohlfarth, E. P. (1979). *Phys. Lett.*, **40A**, 489.

Wohlfarth, E. P. (1980a). *J. Magn. Magn Mater.*, **20**, 77.

Wohlfarth, E. P. (1980b) (Ed.). *Ferromagnetic Materials*, Vol. 2. North-Holland, Amsterdam.

Woirgard, J. & J. de Fouquet (1972). *Scripta Metall.*, **6**, 21.

de Wolff, P. M., T. Janssen & A. Janner (1981). *Acta Cryst. A*, **37**, 625

Wondratschek, H. & W. Jeitschko (1976). *Acta Cryst.*, **A32**, 664.

Wood, E. A. (1964). *Bell System Tech. Publ.*, Monograph No. 4680.

Wooster, W. A. (1973). *Tensors and Group Theory for the Physical Properties of Crystals.* Clarendon Press, Oxford.

Wooster, W. A. & N. Wooster (1946). *Nature*, **157**, 405.

Worlock, J. M. (1971). *Proc. NATO Advanced Study Institute: Structural Phase Transitions and Soft Modes*, Universiteforlaget, Oslo.

Worlton, T. G. & R. A. Beyerlein (1975). *Phys. Rev. B*, **12**, 1899.

Wruck, B., E. K. H. Salje, M. Zhang, T. Abraham & U. Bismayer (1994). *Phase Transitions*, **48**, 135.

Wu, F. Y. (1982). *Rev. Mod. Phys.*, **54**, 235.

Wul, B. (1945). em Nature, **156**, 480.

Wul, B. (1946). *Nature*, **157**, 808.

Wul, B. & I. M. Goldman (1945a). *C. R. Acad. Sci. URSS*, **46**, 139.

Wul, B. & I. M. Goldman (1945b). *C. R. Acad. Sci. URSS*, **49**, 177.

Wul, B. & I. M. Goldman (1946). *C. R. Acad. Sci. URSS*, **51**, 21.

Xu, Y. (1991). *Ferroelectric Materials and Their Applications.* North-Holland, Amsterdam.

Yariv, A. (1985). *Optical Electronics.* Holt, Rinehart & Winston, Holt-Saunders Japan.

Yariv, A. & P. Yeh (1984). *Optical Waves in Crystals.* Wiley, New York.

Ye, Z.-G., E. Burkhhardt, J.-P. Rivera & H. Schmid (1995). *Ferroelectrics*, **172**, 257.

Ye, Z.-G. & H. Schmid (1993). *Ferroelectrics*, **145**, 83.

Yosida, K. (1957). *Phys. Rev.*, **106**, 893.

Yoo, M. H. & M. Wuttig (Eds.) (1984). *Twinning in Advanced Materials.* The Minerals, Metals & Materials Society, Pennsylvania.

Young, A. P. (Ed.) (1998). *Spin Glasses and Random Fields.* World Scientific, Singapore.

Zammit-Mangion, L. J. & G. A. Saunders (1984). *J. Phys. C: Solid State Phys.*, **17**, 2825.

Zamorzaev, A. M. (1957). Dissertation (in Russian). LGU.

Zener, C. Z. (1948a). *Elasticity and Anelasticity of Metals.* University of Chicago Press, Chicago.

Zener, C. Z. (1948b). *Phys. Rev.*, **74**, 639.

Zheludev, I. S. (1971). In H. Ehrenreich, F. Seitz & D. Turnbull (Eds.), *Solid State Physics*, **26**, 429.

Zheludev, I. S. (1978). *Ferroelectrics*, **20**, 51.

Zheludev, I. S. & L. A. Shuvalov (1956). *Sov. Phys. Cryst.*, **1**, 537.

Zheludev, I. S. & L. A. Shuvalov (1957). *Izv. Akad. Nauk SSSR, Ser. Fiz.*, **21**, 264.

Zhirnov, V. A. (1959). *Sov. Phys. - JETP*, **8**, 822.

Zikmund, Z. (1984). *Czech. J. Phys.*, **B34**, 932.

Zikmund, Z. & J. Fousek (1989). *Phys. Stat. Solidi* (a), **112**, 625.

AUTHOR INDEX

SUBJECT INDEX